Rieger · I

R. Rieger · A. Michaelis · M. M. Green

Glossary of Genetics

Classical and Molecular

Fifth Edition

Springer-Verlag
Berlin Heidelberg New York
London Paris Tokyo
Hong Kong Barcelona
Budapest

1st	edition (German)	1954
2nd	edition (German)	1958
	Russian edition	1967
3rd	edition (English)	1968
	Polish edition	1974
4th	edition (English)	1976
	Spanish edition	1982
	Chinese edition	1988
5th	edition (English)	1991

With 13 Tables

ISBN 3-540-52054-6 Springer-Verlag Berlin Heidelberg New York
ISBN 0-387-52054-6 Springer-Verlag New York Berlin Heidelberg

Typesetting: Graphischer Großbetrieb Pößneck
Printing and bookbinding: Clausen & Bosse, Leck
31/3145-543210 – Printed on acid-free paper

Preface

More than a decade has elapsed since the last revision of this Glossary. Circumstances beyond our control necessitated the delay. In the interim, research in genetics has undergone a veritable scientific revolution. The concepts and technology of molecular biology have assumed an integral role in genetics, today and for the foreseeable future. Concomitant with the growth and development of molecular genetics research are new concepts and a new vocabulary. Our goal in this revision has been to incorporate the vocabulary of molecular genetics without unduly increasing the length of the Glossary. To achieve this goal we have scrupulously edited the last edition culling those terms which in our judgement are either little used or obsolete. In addition, the word length of many terms has been condensed without, we trust, loss of comprehension. Wherever possible, synonymy and redundancy have been pointed out, and in the interest of historical accuracy the individual responsible for introducing a particular term or concept listed with the specific paper included in the literature citations. Cross references between related terms are designated by an arrow (→) before each relevant term.

As in the past we solicit comments, corrections and emendations from our readers. We hope this edition of the Glossary will be as useful as earlier editions have been.

We are especially indebted to Mrs. Irma Siegel and Mrs. Heide Trippner who participated directly in the preparation of the manuscript. The cooperation of Dr. D. Czeschlik (Springer-Verlag) and Mrs. J. Schlüter (Fischer Verlag) is gratefully acknowledged.

Gatersleben and Davis
January 1991

Rigomar Rieger
Arnd Michaels
Melvin M. Green

Authors

Professor Dr. Rigomar Rieger
Institut für Genetik
und Kulturpflanzenforschung
Correnstraße 3
O-4325 Gatersleben, FRG

Dr. Arnd Michaelis
Institut für Genetik
und Kulturpflanzenforschung
Correnstraße 3
O-4325 Gatersleben, FRG

Professor Dr. Melvin M. Green
Department of Genetics
University of California
Davis, CA 95616, USA

A

AI, AII — abbreviations for the anaphase of the first and second meiotic division, resp. (→ meiosis).

abbreviation — the successive shortening of ontogenesis through cessation of individual stages.

ABC exinuclease (Seeberg and Steinum 1982) — a → DNA repair enzyme which removes DNA adducts by hydrolyzing phosphodiester bonds on both sides of the modified nucleotides.

aberrant ratio (Sprague and McKinney 1966) — any departure from Mendelian expectations in F_2 generations of maize plants infected with certain viruses.

aberration rate — → chromosome mutation.

abortive infection (Lwoff 1953) — infection of prokaryotic (bacterial) or eukaryotic cells by bacteriophages or viruses, respectively, which does not result in the formation of infectious virus although one or more viral components are synthesized (→ productive infection; reductive infection; lysogenization).

abortive transduction — the failure of a transducing DNA segment to be incorporated into the recipient chromosome.

abortive transfer — any DNA transfer from bacterial donor to recipient cells that fails to establish the incoming DNA as part of the hereditary material of the recipient. A.t. has been observed following → transduction, → transformation, and → conjugation. In all cases the transferred fragment is diluted out as the culture grows. Failure of integration of transferred DNA into the hereditary material of the recipient cell may be due to: (1) The failure of incoming DNA to form circular molecules; (2) circularization takes place, but the circular molecule fails to take up its maintenance system. A.t. of extrachromosomal elements (→ plasmids), as opposed to chromosomal fragments, is relatively uncommon since plasmids are genetic elements of autonomous survival in a bacterial cell. It is only when a mutation in the recipient or a resident plasmid inactivates the host component of the plasmid maintenance system that a.t. of a plasmid occurs. Genes carried on abortive pieces of DNA may be expressed in the recipient cells.

acceptor identity (Normanly et al. 1986) — the specificity (= a. specificity) of a → transfer RNA for an amino acid, as determined by multiple nucleotides whose locations are not conserved between different molecules. The a.i. is read by one of 20 → aminoacyl-tRNA synthetases, which attach the proper amino acid only to their cognate tRNA species.

acceptor splicing site — the junction between the right end of an → intron and the left end of an exon (→ donor site), which exhibits a highly conserved sequence of four nucleotides preceded by a pyrimidine-rich region. A. sites and donor sites are essential parts of the process of exon-splicing and provide a specific molecular signal by which the → RNA splicing machinery can identify the precise splice points.

accessory chromosome (McClung 1900) — → sex and → B-chromosome.

accessory DNA — surplus DNA present in certain cells or cell stages due to, e.g., → gene amplification.

accessory nucleus — any of those cytoplasmic vesicles whose walls, like those of the → nucleus, consist of an inner and outer membrane layer. These walls are pierced by pores identical to those of the → nuclear envelope (annulate membrane structure). A. nuclei arise from the nuclear membrane, to which they often adhere.

accessory plate (Darlington 1936) — a supplementary metaphase plate which owes its origin to bivalents which, during the metaphase, either remain outside the equatorial plate (because of a lack of → centromere orientation) or have not yet attained this (noncongression) (→ congression).

accessory transcription factor (Yura and Ishihama 1979) — any of a number of elements which aid → RNA polymerase in the regulation of → genetic transcription.

accommodation (Thach and Thach 1971) — a GTP-dependent reaction in bacteria which occurs during → genetic translation. → Messenger RNA (mRNA) is moved a distance of about three nucleotides in the 5′ direction relative to the → ribosome during the translocation of peptidyl-tRNA from the A to the P site of the ribosome. This reaction is catalyzed by the → G factor (→ translocation factor) and is dependent on the hydrolysis of GTP. In contrast, mRNA is not moved during the IF 2-catalyzed (→ initiation factor) hydrolysis of GTP which is involved in the activation of ribosome-bound fMet-tRNA (→ initiator tRNA). This second type of GTP-dependent reaction is called a. of fMet-tRNA.

Ac-Ds transposable element system (McClintock 1978) — in maize, a two-element system in which Ac (Activator) is the self-mobile or autonomous element and Ds (Dissociation) the nonautonomous element (→ controlling element). Although both elements can transpose, only Ac can induce transposition of both itself and Ds, i.e., only Ac has a transacting transposition function.

acentric — of → chromosomes or chromosome segments without a → centromere.

achiasmate — of a → meiosis without → crossing over and chiasmata. Where a. meiosis occurs, it is usually confined to one of the two sexes. Generally, the other sex has a chiasmate type of meiosis. A common feature of a. meiosis is the absence of any opening-out of paired chromosomes (bivalents) at diplotene; the four chromatids remain parallel until the beginning of first metaphase. A. meiosis is a distinct and characteristic mechanism which has arisen independently in a large number of organisms. In higher plants it is rare and not yet observed in vertebrates (→ cryptochiasmate).

achromatic (Flemming 1879) — those parts of the → nucleus which may not be stained with dyes characteristic for chromosomes (→ chromatin).

achromatic figure — the → spindle during → mitosis and → meiosis (→ mitotic apparatus).

achromatic lesion — → gap.

A-chromosome (Randolph 1928) — any of the standard chromosomes of the → chromosome complements of all eukaryotic organisms. A-chromosomes form a delicately balanced system and all usually have to be present to secure normal viability. In many species, the standard complement of A-chromosomes may be supplemented by varying numbers of supernumerary or → B-chromosomes. These are derived from the A-c. set and have become functionally subordinate.

acquired character — a → modification impressed on an organism by environmental influences during development.

acrocentric (White 1945) — of chromosomes whose → centromere is very close to one end so that one → chromosome arm is short and the other much longer (→ metacentric).

acrocentric association (Zang and Back 1968) — the nonrandom participation of human acrocentric chromosomes in satellite associations. The frequency of association appears to be a characteristic property of a given individual chromosome.

acrosome (Lenhossek 1897) — a cap-like structure which invests the front part of the sperm head. It is limited externally by an outer membrane that is reflected forwards at the posterior boundary of the a. to form an inner membrane contiguous with the nuclear membrane of the spermatozoon.

The details of a. formation vary but in general follow one of two paths (Nath 1956): either the Golgi elements (→ dictyosomes) are transformed directly into the a. or they "secrete" the a. and then shed with the residual cytoplasm.

In function the a. is concerned with the enzymatic penetration of the protective cover of the egg cell after formation of a surface layer. When sperm draw near to recently ovulated eggs, the outer acrosomal membrane and overlying plasma membran fragment, creating openings to release enzymes of the a. This process is called "acrosomal reaction" (Bedford 1970). Genetically controlled defects of the a. may impair zygote formation.

acrosyndesis (Percival 1932) — incomplete end-to-end pairing of two chromosomes during → meiosis.

actin — any of a group of structurally related isoproteins in higher organisms that play an essential role in cell structure and cell mobility. The functions of actins are diverse, ranging from muscle contraction to → chromosome movement. The structure of the 42000 Da protein is highly conserved, both among various types of actins and in evolution. In the form of F-a., it is the major component of the 5 to 7 μm diameter filaments (the → microfilaments) inside many types of cells.

action system (Hamburger) — in embryology, a system consisting of the → organizer and its organization field, which, in the → reaction system, brings about the realization of one or more developmental potencies (→ induction system).

activation — a chain of responses made by an unfertilized egg to fertilization or to artificial, e.g., penetration by a pipet. These responses include the rupture of cortical granules, the lifting off of the vitelline membrane, and the free rotation of the egg nucleus under the influence of gravity (Gurdon 1974).

activator (Huxley 1935) — (1) a protein cofactor in the presence of which the → promoters of positively controlled genes or operons function. The a. binds to a DNA site upstream of a gene and activates transcription from that gene. The a. is encoded by a → regulator gene. (2) A gene-dependent protein which stimulates the development of a certain embryonic tissue or organ, i.e., is morphologically active. A local a. is only active in the cell in which it is produced, a distant a. or hormone is active beyond the region of its formation and its transport may proceed by diffusion (diffusion hormone) or by body fluids (circulation hormone).

activator RNA (Britton and Davidson 1969) — a regulator molecule capable of recognizing specific sites on DNA. A. RNA is assumed to play a role in the regulation of → gene expression in eukaryotes. According to the Britton-Davidson model of → genetic regulation in eukaryotes, the genetic material is in its normal state repressed. Particular control systems function specifically to overcome this repression and to switch on appropriate genes by the functioning of a. RNA. A. RNA is assumed to be coded for by "integrator" genes which are active only when an adjacent segment of DNA (= "sensor" gene) interacts with a specific protein.

adaptation — any change in an organism's structure or function that allows it to better cope with conditions in the environment. Harmonious adjustment to environmental conditions is the result of a. Adaptation as a process or the result of a process taking a different course in different individuals, signifies the build-up or the possession of → characters which prove advantageous for the individual or the population under the environmental conditions in which it lives and through which the organism acquires → adaptive value or fitness in a certain environment.

A. can be achieved in two different ways, as pure phenotypic a. or as genotypic a. In the first case, the → reaction norm of the genotype can be attuned to those environmental conditions prevalent under natural circumstances. In the second case, genotypic specialization leads to a. in such a way that a change of genotype results in the formation of a new reaction norm which makes it possible for the genotype to harmonize where the original form fails to do so.

Adaptations consist of harmoniously working combinations of genes which are built up and preserved by the process of natural → selection. The adaptation presupposes in each case the possession of a genotype with a favorable reaction norm. The reaction is then favorable if the frequently recurrent environmental influences give rise to phenotypic modifications which permit the organism to remain viable and produce offspring (→ flexibility).

An adaptation process which furthers the group at the cost of the individual, i.e., the chance of the survival of the individual is lowered while the survival rate of the group increases, is designated **"altruistic adaptation"** (Haldane 1932).

The possibility of certain genotypes adapting immediately to changed environmental conditions by means of prospective functions is so-called **"prospective adaptation"** (Simpson 1953) and means the formation of characters which are without adaptive value at the time of formation but which prove to be adaptive in a new environment (= **preadaptation**).

"Pseudoexogenous" (Waddington 1953) is an a. which appears to be occasioned directly by environmental influences but actually results without environmental stimulus or is independent thereof.

adaptedness — the state of being adapted, i.e., the ability of an organism, a genotype, or a population to survive and reproduce in a given environment. The a. of a genotype is a function of its → norm of reaction and of the range of environments in which it occurs (→ selection).

adaptibility – the potentiality for → adaptation. Physiological a. depends upon a → norm of reaction yielding a certain adaptedness in the environments which individuals or populations meet. Genetic a. (= evolutionary → plasticity) occurs by changing the norm of reaction to produce an improved adaptedness in some or all, old or new environments. A genotype that gives rise to adaptive traits in some or all of the environments it meets may be said to possess high a. which may depend upon physiological → homeostasis or upon developmental homeostasis (Dobzhansky 1968).

adaptive — of those changes and traits of an organism which act to preserve life, i.e., tend to increase viability, survival rate, and reproductive rate. An a. trait or character is a structural or functional characteristic of the organism which enables or enhances the probability of survival and reproduction (Dobzhansky 1956).

adaptive gene complex — a suite of genes that confer high → fitness together, but are of much lower fitness when combined with other genes. Such complexes can arise only if there are nonadditive interactions between the loci.

adaptive landscape (*Wright* 1932) — a metaphor used for representation of the interactions between organism and environment: The surface is plotted in a three-dimensional graph, with all possible combinations of allele frequencies for different loci plotted in the plane, and mean fitness for each combination plotted in the third dimension. Populations move through the landscape in the course of evolution. Hills and valleys symbolize the distribution of adaptive fields. The hill tops are → adaptive peaks, the valleys between them symbolize a "no-man's-land" for organisms, i.e., a series of adaptive valleys. Evolutionary trends are the pathways to the adaptive peaks, evolutionary rates correspond to the rate of ascent (Grant 1985).

adaptive norm (Schmalhausen 1949) — a well-adapted more or less stable complex of genetic diversity within the population.

adaptive peak (Wright 1932) — a symbolic representation (from an evolutionary point of view) of the relations between organism and environment in the form of a topographic map. Groups of related genotypes which make their carriers able to occupy certain ecological niches are said to occupy "adaptive peaks" situated in different parts of the map and separated from each other by "adaptive valleys" comprising unfit gene combinations. Any such peak represents a particular configuration of → gene (allele) frequencies and → genotype frequencies characteristic for a population equilibrium and (relatively) stabilized by a complex interaction of opposing forces. If one of these interacting factors becomes modified in its action, three reactions may occur (Dobzhansky 1951; Lerner 1958): (1) compensating adjustments in the responses of the population to the other forces; (2) movement of the population from one such equilibrium across a "valley" to another new a.p. which may be presented by a completly different array of gene frequencies and may require a far reaching rebuilding of the → gene pool. A shift from one to another a.p. is connected with the formation of stages intermediate between the two which may be more or less unbalanced; (3) disappearance of an original a.p. and extinction of the particular gene combinations in those cases in which no mutations occur creating new adaptive gene combinations or if the new combinations do not occur at the right time.

The construction of adaptive peaks is as follows: all genotypes are plotted as points on a plane — the more closely they lie the less they differ from each other — and as ordinate the prevailing → adaptive value of these genotypes within the appropriate environment is used. In this way a "mountain range" is set up with „peaks" separated by "valleys" and "saddles". Each of the various genotypes is then represented by a point on this "mountain surface" and each population occupies a region on the same.

adaptive radiation — the evolutionary diversification of a group of organisms (of a single phyletic line) which leads, often within a relatively short period of time, to the formation by natural → selection of a variety of types from a single ancestral species. These types are adapted to certain environmental conditions. A. r. results in a ramification of an adaptive type in a series of new adaptive zones (Simpson 1953).

adaptive response (Jeggo et al. 1977; Samson and Cairns 1977) — an error-free pathway of inducible → DNA repair that protects against damaging effects of alkylating agents. Two different types of DNA repair enzymes are involved in the a.r.: (1) a broad spectrum DNA glycosylase (3-methyladenine DNA glycosylase) which initiates → excision repair of several methylated bases; (2) 0^6-methylguanine DNA methyltransferase able to directly remove particular methyl and ethyl groups from DNA molecules.

Induction of these repair enzymes follows from treatment with nontoxic levels of alkylating agents and results in a reduced effectiveness of a subsequently applied (higher) challenge concentration of the mutagens.

The regulation of the a.r. is independent of the bacterial → SOS response and is controlled by the product of the ada gene. Cells induced for the a.r. have a reduced ability to express three of the SOS functions: → W-reactivation, W-mutagenesis, and phage lambda induction.

adaptive strategy — the sum total of the adaptations to the environment by a → species.

adaptive topography — = → adaptive landscape.

adaptive value — the comparative → fitness of different genotypes in a given environment; the survival and reproductive value of one genotype relative to the other genotypes in the population. The a.v. includes fertility and functional usefulness of morphological characters as well as physiological variability.

adaptive zone — any major kind of environment to which an organism can become adapted. Niches describe the partitioning of resources within an a.z.

adaptor hypothesis (Crick 1958; Hoagland 1959) — a hypothesis (now experimentally proved) which states that the amino acid sequence of a protein is determined, during the course of → genetic translation, by the alignment of aminoacyl-transfer RNA's at corresponding nucleotide → codons in → messenger RNA. The specificity of the translation mechanism depends on base pairing between a nucleotide region of → transfer RNA (the → anticodon) and nucleotides in the → codon.

adaptor modification hypothesis (Sueoka and Kano-Sueoka 1964) — a hypothesis proposed to explain the regulation of protein synthesis by modification of tRNA molecules at a site which affects recognition of either the mRNA codon, the enzyme (→ aminoacyl-tRNA-synthetase), or the ribosome. When multiple species of tRNA correspond to the same amino acid, modification of one tRNA species may prevent the → genetic translation of mRNA with the corresponding condon(s).

adaptor molecules — → genetic translation.

additive genes — genes interacting and showing no → dominance if → alleles or showing no → epistasis if nonalleles (→ gene interaction).

additive genetic variance — → genetic variance associated with the average effects of substituting one allele for another (= genic variance).

additive theorem — the a.t. of → exchange percentages states: the exchange percentage occurring on → crossing over between the loci A and C of a linkage group is equal to the sum of the values AB and BC, if the locus B lies between A and C, or equals the difference if locus B is localized outside the stretch AC.

adelphogamy — sib pollination. Pollen and stigma belong to two different individuals which derive vegetatively from the same mother plant.

adjacent distribution (McClintock 1945) — the orientation and distribution of chromosomes lying adjacent in the ring or chain configuration of translocation heterozygotes (→ translocation) to the same pole in the first meiotic division (→ meiosis) in contrast to → alternative distribution, in the course of which alternating chromosomes of the configuration are distributed to the same pole. It is a consequence of a.d. that either all or a part of the products of meiosis (genes or gametes) are genetically unbalanced and contain → duplications or → deletions. The proportion of unbalanced meiotic products is determined by the position and number of the chiasmate (in the → pairing segments or → interstitial segments). A.d. is responsible for partial sterility of translocation heterozygotes (→ semi-sterility). Two types of a.d. may be distinguished: (1) **adjacent-1 distribution** (=nonhomologous adjacent distribution; nondisjunctional distribution): chromosomes adjacent in the configuration with nonhomologous centromeres are distributed to the same pole in meiosis I. In such a case the translocated segments and the homologous segments of the structurally unchanged chromosomes are not separated from one another. For them it is a case of "nondisjunction". (2) **adjacent-2 distribution** (= homologous adjacent distribution; disjunctional distribution): chromosomes adjacent in the configuration with homologous centromeres are distributed to the same pole in meiosis I. The translocated segments and the homologous segments of the structurally unchanged chromosomes are separated from one another. For them it is a case of "disjunction".

ADP-ribosylation — a post-translational protein modification during which ADP-ribose is attached to the acceptor proteins as a monomer, oligomer, or polymer. Poly(ADP-ribosyl)ation is found mainly in the nuclei of eukaryotic cells, mono(ADP-ribosyl)ation is catalyzed mainly by prokaryotic enzymes.

ADP-r. utilizes NAD as a donor of the modification group. The ADP-ribose moiety of NAD is transferred to either an acceptor protein or amino acid or to another ADP-ribose moiety to yield an acceptor-linked monomer or polymer. The synthetic reactions are catalyzed by two classes of enzymes: (1) mono(ADP-ribose) transferases, and (2) poly(ADP-ribose) synthetases, which catalyze the initial ADP-ribosylation of protein and subsequent chain elongation.

ADP-r. is a mechanism for alteration of chromosomal protein structure leading to control of a variety of cellular functions, such as → DNA repair, gene expression, and cell differentiation.

ADP-ribosyl transferase — (ADPRT): a nuclear enzyme which catalyzes the formation of poly(ADP-ribose)-modified chromatin proteins by transfer of the ADP-ribose portion of NAD^+ to chromatin proteins, thereby synthesizing mono-, oligo-, and poly(ADP-ribose) which may be linear or branched. ADPRT is dependent on DNA for activity and is activated by breaks in the DNA.

adventitious embryony (Strasburger 1878) — a form of → apomixis (agamospermy); the production of seeds without a sexual process.

affinity — (1) nonrandom → assortment of unlinked markers by some sort of a. between nonhomologous chromosomes causing them to pass preferentially to the same pole of the → spindle in anaphase of the first or second meiotic division. A. of this type has been inferred from genetic studies in mice and yeast (Michie and Wallace 1953). (2) In the case of selective → fertilization, the genetically controlled mutual attraction of male and female gamete types differing as to their genetic constitution. A. is a measure of strength of attraction, while the velocity with which this attraction proceeds is called "reaction velocity" (Haustein 1955). (3) → differential affinity. (4) → terminal affinity.

agameon (Camp & Gilly 1942) — a species reproducing exclusively by → apomixis.

agamete — any noncopulating → germ cell (= spore). Agametes are formed as products either of meiosis (meio.-a., meiospores, tetraspores, gonia) or mitosis (mito.-a., gonidia). Reproduction accomplished by a. is designated as → agamogony, monocytogony, monogenic, or monocytogenic reproduction, agamic or agametic reproduction, agamogenesis, or sporogony (→ gamete).

agamic — reproducing (→ reproduction) asexually.

agamic complex — a plant complex typically including a series of → polyploid sexual, facultatively or obligatory agamospermous microspecies. These are largely the result of various hybridizations among sexual diploid and sexual polyploid members of the complex.

agamogenesis — asexual → reproduction (→ gamogenesis).

agamogony (Hartmann 1904) — asexual → reproduction. The development of a new individual from a single cell. A. in unicellular organisms proceeds in principle in three different ways: (1) Simple fission of the cell the result of which are two roughly equal fission (division) products. (2) "Budding", in which only a small part of the cell is constricted off. (3) "Multiple fission", in which the → nucleus divides several times before the cytoplasm disrupts into as many parts as there are nuclei present.

A. is the sole mode of reproduction of some amoeba and one phase of the complicated (sexual) reproduction cycle of sporozoa. Widely differentiated processes of a. lead to specialized germ cells which are called → agametes and lead without fertilization to the formation of a new individual. These agametes can be formed as the immediate result of meiosis (in **Ectocarpus**, the mosses, the basidiomycetes, and angiosperms) and give rise to haploid organisms or stages of the life cycle as constant elements of the sexual cycle. Agametes can also be produced mitotically by haploid or diploid organisms and in this case yield individuals with the same chromosome complement as the original form.

agamont (Hartmann 1904) — the asexual individual, or the asexual generation, in whose agametangia the agametes are formed (→ gamont).

agamospecies (Turesson 1929) — an apomictic population whose members are of common origin (→ apomixis).

agamospermy — a form of → apomixis: the formation of seeds without sexual processes through → adventitious embryony, diplospory, or apospory. The process can begin autonomously or after pollination (→ pseudogamy). In pseudogamous apomixis the pollen does not make any genetic contribution to embryo formation, but is necessary to initiate the growth of the ovary or to fertilize the endosperm nucleus.

The various embryological pathways of a. are alike in bypassing both meiosis and fertilization in the cell leading to the new embryo. This results normally in the formation of seeds containing embryos which are genetically identical with the maternal parent.

In some plant groups, a. replaces sexual reproduction completely (obligate a.); in other plants some seeds form by agamospermous processes and some by sexual processes (facultative a.).

AG complex (Correns 1928) — the complete set of factors assumed to be responsible for the formation of sexual organs and expression of sexual differences which are not sex-determining. Those factors acting to produce male sex organs are designated by A, those acting to produce female sex organs by G. A and G are carried by the autosomes (or the cytoplasm).

In a diploid cell, A and G each occur twice as AAGG, in a haploid cell only once each. Thus each cell possesses the possibility of development in both directions. Which of the two directions is chosen depends on the specific → sex realizers which are designated M and F. The sex realizers act on the alternative → reaction norm evolved through the → bisexual potency and the AG-system and this is fixed in a male or female direction (→ sex determination).

aggregation chimera — a → chimera in animals produced experimentally by aggregation (or fusion) of cleaving embryos (→ injection chimera).

agmatoploidy (Malheiros-Gardé 1950) — increase of chromosome number by → fragmentation of chromosomes with diffuse or multiple → centromere organization leading to → pseudopolyploidy or → pseudoaneuploidy.

agmato-pseudopolyploidy (Battaglia 1956) — → pseudopolyploidy.

akaryotic — without a → nucleus (→ eukaryotic; prokaryotic).

akinetic — = → acentric.

akinetoplastic — → kinetoplast.

alarmone (Stephens et al. 1975) — any of a variety of signal metabolites synthesized in response to a specific metabolic stress. Alarmones act homeostatically to increase the probability of a cell's survival. The current candidate for such a signal is diadenosine 5′, 5‴-P^1, P^4-tetraphosphate (ApppppA), a compound made by some bacterial → aminoacyl-tRNA synthetases at low tRNA concentrations or in the presence of micromolar concentrations of Zn^{2+}. In eukaryotic cells, this compound has been implicated in the regulation of → DNA replication and as a general a. Alarmones are made in response to heat, ethanol, and a wide variety of oxidants.

albomaculatus (Correns 1904) — ref. to a. → variegation or mottling ("status albomaculatus") in plants caused by genes or extrachromosomal hereditary determinants and consisting of an irregular distribution of white and green regions (white-green variegation), (→ paralbomaculatus).

aleuroplast — a → leucoplast in which protein granules predominate as a storage product.

algeny (Lederberg 1966) — the (purposeful) altering of genes in the body cells or in germinal tissues, or the introduction of desired genes from outside (= "genetic engineering" or "genetic surgery").

alien addition line (Leighty and Taylor 1924; O'Mare 1940) — a line with an extra chromosome (monosomic addition) or an extra chromosome pair (disomic addition) from a related species (→ substitution line; alien substitution line).

alien substitution line (Kattermann 1938; Unrau et al. 1956) — a line in which an alien chromosome or a pair of alien chromosomes from a donor species replaces one chromosome or a pair of chromosomes of the recipient species (→ alien addition line; substitution line). If a pair of alien chromosomes compensates for the missing chromosomes they are considered → homeologous to the pair they replace.

allele (Johannsen 1909) — one of two or more alternate forms of a → gene occupying the same → locus in a particular chromosome or linkage structure and differing from other alleles of the locus at one or more → mutational sites. Members of a set of alleles are mutually exclusive → genetic markers and arise by → gene mutation. Their activity is concerned with the same biochemical and developmental process. A haploid organism or phase of the life cycle has a single representative of each a., a diploid two, a polyploid more than two from the total number existing in the population as a whole.

With respect to a given locus a diploid may be homo- (two identical alleles in the pair of homologous chromosomes) or heterozygous (two different alleles). If the phenotype of a heterozygous allele pair (A/a) resembles that of the homozygous pair (A/A), A is said to be dominant, a recessive. Any new allele may be characterized by its genotypic effect when it is homozygous and by its phenotypic effect when combined in heterozygotes with a sister allele, that is, they are known by their actions and interactions (→ gene interaction). On this basis the following classes of alleles may be distinguished: (1) **Amorphs** (Muller 1932): inactive alleles which act as → genetic blocks to normal biosynthesis. In the broadest sense an amorph may be an a. that is inoperative, one that fails to produce a measurable effect, or even the absence (→ deletion) of a gene. (2) **Hypomorphs** (Muller 1932): alleles that function imperfectly in comparison with → wild-type alleles; sometimes known as → leaky genes. They are prevalent among induced gene mutations. (3) **Hypermorphs** (Muller 1932): alleles that produce an excess amount of a product (the antithesis of hypomorphs), whereby excess is defined in terms of the wild-type alleles. (4) **Antimorphs** (Muller 1932): alleles with an action opposite to that of the wild-type; this type is rare and poorly defined. (5) **Neomorphs** (Muller 1932): alleles whose action differs qualitatively from that of the wild-type allele. Heterozygotes of neomorphs commonly display the allelic products of both alleles. (6) **Isoalleles** (Stern and Schaeffer 1943): alleles producing only very slight differences in phenotypic expression making their identification bound to the use of special methods. (7) **Unstable alleles** are those that mutate at unusually high frequencies somatically or in the germ line. Chromosome breaks and rearrangements often occur in the vicinity of an unstable a. Unstable alleles may result from the insertion of → transposable genetic elements at or near the affected loci.

The individual alleles of a group consisting of more than two are called a **series of "multiple alleles"** (Morgan 1914). A diploid individual or phase of the life cycle may be homozygous for any one allele or heterozygous for a combination of any two. → Segregation in meiosis results in gametes with only a single a. The dominance relations among multiple alleles vary from one group to another. For some groups of alleles every homo- and heterozygous → genotype produces a different → phenotype. In others, the alleles may be arranged in a descending series in which every a. is dominant over all alleles below it. The number of genotypes in a series of multiple alleles is $\frac{1}{2}$ [n×1 (n+1)] where n is the number of alleles of the group [for four alleles: $\frac{1}{2}(4\times5) = 10$]. Multiple alleles of a gene affect similar parts or processes in an organism and are indicated by attaching distinguishing letters or numbers as superscripts to the base symbol (→ genetic nomenclature).

Different alleles may produce detectable effects upon the variability at any stage in the life history of an individual and may modify the expected phenotypic segregation ratio so that certain classes of offspring are in excess, in reduced frequency, or are totally absent (→ lethals).

Two recessive mutations of independent origin are allelic if the heterozygote of m/m′ has a mutant phenotype. If among the products of meiosis in heterozygotes — but not in homozygotes — an occasional revertant or an a. different from the two originals arises, the two alleles are different. Different alleles are called "nonidentical" (Demerec 1956) (in contrast to identical alleles) when recombination has been obtained between them. The "nonidentity" refers to the fact that each represents a mutation at a different mutational site. Another term is "heteroallelic" (Roman 1956) for combinations of two alleles of different mutational origin which yield the wild type by recombination or other mechanisms (in contrast to "homoallelic" combinations). (→ genetic complementation; genetic recombination; pseudoalleles).

Strictly neutral allele pairs are those that under no circumstances make any relevant difference to → fitness and therefore can not contribute to adaptive or progressive evolution. They may include redundant → codon differences and allele pairs that are meaningful codon differences but lead to redundant amino acid substitutions in parts of a protein that do not affect the protein's function.

Quasi-neutral allele pairs affect fitness, but do so to such a small extent that their fate can be accurately predicted by the "neutral theory".

allele center (Reinig 1938) — = → gene center.

allele frequency — a measure of the commonness of an → allele in a population, i.e., the proportion of all alleles of that gene in the population that are of this specific type (= → gene frequency). Allele frequency cline is the term used to describe the a. frequencies of polymorphic genes statistically correlated with spatial or temporal variation of environmental factors.

allele shift — an alteration of allele frequency caused by selection which can lead in extreme cases to complete (± rapidly occurring) loss of an allele depending on its initial frequency. Similar effects may be a consequence of genetic drift.

allele trend — a directed change in allele frequency within a given time.

allelic (Johannsen 1909) —ref. to the relationships between → alleles for the same locus.

allelic complementation — the production of a nonmutant phenotype when two independent mutations at the same gene locus, but on different homologous chromosomes, are introduced into the same cell. A.c. can be accounted for in terms of the subunit structure of multimeric protein molecules and occurs in a wide variety of microorganisms and in some higher organisms (→ genetic complementation).

allelic exclusion (Pernis et al. 1965) — the phenomenon that only one member of an allelic pair of genes controlling the production of functional → immunoglobulin light and heavy chains is active in B cells, even in animals doubly heterozygous for antibody genes. A.e. ensures that all of the antibody molecules produced by a cell have the same specificity (→ isotypic exclusion).

allelic interaction — → allele; gene interaction.

allelism (Johannsen 1909) — the relationship between → alleles.

allelobrachial — of → chromosome structural changes in which the arms of homologous chromosomes (→ heterobrachial; homobrachial) are involved.

allelogenous (Vandel 1938) — of females producing only males or only females in different broods (→ arrhenogeny; thelygeny).

allelomorph (Bateson and Saunders 1902) — = → allele.

allelosomal — of → chromosome structural changes in which both homologs of a chromosome pair are involved (→ heterosomal; homosomal).

allelotype (Strandskov 1950) — a term which corresponds on a population basis to the → genotype (which always relates to a single individual) and refers to the genetic composition — allele frequency — of a → breeding population. Changes in the a. of a population can come about by → mutation, → selection, sampling error (→ genetic drift), and → migration processes in case of incomplete → isolation of populations. With incomplete isolating mechanisms immigrant individuals may increase the frequency of particular alleles or introduce new ones, while emigrating individuals may decrease the frequency of particular alleles or, in extreme cases, produce total loss of a particular allele.

allelozyme — any of genetically variant enzymes (→ isoenzyme).

Allen's rule — the generalization that, in a widely distributed species of warm-blooded animals, populations in colder regions have shorter extremities than those in warmer regions.

allocycly (Wenrich 1916; Darlington and La Cour 1940) — ref. to differences in → chromosome coiling which are shown by chromosomes, or even whole chromosome complements, and may be environmental, genotypic, or cellular in origin.

Regions most frequently subject to allocyclic behavior are the centromeres, telomeres and → nucleolus organizing regions, while the chromosomes most often allocyclic are the → sex chromosomes, → B-chromosomes, and → sex-limited chromosomes of certain Diptera.

The commonest type of a. is that in which the concerned segments appear overcondensed in comparison to regions which follow the standard cycle. This so-called **positive → heteropycnosis** may manifest itself during interphase or during nuclear division with the exception of the meta-anaphase in which the regions following the standard cycle are also maximally condensed. The opposite allocyclic state, **negative heteropycnosis**, is less frequent and is detectable at all times during cell division. Different regions of the chromosomes can follow different types of allocyclic behavior and the same region can behave differently in different cycles of nuclear division.

allodiploid (Serra 1948) — of cells or individuals in which one or more chromosome pairs are exchanged for one or more pairs from another species. In the extreme case, two → genomes, one from either species, are combined in the zygote. (1) Allodiplomonosomes: Allodiploids in which two foreign single chromosomes are present in the chromosome complement. (2) Allomonodiplosomes: Allodiploids which have exchanged one pair of chromosomes for a pair of foreign chromosomes. If there are two exchanged pairs, the individuals are called "allodidiploids", etc.

allogamy — = → cross-fertilization or exogamy.

allogeneic — of different genetic constitution.

allogenes (Leung et al. 1985) — families of related DNA sequences that code for variant forms of a standard tRNA isoaccepting species; the functionally specialized tRNA's may be required only transiently during differentiation.

allogenic — (1) of bacterial → transformation if the induced change produces a new character different from the characters of either the strain being transformed or of the strain from which the inducing DNA was isolated (Ephrussi-Taylor 1951). A. transformations are distinguished from "autogenic" transformations in which the induced change corresponds exactly with the source of the transforming DNA. (2) Of recombination by → crossing over within a → codon which may occur in heterozygotes bearing at the same site codons for different amino acids or different codons for the same amino acid and give rise to a codon for a third, different amino acid (Zamenhof 1967). Allogenic recombination closely resembles → gene mutation in its effects. (3) Of two alleles or chromosome segments of independent origin.

allogenotope (Cohen et al. 1983) — a polymorphic restriction fragment, i.e., one that is generated by action of → restriction endonuclease from the cellular DNA of some individuals but not from that of others.

allograft — a tissue graft (= allogeneic graft; homograft) from a donor of one genotype to a host of another genotype, host and donor being members of the same species. Host and donor are called allogeneic.

allohaploid (Ivanov 1938) — → haploid (= allopolyhaploid).

alloheteroploid (Sharp 1934) — (1) of → heteroploid strains, individuals, or cells whose chromosomes derive from various chromosome sets (→ autoheteroploid); (2) of strains, individuals, or cells which contain additional chromosomes of another species (Serra 1948). Those individuals with one extra foreign chromosome are allomonoheteroploid (→ allodiploid).

alloiogenesis — → alternation of generations.

allolysogenic — → lysogenic.

allometric —ref. to growth, whereby the growth rate of one part of the organism differs from that of another part or of the rest of the body.

allomixis — = → cross-fertilization.

allopatric (Poulton 1903; Mayr 1942) — of populations or species which are mutually exclusive, but usually inhabit adjacent geographic regions. Genotypically distinct allopatric populations of a species are called geographic races, subspecies, or local varieties. Gene exchange between allopatric populations is either limited or completely absent (→ sympatric).

A. speciation is the term used to denote the differentiation of geographically isolated a. populations to the point where they are considered to be taxonomically discrete → species (→ evolution).

allophene (Hadorn 1955) — → allophenic.

allophenic (Hadorn 1955; Mintz 1962, 1967) — (1) of → characters (allophenes) which in a given cell system arise by intercellular gene action, i.e., through the action of genes manifest in other cell systems of the organism (Hadorn 1955). In such a case, a gene has produced an → "autophene" in one cell system and the allophene characteristic of another system is the direct or indirect result of the gene which is manifest in the first cell system (→ gene activation). For allophenes a nonautonomous behavior on transplantation is characteristic. If an autophene secondarily gives rise to one or a series of allophenes, this is called "relational pleiotropy" (Hadorn 1945, 1955) and does not represent a true case of → pleiotropy. (2) Of individuals who display an orderly arrangement of two or more concurrent, allelically alternative cellular → phenotypes, or allophenes (Mintz 1962, 1967). A. individuals (mammals) have been artifically formed by aggregating cleavage-stage blastomeres in vitro from embryos of different genotypes and tranfer of the composites to an "incubator" mother for further development. Such mosaic embryos may develop to form healthy and long-lived adults.

alloplasm (Meyer 1896) —those "cell organelles" which serve a special purpose and are not of regular occurrence such as the neuro- and myofibrils, the cilia, flagella, contractile vacuoles, and Nessel capsules (→ paraplasm).

alloplasmic male sterility (Lacadena 1968) — a type of → male sterility which results from combining artificially the cytoplasm of one species with the nucleus of another.

alloploid (Clausen et al. 1945) — of individuals (called "alloploids" or "amphiploids") which arise after natural or experimental ("synthetic alloploids") crossing of two (or more) species or genera and contain the (structurally and genetically different) chromosome sets of the particular parents, each set being present either once ("allodiploids") or in greater number ("allopolyploids").

In allopolyploids or amphiploids, each chromosome of the original allodiploid or amphihaploid hybrid is represented twice and such forms generally display characteristics of both the contributing parents. By formation of allopolyploids, fertile derivatives may be obtained from hybrids in which sterility was caused by chromosomal imbalance and meiotic irregularity. Meiotic behavior of allopolyploids is dictated by the homology relationships of the chromosome sets they contain. Some form only → bivalents and no → multivalents at → meiosis. In this case every chromosome finds one identical partner with which it pairs (→ chromosome pairing). There is no segregation; the allopolyploid breeds true for its original genotype. Other allopolyploids are multivalent forming. This may result from: (1) the presence of homologous chromosome sets in more than double dose (combination of, e.g., diploid and tetraploid species), or (2) the introduction of partly homologous (= homeologous) chromosome sets.

Segregation of the allopolyploids is the consequence in both cases. Generally, extreme genetical imbalance occurs among segregants of allopolyploids with numerous or complex multivalents in meiosis resulting in reduced fertility.

In conformity with these variations in their meiotic behavior, allopolyploids are classified as follows (Stebbins 1945): (1) **Genome allopolyploids:** Chromosome pairing in meiosis is exclusively in the form of bivalents. Pairing restricted to exactly homologous chromosomes ("homogenetic pairing") may be the consequence of the striking structural dissimilarity between the chromosome sets of the parents or may result from the activity of specific genes which prevent the pairing of partially homologous chromosomes. The sterility of the diploid hybrid from which the allopolyploid arose is usually complete in this category; fertility is fully restored with the production of the genome allopolyploid derivative. Once formed, this derivative is usually completely isolated by sterility barriers from its nearest relatives and may show all the criteria of a new species. (2) **Segmental allopolyploids:** Chromosome pairing in meiosis is characterized usually by bi- and multivalent formation. The parental chromosome sets combined in the diploid hybrid are in this case partially homologous (homeologous) and correspond in a considerable number of segments permitting "heterogenetic pairing". The sterility of the diploid hybrid is thus not pronounced. In contrast to genome allopolyploids, segmental allopolyploids show more or less extensive segregation, the products of which are markedly imbalanced genetically and infertile. Hence segmental allopolyploids are usually unstable. Through segregation they may give rise to stable segregants, which in regard to their chromosome constitution are either autoploid, genome alloploid, of stably segmental alloploid. (3) **Autoallopolyploids:** Chromosome pairing in meiosis is in form of bi- and multivalents as in (2) because homologous chromosome sets are present in more than double dose. Polyploids of this type are possible from the hexaploid level upwards and combine characteristics of auto- and allopolyploidy.

allopolyploid (Kihara and Ono 1926) — → alloploid.

allosome (Montgomery 1906) — a chromosome (= sex chromosome) deviating in size, form, or behavior from the other chromosomes (→ autosome).

allosteric (Monod and Jacob 1961) — of an enzyme whose activity is altered when its structure is distorted by an organic compound at a nonsubstrate site. A. enzymes possess at least two distinct nonoverlapping receptor sites. One of these, the **active** (or catalytic) **site**, binds the substrate and is responsible for the biological activity of the protein. The other, the **allosteric site**, is complementary to the structure of another low-molecular molecule, the so-called **"allosteric effector"**, which binds specifically and reversibly. This bond modifies the catalytic activity of the active site (selective increase or decrease) without the allosteric effect relying on a direct chemical connection between the substrate and the allosteric effector, and operates through introduction of conformational changes in the enzyme molecule which in turn lead to changes in the catalytic site (allosteric transition).

Two classes of → a. effects are distinguished: (1) homotropic effects, i.e., interactions between identical ligands. (2) heterotropic effects, i.e., interactions between different ligands.

The general properties of a. systems may be stated as follows (Monod et al. 1969): (1) Most a. proteins are polymers rather than oligomers involving several identical subunits. (2) A. interactions frequently appear to be correlated with alterations of the quaternary structure of the proteins, i.e., alterations of the bonding between subunits. (3) While heterotropic effects may be either positive or negative (i.e., cooperative or antagonistic), homotropic effects appear to be always cooperative. (4) Few, if any, a. systems exhibiting exclusively heterotropic effects are known. Cooperative, homotropic effects are almost invariably observed with at least one of the two or more ligands of the system. (5) Conditions, treatments, or mutations which alter the heterotropic interactions simultaneously alter the homotropic interactions.

A. effects are assumed to play a role as elements of the negative feedback control and the end-product activation of enzyme activity (positive feed-back control). The allosteric sites of control point enzymes are genetically determined. Mutations can lead to breakdown of the allosteric site and this type of mutations conditions the collapse of the control of the metabolic path in question. The concept of the a. enzyme was developed to interpret a large number of cellular control systems including gene → repression (→ end-product inhibition).

allosteric effect — the effect of a modifier bound at a site of an → allosteric enzyme which is topologically distinct from the active site at which reaction is occurring and which influences the binding or activity at this site by indirect means.

allosteric effector — → allosterism, allosteric, autosteric effector.

allosteric inhibition — → isosteric inhibition.

allosteric preconditioning (Alpers and Paulus 1971) — → preconditioning.

allosteric site (Monod et al. 1963) — a topologically separate site of an enzyme which lacks catalytic activity but which can exert regulatory control over the active site.

allosterism — a mechanism which regulates enzyme activity by means of specific activators or inhibitors (allosteric effectors). These combine with the → allosteric enzyme at sites other than the substrate catalytic site. Such allosteric effectors are usually compounds of low molecular weight, and are believed to exert their influence by changing the conformation of the protein.

allosubstitution (Karpetschenko 1935) — → chromosome substitution.

allosuppressor (Young and Cox 1971) — any mutation which increases the efficiency of suppression exerted by a → nonsense suppressor (→ antisuppressor).

allosynapsis (Sharp 1943) — = → allosyndesis.

allosyndesis (Ljungdahl 1924) — in → polyploids and → aneuploids, meiotic → chromosome pairing of completely or partially homologous (homeologous) chromosomes which have been introduced into the zygote by different parental gametes at → fertilization (→ allosyndesis).

A. can be classified as: (1) Complete a.: All chromosomes are paired allosyndetically; (2) one-sided a.: The chromosomes derived from one gamete pair allosyndetically, those from the other remain unpaired; (3) complete auto-allosyndesis: All chromosomes occur part auto-, part allosyndetically paired.

allotetraploidization — the creation of artificial → allotetraploids.

allotopic (Krooth 1969) — of a mutation which imparts to one tissue a characteristic metabolic feature normally found in another tissue.

allotype — any of the antigenic specificities present in the immunoglobulins and other proteins of different groups of individuals of the same species and determined by different alleles of the same gene.

allotypic nuclear division (Strasburger 1905) — = → meiosis.

allozygous (Crow and Kimura 1970) — of an individual whose two homologous genes are → identical by descent, as opposed to autozygous individuals where the two alleles are of independent origin.

allozygote (McKusick 1973) — a genetic compound, i.e., → heterozygous for different mutant alleles (→ allele).

allozyme — any of the multiple forms of a multimeric enzyme the subunits of which are coded by alleles of the same gene (→ isozyme). Individuals homozygous for a particular allele produce a single protein species, but this enzyme will differ from one individual to another depending on the particular allele carried at the locus in question. Heterozygous individuals will, in general, show additional forms of the enzyme since they carry two different alleles.

alphoid DNA (Maio 1971) — a family of tandemly repeated simple DNA sequences (→ satellite DNA) found mainly at the → centromere of the → chromosomes of many primates. The fundamental monomer repeat length is ~171 base pairs. Within a species, the a. DNA families are complexed and may comprise up to 24% of the → genome.

alternation of generations (Hofmeister 1851) — the alternation of two or more → generations, reproducing themselves in different ways. By generation is understood that phase of the life cycle which lasts from one reproduction process to the next. Most often there is an alternation of sexual and asexual, in rare cases (Protozoa) of different asexual modes of reproduction. If the alternation between the modes of reproduction is not firmly established the term "facultative a. of g." is used; if the alternation proceeds according to an exactly determined sequence the phenomenon is spoken of as "obligatory a. of g.". The following types are distinguished (Hartmann 1939): (1) **Primary a. of g.**, i.e., the alternation between generations which produce → gametes and those which reproduce through → agametes. (a) Homosphasic a. of g.: all generations are equal in respect to their degree of ploidy (number of chromosome sets per nucleus) (= monophasic a. of g., homogeneous a. of g., homologous a. of g.). (b) Heterophasic a. of g.: alternation between generations which produce gametes and those which produce agametes, exhibiting various degrees of ploidy. A. of g. and → alternation of nuclear phases are coupled with one another (= diphasic a. of g., heterogeneous a. of g., antithetic a. of g., botanic a. of g.). (2) **Secondary a. of g.**, i.e., the alternation between generations which produce gametes and those which reproduce by parthenogenesis or by vegetative reproduction. (a) metagenesis: generations producing gametes alternate with generations which reproduce vegetatively. (b) heterogony: generations producing gametes alternate with generations which reproduce by parthenogenesis (= cyclic parthenogenesis, alloiogenesis, heterogamy). (3) **Asexual a. of g.**, i.e., the alternation of two asexual modes of reproduction differing in the products of reproduction (Grell 1956).

alternation of nuclear phases — the regular change of chromosome number connected in the case of sexual → reproduction with the fusion of the → gametes (zygote formation) and the subsequent → meiosis. After meiosis and before → fertilization of the ovum, the organism is in the haplophase (gametic chromosome number), after fertilization and before meiosis in the diplophase (zygotic chromosome number). (1) **Zygotic a. of n.p.**: meiosis takes place in the zygote ("zygotic meiosis"). This type of a. of n.p. is restricted to → haplontic organisms. (2) **Gametic a. of n.p.**: meiosis takes place at the time of → gametogenesis ("gametic meiosis"). This type occurs with → diplontic organisms. (3) **Intermediary a. of n.p.**: meiosis occurs during → sporogenesis of a diploid generation in the case of → diplohaplontic organisms.

alternative distribution — adjacent → centromeres of a ring or chain configuration (composed of four or more chromosomes) in → translocation heterozygotes are orientated towards opposite spindle poles and alternative chromosomes are distributed to the same cell pole in the first meiotic division. With ring formation, a.d. is more frequent than → adjacent distribution; in the case of chain formation the two types of distribution are of about the same frequency. (→ adjacent distribution).

alternative RNA splicing (Tamkun 1984) — the production of more than one mature mRNA by alternative processing of a common → transcript. In alternative RNA s. different specific pairs of splice sites are joined while other functional splice sites are bypassed, according to cell type and developmental stage, i.e., it generates diversity of → gene expression (→ RNA splicing) since it allows a gene to encode multiple protein isoforms.

altruism — in biological terms, a type of behavior that benefits another individual at some cost to the altruist; costs and benefits are measured in terms of individual → fitness. Reciprocal a. implies the exchange of altruistic acts between unrelated individuals as well as between relatives. A. is a group phenomenon and requires at least a group of two. Genes inducing such behavior in their bearers will be positively selected if the recipient of the a. is sufficiently closely related so that the genes themselves enjoy a net benefit. When a. is considered in an evolutionary context, cost and benefit are evaluated in terms of reproductive success.

A trait which is detrimental to the fitness of the individual who expresses it but which confers an advantage on the group is called an "altruistic trait".

alu-equivalent family — a set of DNA sequences in any mammalian → genome, related to the human → Alu sequence family.

alu-like sequence — any of a class of short repetitive DNA sequences dispersed throughout the mammalian genome and distinguished from other repetitive sequences in that they are substrates for the Alu restriction enzyme. These elements range in length from 135 bp (mouse) to 300 bp (man) and have long consecutive runs of deoxyadenosine [poly(dA)] at their 3′ ends. Most elements are flanked by direct repeats of sequences 7 to 20 nucleotides in length. Alu-like sequences are transcribed by RNA polymerase III which initiates transcription at a site immediately downstream from the direct repeat and reads through the second direct repeat, yielding a product co-linear with the integrated element. A region of the Alu consensus sequence is similar to the sequence at the origin of DNA replication of simian virus 40, polyoma, and BK viral DNA's. This region contains the pentanucleotide GAGGC. Roles in regulation of gene expression, transcription, and recombination events have been proposed, but little is known about the functions of Alu-like sequences.

alu sequence family — a family of interspersed repetitive dimeric DNA sequences in the human genome. It is composed of about 300000 members, most of which are 300 base pairs long, end in a dAMP-rich sequence (3′ end of each monomer) and are usually flanked by short (7 to 20 bp), direct repeat sequences. The → promoter is located within the repeat suggesting that complementary DNA copies, synthesized by → reverse transcriptase using the Alu RNA transcripts are inserted into various loci of the genome at staggered DNA breaks (→ retroposon). The Alu sequences are transcribed by → RNA polymerase III and the dAMP-rich sequence is thought to help mediate the transposition event. Alu-sequences have been implicated in the genesis of unequal → crossing over leading to → deletions in specific genes.

alysogenic — ref. to prophage-free sensitive mutants which occur in → lysogenic bacteria.

amber mutation (Epstein et al. 1963) — a suppressible genetic change which results in the → terminator codon UAG in messenger RNA (→ ochre mutation; opal mutation). The corresponding polypeptide chain is terminated at the site of the amber codon if no phenotypic or genotypic suppression (→ amber suppressor) occurs.

A. mutations in bacteria and bacteriophages can be phenotypically suppressed by 5-fluorouracil (FU). This may be due to incorporation of FU into messenger RNA, where it induces coding errors. When incorporated into the amber triplet UAG in the mRNA, FU will occasionally cause misreading of the amber codon as CAG leading to insertion of glutamine at this site and completion of the peptide chain (FU rescue of amber mutants).

amber suppressor — a → suppressor mutation which allows amino acid insertion into the polypeptide at the site of an amber codon (UAG) in messenger RNA. (→ nonsense suppressor).

ambiguity — the phenomenon in which a given → codon can be translated in more than one way (→ genetic code; coding ambiguity; translational ambiguity).

ambivalent (Huxley 1955) — ref. to genes with both advantageous and disadvantageous action for the organism. A distinction is made between genes with "full ambivalence" and genes with "dosage ambivalence". In the first case, the positive and negative effects occur both with heterozygotes (Aa) and with homozygotes (aa); in the second, the deleterious effects are confined to the homozygotes (aa) or are more strongly marked there than with the heterozygotes.

amDNA — acronym for → antimessenger DNA.

ameboid motility — the crawling of cells over a substrate by means of pseudopods as observed throughout the animal kindom.

ameiosis — the failure of meiosis and its replacement by a mode of nuclear division which does not involve a reduction of the chromosome number.

ameiotic parthenogenesis (White 1945) — → parthenogenesis in which → meiosis has been entirely suppressed.

Ames mutagenicity test — a widely used short-term testing system (→ mutagenicity testing) in which the induction of reversion to histidine prototrophy is measured in several special mutants of **Salmonella typhimurium**. The sensitivity of this test was greatly enhanced by the incorporation of plasmid pKM 101 into the tester strains, because many compounds are mutagenic only in the presence of this plasmid.

amino acid activation — the first step of protein synthesis (→ genetic translation), whereby an amino acid reacts with adenosine triphosphate. The reaction is catalyzed by a specific → aminoacyl-tRNA synthetase. The first reaction product is an amino acid adenylate, in which the necessary energy is conserved for the second stage — the attachment of the amino acid to a specific → transfer RNA molecule. The adenylate remains bonded to the enzyme molecule. The same enzyme also catalyzes the transfer reaction.

amino acid attachment site — → transfer RNA.

amino acid replacement — the presence of different → amino acids in equivalent positions of homologous proteins.

amino acids — the building blocks of the → proteins. There are 20 common a.a. of natural occurrence, as L-stereoisomers. They all have the same basic structure but differ from one another according to the side group (R).

The linear order of the a.a. in a peptide or protein is called an a.a. sequence and is genetically determined (→ genetic translation).

aminoacyl adenylate — in protein synthesis an activated compound (AA →AMP) which is an intermediate in the formation of a covalent bond between an amino acid and its tRNA acceptor (→ aminoacyl-tRNA synthetase).

aminoacylation — a mechanism catalyzed by → aminoacyl-tRNA synthetases and involving the formation of enzyme-bound aminoacyl adenylates followed by tranfer (charging) of the corresponding amino acid onto its cognate tRNA. The specificity of the a. is one of the essential requirements for the correctness of protein synthesis. Incorrect a. reactions are characterized by very low rates and strongly depend upon experimental conditions (ionic strength, presence or absence of organic solvents, and pH). A. of noncognate tRNA's is called mis-aminoacylation.

aminoacyl transferase — two soluble protein factors (TI and TII) which are involved in the transfer of amino acids from tRNA to ribosome-bound polypeptides in mammalian systems. **Aminoacyl transferase I** catalyzes the interaction between aminoacyl-tRNA's and 80S ribosomes, placing the aminoacyl-tRNA at the A site of the ribosome. **Aminoacyl transferase II** aids in the translocation of the peptidyl-tRNA at the ribosomal A site to the P site (→ peptidyl transferase).

aminoacyl-tRNA — the product of the reaction between aminoacyl adenylates and the appropriate → transfer RNA. The reaction is catalyzed by specific → aminoacyl-tRNA synthetases. The selection of aa-tRNA by a mRNA-programmed ribosome is one of the key reactions determining the fidelity of protein biosynthesis which exceeds 99,95%. Binding of aa-tRNA to → polysomes is catalyzed by protein → elongation factors in both prokaryotes and eukaryotes.

aminoacyl-tRNA synthetase — any of about 20 specific synthetases which activate the 20 amino acids to form bound aminoacyl-adenylates as the initial step in the synthesis of proteins (→ genetic translation). The activated amino acid is then transferred to a specific → transfer RNA (tRNA) to form the aminoacyl-tRNA. The specific aminoacyl-tRNA's are recognized by a triplet base sequence in the → messenger RNA and the activated amino acid is then incorporated into a polypeptide.

The reactions catalyzed by an aminoacyl tRNA synthetase are: (1) AA (amino acid) + ATP + E (enzyme) $\overset{Mg^{2+}}{\rightleftarrows}$ E · AA − AMP + PP. (2) E · AA − AMP + tRNA ⇄ E + AA − tRNA + AMP.

In addition, the aminoacyl-tRNA synthetases are important in the regulation of → RNA synthesis, enzyme → repression, and intergenic → suppression.

Aminoacyl-tRNA synthetases usually exhibit molecular weights between 90,000 and 180,000 Da. They are either single polypeptide chains or oligomeric proteins of apparently identical subunits. There are indications that these enzymes are present in the microsomal or membrane fractions of the cell, i.e., do not float free in the cytoplasm.

All the enzymes differ from each other in the rate of reaction (PPi exchange as well as aminoacylation of tRNA), composition, or electrophoretic mobility, in addition to recognizing the cognate tRNA.

The majority of a.-tRNA synthetases recognize two or more tRNA species. The specificity of the various synthetases for their substrate tRNA's could be related to the base sequence of the tRNA, to its overall geometry, or to both factors.

aminoacyl-tRNA transferase (Rich and Rajbhandary 1979) — any of a special class of enzymes which transfer amino acids from charged tRNA's to a variety of acceptor molecules without the involvement of a decoding mechanism and ribosomes. The acceptor molecules can be divided into three classes: (1) An intact protein, in which case the amino acid is added to the N-terminus of the protein; (2) a phosphatidyl glycerol molecule, in which case the enzyme catalyzes the formation of aminoacyl esters of phosphatidyl glycerols that are components of → cell membranes; (3) an N-acetyl muramyl peptide, an intermediate in the synthesis of interpeptide bridges in bacterial cell walls.

amitosis (Flemming 1882) — nuclear division by a process other than → mitosis ("direct nuclear division") which in typical cases involves a dumbbell-shaped cleavage of the cell nucleus, in the course of which no chromosomes are recognizable, nor is a → spindle formed.

A. occurs inter alia in ciliates, certain protists, and in specialized animal tissues. Part of the phenomenon described by a. involves so-called → nuclear fragmentation.

amixis (Burnett 1956) — a mode of reproduction (→ heteromixis; homomixis) in haploid organisms where the essential events of sexual reproduction are absent but the pre-conjugation and post-meiotic events normally associated with sexual reproduction may occur (for the same situation in higher plants the term apomixis is applied).

amniocentesis — a procedure for diagnosing genetic abnormalities in utero. By piercing the abdominal and uterine walls with a needle a little amniotic fluid, containing fetal cells, is taken from a pregnant woman. From these cells it is possible to definitely establish whether or not the developing fetus is affected with certain specific genetic and chromosomal defects. A. has made for a significant advance in → genetic counselling.

Three types of disorders are suitable for intrauterine diagnosis by a.: (1) Chromosomal disorders (culturing of cells and determining their karyotype). (2) Autosomal or X-linked recessive diseases due to deficiencies of a specific enzyme that is normally present in cultured amniotic cells. (3) X-linked recessive conditions for which no biochemical lesion can be detected in amniotic fluid, but for which determination of fetal sex will distinguish between girls (who will be unaffected) and boys (who have a high risk of being affected).

If a defect is present, the pregnancy can be terminated. A. is generally not recommended in pregnancies with less than a 1% recurrence risk for a certain disease.

amorph (Muller 1932) — → allele.

amphiagamospecies (Turesson 1929) — a → species consisting of a group of facultative sexual biotypes whose reproduction is essentially apomictic (→ agamospecies; apomixis).

amphiapomictic (Turesson 1926) — of biotypes which propagate partly by sexual → reproduction and partly apomictically (→ apomixis).

amphidiploid (Navashin 1927) — of tetraploid (genome allotetraploid) species hybrids (amphidiploids) whose somatic cells contain the diploid chromosome complements of both parent species (syn. allotetraploid, didiploid, double diploid). The diploid hybrid from which the amphidiploids originate is generally sterile owing to the significant nonhomology between the chromosome sets and the consequent difficulty with → chromosome pairing during meiosis. With the doubling of the chromosome number the sterility barrier is removed. Amphidiploids commonly possess a better viability and competitive ability than the original species and the diploid hybrid. These properties frequently result in the establishment of amphidiploids as new plant species (→ alloploid).

Table 1. Pattern of segregation in the case of amphidiploidy

Locus 1	Locus 2	
$+/+$	$+/+$	Without segregation
$+/+$	$+/a_a$	Segregation for a_a without phenotypic effect
$+/+$	a_a/a_a	Without segregation
$+/a_1$	$+/+$	Segregation for a_1 without phenotypic effect
$+/a_1$	$+/a_2$	Segregation for a_1 a_2 with phenotypic segregation ratio of 15:1
$+/a_1$	a_2/a_2	Segregation for a_1 with phenotypic segregation ratio of 3:1
a_1/a_1	$+/+$	Without segregation
a_1/a_1	$+/a_2$	Segregation for a_2 with phenotypic segregation ratio of 3:1
a_1/a_1	a_2/a_2	Mutant-phenotype, without segregation

Each genetic locus of an a. form which was represented in both parent species occurs four times in the amphidiploids and shows a corresponding → segregation of the → alleles. If a dominant allele (+) is present, a recessive mutation shows no phenotypic effect, provided a second does not occur, since in "amphiploidy" → multivalents are not formed in meiosis (see Table 1). For a recessive gene to manifest itself phenotypically, two independent mutations of the locus are necessary. No segregation results for genes which do not occur in both parent species.

amphigamy (Renner 1916) — (1) the fusion of two sex cells and the formation of conjugated pairs of nuclei (→ "dikaryophase"). If a. immediately follows a → karyogamy, the process is referred to as amphimixis (Renner 1916); (2) the normal fertilization process (Battaglia 1947).

amphigony (Haeckel) — → reproduction by → fertilization, as opposed to → parthenogenesis.

amphihaploid (Zukov 1941) — of "haploid" (allodiploid) types produced from amphidiploids (allotetraploids) C Olsson and Hagberg 1955).

amphikaryon (Boveri 1905) — the nucleus of the zygote produced in the course of → fertilization.

amphimixis (Weismann 1891) — the fusion of two gametes in sexual → reproduction in contrast to → apomixis. A. is cross-fertilization if the two gametes are produced by dioecious organisms (→ automixis).

amphiorientation — → centromere orientation.

amphiplasty (Navashin 1928) — the phenomenon that → nucleolar organizers of one species suppress the activity of the organizers of another species in hybrids (= nucleolar dominance).

amphiploid (Clausen et al. 1945) — of individuals (called "amphiploids") which are either segmental allopolyploid, genome allopolyploid, autoallopolyploid or aneuploid (by addition or substitution of alien single chromosomes) in type (→ alloploid, allodiploid).

amphitelic (Bauer et al. 1961) — → syntelic.

amphitene (Janssens 1905) — = → zygotene.

amphithallic (Lange 1952) — → heteromixis.

amphitoky (Leuckart 1857) — → parthenogenesis.

amphogenous(Vandel 1945) — female individuals which give rise to male and female offspring in the ratio of about 1:1 (→ monogenous; thelygenous).

amphoterotoky (Taschenberg 1892) = → amphitoky (→ parthenogenesis).

amplicon (Lima-de-Faria et al. 1973) — (1) a chromosomal (DNA) region which takes part in → DNA amplification and becomes amplified; (2) a cloning-amplifying vector (Frenkel 1981) which can be used for the cloning and amplifying of foreign genes. Such amplicons can replicate in eukaryotic cells in the presence of standard Herpes simplex helper virus and consist of repeat units of HSV defective genomes. Cotransfection of cells with helper virus DNA and monomeric repeat units of HSV defective genomes results in the regeneration of concatameric defective genomes composed of multiple reiterations of the repeats. Foreign DNA sequences can be introduced into the concatameric chimeric defective genomes that are efficiently packaged and can be stably propagated in serially passaged virus stocks.

amplification — the production of additional copies of a chromosomal DNA sequence, found as either intrachromosomal or extrachromosomal DNA (→ double minutes; DNA amplification; gene amplification).

amyloplast (Errera 1888) — in plant cells, a → leucoplast specialized for the synthesis and storage of starch (→ plastid). Amyloplasts commonly have a less highly developed internal system of membranes than → chloroplasts. The starch is always stored in the stroma, either without or accompanied by only a few short → thylakoids. Internal membranes, together with the stroma, become progressively reduced as starch deposits increase in size.

anagenesis (Rensch 1947) — a mode of → evolution characterized by cumulative changes (production of new organs, structural types, and building plans) in an evolutionary lineage in the course of time without subdivision in the lineage (= phyletic evolution).

According to Huxley (1957) a. is a comprehensive term for all forms and degrees of biological improvements in the course of evolution from specific adaptation to general organizational advances (→ cladogenesis, stasigenesis).

anaphase (Strasburger 1884) — → mitosis, meiosis.

anaphase movement — → chromosome movement.

anaphase separation — during anaphase of → mitosis and → meiosis II, the disjunction of the chromatids of each → chromosome; during anaphase of meiosis I the separation of chromosomes paired as → bi- and multivalents. In all cases the disjunction results in causal participation of the → spindle.

anaphragmic — ref. to mutations which by removal of an inhibiting influence lead to an increased enzyme activity in the mutant.

anauxotrophic — = → prototrophic.

anchorage dependent (McPherson and Montagnier 1964) — of any cell in culture which requires attachment to a solid substrate for growth.

aneucentric — ref. to → chromosome mutations in the course of which chromosomes with more than one → centromere (dicentric, tricentric chromosomes) are formed from monocentric (= eucentric) chromosomes.

aneuploidogen (Oshimura and Barrett 1986) — any agent that induces → aneuploidy (= aneugen, trisomigen).

andro-autosome (Yamamoto 1938) — an → autosome carrying → sex realizers for male character (→ gyno-autosomes).

androdioecious (Darwin 1877) — → dioecious.

androecious (v. Uexküll-Syllenband 1901) — of plants having only male flowers (→ gynoecious).

androecium (Roeper) — the male reproductive organs of a plant; the stamens of a flower taken collectively.

androgenesis (Verworn 1891) — male → parthenogenesis, i.e., the development of a haploid embryo from a male nucleus (= ephebogenesis; merogony) as opposed to maternal haploidy (gynogenesis) originating from the egg by incomplete fertilization, from other nuclei of the embryo sac in plants, by degeneration of the fertilized or unfertilized egg, or from other nuclei or other embryo sacs in connection with → polyembryony.

In the case of a. the maternal nucleus is eliminated or inactivated subsequent to fertilization of the egg cell and the haploid individual (referred to as "androgenetic") contains in its cells the chromosome set of the male gamete only. A. is easily detected by the use of suitable cytological or genetic markers.

androgenous — producing only male offspring (→ gynogenous).

androgynodioecious (Darwin 1877) — → dioecious.

androgynous — ref. to plants bearing both stamens and pistillate flowers in the same inflorescence (= hermaphroditic).

andromerogony — the development of an egg fragment with chromosomes exclusively from the father.

andromonoecious (Darwin 1877) — → monoecious.

androsome — any chromosome occurring exclusively in the nuclei of the male → germ line. Androsomes are present neither in the somatic cells of either sex nor in the female germ line (= male limited chromosomes).

androsporogenesis (Battaglia 1955) — = → microsporogenesis.

aneugamy (Austin 1960) — an abnormal fertilization process, in which a polyploid ovum is fertilized by a normal sperm, or a normal ovum by a diploid sperm (→ polyandry, polygyny).

aneuhaploid (Kimber and Riley 1963) — → haploid.

aneuploid (Täckholm 1922) — (1) of cells, individuals, or populations having one or more whole chromosomes of a → euploid complement absent from or in addition to that complement. In the case of diploidy this type of chromosomal aneuploidy (Dyer et al. 1970) includes inter alia → nullisomics (loss of a chromosome pair, 2n − 2), → monosomics (loss of a single chromosome, 2n − 1), → trisomics (presence of a single extra chromosome of the standard set, 2n + 1), and tetrasomics (occurrence of an extra chromosome pair, 2n + 2). If the loss or addition concerns more than one specific chromosome or chromosome pair, the situation may be characterized by the terms "doubly monosomic" (2n − 1 − 1) or "doubly trisomic" (2n + 1 + 1), etc. (→ pseudoaneuploid).

Chromosomal aneuploidy represents a form of → heteroploidy and occurs in somatic or generative cells either spontaneously or following the action of physical and chemical factors by: (a) Loss of individual chromosomes in mitosis or meiosis leading to the formation of nuclei with hypoploid chromosome numbers. (b) Nondisjunction during mitosis and meiosis, whereby hypo- and hyperploid nuclei are formed. (c) Irregularities of the chromosome distribution in meiosis of polyploids, especially those with an uneven number of chromosome sets (triploids, pentaploids, etc.). (d) Multipolar mitosis with irregular chromosome distribution to the daughter cells; this can result in so-called "multiform aneuploidy" (Böök 1945) characterized by cells with various aneuploid chromosome numbers occurring in the same tissue.

The prefixes "hypo-" and "hyper-" applied to the haploid, diploid and polyploid number allow a classification of aneuploid chromosome number. (2) Of cells or individuals having one or more chromosome segments of a basic chromosome set absent from or in addition to one or more basic sets. This type of segmental aneuploidy (Dyer et al. 1970) includes heterozygosity for → deletions or → duplications where one or more acentric segments of a basic set have been gained or lost, or where centric fragments (sometimes forming new chromosome arm combinations, as in secondary and tertiary → trisomics), have been gained (→ chromosome mutation).

aneuploid reduction — the reduction of → genetic variability by decreasing the number of → linkage groups. A.r. is supposed to play a role in the evolution of plant species in which outcrossing prevails and which occupy pioneer habitats. Such conditions produce a high degree of heterozygosity on the one hand and require repeated colonization of new areas on the other hand (Stebbins 1966).

aneuploidy effect — the phenomenon whereby organisms with abnormal (aneuploid) dosages for chromosomal segments or whole chromosomes are usually adversely affected. A major implication of the a.e. is that chromosomal balance is necessary for normal development. Chromosomal balance also applies to → sex chromosomes, because X-linked genes are → dosage compensated, at least in forms with → XY systems of sex determination.

aneusomatic — ref. to individuals whose cells exhibit variable numbers of individual (mostly heterochromatic) chromosomes (B-chromosomes), or, generally, individuals in which both euploid and aneuploid cells occur.

anisogamete — → anisogamy; gamete.

anisogamy — the → gametes fusing during fertilization vary in respect to size, shape, and behavior. The distinction may include (a) the behavior at fertilization only, (b) size only, or (c) both size and shape (the most extreme case: oogamy). A. can occur in the form of → gametogamy, → gametangiogamy, or → somatogamy.

anisomeric (Sirks 1933) — ref. to nonequivalent genes which by interacting (→ gene interaction) produce a particular phenotype (→ isomeric).

Anlage — (1) in genetics, originally a "hereditary factor" (now replaced by → "gene"); (2) in embryology, the primordium or cell group from which a part of the organism develops.

annealing — the „incubation" of a mixture of DNA's in single-stranded form and the quantitative determination of the helical material formed. The method permits the detection of homologous regions in any two species of DNA, provided the genetic complement is relatively small as in bacteria and viruses.

annidation (Ludwig 1948) — adaptation of genotypes to different ecological niches. By a.a mutant capable of utilizing an ecological niche not available to the parental form may be maintained in the population even if it happens to be inferior to the parent in the ecological niche of the latter. A. is a factor in → evolution.

annulate lamella (Swift 1956) — a class of cytomembranes which occur as stacks of parallel, double-membraned elements each having the same dimensions as the → nuclear envelope and containing pore complexes identical to those of the nuclear envelope. The elements are arranged in highly ordered fashion with a regular spacing between them. A. lamellae occur in oocytes, spermatocytes, a variety of somatic cells, and some cancer cells; generally, they are indicative of actively growing cells and many function in the transmission and storage of genetic information from the nucleus. It seems certain that a. lamellae are directly or indirectly derived from the nuclear envelope. Such membranes occur in both the cytoplasm and the nucleoplasm. Those of the nucleoplasm are small, irregular, and also short-lived. In some instances ribosomes have been observed as being directly attached to the membranes of a. lamellae, indicating a role in protein synthesis.

annulus (Callan and Tomlin 1950) — nonmembraneous material associated with both the inner (nucleoplasmic) and outer (cytoplasmic) rims of the → nuclear pores. This material accentuates the pore perimeter and appears as rings lying on top of either pore margin. The electronopaque material constituting these rings is arranged in eight symmetrically distributed granular subunits with diameters between 100 and 200 nm (annular granules).

anormogenesis — a process of → development outside the realm of variability of the normal development, deviating from normal development (normogenesis) and resulting in malformation or lethality.

anorthogenesis — adaptive changes of evolutionary significance based on preadaptations (→ adaptation).

anorthoploid (Winkler 1916) — = → aneuploid.

anorthospiral — → chromosome coiling.

antephase (Bullough 1963) — → prosphase.

anther — that part of the stamen in which the pollen grains are produced (→ microsporogenesis).

anther culture — the cultivation of immature pollen grains formed by the meiotic division of pollen mother cells and regeneration of haploid plants.

antheridium (Bischoff 1835) — the male gametangium (microgametangium), within which in ferns, mosses, fungi, and gymnosperms, the male gametes are formed (→ archegonium).

anti-association factor — a protein which prevents the joining together of the two ribosomal subunits (→ ribosome). In bacteria, the a.-a.f. is → initiation factor IF-3, in eukaryotes, eIF-3. This factor binds to the smaller ribosomal subunit and prevents the larger subunit from binding.

antibody — any of the bifunctional protein molecules (immunoglobulins; Ig) which have evolved diverse recognition functions for foreign substances (antigens) and effector functions to clear these antigens from the (vertebrate) host. All a. are oligomers of a basic four-chain structure. The monomer consists of two sets of identical polypeptide chains. One chain is shorter (about 220 amino acids) and is called the light chain (L). The other (about 450 or 475 residues) is called the heavy chain (H). The L chain is associated with the amino terminal half of the H chain. The amino terminal regions are highly variable, while the carboxyterminal regions are one of several types within a given species. The variable (V) region is primarily responsible for antigen recognition. The constant regions (C) are responsible for a variety of effector functions such as complement fixation and receptor binding and define the chain types or classes.

A multigene system encodes antibodies. The formation of a. genes requires site-specific recombination of DNA (gene segments) that occurs during maturation of a.-producing cells and brings together the gene segments encoding the variable (V) and joining (J) regions of light chains and V, diversity (D), and J regions of heavy chains.

Known mechanisms of a. diversification are: (1) increase in diversity of a set of germ line V, D, J elements by accumulation of germ line mutations during phylogeny; (2) mechanism of diversification during ontogeny by combining the various V, D, and J elements at random; and (3) a mechanism of diversification by somatic mutation, also during ontogeny.

A. receptor sites are sites which bind antigens in a highly specific manner and may themselves bear antigenic determinants called idiotypic determinants or idiotypes. These can evoke a. production in animals belonging to the same species.

anticipation — the apparent tendency of certain diseases to appear at earlier onset ages and with increasing severity in successive generations.

anticlastogenic — of agents which under defined conditions can decrease the frequency of chromosome damage induced by → clastogens, i.e., chromosome-damaging agents.

anticodon — a contiguous sequence of three nucleotides in → transfer RNA (Table 2) next in sequence to an invariant uridine in the a. loop of the molecule. The a.serves to recognize the appropriate triplet that constitutes the → codon in → messenger RNA (→ genetic code). Mutations in the a. occur that change codon-anticodon pairing without affecting amino acid recognition by the tRNA molecule. Such mutations result in the "wrong" amino acid being placed in many polypeptide sequences during → genetic translation and may, therefore, be expected to be deleterious or lethal in many cases (→ suppressor tRNA).

Table 2. The 45 anticodons in the "universal" genetic code[a] of eubacteria, halobacteria, and eukaryotes. (Osawa and Jukes 1989)

GAA Phe	GGA or IGA Ser	GUA Tyr	GCA Cys
UAA Leu	UGA Ser		
CAA Leu	CGA Ser		CCA Trp
GAG or IAG Leu	GGG or IGG Pro	GUG His	GCG or ICG Arg
UAG Leu	UGG Pro	UUG Gln	UCG Arg
CAG Leu	CGG Pro	CUG Gln	CCG Arg
GAU or IAU Ile	GGU or IGU Thr	GUU Asn	GCU Ser
UAU Ile or $^{+}$CAU Ile	UGU Thr	UUU Lys	UCU Arg
CAU Met	CGU Thr	CUU Lys	CCU Arg
GAC or IAC Val	GGC or IGC Ala	GUC Asp	GCC Gly
UAC Val	UGC Ala	UUC Glu	UCC Gly
CAC Val	CGC Ala	CUC Glu	CCC Gly

[a]Not including differences from the universal code found in mitochondria, Mycoplasma, and ciliated protozoa, and "exceptional" ANN anticodons. Except for $^{+}$CAU ($^{+}$C is lysidine), modifications of the first anticodon nucleotide are not indicated. UAU is not present in prokaryotes and $^{+}$CAU is not present in eukaryotes.

→ Ambiguity in the pairing between the first base of the anticodons and the third base of codons is responsible for a definite pattern of the → genetic code (→ wobble). There are indications that replacement of uridine by 2-thiouridine in the a. may be an evolutionary device for preventing → mistranslation; 2-thiouridine is much less likely to form wobble base pairs.

anticodon-adjacent hypermodification (Jukes 1973) — hypermodification of the (purine) nucleotide immediately 3′ to the → anticodon in → transfer RNA which serves to ensure the fidelity of protein synthesis by preventing → wobble base-pairing interactions between the third anticodon and first codon position nucleotides.

antigen — any agent (substance) which when introduced into an animal elicits the production of → antibodies which bind specifically to the inducing substance (→ immunogenetics).

antigenic determinant — the portion of an → antigen which is recognized by the active site of an → antibody.

antigenic drift — in influenza A viruses, a type of antigenic variation associated with mutations resulting in minor alterations of the viral surface proteins (→ antigenic shift).

antigenic shift — in influenza A viruses, a type of antigenic variation associated with the emergence of pandemic virus strains (subtypes) bearing novel surface antigens. These viral subtypes are antigenically unrelated to previous strains. This change has been speculated to be the result of a recombinative event between a human virus and an animal virus which possesses different surface protein(s).

Within a particular subtype, smaller progressive changes in the properties of surface antigens, referred to as → antigenic drift, are also observed.

antigenic variation (Vickerman 1978) — in African trypanosomes, the sequential expression of genes coding for different variant surface glycoproteins (VSG's). When expression of some VSG genes is switched on, a newly duplicated copy of the expressed gene has been observed within the trypanosome genome, which is not found after the gene's expression is switched off again (antigenic switching). The duplicated copy is referred to as an expression-linked copy (ELC). It is speculated that the duplicative transposition (into a telomeric site) generating the ELC may itself be responsible for the control of VSG expression.

A.v. allows at least a segment of the parasite population to escape the host immune response.

anti-inducer — a compound that inhibits → operon induction caused by → inducers. Anti-inducers compete for inducer binding to free → repressors and also stabilize the repressor-operator complex. Both anti-inducers and inducers are able to bind either to free repressor or to operator-bound repressor.

anti-messenger DNA — a DNA strand (symbol: am DNA) copied from → messenger RNA by RNA-dependent → DNA polymerase (reverse transscriptase, reverse polymerase, revertase) by a process called → reverse transcription (= cDNA).

antimetabolite — a compound which is similar to an essential metabolite and binds to replace the metabolite.

anti-microtubule agent — any agent which favors the disassembly of → microtubules. These include physical agents, low temperature, and high hydrostatic pressure in addition to → tubulin binding drugs, such as colchicine and podophyllotoxin.

antimitotic agent — of agents that lead to the cessation of → mitosis either by decreasing the number of cells entering mitosis **(mitodepressive effect)**, by disturbing the → mitotic apparaturs **(mitoclastic effect)**, by stopping cell proliferation **(mitostatic effect)**, or by causing immediate cell death (→ spindle poison).

antimorph (Muller 1932) — → allele.

antimutagenic (Novick 1956) — of agents (antimutagens) that reduce spontaneous or induced mutation rates. Three types of a. agents are known: (1) agents that prevent the reaction of mutagens with the target molecule, DNA (desmutagenic agents); (2) agents that prevent the formation of mutagens from precursors; (3) agents that act after the induction of DNA damage and interfere with cellular functions.

The main mechanism by which prevention of the primary mutagenic lesions in DNA may be obtained are: (1) direkt interaction of the antimutagen and the mutagen; (2) increased detoxification by antimutagen-inducible enzymes; (3) inhibition of the metabolic activation of promutagens; and (4) adsorption of mutagens on high molecular compounds. (→ DNA repair; gene mutation).

antimutator DNA polymerase — a → DNA polymerase in organisms which are characterized by abnormally low spontaneous → mutation rates when compared with the wild type.

Two major differences characterize a. DNA p. and the normal enzyme: (1) A very high exonuclease activity associated with the mutant enzyme, and (2) a difference with respect to the higher K_m for triphosphates of the mutant as compared to normal enzyme. The reduced spontaneous mutation rate of organisms with the mutant enzyme is probably a consequence of a more effective monitoring system during DNA synthesis.

antimutator gene (Drake and Allen 1968) — a gene that causes a decrease in spontaneous and/or induced mutation rate (→ mutator gene). Enhancement of error-free → DNA repair pathways (→ adaptive response), or inhibition of an error-prone DNA repair pathway (→ SOS response) are expected to result in a. effects. A. alleles have been found which suppress both errors of replication and errors of incorporation (→ gene mutation). → Mutator and a. genes both affecting → DNA polymerase demonstrate the profound role of this enzyme in determining the spontaneous mutation rate (→ antimutator DNA polymerase).

antiparallel — term used to describe the opposite orientation of the two strands of a DNA double helix; the 5′ end of one strand aligns with the 3′ end of the other.

antipolarity (Ito and Crawford 1965) — the lowering by → polar mutations of enzyme production of the genes located on the side of a mutated structural gene proximal to the operator of an → operon (→ polarity).

antirepressor (Oppenheim et al. 1970) — a gene product capable of abolishing repression due to a → repressor (in lambda phage). The a. may inactivate the repressor, compete for the repressor binding site on the DNA, or allow the tanscription of the operons repressed by the repressor.

antirestriction — in bacteriophages, any of various mechanisms that allow overcoming host-induced modification (→ DNA modification-restriction) by, e.g., inhibition of the activity of → restriction endonucleases.

antisense DNA strand — of a strand of DNA that has the same sequence as the → messenger RNA (→ coding DNA strand).

antisense gene — a gene constructed in vitro by reversing the orientation of a portion of transcribed DNA, usually including the region encoding a protein, and placing this next to a transcription-control sequence (a → promoter). This gene cassette can be delivered to the target cell, resulting in genetic transformation and inhibition of expression of the target gene (→ antisense RNA).

antisense RNA (Itoh and Tomizawa 1980; Simons and Kleckner 1983) — any of the small, diffusible transcripts (antisense transcripts) that pair to specific complementary target RNA's and control their function or expression. Duplex formation between a. RNA and target RNA functions at the levels of → genetic transcription, → genetic translation (in both cases blocking expression of the protein gene product), or at the level of → DNA replication (regulating the frequency of replication at the origin when complementary to a given primer RNA). In most cases, the a. RNA and the target RNA are transcribed from opposite strands of the same DNA template by opposing → promoters and the a. RNA's function as negative factors (repressors). The a. RNA's are of varying length (70 to 110 nucleotides), and each has one or more stable stems topped by single-stranded loops (6 to 8 nucleotides in size). The stems are important for RNA stability, the loops are critical for initiation of pairing to the target RNA.

In prokaryotes, a. RNA's are natural components of regulatory systems and involved in, e.g., regulation of → DNA replication and → plasmid incompatibility, → transposition, regulation of phage reproduction, and autoregulation of → cAMP-receptor protein synthesis.

Artificially constructed a. RNA's are able to regulate → gene expression in prokaryotic and eukaryotic systems.

antisense transcript — → antisense RNA.

anti-Shine-Dalgarno sequence — (ASD); a sequence which is part of highly conserved 3′ end of the 16S → ribosomal RNA found in prokaryotes. The → Shine-Dalgarno sequence in messenger RNA interacts with the ASD region of 16S rRNA and this mRNA-rRNA base pairing plays an important role in the inhibition of protein synthesis.

anti-sigma factor — a protein that prevents recognition of initiation sites by the → sigma factor of → RNA polymerase.

anti-specificity factor — a protein synthesized during phage T4 infection that prevents normal termination of RNA synthesis by the specificity factor of → RNA polymerase (→ sigma factor).

antisuppression — the inactivation of allele-specific → suppressors by → antisuppressor mutations. A. may result from various mechanisms: (1) from mutations in ribosomal protein genes inactivating suppressors that are themselves due to a modification in the ribosomal structure, and (2) from mutations in genes that code for enzymes of → transfer RNA metabolism.

antisuppressor (McCready and Cox 1973) — any mutation (genetic modifier) that affects (antagonizes) the expression of → suppressor genes (nonsense suppressors), i.e., diminishes suppressor efficiency rather than to completely abolish it. The phenotypic effects of antisuppressors may be due to any of several types of mutations: (1) a mutant ribosomal or other component affecting codon recognition, tRNA binding, translation or polypeptide chain release; (2) an enzyme affecting → suppressor tRNA structure, e.g., one affecting the modification of bases or the maturation of the precursor tRNA; (3) a mutant → aminoacyl-tRNA synthetase.

antitermination — prevention of termination of → genetic transcription at particular → termination signals. A. allows synthesis of → polycistronic messenger RNA molecules by transcriptional readthrough. Control of the level of a. is a means of genetic regulation. It involves interactions between → RNA polymerase and accessory factors that lead to a modified transcription complex that no longer recognizes transcription termination signals (→ attenuation).

antitermination factor — any of a class of specific proteins that can reduce or prevent → transcription termination by modifying either the transcribing polymerase or sequestering the termination signal.

antiterminator (Roberts 1970) — a positive regulatory element (protein) involved in the regulation of → gene expression in prokaryotes. The a. can block → rho-dependent transcription termination and enables the modified → RNA polymerase to ignore certain transcriptional stop signals and to read through them to produce longer mRNA transcripts. Although the modified RNA polymerase is rendered insensitive to rho action, it still stops transcription at the first strong, rho-independent terminator it meets (→ attenuation).

antitubulin — any substance that binds to the subunits of → microtubules and interferes with their function. Colchicine, colcemid, podophyllotoxin, and vinblastine are antitubulins.

anucleolate — of mutants lacking → nuceloli.

ap DNA — depurinated DNA, i.e., damaged DNA in which purines have been removed from the sugar-phosphate backbone. Depurinization may occur via three known pathways: (1) spontaneous hydrolysis of the purine- sugar glycosylic bond; (2) enhanced spontaneous hydrolysis of alkylated purines in DNA due to labilization of the glycosylic bond caused by alkylation; and (3) removal of alkylated purines from DNA catalyzed by specific → DNA glycosylases (→ AP endodeoxyribonuclease; base-insertion repair).

AP endodeoxyribonuclease — any endonuclease that acts in an early step of the → excision repair pathway and introduces breaks at the 5′ side of apurinic and apyrimidinic residues (→ apurinic/apyrimidinic site). These sites originate either spontaneously or as the result of the action of specific → DNA glycosylases. AP endodeoxyribonucleases catalyze the hydrolysis of the glycosyl bond between the deoxyribose moiety and an altered or incorrect base in DNA.

Some of the AP endodeoxyribonucleases show intrinsic 3′ → 5′ exonuclease activity which would have an "antiligase" function in → DNA repair if incision and generation of a small gap next to the lesion are coupled events occurring prior to excision of the sugar-phosphate residue.

apoamphimict (Turesson 1926) — a biotype which reproduces predominantly by → apomixis but also sexually.

apocyte — a multinucleate cell or mass of cytoplasm.

apoenzyme — the protein part of an → enzyme to which a coenzyme can attach.

apogamety (Renner 1916) — → apomixis.

apogamògony (Fagerlind 1944) — → apomixis.

apogamy (de Bary 1877; Winkler 1908) — (1) = → apomixis (de Bary 1877). (2) The production of a sporophyte, not from the ovum but from other cells of the gametophyte ("euapogamy" according to Farmer and Digby 1907) or from cells of the sporophyte. A distinction is made between somatic (diploid) and generative (haploid) a.

apoinducer — a DNA-binding protein that switches on → genetic transcription by → RNA polymerase.

apomeiosis (Renner 1916) — sporogenesis without reduction of chromosome number during meiosis and giving rise to → apomixis.

apomixis (Winkler 1906) — in plants, the replacement of sexual → reproduction (= amphimixis) by various types of asexual reproduction which do not result in fusion of the → gametes. Owing to failure or suppression of → fertilization and → meiosis, no alternation of nuclear phases is coupled with a. provided it occurs regularly in successive generations as obligatory a. Obligate apomicts have a completely closed → recombination system and heterozygotic genotypes are preserved at the cost of evolutionary flexibility. In the case of facultative apomicts, apomictic and sexual modes of reproduction coexist. The mode of reproduction in animals equivalent to a. is → parthenogenesis.

A., like amphimixis, is a genetically controlled reproduction system and is brought about by mutations which modify the course of sexual reproduction to the point of nonfunction. This occurs inter alia by transformation of meiosis into an → apomeiosis, through degeneration of the megaspore mother cells, the spores or the embryo sacs, through hindrance of gamete fusion, the introduction of parthenogenetic egg cell development, or development of synergides.

Apomictic reproduction is divided into two main types (Gustafsson 1935):

(1) **Agamospermy:** All forms of a., in which reproduction proceeds via asexually produced embryos and seeds. Fertilization is always by-passed, meiosis usually. The embryo generally has the same chromosome number and the same genotype as the mother individual provided there is no → autosegregation. Mechanisms of this type are termed "agamogony" (Fagerlind 1940; Stebbins 1941), "apogamogony" (Fagerlind 1944), "gametophytic apomixis" (Stebbins 1950), and "apospory and displospory with following parthenogenesis" (Gustafsson 1935). In **"apospory"**, a diploid embryo sac is formed directly from a nucellus or integument cell; in **"diplospory"**, the embryo sac forms from an archespore cell, but meiosis is either omitted or does not lead to reduction of the chromosome number. From the diploid gametophyte produced by apo- or diplospory, embryos are produced either by division of the ovum (parthenogenesis) or another cell (apogamety). The simplest mode of agamospermy is adventitious embryony, whereby the embryos develop directly from the diploid sporophyte tissue and the gametophyte stage is omitted.

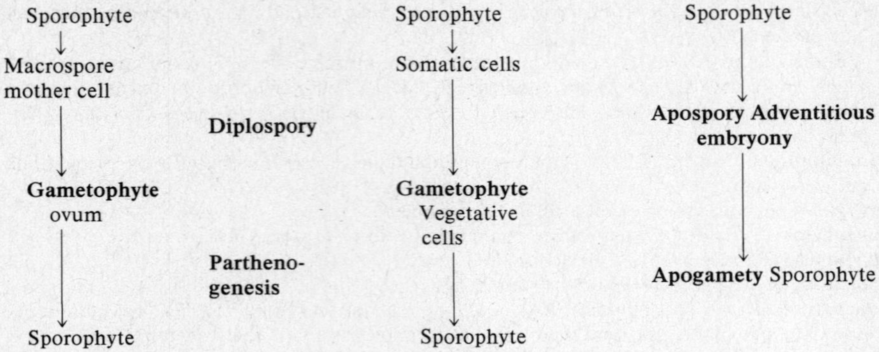

(2) **Vegetative reproduction:** the new individual arises from a group of undifferentiated or differentiated cells, in which case neither embryo nor seed are produced.

A. also occurs in animals (forms usually not higher than insects and Crustaceae), almost exclusively as → parthenogenesis.

apophase (Bullough 1965) — the period of postmitotic reconstruction, i.e., the time when the cell machinery is rebuilt and when the cell grows in mass to recover the size typical of its kind (→ prophase).

aporepressor (Jochimiak 1983) — a protein (the product of the trp R gene) that upon interacting with tryptophan forms a "repressor" with increased affinity for its → operators.

apospory (Druery 1886; Bower 1886) — → apomixis.

AP-site — any of the apurinic and apyrimidinic positions generated by loss of purine and pyrimidine bases from DNA. Such sites represent a substantial fraction of the lesions induced by X-rays, alkylating agents, and to a lesser extent UV irradiation. In addition, spontaneous release of bases from DNA occurs with an appreciable frequency. Base → excision repair has also been identified as a source of AP-sites. AP-sites provide substrates for → DNA repair systems. Endonucleases specific for AP-sites have been identified in a variety of prokaryotes and eukaryotes. → AP-endodeoxyribonucleases, in concert with → DNA polymerase and → DNA ligase activities, can effect the repair of depurinated DNA. Moreover, AP-sites may be substrates also for alternative repair systems which catalyze a direct insertion of bases at apurinic sites.

AP-sites impede chain elongation during → DNA replication; they can be bypassed by the insertion of a single adenine nucleotide followed by chain elongation with complementary deoxynucleotides.

aptitude — a specific physiological state of → lysogenic bacterial strains in which, under the action of inducing agents (→ induction), they react to produce infectious bacteriophages.

apurinic/apyrimidinic endodeoxyribonuclease (Verly and Paquette 1970) — = → AP endodeoxyribonuclease.

apurinic/apyrimidinic site — = → AP site

archebacteria (Woese and Fox 1977) — a diverse group of → prokaryotic organisms that represent a third line of evolutionary descent (a third primary kingdom) separate from the eubacterial and → eukaryotic kingdoms.

The a. consist of three main orders: (1) The extreme halophiles; (2) the methanogens that metabolize carbon dioxide to produce methane; and (3) the sulfur-dependent extreme thermophiles and the sulfolobales that grow under very acid conditions.

Among the archebacterial characteristics are uniquely modified nucleotides in → transfer RNA's, unique features of the secondary structures of each of the → ribosomal RNA's, different → ribosome shapes and sensitivities to antibiotics, DNA-dependent → RNA polymerase subunits that seem to be like those of eukaryotes, unique reverse gyrases that produce positive super-coils in → closed circular DNA and → nucleosome-like structures, unusual modes of gene transfer. A. show similarities to eukaryotes, such as → introns, unlinked 5S rRNA genes, and DNA-dependent RNA polymerases antigenically related to yeast DNA-dependent RNA polymerases.

archegonium — the female gametangium (macrogametangium) of ferns, mosses, fungi, and gymnosperms, within which the ovum is contained (→ antheridium).

archesporium (Goebel 1880) — a cell or group of cells formed by mitosis of the micro- and macrospore mother cells (→ microsporogenesis, macrosporogenesis).

arrhenotoky (Leuckart 1857) — → parthenogenesis.

ARS element (Struhl et al. 1979) — = → autonomous replicating sequence.

artioploid — of even-order polyploids (4n, 6n, etc.) as distinguished from odd-order polyploids (3n, 5n, etc.) or perissoploids.

A-site — → ribosome.

ascertainment — in human genetics, the finding or selection of families for inclusion in a genetic study.

ascogonium — the female → gametangium (oogonium) of anisogamous fungi.

ascospore — one of the spores (regularly eight are formed per → ascus) of ascomycetes, which are produced in the course of meiosis and a subsequent mitosis.

ascus — a membrane-enclosed sporesac of ascomycetes which usually contains eight ascospores. The a. is produced from an ascogenic hypha in which haploid nuclei, previously present as → dikaryons, fuse to form a diploid nucleus (karyogamy). Meiosis follows immediately and then, as a rule, one mitosis, so that from the four meiotic products, eight haploid nuclei are formed, each of which gives rise to an ascospore (→ basidium).

A-selection (Whittaker 1975) — a type of → selection process which favors conservation of → adaptation to consistently adverse environments (= adversity selection).

asexual — → reproduction.

assignment test (Ruddle et al. 1972) — in human genetics, a test with which the location of a particular gene on a specific chromosome is demonstrated by the concordance (association) between the presence or absence of the specific chromosome and the specific phenotype in many hybrid cell clones. (→ synteny test). A modified type is the regional a.t. which can be carried out in a number of ways all based upon the presence of chromosome rearrangements such as → translocations or → deletions (→ clone panel).

association — (1) = → chromosome pairing: (2) in human genetics, the nonrandom occurrence of two genetically separate traits in a population; association has its basis in mechanisms other than genetic → linkage.

assortative mating — → mating system.

assortment — (1) preference for → mating between similar individuals (= assortative mating; → disassortment); (2) in → meiosis, the normally random, in certain cases, nonrandom ("preferential") distribution to the cell poles of whole chromosomes (during anaphase I) contained in pairing configurations (→ chromosome pairing) and of chromatids (during anaphase II) resulting in random or nonrandom → segregation and → genetic recombination of genes.

(a) **Random or independent a.** is the result of the random orientation of the chromosomes with respect to one another and to the spindle poles in anaphase (→ chromosome movement) I and II and their corresponding distribution to the cell poles and meiotic products. It is prerequisite for random → segregation of the gene pairs. (b) **Nonrandom a.** is the preferential distribution of certain chromosomes to one of the cell poles and leads to distorted segregation ratios. It can be the result of different processes and in all cases is caused by a preferential → centromere orientation (affinity, meiotic drive, automatic frequency response). Pairing disturbances, heterozygosity for certain chromosome structural changes (→ complex heterozygosity), aneuploidy, and autopolyploidy are frequently associated with nonrandom a. of the chromosomes.

aster (Fol 1877) — a symmetrical star-shaped figure which surrounds the → centrosome during mitosis and meiosis.

asynapsis (Beadle 1931) — → desynaptic; asynaptic.

asynaptic (Beadle 1931) — ref. to chromosomes in the first meiotic division in which pairing (→ chromosome pairing) either fails completely or is incomplete. Asynapsis is recognizable by the presence of a number (varying from case to case) of → univalents and can be caused by incomplete homology of the chromosomes (e.g., in species hybrids), by gene mutations (asynaptic genes), or environmental influences. As a rule, asynapsis leads to the formation of hypo- and hyperploid meiotic products and frequently to fertility disturbances (→ desynaptic).

asyntenic (Renwick 1971) — → syntenic.

atelocentric (Levan et al. 1964) — of chromosomes whose → centromere is not localized at the end point (terminally) in contrast to telocentric chromosomes.

A + T/G + C ratio — = → base pair ratio.

A/T method — a → molecular cloning method for random DNA fragments into a vector molecule by generating complementary 3′ tails which provide the possibility to anneal foreign and vector DNA. The fragments are treated with lambda-exonuclease in order to generate 3′ single-strand tails; these are catenated by addition of deoxyadenosine residues by terminal transferase. The vector is cut at a unique → restriction site to generate 3′ tails again and these are catenated with deoxyadenosine and terminal transferase (→ dG · dC tailing).

ATP — adenosine triphosphate; a major carrier of phosphate and energy in biological systems which consists of the purine (adenine), the five-carbon sugar, ribose, and three phosphate groups.

ATP-protein phosphotransferase — a protein kinase which catalyzes the covalent linkage of the terminal phosphoryl group of ATP to various heterologous acceptor proteins such as egg white lysozyme or → histone. In vivo, endogenous **E. coli** proteins serve as phosphate acceptor after infection with phage T7, which codes for ATP-p.p.

AT-rich linker (Scherrer 1971) — in eukaryotic DNA, any of a large number of segments rich in adenosine and thymidine. Such linkers are of homogenous size (800 ± 200 bp) and may occur as AT-rich clusters consisting of three to ten AT-rich segments. They are probably of significance with regard to genome organization and function.

attached X chromosome (Morgan 1922) — a metacentric → X chromosome in **Drosophila**, which is an isochromosome and consists of two, normally acrocentric X chromosomes with a single centromere (→ compound X chromosomes; detached-X).

attachment plate (Kaufmann 1925) — a restricted area of the inner membrane of the → nuclear envelope to which the → chromosomes are attached during the → bouquet stage (leptotene of → meiosis). In most cases, this grouping seems to involve the → telomeres.

attachment site (Campbell 1962) — particular sites (att B and att P) in both bacterial and phage DNA molecules (lambda phage). The phage DNA integrates into the bacterial DNA by site-specific recombination between these sites. Both host DNA and phage DNA molecules are broken in these regions; the genomes then fuse to form a continuous structure in which the phage genome (→ prophage) is stably integrated into the bacterial genome. Reversal of integration results in the excision of the phage genome. Integration and excision are cases of → genetic recombination in which → crossing over is restricted to special (attachment) sites. A special phage recombination system, called INT, promotes recombination only at these sites. The recombination system is site-specific and the INT-allele is required for its function. Site specificity is due to the capability of the INT system to recognize structural features, presumably specific nucleotide sequences, of the a. sites. If the a. s. is missing or altered, efficient integration of phage DNA does not occur.

attenuation (Kasai 1974) — a regulatory mechanism by which the full potential for the transcription of bacterial → operons encoding amino acid biosynthetic enzymes is translationally coupled to specific control regions (attenuators) near the 5′-termini of the appropriate mRNA species (→ transcription attenuation).

attenuator (Jackson and Yanofsky 1973) — a variably efficient rho-independent transcription termination site located between the → promoter and the structural genes of bacterial amino acid biosynthetic → operons (other bacterial operons, including those for rRNA and for RNA polymerase may also be controlled by attenuators). The a. contains a region of hyphenated dyad symmetry with a high number of GC base pairs followed by an A+T-rich region. When these regions are transcribed, the RNA transcript (leader peptide RNA) can form regions of secondary structure containing a G+C-rich stem and loop followed by a run of uridine residues at the 3′ end of the transcript. Both RNA structures and the run of uridine residues are important for termination. The a. allows the cell to respond to the level of some metabolites in its environment by allowing transcription of the entire operon or terminating transcription before the structural genes (→ transcription attenuation).

Attenuators in eukaryotes may be regions of DNA that limit the frequency of transcription of a complete primary transcription unit through a process of premature termination.

autoallopolyploid (Kostoff 1939) — of cells or individuals whose → chromosome complement shows the characteristics of auto- and alloploidy (→ alloploid, autoploid). Generally, autoallopolyploids are hexaploid or of even higher ploidy and contain chromosome sets of different species, e.g., AAAABB or AAAABBBB, where each capital represents a chromosome set.

autobivalent (Hakansson and Levan 1957) — a → bivalent in first meiotic division, formed in each case from two structurally and genetically completely identical sister chromosomes produced in the course of a supernumerary reduplication. Thus, autobivalents can be formed if a supplementary reduplication of the chromosomes takes place in a premeiotic interphase nucleus or in early meiotic prophase. The number of autobivalents then corresponds with the somatic chromosome number of the species. The chromosomes formed by supernumerary reduplication are positionally associated with one another from the outset and preferentially pair together. In this way, their pairing forces are completely neutralized, so that for the special case of a meiosis with autotetraploid chromosome number, → multivalents are not generally formed (→ chromosome pairing; autoploid).

autogamy — obligatory → self-fertilization (automixis) within one undivided cell (gamont) by pairwise fusion of haploid nuclei; cell fusion is entirely lacking (→ pedogamy; parthenogamy).

A. is a particular mode of nuclear reorganization in **Paramecium** which takes place in single, unpaired individuals. The two → micronuclei divide twice by mitosis and seven of the eight resulting nuclei disintegrate. The remaining nucleus divides again mitotically and gives rise to a pair of identical haploid nuclei. In the absence of → conjugation the members of the pair of nuclei fuse. Fusion results in homozygosity of each of the newly formed diploid nuclei (→ killer).

autogenic (Ephrussi-Taylor 1951) — of → transformations in which the induced change corresponds exactly with the source of the transforming DNA (→ allogenic).

autogenomatic (Levan 1937) — of → chromosome sets which are strictly homologous and pair normally in → meiosis (→ heterogenomatic).

autogenous control — a gene product exerts a.c. when it either inhibits (negative a.c.) or activates (positive a.c.) expression of the gene coding for it.

autogenous protein (Goldberger 1974) — = → autoregulatory protein.

autogenous regulation (Goldberger 1974) — = → autoregulation.

autograft — a tissue transplant from one site to another in the same individual (→ homograft).

autolysis — digestion of a cell or cell material by some of the cellular enzymes (→ lysosome).

autointerference (Magnus 1954) — the phenomenon that undiluted passages of many of the major groups of animal viruses in cell culture result in a decrease in production of infectious particles. A. is due to the generation of → defective interfering particles which contain only part of the viral genome (deletions) and are dependent in their replication on the presence of homologous infectious virions as helpers (→ helper virus).

automatic frequency response (Brown 1963) — the automatic increase or decrease in the frequency of genes which are responsible for a change in the life cycle (including meiosis). Selection and random fluctuations are without effect on the a.f.r. which is a phenomenon parallel to → meiotic drive. Both the latter and a.f.r. affect the → gene frequency, but usually in different ways. While, in the case of meiotic drive there is a direct gene influence on the behavior of chromosomes in meiosis, the influence proceeds indirectly in the case of a.f.r. (Brown 1964).

automixis — obligatory self-fertilization by → autogamy, → paedogamy, or → parthenogamy, as opposed to amphimixis (cross-fertilization).

automutagen — any → mutagen, formed in the organism as a normal or abnormal product of metabolism, which may induce → gene and → chromosome mutations.

automutation (Imai 1936) — → exomutation.

autonomous replicating sequence (Struhl et al. 1979) — (ARS); any specific DNA sequence responsible for the initiation of → DNA replication, first observed in yeast and then in other eukaryotes. ARS elements support independent replication of a plasmid in host yeast cells and are thought to represent origins of → DNA replication. Several → mitochondrial and → chloroplast DNA's also contain ARS elements. No sequence in **E. coli** DNA can provide ARS activity.

The ARS-core and ARS-box, as well as a 14-bp sequence containing an 11-bp consensus (A/TTTTATPuTTTA/T) has been found for all ARS's and is essential for ARS function; the consensus sequence is a recognition site for a protein that activates the neighboring sequences for replication.

auto-orientation (Darlington 1936) — → centromere orientation.

autophagic vacuole — a type of digestive → vacuole in which some of the cell's own organelles (mitochondria, endoplasmatic reticulum, etc.) are destroyed. An a.v. (= cytolysosome, cytosegresome, composite body, autolysosome) is formed by the development of a membrane around a small area of cytoplasm containing one or more organelles. Like the → phagosomes containing material from outside the cell (→ endocytosis), the a. vacuoles fuse with primary → lysosomes and thus acquire hydrolases which digest the structure inside the a.v. Autophagy occurs in various types of cells under physiologic as well as pathologic conditions.

autophagocytosis — the segregation of cell organelles, such as → mitochondria and → endoplasmic reticulum, in membrane-bounded cavities which acquire acid hydrolase activity by fusion with preexisting secondary and/or primary → lysosomes within the → cytoplasm. Following acquisition of acid hydrolase, digestion of sequestered organelles occurs.

autophene (Hadorn 1955) — a genetically controlled character which is manifested by the cell's own genotype and which shows "autonomous" behavior in transplants and explants (→ allophene). Gene action leading to an a. is designated "intracellular"; an a. is the direct or indirect result of such action confined to the single cell. The term "mosaic pleiotropy" is used to describe the case of a gene influencing or controlling more than one a. (→ pleiotropy, gene action).

autoploid (Clausen et al. 1945) — of cells or individuals with chromosome sets characteristic of the species in question: the chromosome sets are homologous to each other and thus pair fully in meiosis (→ alloploid).

The terms autodiploidy, autotriploidy, autotetraploidy, etc., are used when there are two, three, four, etc., homologous chromosome sets per cell. The presence of more than two (diploidy) of the monoploid chromosome sets characteristic of the species is generally referred to as **autopolyploidy** (Kihara and Ono 1926). It is associated with a corresponding increase in the number of → genomes and → genetic information content.

Owing to the structural identity of the chromosome sets of autopolyploid forms, the homologous chromosomes can pair during meiosis and participate in → crossing over. Since each chromosome has more than one pairing partner, crossing over produces → multivalents (depending on the degree of ploidy, chiasma frequency and → chiasma distribution), i.e., paired groups consisting of more than two chromosomes, referred to as tri- (three paired chromosomes), quadri- (four paired chromosomes), pentavalents (five paired chromosomes per multivalent), etc.

The polyploidization of a diploid species may occur spontaneously or be induced by spindle poisons; it represents a → genome mutation and, like allopolyploidy, it is the result of the formation of a → restitution nucleus in somatic (somatic polyploidisation) or generative (generative or gametic polyploidization) tissue.

The autopolyploid derivative of a diploid species generally only differs quantitatively in → phenotype from its parent. If qualitative differences are present they can be due to the following causes: (1) An increased size of nuclei and cells can differentially influence the gene activity (Goldschmidt 1937); (2) the increase in the number of → alleles per locus associated with autopolyploidy can modify the → expressivity of individual genes; (3) if the diploid form was heterozygous for the allele pairs concerned, an increase in the number of chromosome sets may lead to modifications of the dominance relationships of the alleles.

As a rule, autopolyploids show fertility disturbances, for several reasons: (1) During meiosis, depending on the ploidylevel, the multivalent type, and the type of centromere orientation, there is an irregular distribution of the chromosomes, leading to the formation of unbalanced gametes and zygotes; (2) autopolyploidy is frequently associated with genetic-physiological disturbances which can adversely affect the functioning of the meiotic products independently of and in addition to the irregularities in chromosome distribution; (3) these physiological disturbances associated with autopolyploidy may of themselves cause irregularities in chromosome distribution.

The meiotic → **segregation of genes** of autopolyploid forms differs in many respects from the segregation of the allele pairs of diploid organisms. With autopolyploidy, the number of alleles of each gene is increased according to the ploidy level and gametes are formed which, in contrast to diploid forms, contain more than one allele per gene (homo- or heterozygotic). According to the number of dominant and recessive alleles of the particular locus, the genotypes are distinguished as "quadruplex and monoallelic" (AAAA or A^4), "triplex and biallelic" (AAAa or A^3a), "duplex and biallelic" (AAaa or A^2a^2), "simplex and biallelic" (Aaaa or Aa^3), "nulliplex and monoallelic" (aaaa or a^4), respectively, and in the case of autotetraploidy have the genetic constitution given in brackets. Autopolyploids show a so-called **"polysomic inheritance"**. Further, the segregation of genes of autopolyploids is affected by factors which play no essential part in diploids. Amongst these factors are the number and position of chiasmata in the multivalents, the distance of the particular locus from the → centromere, the behavior in distribution of the homologues paired as multivalents and the presence of → univalents.

(1) **Segregation in the case of autotriploidy:** In this case (as in all other ploidy levels with uneven numbers of chromosome sets) meiosis takes a very irregular course. During meiosis I, the chromosomes are present as tri-, bi-, and univalents. Since the distribution of each homologous group proceeds independently of other such groups, all possible gamete types between haploid to diploid chromosome numbers are formed. These two extreme types are the rarest, the unbalanced intermediate types most common. The actual ratios correspond in principle with those of → trisomics, but as a rule, are strongly perturbed by irregularities of the chromosome distribution and by chromosome elimination; they are thus difficult to predict. (2) **Segregation in the case of autotetraploidy:** In this case, i.e., for the occurrence of four homologous chromosome sets, the prediction of segregation ra-

tios is easier. The chromosomes of each group are present as → quadrivalents or, otherwise, as two bivalents, one tri- and one univalent, or four univalents. Assuming that the four particular homologous chromosomes are distributed to the poles in the ratio 2:2 in anaphase I, the theoretical segregation ratios for the various autotetraploid genotypes of a locus may be calculated. Theoretically calculated segregation ratios which do not take into consideration → crossing over between the concerned locus and the centromere are referred to as "chromosome segregation", those which consider a regularly formed crossing over in this region of the chromosomes and the thereby modified segregation of heterozygous loci, as → "chromatid segregation" (= double reduction). In the first case the segregation expectations result from the summation of the chance distribution combinations of the four homologs into in each case two to the poles. Table 3 shows the relative frequencies of the gamete types and zygote types of autotetraploid genotypes after selfing or, resp., crossing identical genotypes, calculated for pure "chromosome segregation".

Table 3. Segregation ratios (pure chromosome segregation) for the selfing of autotetraploid genotypes

Parent genotype	Gametes				Zygotes					
	AA	Aa	aa	Divisor	A^4	A^3a	A^2a^2	Aa^3	a^4	Divisor
Quadruplex A^4	1	—	—	1	1	—	—	—	—	1
Triplex A^3a	1	1	—	2	1	2	1	—	—	4
Duplex A^2a^2	1	4	1	6	1	8	18	8	1	36
Simplex Aa^3	—	1	1	2	—	—	1	2	1	4
Nulliplex a^4	—	—	1	1	—	—	—	—	1	1

On the other hand, if pure "chromatid segregation" is assumed and two sister chromatid segments with identical alleles of one chromosome become distributed after crossing over and corresponding chromatid distribution into the same meiotic product, modified expectations as to the gamete and zygote classes are realized (Table 4).

Table 4. Segregation ratios (pure chromatid segregation) of autotetraploid genotypes

Parent genotype	Gametes				Zygotes					
	AA	Aa	aa	Divisor	A^4	A^3a	A^2a^2	Aa^3	a^4	Divisor
A^4	1	—	—	1	1	—	—	—	—	1
A^3a	15	12	1	28	225	360	174	24	1	784
A^2a^2	3	8	3	14	9	48	82	48	9	196
Aa^3	1	12	15	28	1	24	174	360	225	784
a^4	—	—	1	1	—	—	—	—	1	1

The presence of "chromatid segregation" is implicit in cases where an unexpected excess of recessive and homozygous genotypes relative to "chromosome segregation" expectancies occur. The extent of "chromatid segregation" actually observed varies from locus to locus between certain limits and depends inter alia upon the crossing over frequency between the locus and the centromere, on the frequency of multivalent formation, and on the types of orientation and distribution of the multivalents.

Segregation ratios of autotetraploid forms generally lie between those theoretically expected for pure "chromosome segregation" and pure "chromatid segregation". The different phenotype ratios in the offspring produced by crossing autotetraploid genotypes with complete dominance of the allele A over a and pure "chromosome segregation", respectively pure "chromatid segregation" are summarized in Table 5.

Table 5. The phenotypic segregation ratios of autotetraploids (complete dominance of A and pure chromosome (I), or chromatid segregation (II)

Type of cross	Phenotypic segregation ratio in the offspring I	II
$A^4 \times A^4$	All A	All A
$A^3a \times A^3a$	All A	783A : 1a
$A^2a^2 \times A^2a^2$	35A : 1a	20.8A : 1a
$Aa^3 \times Aa^3$	3A : 1a	2.5A : 1a
$a^4 \times a^4$	All a	All a
$A^3a \times A^2a^2$	All A	130A : 1a
$A^3a \times Aa^3$	All A	51.3A : 1a
$A^3a \times a^4$	All A	27A : 1a
$A^2a^2 \times Aa^3$	11A : 1a	7.7A : 1a
$A^2a^2 \times a^4$	5A : 1a	3.7A : 1a
$Aa^3 \times a^4$	1A : 1a	0.87A : 1a

The special features of the segregation of autopolyploids compared with the segregation of diploids may be summarized as follows: (1) The occurrence of "chromosome" and "chromatid segregation". (2) The predominance of heterozygous genotypes. (3) The occurrence of several types of heterozygotes in the offspring of biallelic genotypes (e.g., A^3a, A^2a^2, and Aa^3). (4) The occurrence of heterozygous genotypes (e.g., A^3a) which lead to phenotypic segregation only after two generations.

autopolyploid (Kihara and Ono 1926) — → autoploid.

autoreduplication — applies to biological systems (organisms, cells, or subcellular units such as chromosomes, plastids, viruses, or genes) which generate the template for their own reproduction and duplicate themselves or produce mutant systems which are autoreduplicative in the same manner (syn. autoduplication, autoreproduction, identical reduplication).

autoregulation (Calhoun and Hatfield 1973) — a regulatory mechanism of gene expression (= autogenous regulation; self-regulation) in prokaryotic and eukaryotic organisms in which the product (protein) of a → structural gene modulates its own expression or the expression of the operon to which the structural gene belongs. The regulatory gene product has several functions; it may act as a regulatory protein and also as an enzyme, structural protein, or antibody. Often the regulatory protein is the first enzyme in the pathway and participates in control by → feedback inhibitions; its regulatory function is not dependent upon any catalytic or other function this protein may have and it is not involved in the transport or metabolism of an → inducer or → corepressor.

A. may provide a mechanism for amplification of gene expression, for severe and prolonged inactivation of gene expression, for buffering the response of structural genes to changes in the environment, and for maintaining a constant intracellular concentration of a protein, independent of cell size or growth rate (Goldberger 1974). (→ closed regulatory loop).

autorepressor system — a gene system (operon) one member of which does not code for an enzyme but represents a → regulator gene which encodes a → repressor protein. The repressor binds to the operator of the operon and blocks → genetic transcription (→ autoregulation; exorepressor system).

autosegregation (Gustafsson 1935) — the occurrence of changes in the chromosome complement during the formation of the egg cell by plants with gametophytic → apomixis. A. can lead to phenotypically changed, usually → aneuploid individuals and is a result of → subsexual recombination in strongly heterozygous but normally true-breeding apomicts.

autosome (Montgomery 1904) — any chromosome with the exception of the → sex chromosomes which are referred to as allosomes or heterochromosomes. The genes localized in the autosomes follow the mode of distribution of these chromosomes to the gametes in the course of meiosis; they show autosomal inheritance, those localized in the sex chromosomes show "sex-linked" inheritance.

autosteric effector — an effector which is part of or immediately adjacent to the active site of an → enzyme, in contrast to an allosteric effector localized at a distant site (→ allosteric).

autosyndesis (Ljungdahl 1924) — in contrast to → allosyndesis, the pairing (→ chromosome pairing) of complete or partial homologs (→ homologous chromosomes) contained in the same gametes at → fertilization, during the prophase of the first meiotic division (Stebbins 1947). A can occur as → homo- or heterogenetic pairing and is associated with the presence of polyploidy or aneuploidy. Three types of a. can be distinguished: (1) Complete a.: all chromosomes are paired autosyndetically; (2) one-sided a.: the chromosomes derived from one gamete pair autosyndetically, those from the other remain unpaired; (3) complete auto-allosyndesis: all chromosomes occur part auto-, part allosyndetically paired.

autotrophic — of cells or organisms that synthesize their cell components from simple chemical substances (e.g., ammonia, carbon dioxide, etc.) (= prototrophic). Obligatory autotrophs are limited to this way of life; facultative autotrophs are adaptable. Obligatory autotrophs always require carbon dioxide as a carbon source and either light or reduced inorganic compounds as a source of energy (→ auxotrophic; heterotrophic; prototrophic).

autozygous (Crow and Kimura 1970) — → allozygous.

auxocyte — (1) each cell in which → meiosis is incipient (= meiocyte or gonotocont); (2) the androcyte, sporocyte, or primary spermatocyte during the growth period.

auxotrophic (Ryan and Lederberg 1946) — of nutritionally dependent cells, individuals, or strains ("auxotrophs") whose growth depends on a supplement to the basic food medium (minimal synthetic medium) in which → prototrophs are able to grow freely. Auxotrophs arise from prototrophs by → mutations causing one ("monoauxotrophs") or more ("polyauxotrophs") → genetic blocks.

axial core — any of the linear electron-dense cores in leptotene chromosomes (→ meiosis) developed before the onset of → chromosome pairing. The a. cores later become the lateral element of the → synaptonemal complex.

axial filament (Haapala and Nokkala 1982) — a (protein-like) fiber which attaches along the entire length of each chromosome during the leptotene stage of → meiosis. The chromosomes are then zippered together during zygotene to form the → synaptonemal complex (→ chromatid core).

axoneme — (1) the basic microtubular structure (→ microtubule) of → cilia and flagella. The a. is surrounded by the ciliary membrane and is embedded in the ciliary matrix. The fundamental structure is 9 + 2, i.e., nine pairs of peripheral microtubules and two single and central microtubules. (2) The axial filament complex of the mammalian spermatozoon which represents the motor apparatus of the sperm tail. The sperm a. consists of two central microtubules surrounded by a row of nine evenly spaced doublet microtubules, i.e., the same 9 + 2 pattern characteristic of cilia and flagella throughout the plant and animal kingdom.

axostyl — in flagellates, an intracellular bundle of several thousand cross-bridged → microtubules which propagates a wave of bending from its anterior to its posterior end. Motility of flagellates results from undulations in the a.

B

backcross — the crossing of a heterozygote (→ heterozygous) with an individual with a genotype like one of its parents. The first, second, etc., b. generations are symbolized by B_1, B_2, etc. The second and following backcrosses are made by crossing of individuals of the B_1-, B_2-generations, etc. with the same parent.

background genotype — → genotype.

background radiation — the radiation dose to which the organism is subject under natural conditions. It is a mutagenic factor (→ mutagen) and consists of two compenents: (1) **Cosmic radiation:** this consists of elementary particles (protons, alpha-particles, and heavy particles) which come form outer space, meet the layers of the atmosphere and there give rise to photons and various types of elementary particles. The total dose per unit time is dependent upon the height above sea level, the atmospheric pressure, and the geomagnetic latitude. (2) **Terrestrial radiation** from the natural radioactivity of the ground and the environment including the radioisotopes taken into the body. It is subject to great variation in different geographical zones and is essentially dependent upon the rock formation.

The radiation dose received by humans due to b.r. lies between 2−4 milliroentgens per week. At present, 300 milliroentgens per week is accepted as the maximum permissible dose.

back mutation — a heritable change in a mutant → gene (→ gene mutation) resulting in a → revertant which has regained the enzyme or function that was lost with a so-called "forward mutation" (→ transition, transversion). A true back mutation restores the original nucleotide sequence which had been changed by forward mutation. The observation of a wild type phenotype in a mutant line does not necessarily mean that this reversion has occurred by an actual "reversal" of the original mutational event, that is, by true back mutation.

True back mutation can be simulated by (Green 1959): (1) Unequal → crossing over in the presence of a tandem duplication or by crossing-over between → pseudoalleles with positive or negative interference; (2) → gene conversion; (3) → position effects in consequence of chromosomal structural changes; (4) → suppressor mutations.

back selection — an experimental procedure, usually based on drug selection, by which one selects for hybrid segregants that have eliminated a marker that had previously been used to isolate → hybrid cells.

bacterial conjugation — the pairwise joining of male (donor) and female (recipient) bacteria. B.c. encompasses a wide variety of DNA (undirectional) transmission systems which may subsequently alter the genotype of recipient cells as a result of recombination (→ F plasmid).

bacterial sporulation — the conversion of a bacterial cell from the vegetative to the sporulating state as a result of cellular differentiation. Sporulating cells undergo significant changes in morphology and physiology. They synthesize enzymes and → messenger RNA not found in vegetative cells and many vegetative enzymes and mRNA molecules disappear during early b.s. These changes are the result of expression of a new set of genes and the turn-off of certain genes active in the vegetative cell.

bacterial transformation (Griffith 1928; Avery et al. 1944) – the intraspecific (homospecific) and interspecific (heterospecific) transfer of → genetic information by means of "naked" extracellular DNA in bacteria. The donor DNA is bound to the competent recipient cell by interaction with a DNA-binding protein associated with a DNA entry nuclease in a membrane-bound protein complex consisting of two DNA-binding monomers and two nuclease monomers (in **Bacillus subtilis**). During entry, one strand of the donor duplex DNA is degraded, the complementary strand enters the recipient cell as single-stranded fragments. Inside the cell, these fragments interact with the recipient genome and may become integrated into the recipient's chromosomal DNA. Replication of the integrated new information may result in segregation of the transformants.

Heterospecific b.t., i.e., transformation by DNA from closely related species, is of very low efficiency as compared to homospecific b.t.

bacteriocin (Grazia 1925; Jacob et al. 1953) — any substance (protein) produced by bacterial strains which, after adsorption at specific receptors on the surface of sensitive bacterial cells, is capable of killing these cells.

Bacteriocins represent a heterogeneous group of substances ranging from simple protein molecules to high molecular weight particles resembling those of phages or parts of phages. In every case, however, the part responsible for bacteriocidal action is the protein. The resistance and the ability to synthesize bacteriocins are genetically controlled by an accessory genetic element called → bacteriocinogenic factor which replicates in the bacteria as an autonomous → plasmid without physical association with the bacterial chromosome. Transfer of "bacteriocinogenesis" between bacteria of different systematic units takes place via → conjugation and → transduction.

Bacteriocins produced by **Escherichia coli** are called → colicins, those of **Pseudomonas aeruginosa**, pyocins; **Pasteurella pestis** produces pesticins. The mechanism of killing action differs from one b. to another. All activities of bacteriocins are exerted by b. molecules attached externally to the bacterial surface at specific receptors on the bacterial cell wall. Phage receptors and b. receptors are often the same and a mutation can render a bacterium resistant to a phage and one or more bacteriocins.

bacteriocinogenic plasmid — any of a class of → plasmids (= bacteriocinogenic factor) which produce → bacteriocins in many bacterial species. B. plasmids are extrachromosomal, covalently closed circular DNA molecules carrying the genetic information for the production of bacteriocidal proteins which inhibit one or more essential processes such as DNA replication, genetic transcription, or energy metabolism.

bacteriophage — a virus whose host is a bacterial cell. Bacteriophages (= phages) may have a complex morphology. They consist essentially of a core of genetic material (→ DNA or RNA) carrying the particle's → genetic information which is surrounded by a protein coat. Phages exhibit differences in their life cycles and may be grouped as either **virulent** or **temperate**. The introduction by infection of the genetic material of virulent phages into a susceptible host results invariably in the death and dissolution of the cell with the release of (100 – 10.000) new phage particles (**lytic cycle**). Temperate phages can become integrated into the bacterial chromosome at specific points and replicate with it as a unit, the viral functions then being indefinitely unexpressed (**lysogenic cycle**). In the integrated stage the viral material is called a → "prophage" and each bacterium of the "lysogenic" clone carries a copy of the prophage in question. Once integrated into the bacterial chromosome, the viral genes can be considered additional bacterial genes and studied by the usual methods of bacterial genetics. The lysogenic cycle can become converted into a lytic one spontaneously or by → induction. In that case the prophage (→ episome) is released from the bacterial chromosome, multiplies independently, and the cycle ends with the liberation of infectious phages by lysis of the host cell.

During the lytic cycle one pool of "vegetative" nucleic acid and another pool of structural units (such as head, sheath, and tail proteins) are produced. By withdrawing nucleic acid and proteins from both pools essentially at random (→ host-induced modification) mature phage particles are assembled. Within the pool, nucleic acid molecules with different contents of genetic information can give rise by interaction and → genetic recombination to recombinant molecules. Mature phages may incorporate genetic material from the bacterial host cell and transfer it to another cell that they infect after their release from the lysed cell. This process is called → transduction.

The genetic information of a b. is usually contained within one linkage structure or phage chromosome (→ segmented genome) which may be partly → heterozygous in some cases. Infection of a host cell with phages carrying different → genetic markers at homologous loci of the linkage structure constitutes "phage crossing".

The susceptibility of a bacterium to phage infection is primarily dependent on whether or not the phage can attach to specific attachment sites, receptors, on the cell surface. Male-specific DNA phages attach to the tips of → sex pili (F- or I-pili). For male-specific RNA phages nothing is known about their receptors other than they attach to pili.

The **host range** of phages is normally limited to a single bacterial species or at most to very few related species. This limitation is the result of one or both of two general barriers: (1) An adsorption barrier (the b. fails to recognize adsorption sites on the bacterial cell surface), and (2) a barrier that interferes with normal b. development and can be expressed at several levels and/or in several fashions (a common one is that of nuclease-mediated restriction).

Defective bacteriophages are particles which contain either all or some of the normal b. components but fail to form → plaques on any known host. Defective bacteriophages may be divided into three groups: (1) Particles which contain almost exclusively DNA of the bacterial host; (2) particles containing phage-specific DNA; and (3) particles which may not contain DNA and which resemble b. tails without a head.

Defective bacteriophages (prophages) may have arisen by progressive addition or deletion of phage-specific genetic information. Defective prophages, as their intact counterparts, may help to protect the bacterial cell against virus infection. The presence of a defective prophage can be detected: (1) If it confers on its host immunity to superinfecting phages of the homologous type, and/or (2) if it recombines with a genetically marked superinfecting phage.

bacteriophage MU — a very efficient → transposon that replicates by an active process of → transposition. Two phage-coded proteins A and B are essential for this process. Other phage-coded functions are involved in modulating the efficiency of transposition. The terminal repeats of b. Mu are only two base pairs in lenght (→ mutator phage).

bacteriophage resistant — of bacteria being resistant to bacteriophage infection. The mutation of Gram-negative bacteria to b. resistance often involves a change in some component of the outer membrane, which either is the receptor itself or interacts with it. The components which have been observed to change are either proteinaceous or lipopolysaccharide in nature.

bacteriostasis — inhibition of bacterial multiplication without immediate killing.

balance (Bridges 1922) — → genetic balance.

balanced lethals (Muller 1917) — → lethal factor.

balanced load (Muller 1950) — → genetic load.

balancer chromosome — any chromosome that is structurally different from normal-sequence chromosomes. In crosses involving b. chromosomes and their normal sequence homologs cross over products are inviable, crossing over appears to be suppressed, and the chromosomes involved are transmitted to the progeny intact (→ CLB-method). Large → inversions or appropriately constructed combinations of reciprocal → translocation may be used as b. chromosomes.

Balbiani ring (Beermann 1952) — a large RNA → puff of → polytene chromosomes which is present during a greater portion of larval development in the chironomid family of **Diptera**. This structural modification of specific loci is remarkable for its large size and characteristic ring shape. The B.r. structure and its extreme degree of puffing can be understood as the result of the unraveling of a thick, folded cable into individual filamentous elements, probably in the form of extended loops.

Balbiani ring granule (Beermann and Bahr 1954) — any of the numerous ribonucleoprotein (RNP) particles in → Balbiani rings. These particles are assembled from nascent mRNA and nuclear proteins. Similar particles have been found in some → puffs. The finished B.r. particles apparently arise from filamentous precursor particles and are released into the nuclear sap.

Baldwin-effect (Simpson 1953) — the exchange of "nonhereditary" by "hereditary" → characters via → mutation and → selection (→ genetic assimilation). The B.e. is due to an indirect connection between phenotypic modifications and evolution of new adaptive characters. Phenotypic plasticity permits a population to persist in an unfavorably changed environment; this results in postponement of selective elimination and gives the population better chances of acquiring new genetic variations by → mutation, → gene flow, and → genetic recombination. A genotypic response can be build up from these to the new environment.

ball metaphase (Barber and Callan 1943) — a form of → C-mitosis with characteristically clumped chromosomes. The b.m. is followed by either a complete degeneration of the cell or a state similar to interphase.

band (Painter 1939) — (1) the specific association of a large number of → chromomeres in → polytene chromosomes of **Diptera** (the polytene equivalent of "ordinary" chromomeres). Chromosome bands usually have clearcut outlines and characteristic dimensions. Each b. is well defined in relation to neighboring bands and separated from these by regions called interbands (banding pattern).

Bands contain about 95% of the chromosomal DNA. They (or band + interband) are likely units of genetic function and replication and show a selective transcriptional activity which is morphologically manifested as → puffing.

One band contains DNA in the range between 3.000 and 100.000 nucleotide pairs (→ service unit; transcription). (2) A part of the chromosome which is clearly distinguishable from its adjacent segments (→ chromosome banding).

banding pattern — in → polytene chromosomes of **Diptera.** the linear pattern of deeply staining → bands and weakly staining interbands which results from defined local differences in the degree of compaction along the nucleoprotein elements composing the polytene chromosomes. The b.p. is thought to reflect the structural oranization of these elements in the diploid interphase nucleus and represents, for any particular species of **Diptera**, a constant character, disturbed only by chromosome structural changes or temporary modifications such as those in → puffs and → Balbiani rings. If distinct species show precisely the same b.p. they are called homosequential species, as opposed to heterosequential species.

Barr body — → sex chromatin.

basal apparatus (Wilson 1925) — the complex of → basal bodies, bridging material, and rootlets in epithelial cells with multiple → cilia. The multiple cilia are interconnected beneath the → cell membrane by amorphous material bridging the adjacent basal bodies; these in turn are anchored into the cytoplasm by means of long, often bifurcating striated rootlets which arise from the proximal end of each basal body.

basal body (Fawcett 1961) — a cell structure composed of nine triplet → microtubules which nucleates the formation of ciliary and flagellar → microtubules. The b.b. precursors are called probasal bodies. The b.b. is usually assembled adjacent and perpendicular to a preexisting b.b.; it is believed to be homologous to the → centrioles.

In various species, the b.b. varies from 120 to 150 nm in diameter and from 300 to 2000 nm in length.

base analog — a chemical whose molecular structure mimics that of a → DNA base; because of the mimicry, the b.a. may act as a mutagen.

base excision repair — → excision repair.

base-insertion repair (Livneh et al. 1979) — a mechanism for repair (→ DNA repair) of apurinic sites in DNA. An enzymatic activity directly and specifically inserts the correct missing adenine or guanine into the appropriate apurinic sites.

basement membrane — a thin extracellular structure (composed of collagens, noncollagenous glycoproteins, and proteoglycans) that forms a substratum for orderly growing cells and plays a role in cellular growth and → cell differentiation, tissue regeneration and in filtration of macromolecules between tissue compartments.

base pair — two nitrogenous bases that pair in double-stranded → DNA or → RNA molecules. The relative amount of adenine-thymine to guanine-cytosine pairs, wich are those to which pairing in double-stranded DNA is restricted, is called the base pair ratio (A-T/G-C ratio). In RNA, thymine is replaced by uracil.

base pair change — any change of a base pair in DNA; a major source of spontaneous b.p.c. during → DNA replication are the mispairing base modifications which (due to their correct Watson-Crick base-pairing) leak through the replication machinery. Tautomers and isomers of normal DNA bases are thought to be the predominant source of mispairings in spontaneous mutagenesis.

base pair matching — the accurate base pairing in duplex DNA which assures fidelity of → DNA replication. Prior to the polymerization step, the incoming nucleoside triphosphate must be paired with its template partner. If the correct base pair has not been formed, → DNA polymerase rejects the mispaired nucleotide.

base pair mismatch — the presence of a → heteroduplex region in double-stranded DNA. B.p.m. within the DNA double helix may arise in several ways: (1) Spontaneous deamination of cytosine or adenine leads to G-U or I-T mispairs, respectively (such lesions are thought to be repaired via action of → DNA glycosylases); (2) mismatched base pairs may also occur during homologous → genetic recombination if allelic differences are included within the heteroduplex region formed by breakage and rejoining of parental molecules (→ gene conversion); (3) DNA replication errors may also contribute to the generation of mispaired bases (error rate in **E. coli** 10^{-8} to 10^{-11} per bp replicated);

Mismatched base pairs may be corrected prior to replication (→ mismatch repair) with single-stranded breakdown and resynthesis of a section of DNA.

base pair substitution — a class of "lesions" in DNA molecules which may give rise to → gene mutations. Base pair substitution consist of transitions, which preserve the purine-pyrimidine axis (AT ↔ GC), and of transversions, which reverse it (AT ↔ TA ↔ GC ↔ CG ↔ AT). Operationally, b.p. substitutions are recognized either by the amino acid substitutions in the corresponding polypeptide (missense mutations) and suppressible chain terminating signals (nonsense mutations) which they frequently produce, or by their patterns of chemically induced → reversion. Only very rarely are b.p. substitutions measured directly, or within RNA transcribed from DNA.

B.p. substitutions often escape detection when scored on the basis of phenotypic change for the following reasons: (1) For many → codons, the third base can often be substituted by one or more other bases without modifying the coding properties of the reading frame; (2) many transitions give rise to amino acid interchanges which do not significantly alter the functional integrity of the corresponding protein (silent substitutions).

B.p. substitutions are usually reverted by further b.p. substitutions.

basic chromosome set (Dyer et al. 1970) — → chromosome set.

basic defect — a uniform defect of a structural or chemical nature which is the basis of the manifold syndromes of a → genetic disease.

basic number — the number of → chromosomes in a basic → chromosome set. The b.n. represents the smallest → monoploid chromosome number (symbolized by x) of a polyploid series. All chromosome numbers being exact multiples of the b.n. are called → euploid. All numbers deviating by single chromosomes from x and from multiples of it are called → aneuploid. The specific → karyotype comprising the b.n. is the basikaryotype (Sinoto and Sato 1940).

basidium — in Basidiomycetes, a swollen terminal cell which shows a development similar to that of an → ascus until meiosis is over. Basidia are usually club-shaped and towards their distal ends, narrow, pointed stigmata are formed. Their tips enlarge to form → basidiospores and the nuclei, derived from meiosis, migrate from the b. via the stigmata into the basidiospores. Initially each basidiospore is uninucleate but a further mitosis often occurs and results in binucleate basidiospores. Other variants are known.

basikaryotype (Sinoto and Sato 1940) — → basic number.

B-chromosome (Randolph 1928) — any chromosome of a heterogeneous category of → chromosomes (also called supernumerary, accessory, or extra chromosomes) present in many species of both plants and animals which differ in their behavior from normal or A-chromosomes in the following characteristics (some of them may be missing in the special case under observation): (1) Their morphology (they are usually smaller in size than A-chromosomes, frequently but not exclusively heterochromatic and telocentric). (2) Their genetic effectiveness (they normally do not strongly influence the viability and phenotype of their carriers). (3) Their numerical variation (between different cells, tissues, individuals, and populations). (4) Their meiotic behavior (no chiasmate pairing with the A-chromosomes, lower degree of pairing among themselves, lagging and elimination by preferential nondisjunction during meiosis and post-meiotic divisions). (5) Their mitotic behavior (lagging and elimination, polymitosis or preferential distribution).

According to White (1945), two main groups of B-chromosomes may be distinguished: those which are mitotically stable, where all cells of the individual have the same number of Bs. and those which are mitotically unstable, giving rise to cells with different numbers of Bs within the same individual.

Accumulation systems of B-chromosomes in populations are various but normally fall into two categories: (1) Those in which the B. undergoes distorted segregation at meiosis and is thereby preferentially transmitted to the gametes, and (2) those in which the B. undergoes mitotic → nondisjunction in some particular divisions or tissues.

Special types of accumulation systems result in and tendency for the B-chromosomes to increase in frequency in males and to decrease correspondingly in females.

B-chromosomes in both plants and animals may affect growth, development of the external phenotype, genetic recombination, and chiasma formation during meiosis.

The idea is current that supernumerary chromosomes might have originated from small fragments caused by structural rearrangement and might eventually become as large as other chromosomes by repeated duplication.

It is generally held that B-chromosomes are of adaptive importance since their occurrence and frequency within the species varies between populations of different origin and habitat. A minority view is that B-chromosomes are parasitic and maintained more or less independently of any adaptive benefit or handicap they confer upon individuals or populations.

B-DNA — the right-handed conformation of the DNA double helix. B-DNA is in equilibrium with the left-handed → Z-DNA conformation. The equilibrium depends on a large number of parameters such as the presence of various cations in the environment and covalent modifications of DNA.

behavior flexibility (Thoday 1953) — all means of behavior permitting temporary → adaptation to environmental conditions such as habitat selection and measures for controlling the habitat or providing protection against environmental influences (→ phenotype).

behavior genetics — a branch of genetics which deals with the inheritance of different types or forms of behavior, such as intelligence, personality traits, etc.

bicentriole (Perkins 1970) — any coaxial → centriole pair.

bifunctional vector — = → shuttle vector.

bimitosis (Gonzales 1967) — the simultaneous occurrence of two → mitoses in binucleate cell which arise by interference with → cytokinesis.

binary Ti plasmid vector system – a vector system (→ cloning vector) consisting of two plasmids in **Agrobacterium**; one plasmid contains the foreign DNA that can be transferred to plant cells, the other contains the virulence (vir) genes which are necessary for the DNA transfer but are not themselves stably transferred (→ T-DNA).

binding protein (Heppel 1969) — any of a class of bacterial proteins implicated in the transport of compounds (e.g., sugars, amino acids, and ions) across the plasma (cell) membranes.

binom (Grant 1957) — → agamospecies, agameon.

binucleate — of cells with two nuclei.

bioantimutagenic (Kada et al. 1981) — of → antimutagenic agents that act after induction of DNA damage (→ desmutagenic) and prevent the manifestation of the induced DNA damage. Such agents may comprise: (1) Agents that increase the fidelity of → DNA replication; (2) agents that promote repair of DNA damage (→ DNA repair); and (3) inhibitors of error-prone repair.

bioassay — determination of the amount or effectiveness of a biologically active substance by measuring its effectivity on a living organism (→ mutagenicity screening).

biochemical genetics — a branch of genetics dealing with the chemical nature of hereditary determinants and the manner of their action in both development and function (→ gene; gene action).

biocoenosis (Möbius 1877) — a community of organisms which consists of a number of different kinds of plants and animals, living together in a particular habitat or biotope.

bioid (Decker and Speidel 1972) — any of the simplest chemically open systems capable of Darwinian → evolution by mutation from one steady state into a more stable one.

biological altruism — all forms of behavior by which some individuals benefit other individuals of the same species, without benefiting themselves. The benefits may be voluntary or involuntary, direct or indirect. B. a. is involved in the organization and evolution of social systems of insects and other animals including, of course, man (P. I. Darlington 1972).

biological membrane — → cell membrane; endomembrane system; unit membrane.

biome — a community of plants and animals (→ ecosystem). Adjacent biomes usually blend into one another (no sharp demarcation line) through a fairly broad transition region called an ecotone.

biometrical genetics — a branch of genetics concerned with the study of quantitative → inheritance.

biometry — statistics applied to biological problems.

biopoesis (Pirie 1937) — the origin of the first living thing, including the chemical history that preceded it. The first "organism" is called an eobiont (→ eobiogenesis; neobiogenesis). B. can be divided arbitrarily into two phases (Keosian 1965): (1) The phase that covers the period of physical and chemical processes in the abiotic synthesis of macromolecular systems; (2) the phase (eobiogenesis) that includes the transformation of macromolecular systems into the first living things.

biosome (Lehmann 1947) — any autonomous cell constituent multiplying or presumed to multiply by autoreduplication.

biosynthesis — the synthesis of the chemical components of the → cell from simple precursors and their assembly into structures such as, e.g., → membrane systems, → mitochondria, → nuclei, and → ribosomes. The b. of the cell components and the cell's functional morphology are genetically programmed.

biota — all the plants and animals that live together at the same time and in the same place.

biotechnology — the technical and applied aspects of research into biological materials and, in particular, into biological processing.

biotic community — an ecological unit being self-sustaining in energetics and consisting of → sympatric species playing complementary roles in nutrition and energetics (→ interspecific competition).

biotic potential — the inherent ability of a → population to increase in numbers when the age ratio is stable and all environment conditions are optimal (→ reproductive potential).

biotope (Dahl 1921) — → biocoenosis.

biotype (Johannsen 1903) — a group of genetically identical individuals.

biparental cross — → triparental cross.

biparental inheritance — → uniparental i.

bipolarity (Margolin 1965) — the phenomenon wherein both strands of a DNA helix (chromosome) serve in → genetic transcription, although at different regions.

bipolaron (Piperno et al. 1966; Sermonti and Carere 1968) — a nucleic acid segment whose polarity in → genetic recombination between pairs of mutations is opposite at the two ends of the → gene cluster and changes gradually as a function of the position of the selected → crossing over.

bisexual — (1) of organisms having both male and female reproductive organs and producing male and female → gametes within the same individual, which is either monoecious or hermaphroditic; (2) of populations or generations composed of male and female individuals as opposed to a unisexual parthenogenetic population or generation (→ parthenogenesis).

bit — the basic unit ("binary unit") of information content which represents the answer to a binary choice, i.e., describes the possible conditions of which the given condition represents one of two mutually exclusive possibilities (such as "yes" or "no").

bithallic (Ahmad 1954) — → heteromixis.

bivalent (Haecker 1892) — a pairing configuration (→ chromosome pairing) during the first meiotic division which consists of two completely or partially homologous chromosomes (→ pseudobivalent; quasibivalent). The number of bivalents per cell (meiocyte) normally corresponds to half the somatic chromosome number of diploid and genome-allopolyploid species. In chiasmate meiosis the pairing partners are held together after → crossing over by → chiasmata, in achiasmate meiosis by functionally equivalent structures (→ collochores) until the beginning of anaphase I. The formation of bivalents is prerequisite for the reduction of the somatic chromosome number and the random → assortment of the pairing partners (→ segregation). The following types of b. are to be distinguished: (1) **Homomorphic bivalents** in which the two pairing partners are in complete structural agreement. (a) ring b.: Ab. appearing in the form of a ring during meta-anaphase I following chiasma formation in both arms and complete or almost complete → chiasma terminalization; (b) rod b.: Ab. appearing as a rod during meta-anaphase I following chiasma formation and chiasma terminalization in only one arm. (2) **Heteromorphic bivalents** in which (due to heterozygous → chromosome mutations) the two partners of the pair differ structurally and are only partially homologous. (a) unequal b.: the partners are of unequal size, as a consequence of → deletions or various types of → translocation; (b) asymmetrical b.: the partners are of equal size but the → centromeres are located differently. (3) **Precentric and precocious bivalents** in which the centromere regions of the partners are already separated in metaphase I (a departure form normal behavior). (4) **C-bivalents,** i.e., two → C-pairs, lying close together, whose chromatids are separated from one another, apart from the centromere region. C-bivalents are characteristic of → C-meiosis and are produced by the action of colchicine and other spindle poisons.

bivalent interlocking — → interlocking.

Bkm sequence — any of the evolutionary conserved sex associated → satellite DNA sequences containing mainly GATA/GACA tetranucleotides as major repeats. Bkm sequences are located in vertebrate autosomes, in → W chromosomes of birds and reptiles and in the → chromosomes of some mammals. They are absent from → X chromosomes.

blastocyst — an early (pre-implantation) embryonic state, usually of mammals. The b. is a ball-shaped structure consisting of an inner cell mass (100 to 1000 cells according to species), a central cavity (the blastocoele), and trophoblast. The b. is the state at which the embryo is implanted into the uterine wall.

blastocyte — any of the undifferentiated embryonic cells (→ cytodifferentiation).

blastomere — a single large (undifferentiated) cell of a very early multicellular embryo. Blastomeres originate from the first few cleavages of a → zygote in animal embryogenesis (→ cytodifferentiation).

blastulation — → embryonic development.

blending inheritance — → inheritance.

blepharoplast (Webber 1897) — = → basal body.

blotting procedure — → nucleic acid blotting.

blunt end — a perfectly base-paired terminus of double-stranded DNA fragments (= flush end). The process of joining two DNA molecules with b. ends using → DNA ligase is called b.e. ligation (→ cohesive end).

bordered pit — a pit in tracheids or other plant cells involved in water conduction having a distinct rim of the cell wall overarching the pit membrane (Frey-Wyssling and Mühlethaler 1965).

bouquet stage (Eisen 1900) — a meiotic prophase stage (leptotene to pachytene) in those species where the chromosomes are oriented by one or both ends towards one point of the → nuclear envelope. The polarization process responsible for this arrangement originates in the cytoplasm.

brachymeiosis — an abnormal type of → meiosis characterized by the omission of the second meiotic division.

bradytelic (Simpson 1944) — of very slow rates of → evolution, approaching constancy over long periods of time (→ horotelic; tachytelic).

branch migration (Lee et al. 1970) — the ability of a DNA strand, partially paired with its complement in a duplex, to extend its pairing by displacing the resident strand with which it is homologous, i.e., a change of pairing partner in duplex DNA as a result of the release of hydrogen bonding between base pairs and the formation of new bonds (→ displacement loop).

branch site — the site where the 5′ end of the → intron becomes covalently attached near the 3′ end of the intron during → RNA splicing. The surrounding sequences in yeast conform to the specific sequence UACUAAC; in animals the b. sites conform to the less stringent consensus YNYURAY (R, purine; U, uracil; Y, pyrimidine; A, adenine; N, any nucleotide).

break through (Hadorn 1945) — a genotype which in spite of the presence of a → lethal factor in effective dose, survives the lethal crisis and continues to develop despite its autolethal constitution. Having overcome a first critical life period resulting from the → phase specificity of lethals, the b.t. can reach, under certain circumstances, a second sensitive stage in which it perishes.

breeding size — the number of individuals in a population that are actually involved in → reproduction in a given generation (symbolized by N).

Effective b.s. (symbolized by N_e) is the b.s. mathematically adjusted so that populations with varying → sex ratios, degrees of → inbreeding, etc., may be compared. N_e is equivalent to N only in an "ideal population" of large size, with equal numbers of the two sexes, with random → mating and random fusion of the gametes. Normally the effective b.s. is smaller than the b.s., since uneven sex ratios, inbreeding, cyclic changes of b.s. as well as nonrandom sampling of the gametes all reduce the effective b.s.

breeding system — a term used to cover all those variables apart from → mutation which affect the genetic relations of the gametes that fuse in sexual → reproduction. The variable components of the b.s. are numerous and often not easy to estimate. Two main groups may be distinguished (Lewis and John 1964): (1) Those which affect the ability of particular gametes to fuse or parents to mate, and (2) those which affect their probability within the limits set by the first. The variables under (1) comprise an aspect of the b.s. known as the → mating system.

The b.s. may be an essential factor controlling population structure and evolutionary divergence (→ evolutionary rate).

breeding value — the value of an individual as defined by the mean value of its progeny, i.e., the → heritability of anatomical and physiological traits of plants and animals. The b.v. is used for the choice of breeding methods in plant and animal improvement.

bridge-breakage-fusion-bridge cycle (McClintock 1951) — a process which may arise from the formation of → dicentric chromosomes (chromosome type of the BBFB-cycle) or chromatids (chromatid type of the BBFB-cycle), whereby daughter cells are formed (via breakage of the dicentric bridge between the cell poles during the anaphase of meiosis and/or mitosis) differing in their content of → genetic information due to → duplications and → deletions in the chromosomes. A prerequisite for the maintenance of such a cycle is that the daughter cells produced be viable and that the chromosome ends formed by breakage of the bridge should not be stabilized but should fuse. In the course of a BBFB-cycle there may be a repeated regrouping of the genetic loci in the chromosomes or chromatids involved since the breakage of the bridge may occur in different positions. The cycle is initiated by primary → chromosome mutations and results in secondary changes of the chromosomes taking part.

brights (Preer et al. 1953) — → killer.

budding — → cell budding.

bud sport — a somatic → mutation expressed in a bud or branch of plants (= bud mutation).

bulbifery — in plants, a common form of asexual reproduction by small bulbils formed in the axils of leaves, in the floral axis, or in floral branches. In many cases, bulbils fully supplant flower formation and the sexual process; in others bulbils are formed in addition to normal flower formation.

burst size — the mean number of → bacteriophage particles liberated per infected bacterium, or the ratio of phage titers before and after lysis of phage-infected bacterial cells. The interval preceding the increase in titer is called the latent period; the interval of increasing titer is called the rise period or the release period. The b.s. varies widely with strain of phage and bacterium used and is dependent upon cultural conditions.

bypass repair — a → DNA repair pathway that does not remove the → DNA lesion but nevertheless permits → DNA replication to proceed. B.r. preserves continuity of the genome through error-prone replication or recombination.

bypass replication (Radman et al. 1977) — the replication by eukaryotic → DNA polymerases past a → DNA lesion, such as UV-induced dimers, with only a brief delay. B.r. does not require recombination to circumvent the DNA damage blocking replication.

C

CAAC box — in most of the β-globin gene → promoters, a sequence located six bases upstream from the → CAAT box (5′ GCC$\begin{matrix}C\\A\\T\end{matrix}$CACCCT 3′). It is required for efficient in vivo expression of β-globin genes and is often duplicated.

CAAT box — part of a conserved sequence found in most, but not all, eukaryotic genes at about — 75 (75 bp before the initiation of transcription). The CAAT b. is probably involved in binding → DNA polymerase II but its involvement as a transcriptional signal is controversial (→ TATA box).

CA/GT$_n$ repeat — any of the poly (dC-dA/dG-dT) repeat sequences in the eukaryotic → genome that are probably involved in the control of transcription. They harbor the potential to change the conformation between the → B-DNA and → Z-DNA conformation.

callus — a disorganized mass of undifferentiated (parenchyma) plant cells which may accumulate at a wound surface, or grow out from pieces of plant tissue placed in culture.

cAMP — a cyclic nucleotide (cyclic adenosine-3′, 5′-monophosphate) which plays a significant role in regulating → gene expression in prokaryotes (**E. coli**). In the presence of cAMP receptor protein, CAP, cAMP can modulate the transcriptional initiation from many → promoters.

cAMP receptor protein — (CRP); in **E. coli**, a protein (allosteric effector) involved in the regulation of at least 20 genes of catabolite sensitive → operons. It binds in the presence of → cAMP to specific DNA target sites near → promoters at the 5′ end of the respective gene. In some cases, this interaction results in stimulation of mRNA transcription, in others it represses transcription.

Campbell model — the model that explains the integration of → bacteriophage DNA into the bacterial chromosome to form a → prophage. According to the C.m., DNA is circularized first and then prophage integration occurs by physical breakage and rejoining of phage and host DNA. A phage protein, integrase, recognizes the phage DNA and bacterial DNA → attachment sites and catalyzes physical exchange.

canalization (Waddington 1940) — the property of developmental pathways of achieving a standard → phenotype in spite of genetic or environmental disturbances (→ homeostasis). C. is a developmental buffering system reducing potential variation that may act at the gene level or on variation due to either genotype or environment. Phenotypic characters whose development is normally unaffected by environmental stresses or by underlying genetic variability are called canalized characters.

canalizing selection (Waddington 1940) — → selection.

cannonical sequence — an archetypical (prescribed) nucleotide or amino acid sequence to which all variants are compared.

cap (Rottman et al. 1974) — the sequence of methylated, terminal oligo nucleotides in most eukaryotic (not prokaryotic) → messenger RNA; the general structure is $^{7Me}G^{5'}ppp^{5'}Np$ … (N, first nucleotide encoded in the DNA). The cap is added post-transcriptionally near the → TATA box by linking the 5′ end of a guanine nucleotide to the terminal base of the mRNA and methylating at least the additional G- (→ cap-binding protein).

cap-binding protein (Sonnenberg et al. 1978) — any of the proteins that specifically recognize the → cap structure of eukaryotic → messenger RNA. The proteins are believed to promote unwinding of the → ribosome-binding site and are essential in the regulation of eukaryotic mRNA translation (polypeptide-chain initiation).

capping group — a group of methylated nucleotides with the general structure $m^7G^{5'}ppp^{5'}N$

(m) pUp (7-methylguanine linked to the adjacent nucleotide by a 5′-5′-triphosphate bridge) found in most eukaryotic → messenger RNA's (→ cap).

CAP (Zubay et al. 1970) — abbreviation for → catabolite activator protein (= → CRP).

capsid (Lwoff et al. 1959) — the protein coat of a virus particle or → virion which may take different forms (either helical or isometric), depending on the nature of the protein subunits from which the c. is formed. Components (protein subunits) of capsids observed in the electron microscope on axes of symmetry are referred to as capsomeres. Composite capsids may consist of more than one kind of capsomere.

The shape or design of the c. of a virion is determind by the specific bonding properties of its capsomeres, which may be composed of one or more proteins.

capsomere (Lwoff et al. 1959) — → capsid.

capsule — in a wide variety of bacteria, an extracellular layer of polysaccharides or, less frequently, polypeptides surrounding and attached to the bacteria. The biological role of the c. is manifold: (1) It prevents → bacteriophages specific for cell wall receptors from infecting the celles; (2) it protects from → phagocytosis by macrophages. The c. has, on the other hand, also become the receptor for c.-specific bacteriophages.

carcinogenesis — a multistep process which results in a malignant cell. Several specific heritable changes are necessary to achieve this.

carcinogenic — of any agent or process that significantly increases the incidence of malignant neoplasms, irrespective of the mechanism of action. Genotoxic carcinogens are agents that significantly increase the incidence of tumors in a population and have the ability to alter genetic information in test systems. Nongenotoxic carcinogens are those agents that increase the incidence of tumors in a population but do not have the ability to alter genetic information in test systems.

CA repeat — in higher eukaryotes, any of the CA dinucleotides which are part of sequences with putative → Z-DNA properties. Such repetitive sequences occur in the vicinity of a number of expressed genes and may play a role in the evolution of → genomes.

carrier — → genetic carrier.

carrier cell — any virus-infected cell which continuously divides and replicates virus.

caryonide (Sonneborn 1938) — any cell line in ciliates derived from the products of the first cell division after conjugation. Cells in such lineages possess → macronuclei descended from a common ancestral macronucleus. Two caryonides from one exconjugant form a clone, and the four caryonides of a pair represent a synclone. The synclone is the unit of Mendelian inheritance in ciliates.

cascade regulation (Pontecorvo 1963; Scherrer and Marcaud 1968) — (1) in prokaryotes, a form of → genetic regulation on the transcriptional level where one of several → operons codes for the production of an internal inducer which turns on one or more of the other prokaryotic operons. (2) A multistep model of genetic regulation in eukaryotic cells according to which regulation operates at a multiplicity of levels by mechanisms interfering with messenger RNA formation (transcriptional control), mRNA transport, and mRNA translation (translational control).

cassette model — a model to explain → mating-type interconversion in yeast. Information for both the a and α → mating type is assumed to be present as silent "cassettes"; a copy of either type of a cassette may be transposed to the mating type locus, where it is transcribed.

cassette mutagenesis (Wells et al. 1985) — a procedure of → site-directed mutagenesis that can create specific mutants or families of mutants. C.m. involves the following steps: (1) Removal of a stretch of DNA flanked on either end by a → restriction site; (2) insertion of a new cassette of synthetic DNA in its place which directs various amino acid replacements.

The inserted DNA can be long or short, contain many substitutions or have mixtures of bases at different sites.

catabolic — of metabolic pathways fulfilling two functions: (1) They produce a variety of small molecules that can be used as intermediates for the biosynthesis of proteins, nucleic acids, and other essential biological molecules; (2) they allow release of the chemical energy of organic molecules in a biologically useful form (→ catabolite repression).

catabolite activator protein (Zubay et al. 1970) — both a positive and a negative regulator of transcription (= catabolite receptor protein) of bacterial → operons which are subject to → catabolite repression (bio-degradative operons). When complexed with cAMP, CAP binds to specific DNA sites at or near the → promoters, where it stimulates the initiation of RNA synthesis. CAP is a dimer of two identical subunits, each consisting of 209 amino acids. The consensus DNA site for CAP is 5′AA-TGTGA ...TCACA-TT3′ (each hyphen represents any nucleotide). It is 22 bp in length and exhibits two fold sequence symmetry.

catabolite repression (Epps and Gale 1942; Magasanik 1961) — the decreased expression of many bacterial → operons (catabolite operons) resulting from addition of glucose or other carbohydrates. C.r. adapts the carbon catabolic enzyme machinery to the energetically most advantageous use of carbon sources.

C.r. is caused by a decrease of the cellular level of cyclic AMP that inactivates → CRP. The positive control of catabolite-sensitive operons is mediated by the binding of the cyclic AMP receptor protein (CRP) to the → promoter in the presence of cAMP. A conformational change in CRP, induced by binding of cAMP, is required for DNA binding by the protein. Binding of CRP-cAMP greatly increases the productive binding of → RNA polymerase to the promoter and, due to this, enhances initiation of transcription.

catabolite sensitive (Zubay et al. 1970) — of genes whose activities are subject to → catabolite repression.

catalase reactivation (Monod et al. 1949) — the increased survival of UV-irradiated bacteria by treatment of the cells with catalase or peroxidase (→ reactivation). C.r. is enhanced by small doses of visible light and in presence of oxidizable substances (Rupert and Harm 1966).

catenane — → DNA catenane.

catonated — of topologically interlinked, circular DNA molecules.

caveola (Yamada 1955) — any of the cell surface invaginations which (like → microvilli) serve to increase the surface area. (= caveola intracellulare, micropinocytotic vesicle, surface vesicle, surface pit).

C-bivalent — → C-meiosis; bivalent.

CCAAT box (Benoist et al. 1980) — a common sequence element (5′ TTGGPyCAAT 3′) in the → promoters of many genes transcribed by → RNA polymerase II (located at about −75). The CAAT b. is believed to play a part in promoter function and is necessary for efficient initiation of transcription. CCAAT box-binding proteins bind to double-stranded sequences containing TGG and its complement (CCA), making contact to the adjacent guanosines.

cDNA — single-stranded complementary DNA, i.e., a DNA molecule synthesized from a RNA template as an exact template by → reverse transcriptase; used for → molecular cloning or as a specific hybridization probe.

cDNA cloning — molecular cloning of the coding sequence of a gene starting with its → messenger RNA transcript. The cDNA copy will not contain any → intron and flanking sequences of eukaryotic genes since it is a copy of mature (processed) mRNA molecules.

cDNA gene — a → processed gene in the normal genome lacking the → introns that are present in otherwise homologous genes; cDNA genes probably arise by reverse transcription of the spliced RNA transcript (→ RNA splicing) of the homologous genes.

cDNA library — a library obtained by a technique for obtaining DNA molecules corresponding to individual mRNA species by cloning of → cDNA. The cDNA l. is constructed by synthesizing a double-stranded DNA copy of the mRNA population and integrating these cDNA molecules into a restriction enzyme site of a → cloning vector. The cDNA l. is representative of the mRNA sequences expressed in the cell type from which it is derived. It is important for the study of tissue or stage-specific → gene expression.

Compared with a → genome DNA l., a cDNA l. contains very few repetitive DNA sequences and therefore provides a source of single-copy gene-specific probes for the study of chromosome organization and genetic linkage.

cell (Hooke 1665) — the basic structural and functional unit of all living matter which represents a multimembraned system, compartmentalized into specific functional areas or organelles with discrete but complexly interrelated and coordinated roles.

Cells may exist as independent units of life; they may associate with one another to form cell colonies (facultative association) or tissues (obligatory association).

Cells are the building elements of all organisms. In the probionta the cells posses specific organelles to accomplish all the important life functions; in the metabionta, the cells are specialized, resulting in a division of labor whereby particular cells adopt specific functions (→ cytodifferentiation).

Each cell is a strictly delimited system which may be easily identified by its material nature, its structural organization and its behavioral expressions. Cells show manifold variation in size and form, but in all organisms they conform to a single fundamental plan: the protoplasm, i.e., the cellular substance as a whole, is differentiated into the → cytoplasm and the cell → nucleus (in eukaryotes) or → nucleoid (in prokaryotes). Each cell is surrounded by a → plasma membrane and in certain species by a → cell wall as well. The various cell organelles are situated in the → ground plasm of the cell (the "matrix"), i.e., the soluble phase composed of water, salts, and free molecules (including many enzymes). In addition to the cell nucleus or the nucleoid, such cell organelles are the → mitochondria and → chondrioids, the → plastids, the → ribosomes, the → Golgi apparatus, the → centrioles, the → lysosomes, the → cortex, the → vacuoles, and the → endoplasmic reticulum.

Cells of eukaryotes (higher plants and animals, protozoa, fungi and most algae) and prokaryotes (bacteria. blue-green algae) are of different complexity.

(1) **The eukaryotic cell** (Chatton 1925) may be understood as a three-phase system (Morrison 1966): One phase is represented by the nucleo-cytoplasmic matrix in which → nucleoplasm and cytoplasm have continuity via the pores of the → nuclear envelope. The second phase is formed by the membraneous organelles of the cell and the third consists of the contents of these membrane-enclosed structures. Membranes (→ unit membrane) subdivide the cells of eukaryotes into the two major compartments, nucleus and cytoplasm, into a system of interconnected cavities and various separate compartments. The internal skeleton (the cytoskeleton) is made up of networks of fibrous protein. The four major types of subcellular fibers are the → microtubules, the → microfilaments, the → intermediate filaments, and the → microtrabecular system. The discrete, membrane-bounded compartments represent the mitochondria, plastids, and lysosomes, while the system of continuous cavities forms the endoplasmic reticulum and the Golgi apparatus. The chemical constituents (enzymes) by which cells perform their functions are localized mostly in organized membrane structures.

Unlike animal cells, plant cells usually possess → plastids and vacuoles, and are covered by a → cell wall. Mitochondria, ribosomes and a nucleus are present as in animal cells. Plant cells have structures termed → dictyosomes which are similar to the → Golgi complex of animal cells.

(2) The **prokaryotic cell** (Chatton 1925) is the unit of structure in bacteria and blue-green algae and shows less structural components classifiable as organelles than the eukaryotic cell. These organelles include the nuclear equivalent or nucleoid, the cell membrane and a few membraneous structures regularly present in the cytoplasm. Localized invaginations (→ mesosome) of the cell membrane give rise to tube-like structures. Particles equivalent to the ribosomes and mitochondria (→ chondrioid) of eukaryotes are present in bacterial cells.

The cell membrane of prokaryotes seems to perform more complex functions than that of eukaryotes and is engaged in the differentiation of cell structure. In nonphotosynthetic bacteria, the cell membrane contains the enzymes mediating the oxidative degradation of organic materials and the synthesis of energy-rich compounds. In photosynthetic bacteria and blue-green algae, the molecular equipment for photosynthesis is assumed to reside in the cell membrane and its intrusions (vesicles and lamellar structures called → chromatophores).

Cells of the multicellular metabionta are designated somatic if they are body cells. The → gametes and their immediate progenitors are the germ cells, in contrast to the somatic cells.

Reproduction of the cell is accomplished by cell division. In the eukaryotes, this process is subdivided into → cytokinesis (cell cleavage) and karyokinesis (→ mitosis, meiosis).

cell adhesion — the process that keeps cells together in multicellular organisms; cells adhere to each other, to extracellular matrices, and to the substratum. The adhesion between adjoining cells is a fundamental biological event, preceding both → morphogenesis and organogenesis, and also providing the basis for cellular recognition and membrane fusion.

It may be discriminated between molecules carrying out cell-cell interaction (cell adhesion molecules, CAM's), those carrying out cell-substrate interaction (substrate adhesion molecules, SAM's), and those appearing at junctions (cell junctional molecules, CJM's). A family of structurally and functionally related adhesion receptors has been identified (the integrins) which share a common subunit (β) that combines with unique α subunits to create functionally distinct receptors. Many of the adhesive proteins that serve as ligands for the integrin receptors contain the tripeptide sequence, Arg-Gly-Asp (RGD). This sequence is directly involved in the recognition of the receptors.

cell adhesion molecule — any of a class of large cell surface glycoproteins involved in cell-cell interaction during embryogenesis and histogenesis and perhaps in the stabilization of certain tissues in adult life (→ cell adhesion). Primary c.a. molecules (CAM's) are those that appear in early embryogenesis; secondary CAM's appear during later histogenesis to link cells of different types following → cytodifferentiation.

cell age — term used to describe two separate phenomena: (1) In synchronously dividing cultures, age may refer to the position of a culture in the → cell cycle; (2) a second usage refers to the stage of a culture which has been allowed to grow from initial inoculation into fresh growth medium until all the nutrients have been exhausted and waste products accumulate (→ cell culture).

cell aggregation — histogenetic attachment and aggregation of dissociated cells. C.a. is mediated by specific macromolecular products localized at the cell surface and between cells (→ cell ligand).

cell ancestry — → cell lineage.

cell-attachment recognition signal — the amino acid triplet Arg-Gly-Asp that forms part of a widespread cell-extracellular matrix recognition system. Peptides containing this sequence inhibit certain developmental events such as cell-matrix adhesion or cellular migration in vitro and in vivo.

cell autophagy — → cytolysosome.

cell budding — a form of unequal → cytokinesis whereby the mother cell is not halved; instead, bud-like growths form at one or more narrowly restricted sites and one daughter nucleus migrates to each such growth, followed by the separation of the daughter cells from the mother cell.

cell-cell adhesion — the tissue-selective adhesion of cells mediated by cell-specific adhesion molecules on the cell surface. Cells are connected with multiple types of junctions, such as → tight junctions, adherens junctions, gab junctions and desmosomes. C.-c. a. mechanisms are functionally divided into two systems: (1) Ca^{2+}-dependent and (2) Ca^{2+}-independent systems. Both coexist on single cells and for both → cell adhesion molecules have been identified.

cell center — the cell structure which determines the cell poles during the nuclear division.

The c.c. consists of two → centrioles differing in structure and connections with → microtubules. The c.c. is the main locus of microtubule formation in the cell.

cell cleavage — → cleavage.

cell clone — a population of cells derived from a single progenitor cell.

cell colony — → cell.

cell commitment — the commitment of an embryonic rudiment to form a certain differentiated cell type some time before any visible change can be detected (→ cytodifferentiation).

cell communication — long-range and short-range interaction between cells in both excitable and nonexcitable tissues. Short-range c.c. requires direct physical contact between cells and is frequently accompanied by distinct intercellular membrane specializations (→ gap junction; tight junction; septate junction; synaptic junction; desmosome).

cell compartment — any of the membrane-bound units or parts within cells. Presumably each c.c. is able to regulate its contents by virtue of the selective permeability of its double or single → unit membrane and may serve as a regulatory control in the cell. Main compartments of eukaryotic cells are: (1) the cell → nucleus, (2) → mitochondria, (3) → plastids, (4) → cytosol, (5) → endoplasmic reticulum, (6) → ribosomes, (7) → Golgi apparatus, (8) → lysosomes, (9) → peroxisomes, and (10) glycogen granules (→ compartmentalization).

cell cortex (Chambers 1940) — a peripheral layer found in many cell types that includes the plasma membrane and tightly associated intracellular (cytoskeletal) and extracellular (glycocalyx) components. The c.c. may be a repository of developmental information (cortical information) and can be several hundred nanometers deep.

cell culture — the growing of dispersed cells in vitro, including the culture of single cells. The terms **tissue culture** and **organ culture** denote the maintenance or growth of tissues, organ primordia, or the whole or part of an organ in vitro in a way that may allow differentiation and preservation of their architecture and/or function (→ culture alteration). A culture started from cells, tissues, or organs taken directly from organisms is called a **primary culture**. The first subculture of a primary culture gives rise to a → cell line (Feodoroff 1967).

cell cycle — a sequence of events that involves the periodic replication of DNA and segregation of replicated DNA with cellular constituents to progeny cells. Its orderly progression depends on consecutive and/or simultaneous action of a series of genes (→ cell cycle mutant). The c.c. of eukaryotes consists of three periods — G1, S, and G2. Cells become irreversibly committed to DNA synthesis and subsequent → mitosis during G1. Cells in the G1 period have a constant amount of DNA. During S the cells synthesize DNA at some rate until the DNA content is doubled. The cell is then defined to be in the G2 period where it remains until it divides into two G1-phase cells (→ cell maturity). Many nondividing plant and animal cells in vivo and in vitro arrest at some early point of G1. This stage is termed G_0 to distinguish it from an extended G1.

A new concept of the c.c. divides it into two parts. One, termed B phase, is regarded as a deterministic and highly coordinated process that corresponds to the conventional S, G2, mitosis (M), and part of G1 phase. The other part (contained within G1) is the A state and is a sort of limbo in which the cell is not progressing towards mitosis; the cell may remain in this state for any lenth of time subject to the conditions. The transition probability model of the c.c. claims that the initiation of each c.c. depends upon a single critical event, or transition, which occurs at random in G1. A cell waiting to undergo the transition is said to be in the A state, and the rest of the intermitotic period is called B phase.

The c.c. of prokaryotes consists of three periods: (1) preparation for initiation of DNA replication; (2) DNA replication, and (3) a period between termination of DNA replication and cell division. At rapid growth rates, these periods overlap so that synthesis of DNA is discontinuous.

cell cycle-specific mutant (Chu 1978) — operationally, any growth mutant which manifests its defect at specific stages of the → cell cycle.

cell death — loss of function or morphological integrity of differentiated cells; loss ot the ability to undergo unlimited proliferation of rapidly dividing cells as measured by macrocolony formation, physiology, and morphology.

cell determination — the process by which multipotential cells become committed to a particular developmental pathway (cellular d.). Classically, c.d. is said to have occurred at the time when cells acquire the ability to differentiate along a particular pathway even if they are removed from their normal environment.

cell differentiation — = → cytodifferentiation.

cell division — a two-process event consisting of → karyokinesis (mitosis) and → cytokinesis and resulting in the reproduction of cells. The → mitotic apparatus is responsible for → chromosome movement. Mitosis and cell cleavage are linked spatially and temporally. The cleavage plane coincides with the plane of the metaphase chromosomes, but cleavage constriction commences only after chromosomes separate. Transformation of a precursor cell capable of self-replication to a terminally differentiated cell involves modulation of the expression of a number of genes. These include genes responsible for products determining progression through the → cell cycle, for products involved in metabolic processes, and for products determined by the terminal differentiated → phenotype.

In bacteria, c.d. follows an orderly series of events coupling the completion of a round of DNA synthesis with the biosynthesis of cell membrane and septa.

cell division lag (Ryan 1954) — delayed manifestation of induced → mutations which may occur after heavy doses of some → mutagens and is the result of the delayed onset of renewed division (→ mutational lag; phenotypic lag).

cell-division sequence motif — in the majority of proteins that carry out one of the required steps in the cell division competence cascade of deactivating a cell division repressor or activating a cell division activator, a common sequence motif responsible for their activity. The motif consists of a beta turn, a spacer of one to five amino acids, and an alpha helix.

cell envelope — the sum of all surface components of the cell that are located outside the cytoplasmic membrane.

cell fate — the destiny of a cell; the fate of a cell produced by terminal division is either to differentiate into a cell of a specific type or to undergo programmed → cell death, whereas the fate of a cell produced at an intermediary point in a → cell lineage is to generate a specific pattern of cell divisions and a set of descendants (→ fate map).

cell-free extract — a fluid derived from broken cells which contains most of the soluble components of the → cell; capable of performing certain biological functions, such as protein synthesis (→ cell-free translation system), when provided with the necessary precursors.

cell-free translation system — any in vitro system capable of translating natural → messenger RNAs into proteins. A c.-f.t.s. should (1) stringently depend on the addition of exogenous mRNA, (2) faithfully and efficiently translate purified natural mRNA's coding for defined polypeptides, (3) translate de novo a collection of natural homologous mRNA's with relative efficiencies similar to those observed in vivo. C.-f. translation provides evidence for the expression of specific genes. Characterization of purified mRNA's by c.-f. translation, in combination with other techniques, can yield information about the spatial and topological arrangement of transcripts along the → genome, the location of protein coding regions within the mRNA, mRNA structure, and mRNA abundance. The three principal eukaryotic c.-f.t. systems currently in use are from wheat germ, ascites tumor cells, and nuclease-treated reticulocytes.

cell fusion — the fusion of somatic cells to produce viable somatic → cell hybrids (= cell hybridization). Three stages are involved in cell fusion mediated by viruses, chemicals and

→ electroporation: (1) Agglutinations; (2) membrane-fusion between adjacent cells; and (3) osmotic swelling (→ fusogenic agent).

Somatic c.f. between phylogenetically distant species has important applications in → gene mapping and production of homogenous cell products encoded by the genome of one of the partners.

In interspecific somatic c.-f. hybrids, segregation occurs by species-dominant loss of the chromosomes of one parental cell (→ nucleolar dominance). This makes these cells particularly suitable for genetic analysis (→ synteny; synkaryon).

cell generation time — the interval between consecutive divisions of a cell.

cell growth — the doubling, during one → cell cycle, of the DNA and all other cellular components to produce two daughter cells identical to the parent. C.g. regulation is a complex chain of events, beginning with extracellular regulators of many kinds, probably interacting with the cell's plasma membrane. Cytoplasmic mediators then appear to transmit regulatory signals from the plasma membrane to the nucleus, where they perhaps control → DNA-binding proteins. Different cell types seem to recognize different sets of extracellular regulators and may have different mediators and DNA-binding proteins.

cell heredity — heredity at the cellular level. In eukaryotes, the informational patterns of c.h. consist of nuclear DNA, organelle DNA, and priming configurations (Sonneborn 1967) whose presence is required to direct the assembly of molecular species into functional two- and three-dimensional structures (→ cortex).

cell hybridization (Barski et al. 1960) — = → cell fusion.

cell interaction — communication between cells due to cell-to-cell contact (contact-mediated c.i.) or indirectly by factors that are produced by other cells and constitute the "microenvironment" of a group of cells (non-contact-mediated c.i.). C.i. plays an important role in embryonic development, wound healing, and tissue culture (e.g., → contact inhibition of division and movement).

cell junction (Farquhar and Palade 1963) — in animals, any type of specialization of the cell periphery related to → cell adhesion and attachment (→ desmosome). C. junctions represent a variety of morphological forms and all types may exist in various geometrical forms. They arise at characteristic times during ontogeny. The location and specificity of cell junctions play a decisive role in specificity of many histological and cytological characteristics, such as tissue permeability, transmission of electrical signals, recognition of cells during embryogenesis, tissue differentiation and the shape, polarity and internal organization of cells (→ zonula occludens; macula adherens).

cell lethality — the death of individual cells which contain "cell lethals". Cell lethals (Demerec and Hoover 1936) are → mutations (mostly → deletions) which normally bring about the death of the cell in which they occur (→ lethals). A distinction is made between primary and secondary cell lethality (Hadorn 1949): (1) **Primary c.l.**: cells with cell lethals perish even as mosaic spots or implants in association with genetically "normal" tissue. (2) **Secondary c.l.**: following appropriate transplantation into genetically "normal" tissue, the cells with cell lethals circumvent a lethal crisis and survive; their death results from the natural death of the carrier organism.

cell lethals (Demerec and Hoover 1936) — → cell lethality.

cell ligand (Moscona 1968) — any factor associated with the cell surface and intercellular spaces which mediates histogenetic attachment and aggregation of cells. The molecular features, differences of complementarities and distribution patterns of c. ligands on and between cells are assumed to be responsible for differential adhesiveness of the cells, cell recognition, affinities, and preferential association.

cell line — a population of cells which arises from a primary → cell culture at the time of the first subculture (→ cell strain). A **cloned line** is a c.l. descended directly from a clone. A **diploid c.l.** is one in which, arbitrarily, at least 75% of the cells have the same → karyotype as the normal cells of the species from which the line was derived. If the karyotype has changed, but the diploid number of chromosomes remains the same, such cells may be referred to as "pseudodiploid". A **heteroploid c.l.** is one having less than 75% of cells with diploid chromosome constitution (Feodoroff 1967).

An established c.l. consists of cultured cells of unique origin capable of stable growth for many generations.

cell lineage (Wilson 1882) — a pedigree of cells related through asexual division, i.e., the pattern of cell divisions and the set of specific cells generated by those divisions. The term cell ancestry refers to a cell's lineage history, i.e., the pattern of divisions and the set of cells from which it is derived (→ cell fate; regulatory cell).

cell locomotion — the movement of cells which is a vital component of many biological processes during embryogenesis and organogenesis. During locomotion, cells must use either internal "programs" or must respond to signals from their environment, or both. The motile force within the cell is provided by the actin → cytoskeleton, which is contractile and linked to the plasma membrane.

cell mediated immunity — any immune response which is directly mediated by T lymphocytes rather than by circulating (humoral) → antibody molecules.

cell membrane — a component of the → cell surface (= plasma membrane) enclosing every living cell and representing an envelope (usually 8 to 10 nm thick) that selectively regulates the flow of nutrients and ions between the cell's interior and its external milieu. The c.m. of eucells consists of phospholipid bilayers punctuated by globular transmembrane proteins, parts of which span the hydrophobic regions of the bilayer as α-helically coiled polypeptides. Intramembraneous segments create channels (pores, junctions) across the bilayer. A complex set of proteins is attached to the under surfaces of the c.m. and is responsible for stabilizing the membrane and regulating the topography and mobility of the different membrane proteins. The submembrane network (the membrane skeleton) is composed of filaments of actin, actin-binding proteins, and a set of connecting proteins which link the stabilizing infrastructure to the overlying membrane.

The c.m. components are not synthesized in situ; synthesis is a multistep process, associated with the rough and smooth → endoplasmic reticulum and the → Golgi complex.

The c.m. acts as a passive diffusion barrier to charged and large molecules, and in addition carries out facile diffusion (passive transport), active transport, endocytosis (pino- and phagocytosis), and exocytosis. It is further involved in locomotion and chemotaxis, secretion and reception of hormones, conduction of nerve impulses, and direct cellular interactions such as → cell adhesion and → contact inhibition. Differentiated functional sites of the c.m. are involved in these various functions (→ mesosome).

The c.m. and internal membranes (→ membrane system; mesosome; endoplasmic reticulum) of the cell are essentially alike. Both are composed of proteins and lipids; in mammalian cells, small amounts of carbohydrate are also present, associated either with protein (glycoproteins) or with lipid (glycolipids). There is likely to be an interrelation between the protein array and membrane function in cells with a variable distribution of proteins.

cell migration — the active or passive movement of (animal) cells during defined periods of development. Directionality of c.m. may be provided by contact guidance, contact inhibition of movement, chemotaxis, population pressure, as well as other mechanisms.

C.m. plays a important role in, e.g., embryonic development, defense of the body against infections, wound healing, and local spread of tumor cells.

cell motility — the inherent ability of cells to move. Normal tissues are presumed to be stable arrangements of inherently motile cells (→ contact inhibition). Three mechanisms have evolved for c.m. Two of these, bacterial → flagella and eukaryotic flagellae, serve to move cells through fluid media, but function differently. The third motile mechanism is based on the cytoplasmic → microfilaments and is responsible for the ameboid-type movement of a cell on a solid substratum.

cell organelle — a subcellular structure having a specialized function, such as the → mitochondrion, the → chloroplast, or the → spindle apparatus (→ organelle DNA; organelle gene).

cell plate (Strasburger 1882) — the precursor of the new plant → cell wall which appears during early telophase of → mitosis as a thin lamella in the → equatorial plane of the → mitotic apparatus (= phragmoplast). C.p. formation begins with a linear granulation in the equatorial region. The granules are small vesicles produced by the → Golgi apparatus. An additional component are fibrillar elements (microtubules), which arise at the plate and are oriented at right angles to its plane.

cell population — a group of cells designated as either static, expanding, or renewing (Leblond 1964). (1) **Static c. populations** represent homogeneous groups of cells in which no mitotic activity can be detected; the total DNA content remains constant. (2) **Expanding c. populations** are homogeneous groups of cells showing scattered mitoses in numbers which account for the increase in total DNA content. (3) **Renewing c. populations** are homogeneous groups of cells in which mitosis is abundant and exceeds that required for the total increase in DNA content.

cell receptor — any molecule, in the outer membrane of a cell or within a cell's cytoplasm, that recognizes and binds specific messengers (either neurotransmitters or hormones) and passes their message on to other molecules that respond to it in a specific way.

cell recognition — the mutual recognition of cells as expressed by specific → cell-to-cell adhesion. C.r. is due to a specific complementary type of interaction between molecules on adjacent cell surfaces.

cell sap — the interparticular fluid of the cell (= hyaloplasm or soluble phase of the cell). The c.s. is the milieu in which cellular formed elements persist and function and the main processes of cellular metabolism are performed. The c.s. is defined operationally as the supernatant that fails to sediment in a centrifugal field strong enough to deposit all particular material (as, e.g., → ribosomes, → mitochondria, → plastids, etc.).

The c.s. or soluble phase may consist of at least three general subfractions: (1) Micromolecular subfraction, including water, inorganic ions, and dissolved gases; (2) mesomolecular subfraction, including all metabolic intermediates, lipids, sugars, nucleotides, nucleotide derivates, and other low-information content molecules which are not integrated into a preformed structural element; (3) free macromolecules, chiefly proteins and RNA (including → transfer RNA).

cell secretion — the process by which cells synthesize products that will be utilized by other cells or eliminated from the organism. Secretions are either (1) external (or exocrine), i.e., they are expelled into the outer environment (→ signal sequence) or more frequently into natural cavities, or (2) internal (or endocrine), i.e., the secretions enter directly into the circulation to act on other tissues. Secretion may be continuous when the secretion products are discharged without being stored. It is discontinuous when the synthesis and intracellular transport of the product is stored in special granules which are then released to the extracellular space.

cell selection (Darlington 1937) — → selection within a population of genetically different cells. C.s. involves competition between genetically balanced or wild-type cells and unbalanced or mutant cells with the result that the latter are usually eliminated.

cell sensory system — any of the cellular systems that mediate a wide variety of adaptive behaviors involving changes in → gene expression or cell movement. Their information-processing capabilities include signal amplification, integration of multiple inputs, sensory adaptation, and excitatory and inhibitory interactions with other signalling pathways.

cell separation — any technique of sorting cells, i.e., preparing suspensions of dissociated cells from a given organ. Such techniques can take advantage of differences in the physical properties of the cells (such as size or density) or they can exploit differences in the properties of the surface membranes of the cells (such as their electric charge or their ability to adhere to glass or plastics).

cell strain — a population of cells derived either from a primary culture (started from cells, tissues, or organs taken directly from organisms), or a → cell line by the selection or cloning of cells having specific properties or markers. A c.s. derived by isolation of a single cell or groups of cells having properties or markers not shared by all cells of the c.s. is called a **substrain**. **A. cloned strain** is a c.s. descended directly from a → clone, i.e., a cell population derived from a single cell by mitosis (Feodoroff 1967).

cell surface — a two-component structure surrounding the → cell. The inner component is the → cell membrane or plasma membrane. A second component can usually be demonstrated outside the cell membrane. This component is rich in saccharides and always contains other compounds as well. It is called → cell wall, glycocalyx, cell coat, mucous coat, extraneous coat, or zona pellucida.

Specialized areas of the c.s. serving to bind that surface to another structure are termed → desmosomes.

The c.s. plays an important role in the control of cell proliferation, growth, behavior, and differentiation. Cellular response to external factors is often mediated through the binding of a ligand to its specific → cell surface receptor followed by → endocytosis (→ cell surface modulation).

cell surface modulation (Edelman 1976) — any change in the structure, pattern, or dynamic state of → cell surface receptors.

cell surface receptor — any of a class of multifunctional proteins with binding sites that face the external environment and effector sites that couple the binding to an intracellular event. Additionally, many c.s. receptors transport bound ligands into cells (receptor-mediated endocytosis). C.s. receptors receive and transduce a variety of environmental signals across the → cell membrane into the cell interior.

cell synchrony — the situation where a population of cells proceeds through the sequences of the → cell cycle with synchrony, i.e., divides at one time.

C.s. is characteristic for some tissues and may be obtained experimentally by two general procedures: (1) Enforcement, i.e., by a metabolic block which causes a temporary arrest of a particular process (e.g., DNA replication or mitosis). Enforcement permits a high recovery of synchronous cells. (2) Selection, i.e., by a procedure which exploits properties of the cell system so that stage-specific cells may be physically separated from an exponentially growing population in in vitro culture (→ cell culture).

There is evidence that a high degree of c.s. exists in systems (multinucleate cells, syncytia, somatic cell hybrids) where one nucleus in the process of synthesizing DNA transmits inducers through the cytoplasm to stimulate DNA synthesis in an adjacent nucleus.

cell tetrad — → tetrad.

cell theory (Schwann 1838) — the theory, based upon unequivocal experiments, originally put forward by Dutrochet in France and by Schleiden and Schwann in Germany and refined by Schwann (1838/1839) according to which: (1) The → cell is the smallest building element of a multicellular organism and as a unit is itself an elementary organism. (2) Each cell in a multicellular organism has a specific task to accomplish and represents a working unit. (3) A cell can only be produced from another cell (or a → syncytium) by cell division.

cell transformation — the transformation of a normal cell into a neoplastic cell by a multistep cascade of events which are under the control of different classes of → oncogenes.

cellular affinity — the tendency of cells to adhere specifically to cells of the same type, but not to different types. C.a. is lost in cancer cells.

cellular autophagy — the ability of a cell to segregate and to degrade portions of the cytoplasm. Autophagic vacuoles represent a transitional stage between the phase of segregation and the phase of destruction of segregated contents.

cellular cloning — the production from daughter cells, ultimately derived from a single cell, of large, genetically uniform populations of cells.

cellular determination — a process during which the potential of a cell becomes restricted to expressing only a specialized → cellular phenotype.

cellular junction — any of the specialized regions of close contact between cells (→ gap junction; tight junction; zonulae occludentes).

cellular phenotype — the → phenotype of the cell which is the direct consequence of the concentration and spectrum of all the proteins expressed by the cell (→ gene expression).

cellular reprogramming (Tsanev 1975) — of all phenomena in which the protein pattern of a cell is altered without changes in its DNA.

cellular senescence — the loss of proliferative capacity of cells resulting in the inability to increase cell number. C.s. is associated with, e.g., cell-cycle arrest, increase in cell size and size heterogeneity, increase of chromosome aberrations, including polyploidy.

cellular skeleton — → cytoskeleton.

cell wall — a rigid external coat which surrounds plant cells and some animal cells (one of the variants of the → glycocalyx). It is formed outside the → cell membrane and contributes to the mechanical strength of plants and to the ability of cells to withstrand turgor and osmotic variations.

C.walls normally contain carbohydrates, lipids, proteins, and minerals; usually carbohydrates predominate in the form of polysaccharides (cellulose, chitin, xylans, mannans, and lignin). Ultrastructure of the c.w. nearly always comprises long, thin threads (microfibrils 10 to 20 nm wide) embedded in amorphous material. Lipid substances (suberin, cutin, waxes) are deposited either on the inside (suberin) or mainly on the outside (cutin, waxes) of the microfibrillar wall layers (adcrusting layers). The c.w. of plants is perforated in many places by → bordered pits and plasmodesmata which provide direct contact between neighboring cells of a tissue.

CEN — symbol for cloned eukaryotic → centromere DNA, usually followed by an arabic number referring to the chromosome from which the centromere is derived. All CEN sequences contain a high (91 to 95%) A+T region of nearly uniform length (78–86 bp), composed primarily of separate short blocks of As and Ts, rather than alternating or random AT arrays. The AT-rich region is bounded on one side by a partially conserved 25 bp sequence and on the other side by a different conserved 8 bp sequence. The generalized sequence of the 25 bp sequence is

TGTT$^{T}_{A}$TG · TTTCCGAAA···AAA.

CEN's are not chromosome-specific and function in either orientation in the chromosome.

CEN-ARA vector — a yeast → cloning vector that replicates by virtue of its → ARS sequence; due to its CEN sequence it is propagated as an extrachromosome but with lower mitotic stability than natural chromosomes (→ integration vector).

center — → mitotic center.

centimorgan — (cM); a measure of genetic distance, i.e., the distance that separates two → genes between which there is a 1% chance of → genetic recombination. In dependence on the genome size of the organism, one cM represents, e.g., 139 kb (Arabidopsis), 510 kb (tomato), and 1108 kb (human). (→ Morgan unit; map unit).

central body — = → centriole.

central dogma — the hypothesis that information flows only from DNA to RNA to protein; some exceptions are known, but the rule is generally valid.

central spindle (Hermann 1891) — → spindle.

centric — of chromosomes or chromosome segments having a → centromere, as opposed to acentric.

centric fission — a chromosome structural change (→ chromosome mutation) converse to → centric fusion or Robertsonian translocation, i.e., the origin of two → telocentric chromosomes from one metacentric chromosome (= dissociation). C.f. may be due to → centromere misdivision or may involve a donor chromosome providing one centromere and two telomeres. Evolutionary replacement of a metacentric chromosome by two telocentrics has occurred in some animal species.

centric fusion (Robertson 1916) — the fusion by reciprocal → translocation of two → acrocentric chromosomes to give a → metacentric chromosome and a small fragment which is usually lost. C. fusions are the main way by which chromosome numbers have been changed in (animal) → karyotype evolution. In the heterozygous condition, c. fusions show quite regular → meiosis with the two acrocentrics invariably passing to one pole and the metacentric to the other. Aneuploid gametes may result if the three centromeres orientate in a linear manner (→ centromere orientation) on the meiotic spindle instead of in an alternate configuration.

In the case of c.f. the number of large chromosome arms is maintained (no change of the so-called "fundamental number of chromosome arms").

C. fusions may be due to: (1) Two breaks very close to the centromere of acrocentric chromosomes, one in the long arm, the other in the short arm. Reciprocal translocation may then give rise to a large metacentric and a very small chromosome which will usually be lost (usual mechanism of c.f.). (2) Two breaks in the centromere regions of the acrocentrics or in their short arms giving rise to a metacentric with two centromeres so close together that they function as one and an acentric which is immediately lost (→ tandem fusion).

centric region — = → centromere.

centric shift — a shift of the → centromere from its usual position in the chromosome in question. C. shifts may arise in two ways: (1) By pericentric → inversion; (2) by centric → transposition.

centrifugal selection — → selection.

centriolar pinwheel (Fulton 1971) — a structure which, in the electron microscope, resembles a → centriole, i.e., a cylinder about 0.2 nm in diameter with a wall composed of nine equally spaced groups of → microtubules, usually triplets.

centriolar region (Berns et al. 1977) — in many cells, a region containing a pair of → centrioles and → microtubule organizing centers from which the cytoplasmic microtubules emanate (→ pericentriolar cloud).

centriole (Boveri 1895) — a cylinder some 120 to 150 nm wide and of variable length in the cells of animals, some algae and fungi. The cylinder wall is composed of 27 → microtubules arranged in triplets parallel to the long axis of the c. The c. is a highly specialized organelle for → tubulin assembly and serves in intracytoplasmic anchoring for the outgrowth of ciliar and flagellar microtubules (→ basal body; cilium) and as organization center of → spindle microtubules.

Centrioles involved in spindle organization are most often found associated in orthogonal pairs (diplosomes) next to the → nuclear envelope of interphase nuclei. At the beginning of karyokinesis these two separate to the poles. After karyokinesis, new centrioles are assembled orthogonal (at right angles) to the old ones.

Centriolar assembly appears to be analogous to the spontaneous assembly of a virion and coincides with nuclear DNA replication (S period of interphase). There are indications that some centrioles may contain DNA (→ centriole satellite).

centriole cycle — a cycle which includes duplication, elongation, and polar migration of centrioles.

centriole satellite (de Harven and Bernhard 1956) — any of the appendages which may surround the nine centriolar triplets of microtubules.

centripetal selection — → selection.

centrodesmose (Heidenhain 1894) — a set of fibrous connections between the cell centers during the early stages of their separation (→ spindle).

centrolecithal (Balfour 1880) — of egg cells the yolk of which is accumulated towards the middle instead of at one end (→ telolecithal ova) or scattered more or less evently throughout the cytoplasm (oligolecithal ova).

centromere (Waldeyer 1903) — the site (or sites) of attachment of chromosomes to the mitotic or meiotic → spindle. The → microtubules are attached to a trilaminar disk-like plate structure (the kinetochore) on either side of the centromeric constriction which consists of two electron-dense layers of → chromatin and an electron-translucent middle layer. The outer layer of this stratified structure (about 30–40 nm in thickness) is often resolvable into two parallel or plectonemically coiled 9–10 nm fiber sublayers to which microtubules are attached.

Yeast centromeric DNA shares three sequence elements, referred to as conserved centromeric DNA elements (CDE). The A+T-rich central core element (CDE II) is flanked by two different short conserved sequences CDE I (8 bp long) and CDE III (25 bp long).

CDE II regions share three features: (1) They are similar in length (78 to 86 bp); (2) the base composition is always more than 90% adenine plus thymine; (3) they contain stretches of A and T residues 5 to 7 bp in length.

The c. is subject to its own cycle of functional reproduction: It divides, ordinarily, quite regularly at mitosis, and at the second meiotic division (meiosis II). In mitosis and meiosis II, the c. of each chromosome behaves as two units (one for each chromatid), in first meiotic division (meiosis I) as an undivided unit (→ centromere orientation). Transverse instead of longitudinal splitting of the c. at meiosis or mitosis is called → centromere misdivision and may give rise to telocentric chromosomes and → isochromosomes.

The following main types of centromeres (kinetochores) may be distinguished: (1) **Localized centromere:** the → chromosome possesses a permanently localized c. region which, provided the c. is not absolutely terminal, divides it into two arms of equal or unequal length (→ idiogram). Such chromosomes are called "monocentric" ("eucentric"). Chromosomes with two (or more) localized centromeres, dicentric (or "aneucentric") chromosomes, may be produced from them as a result of certain → chromosome mutations. The behavior of such chromosomes during the nuclear division is generally abnormal however (→ bridge-breakage-fusion-bridge cycle). The loss of the localized c. gives rise to acentric chromosomes which are incapable of moving in the → spindle and as a rule are rapidly eliminated.

During metaphase of mitosis, the c. generally appears as a relatively despiralized, negatively → heteropycnotic constriction (→ allocycly) of the chromosomes ("primary constriction"); these constrictions are generally not detectable ("isopycnosis") during the corresponding stage in meiosis I because of a different type of → chromosome coiling. (2) **Neocentromere** (Rhoades 1952): Chromosome ends (→ telomere) which under certain conditions, show movement on the spindle during mitosis and meiosis, just as do the localized centromeres of these chromosomes. They are secondary centromeres whose activity results in the chromosome ends moving first during anaphase movement. (3) **Nonlocalized centromere:** In this case the attachment of the spindle fibers is not confined to a strictly localized segment of the chromosome but takes place along the entire length of the chromosome. As a rule all parts of the chromosome have active spindle — mobility and acentric fragments are lacking. Distinction is made between two variants of this type: (a) The multiple or polycentric (polykinetic) condition: Each chromosome is furnished with many c., separated by small noncentric segments. Chromosomes of this type are actually multiple or compound structures. The only certain example of this organizational type is that of the germline chromosomes of some ascarid nematodes (→ nuclear differentiation). (b) The diffuse or holocentric (holokinetic) condition (Hughes-Schrader and Ris 1941): Every point along the entire chromosome shows active mobility on the spindle. This type of chromosome occurs inter alia in Hemiptera, Homoptera, Lepidoptera, a few Protista and, among higher plants in the genus **Luzula**. (4) **Semi-localized centromere** (Vaarama 1954): The active spindle mobility is coupled to a localized primary c. in mitosis but in meiosis is displaced to another localized site (secondary, tertiary c.). Chromosomes with semi-localized c., as are found in **Pleurozium**, are chromosomes with potentially multiple centromeres and behave functionally like those with localized c. but the position of spindle activity undergoes a regular shift (→ centromere shift). Semi-localized centromeres are regarded as intermediate between the multiple and the strongly localized centromere types.

centromere distance — the distance of a gene from → centromere, measured in terms of → recombination frequency.

centromere effect (Beadle 1932; Mather 1939) — the phenomenon that genetic loci located close to the → centromere show reduced levels of → genetic recombination.

centromere interference — an inhibitory influence (positive → interference) exerted by the → centromere on → crossing over and the distribution of → chiasmata in its vicinity. For each individual chromosome there is a characteristic distance between the centromere and the point at which the first (proximal) crossing over takes place; this is the so-called "differential distance" (Mather 1936).

centromere mapping — the localization of → centromeres on gene maps. Genotypes with a telocentric chromosome additional to the diploid complement (telotrisomics) are especially useful for this purpose. When one of two linked genes is associated with the telocentric arm and the other not, the centromere is assumed to be situated between these two marker genes. The more closely these genes are linked, the more accurately is the position of the centromere determined.

centromere misdivision (Darlington 1939) — an anomalous transverse — instead of lengthwise — division of the → centromere region in chromosomes with localized centromeres. It corresponds to a break in the centromere and can occur in each of the structurally distinguishable zones. If the break occurs in the innermost zone, → isochromosomes with identical arms may be produced. Breaks in other zones lead to chromosomes with terminally localized centromeres, known as telocentric chromosomes.

centromere orientation — the process of orientation of → centromeres during prometaphase of mitosis and meiosis. C.o. is achieved as follows: Each chromosome (mitosis) or → bivalent (meiosis I) exploits two independently acting centromeres, each of which is orientated towards one of the two spindle poles. A stable orientation (→ congression) of the centromeres of all the chromosomes along the spindle equator ("metaphase plate") results from the interaction between the spindle poles and centromeres via the spindle fibers.

Sister centromeres orient to opposite spindle poles (amphiorientation of Bauer et al. 1961) in mitosis, but to the same pole (synorientation) in one (usually the first) of the two meiotic divisions. Mitotic chromosomes with one oriented and one unoriented c. are called mono-oriented (monotelic orientation). If both centromeres are oriented towards the same pole in mitosis, the chromosome is maloriented. During metaphase an equilibrium position is established and is maintained until the beginning of the anaphase movement (→ chromosome movement).

Initial orientation established at the start of prometaphase is often inappropriate and, therefore, re-orientation is of critical importance for controlled chromosome distribution. Re-orientation means the loss of chromosomal spindle fiber connection to one pole followed by the formation of a new connection to the opposite pole. Without re-orientation abnormal chromosome complements would result.

The terms auto- and co-orientation of centromeres (Darlington 1936) are used in describing the orientation process.

(1) **Auto-orientation of centromeres:** The orientation of the two sister centromeres of mitotic chromosomes and (in meiosis) of → univalents towards opposite cell poles ("amphiorientation"); the centromeres do not orientate themselves in any direct relation to one another but lie directly on the equator.

(2) **Co-orientation of centromeres:** The process of mutual orientation of centromeres of → bi- and multivalents during one meiotic division (usually meiosis I), leading to the arrangement of pairing associations on the spindle equator (in this case sister centromeres

orient to the same pole: "synorientation"). For bivalents the arrangement is such that one centromere is "above", the other "below" the spindle equator. In cases where the pairing associations are composed of more than two chromosomes ("multivalents") several types of co-orientation are distinguished. The type of orientation to which a multivalent conforms depends on the location of the → chiasmata: (a) **Linear co-orientation:** all the centromeres of the multivalents lie one behind the other in the spindle; as a rule the chromosomes in the middle of the chain give rise to "false univalents"; (b) **convergent co-orientation:** Alternate centromeres are aligned to the same pole; (c) **indifferent co-orientation:** individual centromeres of the multivalents appear to have no special bearing on the orientation of each other (this also generally leads to the formation of "false univalents"; (d) **parallel co-orientation:** Neighboring centromeres are aligned to the same pole (this type of orientation first becomes possible in quadrivalents which consist of four paired chromosomes).

The various types of orientation of multivalents give rise to various forms of chromosome distribution to the poles and influence the relative propoprtions of genetically balanced and unbalanced meiotic products.

centromere plasmid — a → plasmid containing a cloned → centromere (CEN) from yeast and functioning as a minichromosome in yeast cells. Functionally, CEN plasmids exhibit three characteristics of chromosomes: (1) They are mitotically stable in the absence of selective pressure; (2) they segregate during → meiosis in a Mendelian manner; (3) they are found at low copy number in the host cell.

centromere repulsion — the mutual repulsion of the centromeres of the paired chromosomes towards the end of the meiotic prophase (in diplotene and diakinesis). The extent of this repulsion is possibly wholly or partially responsible for the particular degree of → chiasma terminalization.

centromere shift — (1) the displacement of the centromere region by structural changes of chromosomes (→ chromosome mutation) such as → inversions, → transpositions, and → translocations (= true c.s. according to White 1949). (2) Change of centromere position on the chromosome without detectable structural changes during meiosis of certain species (Coleman 1948). It results in the occurrence of homologous chromosomes (within a cell or in different cells) with differently located centromeres; their behavior corresponds to that associated with semi-localized → centromeres.

centromeric chromomere (Östergren 1947) — → chromomere.

centromeric fusion — a type of → centric fusion.

centromeric granule — a specific part of the → centromere associated with the → spindle microtubules.

centromeric heterochromatin — → heterochromatin associated with the → centromere. There is evidence that c.h. contains DNA with a specific and relatively uniform base composition, so-called → satellite DNA or → repetitious DNA.

centromeric vector — any of the yeast → cloning vectors (shuttle vectors) that can replicate autonomously in **S. cerevisiae** due to the presence of an → ARS element and are stably maintained in low copy number due to the presence of a → CEN element (= centromeric plasmid). The basic feature of c. vectors are: (1) Small size (4–6 kb); (2) high copy number in **E. coli** due to the presence of the pUC origin of replication; (3) ability to screen for recombinants via inactivation of lac $Z'\alpha$ complementation; (4) availability of multiple → restriction sites for easy subcloning; (5) low copy, stable maintenance in yeast; (6) availability of multiple selectable and visual markers expressed in yeast.

centronucleus (Boveri 1901) — a cell nucleus (→ nucleus) with an intranuclearly localized → centrosome.

centroplasm (Erlanger 1897) — (1) = → centrosome. (2) The central region of the protoplasts of blue algae which remains free of the lamellar packets of the → chromatoplasm and in which DNA-containing structures exist.

centrosome (Van Beneden 1883; Boveri 1888) — a major organizer of the → cytoskeleton and one of several → microtubule organizing centers of most animal cells. The c. is found at each pole of the mitotic → spindle in dividing cells and at the focus of microtubule asters during interphase. The c. is composed of two components: (1) A pair of → centrioles, and (2) an associated electron-dense matrix, the pericentriolar material (PCM), that surrounds the centrioles.

At the end of → mitosis each daughter cell receives one c. and by the start of next mitosis each daughter contains two, i.e., the number of centrosomes formed before each division is precisely controlled and c. reproduction must be properly coordinated with nuclear events. Higher plant cells usualy lack a c. and centrioles; instead they display large, more dispersed membraneous vesicles with microtubule organizing capacity.

centrosphere (Strasburger 1893) — the area around → centrioles (= → centrosome).

cephalobrachial (Levitsky 1931) — of → heterobrachial chromosomes whose short arm is represented by a small chromosome segment (= acrocentric).

certation (Nilsson 1915) — competition in growth rate between pollen tubes (= pollen tube competition) of different genotypes, resulting in unequal chances of accomplishing → fertilization (Darlington and Mather 1949). Competition takes place in the style of the female flower and results in a greater percentage of fertilization by pollen of the more compatible genotype, which reach the ovules first and leave insufficient egg cells for fertilization by genotypically less compatible pollen. As a result of c. segregation ratios may be distorted or the → sex ratio may shift in those cases where male and female determining pollen (male heterogamy: → sex detemrination) have a differential chance of fertilization.

Cf — = → colicinogenic factor.

C-factor (Muller 1926) — any gene or heterozygous chromosome structural change (in most cases → inversions) reducing the frequency of → crossing over.

CG suppression — in genomic DNA of higher organisms, the dinucleotide CG is present as about 1% of all dinucleotides, compared with a frequency of about 4% expected from the content of C and G in the same DNA. Thus the ratio (CG observed/CG expected) is suppressed to about 0.25%. CG s. is supposed to result from the fact that cytosine residues within the CG dinucleotide are largely methylated (→ DNA methylation) as 5-methylcytosine with the dinucleotide^{me}CG accounting for about 90% of the 5-methylcytosine in mammalian DNA. Over time, 5-methylcytosine tends to deaminate to thymidine, resulting in the depletion of CG s. by conversion of ^{me}CG to either TG or CA.

CG value — → GC value.

chain-initiating codon — → initiator codon.

chain initiating mutation — a mutation which gives rise to an → initiator codon.

chain-terminating codon — → terminator codon.

chain-terminating mutation (Sarabhai et al. 1964) — a → nonsense mutatation which gives rise to a → terminator codon. Certain c.-t.m. abolish the expression of distal genes in multigenic → operons and are therefore → polar mutations. The degree of polarity exhibited depends on the position of the c.-t. mutation in the proximal gene. If the mutations are arranged in genetic order, several gradients of polarity may be found for the expression of a distal gene. The transition between one gradient and the next appears to coincide with the position of regions which allow the resumption of translation following chain termination.

chalone (Shaver 1916; Bullough 1967) — any of the tissue-specific mitotic inhibitors that control cell proliferation and tissue growth in vertebrates by negative feedback inhibition. Chalones are characterized as follows: (1) They inhibit mitosis in vitro and in vivo; (2) their action is reversible and they are not cytotoxic; (3) they are synthesized by mature cells of the tissue upon which they exert influence; (4) they are released from cells and circulate in blood and humoral fluids (chemical messengers); (5) they are tissue-specific, but not species-specific.

Many chalones appear to be glycoproteins with a mass of about 30,000 to 50,000. A smaller group of such agents has a mass of only 1,000 to 3,000.

The mode of action of chalones is largely conjectural but their activitiy may depend on the rate of cellular membranes which may serve to concentrate c. within the cell. Membrane-chalone interaction is the probable source of the tissue specificity of a c. Stress hormones commonly act as cofactors.

One suggestion is that the "mitotic genes" are activated if the c. action is weak, while the genes for tissue function are activated if the c. action is strong.

Chambon's rule — the base sequence of an → intron begins with GT and ends with AG.

character (Bateson 1907) —a c. (= trait, phene) is any observable (phenotypic) feature of the developing or fully developed individual: a biochemical property, a cellular form or process, an anatomical structure, an organ function, or a mental characteristic (Stern 1960). Genetic characters are in most cases advanced stages, often end points of developmental sequences, the individual steps of which may be governed by different genes. They are produced by → gene action, are removed from the genes by at least one and, often, numerous steps and may be mainly controlled by the → alleles of one or more gene loci in cooperation with the environmental conditions (→ expressivity; penetrance). It is not the character per se which is inherited, but the genetically determind → reaction norm of the → genotype. Genes merely provide the potentiality to produce or contribute to the production of a given c. Whether or not the c. is realized depends upon the influence of the external as well as the genetic environments (→ genetic background).

Characters may be controlled by autosomal or sex-linked genes; they may be either → dominant, → recessive, intermediate, → sex-linked, → sex-limited, or → sex-controlled. They are determined autonomously (→ autophene) if the c. in question is produced as a consequence of the allele being in the cell or are dependent for their expression on the type of cell or organ (→ allophene) independent of the gene (intra- versus intercellular gene action). (1) **Discontinuous, qualitative, or oligogenic characters:** Characters controlled primarily by a few genes each of which has a large contribution to the process of character formation as compared with the nonhereditary fluctuations of the character concerned. (2) **Continuous, quantitative, or polygenic characters:** In contrast to (1) these characters are controlled by many genes with small individual contributions, and show continuous variation. Their study depends on measurements rather than counts. They almost invariably show a large environmental component of variation (→ heritability). (3) **Acquired characters:** Phenotypic modifications arising purely by environmental influences during the developmental process of an organism.

If correlated variations between two or more characters are evident and the basis of the correlation is genetic, these are referred to as "genetic character correlations" (Simpson 1953). Such correlations may be the result of genetically controlled gradients of growth (including allometries), → pleiotropy, → linkage of the genes responsible for the formation of the characters, and correlated or random incorporation of genes into a genetic system which is integrated on another basis.

character association (Anderson 1949) — = → character coherence.

character coherence (Grant 1979) — the tendency of parental species character combinations to stick together in the progenies of hybrids (= character association). C.c. is a property of many natural hybrid populations in plants and may be due to → pleiotropy, resistance to the breakup of parental species character combinations by → recombination, and selection against recombination types.

character convergence — a process whereby two relatively evolved species interact in such a way that either or both converge towards the other with respect to one or more traits (→ character displacement).

character displacement (Brown and Wilson 1956) — in the case of two populations or species whose distributions geographically overlap the accentuation of differences between them in the zone of overlap (→ sympatric) and the weakening or entire loss of differences in the parts of their range outside this zone. The differences may involve morphological, physiological, ecological, or behavior characters (= character divergence).

character divergence (Darwin 1859) — = → character displacement.

character pair — two more or less markedly differing → characters, the origins of which may be of variable complexity. In the simplest case two different → alleles of a genetic locus are responsible for the difference in character.

character progression — the geographic gradation of the expression of specific characters over the range of distribution of a race or species; the gradation may be considered from the center to the border or from one end of the distribution to the other.

character stasis — long-term constancy in a phenotypic character within a lineage.

charomid (Saito and Stark 1987) — a → cosmid vector up to 52 kb long, bearing 1 to 23 copies of a 2-kb spacer fragment linked in head-to-tail tandem arrays. Like cosmids and lambda phage, charomids can be packaged in vitro for efficient introduction into bacteria and allow cloning of restriction fragments of any size up to ~45 kb. The number of spacer fragments is variable and determines the lengths of inserts that can be accommodated within the total length of up to 52 kb.

chasmogamy — in plants, fertilization after opening of flower or floret, as opposed to → cleistogamy.

check cross — the crossing of an unknown → genotype with a phenotypically similar individual of known genotype. The F_2 segregation serves to establish whether the same phenotypes result from the action of identical or nonidentical alleles of the same genetic locus or from the action of nonallelic genes.

chemodifferentiation — the chemical changes which precede visible differentiation in the process of → cytodifferentiation during embryonic development.

chemotaxis — the directed movement of motile cells or organisms towards (positive c.) or away from (negative c.) a chemical source in the surrounding environment.

chiasma (Janssens 1909) — that place in a → bivalent, → multivalent, or in the case of intrachromosomal (internal) pairing within a chromosome (→ isochromosome), from late prophase of meiosis up to first anaphase where (a) pairs of homologous → chromatids remain in contact although they are separated in other parts, and (b) an exchange of homologous parts between nonsister chromatids of the chromosomes has taken place by → crossing over ("chiasmatype theory"). In achiasmate meiosis so-called → collochores may take over function (a) of chiasmata.

With normal chiasmate meiosis, at least one c. is necessary to hold the pairing partners in contact after repulsion ("chiasma theory of → chromosome pairing", Darlingston 1929) between them begins in diplotene. Premature separation of the pairing partners is the consequence of failure of chiasma formation bringing → univalents into being (→ asynapsis; desynapsis). In organisms with large chromosomes two chromatids are seen to cross reciprocally from one chromosome to the other at each c.

Classes of single chiasmata (Darlington 1929) are: (1) **Interstitial chiasma:** on each side of the c. is a length of chromatid. (2) **Lateral chiasma:** the chiasma is terminal as to two chromatids and interstitial as to two others (two kinds, symmetrical and asymmetrical). (3) **Terminal chiasma:** the chiasma is localized at the ends of the pairing partners following → chiasma terminalization. (4) **Multiple chiasma:** a terminal chiasma in which three or four pairs of chromatids participate (as in → multivalents). (5) **Imperfect chiasma:** a chiasma in which one of the four crosswise associations is broken prior to anaphase (= "incomplete chiasma").

If two chiasmata are present in one bivalent (its four chromatids represented by A, A′, B, B′) there may be three different relationships between them (crossing over) (Darlington 1937): (1) **Reciprocal (comparate) chiasmata:** The chromatid A and B or A′ and B′ participate in both chiasmata (= "regressive chiasmata", "two-strand double crossing over"). (2) **Complementary (comparate) chiasmata:** chromatids A and B participate in one chiasma, chromatids A′ and B in the other ("digressive chiasmata", "four-strand double crossing over"). (3) **Diagonal (disparate) chiasmata:** the chromatids involved in one chiasma are, e.g., A and B and those in the other A and B′ or A′ and B (= “progressive chiasmata”, “three-strand double crossing over”).

The formation of a c. is understood to be the consequence of a process of breakage and reunion at homologous sites of both chromatids during → crossing over in which normally only the nonsister chromatids of the paired chromosomes are involved. The reunion process is directed such that no dicentric chromatids are produced by asymmetric reunion.

The formation of a c. represents the culmination of an interrelated sequence of processes each presumably with its own control mechanism. C. formation is temporally associated with a repair type replication (→ repair replication) and there is increasing evidence that both nucleic acid synthesis and protein synthesis are involved (→ nicking enzyme). The crossing over frequency is one half of the chiasma frequency per pairing configuration since only two chromatids of the four are engaged in crossing over at one point. The “map length” of a chromosome is the average chiasma frequency of that pair of homologs, times 50 (→ chiasma map). The average frequency of chiasmata per pairing configuration per meiocyte, or per karyotype, is controlled both by genetic and environmental factors.

Differential distance (Mather 1936) is the interval between the first chiasma and a fixed point (usually the → centromere or → telomere) in a bivalent relative to which this c. is believed to originate. The interval between the centromere and the nearest chiasma may be called the **residual centric distance** (Southern 1967), the distance between the telomeric end and the nearest chiasma may be called **residual telomeric distance** (Southern 1967). Both residual distances are determined by an interaction between the length of a chromosome arm and its chiasma frequency.

Interstitial distance (Mather 1936) is the distance between successive chiasmata in those cases where more than one is present in a bivalent.

chiasma centralization (Resende 1953) — the movement of the chiasmata to the chromosome center in chromosomes with nonlocalized → centromeres or in dicentric chromosomes (→ chiasma terminalization).

chiasma frequency — the mean number of → chiasmata of the → bivalent, or of the whole → karyotype.

chiasma interference (Mather 1933) — the occurrence less often (positive c.i.) or more often (negative c.i.) than expected by chance of two or more crossovers and chiasmata in one segment of a chromosomal pairing configuration in individual cells at meiosis (→ interference). The c.i. normally does not, but in some cases may extend across the centromere region (= chiasma position interference).

chiasma localization (Darlington 1931) — → chiasma terminalization.

chiasma map — a map which gives the distribution of → chiasmata in relation to cytological markers. Plotting the frequency distribution of chiasmata along the individual chromosome arms produces chiasma maps from which genetic maps may be directly constructed. Each chiasma, representing the result of one → crossing over event, gives rise to two recombinants and two nonrecombinants and is taken to correspond to a genetic length of 0.5 Morgan (→ Morgan unit). The standard total genetic map length and the genetic map length of individual chromosomes and chromosome arms are then obtained by halving the corresponding mean number of chisamata in → meiocytes from representative individuals.

chiasma movement index (Slizynski 1955) — a measure of the degree of → chiasma terminalization derived from the average number of chiasmata per → bivalent in an early and in a later stage of prophase of first meiotic division. A high c.m.i. results from quick movement of chiasmata between the two prophase stages compared. The c.m.i. differs from the “terminalization coefficient” in that its determination requires scoring the number of chiasmata in two succeeding stages of prophase.

chiasmata position interference (Carter and Robertson 1952) — = → chiasma interference.

chiasmate — of a → meiosis with normal → chiasma formation (→ achiasmate, cryptochiasmate).

chiasma terminalization (Darlington 1929) — a progressive shift between diplotene and metaphase of the first meiotic division in the distribution of chiasmata along the arms of paired chromosomes from their points of origin to more distal positions. The terminalization process may be complete, partial, or entirely lacking. In the last case the chiasmata are called localized (Darlington 1931).

The "terminalization coefficient" is a measure of the degree and progress of c.t. and denotes the fraction of the chiasmata that are terminal at a given stage between diplotene and metaphase I of meiosis (the coefficient is 1 when all chiasmata are terminal). Generally terminalization coefficients seem to be higher in organisms with small chromosomes and in those showing some distal chiasma localization initially.

chiasmatype theory (Janssens 1909, 1924; Darlington 1929) – a theory postulating that true chiasmata are the direct result of → crossing over, being formed at precisely the points where the exchange of segments between (nonsister) chromatids took place. In contrast to this "one-plane theory" the "two-plane theory" postulates (Sharp 1934) that the cross-figures appear because the chromosomes of the pairing configurations split in contiguous regions along perpendicular cleavage planes. Both the "two-plane" and its variant, the "neo-two-plane theory" (Matsuura 1937), are only of historical interest now since all experimental evidence favors the chiasmatype theory.

chi element (Lam et al. 1974) — any of the special sites (recombinators) near which enhancement of → genetic recombination occurs. Chi activity is determined by a unique nucleotide sequence (5′G-C-T-G-G-T-G-G3′) and stimulates the Rec ^{BC} pathway of bacterial recombination.

chi form (Potter and Dressler 1976) — a recombination intermediate consisting of double-length DNA molecules held together at a region of DNA homology and showing a special symmetry: The point of contact between the unit-size molecules occurs so as to divide the structure into pairs of equal-length arms. The molecules are involved in → genetic recombination.

chimera (Winkler 1907) — in most cases a plant, more rarely an animal, composed of tissues of two or more → idiotypes owing to → mutation, somatic → segregation, or grafting (= mosaic).

Chimeras form a heterogeneous group and in plants may be classified in three different ways (Tilney-Basset 1963): (1) **According to their origin** into those arising by: (a) Spontaneous mutation, (b) induced mutation, (c) sorting out from variegated seedlings (→ variegation), or (d) grafting. (2) **According to their structure** into: (a) Sectorial (different tissues grow side by side and occupy distinct sectors of varying size), (b) periclinal (different tissues are disposed one with the other, one occupies the center of the plant structure as a core, the second tissue grows around it as covering layer), (c) mericlinal (actually an interrupted periclinal c. with one kind of tissue of limited extent placed in a peripheral position and covering a central core). (3) **According to their behavior** into: (a) Species c. or "graft c.", produced artificially by grafting together different species or genera. At the place of union a callus tissue is formed from the stock and scion and this gives rise to adventitious buds of chimerical structure. A.c. of this type is recognized by the general differences between the cells of the species concerned. (b) Chromosomal c., the tissue layers of which differ in their chromosome number. They are artificially produced by treatment of germ layers with → spindle poisons, e.g. colchicine, and sometimes arise spontaneously. (c) Gene-differential c., which arises by somatic mutation of a gene to its recessive allele, or back-mutation to its dominant allele, in one of the germ layers. This type of c. occurs spontaneously and may be induced by → mutagens. (d) Plastid-differential c., which arises either by spontaneous → plastid mutation or by sorting out of two kinds of plastids from a mixed egg or zygote. This type of c. is recognized by the → variegation of the leaves and distinguished from a similar gene-differential c. by the different, non-Mendelian mode of inheritance.

In animals, the following types of chimerism (→ aggregation c.; injection c.) may be distinguished: (1) Zygotic chimerism: Fertilization by one sperm of an ovum and by another sperm of the first or second polar body derived from the same primary oocyte. A single individual results in both cases. (2) Postzygotic chimerism: Fertilization by two sperms of two separate ova and fusion of the two zygotes at a very early stage of development. A single individual results. (3) Twin chimerism: Fertilization by two sperms of two separate ova. Two individuals result.

chimeric DNA — a → recombinant DNA molecule which contains sequences from more than one organism (→ recombinant DNA technology).

chimeric gene — a → hybrid gene.

chimeric plasmid — a → plasmid containing a cloned DNA fragment (→ cloning vector; molecular cloning).

chlamydospore — in fungi, any of the multinucleate intercalary structures developed within hyphae; they usually acquire thick walls and their cytoplasm accumulates lipid material and glycogen (→ conidiospore; sporangiospore).

chloroplast (Schimper 1885) — in plants, a cell organelle characterized by a limiting double membrane and an internal, chlorophyll-containing lamellar structure embedded in a protein-rich stroma. The chloroplasts are capable of light-dependent photophosphorylation and of photosynthetic CO_2 fixation and represent → plastids specialized for photosynthesis.

Chloroplasts are endowed with their own DNA complement and are capable of DNA-directed RNA and protein synthesis. They contain DNA polymerase and a unique class of 70S → ribosomes (forming → polysomes) that are distinct from the 80S particles of the surrounding cytoplasm. The subunits of the 70S ribosomes are 50S and 30S, respectively, and contain → ribosomal RNA molecules of 23S and 16S; the large subunit contains also a 5S rRNA and additional small RNA molecules (4S to 7S, depending on the species). The c. ribosomes contain 50 to 55 proteins. Only about one third of these are encoded in the chloroplast DNA and synthesized within the organelle.

Chloroplasts contain about 56% protein, 32% lipids, and 8% chlorophyll; they are of various shapes (diameter 4 to 6 μm), spherical or ovoid. Their number in a cell varies from one up to 500. The biogenesis and maintenance of a functional c. is under the control of both nuclear genes and genes of the c. genome (→ chloroplast DNA). Nuclear DNA is involved in many steps leading to the synthesis of chlorophyll, of carotenoids, and of the photosynthetic electron transport system.

Chloroplasts originate from submicroscopic ameboid proplastids which have double membranes. When a proplastid reaches a diameter of about 1 μm, its inner membrane invaginates to form lamellae which in light continue developing and differentiating to the mature condition found in the c. (→ plastid).

chloroplast DNA (Chun et al. 1963; Sager and Ishida 1963) — (ct or cp DNA); the genetic material of → chloroplasts; ct DNA is in a covalently closed circular conformation (37 to 62 μm in length) and ranges in molecular sizes from 120 kb to 180 kb. The molecules show remarkable structural homogeneity. An inverted repeat of about 22 kb divides the molecule into a larger (→80 kb) and a smaller (→19 kb) unique sequence. The repeated region contains the genes for chloroplast ribosomal RNA. Each chloroplast contains some 20 copies of its circular DNA molecule.

The replication of ct DNA initiates with the formation of two → displacement loops. The two displacing strands are hydrogen-bonded to the opposite parental DNA strand. They expand towards each other, forming a structure like a Cairns-type of replicative intermediate when the two displacing strands elongate past each other. Replication is by a → rolling-circle type of structure. ct DNA molecules present in each chloroplast can recombine with each other.

Chloroplast DNA sequences have been discovered in the nuclear and mitochondrial genomes (→ promiscuous DNA).

chloroplast genome — The circular → genome of → chloroplasts. The double-stranded, covalently closed circular DNA molecule (120 to 180 kb) codes for about 50 polypeptides in addition to a partial set of ribosomal proteins, a full complement of → ribosomal RNA's and probably a complete set of → transfer RNA's. Polycistronic transcripts have been observed and many c. transcripts are processed posttranscriptionally (→ chloroplast DNA). The c.g. exists as multiple copies per chloroplast.

The → promotors of the c.g. have prokaryotic-like −10 and −35 promoter elements. → Introns analogous to class I and II mitochondrial introns have been found.

chondriamone — = → mitochondrial genome.

chondrioid (Kellenberger and Huber 1953) — in bacteria, a cell organelle functionally equivalent to the → mitochondrion of higher organisms. Chondrioids originate by a process of invagination from the cell membrane surrounding the bacterial cell and in addition are assumed to be capable of reproduction by division.

In morphology chondrioids resemble mitochondria of the tubular type and consist of a covering membrane which surrounds numerous tubular and vesicular interior structures. They contain the same enzymes (inter alia: dehydrogenases, phosphorylases, cytochromes) as the mitochondria of higher organisms. In addition to the "normal" mitochondrial functions, they are involved in the reduplication of the nuclear equivalents (→ nucleoid; endosome) and in the formation of endospores.

chondriokinesis (Nicoloci-Roncati 1912) — the reproduction of → mitochondria and their distribution during mitosis and meiosis.

chondriome (Meves 1907) — (1) collectively the → mitochondria of the cell; (2) all hereditary determinants localized in the mitochondria; the → mitochondrial genome.

chondriosome (Benda 1902) — = → mitochondrion.

choriheterosis (Dodge 1945) — → heterosis effects connected with the → heterokaryotic state.

chreod (Waddington 1961) — a stabilized pathway of change, i.e., a time-extended trajectory of a system that, if it at some time diverts from its path tends to change in such a way as to return to a later section of the trajectory. Systems with this capacity are called homeorhetic.

chromatid (McClung 1900) — one of the two visibly distinct longitudinal subunits of all reduplicated → chromosomes (a half-chromosome) which become visible between early prophase and metaphase of → mitosis and between diplotene and the second metaphase of → meiosis. After these stages the chromatids are known as "daughter chromosomes". The chromatids are those longitudinal units of the chromosome which become separated during anaphase of mitosis and anaphase II of meiosis. Sister chromatids are the two chromatids derived from one and the same chromosome during its replication in interphase, as opposed to nonsister chromatids derived from homologous chromosomes. Normally, chromatids of homologous chromosomes carry identical → genetic information (identical linkage groups), apart from allelic differences between nonsister chromatids, and are similar morphologically. Each c. contains a continuous DNA double helix.

chromatid aberration — a category of chromosome structural changes (→ chromosome mutation) produced in one and not the other → chromatid of a chromosome as opposed to chromosome aberrations in which both chromatids are involved. Chromatid aberrations arise spontaneously or are induced experimentally by → mutagens during and after chromosome reduplication in the interphase nuclei.

chromatid break — a discontinuity in only one of the two chromatids of a chromosome (→ isochromatid break).

chromatid bridge (Smith 1935) — a dicentric chromatid with the two centromeres passing to opposite poles during anaphase (→ bridge-breakage-fusion-bridge cycle). Chromatid bridges arise primarily from spontaneous or experimentally induced chromatid aberrations and secondarily (secondary structural change) through crossing over between the chromosomes of → structural heterozygotes in meiosis.

chromatid conversion — a type of → gene conversion that is inferred from the existence of identical sister-spore pairs in a fungal octade that shows a non-Mendelian allele ratio.

chromatid core (Howell and Hsu 1979) — the axial structure (= axial filament or scaffold) of the → chromatid.

chromatid interference (Mather 1933) — → interference.

chromatid non-disjunction (Sansome 1933) — (1) the distribution of both sister chromatids of a chromosome to one cell pole during anaphase of mitosis or in anaphase II of meiosis; (2) the passing of homologous segments of chromatids to the same pole in → translocation heterozygotes (→ adjacent distribution).

chromatid segregation — → autoploid.

chromatid tetrad — a → bivalent in first meiotic division consisting of two paired chromosomes and thus of four chromatids.

chromatid translocation — → translocation.

chromatin (Flemming 1882) — the material of which both interphase and mitotic/meiotic chromosomes consist. C. is a complex ensemble of (1) about equal parts by weight of DNA and a group of low molecular weight basic proteins, the → histones, and (2) varying proportions of a diverse group of other proteins which are called nonhistone chromosomal proteins.

The fundamental building block of eukaryotic c. is the → nucleosome; each nucleosome consists of eight histone molecules (two each of H2A, H2B, H3, and H4) with 146 bp wrapped in 1.75 turns around the protein core. The fifth histone, H1, interacts with an additional 20 bp of DNA (linker DNA) to complete two turns of DNA around the histone core. The nucleosomes contain specific binding sites for nonhistone proteins the best known of which are the → high mobility group (HMG) proteins.

The nucleosomes are arranged into supranucleosomal clusters (10 nm fiber) and these are packaged into loops (each containing about 30 to 90 kb of DNA) by supercoiling to form the 30 nm c. fiber. The loops (domains) are anchored to a central → chromosome scaffold and to a protein framework called the nuclear matrix. The 30 nm fiber can undergo further coiling and condensation and results in structures observed by light microscopy as a → chromosome.

C. is generally divided into two fractions: (1) → Euchromatin (dynamic c.), which is actively transcribed, and (2) condensed → heterochromatin (static c.), which is usually inactive in supporting transcription. Domains of DNA that contain transcribed sequences show an increase in sensitivity to nucleases (→ nuclease hypersensitive sites) near the 5′ end of active or potentially active genes.

Replication of the eukaryotic genome involves a drastic change in c. structure at or near the → replication fork (→ DNA replication).

chromatin body (Giardine 1901) — in early oogenesis of dytiscid water beetles, a special → rDNA-containing body in the oocyte nucleus (Body of Giardini). It disappears at about → vitellogenesis and its material becomes dispersed in the nuclear interior. The c.b. is assumed to contain extra copies of genes coding for → ribosomal RNA. The repeating units of rDNA in the c.b. are circular.

chromatin diminution (Boveri 1887) — in some nematodes, copepods, insects, and some ciliates, a process by which portions of → chromatin are eliminated from the presumptive somatic cells, thus leading to somatic tissues containing less genetic material than germ line cells. The eliminated material usually consists of highly repetitive (satellite) DNA, but some middle-repetitive and nonrepetitive sequences may also be expelled (→ chromatin elimination). The diminution mechanism must, at least in the case of elimination of intercalary heterochromatin, involve excision of the DNA in heterochromatic sections, e.g., by site-specific recombination and subsequent release of circular DNA.

chromatin elimination (Seiler 1914) — during oogenesis of butterflies and moths, the elimination of ribonucleoprotein (elimination chromatin) which originats from the chromosomes. C.e. may represent a specific type of transcriptive activity of the female genome during oogenesis.

chromatin fiber — the → deoxyribonucleoprotein fiber (→ chromatin; chromosome; nucleosome).

chromatin reconstitution (Dahmus and Bonner 1970) — the reconstruction of → chromatin with chromosomal constituents (histones, nonhistone chromosomal protein, chromosomal RNA) previously removed by chromatin dissociation. It is the source of the → nonhistone chromosomal proteins which during c.r. determine the specific pattern of in vitro RNA synthesis.

chromatinomere (Haapala and Nienstedt 1976) — any of the repeating → chromatin subparticles (→ nucleosomes) along a → chromatin fiber.

chromatoid body (Benda 1891) — a cytoplasmic juxtanuclear clump characteristic for spermiogenetic cells in a wide range of organisms. It is either formed by extrusion of nucleolar material or represents an aggregate of cytoplasmic material. The c.b. contains ribonucleoprotein material.

chromatophore (Schmitz 1882) — any particle (vesicular structures; diameter about 15 nm) isolated from photosynthetic bacteria and containing photosynthetic pigments. Such structures whose pattern may vary with the physiological state of the cell are surrounded by a membrane. They are scattered throughout the cytoplasm.

chromatoplasm — in Cyanophyceae which possess no discrete → plastids, a cell region of lamellar structure in which the → thylakoids are located. The membranes, to which the assimilation pigments of these cells are bound, arise through the invagination of the → plasma membranes which surround the protoplasts.

chromatosome (Simpson 1978) — a → nucleosome core particle containing 166 bp of DNA, a → histone octamer and one molecule of lysine-rich histone H1 (or H5). In → chromatin, chromatosomes are separated from their neighbors by 0–80 bp of spacer (linker) DNA, the exact length depending upon the chromatin source.

chromidia (Hertwig 1902) — basophilic fibrils in the → cytoplasm. They are composed of RNA-containing and RNA-free parts. Monné (1948) referred to the latter as "interchromidia".

chromidiome (Bernhard et al. 1952) — the complete set of all basophilic cell elements (= chromidial substance according to Cowdry). (→ ribosome).

chromocenter (Baccarini 1908; Tischler 1920) — any of the tightly coiled (→ chromosome coiling) chromosomes or chromosome segments within the interphase nuclei. Chromocenters consist of → heterochromatin and are positively → heteropycnotic (→ allocycly). Their number per nucleus may vary since heterochromatin has a tendency to fuse (→ chromosome pairing).

chromofibril (Yasuzumi 1951) — any of the fibrillar subunits of the → chromosome identified by electron microscopy.

chromomere (Fol 1891; Wilson 1896) — any of the linearly arranged, bead-like and compacted segments in mitotic and meiotic prophase → chromosomes and in → polytene chromosomes (→ band). The alternation of chromomeres and → interchromomeres results in chromosome-specific c. patterns (→ chromomere size gradient) which may serve to differentiate between the chromosomes of a → chromosome set.

Evidence is accumulating that the c. represents a unit of DNA replication (a → replication unit), chromosome coiling, RNA synthesis (→ genetic transcription), and RNA processing. The transcription products are molecules of → heterogeneous nuclear RNA. These transcripts are supposed to contain the equivalent of one or a few structural → gene sequences at the 3′ end and regulatory information (utilized for the processes of transcription, maturation and translation of the transcripts of structural genes). The DNA content of a. c. is sufficient to specify from ten to a hundred average-sized proteins. Regions within which

→ repetitious DNA sequences are densely crowded are also suggested to be components of the c. (→ g-region).

The chromomeres localized within the → centromere region are called "centromeric chromomeres" (Östergren 1947) or "kinetomeres" (Lima-de-Faria 1949).

chromomere pattern — the linear order of → chromomeres in chromosomes.

chromomere size gradient (Lima-de-Faria 1952) — the phenomenon that the chromomeres on both sides of the → centromere are large and become progressively smaller towards the chromosome ends (the c.s.g. constitutes one component of the so-called → chromosome gradient).

The c.s.g. represents a system of both structural and replicating order in the chromosomes of eukaryotes. Larger chromomeres which contain more DNA not only replicate later but also require more time for replication than smaller ones (Lima-de-Faria 1972).

chromonema (Wilson 1896) — The smallest light microscopic strand in → chromosomes and chromatids.

chromoplast — a → plastid containing (yellow or orange) pigments other than chlorophyll (→ chloroplast).

chromosite — → lysogenization.

chromosomal — pertaining or referring to the structure, constituents, and function of chromosomes.

chromosomal aberration — = → chromosome aberration.

chromosomal aneuploid (Dyer et al. 1970) — → aneuploid.

chromosomal inheritance — → inheritance of → genetic information contained in the chromosomes (= Mendelian inheritance) as opposed to extrachromosomal inheritance.

chromosomal in situ supression hybridization (Lichter et al. 1988) — (CISS); a method allowing efficient and specific labeling of entire individual (human) metaphase and interphase chromosomes by → in situ hybridization with DNA probes from flow-sorted chromosome libraries. Labeling specificity is due to suppression of the signals from the fast reassociating fraction of (human) genomic DNA: Biotin-labeled DNA inserts from the chromosomal library are partially preannealed with a titrated amount of total genomic DNA prior to hybridization with cellular or chromosomal preparations. Cross-hybridization of repetitive sequences to nontargeted chromosomes can be markedly suppressed under appropriate preannealing conditions. The remaining single-stranded DNA is hybridized to specimens of interest and detected with fluorescent or enzyme-labeled avidin conjugates following posthybridization washes.

chromosomal mobilization activity (Haas and Holloway 1977) — the promotion by → plasmids of the transfer of chromosomal sequences via recombinant excision plasmids. Bacteria which have evolved mechanisms for c.m.a. show a greatly enhanced potential for genetic variability in natural populations.

chromosomal mosaic — an individual showing at least two cell lines with different (either numerical or structural) → karyotypes, derived from a single zygote by post-fertilization events.

chromosomal protein — any protein associated with the eukaryotic chromosome (→ histone; nonhistone chromosomal protein).

chromosomal RNA — RNA transcripts (→ ribonucleic acid; genetic transcription) of chromosomal DNA, i.e., RNA complementary (cRNA) to specific chromosomal DNA sequences. A fraction of cRNA represents precursor molecules of → messenger RNA, → transfer RNA, or → ribosomal RNA (→ heterogeneous nuclear RNA).

chromosomal structural change — a change in chromosome structure (→ chromosome mutation) arising spontaneously or induced by → mutagens.

chromosomal walking (Bender et al. 1983) — a procedure used for the isolation and cloning of an ordered array of overlapping, contiguous regions of chromosomal DNA. C.w. allows mapping of entire chromosomes or even small eukaryotic genomes by overlap hybridization screening of → plasmid and → cosmid DNA libraries. C.w. is based on the systematic isolation of DNA from one end of a recombinant and its use to screen libraries of cloned DNA (→ molecular cloning) to isolate the adjacent DNA sequence (a "step"); the walk proceeds as a linear sequence of steps. Two characteristics are necessary for the hybridization probes: (1) They should reside in terminal positions of the genomic clone from which they are derived, and (2) they must be free of repeated elements.

chromosome (Waldeyer 1888) — a linkage structure (→ linkage) consisting of a c.-specific linear sequence of → genetic information (→ gene). Chromosomes of eukaryotes and prokaryotes are essentially involved in two main activities: (1) Those concerned with the transmission of genetic information from cell to cell and generation to generation. (2) Those concerned with the ordered release of this information to control cellular function (→ gene expression) and development.

The principal synthetic activities related to these functions are the synthesis of DNA and RNA, respectively.

Chromosomes are autoreduplicating structures of differential complexity in eukaryotes and prokaryotes whose number (per cell), morphology and organization are organism-specific characteristics. (1) The chromosomes of eukaryote organisms: Eukaryote chromosomes (→ chromatin) are characterized by their complex structure, their ability to produce a color reaction with basic dyes and their characteristic behavior during → mitosis and → meiosis, in which they undergo regular changes in their degree of compaction (→ chromosome coiling). During interphase of the → cell cycle they are included in the membrane-bounded cell → nucleus. The differentiation of chromosomes into → chromomeres, → heterochromatin, → euchromatin, → constrictions (→ centromere; → nucleolus organizer) are expressions of the differential coiling along the lenght of → chromatids and diagnostic tools for the identification of individual chromosomes of the → chromosome set. Replicated chromosomes consist of two such chromatids which are the functional subunits of chromosomes, i.e., the smallest distributable units.

Number and morphology of chromosomes are species-specific. If the centromere is not strictly terminal (telocentric chromosomes), each c. is divided by the centromere region into two equal (metacentric chromosomes) or unequal (acrocentric chromosomes) arms. The individual chromosomes of a species-specific set are mutually nonhomologous (each bears a different → linkage group). In → diploids each c. finds in the other set a structurally similar, homologous partner (provided no heterozygous → chromosome mutations are present in the diploid complement) with which it may pair in meiosis (→ chromosome pairing). Cells produced from diploid cells in the course of meiosis (germ cells, gametes, or gonia) contain only one chromosome set; they are → haploid. Somatic cells of a → polyploid species contain more than two chromosome sets.

The → sex chromosomes are causally related to → sex determination and are called X- and Y-chromosomes, in contrast to → autosomes. The standard chromosomes of a chromosome complement are called → A-chromosomes, as distinguished from → B-chromosomes or supernumeraries, which in certin species are present in addition to the A-chromosomes.
(a) **Chemical components of the chromosomes of eukaryotes.** The chromosomes contain almost equal amounts of DNA and → histones, a variable amount of → nonhistone chromosomal proteins, and a small quantity of chromosomal RNA. In addition, lipids, polysaccharids and metal ions have been reported to be components of the c.
Chromosomal DNA forms the backbone of the chromosome fibers. Each chromatid contains a single DNA double helix associated with chromosomal proteins and condensed or coiled along its length. Nonrepetitive (unique DNA) and → repetitious DNA sequences (serially repeated sequences) are interspersed in a generally ordered manner (→ g-region). This is reflected by the fact that eukaryotic chromosomal DNA after denaturation reassociates in a heterogeneous manner over a wide range of rates: some fractions associate rapidly (→ satellite DNA is represented in this fraction), others at an intermediate rate; a fi-

nal portion reassociates very slowly and is called unique or single-copy DNA (→ Cot value), since it associates at a rate expected for nonrepetitious DNA. The other portions represent repetitious DNA. Frequently part of the repetitive DNA differs from the bulk of chromosomal DNA in G + C value and hence appears as "satellites" (satellite DNA) when the DNA is banded in a CsCl density gradient. Such satellite sequences are located predominantly in the centromeric → heterochromatin.

Chromosomal DNA (→ deoxyribonucleic acid) is a multi-replicon structure. Each replicon (average size about 30 μm) has its own starting point (origin) at which → DNA replication is initiated at different times during the S period of interphase. Replication is semiconservative and bidirectional with two forks moving in opposite directions to create a serial array of replication bubbles. Finally the intermediate pieces in adjacent replicons fuse (fork movement at a rate of 2.6 kb/min /fork). There is evidence that protein synthesis is required for the initiation and maintenance of DNA replication throughout the S period of → interphase. The units of replication of DNA are activated asynchronously (asynchronous replication). Their number is about 2 to 10 $\times 10^4$ per cell. Early replicating DNA is relatively rich in G + C and late replicating DNA (usually localized in constitutive → heterochromatin) relatively rich in A-T base pairs. There are indications that the replication of chromosomal DNA might occur in association with the inner membrane of the → nuclear envelope.

Chromosomal proteins include two classes, the basic chromosomal proteins or → histones, and the acidic or → nonhistone chromosomal proteins (→ hertones).

The nonhistone proteins vary in quantity and there appears to be a limited heterogeneity between different tissues and species. Histones (which are absent in some fungi and dinoflagellates, and replaced by protamines in the sperm chromosomes of some animals) may be separated into five major fractions which do not vary significantly in type or quantity over a large range of organisms.

The newly synthesized chromosomal DNA becomes quickly associated with newly synthesized histone. Possibly, histone enters the chromosome very near to the replication fork.

(b) **The ultrastructure of the chromosomes.** Chromosomes have before replication two longitudinal subunits of DNA which are conserved during replication (→ polytene chromosome). They are uninemic, i.e., the DNA is one continuous double helix before replication (from anaphase through G_1 of interphase). Evidence for non-DNA linkers is absent. The basic → nucleosome fiber (diameter 10 nm) is folded to form a fiber 30 nm in diameter which, in turn, is folded into a series of loops arranged within a fiber 200–300 nm in diameter. Absence of this helical conformation in the centromere accounts for the typical constricted appearance of the centromeric region in metaphase chromosomes (→ chromosome scaffold).

The chromosomes of eukaryotes appear to be arranged in large domains and the nuclear substructure involved in the organization of the chromatin fiber in domains or loops is the nuclear matrix or scaffold. The DNA loops are anchored to the nuclear matrix at specific sequences (scaffold attachment regions).

(c) **Functional structures of chromosomes.** In addition to the regularly repeated changes of structure by chromosome coiling (compaction and decompaction), interphase chromosomes undergo changes of structure associated with chromosome reduplication and DNA transcription via differential gene activation (→ gene expression). These are reversible structural changes which involve individual replication and transcription units (chromomeres). In → lampbrush and → polytene chromosomes such changes associated with DNA-dependent RNA synthesis are clearly recognizable (→ puff; Balbiani ring).

(d) **Pairing structures of chromosomes.** In → chromosome pairing during meiosis, a special pairing structure is formed between pairing partners. This is called the → synaptonemal complex.

(2) **The chromosomes of prokaryotic systems and cell organelles in eukaryotes:** The chromosomes of prokaryotic systems differ in structure and complexity from those of euka-

ryotes. Usually, the entire genetic information (→ plasmid) is contained in a single circular structure which is double- or single-stranded DNA (in several → viruses RNA takes the place of DNA) of various lengths and molecular weights. Each of these molecules is a single → replicon.

In bacteria, chromosomal DNA is folded and condensed in a compact conformation as a superhelical structure with integrated particular → DNA-binding proteins.

The genomes (chromosomes) of viruses and bacteriophages occur in several geometric forms. Genetic information may be encoded in either DNA or RNA and these linear or circular molecules may be single- or double-stranded dependent upon the type of virus. Single-stranded DNA or RNA is rearranged into a double-stranded → replicative form (→ replicative intermediate) during replication, which is semi-conservative as in the case of chromosomal DNA of eukaryotes. Some viruses (mostly RNA viruses) contain → segmented genomes.

Phages with linear duplex chromosomes may be divided into three broad classes on the basis of their chromosome type: (1) One class has chromosomes with unique ends and short, single-stranded → cohesive extensions (coliphages lambda, P2, 186, and phi 80). The cohesive extensions are complementary in nucleotide sequence and may give rise by pairing to circular molecules. (2) A second class has circularly permuted ends (→ circular permutation) and substantial duplex terminal repetition (T-even coliphages, coliphage P1 and 15, *Salmonella* phage P22). (3) Phage chromosomes exemplified by the T-odd coliphages (T1, 3, 5, and 7) where interruptions occur in the polynucleotide chain at specific points.

chromosome aberration — (1) In the broadest sense, all types of changes in chromosome structure and chromosome number. A standard → karyotype is used for comparison and identification of the type of change. (2) All types of chromosome structural changes or → chromosome mutations, collectively (= chromosomal aberrations). (3) A category of chromosome structural changes in which both chromatids of the chromosomes are involved (= presplit aberrations), as opposed to → chromatid aberrations. Presplit aberrations normally arise (spontaneously or by application of → mutagens) in the not yet reduplicated chromosomes of the presynthetic (G_1) interphase nucleus.

chromosome addition — the addition of an extra chromosome derived from a different species.

chromosome arm (Navashin 1912) — one of the two main segments (= chromosome limbs) of the → chromosome; they are separated by the → centromere and their lenght depends upon the centromere position. If the two arms are of the same length, i.e., the centromere occupies a median position, the chromosome is said to be "isobrachial" (Sorokin 1929) or metacentric, otherwise (arms of unequal length) the chromosome is described as "heterobrachial" (Sorokin 1929), "cephalobrachial" (Levitsky 1931) or acrocentric (White 1945). Chromosomes with terminal contromeres are "monobrachial" (Sorokin 1929) or telocentric.

chromosome assignment — the assignment of genes to specific chromosomes by examining sets of hybrid cell clones with different chromosome constitution for expression of the phenotype with a particular chromosome (concordant segregation of phenotype and chromosome).

chromosome assortment — → assortment; chromosome movement.

chromosome banding — the experimental production in → metaphase chromosomes of differentially stained or fluorescing regions which serve in chromosome identification. Banding patterns can be divided into several major categories: (1) C-banding with Giemsa staining after specific pretreatment which indicates regions of constitutive → heterochromatin on both sides of the → centromere.

(2) Banding of certain sites along the chromosome arms due to quinacrine fluorescence (Q-banding) or Giemsa and other staining after various pretreatments (G-banding). A specific variant of this technique reveals c.b. which is the reciprocal of Q- and C-banding and is called R-banding. (3) T-banding by prolonged pretreatment which results in strong stai-

ning, both with Giemsa and Quinacrine, of chromosome ends (telomeric bands). (4) Restriction endonuclease banding (R-banding) produced by restrictase treatment of fixed metaphase chromosomes followed by Giemsa or fluorochrome staining. Banding is correlated with the specificity of the → restriction endonuclease and with the length of the DNA fragments produced in isolated DNA.

chromosome break — any discontinuity in chromosome structure occurring across the entire cross-section of the chromosome; chromosome breaks are either of spontaneous occurrence or are produced by the action of → mutagens. They are recognizable during nuclear division (mitosis, meiosis) and lead to the formation of centric and acentric fragments. Discontinuities which involve only one of the two chromatids are referred to as chromatid breaks.

According to the "breakage-reunion model" (which serves to interpret the production of chromosome structural changes), spontaneous or experimentally induced breaks of the chromosomes are assumed to be the primary lesions which ultimately give rise to → chromosome mutations.

chromosome breakage syndrome (German 1969) — any of a number of genetically determined human disorders (such as Bloom's syndrome, Ataxia telangiectasia, Fanconi's anemia) characterized by chromosome instability in the form of increased frequencies of broken and rearranged chromosomes occurring either spontaneously or following treatment with various DNA-damaging agents.

chromosome bridge — a double bridge linking the cell poles, formed by dicentric chromosomes (with two → centromeres) during anaphase of mitosis and meiosis (→ bridge-breakage-fusion-bridge cycle). The dicentric chromosomes are produced as a result of primary or secondary → chromosome mutations. If only one of the two chromatids participates in bridge formation, the term chromatid bridge is used.

chromosome chimerism (Chu et al. 1964) — the presence of cell populations of various → karyotypes in the same individual. C.c. is the result of transplantation or of double fertilization, if both fertilized meiotic products share in the development of the embryo (→ chromosome mosaicism).

chromosome coiling — the cyclic compaction and decompaction of eukaryotic → chromatin during → mitosis and meiosis as accomplished by the hierarchical folding of the DNA filament of chromosomes by chromosomal proteins. Four levels of chromosome compaction may be distinguished: (1) The primary coiling of the DNA duplex; (2) the secondary coiling around a → nucleosome to give a nucleofilament (10 nm in diameter); (3) formation of a coiled structure 30 nm in diameter with 5 to 7 nucleosomes per turn (total packaging ratio of DNA in the 30 nm fiber around 40:1); (4) further folding into loops of the 30-nm fiber of interphase and metaphase chromosomes with each loop being an independent domain of supercoiling due to topological contraints placed on the loops by nonhistone proteins.

The coiling of the axial fiber results eventually in the formation of the compacted → chromosome; freeing of → sister chromatids in later stages of prometaphase of mitosis occurs via spiralization or macrocoiling of chromonemata to become coiled metaphase → chromatids. The coiling cycle is reversible and characterized by: (1) Chromosome decondensation (gradual) throughout the G1 period of → interphase (→ cell cycle); (2) maximal level of dispersion at the time of → DNA replication during the S period; (3) beginning of recondensation which culminates in the formation of maximally condensed metaphase chromosomes.

Relational coiling: The twisting of one strand (chromatid or individual chromosome) about another strand ("external coiling") as in a rope. The coils interlock in such a way that the strands are not separable without unwinding the coils. Relational coiling is characteristic of sister chromatids during mitosis and paired homologous chromosomes in the meiotic prophase.

Relic coiling: This concerns relaxed spirals recognizable in the chromosomes of mitotic and meiotic prophase. They are regarded as the incompletely resolved internal coils of the preceding nuclear division ("supercoils").

The mitotic and meiotic coiling cycles show a series of differences. In mitotic prophase, the chromosomes occur as irregular, undulating threads; as prophase continues, these threads become coiled. The fine coils increase in diameter and the chromosomes become shorter and thicker to give the characteristic maximally spiralized metaphase chromosomes. There is no change in coil structure during anaphase. During telophase coiling is relaxed, the coils pull away from one another and despiralization commences. In interphase, the chromosomes are extensively despiralized and in this state are capable of the DNA-dependent RNA synthesis (functional form of the chromosome).

As compared to mitosis, the relic coils of the last premeiotic mitosis are much more completely smoothed out in meiotic prophase (leptotene and zygotene). In pachytene, chromosome contraction begins by formation of major and minor coils. The coils of the meiotic metaphase I have a greater diameter than those in mitotic metaphase and occur in smaller number. Maximum contraction is generally in diakinesis.

The causes and mechanisms of chromosome coiling are still but little understood.

chromosome complement (Darlington 1932) — the group of chromosomes derived from a particular gametic or zygotic → nucleus. It may be composed of one (haploid or monoploid nucleus), two (diploid nucleus); or more (polyploid nucleus) → chromosome sets.

chromosome complex — a group of chromosomes which in the extreme case may consist of an entire → chromosome set and which is distributed as a unit in meiosis. Chromosome complexes occur in → complex heterozygous species (which are heterozygous for several reciprocal → translocations) if alternative chromosomes of the configuration are regularly distributed to the same pole after ring or chain formation by meiotic → chromosome pairing between the chromosomes involved in the translocations and their structurally unchanged homologues (→ alternative distribution).

chromosome configuration (Darlington 1929) — any association by → chromosome pairing of chromosomes at → meiosis which segregates independently of other such configurations at anaphase I.

chromosome congression (Darlington 1937) — the movement of mitotic chromosomes to the spindle equator half way between the spindle poles where they reach an equilibrium position. Its consequence is auto-orientation of centromeres (→ centromere orientation). The equilibrium position for meiotic chromosomes (bivalents) is different from that of single mitotic chromosomes in so far as their centromeres (centromere co-orientation) do not come to lie at the spindle equator but in the longitudinal axis of the spindle, equidistant from the equator and their respective poles.

chromosome contraction — the thickening and shortening of chromosomes during → mitosis and → meiosis (= chromosome condensation) by → chromosome coiling.

chromosome core — an axial structure or scaffold forming the backbone of the eukaryotic → chromosome. The c.c. is composed of nonhistone proteins and extends from one end of a chromatid to the other (= chromosome scaffold).

chromosome diminution — the elimination of whole chromosomes or heterochromatic chromosome segments in mitosis, meiosis, or interphase (e.g., in Ascaridae, in certain species of the genus *Cyclops* and various *Diptera*). The process leads to differences in the → chromosome complements of the germ line and the soma and, in few cases, within the gonads themselves, to differences between the gametes and the remaining cells. There are various mechanisms leading to c.d. (→ nuclear differentiation).

chromosome disjunction — the → assortment of chromosomes during the first meiotic division.

chromosome elimination (Seiler and Haniel 1921) — the loss of chromosomes from nuclei during certain early → cleavage divisions in dipteran and hemipteran insects. Only the germ-line cells retain the full → chromosome complement; all somatic cells have a "reduced" number. C.e. is a mechanism of somatic versus germ-line differentiation (→ chromosome diminution).

chromosome engineering (Sears 1972) — the incorporation of alien chromosomes, or parts of alien chromosomes, into a given chromosome set. The alien material may be present as translocations, additions, or substitutions within the → chromosome complement.

chromosome erosion (Levan 1948) — the occurrence of a number (variable but often very high) of constrictions in the chromosomes due to the action of C-mitotic or mutagenic agents (= chromosome mottling). Characteristic of this condition are sharply defined alternatively stained and unstained segments.

chromosome field (Lima-de-Faria 1954) — → chromosome gradient.

chromosome fission — the splitting into two functional halves of a → metacentric or → submetacentric chromosome with one functional → centromere resulting in two stable → telocentrics (→ centromere misdivision; chromosome fusion).

chromosome flow sorting (Carrano et al. 1979) — a method for isolation of individual eukaryotic → chromosomes. It typically employs DNA-specific fluorochromes as stains. Instruments equipped with lasers capable of excitation in the near UV are used to quantitate the fluoroscence of individual chromosomes isolated in buffers containing organic solvents and/or detergents and to analyze them at rates exceeding 1000 per second. Individual chromosomes can thus be sorted, at purities of about 50% (10- to 20-fold enrichment) at rates of 100,000 per hour. The chromosomal DNA can then be cloned to obtain minilibraries of 50,000–10,000 plaque-forming units (pfu). This amount allows, at an average insert size of ≥ 5 kb, to span an entire (human) chromosome.

chromosome fusion (Seiler and Haniel 1921) — the union of two (or more) chromosomes by means of chromosome structural changes to form a single chromosome (→ chromosome mutation). Among chromosomes with localized → centromeres, stable unions are possible only if the fusion takes place in the centromere region (→ centric fusion).

Otherwise di- or polycentric chromosomes are formed; these behave anomalously during nuclear division, resulting in the formation of anaphase bridges (→ chromosome bridge) which are broken. Chromosome fusions of the centric fusion type play a part in the evolution of karyotypes.

chromosome gradient (Lima-de-Faria 1954) — the gradual decrease (or increase) in → chromomere size, stainability, and thickness of the chromosomal fibrillae observed in specific regions of the chromosome. The gradients exhibit a variation of defined form which bears definite relations to particular regions of the chromosomes: (1) The origin of the gradients is related to the → centromere. The chromomeres are generally larger and the fibrillae more stainable near the contromere. (2) The rate of decrease of the constituents of the gradients bears a definite relation to the position of the chromosome ends. Size of chromomeres and stainability of fibrillae decrease gradually towards the chromosome ends. The slope of the gradients is a function of the length of the chromosome arms. (3) The shape of the gradients is influenced by intercalary knob formations and their adjacent regions. Particular seriations located at regular intervals in the chromosome body represent regions where the gradients are locally disturbed. This means that the chromosome gradients may vary in a defined way under the influence of specialized chromosome regions.

The term "chromosome field" refers to the fact that the variations observed within the gradients take place in such a way that the properties exhibited by certain entities in one part of the chromosome bear a definite relation to the properties exhibited by other entities in other parts of the chromosome.

chromosome hybridity — → structural heterozygosity.

chromosome imprinting (Crouse 1960) — the process by which one of two genetically homologous chromosomes becomes altered, i.e., predetermined to function differently from its homolog at a subsequent stage of development. C.i. may occur several to many cell generations before the stage at which behavioral differences become obvious (Chandra and Brown 1975).

chromosome interference — → interference.

chromosome loss — failure of a chromosome to become incorporated into a daughter nucleus at cell division.

chromosome loss technique — in man-mouse somatic cell hybrids, a technique for linking genes to specific chromosomes. In this system the human chromosomes are lost one or a few at a time and the rodent chromosomes are retained. Two human chromosome loci are called syntenic if they are either lost or retained together in a series of hybrid clones (Pontecorvo 1975). The initial discovery of the loss of human chromosomes in man-mouse somatic cell hybrids was made by Weiss and Green (1967).

chromosome map — the graphic representation (in linear form) of a chromosome in which the genes (genetic markers) belonging to a particular → linkage group are plotted according to their relative distances. A distinction is made between → genetic maps and cytological maps. (1) **Genetic maps:** The frequency of intergenic → crossing over between any two linked markers is used as a measure of their distance (→ genetic recombination) which is expressed in → map units. The accuracy of mapping is dependent upon the precision with which the crossing over frequency can be estimated from the recombination frequency. (2) **Cytological maps:** Genes are located on the basis of cytological findings obtained with the aid of → chromosome mutations. Detailed cytological maps can only be prepared if the species concerned possesses large and visibly differentiated chromosomes which can be thoroughly analyzed genetically. To prepare a map, → deletions, → inversions, or → translocations are experimentally induced by → mutagens. They result in specific changes in the linkage relationships which may be defined precisely by genetic analysis. In addition, the location and extent of such structural changes may be determined microscopically (usually during meiosis, sometimes during mitosis) and in this way it is possible to correlate the genetic and structural findings.

Another method of preparing cytological maps is based upon microscopic analysis of the → polytene chromosomes of Diptera in which the position and extent of the chromosomal structural changes may be determined exactly. Here again, the genes localized by recombination analysis may be correlated with chromosome structure.

The linear order of genes in cytological maps constructed in the above manner agrees completely with that in genetic maps. However, the distances between the genes under consideration may differ appreciably in the two types of c.m. In general, gene distribution is more regular on the cytological map than on the genetic map. The explanation lies in the nonrandom distribution of crossing over and the effects of double crossing over and → interference.

chromosome-mediated gene transfer (McBride and Ozer 1973) — (CMGT); The use of isolated (mammalian) metaphase chromosomes as a vehicle for the transfer of genes between cultured cells. The technique is closely analogous to → DNA-mediated g.t., or DNA transfection. With an appropriate selection system, stable cell hybrids expressing the selected marker can be derived. The transferred chromosomal material is associated with recipient cell chromosomes and called a → transgenome and the cells expressing these genes are transformants. Transformants usually contain a subchromosomal and cytologically undetectable amount of the donor genome (microtransgenome); in a minority of transformants, a donor chromosome fragment (macrotransgenome) is visible. Initially, transformed cell lines are unstable, losing the transferred gene at a constant rate per cell generation. Upon prolonged culture, transformants may convert to stable expression.

CMGT provides a method for fine structural mapping of chromosomes (regional mapping of closely linked genes).

chromosome mobilization (Haas and Holloway 1978) — operationally, the production of a donor chromosome in bacterial Hfr cells which is ready for linear sequential transfer to a recipient cell during → conjugation. C.m. is under → sex factor control. C.m. requires protein synthesis for the production of an F-coded initiator followed by obligate requirement for DNA synthesis. DNA synthesis in the Hfr parent acts as a driving force for the initiation of chromosome transfer into the recipient parent. Effective chromosome transfer (more than serveral percent of the leading extremity of the donor chromosome) requires tight homologous pairing between the lead region of the donor chromosomes and the appropriate segment of the recipient chromosome. This allows pairing of homologous regions of the donor and recipient chromosomes and provides for the orderly and efficient integration of genetic material from the donor into recombinants.

During c.m. the fully replicated circular donor chromosome becomes linear.

chromosome mosaic — → chromosome mosaicism.

chromosome mosaicism — the presence of cell populations of various karyotypes in the same individual. The different karyotypes are produced from an originally "pure" karyotype by → chromosome mutations, mitotic chromosome loss, → nondisjunction, etc. (→ chromosome chimerism). Such individuals (chimeras) are termed "chromosome mosaics". The various karyotypes which they exhibit may differ both in chromosome number and chromosome structure (→ gynandromorph).

chromosome mottling (Tjio and Levan 1948) — = → chromosome erosion.

chromosome movement — the movements of chromosomes during mitosis and meiosis which are a prerequisite for the anaphase separation of chromatids (mitosis and meiosis II) or chromosomes (meiosis I).

Three basic mechanisms, which are dependent upon the integrity of → microtubules, have been proposed for c.m. during mitosis and meiosis: (1) The chromosomes are moved by the controlled assembly and disassembly of microtubules; (2) c.m. is due to the sliding of microtubules past one another by dynein-like bridges; (3) actin filaments and microtubules interact in some way to provide movement.

During mitosis, the chromosomes undergo orientation (→ centromere orientation) relative to the poles (auto-orientation) leading to the "congression" of single chromosomes on to the metaphase plate. After that the polar movement of the chromatids takes place.

During meiosis the pairing of homologous chromosomes (→ chromosome pairing) involves a certain amount of movement. Furthermore, there is a twisting of homologs about one another during pairing and at diplotene a segregation of the pairing patterns takes place. The latter process begins at the centromeres and has been regarded as a repulsion. The movements which the paired chromosomes undergo on the spindle prior to metaphase I of meiosis are of three kinds (John and Lewis 1965): (1) orientation of the chromosomes relative to their partners and the poles (co-orientation); (2) → pre-metaphase stretching and the subsequent reapproach of cooriented centromeres; (3) congression movement determining the distribution of the pairing configurations on to the metaphase plate.

Only after the completion of these movements are the meiotic chromosomes able to undergo anaphase separation leading to their → assortment to the poles.

The congression and polar movements during mitosis and meiosis depend upon the joint action and interaction of the centromeres and spindle fibers. The mechanisms of c.m. and the forces underlying them are not yet accurately understood. The numerous hypotheses advanced to explain pre-metaphase and anaphase movements embrace traction fibers, electrostatic forces, tactoid, and sol-gel transformations as well as jet-propulsion mechanisms.

chromosome multiformity (Tobias 1953) → the existence of widely different → chromosome complements within the members of a given taxon.

chromosome mutation — any structural change (= intergenic mutation or chromosome aberration) involving the gain, loss, or relocation of chromosome segments. Chromosome mutations arise spontaneously or are induced experimentally by chemical or physical → mutagens (→ mutation) and can be detected cytologically during mitosis, meiosis or in polytene chromosomes of Diptera. The ability to detect different kinds of chromosome mutations will depend on the size, structure, and number of chromosomes concerned and the ease with which they can be handled.

Chromosome mutations are either intrachromosomal (= homosomal) and then called "intrachanges", or interchromosomal and then called "interchanges" (allelosomal when in homologous, or heterosomal, when in nonhomologous chromosomes). Intrachanges are either in one arm (= paracentric or homobrachial) or involve both chromosome arms (= pericentric or heterobrachial).

Descriptively, all such structural changes include → **deletions** or deficiencies (the actual loss of terminal or intercalary portions of the chromosome and the included genes), → **duplications** (the presence of a group of genes more than once in the genome), → **inversions** (a chromosome segment becomes inverted and reinserted into its original chromosomal position), and → **translocations** (relocations of chromosome segments within one or between chromosomes). Deletions and inversions are intrachanges, duplications and translocations may be intra- or interchanges. All these types of structural changes may be basically of two sorts: (1) **Chromosome type structural changes:** Both chromatids at identical loci are engaged in the aberration formation. The whole chromosome is the unit of aberration production ("chromosome aberrations"). (2) **Chromatid type structural changes:** The single chromatid is the unit of aberration formation ("chromatid aberrations").

In all these kinds of structural changes, rejoining must take place in such a way that the polarity of the two strands of the DNA (the structural backbone of eukaryotic chromosomes) is conserved, i.e., if the two strands of double-stranded DNA are called C and W and, e.g., an inversion occurs, then the segment of strand C between the two inversion points must be inserted into strand W and vice versa. Thus neither strand will have the same base composition as it did before rearrangement.

With heterozygosity for the chromosomal structural changes (= structural hybridity) "secondary chromosome mutations" may arise from the primary ones by → crossing over between the normal chromosome and the structurally changed pairing partner during first meiotic division.

The chromosome mutations are the final result of complicated multiphase processes arising as a consequence of mis-repair (S-independent or ionizing radiation type) or mis-replication (S-dependent or alkylation type) of damaged DNA (→ DNA repair; DNA replication). The ultimate lesions responsible for the formation of c.m. seem to be DNA strand breaks (double-strand breaks).

C. mutations can have three effects: (1) They can bring genes into new regulatory relationships with one another so that altered pathways of → gene expression can result (→ position effect); (2) by linking certain genes they can produce particular new combinations of → genes; (3) The c.m. can act as a → sterility barrier because, in → heterozygotes, the mutant chromosome cannot pair properly with the wild-type homolog and, in consequence, meiosis may be defective and the heterozygote of low fertility (→ speciation).

As a rule, chromosome structural changes which produce acentric fragments and dicentric reunion products rapidly succumb to cell selection and are eliminated. Other reconstruction types (duplications, inversions, eucentric translocations) are of fundamental importance in the evolution of karyotypes.

Chromosome rearrangements analogous to eukaryotic c. mutations (deletions, duplications, inversions, and → transpositions) also occur in prokaryotes. An additional class is the fusion of two (bacterial, plasmid, or phage) chromosomes without any loss of genetic material during this event.

chromosome orientation — → centromere orientation.

chromosome pairing — (1) the highly specific side-by-side association of homologous chromosomes. **Homologous c.p.** can be classified into four types defined as follows (Riley and Law 1965): (a) **Meiotic synapsis or pairing:** The intimate association of homologues easily recognized during first prophase of meiosis and leading to → bivalent (pairing of two homologs in diploids) or → multivalent formation (pairing of more than two homologues as in poly- or aneuploids). The bi- or multivalents resulting from pairing of zygotene chromosomes are maintained to metaphase I if → crossing over and → chiasma formation took place (→ desynapsis). Meiotic pairing is the prerequisite for the regular anaphase separation of meiotic chromosomes. When the pairing process is viewed under the electron microscope the visibly single zygotene chromosomes form a tripartite ribbon called → "synaptonemal complex" (Moses 1958), which is absent in achiasmate meiosis.

Meiotic pairing begins at "contact points" (→ zygomere) from which it runs along the chromosomes in zipper-fashion. The contact points are at the chromosome ends, at the centromeres, or at both concurrently. "Localized pairing" is a form of pairing confined to those chromosome segments near the first contact point. In all cases, the pairing affinities of chromosomes are satisfied by associations in twos. For multivalent production between more than two homologues a regional change of pairing partners takes place.

According to Grell (1962) two pairing phases may be distinguished in meiosis: "Exchange pairing" and "distributive paring". Exchange pairing is specific pairing leading to crossing over between homologs. If exchange has taken place, exchange pairing is directly followed by distributive pairing and the chromosomes with chiasmata remain associated up to the time of anaphase separation. Those chromosomes which have not undergone exchange by crossing over may take part in associations during the phase of distributive pairing and under these circumstances even nonhomologs may pair.

The mechanism of meiotic pairing is still unclear. Two distinct aspects are under genetic control. The extent to which pairing is realized within the chromosome complement is controlled by the activities of major and polygenes. Mutants of major synaptic genes can so disorganize the pairing process that no pairing occurs and quantitative modifications of a wide magnitude can result from other genetic conditions. Furthermore, pairing specificity can be increased or decreased by the action of specific genes permitting or preventing the pairing of chromosomes distantly or closely related genetically and phylogenetically.

Meiotic c.p. possibly occurs in two steps (Hotta and Stern 1971): During zygotene relatively small sequences of DNA interspersed along the homologous chromosomes begin to pair ("gross pairing") and chromosome pairs may be stabilized by the synaptonemal complex. A second step follows gross alignment of chromosomes and consists in the formation of single-stranded DNA regions by action of an endonuclease and possibly other enzymes. These could pair and recombine ("molecular pairing"). Duplex formation between specialized zygotene regions or single-stranded regions produced by endonuclease might be facilitated by a specific DNA-binding enzyme. There is evidence that meiotic c.p. is temporally associated with a semi-conservative replication of DNA (→ zygotene DNA). This means that a distinctive component of DNA evidently does not replicate during premeiotic S period but does so at the zygotene stage (→ colchicine-binding protein; distributive pairing).

(b) **Somatic pairing:** The close juxtaposition and coorientation of homologous chromosomes as, for example, during the prophase and metaphase of somatic divisions in *Drosophila* and other Diptera.

(c) **Salivary gland pairing:** The intimate association of homologous → polytene chromosomes in the salivary glands and other tissues of species of Diptera. This pairing process may be a special expression of the somatic pairing seen in other cells of these organisms.

(d) **Secondary association or pairing of bivalents:** The nonrandom distribution of bivalents at first metaphase of meiosis in some polyploid species. The bivalents occur in pairs or groups if they are related by genetic and evolutionary factors.

(2) **Nonspecific pairing** of chromosomes which can be classified into three types defined as follows (Riley and Law 1965):

(a) **Heterochromatic fusions:** Associations of chromosomes, leading to the formation of → chromocenters or to the random association of heterochromatic chromosome regions.

(b) **Terminal nonspecific associations:** Terminal associations between univalent chromosomes at first metaphase of meiosis; these may also be a consequence of heterochromatic fusions.

(c) **Nonspecific pairing at prophase of meiosis:** The pairing of nonhomologous chromosomes or chromosome regions if homologous synapsis is interrupted by → structural heterozygosity or in the absence of homologous partner chromosomes (as in some monoploid individuals: "intra-haploid pairing").

chromosome polymorphism — the occurrence of one or several chromosomes in two or more alternative structural forms within a population. The structurally changed chromosomes are the result of → chromosome mutations which may be fixed, homo- and heterozygous, in a certain percentage in the population ("balanced chromosomal polymorphism"). The chromosomal structural types then fluctuate around a mean, roughly constant from generation to generation and possess → adaptive value. The various chromosomal structural types present in c.p. are, in a sense, homologous chromosomes, as they can mutually replace the alternative structural form in the karyotype. The c.p. may be eliminated if one of the structural variants proves to be superior under the prevailing environmental conditions and its proliferation is strongly furthered by natural selection at the expense of the alternative type (→ genetic polymorphism).

chromosome puffing — → puff; polytene chromosome.

chromosome pulverization (Levan and Tjio 1948; Nichols et al. 1964) — the destruction of chromosome structure, varying from an apparently total fragmentation of the chromatin continuity to various degrees of defective condensation and erosion (→ premature chromosome condensation; prophasing).

chromosome rearrangement — = → chromosome mutation.

chromosome reduplication — the synthesis of all those compounds which give the chromosome form, control its mechanical activity during the cell cycle, and regulate its metabolism. The primary event in c.r. in pro- and eukaryotes is the replication of DNA.

chromosome reinitiation — in bacteria, the beginning of a round of chromosome replication in an already replicating chromosome.

chromosome scaffold (Adolph et al. 1977; Poulson and Laemmli 1977) — in eukaryotic metaphase chromosomes, an axial organizing structure of nonhistone proteins (the central core) which folds the 30 nm chromatin fiber into loops that project outward radially from the chromatid axis. The c.s. is a fibrous network without prominent longitudinal fibers. The DNA is organized in loops (45,000 to 90,000 bp in length) anchored to the c.s.

chromosome segregation — (1) the orderly segregation of chromosomes during → mitosis and → meiosis on which the faithful inheritance of genetic information depends. (2) The process of preferential elimination of chromosomes of one of the parental cells in interspecific cell hybrids.

chromosome set — in eukaryotes, a minimum viable complement of indispensable (→ B-chromosome) chromosomes, each represented once. The number of chromosomes in a basic c.s. is called the **basic number**. The gametic complement of a fertile, functionally diploid, polyploid is a secondary basic c.s.; the number of chromosomes in such a set is called the secondary basic number (Dyer et al. 1970).

The genes contained in the chromosomes of one c.s. represent one → genome; each chromosome of the set is a carrier of a specific → linkage group.

Diploid organisms with sexual reproduction possess two c. sets in the → diplophase of their life cycle, one in the → haplophase. This alternation of nuclear phases is the result of → fertilization and → meiosis. In the case of → polyploid organisms, the number of c. sets is increased proportional to the degree of polyploidy.

chromosome shattering — extensive fragmentation and/or pulverization of all chromosomes of a mitotic cell (generalized c.s.).

chromosome-specific DNA library — a → DNA library representative of a single eukaryotic chromosome.

chromosome sterility — → hybrid sterility.
chromosome stickiness (Beadle 1932) — → stickiness.
chromosome substitution (Karpechenko 1935) — the exchange of single chromosomes or chromosome pairs by other chromosomes of the same chromosome complement ("autosubstitution") or by chromosomes of the complements of other species or genera ("allosubstitution" or "alien substitution").
chromosome theory — the chromosome theory of → Mendelian heredity states that the → chromosomes, as carriers of → genetic information, represent the material basis of "nuclear inheritance". The regularity of the gene distribution from cell to cell (→ mitosis) and from generation to generation (→ meiosis) and the concomitant → genetic recombination depend on the characteristic behavior of the chromosomes in the course of nuclear division.

The first clear formulation of the chromosome theory was made by Sutton (1903). He referred to the relationship between the separation of homologous chromosomes in meiosis and the separation of alleles in gamete formation as postulated by Mendel.
chromosome transfer (McBride and Ozer 1973) — The transfer of isolated metaphase chromosomes into (mammalian) cells cultivated in vitro (→ chromosome-mediated gene transfer).
chromosome walking (Bender et al. 1983) — = → chromosomal walking.
chromosomin (Stedman and Stedman 1943) — = → nonhistone chromosomal protein.
chromotype (Battaglia 1956) — = → chromosome set.
chronocline — a character gradient in the time dimension (→ cline).
chronogenetics (Gedda and Brenci 1973) — a branch of genetics devoted to the study of the fundamental, endogenous genetic time (hereditary biological time) with its theoretical and practical, normal and pathological implications.
cilium — in many algal, protozoan and metazoan cells, any of the whiplike organelles (= undulipodium) with a diameter of the order of 200 nm. Their dry weight composition is roughly 70 to 84% protein, 13 to 23% lipid, 1 to 6% carbohydrate, and 0.2 to 0.4% adenine and uracil nucleotides. Cilia contain ATPase and adenylate kinase and are composed of 11 fibrils, two of which are centrally located and somewhat smaller and more fragile than the nine peripheral ones (diameter 20 to 25 nm, length 35 to 37 nm). The peripheral fibrils have a double tube substructure. Both cilia and → flagella, which have the same chemical constitution, are enclosed by a semipermeable membrane (about 9 nm wide); within this membrane the peripheral fibrils are evenly spaced around a circle some 160 nm in outside diameter.

In addition, cilia contain 9 nontubular fibrils (5 nm in diameter) which lie in the matrix between the tubular peripheral and central fibrils of the 9 + 2 arrangement (→ centriole; basal body). Cilia exert a unilateral beat and a sideways force on the water to move the cell in a direction perpendicular to the organelle. Frequently, cilia are aggregated into compound structures within which the units act in synchrony.

Cilia and flagella derive their motile behavior from the generation of sliding shear force which is simultaneously captured and transformed into rhythmic propagated bands. The distinction between cilia and flagella consists in the general form of their beat cycle.
cin-duction (Ozeki and Howarth 1961) — = → colicinoduction.
circadian rhythm — any cyclic quantitative variation regularly occurring with 24-h periodicity (e.g., mitotic indices of proliferating cells).
circular DNA — any of the covalently closed DNA molecules found in bacteria, many viruses, in mitochondria, plastids, and plasmids. Small, polydisperse circular DNA's have also been observed in a number of eukaryotic organisms and are suggested to have homology with chromosomal DNA and the capacity to be inserted into, and excised from, chromosomal DNA.
circular permutation — of physically linear DNA molecules in bacteriophages with the same

set of genes arranged in the same order which are related to each other as if each had once been in a circle, and the circles had been opened by randomly distributed breaks:

ABCDE...XYZAB
DEFGH...ZABCDE
FGHIK...CDEFG

In addition to their fixed length and circular p., these DNA molecules have a terminal redundancy. Such molecules reveal unusual → recombination frequencies suggesting → genetic circularity.

cis-acting DNA sequence — in eukaryotes, any of a class of DNA sequences consisting of arrays of short (10 to 12 bp) nucleotide sequences found in the 5′-flanking regions or within genes. They represent recognition elements that interact with specific → transcription factors and are required for eukaryotic gene regulation.

cis-acting regulatory protein — any regulatory protein involved in → gene expression and produced by a regulatory gene close to the site (cis) at which its product acts. Cis-specifity can be attributed to limited diffusibility of the active protein in vivo.

cis-configuration (Haldane 1942) — the → alleles of two linked genes or the → heteroalleles of a → cistron are said to be in the cis-configuration (coupling phase) if in doubly → heterozygous or → heterogenotic cells the two (dominant) wild-type alleles are located in one, the two mutant alleles in the other of two homologous chromosomes (AB/ ab or + +/ab).

A double heterozygote in the repulsion phase, with the dominant alleles each in one of the homologous chromosomes (Ab/aB or + b/a +), is termed "trans-configuration" or "trans-heterozygote" (→ cistrans test).

cis dominant (Epstein and Beckwith 1968) — of mutations (genes) that affect adjacent linked genes. Their effect (c.d. effect) cannot be overcome by the wild type allele residing on another DNA molecule, i.e., c.d. mutations cannot be complemented (→ genetic complementation; operon).

cis regulation — → regulatory gene.

cisterna — any of the spaces (sacs or cavities) between the membranes of the → endoplasmic reticulum and filled with fluid contents (→ nuclear envelope).

cis-trans test (Lewis 1951; Benzer 1957) — a genetic test (= complementation test) to decide whether two → gene mutations (m_1 and m_2) occur in the same functional gene (→ cistron) and to establish the limits of this genetically active region. To this end, the heterozygotes (or → heterogenotes) in which the concerned mutations are located in the same chromosome (cis-configuration: $++/m_1m_2$) and those in which they are present in separate chromosomes (trans-configuration: $+m_1/m_2+$) are compared. Two recessive mutations are assigned to the same cistron if their trans-configuration exhibits the mutant phenotype and their cis-configuration the normal phenotype (wild type). If one or both mutations are dominant they are considered to belong to the same cistron if the cis- and trans-configurations differ in phenotype. On the other hand, should the cis- and trans-configurations be phenotypically identical, this is evidence that the mutations occurred in different cistrons. The fact that among mutations of a cistron "intra-cistronic complementation" (→ genetic complementation) can occur whereby independent mutations, in the trans-configuration, interact to produce the normal phenotype demands that the c.t. be used critically. However, in "intra-cistronic complementation" the normal phenotype is often more fully expressed with cis-configuration than with trans. Since cis- and trans-configurations are not equivalent in all cases of complementation, it is reasonable under these conditions to assign the independent mutations to the same cistron.

cistron (Benzer 1957) — a segment of DNA or RNA which codes for a specific gene product, either a protein (messenger RNA) or an RNA involved in → genetic translation (→ ribosomal RNA; transfer RNA); within the cistron, pairs of mutations in trans-configuration are either deficient for a particular protein (enzyme), or produce that protein in a structurally abnormal form (Fincham 1959). The c. may be equated to the → gene as a functional unit and is defined operationally by the results of unequivocal → cis-trans tests (complementation tests). If a protein consists of a single polypeptide chain or a series of identical chains, then it is determined by a single c.; if a protein is composed of two or more nonidentical polypeptides, two or more cistrons are involved in its determination ("one c. — one polypeptide concept"). Within each c. → genetic recombintion (→ recon) and → mutation (→ muton) may take place at many sites (→ genetic transcription; operon).

The eukaryotic c. is a transcription unit containing regions (→ intron) to be excised (→ RNA processing) from the mature → messenger RNA; these alternate with regions which will be expressed (= → exon). The eukaryotic gene is thus a mosaic.

Overlapping cistrons which are translated in different reading frames or become alternatively spliced have been observed.

cis vection effect (Lewis 1955) — = cis-trans position effect (→ cis-trans test; position effect).

clade (Huxley 1957) — a segment of an evolutionary tree, i.e., a monophyletic group of taxa formed by → cladogenesis and sharing a closer common ancestry with one another than with members of any other c.

cladogenesis (Rensch 1947) — evolutionary splitting, i.e., the formation (by branching or divergence at the → species level or higher) of independently evolving lineages from a single ancestral line. A fundamental cladogenetic process is → speciation. C. is assumed to be based on changes of regulatory DNA sequences and karyotype repatterning, while → anagenesis and → adaptation are mainly brought about by → gene mutation and enzyme polymorphism.

cladogram — a diagram that shows the distribution of characteristics within a group of related organisms.

clastogenic (Shaw 1970) — of any agent (clastogen) capable of inducing chromosomal structural changes (→ chromosome mutation). Most c. agents may be classified into two major groups: (1) S-dependent agents which induce damage at any phase of the → cell cycle; expression of the primary lesion as a chromosomal aberrations requires, however, a round of → DNA replication and chromosome replication (agents inducing delayed effects). (2) S-independent agents which produce chromosomal aberrations independently of DNA or chromosome replication (agents inducing nondelayed effects).

cleavage — (1) the furrowing of the cell cytoplasm after completion of nuclear division. The furrow originates in the → cortex immediately peripheral to the equator of the → spindle. The deepening of the cleavage furrow involves actual growth of the cortex in this region, which continues until c. is complete (→ cytokinesis). (2) The division of the egg cell after fertilization (zygote) into a number of cells during a time interval called the "period of c.". It lasts to the moment when a shifting of regions of the egg relative to one another (blastulation, gastrulation) sets in. Henceforth the individual is usually referred to as an embryo.

At the beginning of gastrulation the cleavage cells are arranged into a "blastula", which in its typical form is a hollow sphere. The main physiological function of c. is the restoration of the balance between the nuclear size and the cytoplasmic volume with which it is associated.

Cleavage types may be classified as (1) **equal c.** in eggs with little yolk, (2) **unequal c.** in eggs with somewhat more yolk, and (3) **superficial c.** in extremely yolky eggs in which c. occurs in only a small area of the egg.

cleavage furrow — a temporate organelle arising during cell → cleavage. The c.f. has constituent → microfilaments and first appears at late anaphase of telophase of animal cells. It normally lies in the plane previously occupied by the → metaphase plate. At the time the furrow surface indents, an equatorial ring of circumferentially oriented microfilaments develops. The microfilamentous band is the visible manifestation of the mechanism which exerts tension and accomplishes → cytokinesis.

cleavage nucleolus — a → nucleolus which differs from normal nucleoli in that it lacks the granular outer region. C. nucleoli, observed in the interphase nuclei of cleaving amphibian eggs in the form of nucleolar bodies, are almost devoid of basophilia and do not incorporate appreciable amounts of radioactive uridine. Possibly, they are sites of accumulation of proteins of cytoplasmic origin.

cleavage nucleus — (1) the nucleus of the fertilized egg cell or zygote; (2) the nucleus of egg cells which develop parthenogenetically (→ parthenogenesis).

cleistogamy — fertilization within closed flowers making selfing obligatory.

cline (Huxley 1939) — a gradient (decrease or increase) within a continuous population in the frequencies of different → genotypes in different localities ("genocline") or → phenotypes ("phenocline") (Darlington and Mather 1949).

clonality (Smith 1973) — The probability of gene identity in a → multigene family, i.e., the probability that two randomly chosen gene units within a family belong to the same lineage.

clone (Webber 1903) — (1) a population of cells or organisms derived from a single cell or common ancestor by mitosis. The mode of → reproduction giving rise to a c. is asexual. A c. is not necessarily homogeneous and therefore the term c. or cloned should not be used to indicate homogeneity in a population. (2) A population of → recombinant DNA molecules all carrying the same inserted sequence. The verb "to clone" means to insert by use of in vitro recombination techniques a particular DNA sequence into a → cloning vector.

cloned library — a "library" obtained by → molecular cloning of random fragments of genomic DNA. Provided the number of clones is large enough for complete sequence representation, in principle any → gene can be isolated by screening the c.l. with a specific hybridization probe (→ DNA library).

cloning — → clone; molecular cloning.

cloning efficiency — the efficiency of cloning (in percent) of in vitro cultivated cells.

cloning host — a bacterial cell used for transformation with → recombinant DNA inserted into a → cloning vector. Each transformed cell is grown as a separate colony or plaque. Selection of the desired clone is based on the presence of both the cloning vector and appropriate markers. Subsequently, the isolated clone can be extensively amplified by growing a large-scale culture. The inserted genetic segment can then be separated from the cloning vector with the → restriction nuclease used originally to insert it.

cloning vector — in recombinant DNA technology, any DNA molecule capable of autonomous replication within a host cell (= cloning vehicle) into which other DNA sequences can be inserted and thus amplified (→ molecular cloning). C. vectors are derived from bacterial → plasmids, → bacteriophages, and animal viruses. They usually have a single → restriction site at which foreign DNA can be inserted (insertional vectors) or a pair of restriction sites defining a sequence that can be removed and replaced by foreign DNA (replacement vectors).

A useful c.v. should have the following properties: (1) It is stably inherited (either extrachromosomally without sequence homology to the host cell genome, or intrachromosomally in the host genome); (2) it is genetically stable and capable of conferring selectable phenotypic traits on the → cloning host; (3) it is of low molecular weight for easy handling and can be amplified readily; (4) it contains DNA restriction sites for one or more restriction endonucleases, permitting insertion of foreign DNA at these sites, preferably inside genes of readily scorable phenotype; (5) any vector-borne aids for the expression of inserted DNA.

Most of the c. vectors can be categorized according to purpose, host-range, and type of extrachromosomal element involved: (1) Generalized c. vectors for the unbiased cloning of genomic and → cDNA fragments for genetic and biochemical analysis; (a) *E.coli* vectors, e.g., plasmids, single-stranded bacteriophage vectors, double-stranded bacteriophage vectors, cosmid vectors; (b) non-*E.coli* vectors; (c) eukaryotic c. vectors). (2) Specialized plasmid c. vectors; (a) *E.coli* cloning vectors, e.g., plasmids which represent expression vectors, promoter cloning vectors, terminator cloning vectors; (b) double-stranded bacteriophage vectors;

(c) eukaryotic cloning vectors). Such vectors have been developed for specific needs (→ cosmid, DNA splicing; expression vector).

cloning vehicle — = → cloning vector.

closed reading frame — a → reading frame containing → terminator codons that prevent its translation into protein (→ open reading frame).

closed regulatory loop (Kourilsky and Gros 1974) — a form of → autogenous regulation of an → operon; the co-operation of regulatory proteins which control the rate of synthesis of each other.

cluster gene (Gaertner-Cole 1977) — a → gene coding for a multifunctional enzyme. Each part of the polypeptide, translated from a single → messenger RNA, and bearing a specific enzymatic activity corresponds genetically to a → complon. C. gene are more prevalent in eukaryotes than in prokaryotes.

C-meiosis (Levan 1939) — a modified form of → meiosis which occurs following the action of colchicine ("C-meiosis" = "colchicine meiosis") and other spindle poisons. The spindle poisons cause partial or complete blockage of the spindle mechanism which normally directs chromosome distribution. At the same time, → chromosome coiling and → chiasma formation are inter alia functions which are indirectly affected. The complete suppression of the spindle mechanism, which may involve only the first meiotic division or only the second, or both divisions, gives rise (via the formation of → restitution nuclei) to meiotic products with more than one chromosome set (→ C-mitosis).

C-mitosis (Levan 1938) — a modified form of → mitosis, caused by partial or complete inactivation of the spindle mechanism and concomitant disturbance of the → chromosome movement normally leading to chromatid distribution in anaphase. The modification is produced by the action of colchicine ("C-mitosis" = "colchicine mitosis") and other spindle poisons. Parallel with the effect on chromosome movement is an increased chromosome contraction. The → centromeres divide late and the chromatids associated at the centromere mutually repel one another. In this way cross-shaped, so-called "C-pairs" are formed.

Complete spindle inactivation leads to the formation of → restitution nuclei with doubled chromosome numbers. Partial spindle inactivation hinders the regular anaphase movement of the chromatids and frequently gives rise to multipolar anaphases.

coadaptation (Dobzhansky 1950) — the → selection process for a balanced combination of genes in an individual, and of individuals in a population.

coadapted (Dobzhansky 1950) — of a combination of linked genes maintained by natural → selection that acts to produce a favorable phenotype.

coalescent (Kingman 1982) — a parameter (n_o coalescent) describing the family relationships among a sample of n_o individuals from a particular generation of a population. It is due to a stochastic process resulting from n_o-1 independent random events (coalescence) at each of which n ($2 \leq n \leq n_o$) ancestral lineages of a sample are descended from n-1 distinct ancestors for the first time. The n_o c. plays a central role in describing the genealogical behavior of generations of individuals.

coancestry (Falconer 1960) — the degree of relationship by descent of two parents of a diploid individual. Values of c. are obtained from the probability of occurrence of genes identical by descent in a pair of gametes taken at random from a pair of parents.

coated pit (Roth and Porter 1964) — in eukaryotes, any of the ubiquitous cell surface specializations (depressions or invaginations) that mediate the receptor-dependent → endocytosis of a wide variety of macromolecules. C. pits are characterized by a thick (~20 nm) coat at their cytoplasmic surface and probably bud into a cell to yield → coated vesicles. There are indictions that the vesicle responsible for the internalization of receptor-ligand complexes is a structure named the → receptosome, a smooth-surfaced vesicle budding off from the neck of the c.p.

coated vesicle (Roth and Porter 1964) — any of the cytoplasmic organelles (intracellular vesicles) in all eukaryotic cells that are composed of a membrane surrounded by a polygonal protein lattice containing clathrin and other closely associated polypeptides. C. vesicles form at the cell surface by invagination and bud off into the cytoplasm or pinching off from the cisternae of the → endoplasmic reciculum or → Golgi apparatus. They are involved in receptor-mediated → endocytosis, formation of primary → lysosomes, membrane recycling, and intracellular vesicular transport (→ coated pit).

cocarcinogenic — of any compound which is not carcinogenic itself, but which increases the potency of a carcinogen either by shortening the latency period or by increasing the incidence. Several mechanisms may underlie c. activity: (1) Alternation of the metabolism of a procarcinogen; (2) facilitation of the transport of a carcinogen to its receptor; (3) alteration of binding of carcinogen with its receptor; (4) inhibition of → DNA repair; and (5) promotion of initiated cells.

co-conversion — the linked conversion of two or more genes (→ gene conversion).

code — → genetic code.

coding ambiguity — the phenomenon whereby one nucleotide triplet (→ codon) codes for more than one amino acid. Ambiguity in the → genetic code may result from: (1) A codon may be misread owing to environmental conditions existing around the tRNA-mRNA-ribosome complex which interfere with the codon recognition process (→ mistranslation). (2) A codon may be misread because of a specific change in the primary, secondary, or tertiary structure of either the → transfer RNA or the → ribosome such that it affects the codon-anticodon interaction. (3) An → aminoacyl-tRNA synthetase may misrecognize either an amino acid or a tRNA and form a wrong → aminoacyl tRNA.

coding capacity — the amount of protein (amino acid sequence) which can be specified by a given DNA or RNA sequence. Estimates of c.c. usually require assumptions about the absence of → introns or of → overlapping genes.

coding DNA strand — the strand of duplex DNA (= sense strand) which is transcribed into a complementary RNA strand. The DNA strand that is not transcribed (→ genetic transcription) is referred to as the noncoding or antisense strand.

coding efficiency (Mitra et al. 1979) — the comparative preference for one → transfer RNA isoacceptor over another in reading the same → codon.

codon fidelity — the fidelity of the genetic coding process (→ genetic code) as maintained during → DNA replication and the synthesis of proteins by a series of editing or proof-reading reactions which remove errors. Without these checks, → mutation rates would be unacceptably high and proteins largely heterogeneous.

The error rate in DNA replication (*E.coli*) is one mistake in 10^8-10^{10} nucleotides polymerized, but the overall error rate in transcription of the DNA and translation of the → messenger RNA into protein is only about 1 in 10^4 amino acid residues incorporated. This specificity is due to the evolution of editing or proof-reading mechanisms (→ DNA polymerase).

coding ratio (Crick 1963) — the number of bases in nuclei acids divided by the number of amino acids whose sequence they determine in a particular polypeptide (→ genetic code).

coding recognition site — → transfer RNA.

coding sequence — that part of a gene which directly specifies the amino acid sequence of its protein product. Noncoding sequences of genes include: (1) control regions, such as → promoters, → operators, and → terminators, and (2) the intron sequences of certain eukaryotic genes.

coding strategy — the mode of employment (choice between synonymous codons, choice between degenerate bases in codon position III) of the codon catalog (→ genetic code) by different → genomes. Different c. strategies for different species may be due to different optimizations of the tolerated error level and amount of each base in the pool. The error depends both on the time available for base incorporation and for proof-reading. Incorporation time is a function of the concentration of the base being incorporated; time for correction depends on concentration of the next base in the sequence (if the pool contains an abundance of the next base it will be incorporated rapidly, leaving little time for proof-reading of the first base).

coding triplet — → codon.

codogenic — = → coding DNA strand.

codominant — of two alleles when there are three phenotypes corresponding to the three genotypes (A^1A^1, A^1A^2, and A^2A^2).

codon (Crick 1963) — any triplet of → nucleotides (coding unit) in DNA or RNA (if RNA is the carrier of primary → genetic information as in some viruses) that codes for a particular amino acid or signals the beginning or end of the message (→ anticodon). The term c. is also used for the corresponding (and complementary) sequences of three nucleotides in the → messenger RNA (→ nodoc) into which the original DNA sequence is transcribed (→ genetic transcription). Of the 64 possible codons in the → genetic code, the mRNA triplets UAA (ochre), UAG (amber), and UGA (opal), serve as → terminator codons. AUG and GUG are → initiator codons (→ regulatory codon).

Synonymous (= degenerate) codons are different codons for the same amino acid. Different organisms (both prokaryotes and eukaryotes), different tissues, and different proteins within one organism, demonstrate clear preferences for use of synonyms (→ codon usage). Most changes in the third bases of codons do not change the corresponding amino acid and are hence synonymous.

There are indications that reading of a given c. may be influenced by mRNA sequences external to this c. (effect of the reading context on translation).

Any base triplet which can encode any of two or more amino acids is called an ambiguous c.

codon-anticodon complex — the complex formed by recognitions of a → codon by its complementary → anticodon in → transfer RNA, The c.-a.c. is more stable than two complementary trinucleoside diphosphates; there is evidence that the anticodon loop is flexible and changes conformation on binding the codon.

codon bias (Air et al. 1976) — the nonuniform distribution of → codon usages, i.e., the phenomenon that an organism frequently uses a particular codon, rather than its alternative, to specify a given amino acid.

codon context effect (Akaboshi et al. 1976) — the phenomenon that the efficiency of translation of some → codons may be influenced by the nucleotides preceding or following the translated codon in → messenger RNA. The c.c.e. may help to determine the level of → gene expression via translational activity of the RNA.

codon family (Mitra et al. 1977) — a group of four (synonymous) → codons (a code box) differing only in the third position and all coding for the same amino acid. All codons of a c.f. have the first two nucleotides in common while the third codon position can be occupied by any of the nucleotides U, C, A, and G. Half of the codons of the → genetic code belong to c. families. This kind of codon reading is called "two-out-of-three" reading.

codon misreading — the → mistranslation of a → codon in mRNA that increases errors in protein synthesis by generation of amino acid substitutions. Usually only one base is misread at a time, misread bases are 5′ or internal in the codon; the most typical misreading error is pyrimidine for pyrimidine. The context of the misread base is important and therefore some codons are resistant to misreading.

codon reassignment (Sueoka 1962) — the disappearance, during evolution, of a → codon and its → transfer RNA carrying the anticodon for its translation and, at a later period, the reappearance of the codon together with the anticodon translating it. The anticodon may be assigned to the same amino acid, or to a different amino acid from that of the tRNA that disappeared, i.e., the codon has been reassigned or "captured". C.r. may be due to directional mutation pressure resulting in a change in the GC content of DNA during evolution (GC pressure or AT pressure, respectively).

codon recognition — the relationship between the anticodon of → transfer RNA and the codons in → messenger RNA. A site on the ribosome-mRNA complex (→ initiation complex) at which codon-dependent binding of → aminoacyl-tRNA occurs during → genetic translation is called the codon recognition site (Watson 1964).

codon selection — → codon usage.

codon usage (Grosjean and Fiers 1982) — in prokaryotes and eukaryotes, the intraspecies mode and variation in choice among synonymous → codons, which results in a distinct bias (codon bias) in the frequency with which a particular synonymous codon is used to code for a particular amino acid (codon selection). Correlations exist between the degree of bias and the extent of expression of a gene, with highly expressed genes displaying the most extreme bias (avoidance of codons recognized by minor tRNA's).

One possible mechanism producing this nonrandom use of codons might be the constraint that arises from → transfer RNA availability. C.u. among organisms fluctuates with the G+C content of the organism's genome (→ coding strategy).

coefficient of aberration production (Lea and Catcheside 1942) — the aberration frequency (→ chromosome mutation) per cell per roentgen in the case of experimental induction of aberrations by means of ionizing radiations.

coefficient of coancestry (Malecot 1948) — the probability that two homologous genes, one from individual I and the other from L, are identical by descent, i.e., are descended from the same ancestral gene. The complementary probability, $1\text{-}f_{IL}$, is the probability that these two genes come from unrelated ancestors.

coefficient of coincidence (Muller 1916) — → coincidence.

coefficient of hybridity (Morton et al. 1971) — a measure of the distance between populations estimated from the allele frequencies in the individual populations, i.e., without reference to the general frequencies in the ensemble of populations (→ coefficient of kinship).

coefficient of identity — the probability of identity by descent of particular genes.

coefficient of inbreeding (Wright 1929) — → inbreeding coefficient.

coefficient of integration — relationship between the frequency of transfer of a donor marker to recipients and the frequency of its transmission from recipients to recombinants following F-x Hfr conjugal mating in bacteria (→ F-factor).

coefficient of kinship (Malecot 1948) — the most condensed measure of relationship between two individuals from a population, i.e., the probability that a gene taken at random from individual A is → identical by descent with a gene at the same locus taken at random from individual B. The c. of k. is an index whose value decreases as the relationship between the individuals in question becomes more remote.

The c. of k. of two populations is the probability that two genes at a locus carried by an individual randomly chosen from the F_1 after crossing the populations are identical by descent (→ coefficient of hybridity).

coefficient of relationship (Wright 1922) — the theoretical correlation between relatives X and Y (r_{XY}) for some quantitative trait, assuming additivity of gene effects both between different loci and between different alleles at each locus.

coefficient of reunion (Darlington and Upcott 1941) — the ratio of the total number of identifiable chromatid and chromosome breaks to the number of breaks which participate in a discernable reunion (→ chromosome mutation).

coefficient of selection — the intensity of → selection as measured by the proportional reduction in the gametic contribution of a particular → genotype compared with a standard genotype (symbol s). The contribution of the favored genotype is taken to be 1, the contribution of the genotype selected against is 1-s. This expresses the → fitness or adaptive value of one genotype compared with the other. A coefficient $s = 0.2$ means that for every 100 zygotes produced by the favored genotype, only 80 are produced by the genotype selected against.

coenocyte — a multinucleate cell produced by repeated nuclear division without cytokinesis (= multikaryon).

coenogamodeme (Gilmour and Heslop-Harrison 1954) — = → coenospecies.
coenospecies (Turesson 1922) — a group of individuals of common evolutionary origin comprising more than one taxonomic species. The c. is made up of → ecospecies which, following hybridization, may participate to a limited extent in mutual exchange of genes. Gene exchange between different c. does not occur. Sterile hybrids are produced by hybridization.
coenzyme — usually a nonprotein compound of low molecular weight necessary for the action of some → enzymes.
coevolution — coadaptation, i.e., the reciprocal adaptation of taxa to each other, and cospeciation, i.e., the joint → speciation of associated taxa such as hosts and their parasites.

A coevolutionary trait represents an adaptation evolved specifically in response to natural → selection imposed by close interaction with other species.
cognate transfer RNA's — → transfer RNA species recognized by a particular → aminoacyl-tRNA synthetase.
cog region (Angel et al. 1970) — any group of similar sequences of nucleotides in DNA molecules which may specifically be recognized by endonucleases or other enzymes.
cohesive end (Ris and Chandler 1963; Hershey et al. 1963) — any of the single-stranded (overhanging or sticky) complementary ends in DNA molecules (→ blunt end).

In the DNA of bacteriophage lambda, the 5′ ends of the DNA strands are complementary, single-stranded segments, 12 bases long. Upon injection of lambda DNA into a cell, the c. ends anneal, cyclizing the DNA, and the nicks are sealed by host ligase.
cohort selection (Mertz et al. 1984) — a kind of natural → selection that takes place due to interactions among the age groups in a biological population. The age groups must be genetically differentiated; selection takes place when the age groups modify one another's survival and/or reproduction. C.s. is frequency-dependent and depends upon interaction between random → genetic drift and selection.
coil — one complete revolution of the internal spiral of a chromosome (→ chromosome coiling). (Darlington and Mather 1949).
coiling — → chromosome coiling.
coimmune — of → temperate phages if the → repressor made by each of them can repress lytic growth of the other (heteroimmune), i.e., recognizes the operators of the others.
coincidence (Muller 1916) — the coefficient of c. denotes the proportion of observed double → crossovers to the number expected from the random combination of single crossovers among three or four linked genes (Darlington and Mather 1949).

The coefficient of c. is the most commonly used genetic measure of the → interference intensity. For positive interference it is less than 1, for negative interference greater than 1. In the absence of interference, the coincidence value equals 1. In cases of absolute interference (no double crossing over), the coincidence value equals O.
coincidental evolution (Hood et al. 1975) — the maintenance of sequence homology among nonallelic members of a → multigene family within a single species (= → concerted evolution).
co-inducer — a molecule which interacts with → repressor protein to free the DNA of the → operon from restraints on its transcription into → messenger RNA.
cointegrate — a product originating by fusion of two → replicons, one originally possessing a → transposon, the other lacking it. Copies of the transposon are present at both junctions of the replicons in the c., oriented as direct repeats. The c. is an intermediate in the → DNA transposition pathway for some → transposable genetic elements.
colchicine-binding protein (Hotta and Stern 1973) — a protein found in mitotic and meiotic nuclei which might be involved in meiotic and mitotic → chromosome pairing. An increase of c.-b.p. coincides with the leptotene to pachytene stages of → meiosis.
colchiploidy (Dermen 1953) — polyploidy induced following the application of the spindle poison colchicine (→ polyploid).

colicin (Frederiq and Betz-Bareau 1953) — any of a class of bacterial protein antibiotics which are specified by extrachromosomal → plasmids (→ colicinogenic factor) in certain bacterial strains, can adsorb to the surface of susceptible bacteria and, in the majority without penetrating the surface, kill these cells. The receptor sites on the bacterial surface can also bind certain bacteriophages; hence colicins may have evolved from bacteriophages by a process of gene elimination. Three types of action are known by which different colicins may kill sensitive cells: (1) Arrest of DNA synthesis and DNA degradation; (2) inhibition of protein synthesis which can be traced to a specific alteration of some component (16S → ribosomal RNA) of the ribosomes; (3) inhibition of oxidative phosphorylation and overall arrest of macromolecular syntheses.

Cells producing ac. are themselves immune to its effects due to the formation of a specific c. inhibitor which, probably by reversible association with the c., inhibits its action. Bacterial mutants resistant to c. which nevertheless still adsorb c. are called "mutationally immune", "tolerant", and "refractory". Different colicins have different host specificities and different antigenic properties.

colicin factor — = → colicinogenic factor.

colicinoduction (Ozeki and Howarth 1961) — transfer of bacterial genes from donor (col^+) to recipient (col^-) cells due to the presence of → colicinogenic factors which act as a → conjugon and mediate bacterial → conjugation (→ sex factor).

colicinogenic (Fredericq 1953) — of bacterial strains able to produce → colicins.

colicinogenic factor (Fredericq 1953) — any of the extrachromosomal genetic elements (plasmids, symbolized by Cf) responsible for the production of → colicins in colicinogenic strains of Enterobacteriaceae (= colicin factors; col-factors). Col-factors are transferable from colicinogenic (col^+) to noncolicinogenic (col^-) strains by means of cell contact or by → transduction. Each col-factor represents an independent → replicon and consists of double-stranded (circular) DNA.

Some of these col-factors have been reported to be → conjugons mediating bacterial → conjugation or to interfere with bacerial fertility, while others do not affect the fertility of donor strains.

Col-factors are genetic units of varying complexity in which the structural genes for the synthesis of one or several different colicins may be associated with genes governing other properties (as, e.g., repression, immunity, fertility, susceptibility to male specific phages, inhibition of fertility, restriction of phages, and resistance to UV). As a rule, col-factors are independent of the bacterial chromosome. Some col-factors of *E. coli* K 260 have, however, been found associated with the chromosome or with detached chromosomal genes. Recombination with other plasmids is obtained by transducing col-factors to recipients already carrying the → F factor Fd, F'lac, or some other col-factors. Recombination with → resistance factors occurs when the recipient cell already carries a certain col-factor (Col V) in addition to the R factor. The fertility properties determined by col-factors are due to an F-factor called F_x and closely attached to the colicinogenic determinants. F_x can be physically or functionally separated from these. Like the original col-factor, F_x is transferred with efficieny to F^-(recipient) cells and imports to them a low level of chromosomal fertility. F_x is itself capable of autonomous multiplication and transfer.

The c. factors range in size from a molecular weight of a few million (for those that are not self-transmissible), to more than 60×10^6 (for those that are self-transmissible).

colicinogeny — → colicin.

colicin tolerant (Hill and Holland 1967; Nomura and Witten 1967) — of bacterial mutants that are tolerant to the killing action of a number of → colicins. Such mutants still possess normal colicin receptor sites but the sequence of biochemical changes that normally follow colicin adsorption does not occur. The mutation results in widespread changes in the structure of the cell surface.

colinearity (Crick 1963) — the exact point-by-point relationship between the order of amino acids along the polypeptide chain and the order of the corresponding → codons along the polynucleotide chain of the nuclei acid (→ sequence hypothesis; genetic transcription; genetic translation).

Many eukaryotic genes are discontinuous (→ exon; intron), i.e., the information in the DNA is not colinear with the information found in the corresponding mature functional products that they encode (→ split gene; RNA splicing).

collaterals — individuals of a family not related by direct descent. Inheritance is spoken of as "collateral" if particular → characters (traits) appear in collaterals (e.g., uncle and niece, etc.) as is characteristic for some irregularly manifesting recessive traits.

collochore (Cooper 1941) — a small nonspecific heterochromatic region causing chromosome association in → achiasmate meiosis. Collochores are points of close contract between homologous chromosomes. They fulfill the function of → chiasmata in achiasmate meiosis without being the result of genetic → crossing over and exchange of chromatid segments.

colony hybridization — a technique that uses in situ hybridization for identification of bacteria carrying chimeric vectors (→ cloning vector) whose inserted DNA is homologus with some particular sequence (probe).

Col plasmid — a → colicinogenic factor or colicinogenic plasmid able to produce → colicins.

combinant (Lenz 1938) — (1) = → codominant; (2) of nonallelic genes showing functional interaction in the production of a particular phenotypic character (→ gene interaction).

combining ability (Griffing 1956) — a potential tool for identifying prospective parents of productive hybrids (→ heterosis) from a set of crosses in F_1. (1) General c.a.: In polycrosses, the general ability of a clone to contribute yield or some other character to hybrids in crosses with all other clones. General c.a. results from additive gene action. (2) Specific c.a.: C.a. expressed by clones of certain specific crosses; the progeny perform better or worse than would be expected on the basis of general c.a. of the clones involved. Specific c.a. results from → dominance, → epistasis, and other nonadditive effects.

commiscuum (Danser 1929) — a group of individuals all of which can actually or potentially exchange genes (a species).

commitment — the fixation of a cell to a unique developmental sequence alternative to its previous state (predominantly in microorganisms). Examples of c. are → sporulation in bacilli, meiosis and sporulation in yeast, or meiosis in explanted microsporophytes.

community — an association of co-existing populations bounded functionally by their interactions or spatially by their co-occurrence in a habitat or biotope.

compact X-chromosome — → dosage compensation.

comparate chiasmata (Darlington 1937) — → chiasma.

comparium (Danser 1929) — a group comprising all → coenospecies which may participate in mutual hybridization, whether direct or indirect.

compartmentation — the subdivision of cells into parts (→ cell compartments) by double or single → unit membranes.

competence (Waddington 1932; Thomas 1955) — (1) the state of a (bacterial) cell when it will take up extracellular DNA in → genetic transformation or → transfection. C., a transiently acquired physiological state of recipient cells, confers the ability to bind and, after a series of events, to integrate and express exogenous DNA. (2) Embryonic → induction.

competence factor (Charpak and Dedonder 1965) — → competence (= competence inducing factor; competence provoking factor).

competition — in ecology, genetics, and evolution, the endeavor (which may take various forms and have various, direct or indirect, deleterious results) of two (or more) organisms of the same or of different species to gain the same particular factor or thing, or to gain the measure each wants from the supply of any factor or thing when that supply is not sufficient for both (or all) (Milne 1961).

competitive ability — the success of one species in a mixture which is generally attributed either to the ability of that species to produce a greater proportion of offspring in the next generation at the expense of the other(s) or to its ability to reduce the numbers of the other species without itself changing (the success of one component of the mixture may also depend on its ability to avoid the other(s) without necessarily having any effect on it).

competitive character displacement — in a population, a change of character state that is due to natural → selection arising as a result of competition with individuals of another population in the same environment. The change may be either divergent or convergent.

competitive exclusion (Hardin 1960) — the principle that indefinite coexistence of two or more → species limited by the same resource is impossible.

competive release — = → niche expansion.

complement (Darlington 1932) — → chromosome complement.

complementary — (1) of two structures, each of which defines the other, e.g., the two strands of a → DNA double helix; (2) of four-strand double → crossing over; (3) of genes which by interaction produce an effect qualitatively distinct from the effects of any of them separately (→ gene interaction).

complementation — (1) all cases in which interaction between viral gene products or gene functions in multiply or mixedly infected cells results in an increased yield of infective virus of one (or both) parental types (→ helper virus). (2) The production of a wild-type phenotype when two different mutations are combined in a diploid or a → heterokaryon and tested in → trans-configuration (→ genetic complementation).

complementation analysis — determination of whether cells with the same phenotype have the same genotype (→ genetic complementation).

complementation group — a series of mutations unable to complement (→ genetic complementation) when tested in pairwise combinations in trans. The c.g. defines a genetic unit (the → cistron) that might better be called a "noncomplementation group".

complementation map — a diagrammatic representation (gene map) of the complementation matrix (→ genetic complementation). The normal, fully functional gene can be represented by a line (bar) and the various classes of mutants by shorter lines representing regions for which they are defective in function. Two noncomplementing mutants will be defective in the same region and so their representations will overlap; two complementing mutants are defective in different regions and so their representations are nonoverlapping.

The c.m. indicates which alleles, defective in function, can in concert with other alleles, defective in a different function, restore the full function of the locus.

complementation mapping — a method for obtaining → gene assignments by fusing a cell line, harboring a recessive conditional lethal mutation, with wild-type cells of another species and direct selection of hybrid clones in which the mutant phenotype has been complemented.

complementation test — = → cis-trans test.

complementation unit (Demerec and Hartmann 1959) — a subunit of the complementation map of a → cistron as determined by → cis-trans tests (complementation tests) which show that → genetic complementation may also occur between two mutants of one cistron ("intracistronic complementation"), neither of which complements a third.

This result is evidence that some mutations involve one c.u., other mutations another c.u., while still others affect both complementation units or "complons" simultaneously.

complete digest — cleavage of all of the potential target sites within a DNA preparation by a → restriction endonuclease.

complex heterozygous (Renner 1917) — of a special type of genetic system in plants (called complex heterozygotes) based on heterozygosity for multiple reciprocal translocations (as examplified by *Oenothera*). C.h. species are heterozygous for numerous allele pairs linked to different chromosomes but segregating as a unit during → meiosis.

The best example of "complex heterozygosity" is the plant genus *Oenothera*. Species of this genus have 6, 8, 10, 12, or all 14 of their chromosomes linked in a chain or ring at meiosis as a consequence of reciprocal translocations and the specific type of chromosome pairing in the case of translocation heterozygosity (Cleland 1922, 1962). Within this pairing configuration paternal and maternal chromosomes alternate.

During anaphase I of meiosis, a regular distribution of adjacent chromosomes to opposite cell poles (→ alternative distribution) occurs. Thus all paternal chromosomes and genes of the pairing configuration are carried to one pole, all maternal chromosomes and genes to the other pole. With a ring of 14 chromosomes only two kinds of gametes are produced, identical genetically with the two that united to form the plant. The group of chromosomes and genes distributed as a unit during meiosis and identifiable by a particular group of alleles is called a "complex". It acts as one → linkage group (= Renner complex). Since in different *Oenothera* species different numbers of chromosomes are included in the ring, the size of the complexes is species-specific.

Selfing of a complex heterozygous *Oenothera* does not produce a Mendelian combination of 1:2:1 (→ segregation). The two expected classes of "complex" homozygotes normally fail to appear; only complex heterozygotes are recovered thus representing a system of "enforced heterozygosis". Their hybrid nature becomes evident when "heterozygotes" are crossed to another heterozygous or homozygous species resulting in the production of "twin hybrids". Complex heterozygosity is maintained in most cases by one or more → lethal factors per complex which are lethal in the homozygous state. The lethals are either zygotic or gametophytic. Zygotic lethals cause the death of zygotes; gametophytic lethals kill or inhibit either the male or female gamotophyte. In most *Oenothera*, these lethals are "balanced", i.e., both complexes possess nonallelic lethals. A species with a balanced set of gametophytic lethals gives rise to only one kind of functional egg and one kind of sperm. Plants with zygotic lethals produce two types of functional egs and sperm, but half the zygotes produced by selfing will die, since they receive one or the other of these lethals in double dose.

complex locus (Dunn 1954) — any of the genetic loci characterized, in most cases, by: (1) Different mutations result in different phenotypes and are often → pleiotropic; (2) pairs of mutants often show partial or complete interallelic → complementation.

Many c. loci are considered to be groups of closely linked genes (gene clusters). Others may code for a single multifunctional protein with several domains, each of which is responsible for a particular function.

complon — a unit of allelic → chomplementation, i.e., a complementation subgroup within a linear → complementation map (→ genetic complementation).

composite gene — any gene arising by recombination between two nonallelic genes, located on two nonhomologous chromosomes, and containing portions of both genes. The nonallelic genes are of extensive but not of complete homology. C. genes may also be formed by mitotic → gene conversion.

composite transposon — a bacterial transposon (= compound t.) whose central region is flanked on each side by → insertion sequences, either or both of which may enable the entire element to transpose (→ transposable genetic element).

compound — a genotype heterozygous for two mutant → alleles (m^1/m^2) of the same locus. Together with the wild-type allele, the mutant alleles form a series of multiple alleles (+, m^1, m^2, etc.).

compound chromosome (Robertson 1916; Novitski 1954) — (1) any new monocentric chromosome resulting from structural changes (→ chromosome mutation; Robertsonian translocation) which transformed two rod-shaped chromosomes into one V-shaped chromosome. (2) In *Drosophila*, any monocentric chromosome in which the material from one chromosome arm is represented twice (→ X chromosome). Gametes of the compound-bearing individuals generally carry either two or no doses of the chromosome arm (Lindsley and Grell 1968). The two arms comprising a c.c. may join (1) by attachment of the base of one to the terminus of the other to form an acrocentric chromosome, or (2) by attachment of both proximally to a single → centromere to form a metacentric chromosome; the ends of either an acrocentric or a metacentric chromosome may join to form a compound → ring chromosome.

The compound arms may be in the same sequence or one may be inverted with respect to the other. The elements of a c.c. may, therefore, pair as a spiral (tandem configuration) or as a hairpin (reversed configuration).

Compton effect — in an atom, the interaction of a high energy photon with a single electron resulting in the ejection of the electron and secondary emission of a lower energy (longer wave length) photon.

co-mutagenic (Mukherjee and Beermann 1965) — of any two agents, nonmutagenic in themselves, whose co-operative activity results in gene mutations or chromosome structural changes.

co-mutation (Guerola et al. 1971) — the process of co-ordinate occurrence of clustered mutations.

concatemer (Thomas et al 1968) — a DNA molecule made up from repeated monomers of a single kind to give a linear multimer with all the monomers in the same relative orientation to one another. Concatemeric DNA is the substrate for the in vitro packaging of → recombinant DNA molecules constructed in → cosmid or lambda phage cloning experiments. In some phage, concatemeric DNA is the precursor from which mature phage chromosomes are cut and packaged.

concatenated — of circular DNA molecules interlocked like rings on a chain.

c-onc gene — any of the normal (vertebrate and invertebrate) cellular genes containing sequences homologous to the transforming genes (v-onc genes) of RNA tumor viruses (→ oncogene, proto-oncogene).

concerted evolution (Zimmer et al. 1980; Dover and Coen 1981) — the process which results in internal homogeneity for sequence variants of many → multigene families and noncoding repetitive families of DNA within an organism or a species (= coincidental e.; horizontal e.). → Rectification mechanisms contributing to c.e. of tandemly arranged and of dispersed genes include → gene conversion, unequal → crossing over, and transposition (→ molecular drive). C.e. results in the tendency of a family of repeated DNA sequences to evolve in unison. All the members of a repeat family undergoing c.e. can be considered to be descended from a single member at some point in the past.

concordance — the identity of members of matched groups or pairs (such as twins) in any one or in a given → trait. The opposite of c. is → discordance.

concordant orientation — → translocation.

concordant segregation — in → hybrid cells undergoing chromosome elimination, the simultaneous appearance or disappearance of gene markers.

conditional mutation (Edgar and Lielausis 1964) — any mutation whose manifestation is dependent on one or more parameters, such as temperature, age, and nutrition. Restrictive conditions are those in which the mutant phenotype is observed, whereas permissive conditions in effect suppress the mutant phenotype resulting in wild type or normal gene function. Major categories of c. mutations are: (1) → Nonsense mutations in the presence of a relevant → nonsense suppressor (in prokaryotes and lower eukaryotes; in higher eukaryotes, no nonsense suppressors have been observed); (2) temperature-sensitive mutations resulting in temperature-sensitive gene products (they are under permissive conditions more or less active); (3) auxotrophic mutations (of bacteria and certain fungi) which are only lethal in the absence of a certain growth factor (→ lethal factor).

conditioning — a direct alteration in the development of the → phenotype of an individual resulting from the action of external agents during early, critical stages of development. Examples of c. are → phenocopies, → dauer-modifications, and other changes transmitted partially or wholly through successive asexual or sexual generations in the absence of the inducing stimuli (Hill 1967).

conduction (Clark and Adelberg 1962) — (1) in bacterial → conjugation, the transfer of the bacterial genetic material and of the → conjugon (= "promotor") as linked elements (→ donation). (2) The physical association of two → plasmids (a small nonconjugative one and a conjugative one) whereby the small plasmid is transferred with the large conjugative replicon as a consequence of genetic recombination or transposition between the two plasmids. Recombination results in a single transferable DNA molecule (→ donation).

configuration (Darlington 1929) — = → chromosome configuration.

conformation — of the different spatial arrangements of the atoms that may arise by rotation about bonds in a molecule. The term conformational change refers to a change in the average positions of the atomic nuclei not including covalent bond changes.

congenital — of a → phenotype (character, trait) recognizable at birth or referable to constitution at birth. C. is a descriptive term neither including all inherited conditions, nor excluding those characters which result from environmental influences alone. Malformations evident at birth are therefore c. and may be caused by genetic or extrinsic factors.

congression (Drlington 1937) — → chromosome congression.

conidiospore — an exospore of fungi which, in contrast to the endospores (→ ascospore) formed within a cell or an organ, is constricted singly or in chains. Conidiospores represent "aplanospores" (= conidia).

conjugal DNA transfer — in bacteria, transfer of genes between bacterial cells during → conjugation. In the case of conjugal transmission of the *E. coli* → F plasmid, the passage of a single DNA strand of the F plasmid (with a 5′ terminus at the origin of transfer) from donor (F^+) to recipient (F^-) during conjugation. DNA synthesis produces complementary strands using the strand transferred to the recipient and the strand retained in the donor as templates (conjugal DNA synthesis) (→ plasmid mobilization).

conjugation — the term c. denotes several events that connect, couple, or unite chromosomes, nuclei, cells or individuals. (1) **Conjugation of chromosomes:** = → chromosome pairing. (2) **Conjugation of nuclei:** a) = → karyogamy, b) = → dikaryotization. (3) **Conjugation of cells:** (a) in general, → syngamy, the union of sex cells (gametes) during fertilization; (b) the fusion of a number of cells into a multinucleate plasmodium or syncytium. (4) **Conjugation of individuals:** the union of unicellular organisms or of filaments of equivalent cells; (a) **in certain algae,** the formation of connections between pairs of laterally juxtaposed cells leading to the fusion of the cells and their nuclei; (b) **in ciliated protozoa,** a temporary association of two individuals during meiosis and reciprocal cross-fertilization; (c) **in bacteria** (bacterial conjugation or bacterial mating), a parasexual mode of undirectional (polarized) transfer of genetic information involving direct cellular contact between a donor ("male") and a recipient ("female") cell.

Bacterial c. is a multistage process including: (1) The transfer of one preexisting DNA strand beginning with the 5′ end into the recipient cell (→ conjugal DNA transfer); (2) synthesis of a complementary DNA strand in the recipient to make the transferred DNA sequence double-stranded; (3) recombination between the donor DNA fragment and the cell genome (→ F plasmid; conjugative plasmid).

conjugational DNA synthesis — → conjugal DNA transfer.

conjugative plasmid — any of a number of large → plasmids which allow bacterial cells harboring them to transfer DNA to other cells by a process called bacterial → conjugation. Such donor cells are poor conjugal DNA recipients (surface exclusion). For the conjugative → F plasmid and F-like plasmids, all the plasmid genes required for the expression of the conjugative properties are organized in a continuous block of DNA (~ 35 kb long). At least four chromosomal genes of *E. coli* are also required for DNA donor activity and → surface exclusion.

conjugative transposon — → transposon.

conjugon (Luria 1963) — any of a number of different genetic elements (→ plasmids) in bacterial that promote bacterial → conjugation and gene transfer. These include the → F plasmid, → colicinogenic factors (Col), and → resistance transfer factors (R). Conjugation due to the presence of F and certain R-factors is brought about by the synthesis of a special kind of pilus (→ sex pilus) at the bacterial surface which also acts as receptor for F-specific (male-specific) bacteriophages.

Cells with any one of these factors (donor cells) can conjugate with related cells (usually lacking such a factor) and during conjugation the conjugons are transferred from donor (male) to recipient (female). In some cases, the conjugons may transport from the donor to the recipient male chromosomal material that has become temporarily associated with the conjugons which are DNA elements (→ replicon). Conjugons have three functions: (1) determine the surface properties and synthetic abilities which lead the formation of effective contacts between conjugation partners; (2) mobilize the "genote" to be transferred from donor to recipient; (3) provide the immediate energy source necessary for the transfer of chromosomal material.

Conjugons are tranferable from cell to cell by → transduction via bacteriophage as well as by bacterial conjugation.

conjunction (Gonda and Radding 1983) — the first stage in synapsis of DNA molecules during → genetic recombination (in *E. coli*). C. involves the invasion of duplex DNA by single-stranded DNA (→ conjugal DNA transfer) and is independent of sequence homology. During a second stage of synapsis, homologous alignment of complementary sequences occurs either by facilitated diffusion or by translocation of the DNA molecule in a ternary complex with the Rec A protein.

The initial structure between paired complementary sequences is assumed to be a hydrogen-bonded → heteroduplex molecule. A transient → nick then allows interwinding of the two strands which converts the unstable heteroduplex into a stable joint molecule.

conjunctive segment (Cooper 1944) — =→ collochore.

connexon (Payton et al. 1969) — a communicating channel between eukaryotic cells associated with → gap junctions.

consanguineous — of two or more individuals if they have a common recent ancestor (usually not further back than three to four generations). The effect of c. mating is to increase the probability that the offspring will be homozygous at any pair of genetic loci, i.e., will be on the average less → heterozygous than the offspring of a non-consanguineous mating (→ inbreeding).

consensus junction sequence — → intron; intervening sequence.

consensus sequence — a particular (idealized) nucleotide sequence in which each position represents the base most often found when many actual sequences of a given class of genetic elements are compared. C. sequences have been defined in two ways: (1) by comparison of those naturally occurring DNA (or RNA) sequences that are believed to encode a particular genetic function; (2) by generation of many mutations of an individual genetic element.

conservative substitution — the phenomenon that natural → selection pressures favor changes in protein sequence in which certain physical and chemical properties of residues are conserved, i.e., a substitution of one amino acid residue for another has a significantly greater chance of being accepted if the two residues are similar in general properties.

constitutive — (1) of any gene that encodes a product required in the maintenance of basic cellular processes or cell architecture ("household genes"), as opposed to regulated genes that typically encode products that are required when cells grow under specific physiological conditions, or (in multicellular organisms) that are expressed in particular cell types (→ gene expression). (2) Of mutations resulting in alleles that usually are regulated to be expressed without regulation. (3) Of enzymes (proteins) produced by the cell in fixed amounts independent of need, in contrast to → inducible and → repressible enzymes.

constriction (Agar 1911) — a lightly staining (unspiralized) segment of fixed position in the metaphase chromosome. (1) **Nucleolar constriction** (Darlington 1937): A secondary constriction in the region of which the → nucleolus becomes organized (→ satellite). (2) **Pri-**

mary or centric constriction (Darlington 1937): A constriction associated with, and determined by the → centromere region. (3) **Secondary constriction** (Darlington 1929): Any noncentric constriction either nucleolar or non-nucleolar.

consultand (Murphy 1970) — the person whose genotype is of primary relevance to the → genetic counselling problem at hand. It may or may not be the same person as the counselee (frequently c. is used synonymously with counselee).

contact feeding (Corsaro and Migeon 1975) — = → metabolic cooperation.

contact guidance (Weiss 1959) — the oriented response of isolated tissue cells in vitro according to the topography of their substratum. There are three postulated mechanisms of cell guidance: (1) Alignment by submicroscopic fibers; (2) differential adhesion; (3) changes in the motile apparatus of the cell (→ contact inhibition).

contact inhibition (Abercrombie and Heaysman 1954) — in cell culture, the cessation of cell locomotion and/or growth when freely growing cells from a multicellular organism come into physical contact (on a plane substrate) with each other (→ contact guidance). C.i. requires the presence of actively metabolizing cells and probably involves inhibitory factors associated with the → cell membrane. As a result of ci.i., cells in culture tend to form confluent monolayers with negligible overlap (→ contact retraction).

contact paralysis — the cessation of forward extension of the pseudopods of a cell as result of its collision with another cell (→ intercellular invasion).

contact retraction (Weiss 1958) — in cell culture, intercellular collision in the form of an abrupt retraction of the contacting cells, often causing them to spring apart (→ contact inhibition).

containment — the precautions used to prevent the replication of the products of → recombinant DNA technology outside the laboratory. (1) Biological c.: The use of hosts and → cloning vectors which render the recombinant products less able to survive and multiply outside the laboratory environment. (2) Physical c.: The use of special laboratories which prevent escape of products of recombinant DNA technology.

context effect — → codon context effect.

contig (Staden 1980) — a group of cloned nucleotide sequences that are contiguous (→ molecular cloning).

contracomplementation (Portion and Ruodhonen 1972) — in *Drosophila melanogaster*, the lethality of certain heterozygous combinations of alleles of sex-linked genes, each allele being homozygous viable.

contractile ring (Lewis 1939) — a cell organelle involved in → cleavage and representing a contractile belt of aligned → microfilaments (c.r. filaments) arranged parallel to the plane of cleavage. The c.r. has an overall width of about 10 μm and the layer of microfilaments is a uniform, coherent, and apparently cohesive stratum in the → cell cortex.

control gene — any → gene that regulates the time and rate at which neighboring structural genes are transcribed into → messenger RNA (→ operon).

controlled allele (Peterson 1976) — an → allele under the control of a→ controlling element system.

controller node (Zuckerkandl 1978) — a genetic unit of regulation consisting of a set of regulators, effectors, and receptors (such as → promoters and enhancer sequences) located in the neighborhood of the gene(s) it controls. C. nodes are involved in the control of transcription of a given gene and in the processing of its transcript.

controlling element (McClintock 1956) — any member of the first identified family (in maize) of → transposable genetic elements which have the capacity to control → gene expression at several loci, in addition to the production of unstable mutant target genes. They may cause adjacent chromosome breakage and other structural rearrangements and can regulate the specific expression of nearby genes.

Upon insertion into genes c. elements generate mutated alleles which, after excision of the inserted element, revert to wild type (giving rise to plant or seed variegation). At least six different systems are known in maize. Each is based on a single autonomous element or on two interacting elements (receptor and regulator). In two element systems, a nonautonomous receptor element, cis-located at the controlled locus, is receptive to signals from a translocated regulatory element which mediates the excision of the former.

An allele with an inserted c.e. at the locus is called c.e. allele. An allele under the control of a c.e. system is a controlled allele and includes an inserted receptor or regulatory element.

convection (Hanahan et al. 1980) — the process of → DNA-mediated gene transfer. DNA is introduced into eukaryotic cells and stably established in high molecular weight DNA, as distinguished from → transfection and → transformation.

convergence — in evolutionary biology, the evolution of greater similarity between members of two or more distantly related phyletic lines (whose ancestors were less alike), as opposed to → adaptive radiation. C. is the result of adaptation to similar ecological conditions.

conversion (Winkler 1930) — → gene conversion; phage conversion.

conversion polarity — a gradient in the frequency of → gene conversion from one end of a gene to the other.

convertant — the product of → gene conversion.

convertogenic — of agents able to induce → gene conversion.

convivium — a group of genotypes within one → commiscuum, which is geographically isolated from other such groups.

coordinate repression (Ames and Garry 1959) — → repression.

co-orientation (Darlington 1937) — → centromere orientation.

copia element (Dunsmuir et al. 1980) — any member of a class of repetitious, transposable DNA sequences (in *Drosophila*) that have strong similarities to vertebrate → retroviruses. The copia sequences have widely scattered chromosomal locations (30 to 50); the integrated c.e. has a characteristic structure of a central 4.7-kb coding region flanked by 276-bp direct terminal repeats (LTR's). Two major transcripts are initiated within the 5′ LTR of the element. Insertion of the c.e. generates a five base pair duplication at the site of integration.

Among known copia-like repeated sequence families, three (copia, 412, and 297) are 4.7, 7, and 6.5. kb long, respectively. They are nonhomologous in nucleotide sequence, but share the following properties: (1) Elements of each family occur at widely separated chromosomal sites; (2) their sequences are closely conserved and nonpermuted; (3) all the sequences are terminally redundant, with direct repeats of 0.5 kb, 0.3 kb, and 0.4 kb, respectively; (4) all undergo → transposition; (5) their DNA codes for abundant poly(A)-containing cytoplasmic RNA's.

Both cultured *Drosophila* cells and flies contain small amounts of circular extrachromosomal copia DNA with structures similar to extrachromosomal retrovirus proviral DNA.

copolymer — a polymeric molecule containing more than one kind of monomer unit (→ homopolymer).

copulation — the union of sexual units. In a broader sense, the sexual act. According to Kniep (1928) the following types of c. are to be distinguished: (1) **Gametic c.:** The fusion of differentiated, uninucleate → gametes. (2) **Gametic-gametangial c.:** The fusion of a differentiated uninucleate gamete (male or female) and a differentiated gametangium which produces no differentiated gametes of the opposite sex. (3) **Gametangial c.:** The fusion of two gametangia which can be distinguished morphologically as male or female. The gametangia do not differentiate gametes. One or many pairs of nuclei may participate in gametangial c. (4) **Somatic c.:** The fusion of undifferentiated vegetative cells.

copy-choice recombination (Lederberg 1955) — a model of the mechanism of → genetic recombination which is not regarded as a physical exchange of preformed genetic strands. There is no experimental support for this model.

copy error — an error in the → DNA replication process giving rise to a → gene mutation.

copy error lag (Braun 1965) — in the case of → copy errors giving rise to → gene mutations after incorporation of a fraudulent base into nucleic acid, time will elapse between the base incorporation and the permanent change in base sequence. This is called c.e.l. (→ phenomic lag; segregation lag).

copy number — the number of molecules, per → genome, of a → plasmid or gene contained in a cell.

core nucleosome — → nucleosome.

co-repressor (Jacob and Monod 1961) — an → effector which activates a → repressor (→ operon).

correlated response (Wigan and Mather 1942) — the phenomenon that selection exercising its pressure on some character (or genotype) changes others in concert; the c.r. may be due to linkage or pleiotropy (on the level of the genotype) and may also be a form of coadaptation (on the level of the gene pool). (→ genetic correlation).

correlation coefficient — a coefficient which measures the degree of association between two or more variables.

correndonuclease (Grossmann et al. 1975) — any → endonuclease active in → DNA repair (= correctional e.). These enzymes not only recognize and incise at the → DNA damage but also produce termini suitable for removal of the damage by excision enzymes, reinsertion of nucleotides by polymerases, and final phosphodiester bond formation by → DNA ligase.

Correndonucleases generally are small proteins (mw less than about 30,000). A type I c. is specific for damaged DNA's possessing monoadduct derivatives (base modifications in which an addition reaction, adduct formation, or a chemical transformation leads to the alteration of a single nitrogenous base); a type II c. is specific for diadduct modified regions of DNA (diadducts comprise more than one nitrogenous base involved in the final chemical modification).

cortex (Chambers 1940) — a peripheral cytoplasmic region of animal cells (the "ectoplasm") including the → cell membrane by which the cells are bounded. It mediates the cell's interactions with its environment. In egg cells and *Paramecium*, the cortex has been proved to play an important role in early development and → differentiation (Raven 1961; Sonneborn 1963, 1964); it is a carrier of a system of developmental factors — the "cortical field" — which is of importance for the determination of early steps in the differentiation of the embryo from the fertilized egg cell. The cortical field is the seat of a system of morphogenetic factors which determine the polarity and symmetry of eggs. The developmental information in the fertilized egg is presumably stored in the nucleus, the cytoplasm as well as the cell cortex.

cortical granule (Allen and Hagström 1955) — any of the round to elliptical, membrane-bound bodies (approximately 0.5 to 0.8 μm in diameter) in the → cortex of animal oocytes, expecially just before fertilization. They contain mucopolysaccharides and participate in the formation of the fertilization membrane by the cortical reaction, i.e., the granules are released into the space which lies immediately outside the oocyte → cell membrane. C. granules are probably derived from the → Golgi apparatus and may be formed by coalescence of tiny vesicles.

cosduction (Flock 1983) — transduction of foreign DNA by use of a cosplasmid, i.e., a → plasmid onto which the cohesive ends from, e.g., phage Φ 104 reside. The transducing phage particles are called cosducing particles and the recipients cosductants.

cosmid (Collins and Hohn 1978) — a hybrid-bacteriophage → cloning vector used for cloning of long fragments of DNA in *E. coli.* C. are characterized by the following features: (1) The lambda phage sequence (cohesive end site) which is required for packaging into the lambda capsid; (2) a selectable drug resistance marker; (3) single restriction sites useful for cloning.

Subsequent to restriction and insertion of large fragments of DNA into the c., long concatenated DNA structures are formed which mimic the natural packaging substrate. Packaging (the cos sequence is recognized by the head proteins when lambda DNA is packaged to form complete virus), cleavage of the cos sequence to yield cohesive ends, and transduction impose a size selection: only those hybrid molecules are obtained that fill a phage-head with about 37 to 50 kb long DNA.

After transduction into a lambda-sensitive bacterium, the hybrid c. replicates as a plasmid and is selected for by using the antibiotic resistance carried by it. Cosmids are useful vectors for making a → gene library for the following reasons: (1) The high efficiency of hybrid clone formation; (2) only hybrid plasmids are packaged; and (3) only large fragments of DNA are selected for.

cosmid shuttle vector — a → cosmid vector that can be used as a vehicle for genomic DNA cloning and as → shuttle vector.

cost of natural selection (Haldane 1957) — the amount of selective elimination which accompanies the process of substituting one allele for another in evolution. The process of substitution involves lowering of population → fitness and thus creates a → genetic load (substitutional or evolutionary load).

Cot (Britten and Kohne 1968) — the parameter used in the renaturation analysis of DNA genomes: original concentration of DNA x time (Cot). Cot ½ is the Cot required to proceed to half completion of the reaction. Highly reiterated DNA sequences will renature at low Cot values, unique sequences at high Cot.

cotransduction — the transfer of more than one host gene within the same phage particle (→ transduction). Since the amount of DNA carried in a phage head is small, c. implies that the genes concerned are also close together on the host chromosome.

cotransfection — a method in which a nonselectable gene present in genomic DNA is transfected with a cloned selectable gene (→ transfection).

cotransformation — (1) a technique whereby two or more unrelated genes are moved simultaneously and integrated into eukaryotic (mammalian) cells growing in vitro. One of these genes serves to improve the chances for accompanying genes, whatever they code for, to move successfully into the recipient cell. (2) In → genetic engineering, incubation of host cells with two types of → plasmid, one selectable, the other not. Cells transformed (→ genetic transformation) with the first are selected. If transformation has been carried out at high DNA concentration, then it is probable that these cells will also have been transformed with the second (→ DNA-mediated gene transfer).

countertranscript (Kumar and Novick 1985) — any RNA species that is initiated from a single start within the leader region and transcribed from the opposite strand, one terminating very close to one mRNA start, the other further upstream. One or both of these countertranscripts serve(s) as a negative control element(s).

coupled transcription-translation — formation of the 70S → initiation complex before the → messenger RNA is released from the DNA, i. e., coupling of → genetic transcription and → genetic translation. C. t.-t. is the rule in prokaryotes.

In eukaryotes, no c.t.-t. occurs since the mRNA is synthesized and processed in the → nucleus and later transported to the cytoplasm where translation takes place.

coupling (Bateson et al. 1905) — → cis-configuration.

covarion (Fitch and Markowitz 1970) — in a given → gene, those codons, which are concomitantly variable resulting in favorable mutations, or in mutations leading to amino acid substitutions of little or no effect (→ varion).

C-pair (Levan 1938) — → C-mitosis.

cp DNA — acronym for → chloroplast DNA (= ct DNA).

CpG island — in vertebrate genomes, regions of DNA with a high G+C content and a high frequency of CpG dinucleotides relative to the bulk genome. The CpG islands are nonmethylated and frequently include the 5′ end, the 3′ end, or both ends of a gene. Clusters of CpG dinucleotides are implicated in → nuclease hypersensitivity and probably act as control elements in post-transcriptional regulation of various "housekeeping" genes.

CpG suppression — the phenomenon that the dinucleotide CpG is deficient in vertrebrate genomes (found at only 25% of the expected frequency calculated as the product of the known frequencies of the mononucleotides C and G). This suppression may involve → DNA methylation. Most CpG dinucleotides in vertebrate DNA are methylated as mCpG, in which the mC is 5-methylcytidine. 5-methyl cytidine has a propensity to deaminate to thymidine; thus CpG dinucleotides in DNA tend, over evolutionary time, to be replaced by TpG (CpA on the other strand).

criss-cross inheritance (Bridges 1913) — → inheritance.

cristae (Palade 1953) — folds (crests or ridges) of the inner mitochondrial membrane (cristae mitochondriales) to which ATPase knobs are attached. → Mitochondria with c. are usually found in animal cells with intensive respiratory metabolism and are uncommon in plants.

cRNA — RNA molecules complementary to specific chromosomal DNA sequences (→ chromosomal RNA).

cron (Huxley 1957) — the time unit for evolutionary processes (1 cron = 1,000,000 years; 1 kilocron = 10^9 years; 1 millicron = 1000 years).

cross — in genetics, the bringing together of genetic material from different individuals (genotypes) in order to achieve → genetic recombination. (1) **In eukaryotic organisms** with sexual → reproduction, genetic crossing is achieved by the processes of nuclear fusion (karyogamy) and → meiosis (cross-fertilization). (2) **In bacteria**, genetic crossing is achieved by the → parasexual phenomena of → conjugation, → transduction, → sexduction, and → transformation. (3) **In viruses and bacteriophages,** genetic crossing is achieved by multiple infection of the host cells with particles of differing genotypes. If bacteria are infected simultaneously with two or three different virus particles these crosses are referred to as "biparental" and "triparental", respectively.

crossbreeding — = → outbreeding.

cross-feeding — → syntrophic.

cross-fertilization — → fertilization.

cross-induction (Borek and Ryan 1958) — → induction.

crossing barrier — any of the genetically controlled mechanisms which either entirely prevent or at least significantly reduce the ability of individuals of a population to hybridize with individuals of other populations. The formation of c. barriers is a fundamental by-product of natural → selection altering the genetics of reproduction and setting a population on its own evolutionary pathway.

crossing over (Morgan and Cattell 1912) — a process consisting of a sequence of molecular events that may lead to → genetic recombination between linked markers in both prokaryotes and eukaryotes (and its result: the crossover). C.o. gives rise to reciprocal exchange of segments (by symmetrical breakage and crosswise rejoining) at corresponding positions along pairs of homologous DNA molecules or → chromosomes (linkage groups). It involves the following steps: (1) The formation of recombinogenic sites (→ chi site); (2) formation of recombinational structures at the molecular level (hybrid DNA or → Holliday structures); (3) resolution of the Holliday structures at the molecular level; (4) segregation of the c.o. products (recombinants).

In eukaryotes, c.o. may be meiotic or mitotic. (1) **Meiotic crossing over:** The exchanges are observable microscopically in suitable material by the formation of → chiasmata between homologous chromosomes. In a population of cells going through meiosis, the incidence of exchanges between any two given points in one chromosome pair is highly correlated with the physical distance between these two points (→ genetic map). Genetically, the process of crossing over is inferred from the proportion of recombinant individuals among the progeny of heterozygotes. From the fraction of recombinants one can calculate the fraction of recombinant gametes from which those individuals originated (→ crossing over value; recombination value). In haploid organisms, the fraction of recombinant individuals and the fraction of recombinant gametes coincide.

Any model of meiotic crossing over must satisfy three essential requirements: (a) Recombination is always reciprocal; (b) each event involves only two of the four chromatids (normally nonsister chromatids) of two homologous chromosomes and therefore only two of the four products of meiosis, (c) double or multiple crossing over ("compound crossing over") may involve all four, three, or only two of the four chromatids. If two crossing over events within homologous linkage structures involve the same two chromatids (→ coincidence; interference) they are said to be "regressive" ("two-strand exchange"). If they involve completely different pairs of chromatids, they are said to be "digressive" ("four-strand exchange"), and if they involve one common chromatid they are referred to as "progressive" ("three-strand exchange").

C.-o. involves the breakage and reunion of complementary DNA strands (in eukaryotes, probably during the genesis of the → synaptonemal complex). The enzyme mechanism includes single-strand breakage, strand digestion, end-to-end joining, and repair. These activities are prerequisite to hybrid formation (→ polaron hybrid DNA model) and accomplishment of reciprocal recombination.

Special types of crossing over are called "unequal" and "illegitimate".

Unequal crossing over (Sturtevant 1925) produces one chromatid containing a gene twice (duplication) and another lacking that gene (deletion).

Illegitimate crossing over (Darlington 1932) gives rise to secondary structural changes in heteromorphic (structurally heterozygous) chromosomes and takes place in haploid or polyploid individuals between homologous segments or homeologous (= partially homologous) chromosomes.

(2) **Mitotic or somatic crossing over** (Stern 1936): Crossing over at mitosis in either somatic or gonial tissue which like meiotic crossing over takes place in the four-strand stage of chromosomes.

The consequence of somatic crossing over depend upon: (a) the location of the crossing over site, and (b) the mode of orientation and, thus, distribution of the centromeres of the chromosomes engaged in somatic crossing over. A single cross over between two linked heterozygous loci (A/a and B/b) and distribution of the centromeres 1 + 2 to one and 1′ + 2′ to the other cell pole in anaphase results in homozygosity for genes distal to the point of exchange (B/B and b/b) for which the genotype was originally heterozygous (B/b). This event is detectable phenotypically by the production of a b/b-spot, provided that it is the right part of the body and that the expression of the gene is autonomous.

With a single somatic crossover between the centromeres and the first of the two heterozygous loci a twin patch aa and bb may result if the centromeres 1 + 2 are distributed to one and 1′ + 2′ to the other cell pole.

Patterns of distribution of the type 1 + 2′ (one pole) and 1′ + 2 (other pole) leave the original heterozygosity unchanged.

crossing over homozygote (Renner 1941) — → complex heterozygous.

crossing over map — a → genetic map made by utilizing crossing over frequency as a measure of the relative distances between genes in one linkage group.

crossing over modifier — any gene or chromosomal structural change increasing or decreasing the crossing over frequency, Generally, reduction in crossing over is easier to achieve than its enhancement. With rare exceptions, all structural heterozygotes ("heterokaryotypes") have lower crossing over values for regions included in or immediately adjacent to the structural change. A compensatory increase often takes place in those segments farther removed from the aberration.

crossing over potential (Sybenga 1965) — the basic number of crossovers of the chromosome segment under observation, which is only realized in the absence of → interference.

crossing over suppressor — any gene or heterozygous structural change of chromosomes suppressing or reducing (in most cases locally) the frequency of meiotic crossing over (= C-factor). (→ crossing over modifier).

crossing over unit — a frequency of exchange of 1% between two pairs of linked genes; 1% of crossing over is equal to one unit on a linkage map (= → map unit).

crossing over value — the frequency of crossing over between two linked genes. For a fixed set of genetic and environmental conditions, recombination in a particular region of a linkage structure (chromosome) tends to be constant and the same is then true for the crossing over value which is used in the production of → genetic maps. The crossing over value is equal to the → recombination value or fraction when the distance between the markers in question is short.

cross-link — a covalent linkage between the complementary strands of a DNA duplex (interstrand c.-l.) or between bases of a single strand (intrastrand c.-l.). C.-links are formed by a number of agents inducing → DNA lesions (→ cross-link repair).

cross-link repair (Boyce and Howard-Flanders 1964) — the repair (→ DNA repair) of interstrand → cross-links in (bacterial) DNA. In *E. coli*, the c.-l.r. mechanism is a two-step process which involves uvrA- and uvrB-dependent excisions in one strand from the sides of a crosslink, followed by recA-controlled exchanges between homologous duplexes. The half-excised cross-link (half-excision is the double nick or tandem single-strand break in the same strand, one from each side of the cross-link) attached to one strand in the first repair step results in a gap in the other strand. C.-l.r. in eukaryotic cells may follow a similar two-step mechanism.

crossover — any of the points of physical exchange by → crossing over between homologous DNA molecules or nonsister chromatids. In eukaryotes, the frequency and distribution of c. can be measured directly by observation of → chiasmata at the diplotene stage of → meiosis. It can be measured indirectly from recombinant progeny in the offspring of appropriate crosses.

crossover heteroduplex — the joining of duplex segments of DNA of different parentage by means of a → heteroduplex structure. Flanking markers are of crossover genotype.

crossover position interference — the phenomenon that one crossover affects the probability of another one occurring nearby.

cross-pathway regulation — the involvement of a metabolite in the regulation of enzymes (genes) of an unrelated pathway (= supraoperon control), as opposed to cases in which only those genes are regulated that are concerned with the synthesis of the regulatory metabolite (→ operon; regulon).

cross-reacting material — a protein produced by mutant genes that is enzymatically inactive but shows serological properties similar to the protein of the not mutated (wild-type) gene (symbol CRM). Mutants producing CRM are referred to as CRM-positive (CRM^+), mutants that do not are called CRM-negative (CRM^-).

cross-reactivation (Luria 1947) — the ability of a genetically marked coinfecting bacteriophage to rescue a particular marker (locus) from a UV-irradiated phage (= marker rescue). C.-r. is assumed to occur as a result of recombination between the phage genomes.

cross-sterility — the failure of fertilization because of genetic or cytological conditions (→ incompatibility) in crosses between pairs of individuals, despite the formation of normal, functional gametes.

CRP — symbol for → cAMP receptor protein of *E. coli.*

cruciform DNA — a nondouble helix form of DNA generated as an intermediate in → genetic recombination or from → palindromic sequences under the effect of supercoiling. C. DNA may be selectively recognized by → DNA-binding protein.

cryptic — of (1) → structurally heterozygous individuals not identifiable on the basis of abnormal meiotic → chromosome pairing configurations ("cryptic structural hybrids"); (2) a form of → polymorphism controlled by recessive genes ("cryptic polymorphism"); (3) any base change (in DNA or hereditary RNA) which goes undetected (cryptic mutations; Lamb 1975) since it does not result in abrupt phenotypic changes for several reasons such as the degeneracy of the → genetic code and the origination of synonymous codons; (4) in prokaryotes, any inactive gene (cryptic gene) that can be reactivated by mutation; (5) a → plasmid whose presence causes no discernable change in the host's phenotype; (6) a highly repeated → satellite DNA (cryptic satellite) with a density similar to the bulk of the DNA; (7) phenotypically very similar → species (cryptic species) which do not hybridize under normal conditions.

cryptochiasmate (White 1965) — of a type of → meiosis characterized by delayed opening-out of → bivalents in spermatogenesis. Under such circumstances, the chiasmata are revealed only for a brief period of time just before anaphase I (→ achiasmate).

cryptochimera — a → chimera which is outwardly recognizable as such only after regeneration.

cryptoendomitosis (Bauer 1952) — an → endomitosis which takes place in the interphase nucleus and shows none of the stages comparable to mitosis. The polytene → giant chromosomes of Diptera are produced by c.

cryptopolyploid — of prokaryotic and eukaryotic major phylogenetic groups characterized by logarithmic distributions of DNA contents which tend to form several peaks and are taken to represent intragroup doublings of DNA.

crystal-containing body (Thornton and Thimann 1964) — any of the membrane-bounded organelles in plant cells containing crystalloids. They are predominantly localized in storage cells and other metabolically less active cells and may represent a specialized type of → microbody.

ctDNA — acronym for → chloroplast DNA.

C-terminus — that end of the peptide chain which carries the free alphacarboxyl group of the last amino acid (→ N-terminus; genetic translation).

cultural evolution — the progressive development and accumulation of the → cultural heritage; in combination with organic → evolution c.e. plays a crucial role in the evolution of man.

cultural heritage — the accumulated store of knowledge, understanding, arts, customs, and technological ability available to a human social group at any given time in its history (Grant 1985). C.h. will be transmitted from generation to generation by education (→ cultural evolution).

culture alteration — any peristent change in the properties or behavior of cultured cells (→ cell culture; tissue culture), e.g., altered morphology, chromosome constitution, virus susceptibility, nutritional requirements, proliferative capacity, malignant characters, etc. (Feodoroff 1967).

cumulative genes (Nilsson-Ehle 1911) — = → polymeric nonallelic genes (→ gene interaction).

curing — the partial or total loss of integrated viral or plasmid DNA from the chromosomes of eukaryotic transformed cells (→ cell transformation). C. appears to involve excision of integrated viral/plasmid DNA or chromosome loss. Agents which increase spontaneous c. are called curing agents.

cut — a double-strand scission (double-strand break) in the DNA duplex, as opposed to the single-strand nick (→ double-strand break repair).

cuticle (Brongniart 1834) — a lipophilic layer of material superimposed on to the ordinary epidermal → cell walls in the outer parts of the plant. From the exterior to the interior of the cell wall, a distinction may be made between the epicuticle, the cuticle proper, and the cuticular layer.

C value (Swift 1950) — in eukaryotes, the DNA quantity per → genome, i.e., per chromosome set. The DNA content of diploid nuclei is usually referred to as the 2 C value. In cells which are about to divide by → mitosis, the amount of DNA is gradually doubled (during S-period of → interphase) until it reaches the 4 C value. Haploid nuclei, e.g., those of sperm and egg nuclei or the spore nuclei in higher plants, will contain the 1 C amount of DNA.

C value paradox (Cavalier-Smith 1978) — the absence of a relation between quantity of total DNA per genome and the apparent phenotypic complexities of the organisms and the general gross discrepancies between the → C values observed and those required to account for the numbers of RNA and protein species estimated to be encoded by the genomes.

cyanelle — in the protist *Cyanophora paradoxa*, two to four intracellular structures, each surrounded by a vacuolar membrane. The c. contains unstacked → thylakoids and photosynthetic pigments typical of free-living cyanobacteria. The genome of the c. resembles that of → chloroplasts both in structure (circular DNA containing an inverted repeat encoding rRNA) and in size (molecular weight about 10^8).

cybrid (Bunn et al. 1974) — the fusion product of an enucleated cytoplast with an intact (nucleated) cell (→ protoplast). In cybridization, the nuclear genome of one parent is combined with the organelles of a second parent. Sendai virus or polyethylene glycol may be used as fusing agents.

cyclic adenosine monophosphate — (cAMP); a major positive and/or negative regulator of metabolic processes in prokaryotic and eukaryotic organisms. In most cases, the effect of cAMP (adenosine-3′, 5′-cyclic monophosphate) is mediated by a cAMP-binding protein (CRP or CAP) which is a DNA-binding protein that acts at → promoters to regulate → gene expression. In bacteria, cAMP is involved in → catabolite repression, in eukaryotes it mediates the hormonal induction (transcriptional regulation) of many genes through a conserved cAMP response element (CRE) and requires cAMP protein kinase (A kinase). The 5′flanking region of such genes is identical or similar to the palindromic octamer (TGACGTCA). cAMP is derived from ATP by the action of the enzyme adenyl cyclase.

cyclic AMP receptor protein — (CRP); a protein responsible, in conjunction with → cyclic AMP, for the coordinate regulation of transcription of a large number of bacterial genes. The CRP · cAMP complex binds to specific DNA sequences within the → promoter regions of → operons. CRP has at least two functions: (1) Promotion of efficient transcription of operons sensitive to → catabolite repression by increasing the rate of initiation of RNA synthesis; (2) modulation of the natural polarity of polycistronic gene units.

cyclic permutation — in bacteriophage populations, the phenomenon that the termini of the DNA molecules can be found at many different positions (bases) within an overall sequence (probably at any point in the base sequence).

cyclomere (Engelhardt and Pusa 1972) — any of the inner halves of the → nuclear pore complexes which are assumed to be the regulating units of → chromosome replication and to be permanently associated with the chromosomes of eukaryotes.

cyclosis — cytoplasmic streaming.

cyst (Davis 1908) — (1) a group of → spermatogonia, spermatocytes, or early spermatids of common origin and surrounded by a septal or cyst membrane; (2) any enveloping structure, usually a membrane; (3) a resting spore.

cytaster (Flemming 1882) — one of the aster-like figures containing the → contrioles which is formed in the cytoplasm outside the nucleus prior to the onset of mitosis and meiosis (→ mitotic center).

cytocatalytic (Lewis 1967) — of a mode of → evolution initiated by an abrupt gross mutation, usually doubling of the chromosome complement or, perhaps less commonly, the gain or loss of individual chromosomes.

cytochemistry — the identification and localization of chemical constituents at the level of the cell and subcellular structures (→ histochemistry). The cytochemical studies may be based on four main analytical techniques: (1) Separation of cell fractions by conventional biochemical techniques; (2) isolation of minute amounts of tissues, or even single cells, by micro- and ultramicromethods; (3) direct detection of cell components in the cell by staining procedures; and (4) use of measurements of physical parameters.

cytochimera (Derman 1947) — a → chimera in which different tissue sectors or tissues differ in chromosome number (="chromosomal chimera").

cytodifferentiation — the sequence of progresses (=cell differentiation) by which different cell types acquire the ability to stably express a characteristic set of genes (→ gene expression) for performing special functions. The stability of the differentiated state can be due to various molecular mechanism: (1)→ DNA methylation; (2)→ chromatin structure; (3) DNA-protein interactions; (4)→ DNA rearrangements.

Some genes whose expression is restricted to a specific subset of cells contain cis-acting → transcriptional enhancer seqences required for efficient transcription in the appropriate differentiated cells (→ homeobox).

C. ist the process by which, during the development of the individual from the fertilized ovum, specialized cells, tissues, and organs are formed. C. consists in the progressive restriction of the developmental potential and in the specialization of function. A biologically equivalent progress takes place in some unicellular organism in which part of the cell may undergo morphological and functional spezialization.

Among the important characteristics of c. are (Grobstein 1959): (1) Inititation of change; (2) stabilization of change after cessation of stimulus; (3) the efficacy of many substances, exogenous and endogeneous, as inductive stimuli; (4) a limit of five or six as the number of cell types which may differentiate directly from any cell type; (5) progressive limitation in the number of developmental pathways; (6) restricted periods during which a cell is competent to respond to a inductive stimulus; (7) the discreteness of cell types produced in c.; (8) a requirement for a minimal and preferably heterogeneous cell mass to initiate c. in many instances, and to maintain it in some.

A prerequisite for an orderly c. is an overall regulatory program (→ messenger recognition factor) which insures that the developmental alterations (1) occur in a definite chronological order, (2) are spatially restricted to certain cells of the multicellular assembly, and (3) are of definite extent (→ modulation.)

Two aspects of c., called sequential and branching c., may be distinguished: (1) Sequential c.: One Cell type gives rise to one single but different cell type although the total number of cells may have increased by mitosis. (2) Branching c.: A single cell type yields a multiplicity of cell types.

In all types of c. both the cell's inner controls and influences from its environment play a role. The environment is assumed to trigger the activation of genes or gene products previously inactive, or to inactive genes that were previously active (differential → gene activation.) Branching c. may be due to environmental influences acting unequally on identical cells or due to nonrandom segregation of cytoplasmic elements to daughter cells during mitosis, thought to cause unequal positioning of inducing material to the two daughter cells. In some specific cases, c. has been found to be accompanied by elimination of certain chromosomes or chromosome segments (→ chromosome elimintation; positional information; commitment).

cytoduction (Zakharov et al. 1969) — in yeast, the production of cells with mixed cytoplasm but with the nucleus of one or the other parent. Cytoductants arise when two haploid strains mate and develop a bud befor nuclear fusion occurs within the transient heterokaryon.

cytogamy (Wichtermann, 1939, 1940) — the fusion or → conjugation of cells.

cytogenetic mapping — a mapping technique which seeks to determine the locus of a specific gene of the → chromosome map. Such locus may be detected for a specific chromosome arm, for a fraction of such arm, or for a minute deleted segment of a chromosome (→ deletion mapping; gene mapping).

cytogenetics (Sutton 1903) — a field of investigation which developed from the (originally) separate sciences of → genetics and → cytology and is concerned with problems based on the correlation of genetic and cytological (especially chromosomal) features characterizing a particular genetic sytem under investigation. Essentially, the field of study comprises the behavior of the chromosoms during mitosis and meiosis, their origin and their relation to the transmission and recombination of genes. Modern c. ("molecular cytogenetics") has a distinctly chemical accent.

cytogony (Hartmann 1904) — the → reproduction by single cells (gametes or agametes) (→ somatogony).

cytohet (Sager 1973) — a cell containing two genetically distinct types of a specific organelle (= cytoplasmic heterozygote).

cytokinesis (Whitman 1887) — the process of division or segmentation of the cytoplasmic portion of a cell and the segregation of daughter nuclei to separate cells, as distinct from nuclear division or karyokinesis; the cell becomes divided into two or more (multiple division) cells. During c. the cytoplasmic components are distributed to the daughter cells.

C. represents a basic progress of development and, as a rule, proceeds in conjunction with nuclear division or karyokinesis (→ mitosis, meiosis). In certain cases, c. is omitted after nuclear division and multinucleate plasmodia or syncytia are formed.

The period of c. constitutes only a small interval of the → cell cycle in most organisms. C. is regulated by the → mitotic apparatus. The plane of division is correlated with the equatorial plane, the position of the mitotic apparatus; it is usually perpendicular to the center region of the → spindle.

C. in cells with flexible surfaces (such as animal cells and cells of many green algae) is accomplished by a constriction at the middle of the cell (→ contractile ring) which is achieved by a bundle of membrane-associated → microfilaments. This constriction becomes accentuated and deepens until division is complete. Formation of the new → cell membranes takes places in the stalk at the base of the furrow (Roberts 1961).

The following types of furrowing may be distinguished (Mazia 1956): (1) Furrowing at the equator (egg cells in animals) which may take place simultaneously around the circumference or may begin to one side. (2) Furrowing displaced towards one pole (occasional animal cells) often associated with asymmetry of asters or centrosomes (smaller aster towards the pole where the small cell will form). (3) Furrowing accompanied by pulling away of daughter cells (many types of animal cells, especially ameboid cells) sometimes with a stalk between daughter cells. (4) Furrowing along an irregular path that seems to be determined by surface differentialtions of the cell (some ciliate protozoa).

In higher plant cells and some algae, c. involves → microtubule-based organelles called the → phragmoplast and → phycoplast, respectively and which occur without furrow formation. The movements of the → endoplasmic reticulum and the → dictyosomes, and the fusion of vesicular material at the equator are important components of normal c. (→ phragmosomes).

The term "equal" c. is used when the mother cell divides into two daughter cells which show a close agreement in their sizes. "Unequal" c. describes the process when qualitative and/or quantitative differences in the daughter cells are produced in the course of c.

cytokinin — any of the N-6-isoprenoid-substituted purine → nucleosides which have wide distribution as minor components of the → transfer RNA's of animals, plants, and bacteria.

Cytokinins promote → cytokinesis in plants and play, together with other plant growth substances, a major role in ensuring orderly growth and development.

C. activities include: (1) Induction, promotion and regulation of DNA, RNA, protein, and thiamine biosynthesis; (2) regulation of organ formation, apical dominance, and branching; (3) enhancement of flowering and of seed germination; (4) regulation of phloem transport and mobilization of metabolites; (5) preservation of flowers, fruits, vegetables, and leaves through prevention of senescence.

cytological race — in → species whose populations are cytologically polymorphic, any → race differing in chromosome number and/or chromosome structure from others (= chromosomal race; cytorace).

C. races my be classified (according to White 1973) into (1) microcytoraces, when population A and B differ only in the relative proportions of certain gene sequences; (2) mesocytoraces, when population B contains one or more sequences not present in A; (3) macrocytoraces, when population A is monomorphic for one sequence and population B for an alternative one. If a zone of overlap occurs, the population in that zone will be polymorphic for the rearrangements in question.

cytology — the study of structure, function, development, reproduction, and life history of cells (= cell biology). "Enzyme cytology" is the study of the intracellular localization of enzymes.

cytolysosome((Novikoff 1960) — any enlarged → lysosome which also contains mitochondria or other cell components. Since it has been established that these bodies do not function in cytolysis, the terms "areas of focal degradation" (Hruban et al. 1963) or "autophagic vacuoles" (de Duve 1963) are more appropriate.

cytomatrix — in eukaryotes a fibrous intracellular meshwork composed of → microtubules, → intermediate filaments and → microtrabeculae in aqueous phase. The dense network embraces, at least transiently, almost all of the proteins in the cytoplasm (the microtrabecular lattice; MTL). The large collection of intracellular membranes is an integral part of the c., which has been implicated in many cellular functions including the maintenance of cell form and the translocation of organelles within the cytoplasm.

cytomembrane (Sjöstrand 1955) — the basic unit of the → membrane system of the cell (→ unit membrane). It may consist of paired structures separated by a space called cisterna. Cytomembranes are divided into rough(granular) surface cytomembranes (**alpha cytomembranes,** as examplified by the → cell membrane), and smooth (agranular) cytomembranes (**beta cytomembranes,** as examplified by the → endoplasmic reticulum and the → nuclear envelopes **gamma cytomembranes,** as examplified by the membrane system of the → Golgi apparatus). Beta cytomembranes are membranes which appear to represent invaginations of the cell membrane.

cytomixis (Gates 1911) — the phenomenon of extrusion or passage of → chromatin from one cell into the cytoplasm of an adjoining cell (as in pollen mother cells, meristematic, tapetal, integumental, and ovary cells).

cytoplasm (Strasburger 1882) — all the protoplasm of a → cell excepting the → nucleus. The c. is separated from the external surroundings by the → cell membrane and from the nucleus of eukaryotic cells by the → nuclear envelope. The c. is divided into two phases, in part, through the → microtrabecular lattice: (1) A protein-rich phase containing → cell organelles and organized macromolecular complexes, and (2) a water-rich phase filling the remaining interstices. The filamentous components and their associated proteins may be interconnected into an isotropic network (the → cytomatrix or cytoskeleton) that pervades the cytoplasmic space and confers on c. the properties of a visoclastic, thixotropic, contractile gel.

Most of the metabolic and biosynthetic functions of the cell are completed in the c. It is the substrate for → gene action and exerts a regulatory effect on the expression of genetic information. In the course of → embryonic development it is organized into different, organ-forming regions occupied by equivalent nuclei (→ cytodifferentiation). The nucleus and c. are so interdependent that an isolated nucleus without c. ist usually as incapable of prolonged living as is a c. without nucleus.

cytoplasmic filament — in eukaryotic cells, any of three classes of filaments; (1) actin-containing → mircofilaments; (2) tubulin-containing → microtubules, and (3) → intermediate filaments. Microfilaments and microtubules have both a skeletal (→ cytoskeleton) and a contractile role while intermediate filaments seem to play only a skeletal role, e.g., in the anchorage of the → nucleus at its place.

cytoplasmic inheritance — → inheritance of → characters or traits whose determinants are not located in the nuclear chromosomes (= extranuclear inheritance). (→ cytoplasmon).

cytoplasmic male sterility (Welch and Grimball 1947) — (CMS); the maternally inherited inability of a higher plant to produce viable → pollen. Fertility may be regained either by a cytoplasmic reversion to fertility or by a nuclear restorer gene able to override the effects of the cytoplasm. Malfunction of → mitochondria is probably a general cause of CMS.

cytoplasmon (Renner 1929) — all cytoplasmic hereditary determinants (plasmids, plasmogenes or nonchromosomal genes) excepting those localized in the → plastids (→ plasmon; plastom) and → mitochondria (→ chondriome).

cytoplast — an enculeated eukaryotic cell. Cytoplasts may be viable and can then perform a number of cellular functions up to 24—48 h. The fusion product of a c. and an intact cell is a cybrid.

cytopon (Cone 1969) — a cytoplasmic bridge connecting two or more cells which represent a → syncytium. Cytopons may arise in vitro and in vivo as a result of incomplete → cytokinesis following mitosis.

cytosegresome — an area of cytoplasm which has become screened off by a membrane from the remainder of the cell (= cytolysosome). The segregated portions may either be derived from the cell's own cytoplasm (autophagy), or may stem from other cells (heterophagy). Cytosegresomes are involved in lysosomal digestion (→ lysosome).

cytosis (Novikoff 1961) — a general term for cellular ingestion processes by pinocytosis or phagocytosis. "Endocytosis" for intake and "exocytosis" for extrusion were proposed by de Duve.

cytoskeleton (Goldman et al. 1975) — in eukaryotic cells, the three-dimensional subcellular framework consisting of filamentous contractile and structural proteins (→ cytoplasmis filament) that function in motility-related activities and in maintenance of cell shape. The c. may also provide an essential scaffolding in which metabolic procresses occur.

cytosol — the fluid portion of the → cytoplasm outside the organelles.

cytostatic — of any physical or chemical agent capable of inhibiting cell growth and cell multiplication.

cytotaxis (Sonneborn 1963) — a procress which specifies the arrangement of macromolecules into organelles and these into supra-organellar structures. These arrangements serve as sources of information for the assembly and organization of similar structures.

cytotaxonomy — the study of the natural relationship of organisms by a combination of cytology and taxonomy.

cytotubulus (Ledbetter and Porter 1963) — a delicate, unbranched tube (diameter 23—27 nm) of unspecified length. Cytotubuli are usually evenly spaced in the cortial plasm (→ cortex) of the cell. Their function is still unclear. The outer layer of each c. consists of an annular bundle of lengthwise oriented elementary fibrils. The fine strucure corresponds with that of the → undulipodia. Cytotubuli of somewhat smaller diameter are detectable in the region of the → spindle during nuclear division and are identical with the spindle fibers.

cytotype (Müntzing 1953) — any variety (race) of a species whose chromosome complement differs quantitatively (in chromosome number) or qualitatively (in chromosome structure) form the standard complement of the species.

D

dam — the female parent (applicable to mammals) (→ sire).

Dam-directed mismatch repair (Carraway et al. 1988) — in *E. coli,* a → DNA mismatch repair system which removes errors in the newly snythesized daughter DNA strand. Discrimination between daughter and parental DNA strands resides in the differential state of methylation (→ DNA methylation) of 5'-GATC-3' (Dam recognition) sequences. The dam locus specifies a DNA adenine methyltransferase which modifies the adenine residues in 5'-GATC-3' sequences.

Danforth equilibrium — ref. to a theory which states that the frequency of occurrence of a given mutant allele within a population is equal to the freqency with which the allele arises by → mutation, multiplied by the average number of the generations in which it may manifest itself before it becomes eliminated owing to its unfavorable action.

dark holding recovery (Patrick et al. 1964) — in yeast, an increase in viability of UV-irradiated wild-type cultures due to storage in water or saline for periods of up to 8 days. D.h.r. is assumed to result from the corrective action of intracellular enzymes upon UV-induced lesions prior to DNA replication and cell division in these cells.

dark repair — any repair of DNA lesions which does not require light (= dark recovery). The basic types of d.r. are → excision repair, → postreplication repair, → reinitiation recovery, and replicative repair (→ photoreactivation).

Darlington rule (Darlington 1937) — according to this rule, which governs chromosomal sterility, the fertility of allopolyploid forms (→ alloploid) is inversely proportional to the fertility of the hybrid from which they arose. Sterile diploid hybrids with little or no → chromosome pairing give rise to fertile allopolyploids which form bivalents almost exlusively during meiosis.

Conversely, allopolyploid species produced from diploid forms which form bivalents exhibit an irregular chromosome distribution (→ centromere orientation) due to the formation of → multivalents which often leads to sterility.

Darwin — a rate of phenotypic evolution corresponding to a change by a factor of e per million years.

Darwinian evolution — → evolution.

Darwinian fitness — operationally, the average contribution which the carrier of a → genotype or of a class of genotypes make to the → gene pool of the following generation relative to the contribution of other genotypes. D.f. (= adaptive value) is a function of both the genotype and the environment (Dobzhansky 1962).

dauermodification (Jollos 1921) — a character change usually induced by extreme environmental conditions which survives for a long time in the vegetative or generative offspring in the absence of the inducing stimulus. It gradually weakens, however, and finally disappears. By reciprocal crossing, it has been shown that the persistence of d. is effected by the cytoplasm (= persistent modification).

daughter chromosome — any of the two chromatids (sister chromatids) of which the replicated chromosome consists after mitotic metaphase or meiotic anaphase II.

daughter nucleus — any of the two cell nuclei resulting from a nuclear division (→ mitosis; meiosis).

daugther-strand gap repair — a type of → DNA repair in which strand exchange or damage-bypass snythesis restores duplex DNA integrity at the sites of gaps that occur when a replication fork overcomes a → DNA lesion that blocks chain elongation (→ translesion synthesis). D.-s.g.r. war originally called postreplication repair.

dedetermination — the loss of the → determination of cells which consequently recover their totipotent state.

dedifferentiation (Butler 1933) — the loss of recognizable specializations that define a differentiated cell (→ cytodifferentation) and equip it for a particular array of functional activities. The initial steps in d. involve the following events: (1) Decondensation and dispersal of → chromatin; (2) enlargement of → nucleoli and increase of the complexity and distribution of the fibrilar and granular zones; (3) increase in RNA synthesis, usually preceding DNA synthesis; (4) accquisition of nuclear and cytoplasmic characteristics considered necessary for division; (5) activation of other biochemical processes (→ redifferentiation).

dedikaryotization (Mittwoch 1954) — =→ dediploidization.

dediploidization (Buller 1941) — in Basidio- and Ascomycetes, the formation of haploid cells or hyphae from a dikaryotic diploid mycelium or a dikaryotic diploid cell (→ dikaryon). Mittwoch (1954) used the term "dedikaryotization" in the same sense to avoid the concept "diploid" (→ haploidization).

defective interfering virus particle (von Magnus 1951) — (DI); in numerous groups of DNA and RNA viruses, any of a class of deletion mutants which are produced during serial high-multiplicity passaging of the parental standard virus. DI's are unable to multiply without coinfection with the nondeleted virus; in coinfection they become enriched and inhibit the multiplication of the parental virus. DI particles have the nucleic acid sequences necessary for replication and encapsidation and, in some cases, for translation.

deficiency (Bridges 1917) — a chromosome structural change (→ chromosome mutation) resulting in the loss of a terminal acentric chromosome or chromatid segment and the → genetic information which it contains (symbol Df) (→ deletion).

deficiency-exaggeration — → exaggeration.

degradative plasmid (Chakrabarty 1973) — any of a class of → plasmids each of which specifies a set of genes involved in biodegradation of an organic compound. D. plasmids can be either transmissible or nontransmissible.

degree of genetic determination — the fraction of the total variance (→ genetic variabilty) that is genetically determined.

degree of relatedness — the probabilits that a second individual has a given gene if a related individual is known to have it.

delayed inheritance (Boykott and Diver 1923) — → inheritance.

deleterious — of mutations which in wild species lower the → fitness of their carries to survive or to reproduce and in cultivated species impair the traits considered desirable by the breeder (Dobzhansky 1952).

deletion (Painter and Muller 1929) — in prokaryotes and eukaryotes, a genetic rearrangement through loss of segments of DNA (chromosomes), bringing sequences, which are normally separated, into close proximity. Genetically deletions are defined by two properties: (1) They do not revert to wild type; (2) they fail to recombine in pairwise crosses with at least two nonidenditcal point mutations adjacent to each other (→ deletion mapping).

At the molecular level, in-phase deletions produce fragments with normal ends, but with amino acids deleted internally in the transcript. Internal deletions of a nontriplet number of nucleotides result in termination at an out-of-phase triplet (→ frameshift mutation), and in some cases also generate internal reinitation fragments.

In eukaryotes, the loss of an intercalary chromosomal segment may be distinguished from the loss of a terminal chromosome region by applying the term deletion to the former and deficiency to the latter. In practice, deletion is frequently used for both types of structural change. In both cases a centric and an acentric fragment are produced in those eukaryotes whose chromosomes possess a localized → centromere (→ chromosome mutation).

The size of the deleted material can vary from a single nucleotide to sections containing a number of genes or a whole chromosome region.

The existence of relatively large deletions in the chromosome complement of eukaryotes may be proved cytologically by the occurrence of centric and acentric fragments in mitosis or by the absence of regional pairing (loop formation) in first meiotic prophase and in (somatically paired) → polytene chromosomes of Diptera in that pair of homologs which is heterozygous for the segment lost.

The genetic consequences of deletions are primarily due to the loss of genetic information and secondarily due to quantitative changes in the genotype as well as changes of the genetic balance. Depending upon the size of the deleted chromosome segment, a d. may act as a recessive or dominant → lethal.

deletion loop mutagenesis (Kalderon et al. 1982) — a method for → site-directed mutagenesis which allows point mutations to be introduced into a sequence of DNA defined by a previously isolated deletion mutant. Both wild-type and deletion mutant DNA are cloned (→ molecular cloning) into a bacterial → plasmid and each is cleaved with a different single cut → restriction enzyme. Heteroduplexes are formed between the two DNAs to produce circular molecules, each containing a → nick in each strand and a single-stranded deletion loop. Deletion loops are then treated with a mutagen (sodium bisulphite) and the DNA is transfected directly into a uracil repair-deficient strain of *E.coli.* Up to half of the resultant clones contain DNA produced by replication of the wild-type length strand and bear mutations exclusively within the target area.

deletion mapping — the genetic localization (mapping) of the positions of deletions in the linkage structures (chromosomes) of eu- and prokaryotes. The genetic positions of deletions in eukaryotes may be estimated without extensive linkage studies, simply according to whether or not they phenotypically "disclose" certain recessive mutations. All segmental losses which permit the manifestation of a recessive mutation in the heterozygous state must at a minimum include the locus of this mutation.

Deletion mapping in prokaryotes exploits the following principles: (1) Three mutants (a, b, and c), each differing from the wild type by deletion of a part of the linkage structure are tested for recombination. (2) If two mutants (a and c) mutually recombine and yield the wild type (+), but neither recombines with the mutant b, then (3) these results may be described by a matrix based upon the results of the particular recombination tests (+ = recombination; o = failure of recombination):

	a	b	c
a	o	o	+
b	o	o	o
c	+	o	o

Recombination results permit a unique ordering of the three mutants in the linkage structure. In this case, the deletions fall in the serial order a, b, c such that a and b overlap on one side and b and c on the other:

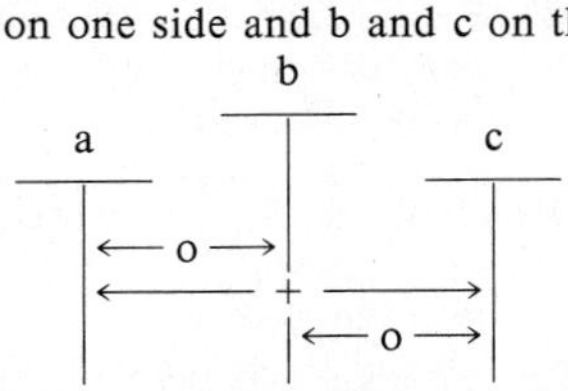

deletion mutagenesis — an in vitro mutagenesis procedure for generation of a family of → deletions by exonucleolytic removal of the ends of a linear DNA fragment.

deme (Gilmour and Gregor 1939) — a local interbreeding group ("panmictic unit") consisting of a series of individual genotypes. The term is always used with a prefix which characterizes the d. more precisely. Among others the following combinations are possible:

(1) with regard to the spatial relationships: **Topodeme:** a group of individuals existing in a certain geographic region (e.g., all individuals of *Ameria maritima* in the region of the Brit-

ish Isles); **ecodeme:** A group of individuals associated with a specific habitat (e.g., a particular mountain peak); (2) with regard to character differences: **Phenodeme:** a d. differing phenotypically from others, assuming that it is not yet experimentally possible to establish whether this is due to genotypic or environmental factors; **genodeme:** a d. differing from others in genotype; it needs not also be characterized by divergent phenotypic characters; **plastodeme:** a d. differing phenotypically from others owing ot the effect of the environment:

(3) With regard to propagation relationships: **Gamodeme:** a d. whose individuals are so distributed spatially and temporally that within the limits of the mating system they can all mate with one another (= "breeding population"). The suffix N% is used for a more precise, quantitative characterization. Thus, N%-gamodeme means that each individual has N% chance of mating within the d.; **autodeme:** a d. composed predominantly of autogamous individuals; **endodeme:** a "gamodeme" of predominantly endogamous (→ dioecious) plants (or bisexual animals); **agamodeme:** a d. of predominantly apomictic plants or asexual animals; **clonodeme:** a. d. with predominantly vegetative reproduction; (4) with regard to trends in variation: **Clinodeme:** several d. which together form a gradient (→ cline).

Secondary derivatives may be made from these primary word forms by assembling a series of prefixes in the order corresponding with the required meaning. This allows a more exact characterization of the d.; thus, e.g., "phenotopodeme" (a "topodeme" which differs phenotypically from others).

demographic genetics (Schaal and Levin 1976) — a branch of → population genetics and ecology studying genetic differences related to age, population size, genetic alteration in competitive ability, and viability.

demography — the study of survival, death, reproduction, and growth of populations.

denaturation — the loss of the native configuration of a macromolecule resulting from specific treatment (e.g., heat treatment in thermal d.). D. is extensively used to study the helix-coil transition in DNA, and as a probe of the stability, composition, and structure of DNA in solution. The same technique is applied to complexes of polypeptides and proteins with DNA (→ denaturation mapping).

denaturation map — a map that reveals base content fluctuations along a stretch of DNA as established by electron microscopic analysis of partially melted DNA molecules. The d.m. may be used for two purposes: (1) As a frame of reference for mapping (→ denaturation mapping) other physical markers on that DNA molecule, and (2) to study DNA metabolic processes, e.g., replication, transcription, or transposition.

As a rule, d. maps are obtained after the kinetics of DNA unwinding by formaldehyde is interrupted at some stage. Peaks in the map correspond to AT → enriched regions of DNA.

denaturation mapping (Inman 1966) — a method employing electron microscopy of DNA partially denatured in the presence of formaldehyd (to keep regions from renaturing), which permits a decision as to whether the gene sequence in a viral genome is linear and unique or circularly permuted, and the characterization of viral double-standed nuclei acids as to the distribution of specific nucleotide sequences rich in adenine-thymine base pairs (these melt a lower temperatures).

The DNA molecules are photographed in the electron microscope and the d.map is constructed by measuring the location of denatured regions along each molecule. The denaturation map thus constructed charts the extent of denaturation along the molecule. The first sequences to denature have the lowest G + C content. Compared with → deletion mapping, d.m. requires no second (mutant) DNA. D.m. identifies genes by using their characteristic denaturation map as a fingerprint.

dense caveola (Fawcett 1965) — = → coated vesicle.

dense fibrillar component — a component of the → nucleolus which lacks granules and stains more intensely than other nucleolar components; it is probably the site where active ribosomal transcription units reside (→ granular component).

dense lamella (Kakifat et al. 1967) — = → nuclear lamina.

deoxynucleotidyltransferase — = → terminal d.

deoxyribonuclease — any of a class of hydrolytic enzymes which are ubiquitous in biological systems, contribute significantly to the biosynthesis of nucleic acids and have a central role in → DNA repair, → genetic recombination, and restriction of foreign DNA in prokaryotic systems. Deoxyribonculeases are phosphodiesterases which either require a terminus for hydrolysis (exonucleases) or do not require a terminus (endonucleases). The endonucleases often show a preference for either single-stranded DNA or duplex DNA and recognize certain base sequences (→ restriction endonuclease). Such sequence recognition is in contrast with the single nucleotide recognition typical of → ribonucleases.

deoxyribonucleic acid — a polymer of subunits called deoxyribonucleotides which is the primary genetic material of all cells, i.e., the material equivalent to → genetic information (symbol DNA). DNA of prokaryotic and eukaryotic organisms is normally found in a double-stranded state (Watson-Crick double helix), yet several important biological processes (replication of DNA, → genetic transcription, → genetic recombination, repair processes, certain stages of the life-cycle of some viruses) transiently involve single-stranded regions.

In eukaryotes, nuclear DNA contained in the chromosomes and organelle DNA (→ mitochondrial DNA, → chloroplast DNA) may be distinguished. Nuclear DNA is composed of three major classes: highly, moderately, and nonrepetitive DNA (→ Cot value).

The content of these classes may vary considerably from organism to organism. In general, highly → repetitious DNA (simple sequence DNA) constitutes 3–30%, nonrepetitive (single copy) or → unique DNA 30–80%, and intermedidate the remainder. Not all highly repetitive DNA is → satellite DNA; a significant portion is dispersed throughout less repetitive sequences in → main band DNA.

(1) **Structure of DNA:** Single- or double-standed DNA molecules can exist as threads, circles, or chains, and may be coiled or supercoiled. Circular DNA has been found in bacteria, bacterial viruses, and some animal viruses; it is characteristic for the DNA of bacterial → plasmids, → mitochondria, → kinetoplasts, and → chloroplasts of eukaryotes. This DNA may have various structural forms including supercoiled and open circles (one or both strands have breaks) that are either monomeric or multiple length (catenated oligomeric molecules) and also interlocked cyclic molecules.

Each nucleotide consists of the pentose sugar 2-deoxy-D-ribose, phosphoric acid (joined to the sugar moiety by an ester linkage and conferring on DNA its acid properties) and one or another of several different nitrogenous bases linked to the deoxyribose at carbon atom number one of the sugar. The glycosidic combination of base and sugar is called a **deoxyribonucleoside** and their phosphate esters are the **deoxyribonucleotides.**

All DNA samples isolated from a variety of organisms contain as the major components four nitrogenous bases: **adenine** and **guanine** (purines with a double ring structure), and **thymine** and **cytosine** (pyrimidines with only a single ring). These four bases predominate but in addition DNA may contain minor components such as 6-methyl-adenine (in bacteria) or 5-methyl-cytosine (in plants and animals). Exceptions to these generalizations exist: DNA of T-even phages of *Escherichia coli* contains 6-hydroxymethylcytosine instead of cytosine, and certain phages of *Bacillus subtilis* have other bases apparently replacing thymine in their DNA. In spite of these exceptions there are generally four types of nucleotides, depending on which of the bases is attached to the sugar-phosphate ester. These are known as deoxyadenylic acid, deoxyguanidylic acid, deoxythmidylic acid, and deoxycytidylic acid.

The DNA macromolecule built up of a long chain of nucleotides is joined together by means of diester links in which one phosphoric acid molecule forms bonds between the 3' and 5' positions of consecutive Nucleotides to form a chain referred to as a **polydeoxyribonucleotide.** The sequence of nucleotides in the chain makes up the **primary structure** of DNA, which can best thought of as having a sugar-phosphate backbone to which is attached at each level one or another of the nitrogenous bases.

DNA usually consists of two polynucleotide chains joined by hydrogen bonds between pairs of bases that occupy the same level on the two strands. The two chains are coiled plectonemically (in an interlocked form) about the same axis. They follow right-handed helices, with ten bases in each for one turn (3.4 nm) of the spiral. The phosphates are on the outside of the double helix; the bases are perpendicular to the long axis of the molecule on the inside and spaced at intervals of 0.034 nm along it. The two chains run in opposite directions in terms of the 3'—5' phosphate-deoxyribose linkages and thus have a definite polarity, which is relevant to the genetic function of the molecule. If the sequence of atoms in the backbone of one chain is ... $-C_3-C_4-C_5-O-P-O-C_3-C_4-C_5-$ ... then it will be ... $--C_5-C_4-C_3-O-P-O-C_5-C_4-C_3-$ ... in the other.

The pairing between bases of the two chains is highly specific. Adenine (a purine) is always hydrogen-bonded to thymine (a pyrimidine) and guanine (another purine) to cytosine (another pyrimidine). The existence of these specific base pairs means that the sequences of nucleotides along the two chains are complementary. A consequence of this **secondary structure** of DNA molecules, first suggested by Watson and Crick (1953), is that the molar amount of adenine (A) is equal to that for thymine (T), and that of guanine (G) is equal to that for cytosine (C). The overall composition of a particular DNA sample may, therefore, be described by the percent of G + C and A + T or by the **base ratio** A + T/G + C.

Conformational heterogeneity exists along the DNA strand. Aside from the common → B-DNA conformation, DNA can adopt left-handed helical conformations (→ Z-DNA). In B-DNA, all the bases are in the anti conformation, in Z-DNA, consecutive bases alternate between the anti and syn conformation more readily. The equilibrium distribution of right- and left-handed conformations by a tract of DNA with a specific sequence depends on: (1) The type of base or sugar-phosphate backbone modification; (2) the general nature of the ions, ligands, and solvents; (3) the torsional stress of → DNA supercoiling; (4) temperature; and (5) the presence of proteins. Z-DNA is widespread and exists at isolated or multiple sites in both prokaryotic and eukaryotic genomes.

In addition to the B and Z conformations, DNA under certain conditions may adopt other shapes, such as single-stranded loops, hairpins, and cruciforms. Some of these structural pertubations could provide recognition elements in DNA-protein interactions and ultimately play an important role in the utilization of genetic information. An important factor promoting structure diversity is negative supercoiling.

Heterogenity in base composition in eukaryotic DNA's is due to four classes of DNA sequence: (1) highly repetitive; (2) moderately repetitive (→ repetitious DNA) (3) unique; and (4). → inverted repetitive DNA.

The chromosomes of viruses (with the exception of some RNA viruses) and bacteria consist of DNA exclusively, whereas those of eukaryotes contain besides DNA other chemical components. DNA with coding properties is present in → mitochondria and → chloroplasts.

In order to fulfil the functional requirements of genetic material DNA must carry → genetic information (instruction), must be able to replicate (to reproduce itself) and to undergo → mutation (to alter the instructions occasionally). The genetic information necessary for directing all the specific activities of the cell resides in the sequence of the four primary bases (→ cistron, gene) in form of a → genetic code which is read by the processes of → genetic transcription and → genetic translation.

The presence of 5-methyl-cytosine is a common but not universal feature of prokaryotic and eukaryotic DNA (→ DNA methylation; DNA-modification-restriction). The 5-methyl cytosine content is species-specific and can vary from 1 to 10%. The dinucleotide CpG is the principal target for methylation.

Eukaryotic DNA is arranged in the nucleus by at least three hierarchical levels of organization: (1) The → nucleosome level; (2) The nucleosomes are arranged into ordered arrays of solenoids or toroids; (3) these supranucleosomal clusters are further packaged into supercoiled loops (about 30,000 to 100,000 bp each) that are anchored at their bases to a → nuclear matrix or nuclear scaffold.

One picogram (gram x 10^{-12}) of DNA corresponds to about 950,000,000 nuleotide pairs or a 31 cm length of double helix.

2. **Replication of DNA:** DNA replication is a complex series of events in which diverse multienzyme systems (→ replisome) function whose operations concern three stages of the process: initiation, elongation, and termination. Replication occurs with the two strands of the double helix undergoing partial separation (breakage of hydrogen bonds and regional unwinding of the chains), thus allowing each strand to act as a template for the synthesis of a complemantary strand by specific base pairing ("semiconservative replication" according to Delbrück and Stent 1957). Separation of strands goes hand in hand with replication and is not complete before the synthesis of the new polynucleotide chains. Deoxyribonucleoside triphosphates are the monomeric precursors for DNA synthesis. They are enzymatically (→ DNA polymerase) joined together in the presence of a single-stranded DNA sequence which acts as template. The biosynthesis of DNA involves three main stages: (1) The synthesis of deoxyribonucleosides; (2) their phosphorylation to triphosphates by appropri kinases, and (3) the polymerization of these triphosphates to form polynucleotide chains.

All DNA replication systems have in common one (prokaryotes) or more (eukaryotes) unique sites (replication origins) for the initiation of replicaction from which replication proceeds uni- or (in most cases) bidirectionally (→ replicon). At least three general features distinguish DNA replication from other macromolecular biosyntheses: (1) DNA precursors (the deoxyribonucleotides) are highly specialized and play rather limited roles outside their function as DNA precursors; (2) DNA replication is localized both in time and space, occurring only during certain periods and at a limited number of intracellular sites; (3) rates of DNA chain growth are one or two orders of magnitude higher than for the chemically similar process of RNA synthesis.

The mechanism of DNA replication involves: (1) Unwinding of the parental strands; (2) synthesis of a short RNA chain along the parental strands (→ RNA primer); (3) extension of the chains by a → DNA polymerase in the 5' → 3' direction, using the RNA as a primer; (4) removal of the RNA by RNase activity; (5) filling in of the gaps between the DNA pieces by a DNA polymerase; (6) joining of the DNA pieces by → DNA ligase (→ Okazaki pieces).

In eukaryotic → chromosomes DNA replication occurs in → replication units (as many as 10.000) arranged in tandem and replication in multiple subchromosomal clusters. From any single initiation site or origin replication progresses by divergent replication forks, i. e., bidirectionally at a one-way rate of some 0,5 to 1,2 nm/min (in bacteria 20—30 nm). The intervals between neighboring origins vary, ranging from 15 to 120 nm with a mean interval of about 50 nm; replication my start from neighboring origins staggered in time. The DNA contained in different portions of chromosomes is replicated during charcateristic parts of the → interphase, the order depending in part on the frequency of repetition of their nucleotide sequences. Some kinds of highly repeated sequences are replicated in either the early or late part of the S phase (→ heterochromatin).

The fidelity of DNA replication can be determined by both the accuracy with which a template is copied during the initial incorporation of a nucleotide, and the efficiency with which incorrect nucleotides are removed (→ DNA mismatch repair; → DNA relaxing enzyme; → small polydisperse circular DNA). In prokaryotes, protein synthesis is required only for the initiation of DNA synthesis. In eukaryotes, protein synthesis seems to be required for both the initiation and completion of DNA synthesis. Both asymmetric and symmetric replication have been found at different stages of the same bacteriophage or under different conditions in the same bacterium. In other cases, only symmetric replication or, in the case of single-stranded DNA synthesis, only asymmetric replication is found. Nuclear and organelle DNA's replicate at different times in the eukaryotic cell cycle, but their replication cycles are tightly coupled.

Single-stranded instead of double-stranded DNA is found in some viruses. The DNA strand present in the infectious virus particle is called the "plus" strand. After its entry into a host cell by infection, this strand serves as a template for the formation by DNA polymerase of a complementary "minus" strand, whereby a double helix (the → "replicative form", symbol RF) is formed. This double helix is·then replicated in the conventional manner using host cell DNA polymerase to produce many new double-helical copies. Synthesis of single-stranded DNA for progeny viruses needs a special device to insure that only one of the two potential replicas is produced.

As expected from its genetic function, the quantity of chromosomal DNA is dependent upon the number of chromosomes present in the cell. This relative constancy of DNA means that both the cycle of DNA replication and chromosome reduplication must keep pace. There are, however, indications of variations in the incorporation rate of DNA precursors in the DNA content in interphase nuclei, in → polytene chromosome loci, or in particular heterochromatic regions (such as "nucleolus-assosciated" chromatin).

The evolution of the higher, multicellular organisms from more primitive ones has been accompanied by al large increase in the total amount of DNA per cell. This probably reflects an increase in the total genetic information content necessary for the development of increasing structural and behavioral complexity. The range of complexity in the → genome, provided that this is represented by the total amount of DNA, increases by about 10^3 from bacteria to mammals. The range becomes even greater if viruses are included. The number of nucleotide pairs ranges from about 5×10^3 (in small viruses), 1.9×10^5 (in *Escherichia coli*), 3.0×10^7 (in *Bacillus megaterium*) to 5.6×10^9 in human cells. (For comparison one → cistron or gene contains about 1500 nucleotide pairs). All the genetic information encoded in chromosomal DNA is contained in one chromosome (linkage structure of a definite number of lineary arranged genes) in viruses and bacteria, and distributed among a species-specific number of chromosomes (the chromosomes set) of eukaryotes.

Only in the RNA viruses has the role of DNA as carrier of genetic information been taken over by → ribonucleic acid.

deoxyribonucleoprotein fiber— (DNP fiber); a → chromatin fiber which is the basic unit of the eukaryotic → chromosome and is composed of DNA, → histones, and variable amounts of RNA and → nonhistone chromosomal proteins.

deoxyribonucleoside— → deoxyribonucleic acid

deoxyribonucleoside kinase— any of the enzymes that catalyze the transfer of a phosphate group from a nucleoside triphosphate to position 5' of a deoxyribonucleoside.

depurination — the removal of purines from the sugar-phosphate backbone of damaged DNA (→ DNA lesion; DNA repair; AP site). Damage caused by d. is repairable by → excision repair initiated by an A/P endonuclease specific for apurinic sites in DNA. It causes a single-strand break in the DNA in the vicinity of the apurinic site. Subsequently, excision, polymerization, and ligation steps occur and DNA integrity is restored (→ base-insertion repair).

depressor effect (Delbrück 1945) — following the mixed infection of a bacterial cell by two bacteriophage types and successful → mutual exclusion, the numerical reduction of the offspring of the reproducing phage type freed by lysis of the bacterial cell.

depression — the release of → repression of enzyme (protein) production after removing a specific metabolite, no → inducer being necessary (→ trans-depression; induction).

dermatoglyphics — the pattern of the ridged skin of palms, fingers, soles, and toes, and the study and the systematic classification of such patterns into a number of different types. The study of d. is important in medical genetics chiefly becaus of the characteristic combinations of pattern types associated with some chromosomal disorders. Furthermore, since these pattern are heritable, d. may be used as one method of measuring resemblance between twins in the determination of twin zygosity.

dermatoplast (Küster 1935) — gymnoplast.

descendants — all the individuals resulting from the sexual reproduction of one pair of individuals.

desmosome (Bizzozero 1871) — any of the specialized areas of the → cell surface that serve to bind the surface to another structure. The latter may be another area of the cell surface (intercellular contact), or some extracellular structure (= cell junction.) In general desmosomes appear to be permanent cell structures and probably remain intact during cell division.

According to Bennet (1969) desmosomes my be classifield in several ways: (1) According to shape: (a) **Macular d.** which have a discrete, spotlike shape (maculae adherentes); (b) **Zonular d.** which encircle a cell like a belt or zone and bind it to its neighbors over its entire circumference at the level of the zonular d. (terminal bars or zonulae adherentes). (2) According to the structure to which the cell surface is bound at the d.: (a) **Autodesmosomes** which bind an area of the cell surface to another d. on another area of the surface of the same cell; (b) **Homodesmosomes** which bind an area of the cell surface of one cell to a d. on the surface of another cell (**iso- or allodesmosomes**); (c) **Heterodesmosomes** which bind an area of the cell surface to some noncellular structure (**hemi-isodesmosomes**).

In addition to their primary function in intercellular adhesion, desmosomes also appear to be foci for the organization of cytoskeletal elements (→ cytoskeleton) within the cell and are thus important in the determination of cell shape and internal organization.

The d. is composed of peripheral and integral membrane proteins that bridge the cell membrane and connect the internal tensile skeletons of adjacent cells. Ultrastructurally, desmosomes posses electron-dense subplasma membrane plaques to which → intermediate filaments appear to attach.

desmotubule (Robards 1968) — a tubule (diameter about 20 nm) traversing each → plasmodesma. The wall of the d. probably consists of eleven subunits arranged around a central cavity through which runs a 4 nm diameter rod. This is connected to the inside of the tubule wall by fine filaments. At the ends of each plasmodesma the plasmalemma and cell wall are closely appressed to the tubule, thus precluding direct continuity between the cytoplasm of adjacent cells.

desmutagenic — (Kada 1974) — of → antimutagenic agents (desmutagens) that act on a mutagen directly before the mutagen acts on target DNA, as opposed to antimutagens that block the processes eventually changing normal cells to mutant cells.

desynapsis (Li et al. 1945) – the falling apart during diplotene or diakinesis of → meiosis of chromosomes which paired during zygotene and remained paired at pachytene (= desyndesis). Due to d. most or all chromosomes may be present as → univalents at metaphase of meiosis I. An accompanying decrease in → chiasma number and chromosome contraction, precocious division of univalents, imbalanced segregation, and reduced fertility may ensue. D. is often genetically controlled. If so, one or a few recessive genes are involved (→ asynapsis). Mutant genes resulting in d. are called desynaptic genes.

desyndesis (Sharp 1934) — = → desynapsis.

detached-X (Kaufmann 1933) — a metacentric chromosome produced from → attached X-chromosomes of **Drosophila melanogaster** by → interchange or crossing over; one arm contains the X-chromosome while the other arm consists either of the long or the short arm of the Y-chromosome. The process leading to d. is termed "detachment".

detachment (Belling 1924) — (1) the loss of bi- or univalents in meta- or anaphase I of meiosis by their exclusion from the → spindle; (2) the production of → detached X-chromosomes.

determinant (Weismann 1891) — any of the hypothetical units of the → germ plasm (= hereditary determinants).

determinated disjunction — → disjunction.

determination (Roux 1905) — a process which initiates a specific pathway of development by singling in out from among various possibilities for which a cellular system is competent. In multicellular organisms it is useful to distinguish between determinations resulting in immediate → differentiation and those resulting in a reproducible cell state propagated in the nondifferentiating phase by cell heredity (Hadorn 1965).

Provided the external conditions are favorable, d. confers an entirely specific differentiation upon each part of the embryo. The d. may be stable (definitive), in which case there is only one developmental pathway open to the cell or embryo part, i.e., an automatic course of differentiations is prescribed; or it may be labile. Different parts of the developing organisms may acquire d. at different periods (→ transdetermination).

During d., an event takes place which decides which of several types of development will actually be realized. This "event" may be regarded as the activation of one or another set of gene actions, i.e., in the course of development, genetic information is selectively transformed into phenotypic effects (→ gene activation; embryonic development).

Three main types of mechanisms resulting in d. may be distinguished (Waddington 1956): (1) **Ooplasmic segregation:** Different regions of the cytoplasm of the egg possess specific properties. Thus, certain cytoplasmic regions called "ooplasms" or "organ-forming substances" (the older name) can only develop in one direction. These cytoplasmic regions may be highly localized, showing definite boundaries, or they may overlap into one another. (2) **Evocation:** Two adjacent regions of an egg or embryo mutually affect each other so as to change the direction of development of one or both. Such progresses are very frequent afther the → cleavage period when parts of the embryo which were previously separated come into contact owing to positional changes and folding during gastrulation. Thus, a region determined very early in the development, e.g. by ooplasmic segregation, may come in contact with an as yet undetermined part of the embryo and so affect it that it develops to form a definite type of tissue. (3) **Field action:** In many embryological processes, the development of a particular point in a region of the egg depends upon its relationship to other adjacent points or to its position within the region. The mutual influence between the parts of an embryonic region is of such a nature that it is possible ot remove parts of the embryonic tissue and replace them in an inverted position. Owing to the field action, the development of every one of the points is so influenced by the points of the rest of the tissue that a normal embryo can be formed ("field phenomena"), i.e., any piece of embryonic tissue is so capable of regulation, that the original and complete field condition may be restored fully after experimental interference. The term "field" is chosen by analogy with physical field theories (as, e.g., magnetic field, gravitational field) although no physical forces operate in the embryonic field; nevertheless, the biological events have general characteristics similar to the physical. In both cases certain activies are effective over the whole field.

detrimental (Muller 1934) — any mutation which lowers the viability (= subvitals). The survival rate of the individuals concerned is above 10% but less than 100%. The great majority of spontaneous or induced mutations which produce visible changes are d. (→ lethals).

deuterohermaphroditic (Correns 1928) — → hermaphroditic.

deuterotoky — → parthenogenesis.

deutoplasm (van Beneden and Bessels 1868) — products elaborated by the cell and stored in the cytoplasm as lipid droplets, yolk bodies, pigment, and secretion granules (= inclusion, paraplasm).

development — a process of regulated growth and differentiation (→ embryonic development) that results from the interaction of the → genome with cytoplasm, internal cellular environment, and external environment.

D. is a programmed sequence of phenotypic changes, under temporal, spatial, and quantitative control which is irreversible or difficultly reversible at least under ordinary environmental conditions. The sum total of these modifications constitutes the life cycle of an organism (Sussman 1965).

Four major processes interacting with one another make up the complex process of d.: (1) Genetic replication, i.e., the means by which (in the course of → mitosis), the sets of genetic information (the genes in the chromosomes) are duplicated; (2) growth, i.e., the increase in mass of the organism which is intimately associated with cellular metabolic activity: (3) cellular differentiation, i.e., the process by which cells of common origin, and thus genetically identical, diverge in their structure and/or function and give rise to morphologically and physiologically distinguishable cell lineages; (4) histo- and organogenesis, i.e., the processes by which differentiated cells aggregate to form tissues of unitary function. Theses tissues then associate to make up organs.

developmental canalization (Waddington 1942) — the phenomenon wherein natural → selection acts on → development to produce pathways which are insensitive to minor genetic or environmental variation. D.c. results in the phenotypic norm of the species.

developmental constraint — any of the nonrandom canalizations of evolutionary direction due to limitations imposed by → epigenetic interactions in the developing organism. D. contraints may influence → fitness and thus restrict evolutionary direction.

developmental cycle — the complex progression of phenotypic alterations in a cell or a multicellular organism during → development. This progression occurs in a fixed chronological order at specific intracellular sites and/or within specific cells of a multicellular organism, and to fixed amounts and extents. Initial stages of development of a multicellular organism commence while nearly all of its nuclear → genes are inactive, i.e., the initial stages are sustained by stable gene products (→ maternal messenger RNA) formed during → oogenesis.

developmental fate map — → embryonic fate map.

developmental flexibility (Thoday 1953) — → flexibility.

developmental genetics — a branch of genetics primarily concerned with questions dealing with the manner in which genes control or modulate the process of development (→ phenogenetics).

developmental genome rearrangement — the rearrangement and/or elimination of DNA sequences in some or all somatic nuclei during some stage of → development. Germ-line-specific sequences are usually eliminated from the soma of certain invertebrates (such as ascarid worms, flies, and copepods). Mammalian immungoglobulin genes undergo rearrangement in B lymphocytes, deleting internal sequences and joining previously separated regions. Extensive d.g.r., including sequence elimination, also occurs in various species of ciliates during macronuclear development.

developmental homeostasis (Lerner 1954) — → homeostasis.

developmental instability — intragenotypic variation of → development under a certain environmental condition. D.i. may be due to local fluctuations in environmental conditions outside and/or inside the organism and also to mistakes in the biochemical steps of the developmental process (→ phenotypic plasticity).

developmental noise — any uncontrollable variation in phenotype due to random events in → development.

devitation (de Beer 1951) — a mutation-controlled (→ mutation) shift of the developmental process in an alternative direction. The extent of the phenotypic effects produced by the mutation in such as case is variable. As a rule, the effect is greater the earlier the shift takes place.

dG·dC tailing — a method used for → molecular cloning of random fragments of DNA by adding complementary homopolymer tails to the → insert and the vector DNA (A's and T's method). Oligo dG and oligo dC tails are added by → terminal transferase.

DH site — a → DNase I hypersensitive site in → chromatin.

diagynic — of genes localized in the → X-chromosome (X-linked genes) which are transferred from mother to sons in the case of female heterogamety (→ diandric).

diakinesis (Haecker 1897) — → meiosis.

diallel cross — the set of all possible matings between several genotypes (individuals, clones, homozygous lines, etc.).

diallelic (Atwood 1944) — → monoallelic.

diandric — of genes localized in the → X-chromosomes (X-linked genes) which are transferred from father to daughters in the case of male heterogamety (→diagynic).

diaphoromixis (Burnett 1956) — → heteromixis.

diauxy (Monod 1942) — the adaptation of microorganisms to culture media which contain two different carbohydrates. Initially, only one of the carbohydrates is used, later the second as well. The organism possesses constitutive → enzymes for the first and requires induced enzyme synthesis befort it can metabolize the second carbohydrate.

dibasic (Darlington and Janaki Ammal 1945) — → polybasic.

dicentric (Darlington 1937) — of chromosomes or → chromatids with two localized → centromeres (usually associated with the formation of bridges at anaphase of mitosis; → bridge-breakage-fusion cycle). D. chromosomes or chromatids are lost after one or more cell divisions unless they get rid of the accessory centromere. They may persist under certain circumstances: (1) Dominance of one centromere over the other; (2) total failure of expression of one centromere; (3) proximity of the two centromeres so that both attach as a single unit to spindle microtubules.

D. derivatives of originally monocentric chromosomes or chromatids are the result of chromosomal structural changes (→ chromosome mutation).

dichogamous (Sprengel 1793) — of flowers or hermaphroditic animals whose male and female sex organs are active at different times. "Dichogamy" is subdivided into: (1) **Protoandrous or proterandrous dichogamy:** the male sex organs are active before the female ones; (2) **Protogynos or proterogynous dichogamy:** the female sex organs are active before the male ones.

Organisms in which both protoandrous and protogynos dichogamy occur are termed "**heterodichogamous**".

dichopatric (Smith 1965) — of → allopatric populations whose geographic ranges are separated to the extent that individuals of the involved populations never meet (→ parapatric).

dichophase (Bullough 1963) — that phase of the mitotic cycle in which a cell is faced with the choice between → differentiation for mitosis and differentiation for aging and tissue function (→ apophase).

diclinous — having male and female germ cells in separate flowers either on the same plant, which is said to be "monoecious", or on different plants, which are said to be "dioecious" (→ monoclinous).

dictyokinesis (Perroncito 1910) — the distribution of → dictyosomes during cell division.

dictyosome (Perroncito 1910) — any of the parts of the → Golgi apparatus (= Golgiosomes) isolated by differential centrifugation.

Dictyosomes consist of 2 —7 (sometimes up to 20 or more) layers of flattened membranes (cisternae) surrounded by small vesicles. The → cisternae are isodiametric in the plane of flattening (diameter between 0.2 and 1.2μm; in conjugates up to 5μm). The vesicles surrounding the stacks of cisternae are formed by localized dilatations which are probably a product of dictyosome activity. The cisternae are joined by some bonding constituent within the intercisternal matrix which contains intercisternal elements (diameter 7 —8 nm). The center-to-center spacing of intercisternal elements is about 15 nm.

Dictyosomes dispersed throughout the cytoplasm make up the Golgi apparatus of plants and invertebrates. At telophase (→ mitosis), the dictyosomes become grouped at the periphery of the → cell plate in plants. In vertebrates, the dictyosomes are either aggregated in al loose or compact juxtanuclear mass or they are scattered through the cytoplasm either as single elements or, more often, as small aggregates. Eventually, they occupy a cortical position (→ cortex).

The component cisternae of a d. are not equivalent. They differ structurally and probably have different functional capabilities.

Dictyosomes are somehow replicated to form new such elements, but the mode of reproduction is not yet known.

dictyotene — a stage that interrupts meiotic prophase in mammalian oocytes. During d., the chromosomes become decompacted; the oocyte grows in size and accumulates cytoplasmic organelles or their precursors to provide structures for the meiotic divisions and a maternal supply of RNA and proteins for the early embryo.

diheterozygous — = → dihybrid.
dihybrid — → heterozygous in respect to two pairs of alleles.
diiso-compensating trisomic (Kimber and Sears 1968) — a compensating → trisomic in which a missing chromosome is compensated for by two → isochromosomes, one for each arm of the missing chromosome (= double-isotrisomic; Kimber and Sears 1968). (→ ditertiary compensating trisomic; telotertiary compensation trisomic; ditelotrisomic; teloisotrisomic).
diisosomic (Kimber and Sears 1968) — of a cell or individual which lacks one chromosome pair but has two homologous isochromosomes for the same arm of the missing pair (→ monoisotrisomic; triisotrisomic).
diisotrisomic (Kimber and Sears 1968) — an individual deficient in one chromsome but which has a pair of homologous → isochromosomes for one arm of the missing chromosome (→ diisocompensation trisomic).
dikaryon (Maire 1902) — a binucleate cell, spore, or mycelium consisting of such cells in Asco- and Basidiomycetes, which may be homokaryotic (containing two genetically identical nuclei) or heteokaryotic (containing two genetically different types of nuclei).

"Dikaryotization" is the process by which a dikaryon is produced: Two sexual cells or organs, each containing one or more haploid nuclei of unlike → mating type fuse and the nuclei from the different strains remain associated in one or more pairs, each called ad., while retaining their individuality during → dikaryophase. Plasmogamy (fusion of cytoplasms) is in this way separated from karyogamy in a kind of "interrupted fertilization" because the two members of each d. may repeatedly undergo mitotic divisions (which are generally simultaneous) before performing karyogamy (→ di-mon matings)
dikaryophase — a third nuclear phase (→ alternation of nuclear phases) in addition to the haploid and diploid phase in Euascomycetes and Basidiomycetes which is interposed between plasmogamy and karyogamy (= dihaplophase). The → dikaryon characteristic for the d. originates (following plasmogamy) by the intracellular association of pairs of compatible nuclei, which may be of identical or different origin in self-fertile and cross-fertile species. The pairs of haploid nuclei divide synchronously during vegetative growth, thus establishing and maintaining an equal ratio of the two nuclear types. In Euascomycetes, the d. occurs as the ascogeneous hyphal system, in Basidiomycetes, as the secondary mycelium. The d. leads to and is terminated by karyogamy and meiosis in homologous specialized cells called asci and basidia (Raper and Esser 1964).
dikaryosis — in higher fungi (Euascomycetes and Basisdiomycetes), a substitute for diploidy. It differs from diploidy primarily in the spatial separation of the two genomes into a genetically balanced system (→ dikaryophase; heterokaryosis).
dikinetic — = → dicentric.
dilution gene — any modifying gene which reduces the action of a major gene and shows no other phenotypic manifestation (= diluting factor).
dimegaly (Henking 1891) — a condition in which two types of female and/or male gametes are produced in the same organism.
dimer — any structure resulting from the association of two identical subunits (→ pyrimidine dimer).
diminution (Herla 1895) — = → chromosome diminution.
dimixis — → heteromixis.
di-mon mating — mating between di- and monokaryons in Basidiomycetes, as first described by Buller (1931) (Buller phenomenon). Three types may be distinguished: (1) **Compatible di-mons:** the monokaryon (e.g., A^1B^1) is compatible with both components of the → dikaryon (e. g., $A^2B^2 + A^3B^3$); (2) **hemicompatible di-mons:** the monokaryon is compatible only with one component of the dikaryon (e.g., $A^1B^1 \times A^1B^1 + A^2B^2$); (3) **noncompatible di-mons:** the monokaryon is compatible with neither component of the dikaryon (e.g., $A^1B^1 \times A^1B^2 + A^2B^1$).

As a rule, the monokaryon changes to a dikaryon in all three cases.

dimorphism — (1) the occurrence of two forms (differing in genotype or chromosome morphology) within a population (→ polymorphism); (2) the occurrence of two different types of flower on one plant.

dinokaryon (Chatton 1920) — a type of nuclear organization in dinoflagellates which differs from that of other eukaryotes. A striking feature is the absence of a → chromosome coiling cycle. Chromosomes maintain the same appearance throughout the cell cycle and are distinctly visible as rod-shaped bodies within the interphase nucleus. The DNA is not associated with protein. Dinoflagellate mitosis (dinomitosis) is without a typical spindle. During division a complex process of invagination of the intact nuclear envelope takes place at one side of the d. and results in the formation of parallel cylindrical channels through the d. which contain bundles of → microtubules. Nuclear constriction occurs perpendicular to these channels which may be involved in the separation of daughter chromosomes. Dinoflagellates thus resemble prokaryotes in the manner of chromosome separation, chemistry, and ultrastructure and the d. may represent an intermediate (mesokaryotic) type in the evolution of the typical eukaryote nucleus.

dioecious — of plants with either male or female sex organs, in contrast to → monoecious organisms. D. plants require a partner of opposite sex for the completion of their sexual cycle. Plants with hermaphroditic and male flowers or hermaphroditic and female flowers on different individuals are called androdioecious and gynodioecious (Darwin 1877), respectively. Species which, in addition to hermaphroditic offspring, also produce monoecious offspring with male and female inflorescenes are referred to as androgynodioecious.

In fungi, individuals which function as either donor or recipient of a nucleus are called d. Dioecious fungi include: (1) All species in which mycelia bear either female or male sex organs, but not both; (2) all species with two mating types, in which zygote formation follows the fusion of isogametes or isogametangia. Since sexual behavior is determined purely through the physiological difference between the two mating types, they are called physiologically d. (Esser 1971).

DI particle — = → defective interfering virus particle.

diphasic — of chromosomes with one arm euchromatic, the other heterochromatic.

diplandroid (Mok and Peloquin 1975) — of 2n pollen in diploid plants; d. pollen may result in male gametophytes with the somatic chromosome number.

diplobivalent (Barber 1940) — a bivalent composed of two → diplochromosomes and hence eight chromatids.

diplochromosome (White 1935) — a → chromosome which has reduplicated twice since the last effective mitosis, the centromere region being undivided (Darlington and Mather 1949). It consists of four chromatids, as opposed to normal chromomomes with two chromatids.

diplo-haplontic (Svedelius 1915) — of organisms ("diplo-haplonts") in whose life cycle a haploid phase exists between meiosis and fertilization, including all higher plants, many algae and fungi (→ diplontic, haplontic). Meiosis in diplo-haplontic organisms is intermediate and its products are not gametes but meiospores. These spores do not fuse to form a zygote but undergo mitodic divisions and develop into haploid individuals. These then eventually contain the gametes and hence they are called gametophytes. The diploid individual arising from the zygote is the site of meiosis, and since it forms the meiospores it is called the sporophyte. In diplo-haplontic life cycles two generation alternate (→ alternation of generations), one reproducing sexually and one reproducing asexually.

diploid (Strasburger 1905) — of cells, phases of life-cycles (→ diplo-haplontic, diplontic), and organisms ("diploids") with two homologous → chromosome sets (one paternal, the other maternal) in which each type of chromosome (autosome) except the → sex chromosomes of the heterogametic sex is always represented twice (symbol 2x or 2n), as opposed to → haploid (one chromosome set; 1x or 1n) and → polyploid (more than two chromosome sets; 3x, 4x, etc).

"Diploidy" is initiated with the fusion of two haploid gametes during → fertilization. By this process al "holozygote" (→ zygote) is formed. The adaptive advantage of a dominant phase of diploidy in higher organisms may result from: (1) Diploidy masks recessive deleterious mutations and (2) allows interactions between alleles to produce heterozygotes superior to either homozygote.

Diploidy for part of the → genome is only known in viruses and bacteria and may result from several mechanisms (→ heterozygous). "Zygotes" in bacteria which are diploid only for part of the genome and haploid for the remainder are referred to as "merozygotes" and are usually unstable.

The process of partial genetic transfer producing a bacterial merozygote by incorporation of a → "merogenote" into a recipient bacterial cell is knows as "meromixis". In the case of heterozygosity for the → alleles in question partial diploids show the phenomena of dominance, recessivity and → genetic complementation.

The term "**functional diploid**" (Darlington 1928) denotes an → alloploid organism behaving in → segregation like a diploid one.

diplontic — of organisms ("diplonts") having a life cyle in which the products of meiosis behave directly as gametes (→ diplo-haplontic, haplontic), including all multicellular animals. Only the gametes of diplonts are haploid. The organism arising from the zygote and all derivatives produced by vegetative reproduction or parthenogenesis are diploid and contain two (in case of polyploidy more than two) chromosome sets in their somatic cells. As opposed to diplo-haplonts, there is no → alternation of generations in diplonts, but only an → alternation of nuclear phases.

diplophase (Buder 1916) — (1) the diploid phase or generation between fertilization (zygote formation) and the beginning of meiosis (→ haplophase); (2) = → diplotene in the prophase of the first meiotic division (Belling 1928).

diplosis (Renner 1916) — the doubling of the gametic chromosome number.

diplosome — a pair of → centrioles.

diplospory — → apomixis.

diplotene (v. Winiwarter 1900) — the fourth stage of meiotic prophase (→ meiosis) where the paired chromatids begin to separate.

diplounivalent — a → diplochromosome in pollen grain mitosis, consisting of four chromatids. Diplounivalents are produced, following the failure of the second meiotic division, by chromosome reduplication preceding pollen grain mitosis.

directional selection (Mather 1948) — → selection.

directed mutagenesis — → site-directed mutagenesis.

direct repeat — any stretch of DNA within a single molecule which has the same nucleotide sequence in the same orientation, such as 5'AGTCT ... AGTCT ... 3'
3'TCAGA ... TCAGA ... 5'

D. repeats can be either adjacent to one another or separated from each other on the same molecule (→ inverted repeat).

disassortment — preference for → mating between dissimilar individuals (→ assortment).

discontinuous coding sequence — the coding sequence in genomic DNA of eukaryotic → split genes consisting of → exons and → introns.

discordance — the nonidentity of matched groups or pairs (such as twins) in any one or in a given → character, as opposed to concordance.

By means of "discordance analysis", which may be used in human genetics, it is possible to establish that a pair of twins originated from the same egg.

discordant clone — a cell clone which in → gene assignment experiments with hybrid cells disagrees with the majority of hybrid clones. The disagreement may be due to chromosome rearrangements or other factors.

discordant orientation — → translocation.

disjunction — the separation of daughter chromosomes (chromatids) during anaphase of mitosis and second meiotic division and of paired chromosomes (→ chromosome pairing) during anaphase of first meiotic division (→ assortment).

dislocation (Navashin 1926) — a structural change of chromosomes (→ chromosome mutation) produced by loss or displacement of chromosome segments. Homologous pairs of chromosome segments differing in their linear sequence with other segments are referred to as "dislocated" segments (Darlington 1937).

dislodgement (Coetzee et al. 1972) — a type of interaction between a compatible resident → plasmid and another plasmid transferred by conjugation into a recipient bacterial strain which leads to the elimination of the resident plasmid, loss of some, but not all, of the phenotypic characteristics determined by that plasmid, or recombination between the two plasmids.

disomic (Blakeslee 1921) — (1) of cells and individuals (2n) with two → chromosome sets whose members are represented by pairs of homologs, as opposed to → monosomic, → nullisomic, → trisomic, and → tetrasomic ones; (2) of cells containing one chromosome set plus a second representative of one member of the set (n + 1).

disparate chiasmata (Darlington 1937) — → chiasma.

dispermy — the entrance of two spermatozoa into an egg cell (= dispermic fertilization) (→ polyspermy).

dispersal — the enforced or voluntary movement of individuals from native areas, in many cases presumably in an effort to increase their reproductive potential. Reproductive enhancement of migrants may be achieved by way of increased access to mates or resources and by the avoidance of → inbreeding within the native area. D. plays a role in the regulation of animal populations and in speciation.

dispersed gene family — → multigene family.

dispersed repetitive DNA sequence — any of a family of repeated DNA sequences (= dispersed repeated gene family) dispersed throughout the → genome of eukaryotes (→ Alu family; copia-like sequence family). D.r. DNA sequences are highly respresented in the poly(A) RNA pool. At least some of these sequences are capable of → transposition and structurally similar to bacterial → transposons and to the integrated genomes of RNA viruses of the retrovirus group; they have short terminal direct repeats and undergo rearrangement, i.e., may change in both number and location. The integration of the sequences occurs at many alternative sites. D.r. DNA sequences may function in the regulation of → gene expression at diverse locations and their integration may result in → insertion mutations.

dispersed repetitive gene family (Rubin et al. 1976) — = → dispersed repetitive DNA sequence.

displacement loop (Kasamatsu et al. 1971) — (D-loop); (1) in covalently closed circular DNA, a region in which a short stretch of RNA is paired with one strand of DNA, displacing the original partner DNA strand in this region. D-loops are early intermediates in mitochondrial DNA replication (displacement replication). The unique region of the mitochondrial genome to which initiation of → mitochondrial DNA synthesis is confined, is called the D-loop origin. Leading strand synthesis commences within this region and proceeds unidirectionally until the single complementary strand origin is unmasked. At this time, lagging strand synthesis begins in the opposite direction (→ R-loop). (2) The displacement of a region of one strand of duplex DNA by a single-stranded invader in the reaction catalysed by the Rec A protein of *E. coli.*

displacement replication (Robberson et al. 1972) — a mode of continuous DNA replication in → mitochondria. DNA synthesis proceeds from the replication origin on one strand of the closed circular DNA molecule until about half the chromosome is partially duplicated. Retrograde synthesis then begins while the first strand continues.

disruptive selection (Mather 1948) — → selection.

dissociation — (1) a type of structural change in eukaryotic chromosomes which can increase the chromosome number (White 1957). D. is a special type of translocation (→ centric fusion gives the opposite result) whereby a metacentric large chromosome and a small supernumerary (supplying one → centromere and two → telomeres) are translocated so that two acrocentric or submetacentric chromosomes are produced. (2) The spontaneous or induced separation of the nuclear components of a heterokaryotic → dikaryon; in fungal colonies this may result in changed sectors (Pontecorvo 1953).

dissociation factor (Kohler et al. 1968; Bade et al. 1969) — a protein factor (in *E. coli*; mw about 9000) which plays a role in the cycling of → ribosomes. By complexing with the 30S subunit the d.f. (DF) splits the free 70S runoff ribosome at the end of polypeptide synthesis into its 30S and 50S subunits. After that DF is cyclically released at a subsequent stage in the formation of a → polyribosome. DF is identical with one of the → initiation factors (IF_3), i.e., it has, in addition to its role in separating ribosomes into subunits, other functions in the initiation of protein synthesis (→ genetic translation).

dissymmetry coefficient (Fogel et al. 1978) — in → gene conversion the radio of 3:1 to 1:3 segregations.

distal (Navashin 1912) — of a part of a eukaryotic chromosome arm, which is farther from the → centromere than another (proximal) part (Darlington and Mather 1949).

distortional segregation — → segregation; B-chromosome.

distributive pairing (Grell 1962) — that phase of the meiotic cycle (→ meiosis) during which chromosomes that had previously failed to undergo → exchange pairing may enter into segregation associations with other non-cross over chromosomes, homologous or nonhomologous.

distylous (Darwin 1877) — of flowering plant species with two types of individuals that bear different forms of flowers: "pin" flowers with long styles and short stamens, and "thrum" flowers with short styles and long stamens. In most cases, this flower dimorphism is associated with a sporophytically controlled diallelic → self-incompatibility mechanism that prevents fertilization after self-pollination or pollen transfer between individuals of the same flower type.

ditelomonotelosomic (Kimber and Sears 1968) — a cell or individual in which one chromosome pair is missing but which has a pair of telocentric chromosomes for one arm and an unpaired telocentric chromosome of the other arm of the missing chromosome pair (→ double-monotelosomic).

ditelosomic (Kimber and Sears 1968) — of a cell or individual deficient in two homologous chromosome arms (→ tritelosomic; diisosomic).

ditelotrisomic (Kimber and Sears 1968) — an individual deficient in one chromosome but which has a pair of homologous telocentric chromosomes for one arm of the missing chromosome (→ diiso-compensating trisomic).

ditertiary compensating trisomic (Kimber and Sears 1968) — a compensating → trisomic in which the missing chromosome is compensated for by two tertiary chromosomes, one having one arm of the missing chromosome and the other the second arm (→ diisocompensating trisomic).

divergence index (Ginsburg 1954) — → intergradation index.

divergent transcription — the initiation of → genetic transcription at two → promoters facing in opposite direction, i.e., transcription proceeds away in both directions from a position internal to the gene cluster (operon).

division septum — in all cells that divide by binary fission, an ingrowth at the midpoint of the cell which is controlled both spatially and temporally. Usually, the formation of the d.s. occurs by the circumferential invagination of the components of the cell envelope and eventually leads to complete separation of the two daughter cells.

dizygotic — of twins produced by two separate ova, fertilized by separate sperm, as opposed to → monozygotic twins arising from a single fertilized ovum which develops into two embryos at some relatively early developmental stage. Mono-and dizygotic twins are also called identical and fraternal twins, respectively. Monozygotic twin pairs normally have identical genotypes, they are like-sexed, and usually closely similar in physical appearance. Dizygotic twins are as similar genetically as ordinary sib pairs, having, on the average, half their genes in common. Zygosity of twins may be investigated in two ways: (1) By examining the fetal membranes, or (2) on the basis of similarities (concordance) and differences (discordance) between the cotwins.

D-loop — → displacement loop.

DNA — a shortland for → deoxyribonucleic acid.

DNA adduct — a covalent adduct between a chemical mutagen and DNA. Such lesions activate → DNA repair processes and, unless repaired prior to DNA replication, may lead to → nucleotide substitutions, → deletions, and → chromosome rearrangements.

DNA amplification — amplicifation (→ gene amplification; sequence amplification) of a variety of DNA sequences (selected genes and flanking DNA) either as a result of selection for drug resistance or as part of the malignant process in animal cells. DNA a. results typically in two types of karyological consequences: (1) In some cell lines, the amplified gene copies are present in one or more chromosomes, originally called → homogeneously staining regions (HSRs); (2) in other cell lines, amplified genes are present on small extrachromosomal elements lacking → centromeres, called → minute chromosomes.

Secondary DNA a. or re-amplification (Flavell et al. 1977) is a process that generates a → DNA sequence family containing a subset of recently amplified, relatively nondiverged sequences against a background of more highly diverged members amplified in an earlier event.

DNA annealing — hybridization of single-stranded DNA to another single-stranded DNA molecule (→ DNA assimilation) due to pairing of the molecules and hybridization of complementary sequences.

DNA assimilation — (Shibata et al. 1979) — the hybridization of single-stranded DNA to a homologous sequence within a duplex DNA molecule (→ DNA annealing).

DNA bending (Marini et al. 1982) — a conformational change of the DNA double helix due to heterogeneities in its local structural composition. DNA b. may result from interaction with proteins (e.g., in → nucleosomes) and smaller ligands. DNA can assume a bent conformation by itself, as in → supercoiled DNA and → kinetoplast DNA. Molecular models for bending include smooth curvature along the axis and localized bend such as kinks or junction bents.

DNA-binding protein (Hotta and Stern 1970) — any of a class of proteins that bind to specific DNA sequences (promoters, enhancers) and play integral roles in → DNA replication, → genetic recombination, strand scission (→ restriction endonuclease), → genetic transcription, and → DNA repair. They include many of the enzymes of DNA metabolism and other proteins, such as → repressors, activators, and proteins binding to → promoters, and → enhancer elements.

Sequence specific DNA binding may result from: (1) Direct atomic interaction between amino acid side chains and base pairs in the DNA; (2) a limited number of structural motifs (→ zinc finger motif; helix-turn-helix motif; leucine zipper) which direct the appropriate amino acid side chains of a protein into the grooves of double helical DNA where they can interact directly with base pairs. The motifs form the threedimensional scaffolds that match the contour of DNA and dictate the appropriate positioning of the interacting protein surface, allowing atomic interaction between amino acid side chains and the base pairs of the specific binding site.

DNA breathing — the phenomenon of random and transient breaking and reforming of hydrogen bonds in duplex DNA (→ branch migration; single-strand inversion).

DNA catenane (Clayton and Vinograd 1967) — any of the interlocked duplex DNA rings in prokaryotic and eukaryotic cells. They are suggested to be intermediates in → genetic recombination or DNA replication. Bacterial → DNA gyrase is representative of a general class of enzymes that facilitate diffusion of duplex DNA segments through each other. All cells contain enzymes called → DNA topoisomerases that can form and resolve DNA catenanes.

DNA circularization — conversion of a linear DNA fragment into a covalently → closed circle by annealing of → sticky ends (5' and 3' extensions) of the DNA fragment generated by digestion with a single → restriction endonuclease.

DNA clone — a sequence of DNA that has been inserted into a → cloning vector and then replicated to form many copies.

DNA cloning — → molecular cloning.

DNA compaction — in eukaryotic nuclei, the packaging of DNA into → nucleosomes and higher order structures of compaction eventually resulting in compact metaphase → chromosomes. DNA c. is achieved through an extensive hierarchical folding of the DNA, mediated by → histones and nonhistone nuclear proteins. The first order of folding is defined by the nucleosomes; at the next level, nucleosomes are arranged into solenoids or superbeads (6 to 10 nucleosomes) to give a linear condensation ratio of DNA of about 40. At the highest level, interphase chromatin and metaphase chromosomes are organized into arrays of supercoiled loops (each containing 50 to 100 kb of DNA).

DNA concatemer — a chain of repeated units of duplex DNA.

DNA conformational microheterogeneity (Klysik et al. 1981) — the phenomenon that neighboring DNA segments may have different secondary structures. Such DNA ist suggested to play an important role in → DNA replication, → genetic recombination, mutagenesis-carcinogenesis, → DNA repair, → genetic transcription, chromosomal organization, and virus packaging.

DNA control element — any of the sites within genes or operons that interact with specific regulatory proteins (→ DNA binding protein) to control → gene expression. Positive-acting sites are called → transcriptional enhancers; negative-acting sites are → operators or → transcriptional silencers.

DNA damage — any of various types of → DNA lesions occurring either spontaneously or induced by → mutagenic, → clastogenic, or → carcinogenic agents. These lesions (single- and double-strand breaks, cross-links, or steric alterations and loss of bases) affect → DNA replication and decrease the rate of DNA synthesis, which results in a decrease in the size of newly synthesized DNA, altered linkage of newly synthesized fragments to higher molecular weight DNA, and a decrease in the number of actively synthesizing → replicons in eukaryotic cells. The lesions may be removed by → DNA repair, replaced by recombination, or retained, leading to mutation and cell death.

DNA damage repair hypothesis (Haynes et al. 1984) — a hypothesis according to which lethal hits in cells exposed to mutagens are assumed to arise from potentially lethal damage (→ DNA damage), left unrepaired in DNA. Mutational hits result from premutational lesions which have not been removed by error-free repair, but have been processed by error-prone repair. Observed levels of induced mutagenesis in rapair-proficient cells are suggested to result from the net effect of error-free and error-prone processes acting on premutational lesions in DNA. → DNA repair systems are seen as causative as well as ameliorative factors in mutagenesis.

DNA damge tolerance mechanism — any mechanism that determines a route around a → DNA lesion that has not been repaired before the approach of a DNA replication fork (= post-replication repair). Such mechanisms aim to ensure the completion of DNA replication by a replicational bypass of unrepaired lesions, i.e., restoration of the double helix structure while the initial lesions remain in the DNA, and a second chance to repair.

DNA denaturing enzyme (Abdel-Monem et al. 1977) — = → DNA unwinding enzyme.

DNA-dependent RNA polymerase — → RNA polymerase.

DNA diagnostics — the pre- or postnatal analysis of → genetic diseases, i.e., the identification of individuals predisposed to conditions and the analysis of disease. Fundamental to

DNA d. is: (1) The definition of nucleic acid sequences that are of diagnostic value; (2) the development of techniques that permit monitoring these sequences in individual patients.

DNA-DNA hybridization — hybridization (annealing) of purified, denatured DNA as used to investigate homologies between DNA molecules from different sources and to examine the relation of particular DNA fractions to the rest of the genome. Reannealing kinetics are used as a measure of genome complexity (→ Cot value).

DNA-driven hybridization — the reaction of an excess of DNA with RNA.

DNA elimination — in sciarid flies, a process in which whole chromosomes are eliminated at specific stages of development.

DNA encapsidation — the packaging of virus DNA into → virions. In the morphogenesis of complex double-stranded DNA-containing → bacteriophages, DNA e. is a three-component system consisting of newly replicated DNA, a functional prohead, and specific phage proteins, which link concatemeric DNA with the prohead as well as act in its subsequent encapsidation (linkage/packaging proteins).

DNA exonuclease — any → deoxynuclease characterized by the following typical features: (1) Determination of the polarity of attack (5' or 3'); (2) the product specificity (5'-nucleotides or 3'-nucleotides); (3) secondary structure specificity (single- or double-stranded DNA preferred) and processivity (enzymes that dissociate from DNA after every catalytic event and those which remain bound to DNA until template hydrolysis of all phosphodiester bonds has occurred).

DNA extension — the single-stranded DNA tail located at the end of a restriction fragment. Different → restriction endonucleases generate DNA e. at either the 5' or 3' end of the DNA strand.

DNA fingerprinting (Jeffreys et al. 1985) — a method using DNA probes that hybridize to hypervariable → minisatellite regions of human (and other) DNA. The probes, constructed from single tandem repeats of a short human core sequence (15 bases long), reveal similar hypervariability in mice, dogs, cats, and birds. Each of the DNA probes is based on a tandem repeat of one of the core sequences. As with conventional → restriction fragment length polymorphism technology, the probe is made radioactive and is applied to gel separation of a restriction enzyme digest of an individual's genomic DNA. It hybridizes to any fragment that contains members of the particular set of minisatellites represented by the core sequence in the probe.

DNA footprinting — a method for determining the sequence specificity of → DNA-binding proteins. DNA f. utilizes a DNA damaging agent which cleaves DNA at every base pair; DNA cleavage is inhibited where the ligand binds to DNA.

DNA fusion — the joining of DNA segments not previously associated and with no appreciable nucleotide sequence homology. DNA f. products have the potential of generating two fused protein products, provided that both components are translated, that they are not separated by translation stop codons, and that they lie within the same → reading frame following the fusion event.

DNA glycosylase (Lindahl 1974) — any of class of repair enzymes (→ DNA repair) that remove certain abnormal bases from DNA leaving an apurinic/apyrimidinic (AP) site and initiating base → excision repair. The repair pathway involves the incision of the phosphodiester bond adjacent to the AP site by an AP endonculease, the excision of the baseless site, and the successive action of → DNA polymerase and → DNA ligase as in nucleotide excision repair.

DNA glycosylases are generally small proteins (mw 18.000–31.000) and have no subunit structure. There is no requirement for cofactors or divalent cations.

DNA gyrase (Gellert et al. 1976) — an enzyme of *E. coli* (composed of two subunits) that catalyzes negative supercoiling of closed (initially relaxed or positively supercoiled) duplex DNA (viral, plasmid, and chromosomal DNA). The enyme is involved in → DNA replication, → genetic transcription, → DNA repair, and → genetic recombination. Six activities are known: (1) Binding to DNA; (2) introduction of supercoils in the presence of ATP; (3) relaxation of supercoils in the absence of ATP; (4) site-specific cleavage of DNA following denaturing of the enzyme-DNA complex; (5) ATP-dependent → catenation and uncatenation of DNA rings; (6) DNA-dependent hydrolysis of ATP to ADP and P_i (→ DNA topoisomerase).

DNA hairpin structure — a DNA sequence capable of forming a h. structure, i.e., a double-stranded DNA caused by a turn back on itself and pairing of complementary base sequences. DNA h. structures are often found near functional loci such as regulation and promotion sites as well as within cruciform structures. DNA hairpin-containing domains may play an important role in control mechanisms of → gene expression as well as providing sites for exogenous drugs and endogenous proteins (→ RNA hairpin structure).

DNA helicase (Kuhn et al. 1978) — any of a class of DNA unwinding enzymes (single-stranded DNA-dependent nucleoside 5'-triphosphatases) which either unwind DNA propressively by ATP-energized migration of the enzymes along one DNA strand (5'→ 3' direction) or nonprogressively by binding in stoichiometric amounts to one of the strands. DNA helicases bind and move on the lagging strand template during → DNA replication and unwind duplex DNA in advance of the replication fork. In addition, some DNA helicases interact with other replication proteins and are involved in the synthesis of → RNA primers on the lagging strand.

DNA helix-destabilizing protein (Hubermann et al. 1971) — (HO-protein); a protein isolated from various organisms by virtuse of its strong preferential affinity for single-stranded DNA. It lowers the melting temperatures of double-stranded DNA and stimulates the activity of the homologous → DNA polymerase in vitro.

DNA homoplasmicity — DNA sequence homogeneity in somatic cells of an individual organism as opposed to DNA heteroplasmicity.

DNA insert — a DNA sequence inserted into a → cloning vector.

DNA insertase (Linn et al. 1979) — an enzyme that directly inserts missing bases into DNA. The enzyme adds adenine at apurinic sites and guanine at apyrimidinic sites (→ AP site).

DNA insertion element — any specific DNA sequence capable of transposing itself from one chromosomal location to another (→ insertion sequence; transposon; transposable genetic element; insertion mutation; insertosome).

DNA intercalation (Lehrman 1961) — a binding mode originally suggested to explain the high affinity of planar aromatic dyes for DNA. Among the molecules that intercalate are the aminoacridine dyes, antimicrobial drugs such as ethidium bromide, and the actinomycin antibiotics.

DNA invertase — → invertable DNA.

DNA in vitro packaging (Kaiser and Masuda 1973) — a technique for → molecular cloning employing pairs of → lysogens that have complementary defects in phage lambda packaging protein genes. While each lysogen is grown and induced, the absence of required proteins prevents prophage genomes from packaging; lambda packaging proteins thus accumulate. When extracts of the two lysogens are combined, both added lambda DNA and endogenous lambda DNA are packaged by the full complement of packaging proteins.

DNA lesion — any of different types of → DNA damage (steric deformations) which may result in → gene mutation or gross structural changes of chromosomes. DNA lesions may occur spontaneously or may be introduced into DNA after exposure to UV light, ionizing radiation or chemical mutagens. The most common DNA lesions are (Lindahl 1979): (1) Monofunctional lesions causing neglegible helix distortion; (2) monofunctional lesions causing minor helix distortion; (3) monofunctional and difunctional lesions causing major helix distortion.

Such lesions are responsible for insertion of an incorrect base in one strand of

DNA, missing bases, altered bases, single-strand breaks, double-strand breaks, and cross-linking.

Each of these defects can be corrected by any of several → DNA repair systems. A classification of DNA lesions based on the repair enzymes involved, rather that on the degree of DNA distortion or the type of adduct formed (monoadducts and diadducts, respectively) is as follows: (1) DNA lesions dependent on uvr gene products for repair; (2) DNA lesions involving repair by → A/P endonucleases; (3) DNA lesions dependent on other repair endonucleases; (4) DNA lesions not regularly recognized by repair enzymes (→ DNA damage tolerance mechanism; DNA damage repair hypothesis).

When not repaired, the DNA lesions produced by some → mutagens alter the coding properties of the bases of DNA directly (miscoding lesion), leading to → mutation via mispairing during DNA replication; others act as blocks to DNA replication and require the active involvement of an error-prone DNA repair process (→ SOS response) for the generation of mutations.

DNA lesion reversal — → DNA repair.

DNA library — a store of cloned DNA (cloned library) in recombinant → cloning vectors containing chromosome-specific DNA fragments. A DNA l. may be either a → cDNA library or a → genomic DNA library.

DNA ligase (Gellert 1967) — any of a number of enzymes (= polynucleotide ligases, DNA sealases) which catalyze the closing of single-strand breaks (nicks) in native double-stranded DNA and the circularization of DNA by the formation of phosphodiester bonds between 3' OH groups and 5' monophosphate groups of successive nucleotides in the DNA chain. DNA ligases from various sources require either ATP or NAD as a cofactor.

DNA l. is essential for → DNA replication (joining of → Okazaki fragments), → DNA repair, and → genetic recombination (→ RNA ligase).

There is one exception to the generalization that DNA ligases only join together polynucleotides held in juxtaposition by a complementary strand. DNA l. from bacteriophage T4-infected cells can join duplex DNA substrates intermolecularly at base-paired ends ("blunt end joining"). This reaction is useful for the construction of chimeric DNA molecules (→ recombinant DNA technology).

DNA-like RNA (Scherrer et al. 1963) — = → pre-messenger RNA.

DNA looping — protein-mediated bridging of two DNA binding sites that results in looping out of the intervening DNA sequence flanked by the two binding sites. DNA l. may play a role in the interactions of eukaryotic → transcription enhancer elements and → promoters and in controlling prokaryotic DNA replication and transcription.

DNA-mediated gene transfer — the introduction, via purified DNA (= DNA transfection), of new genetic information into cultured animal or plant cells. DNA-m.g.t. is a common method for transformation of eukaryotic cells with specific genes (→ gene transfer) employing coprecipitates of DNA and calcium phosphate (DNA-Ca-P_i). Transformed cells expressing exogenous DNA coding for selectable markers simultaneously integrate other DNA sequences at high frequency. Cotransformed cells may thus contain any cloned prokaryotic or eukaryotic gene. During transformation disparate DNA fragments ligate and recombine, associating randomly with the selected locus; sequences longer that 100 kb are rarely transferred intact (→ chromosome-mediated gene transfer; pekalosome; transformation). Stable DNA-m. genetic transformation may occur by either of two mechanisms: (1) Integration into host chromosomes, or (2) independent replication (if the DNA is introduced as part of an autonomously replication → plasmid).

DNA methylation — the specific methylation of DNA bases by DNA methylases. DNA m. occurs in virtually all organisms examined and involves the enzymatic transfer at the nucleotide level of the methyl group of S-adenosyl-methionine to specific bases in DNA. It is the bais of bacterial → DNA modification-restriction systems and has been implicated in the regulation of bacterial DNA replication, genetic recombination, and mutation.

In prokaryotes, the pattern of DNA m. is species-specific. In some prokaryotes, a small fraction of the cytosines is methylated, in others only adenine residues are methylated. A third group of prokaryotes contains both m^5Cyt and m^6-Ade.

In eukaryotes, the most common modified base is 5-methyl-cytosine (m^5Cyt). The dinucleotide CpG is preferentially methylated in eukaryotic DNA. A second modified base is N^6-methyladenosine (m^6Ade). Modification occurs posttranscriptional in most eukaryotic cellular and viral mRNAs with one to three m^6A residues per 100 nucleotides; the methylated A is usually found within a AAC or GAC consensus sequence (the function is unknown). The methylation pattern in particular genes is tissue-specific, indicating that methylation may play a role in the regulation of → gene expression. Hypomethylation of specific sequences in and around certain genes has been correlated with expression of these genes. Sites of cytosine methylation seem to be hot spots for mutation in eukaryotic DNA (→ HTF island).

Maintenance methylation is the postreplicative methylation of hemimethylated sites that maintains preexisting methylation patterns after → DNA replication. A maintenance methylase is specific for the hemimethylated DNA generated during DNA replication.

DNA methyltransferase (Wyatt 1951) — (1) any of the enzymes (= DNA methylases) involved in → DNA methylation, i.e., transfer of methyl residues from S-adenosyl-methionine to cytosine (or adenosine) in double-stranded DNA. The process occurs subsequent to → DNA replication and → DNA repair and, therefore, undermethylated tracts of DNA exist for some period of time after strand synthesis. DNA methyltransferases produce three kinds of methylated nucleotides: 5-methlycytosine (m^5C), N^4-methylcytosine (m^4C), and N^6-methyladenine (mA). (2) A repair enzyme for → DNA damage induced by alkylating agents that removes particular methyl groups from DNA in a suicide reaction inactivating the enzyme (→ adaptive response).

DNA microinjection — a method of direct introduction by microinjection of specific DNA sequences into animal (and plant)cells, transformation of these cells, and production of → transgenic individuals.

DNA mismatch — any noncomplementary base pair in double-stranded DNA. Mismatched base pairs (or unpaired bases) can occur in vivo as a result of: (1) Errors in DNA replication; (2) hybrid DNA formation during recombination between homologous, but nonidentical DNA sequences; (3) chemical modification of bases in duplex DNA.

DNA molecules containing mismatched or unpaired bases are referred to as heteroduplexes. Their replication produces daughter molecules with different sequences at the site where the heteroduplex had a mismatch or an unpaired base (→ DNA mismatch repair).

DNA mismatch repair (Holliday 1964) — a → DNA repair process that recognized and repairs incorrect base pairs in double-stranded DNA. Mismatched bases (due to errors during DNA replication) are excised from the newly synthesized DNA strand and replaced with properly matched bases (preferentially GT in favor of GC). After excision of a polynucleotide from one strand, thereby removing one member of the unmatched base pair, the resulting gap is filled in by → DNA polymerase to form a correct base pair. The seal is then made by → DNA ligase.

For preferential correction of newly synthesized strands, the DNA m. r. system must be able to discriminate between daughter and parental strands. Single-strand discontinuities and → DNA methylation may provide the basis for strand discrimination. The informational transfer process coupled with DNA m.r. is considered to be the basis of nonrecipocal recombination or → gene conversion,

DNA misrepair — → DNA repair.

DNA modification — any post-synthetic chemical change of a DNA molecule, such as → DNA methylation or glucosylation (covalent attachment of methyl groups and glucose moieties). In host-controlled → DNA modification-restriction, the DNA of the host is protected against the action of its own → restriction endonucleases by such DNA modification.

DNA modification-restriction (Luria and Human 1952; Arber and Dussoix 1962) — in bacteria, any of the enzymatic mechanisms for degradation of foreign DNA molecules that penetrate the cell (viral genomes and plasmids). DNA r.-m. systems result in host-specific barriers to interstrain and interspecies transfer of genetic information and comprise two activities: (1) A sequence-specific → DNA methyltransferase (MTase); (2) a → restriction endonuclease (ENase) of identical specificity responsible for restricting unmodified DNA. The molecular basis of DNA m.-r. is the presence of specific nucleotide sequences in bacterial DNA (4 to 8 nucleotides in length) which are recognized by modification and restriction enzymes. If the modification enzyme acts first, it protects the recognition site by DNA methylation from attack by the restriction endonuclease. If the DNA is unmodified both strands can be endonucleolytically cleaved by the restrictase at or near the restriction site. Strains with a given restriction specificity also have the corresponding modification specifity protecting their own restriction activity.

The group of site-specific enzymes that protect bacterial DNA from endogenous type II restriction endonucleases are called modification methylaseses. Type II modification methylases require S-adenosylmethionine (SAM) for their activity. They methylate within the sequence recognized by the corresponding restrictase at either adenine or cytosine.

When a strain possesses more than one DNA m.-r.s., each system functions independently. Restriction and modification activities are probably compenents of a single complex molecule which may be able to dissociate, splitting off a functional modification methylase without restriction activity.

Bacterial mutants with altered restriction and modification properties have been isolated: $r^- m^+$ mutants fail to restrict but can still modify DNA; $r^- m^-$ mutants neither modify nor restrict.

DNA nick — a break introduced in a DNA strand that can be sealed by → DNA ligase. Duplex DNA can be "activated" as a template primer for certain → DNA polymerases by single-strand scissions introduced by an → endonuclease (→ nick translation).

DNA nicking-closing enzyme (Vosberg and Vinograd 1975) — any of a class of enzymes, belonging to the type 1 → DNA topoisomerases, which catalyze the overall process of nicking, swivelling, and resealing DNA; this relieves torsional stress in both linear and closed circular DNA. Such changes in the duplex winding of DNA are involved in → DNA replication, → genetic transcription, → genetic recombination, formation of prokaryotic chromosomes, and encapsidation of virus DNA. DNA n.-c. enzymes act via a single-strand breakage mechanism.

DNA overreplication — a mechanism that might result in → DNA amplification; several successive initiations of DNA replication at one replication origin produce several copies of a gene or DNA sequence in one cell cycle ("onion-skin" model).

DNA packaging — (1) in viruses, the process by which DNA enters the prohead, i.e., the empty capsid structure; (2) in prokaryotes and eukaryotes, the compact folding of the long DNA molecules of the → genome in a manner which allows their accommodation within the cell without hindering their accessibility.

DNA packing ratio — the ratio of the length of → B-DNA (0.34 nm/base pair) to the unit length of the → chromatin fiber containing it.

DNA pairing — the correct alignment of interacting (homologous) DNA molecules. DNA p. is an essential step in homologous → genetic recombination; it may be mediated by the formation of → heteroduplex DNA (h DNA) in which the DNA duplex is derived from one strand of each of the two parental molecules.

DNA palindrome — in a DNA molecule, an inverted duplication with twofold rotational symmetry (the 3' → 5' base sequence is the same on each strand) in which the symmmetry is perfect right up to the axis of symmetry, as distinguished from inverted repeats in which a nonsymmetrical region separates the inverted symmetries. DNA palindromes are probably involved in interactions between DNA and sequence-recognizing proteins (→ DNA-binding protein).

DNA periodicity — the number of base pairs per turn of the DNA double helix.

DNA photolyase — a → DNA repair enzyme (deoxyribodipyrimidine photolyase) which catalyzes photomonomerization of the cyclobutandipyrimidines, or pyrimidine dimers, produced by 200—300 nm UV radiation in DNA. The enzyme acts by binding to the pyrimidine dimer containing regions of DNA through a light-independent reaction, and subsequently produces a photochemical repair of the dimers, after which it detaches from DNA (→ photoreactivation).

DNA plasm (Kellenberger 1962) — any of the areas in prokaryote cells in which the DNA is concentrated (= nucleoid). The shape of such areas is highly variable and covers a range from a delicated network of branches to a compact nuclear body.

DNA polymerase — any of a number of multifunctional enzymes catalyzing the formation of → deoxyribonucleic acid from deoxyribonucleoside triphosphates and using DNA (DNA-dependent DNA polymerases) or RNA (RNA-dependent DNA polymerase) as template. A second type of reaction catalyzed by DNA p. is called → repair synthesis in which a partially single-stranded template is converted to a completely double-stranded form. Thus repair synthesis restores a damaged DNA molecule to its native state (→ excision repair). Mutant alterations of DNA-dependent DNA p. can affect the specificity of DNA replication (→ mutator gene).

All known **DNA-dependent DNA polymerases** require a primer with a free 3' hydroxyl group available to make a phosphodiester bond. Since the two chains of a parental DNA molecule have opposite chemical polarities, synthesis of a least one strand is discontinuous as a series of so-called Okazaki pieces. In bacteria, three DNA p. species may be distinguished: (1) Pol I, which is present in highest molar concentration; it is implicated in → DNA repair and in discontinuous DNA synthesis. (2) Pol II, for which no function is definitely known. (3) Pol III, which is responsible for chromosomal DNA synthesis.

All three enzymes exhibit 3' to 5' exonuclease activity; Pol I and Pol III also have 5' to 3' exonuclease activity. The 3' to 5' exonuclease activity can be called a proofreading or editing activity since it preferentially removes incorrectly inserted nucleotides at the site of DNA replication (primer terminus).

Four distinct molecular species of DNA polymerases have been reported for eukaryotes: (1) DNA p. α (DNA p.I), which is a high molecular weight species (6-8S) and the major replicating polymerase; it contains an associated primase and is recovered from the cytoplasmic fraction. (2) DNA p. β, which is a low molecular weight molecule (usually isolated from purified nuclei); not found in fungi, free-living protozoa, or some plants. (3) DNA p. γ (mitochondrial DNA p.), which is only found in → mitochondria. (4) DNA p. δ (DNA p. II), which is a large molecule containing an associated 3'-exonuclease.

Many DNA polymerases operate with a variety of ancillary protein components. These components form a replication complex with the DNA p. and participate in the translocation of the DNA p. along the template; they may also be necessary for the attachment and removal of ribonucleotide-containing primer sequences (→ primase; primer RNA).

Polymerase fidelity is a function of two enymatic activities; (1) nulecotide selection, which results in a mis-insertion frequency of about 10^{-5} incorrect bases per nucleotide incorporated; (2) 3' → 5' exonuclease proofreading activity, which decreases the misincorporation frequency by another factor of 10 to 100.

RNA-dependent DNA polymerase (→ reverse transcriptase) is a polymerase contained in virions of RNA tumor viruses. It catalyzes the synthesis of a DNA copy (a → provirus) of the resident or endogenous RNA and also the polymerization of deoxyribonucleotides using exogenous polynucleotides as templates in the presence of poly- or oligodeoxyribonucleotides. The enzyme has been used succesfully to obtain DNA copies (cDNA) of → messenger RNA (→ antimessenger; → terminal deoxyribonucleotidyl transferase).

DNA polymerase-primase complex — a multisubunit protein complex containing → DNA polymerase and → DNA primase; involved in DNA replication (→ deoxyribonucleic acid).

DNA polymorphism — any silent or neutral varitation in DNA. Such variation may be detected in one of two ways: (1) Differences in DNA sequence as identified in the course of → DNA sequence analysis of cloned DNA fragments; (2) introduction or removal of cleavage sites for → restricition endonucleases by neutral → nucleotide substitutions.

DNA primase (Scherzinger et al. 1975) — any of a group of → RNA polymerases which synthesize the → RNA primers (oligoribonucleotides of about ten nucleotides that prime the initiation of DNA replication) that are extended by → DNA polymerase to form → Okazaki fragments. In eukaryotes, DNA p. acitivity is associated with DNA polymerase.

DNA p. (of prokaryotes) differs from the usual DNA-dependent RNA polymerases in that the enzyme has a lower molecular weight, the products are short oligoribonucleotides, and it is insensitive to α-amanitin.

DNA-protein interaction — interaction between DNA and proteins that plays a fundamental role in all living cells. D.-p.i. can be subdivided into two categories: (1) Specific recognition, in which a protein recognizes a particular DNA sequence (e.g., → repressor-DNA interaction, → DNA polymerase-DNA-interaction); (2) nonspecific recognition, where the protein recognizes primarily the general features of DNA with little or no specificity for a particular base sequence.

In the case of specific recognition, the protein initially binds to DNA nonspecifically and then finds a cognate DNA sequence by sliding, short hopping and/or interstrand jumping (→ DNA-binding protein).

DNA puff (Breuer and Pavan 1955) — a unique and exclusive type of → puff in sciarid → polytene chromosomes which gives rise (by DNA puffing) to an excessive and disproportionate synthesis of DNA. DNA puffs constitute an unusual mechanism for amplifying (→ gene amplification) the activity of particular chromosomal sites. Disproportionate DNA synthesis in DNA puffs involves a number of additional rounds of replication. The excessive DNA is not shed from the chromosome but remains integrated. DNA p. formation is intimately linked to metabolic changes in salivary gland metamorphosis.

DNA rearrangement — any of the changes of the "normal" DNA arrangement which include homologous events (rearrangements proceeding through a → heteroduplex intermediate) and nonhomologous events (events that do not use such an intermediate). Genetic recombination during meiosis, sister chromatid exchange and interchromosomal recombination during mitosis as well as nonallelic → gene conversion, chromosome translocations, certain → gene amplification events, the movement of → retroviruses and of → transposable genetic elements, the rearrangement of antibody and T cell receptor genes of the immune system, and the formation of processed → pseudogenes represent examples of DNA r.

During evolution, DNA r. can duplicate genes that diverge to provide new functions and can establish novel chromosome combinations during → speciation.

DNA reassociation — the formation of duplex DNA from denatured single-stranded DNA. DNA r. is an important tool for the examination of the organization of the → genome and for studying its evolution. The methods commonly used for measurement of DNA r. kinetics are binding to hydroxyapatite (HAP), measurement of hypochromicity, and assay of resistance to single-strand specific nucleases (measurement of the fraction of nucleotides in duplex at each point in the renaturation reaction.) The kinetics of DNA r. are needed for evaluation of the frequency of occurrence of → repeated DNA sequences and for the design of measurements comparing DNA sequences from different species.

DNA recombinant technology — = → recombinant DNA technology.

DNA reiteration — the presence in eukaryotes of DNA sequences which exist in many similar, probably not identical copies in the → genome (→ gene amplification; repetitious DNA).

DNA relaxing enyzme (Wang 1971) — any of the enzymes that alter the topological winding number of DNA, i.e., catalyze the conversion of superhelical DNA to a nonsuperhelical covalently closed form. DNA r. enzymes have been reported for prokaryotic and eukaryotic organisms. They may act as "swivels" during DNA replication by the catalytic nicking and re-sealing of DNA chains. The enzymes of eukaryotic origin are distinguished from the protein of prokaryotes in that they relax both negatively and positively twisted DNA's and do not require Mg^{2+} for activity (→ DNA topoisomerase).

DNA renaturation — the renaturation of DNA which has been denatured (by heat). Two types are recognized: (1) Rapid reversal of thermal denaturation of parts of the DNA molecules which are incompletely denatured by heat and therefore remain "in register" (G-C-rich regions in the DNA); (2) renaturation of strands which are completely separated by denaturation (= reannealing or reassociation); dependent on collisions between complementary strands (reaction exhibits second order kinetics and depends on the concentration of the two strands of DNA).

DNA denatured by alkali treatment instead of heat does not show type 1 renaturation since G-C-rich and A-T-rich regions are equally vulnerable with alkali.

DNA repair — any (enzyme-mediated) process that removes lesions from DNA (→ DNA lesion) and/or restores an intact DNA molecule. DNA r. is sometimes accompanied by a modification of the base sequence resulting in mutations (error-prone repair; misrepair or incorrect repair) as opposed to correct (error-free) DNA r. that does not alter the original information of the DNA molecule concerned.

DNA r. can be accomplished through different pathways, depending on, e.g., the kind of → DNA damage or the physiological state of the cell: (1) Light-induced repair (→ photoreactivation); (2) excision of the damaged bases → excision repair); (3) reconstruction of a functional DNA molecule from undamaged fragments (= → recombinational repair); (4) disregard of the damage (→ SOS repair).

Excision r. and photoreactivation both repair the template, recombinational r. forms a new template, and SOS r. ignores the damage and forms uninterrupted daughter strands, despite the presence of damaged sites on the template.

DNA r. functions are present in all cells either as corrective activities in → DNA polymerases or as independent enzymes or both. The repair processes play a fundamental role in the maintenance of genetic integrity and fidelity in living systems. The repair capacity of organisms not only determines their survival after exposure to → clastogenic and → mutagenic agents which damage their DNA, but also influences their response to the effects of these agents (→ adaptive response). (→ reinitiation recovery; repair inducer; base-insertion repair).

The observation of preferential DNA r. in active eukaryotic genes provides evidence that there are variations in the repair processes corresponding to different parts of the genome. There are indications that → DNA topoisomerase 1 is involved in the preferential repair, while topoisomerase 2 is involved in the average genome repair.

DNA repair assay — any (bacterial) assay used to study → DNA repair as an indicator of → DNA damage and based on two assumptions: (1) → Mutagenesis (and carcinogenesis) involves an alteration in DNA; (2) since all cells have mechanisms to eliminate such damage, measurement of repair activity can be extrapolated to estimate the extent of such damage in DNA.

DNA r. assays include studies of → DNA repair synthesis, repair of single-strand breaks and crosslinks, and removal of adducts from DNA.

In the DNA r. host-mediated assay (Mohn et al. 1983), genotoxic activity of a given chemical can be measured in mice infected with *E. coli* strains differing vastly in DNA repair capacity. Differential killing of these strains recovered from various organs of the treated animal is used to determine the induction of (repairable) DNA damage.

DNA repair synthesis — the synthesis of new DNA to replace the nucleotides which are removed during the excision of → DNA damage.

DNA replicase (Yagura et al. 1983) — a multienzyme complex (in mice) consisting of → DNA polymerase and → DNA primase. The DNA and → primer RNA synthesized by DNA r., with single-stranded DNA as template, are stimulated by stimulation factors.

DNA replication — the direct copying of → deoxyribonucleic acid by a discontinuous mechanism involving the synthesis of short nascent DNA intermediates (→ Okazaki fragments) on at least the retrograde strand of the replication fork. DNA r. requires the concerted action of many proteins and enzymes, either separately or in a complex (→ replisome). These include, e. g., → DNA polymerases, DNA polymerase accessory proteins, → DNA primase, → DNA topoisomerase, → DNA helicase, → DNA-binding proteins, → RNase, and → DNA ligase. Basic rules of DNA r. are: (1) Semi-conservative mode of replication; (2) → RNA priming; (3) replication direction 5' to 3'; (4) leading and lagging strand replication; (5) usually bidirectional DNA r. from two replication forks (→ replicon) moving in opposite directions (in the case of unidirectional DNA r. the replicon has a single replication fork and there is a termination signal for synthesis of the first fragment).

DNA r. involves at least three stages: (1) Chain initiation; (2) chain elongation; (3) joining of the short chain intermediates. Since DNA polymerase cannot initiate new DNA chains, → RNA primers at the 5' ends are involved in initiation.

DNA replication arrest site — any of the specific nucleotide locations along the single-stranded DNA template at which → DNA replication by some prokaryotic and eukaryotic → DNA polymerases is interrupted. The DNA r. a. sites are particularly prominent in the termination regions.

DNA replication fidelity — the level of accuracy of DNA synthesis in each cell generation. High DNA r.f. ensures accurate transfer of genetic information to progeny cells and thus species identity.

→ DNA replication is an extremely accurate process. Mutation rates are typically 10^{-9} to 10^{-19} per base pair replicated. This high fidelity is thought to be achieved by a multistage mechanism: (1) Base selection in the original incorporation of the complementary nucleotide; (2) exonucleolytic proofreading of the newly added base; (3) postreplicative scanning of the DNA for mismatched bases (→ DNA mismatch repair).

DNA replication fork — an asymmetric structure in replicating DNA, consisting of a leading side where DNA synthesis is continuous and a lagging side where discontinuous replication occurs.

DNA replication unit — in eukaryotic chromosomes, the length of DNA estimated by the distance from the center of one autoradiograph to the center of the next sequence (→ replicon; chromosome).

DNA replication terminus — on → plasmid or chromosomal DNA, the position at which movement of the → DNA replication fork is either arrested or is severaly impeded.

DNA restriction fragment — any fragment of DNA originating by the action of → restriction endonucleases (→ DNA modification-restriction). —

DNA-RNA hybridization — a process based on the fact that isolated RNA on being mixed with denatured (single-stranded) DNA under appropriate conditions forms hydrogen bonds with the complementary DNA sequences (from which they were transcribed) leading to DNA-RNA hybrids. The rate of DNA-RNA hybridization is limited by the collision frequency of complementary nucleotide sequences. DNA-RNA h. is a useful tool for localizing genes in chromosomes (→ DNA-DNA hybridization).

DNase I hypersensitive site — (DH-site); in eukaryotic → chromatin, any of the short regions (50—400 bp) localized near the 5' ends of active genes as well as at internal and 3' regions which are hypersensitive to DNase I, micrococcal nuclease, S_1 nuclease, and specific restriction endonucleases (= → nuclease hypersensitive sites).

DNA sealase — = → DNA ligase.

DNA sequence amplification — = → DNA amplification.

DNA sequence analysis — = → DNA sequencing.

DNA sequence change — any of a variety of changes of nucleotide sequence, such as base substitutions, insertions, deletions, and rearrangements. DNA s. changes can affect genes or their regulation (→ gene mutation; DNA rearrangement). The majority appears to be neutral, i.e., they have little or no effect on the → phenotype. The → mutation rate is the underlying or basal rate of DNA s. changes.

DNA sequence family (Britten and Kohne 1968) — a group of very similar DNA sequences which are present more than once in each (haploid) genome of eukaryotes. DNA s. family members may lie in tandem or dispersed and the family may be divisible into subfamilies or segments according to the presence or absence of certain → restriction sites. They originate by → DNA amplification.

A heterogeneous family is one whose members are diverged from each other to different extents. The members of a homogeneous family are about equally diverged.

Sequence variation within a repeat family may reflect a process of homogenization caught in time such that continual rounds of homogenization establish or eliminate sequence variants (→ concerted evolution; multigene family).

DNA sequence interspersion — The interspersion, in highly ordered patterns, of single copy and middle repetitive DNA sequences in the genome of eukaryotes (→ DNA sequence organization).

DNA sequence organization — in eukaryotes, three general patterns of repetitive DNA sequences alternating with single-copy sequences. The salient structural difference between the patterns is the length of the repetitive and single-copy DNA sequences: (1) **Short period interspersion** (all major animal and plant phyla), i.e., the major fraction of the DNA contains repetitive sequences of average length of about 300 nucleotides; (2) **long period interspersion (Drosophila, Apis,** and some plants), i. e., long single copy regions (several thousand bp in length) are interspersed by long repeat sequences; (3) an intermediate interspersion pattern (chicken), i.e., about half of the genome contains repetitive sequences (about 2000 nucleotides in length) interspersed among single-copy sequences (about 4500 nucleotides in length). The remainder of the genome contains single copy sequences (of 17,000 nucleotides or greater) with no observed interruption by repetitive sequences.

In general, the long period pattern is characteristic of species with relatively small → genomes (~ 0.4 pg) and few repetitive sequences.

DNA sequence repetition — in eukaryotic genomes, the presence of different degrees of various types of → repetitious DNA sequences: (1) Highly repetitious tandem arrays of simple sequences or satellites (→ satellite DNA); (2) tandemly arranged moderately repeated sequences coding for structural RNA's or proteins; (3) interspersed repeated sequences (→ DNA sequence organization).

DNA sequence variation (Grula 1982) — the percent difference in sequence between individual genomes arranged for a population and averaged over all of the single copy DNA. The single copy DNA sequence heterozygosity is defined as the mean percent difference between the two genomes of an individual for all of the single copy DNA. Base substitution events lead to DNA s.v..

DNA sequencing — a routine analytical procedure (nucleotide sequencing) for investigation of the phylogenetic relationship of organisms, the divergence of → multigene families, and the evolution of gene structure (= DNA sequence analysis). The procedure consists of: (1) Selection of DNA clones; (2) treatment of these individually by base-specific reactions; (3) separation of the reaction products by size on sequencing gels; (4) computer-aided reading of separated bands as a sequence of A, C, G. and T bases.

Two widely used nucleotide sequencing methods involve basespecific chemical cleavage (Maxam and Gilbert 1977) or termination of in vitro DNA synthesis (Sanger et al. 1977). The former requires at least partial → restriction mapping and is relatively labor-intensive, the latter is easier and for most applications the method of choice.

DNA silencing (Sager and Kitchin 1975) — in eukaryotes, the selective inactivation or elimination of specific chromosomes, chromosome regions, or DNA molecules in the presence of unaffected homologues (→ chromosome imprinting).

DNA spacer — any extra DNA sequence associated with a → gene or group of genes: (1) Transcribed DNA spacers, i. e., sequences that are represented in the primary transcript, but not in the mature transcription product (tRNA, rRNA); (2) nontranscribed DNA spacers, i.e., sequences associated with the gene in question, which are either not transcribed at all or only infrequently.

DNA splicing — the preparation of foreign DNA for the → cloning vector by fragmenting the → genome or by reverse transcription of → messenger RNA to its duplex → complementary DNA (cDNA). Subsequently, the duplex DNA fragment is covalently joined to the cloning vector by → DNA ligase through one of several procedures: (1) Cohesion of overlapping ends generated by → restriction endonucleases; (2) cohesion of oligo dA and oligo dT or oligo dG and oligo dC tails (12 to 200 nucleotides in length) added to the 3' → flush ends by → terminal deoxynucleotidyl transferase; (3) joining of flush ends directly; (4) cohesion of chemically synthesized → overlapping ends up a restriction endonuclease recognition sequence, added by flush end joining (→ RNA splicing).

DNA steresis (Bloom and Leider 1962) — the loss of parts of the chromosome material after local irradiation with heterochromatic ultraviolet light. The irradiated segments stain but weakly or not at all when treated according to the Feulgen or methyl green-pyronine methods. Only a weak color reaction is obtained with the Alfert-Geschwind reagent for basic proteins. It is assumed that the effect is due to the lack of DNA containing material in the segment concerned.

DNA strand displacement — displacement of one strand from a duplex DNA by concomitant replacement with an equivalent nucleotide sequence. DNA s.d. is an integral aspect of → DNA replication and → genetic recombination. Branch points result in molecules subject to DNA s.d. where two strands compete for base pairing interactions with complementary sequences of the third strand. The movement of branch points in these molecules (branch migration) may proceed with or without the concerted action of specific enzymes and other proteins.

DNA supercoiling (Bauer and Vinograd 1968) — an altered state of DNA by induction of torsional stress in the reverse sense to that of the DNA helix (negative supercoiling). DNA s. is an intrinsic property of the DNA of → plasmids, viruses, mitochondria, plastids, bacteria, and eukaryotes; the level of DNA s. in bacterial cells is controlled by the activity of two enzymes with opposing functions: → DNA gyrase and → DNA topoisomerase 1. Supercoiling in eukaryotes may be introduced through the coiling of duplex DNA around → nucleosomes mediated by the → DNA nicking-closing enzymes. At sufficiently high negative superhelical density, torsional stress promotes untwisting of the DNA helix. This renders supercoiled DNA highly reactive to reagents that prefer single-stranded to double-stranded DNA.

DNA s. greatly affects the substrate properties of DNA molecules and plays an essential role in → DNA replication, → genetic recombination, → genetic transcription, and → transposition.

DNA swivel enzyme — any of the enzymes capable of untwisting supercoiled DNA (→ DNA supercoiling) by nicking and rejoining of a single-strand (= DNA nicking-closing enzyme of eukaryotes). These enzymes provide the swivel (switch point) necessary for the unwinding of parental DNA strands during replication of circular or very long linear DNA molecules.

DNA synaptase (Potter and Dressler 1978) — any of a class of enzymes capable of fusing duplex DNA molecules at a region of homology (genome fusion). In addition, the enzymes are able to catalyze the association of single-stranded DNA with homologous duplex DNA. DNA s. has properties of an enzyme participating in an early step in → genetic recombination.

DNA synthesis inhibition — a usually transient delay in normal de novo → DNA replication in consequence of induced → DNA damage. Depending on the inducing agent, such damage may block initiation, inhibit progression existing at growing points, or both. The normal rate of DNA replication may be restored when the → DNA lesions are repaired (→ DNA repair).

DNA topoisomerase (Liu and Wang 1978) — any of a ubiquitous class of enzymes that catalyze the breaking and rejoining of DNA strands (without altering the primary structure of DNA) in a way that allows the strands to pass through one another, thus altering the topology of DNA. DNA topoisomerases are implicated in processes such as → DNA replication, segregation of replicas, → genetic recombination, → genetic transcription, and → nucleosome assembly. Two classes of enzymes are distinguished on the basis of their mechanism of action: (1) Type I enzymes (nicking-closing enzymes) break and reseal one strand of DNA changing the → linking number in steps of one; (2) type II enzymes break and reseal both strands in a conterted manner changing the linking number in steps of two.

Both enzyme types can perform a variety of similar topological interconversions including relaxation of supercoiled DNA (→ DNA supercoiling), knotting/unknotting, and catenation/decatenation of duplex DNA (→ DNA catenane). Type I enzymes catalyze the removal of superhelical turns, the linking of single-stranded rings of complementary DNA sequences, and the interconversion between single-stranded rings with and without knots. Type II enzymes catalyze removal or introduction of superhelical turns, the linking of double-stranded DNA rings with or without gross sequential homology, and the interconversion between double-stranded rings with or without knots.

DNA t. of prokaryotes is larger than eukaryotic DNA t. and depends on Mg^{2+}.

DNA topology — the topological state of DNA (→ DNA supercoiling) which is a crucial parameter in processes such as DNA replication, genetic transcription, and genetic recombination.

DNA transfection — the introduction of exogenous DNA (genes) into cells by a variety of techniques (→ DNA-mediated gene transfer).

DNA transformation — a complex series of molecular events involving DNA transfer, integration, and expression of exogenous DNA (→ genetic transformation).

DNA transposition (Cohen 1976) — the (homology-independent) change of position of → transposable genetic elements from one site to another in the genetic material of prokaryotes and eukaryotes. The end result of DNA t. is the insertion of a defined DNA sequence between two base pairs in a recipient DNA molecule (chromosome). One feature of the reactions involved in DNA t. is the generation of a short duplication (3 to 12 bp) of the target site sequence at the ends of the transposed element. This is due to a staggered cut made in the target DNA whose protruding ends are joined to the ends of the transposed sequence. Most transposable genetic elements code for a single protein, called transposase, that is essential for DNA t.

DNA t. involves replication and recombination events, i.e., a copy of the transposable sequence remains in its original site, and another copy appears at the new site.

The end products of DNA t. (in bacteria) are of two kinds: (1) Cointegrates which consist of a fused donor and target molecule with duplicated copies of the transposable genetic element at the junctions; (2) simple inserts consisting of one copy of the transposed sequence inserted into the target molecule. Single inserts can be produced from a primary cointegrate structure by site-specific homologous recombination between the two copies of the transposed element, or directly, without extensive DNA replication, by a set of cutting and strand transfer reactions followed by gap repair.

Transposition events generate → insertion mutations which disrupt the integrity of the target DNA. Since transposable elements can carry transcription initiation and/or termination signals, they can alter downstream → gene expression and generate → deletion formation, → inversion formation and → replicon fusion.

The host is known to play a role in DNA t. in three different ways: (1) It provides target sites in recipient DNA; (2) it might provide some of the enzymes involved in DNA t.; (3) it presumably provides the enzymes generating the proper topology for the donor and recipient DNAs.

Two mechanisms are involved in DNA t. regulation: (1) Transposition immunity, in which an element located in a sequence of host DNA prevents the insertion of other elements of the same sequence family from insertion in cis nearby (prokaryotes); (2) transposition repression, in which elements produce transacting → repressors of transposition (prokaryotes and eukaryotes).

The transposition of certain eukaryotic transposable genetic elements, such as the → copia-like elements, the retroviral proviruses, and the dispersed repetitive → Alu sequences, move by a mechanism that involves → reverse transcription of messenger RNA (→ retroposon).

DNA tumor virus — any DNA virus capable of inducing tumors and differing from → retroviruses in some fundamental respects: (1) The transforming information of the DNA t. viruses is essential to the viral replication cycle while the retroviral transmitted → oncogenes represent a burden as far as the viral reproductive cycle is concerned; (2) DNA t. virus replication is incompatible with cell survival, i.e., the transforming information must be an early product of the viral genome, while the productive cycle of the retroviruses does not interfere with cell multiplication.

DNA turnover — any mechanism of nonreciprocal exchanges of DNA sequences (e.g., → gene conversion, → DNA transposition, unequal → crossing over, and RNA-mediated retroposition) that cause continual fluctuations in the copy number of DNA motifs during the lifetime of an individual. Such mechanism are potentially capable of accelerating, retarding, or maintaining constant the rate of DNA differentiation between populations (→ molecular drive).

DNA twisting — the bending of the DNA duplex required in numerous situations. These include supercoiling of covalently closed DNA circles (→ DNA supercoiling), wrapping DNA in a tight helical coil about the → histone core of a → nucleosome, compacting DNA inside the bacteriophage head, and cyclization of DNA molecules with → cohesive ends (→ DNA-untwisting enzyme).

DNA typing — the characterization of genetic variation at the DNA level by various means, such as determination of amplified DNA fragment length differences, hybridization with allele-specific oligonucleotide probes, and direct → DNA sequencing.

DNA-untwisting enzyme (Champoux and Dulbecco 1972) — = → DNA relaxing enzyme.

DNA unwinding — the local unwinding of the DNA helix which plays an important role in many biological processes.

DNA-unwinding protein (Alberts and Frey 1970) — in prokaryotes and eukaryotes, any of a class of proteins (= single-stranded DNA binding protein; DNA-melting protein) which preferentially bind to and stabilize single-stranded DNA regions. They play a role in → DNA replication, → genetic recombination, and → DNA repair.

dNdP transferase (Nazarenko 1978) — an enzyme that catalyzes poly (dG)·poly (dC) synthesis through the formation of complementary homooligonucleotides which serve as templates and primers for → DNA polymerase I.

DNP — acronym for **d**eoxyribo**n**ucleo**p**rotein.

D_{null} — the radiation dose required to reduce the cell population (or other properties of the biological system) by $^1/_e$ (or 37%) on the exponential part of the dose-effect curve.

Dollo's law — the generalization that the direction of → evolution cannot be reversed.

domain — (1) a discrete structural entity of chromosomes, i.e., a region within which supercoiling is independent of other domains; (2) a discrete continuous part of the amino acid sequence of a protein or the nucleotide sequence of a nucleic acid molecule that can be equated with a particular function.

dominant (Mendel 1866) — of genetically controlled → characters and the corresponding → alleles which are manifest in all → heterozygous, heterokaryotic or heterogenotic members of the first filial generation (F_1) from a cross between two homozygous (pure-breeding) strains differing in respect to these characters. The dominant character is evident in three-quarters of the individuals of the second filial generation (F_2) in diploid sexually reproducing organisms (→ segregation). Those characters and alleles being masked by the dominant ones are referred to as "recessive".

"Dominance" and "recessiveness" are not properties of the genes per se, but the result of the action of the genetic locus in question within the total reaction system of the particular → genotype. Both may be modified in degree by environmental influences as well as by other genes ("dominance modifiers") of the genotype. These modifications are brought about by influences on the gene-controlled processes and not on the genetic material proper. In genetic nomenclature, recessive characters and genes (alleles) are generally symbolized by small letters, the corresponding dominant ones by the same letter or letters, but beginning with a capital.

Complete dominance and complete recessiveness are the extreme cases between which all transitional "degrees of dominance" are possible.

Degree of dominance is the ratio of deviation of a heterozygote to that of the dominant homozygote from the midparental value. It is a most important genetic parameter in studying quantitative inheritance concerned with → heterosis and represents a measurement of the mean effects of genes. (1) **Complete dominance:** The heterozygous form (Aa) is phenotypically identical with the homozygous form (AA); (2) **semi-dominance** (= partial dominance; incomplete dominance): the phenotype of the heterozygous form (Aa) lies between that of AA and aa, i.e., it is "intermediate"; (3) **conditioned dominance** (Goldschmidt 1938) (= irregular dominance): the characters and alleles concerned show variable manifestation in the heterozygous state, i.e., in one genotypic environment and under certain environmental conditions they are dominant, under others they are recessive; (4) **conditional dominance:** the dominance of the characters and alleles in question can be observed in the heterozygous state but the homozygous phenotype is not known; (5) **alternating dominance** (= **change of dominance; reversal of dominance):** In the course of ontogenetic development of a heterozygous individual the d. effect of one allele of a heterozygous allele pair may be transferred to the other ($A_1a_2 \rightarrow a_1A_2$). In such a case, the alleles are manifest one after the other at various phases of development of the organism, i.e., first one allele is d., and then the other. Alternating dominance can also occur in the phylogeny of some species so that the dominance relationships change from generation to generation. Numerous external and internal factors may cause this phenomenon, e.g., changes in genotypic environment, → position effects, polyploidy, labile genes, etc.: (6) **delayed dominance:** the delayed manifestation of the dominant allele of a heterozygous allele pair so that the alternative allele only relinquishes its phenotypic action at a later stage of ontogenesis; (7) **shifting dominance:** variable dominance of an allele occasioned by external influences; (8) **sex-influenced dominance:** the dominance of the allele concerned differs in the two sexes.

Dominance modifiers (Fisher 1928) or **domini genes** (Goldschmidt 1935) are genes influencing the degree of dominance of homozygous and/or heterozygous nonalleles and their corresponding phenotypic characters.

dominant lethal — any → mutation (mainly structural and numerical chromosome aberrations) which kills an individual heterozygous for it. It is not possible in any particular case to confirm its genetic nature or to allocate it to a particular chromosome by tests such as are applicable to other kinds of mutant genes (→ lethal factor).

dominant lethal assay (Russell 1951; Bateman 1958) — in → mutagenicity testing, a procedure for the detection of germinal mutations causing death of offspring as early or mid-term embryos. The mutational events include major aneuploidies, especially all autosomal monosomics, large deletions, and duplication-deletion types that act after fertilization but early in embryonic development, i.e., before, at, or shortly after implantation.

dominant negative mutation (Herskowitz 1987) — a mutation (produced by manipulation of a cloned gene) encoding a polypeptide that when overexpressed disrupts the activity of the wild-type gene. Its phenotype is manifested in the presence of the wild-type gene (dominance) and inactivates the wild-type allele (negative). Use of d.n. mutations can provide information on the in vivo function of a diverse set of cloned genes and gene segments.

donation (Clark and Adelberg 1962) — in bacterial → conjugation, a process that involves two different → plasmids, a nonconjugative and a conjugative one, the latter providing (without physical contact) the missing conjugative function to the other (the mobilizable plasmid) so that this one can be transferred (→ conduction).

donor conjugal DNA synthesis — in bacterial → conjugation, DNA synthesis in donor cells; it serves to replace the single-strand being transferred. DNA synthesis in the recipient cell is called recipient conjugal DNA s. and converts the transferred single-stranded to double-stranded DNA. Subsequently, circularization of the molecules occurs.

donor splicing site — the boundary between an → exon and an → intron (the left splice junction). It consists of a highly conserved, nine-nucleotide sequence (→ acceptor splicing site).

dosage — → gene dosage.

dosage compensation (Muller et al. 1931) — a mechanism that ensures that the activity of X-linked or Z-linked genes is made equal in the two sexes of organisms with an XX compared to XY or ZZ compared to ZW basis of → sex determination.

In **Drosophila**, both X chromosomes are simultaneously active in female somatic cells; d.c. operates at the level of transcription. Mammalian females achieve d.c. by inactivating, randomly, one of the two → X chromosomes.

dosage effect (Stern 1929) — (1) the quantitative differential action of the → alleles of a gene on the phenotypic expression of the corresponding characters. The extent of the genetic dosage effects is correlated with the frequency of the particular allele in the genotype. — (2) The deviation of phenotypic expression from → dosage compensation towards an effect proportional to the gene dose.

dose fractionation — the application of a radiation dose in two (or more) fractions separated by a certain minimal time interval. D.f. tends to reduce the total effect compared with application of the entire dose in a single exposure of short duration. This difference is due to → DNA repair processes.

double crossing over (Sturtevant 1914) — → crossing over.

double diploid (Warmke and Blakeslee 1939) — → alloploid cells and organisms with the diploid chromosome complement of two diploid species (= amphidiploid; allotetraploid) (→ double haploid).

double fertilization (Navashin and Guignard 1899) — → fertilization.

double haploid (Warmke and Blakeslee 1939) — → alloploid cells and organisms with two chromosome sets, each from one species (→ double diploid).

double minute (Mark 1967) — (DM); any of the small extrachromosomal fragments associated with particular tumor types and observed only in neoplastic cells. DM's result from → DNA amplification, are acentric structures and highly variable in number from cell to cell (→ homogeneously staining region). At mitosis they appear as paired or unpaired spheres (0.3 —0.5 μm in diameter).

double monoisosomic (Kimber and Sears 1968) — a cell or individual which lacks one chromosome pair but has two isochromosomes, one for each arm of the missing chromosome pair (→ monotelomonoisosomic).

double monotelosomic (Kimber and Sears 1968) — a cell or individual in which one chromosome pair is missing but two telocentrics, one for each arm of the missing pair, are present (→ monotelomonoisosomic; ditelomonotelosomic).

double reduction (Darlington 1929) — → chromatid segregation.

double-strand break — (dsb); loss in continuity of the phosphate sugar backbone in two adjacent strands of the DNA duplex such that these breaks are no more than five to ten base pairs apart.

double-strand break repair — a → DNA repair process that involves recombination functions and repairs → double-strand breaks in DNA (→ recombinational repair; postreplication repair).

doublet — an especially thick → band in → polytene chromosomes, probably formed by a → duplication (one-band tandem repeat). Doublets are presumed to be composed of a number of recombinationally separable subunits with a common or related information content.

double telotrisomic (Kimber and Sears 1968) — an individual deficient in one chromosome but which has two telocentric chromosomes, one for each arm of the missing chromosome (= pseudotrisomic).

doubling dose — the radiation dose that would double spontaneous mutation rates. It is used for estimates of genetic risk from ionizing radiations.

downstream — term used to describe the relative positions of sequences on a nucleic acid or protein molecule. D. means in the direction in which a nucleic acid or protein molecule is synthesized, i.e., on the 3' side of any given site in DNA or RNA, and on the C-terminal side of any site within a polypeptide.

With reference to the direction of → genetic transcription, in the 3' direction on the non-template strand, but in the 5' direction on the template strand.

Dq — the radiation dose required before the dose effect curve becomes exponential – often referred to as the "shoulder" (→ D_{null}).

drift (Wright 1921) — → genetic drift.

drift-region (Blattner et al. 1972) — the area between the start site and the entry site(s) where DNA-bound → RNA polymerase can move without producing RNA chains by → genetic transcription. The start site is the nucleotide sequence which triggers RNA synthesis at the start point nucleotide (→ promoter).

D-RNA (Georgiev 1961) — in eukaryotes, RNA species with nucleotide sequences complementary to nuclear DNA sequences and not representing → ribosomal RNA or → transfer RNA (D-RNA = DNA-like RNA) (→ pre-messenger RNA; heterogeneous nuclear RNA).

drug resistance factor — = → R plasmid.

drumstick (Davidson and Smith 1954) — the → sex chromatin of polymorphonuclear neutrophil leukocytes of human blood which is contained in a specific drumstick-like nuclear appendage that has its head (about 1.5 μm in diameter) attached to the nucleus by a thread-like stalk. Drumsticks differ from the sex chromatin of other cells in being extruded from the nucleus and in being visible in only a relatively small proportion of cells (in about 1 in 40 neutrophiles in normal females and in less than 1 in 500 cells in normal males).

dsRNA — acronym for **d**ouble-**s**tranded **RNA** (ssRNA: single-stranded RNA).

Dubinin effect (Stern and Kodani 1944) — a form of variegated-type → position effect which leads to the reduction or inactivation of the action of a normal allele and is conditioned by translocated → heterochromatin in the immediate vicinity of the allele. This type of position effect was first found for the cubitus interruptus gene in *Drosophila* by Dubinin and Sidorov (1935)

duplex (Blakeslee et al. 1932) — of a polyploid having two dominant alleles for a given genetic locus (as AAa in triploids, AAaa in tetraploids) (→ nulliplex).

duplicate genes (Shull 1914) — (1) two identical genes showing the same phenotypic action but localized in different chromosomes. D.g. are of relatively frequent occurrence and are probably due to secondary polyploidy (→ polyploid). In gene formulae, the d.g. are generally given the same symbol but characterized by different suffixes or subscripts.

For the case of heterozygosity and complete dominance, A_1 is dominant over a_1 and A_2 (identical with A_1), over a_2(identical with a_1). Since A_1 and A_2 have the same phenotypic action, the combinations A_1a_2 and A_2a_1 are phenotypically equal and different from a_1a_2. For these reasons, the normal F_2-segregation (9:3:3:1) does not occur; instead the segregation ratio is 15:1 provided that the dominant alleles of the d.g. do not act additively or cumulatively ("cumulative d.g."). In this case, the 15:1 segregation becomes 9:6:1 [$^-9A_1 \cdot A_2 \cdot$:6($3a_1a_1A_2$. and $3A_1 \cdot a_2a_2$):$1a_1a_1a_2a_2$].

By analogy with d.g., the presence of three identical allele pairs is referred to as triplicate, of four identical allele pairs as quadruplicate genes, and more than four allele pairs generally as polyplicate genes (→ gene interaction).

(2) Homologous genes present in more than one copy per → genome. D. genes originate by → gene duplication (paralogous genes) rather than by speciation. Three categories with respect to → gene expression in a given species and for a given enzyme: (a) Nondivergent expression, i.e., the d.genes are equally expressed in all tissues in which the enzyme is expressed; (b) unidirectionally divergent expression, i.e., the d. genes are expressed unequally in one or more tissues; (c) bidirectionally divergent expression, i.e., there is no consistent predominance of one locus product over another among the various tissues.

duplication (Bridges 1919) — a chromosomal structural change (→ chromosome mutation) resulting in the doubling of a section of the genome of pro- and eukaryotes. The size of the duplicated segment may vary considerably.

D. mutants of prokaryotes contain a repeat of part of the parental genome. They may be direct repeats, in which the duplicated sequence has the same informational polarity as the original, and inverted repeats, in which it does not.

Duplications are considered to play a role in origin of new genes through functional diversification of duplicate members (→ gene duplication).

Cytologically, the existence of relatively small duplications in the chromosome complements of eukaryotes may be proved by the appearance of regional alternations of pairing (→ chromosome pairing) in first meiotic prophase and (somatically paired) → polytene chromosomes of Diptera in that member of a pair of homologous chromosomes which is heterozygous for the duplicated segment. Where whole chromosomes (linkage groups) are doubled in the haploid chromosome set, this constitutes a → genome mutation (→ aneuploid). Duplications may occur, inter alia, as primary structural changes of chromosomes, as a result of disturbances in the crossing over process (unequal → crossing over), or as secondary chromosome mutations due to crossing over in inversion- or translocation heterozygotes.

A distinction is made between "inter- and intrachromosomal duplications" according to whether the duplicated chromosome segment is incorporated in another chromosome of the → chromosome set or whether the duplicate region is present in the same chromosome.

(1) **Interchromosomal duplications:** the duplicated segment is incorporated into a non-homologous chromosome or it occurs as a fragment in the chromosome set. (2) **Intrachromosomal duplications:** the duplicated segments are contained within a single chromosome and may be non-contiguous, i.e., separated by other segments, or contiguous, i.e., immediately adjacent (= repeats or tandem-duplications). Also, they may occur in the same chromosome arm ("intraarm duplications"), or in different ones ("interarm duplications"), with the same order of loci with respect to the centromere as in the original ("direct duplications"), or with the order inverted ("reversed" or "inverse duplications"). In nature, the most frequent type o d. is the tandem d. or repeat.

The genetic consequences of duplications depend on the genetic information which they contain and the change in the gene balance effected by them. When homo- and heterozygous, they may cause an increase or decrease in the viability of their carriers and in extreme cases may act as lethals. Under evolutionary conditions, small duplications may provide a basis for the mutational differentiation of genetic material.

"Complementary duplication-deficiencies" occur in the case of unbalanced gametes (e.g., AB/CB and AD/CD) produced as a result of translocation heterozygosity (→ translocation) and → adjacent distribution if the chromosome segment duplicated in the one gamete is lacking in the other and vice versa. The gametes mutually complement each other and their combination in the zygote $\frac{AB}{AD}\frac{CB}{CD}$ leads to viable offspring.

duplication mapping — like → deletion mapping, a gene dosage method for → genetic mapping. An individual → trisomic for part or all of a chromosome has about 50% more of a particular gene product and this allows to locate some genes.

duplicon (Sibatani and Hiai 1964) — = → replicon.

durability (Dobzhansky 1968) — the probability that a → unit of evolution (i.e., a strain, Mendelian population, or a species) will leave descendants after a given period of time.

dyad (Nemec 1910) — (1) a pair of cells resulting from the first meiotic division or from an aberrant course of both meiotic divisions; (2) the two chromatids connected at the → centromere which make up one chromosome in first meiotic division (*Nemec* 1910).

dyscentric (Darlington 1926) — of → chromosome mutations (e.g., inversions, inverted duplications) which involve the rotation of a chromosome segment and its constituent genes through 180° with reference to the position of the → centromere (→ eucentric).

dysgenic — tending to be harmful to the hereditary qualities of a species (→ eugenic).

dysploid (Tischler 1937) — ref. to differing basic chromosome numbers within populations or species. "Dysploidy" is of relatively frequent occurrence among seed plants such that within a group of individuals under investigation there occur a series of basic chromosome numbers which usually differ by one chromosome (e.g., x = 5; 6; 7; 8; 9; etc.). In contrast to populations with → B-chromosomes and → aneuploid derivatives of normal diploid populations, dysploid basic numbers are stable within the group of individuals. Frequently, dysploidy is accompanied by more or less drastic structural changes of the chromosomes concerned.

E

early enzyme — any of the bacteriophage-induced enzymes (= early protein; phage-induced enzyme) that are synthesized de novo after phage infection under the direction of the invading viral genome. Early enzymes direct the synthesis of phage DNA or its precursors and are of two types: (1) An enzyme activity which was absent in the uninfected bacterial cell; (2) a protein synthesized after infection having an enzyme activity very similar or identical to an enzyme found in uninfected cells, whose rate of reaction is many times greater than that of the host enzyme or with different properties or requirements.

early gene — any of those bacteriophage or virus genes expressed before DNA synthesis in the host cell begins (→ late gene).

echylosis (La Bella and Krass 1968) — the release of nonparticulate material from the cell through an apparently intact → cell membrane (→ inchondriosis; inchylocytosis).

eclipse — the decrease or disappearance of infective virus particles which accompanies the release of the viral genome into a host cell.

ecliptic period (Luria 1950) — the period (= dark period) which elapses between the infection of a bacterium with bacteriophages and the occurrence of the first infectious phage offspring. It corresponds to the first half ot the so-called latent period.

ecoclines — grades of variation within a species. They occur as reactions to different ecological zones occurring within the species distribution (→ cline).

ecogenetics (Brewer 1971) — a branch of → genetics studying genetically determined variation in the response of individuals to environmental agents.

ecological density — the number or mass of individuals per area or volume of habitable space.

ecological genetics — a branch of genetics that analyses evolution in natural population and examines the adjustments and adaptations of wild populations to their environment.

ecological niche (Elton 1927) — a multidimensional space representative of the environment in which a species exhibits positive fitness (Hutchinson 1958), i.e., the position of a species in a community of organisms, based on its adaptive requirements and functions.

ecological race — any of a number of genetically and phenotypically different but interfertile races (= ecotypes) of a species which are adapted to different habitats in the same territory.

ecophene (Turesson 1922) — the range of → phenotype produced by one → genotype within the limits of habitat under which it is found in nature (Darlington and Mather 1949).

ecophenotype — any nongenetic modification of the → phenotype in response to an environmental condition.

ecospecies (Turesson 1922) — a group of → ecotypes capable of interbreeding without loss of fertility or vigor in the offspring.

ecosystem — the complex of living an nonliving components in a specified location that comprise a stable system in which the exchange of material follows a circular path (→ biome).

ecotone — → biome.

ecotopic recombination — recombination between tandem or dispersed non-alleles (= nonallelic recombination) which include members of → multigene families that share homology but not chromosomal location.

ecotropic (Levy 1974) — → xenotropic.

ecotype (Turesson 1922) — a local race (= ecological race) with genotypes adapted to a particular restricted habitat as a result of natural → selection within the local environments. Ecotypes of the same → ecospecies are interfertile and produce fertile offspring.

ectodesma (Lambertz 1954) — any of the → plasmodesma-like structures in the outer epidermal cell walls of many plants. They extend from the cell lumen to the undersurface of the cuticle.

ectogony (Waller 1917) — → xenia.

ectopic pairing (Slizinski 1945) — the tendency of specific regions (→ bands) of *Drosophila* → polytene chromosomes to stick to other regions in different chromosomes.

ectoplasm (Haeckel 1873) — the external, peripheral layer of the → cytoplasm of a cell which is more rigid and devoid of granules (= cortex).

eduction (Sunshine and Kelly 1971) — the loss of host genetic material concomitant with the exit of a → plasmid (episome) integrated into the host chromosome. The resulting deletion-containing cells are called eductants.

effective breeding population (Wright 1931) — the effective number of breeding individuals of a population (= effective population size). (→ breeding population).

effectively dominant (Muller 1950) — → dominant.

effectively neutral mutation — any → mutation for whose fate stochastic events are more important than selection (→ non-Darwinian evolution).

effector (Jacob and Monod 1961) — (1) in genetics, a small molecule (metabolite) that interacts with the → repressor of an → operon, activating or inactivating the repressor molecule with respect to its ability to combine with the → operator. In the case of enzyme induction, "positive effectors" or "inductors" decrease the activity of the repressor. In the case of → repression of enzyme formation, "negative effectors" or "corepressors" enhance the activity of the repressor and inhibit the synthesis of the repressible protein. (2) In enzymology, a regulatory metabolite of an enzyme (= modifier or modulator) that typically modifies the affinity of an enzyme for its substrate and frequently for other reaction components as well. Effectors need not have similarity with the substrate of the enzyme whose activity is influenced. In some cases, the substrate has been found to act as an activator. This situation is referred to as substrate co-operation. A positive e. increases the activity or binding of a second ligand, a negative e. decreases it (→ allosteric).

egg — in the case of oogamy, the female → gamete specialized for the storage of food, as opposed to a sperm (male gamete) which is specialized for motility and fertilizes the e. in case of sexual → reproduction. Its cytoplasm is ordered in the form of a gradient system. Certain cell particles can be enriched and secreted in specific regions and a pattern may be discernible in the → cortex which is decisive for the further course of development (→ predetermination).

egg mother cell — a megasporocyte (symbol EMC) from which an → egg cell is derived during → megasporogenesis in higher plants.

elaioplast (Wakker 1888) — a → plastid in which oil predominantes as the storage product. Elaioplasts are specialized leukoplasts.

electroinjection (Morikawa et al. 1986) — the direct introduction, by means of electric- field impulses, of foreign DNA into intact (plant) cells.

electromorph (King and Ohta 1975) — any of the electrophoretically distinguishable allelic forms in a population or inbred strain.

electron transport particle — a particle (symbolized ETP) derived from → mitochondria, and capable of carrying out electron transport from suitable substrates to oxygen.

electroporation-mediated gene transfer — → gene transfer into cells or protoplasts by means of electroporation (application of high-intensity electric fields of short duration to reversibly permeabilize biomembranes).

elementary membrane (Sitte 1961) — = → unit membrane.

elementary particle (Fernandez-Morán 1963) — (EP or F_1 particle); any of the regularly spaced (intervals of 10 nm) particles covering the inner membrane of the → mitochondrion. The EP's represent a special ATPase involved in the coupling of oxidation and phosphorylation and have a stem linking them with the membrane.

Particles similar in dimensions to the EPs have been isolated from → chloroplasts and are designated → quantasomes.

elimination chromatin (Seiler 1914) — → chromatin elimination.

elimination coefficient — the frequency with which certain genotypes (carrying specific genes or gene combinations possessing deleterious effects) die prematurely or are hindered during reproduction and are genetically eliminated as a consequence (→ genetic death). If, for example, a certain gene has an average e.c. of 5% then this means that of 20 individuals possessing this gene among their genotypes, one will perish before the gene concerned can be transmitted to the offspring. Under these conditions, the gene will survive in 20 generations before it is eliminated. Its → persistence is 20 generations.

elongation factor (Lucas-Lenard and Lippmann 1966) — (EF); any of the specific protein factors involved in the elongation of polypeptides during → genetic translation in prokaryotes and eukaryotes: they associate with → ribosomes cyclically, for addition of each amino acid to the polypeptide chain. The bacterial e. factors EF-G, EF-Tu, and EF-Ts vary in content with growth rate and ribosome content; they are produced in coordination with ribosome synthesis.

The eukaryotic factors that in terms of function roughly correspond to *E. coli* factors are the following:

Eukaryotes	*E. coli* factors
eEF 1α	EF-Tu
eEF1β	EF-Ts
eEF1γ	None
eEF 2	EF-G

The reactions that are catalyzed in eukaryotic protein synthesis do not exactly correspond to those in prokaryotic protein synthesis.

EF-Tu and EF-G interact with the → ribosomes during protein synthesis: EF-Tu presents incoming aminoacyl tRNA to the programmed ribosome as an EF-Tu-GTP-tRNA ternary complex and EF-G promotes translocation of peptidyl-tRNA and its associated mRNA from the A to the P site after peptidyl transfer. Both events are accompanied by ribosome-dependent GTP hydrolysis.

Chloroplasts and mitochondria contain ET's different from those present in the cytoplasm. The chloroplast e. factors (EF-G_{chl} and EF-Tu_{chl}) are synthesized in the organelle, while the mitochondrial EFs are synthesized in the cell cytoplasm and the genetic information for these factors resides in the nuclear DNA.

elongator transfer RNA — any → transfer RNA molecule used for inserting an amino acid into internal peptide linkages, as distinguished from → initiator transfer RNA.

embryo — in animals, the young organism arising from the fertilized → egg cell. In seed plants, the young → sporophyte arising from the union of the female and male sex cells (→ embryonic development). The origin, development and evolution of the e. is referred to as "embryogenesis" or "embryogeny"

embryoid — an embryo-like, well-organized structure originating from immature plant embryos, inflorescences and/or leaves cultivated in vitro (→ somatic embryogenesis).

embryonic determination — the limitation of the developmental capacity that occurs at an early stage of the → embryonic development; once established, e.d. is transmitted through all cell divisions.

embryonic development — the canalized process of histogenesis of multicellular organisms whereby fertilized egg cells (zygotes) are transformed into a restricted number of alternative states among which there are few if any intermediates. During e.d. the cells arising from the zygote by mitosis become different (→ induction) from one another although they are endowed with identical genomes (→ cytodifferentiation). Programming for e.d. begins as early as oogenesis and the cells of an embryo differ from one another at their earliest delimitation. Each of the canalized pathways of e.d. is called a "**creode**" by Waddington (1962), and is the result of essentially all the available gene-reaction systems whose intensities are mutually adjusted by interlocking control systems (→ genetic regulation).

E.d. is a highly complex process during which an ordered sequence of genetically programmed cellular commitments progressively orchestrates the development of a multipotent egg into a complex organism. The expression of the genetic program relies on the availability (in time and space) of adequate environmental → epigenetic factors. The interaction ensures initiation, amplification, and stabilization of each differentiated state.

Although the chromosomal information is similar in both gametes, the oocyte also provides a structural frame (determining embryonic axes), determinants in the → cortex and a metabolic pool to support early embryonic events. In the first phase of e.d., the various parts of the germ come to exhibit clearly apparent differences ("regionalization", "embryonic segregation" or „arealization"). By a second process (histogenesis) any one particular region gradually changes its character. A third major process consists in the molding of a region of the egg into a definite shape and is called "morphogenesis".

E.d. involving all three major processes may be called "individuation" since it confers on the developing material the particular and specific individual character of the organ into which it is developing.

Three major types of developmental systems may be distinguished: (1) Those depending on segregated ooplasms, (2) those depending on gradient systems, and (3) those depending on inductive relationships.

Early e.d. begins with → fertilization and proceeds by a series of stages (cf. Waddington 1956):

(1) **Fertilization:** In this stage, two processes are important for the inception of e.d. in sexually reproducing organisms: The union of the haploid egg nucleus with the sperm and the activation of the egg leading to the start of division. The processes are separate from one another and it may happen that activation occurs without fusion of the nuclei. (2) → **Cleavage:** The egg cell divides into ever smaller descendants by → mitosis. In this process the eggs of each species follow a quite definite cleavage pattern at least for the first three or four cleavages (some species considerably longer). The position of the cleavage planes is largely a consequence of the arrangement of the mitotic spindles which in turn is controlled by local differences in the egg cortex, including the organization of the cytoplasm. (3) **Blastulation:** The newly formed cells migrate to the periphery of the „cell cluster" producing in the simplest case a cavity filled with liquid and surrounded by a cell layer – the blastula. This marks the end of cleavage period. Depending on the amount of yolk, the blastula may be formed asymmetrically. In eggs very rich in yolk, e.g., bird's eggs, it may be reduced to a flat plate of cells – the blastoderm – which swims to the upper pole of the yolk mass. (4) **Gastrulation:** This takes place immediately after the formation of the blastula. During this short and extremely critical period of e.d., the different regions of the blastula are so folded and pushed into each other that the actual embryo is produced, generally consisting of three separate layers (in Coelenterata and lower forms, only two layers – an outer and an inner). These three fundamental layers are called the **ectoderm, endoderm,** and **mesoderm.** The ectoderm is the outermost layer, from which the skin and nervous tissue originate. The inner layer, the endoderm, gives rise to the intestines and their appurtenances, while the middle layer, the mesoderm, builds the muscles, skeleton, etc..

In the different groups of organisms, the folding which brings the individual layers into their correct positional relationship may differ greatly, but apart from this, the formation of an outer, a middle, and an inner layer is common to all animal species.

During gastrulation a new and widely distributed cell differentiation begins. From this point onward the e.d. is organized by the controlled activity of the cell-borne genetic apparatus. Although the genome of the zygote and the blastula cells stemming from it show a slowly increasing level of activity throughout cleavage and blastulation, the indispensible nature of embryonic gene action is not observable prior to gastrulation. Gene activation sets in suddenly in a short period preceding the onset of gastrulation and most of the RNA synthesized during this activation appears to be → messenger RNA (Bacharova et al. 1966). Likewise, ribosomal RNA occurs shortly after the beginning of gastrulation. Nevertheless, the ribosomal RNA necessary during cleavage and blastulation is synthesized very much earlier (mainly during oogenesis, as has been shown in studies with → lampbrush chromosomes). (5) **Formation of basic organs:** Very soon after gastrulation, the fundamental pattern of the embryo begins to emerge. In most cases those organs are produced which remain during subsequent stages of development and form the most important organs of the adult animal. However, in some animals the embryo first develops into a larva in which organs are produced which will undergo further radical changes during the transition to the adult animal, so that it is very difficult to establish a general scheme. The e.d. of the vertebrates proceeds more or less according to the following scheme. First, the ectoderm forms the epiderms covering the entire body. A thickened plate, differentiated from the epidermis, folds itself into a groove and finally becomes enclosed into a tube forming the central nervous system. Cells from the ectodermal sheet inside of the embryo migrate to the juncture between neural and epidermal parts of the ectoderm and there form nerve ganglia and other organs. The mesoderm sheet is divided lengthwise in a series of individual segments. Along the middle of the embryo there forms a long rod-shaped structure, the notochord, the first skeletal element to form. The mesoderm thickens on either side of the neural tube and becomes transversely segmented so that a series of block-shaped tissues (the somites) are formed. These form the main musculature and the innermost layers of the skin. At each side of the somites, nephri or kidneys are produced from a mesoderm segment. Within the embryo the endoderm is folded into a tubular structure from which the gut and the digestive system arise during the further course of development. The formation of this organ always begins earlier at the anterior than at the posterior end of the embryo.

embryonic differentiation — differentiation resulting in embryonic development.

embryonic fate map (Vogt 1929) — a map which basically illustrates the developmental fate of cells in a primordium. The construction of an e.f.m. typically involves marking presumptive migrating cells so that their movements, development and differentiation can be followed during ontogenesis.

embryonic field (Weiss 1939) — a chemo-differentiated region of embryonic tissue during early → embryonic development. The development of any given point of an e.f. depends on its relations with the other points in its neighborhood or on its position within the region as a whole. The term "field" is used to refer to the coordinated and integrated character of a whole complex of processes proceeding within such a region. An e.f. giving rise to the formation of a definite organ with characteristic individual shape may be referred to as „individuation field" (Waddington and Schmidt 1933).

embryonic induction (Spemann 1901) — a phenomenon, due to interaction between one (inducing) tissue and another (responsive) tissue, which gives rise to alternations of the developmental course of the responding tissue.

embryonic lethal — → lethal factor.

embryo sac (Hofmeister 1849) — the mature female gametophyte in angiosperms (→ megasporogenesis).

embryo sac competition — → Renner effect.

embryo sac mothercell — → megasporogenesis.

emiocytosis (Lacy 1961) — the process by which cellular constituents are released from cells (= reverse → pinocytosis or exocytosis).

empiric risk — in human genetics, the risk that a certain child will have the condition or trait under study in those cases where genetics is obviously involved but no definitive pattern of inheritance can be demonstrated.

end labeling — the addition of a radioactively labeled group to one end (5' or 3') of a DNA strand.

endochromocenter (Geitler 1953) — an enlarged → chromocenter produced by → endomitosis and the concomitant increase in chromosome number.

endocytobiosis — intracellular symbiosis in which the symbionts live within host cells. The larger partner is called the host, and the smaller one the symbiont.

endocytosis (de Duve 1963) — the internalization, by a large variety of eukaryotic cells, of extracellular fluid or particles. E. occurs by invagination and pinching off of the plasma membrane and serves an important role in uptake of nutrients, scavenging of extracellular material, and in the internalization of receptor bound ligands such as hormones, growth factors, lipoproteins, and antibodies. Internalization of large particles is called phagocytosis, internalization of small particles, solutes, and fluid is called pinocytosis.

Adsorptive e. is the process by which cells selectively concentrate and internalize ligands that bind to plasma membrane receptors.

Fluid phase or bulk e. is a constitutive process not modulated by extracellular ligands or by other inducers; it allows molecules without affinities to plasma membrane moieties to enter cells.

Receptor-mediated e. is the internalization of a heterogeneous class of physiologically active ligands which bind to a relatively small number of cell surface receptors; the receptor-ligand complexes accumulate in → coated pits of the plasma membrane.

endoduplication (Jorgensen 1928) — the regular doubling of the haploid chromosome number in andro- or gynogenetic development owing to the failure of cell-wall formation in the first division of the zygote and the union of the spindles in the subsequent mitosis.

endogamy all systems of sexual → reproduction (= inbreeding) in which the mating partners are more closely related than in random mating (panmixia) within the population (→ mating system).

endogenote (Morse et al. 1956) — that part of the bacterial chromosome which is homologous to the genome fragment (the "exogenote") transferred from the donor into the recipient cell during the formation of a → merozygote. A bacterial cell which contains an "exogenote" in addition to its genetic material is referred to as a "syngenote" and is hyperploid for the fragment concerned. The syngenote may be → heterogenotic or homogenotic.

endomembrane system (Morré et al 1971) — the functional continuum of membraneous cell components consisting of the → nuclear envelope, → endoplasmic reticulum and → Golgi apparatus as well as vesicles and other structures such as → annulate lamellae derived from the major components. All these are local specializations of a single interassociated membrane system peculiar to eukaryotic cells.

endomitosis (Geitler 1939) — in eukaryotes, a form of somatic polyploidization (→ polyploid) which is fairly common in differentiated or differentiating tissues. It takes place within an intact → nuclear envelope and gives rise to endopolyploidy, i.e., duplication of the chromosomes without cell or nuclear division in the usual sense. E. is characterized by an increase in DNA content.

In some organisms, e. has been found to resemble a normal mitotic cycle. Endoprophase is followed by endometaphase. In endonanaphase the chromatids separate within the nucleus, and in endotelophase an interphase state with double the original chromosome number is attained. The formation of the → mitotic apparatus and of a metaphase plate is omitted. The autonomously separated chromatids normally remain in a typical parallel arrangement.

An e. is termed "masked e." (Resch 1952) or „**endoreduplication**" (Levan and Hauschka 1953), where no stages formally comparable with the normal mitotic cycle are observed. An endoreduplication may be established with certainty if it is followed by a normal mitosis in which the products of endomitotic chromosome reduplication reveal themselves as → diplo- or quadruplo-chromosomes which separate before anaphase producing a doubled or quadrupled number of individual chromosomes, as the case may be.

Endopolyploidization associated with e. may lead to very high degrees of ploidy. In general, fixed degrees of ploidy are characteristic of certain tissues. In most cases, endopolyploidy represents an ultimate state of → cytodifferentiation and is not a factor of cytodifferentiation.

If separation of the reduplicated chromosomes formed by e. fails to occur, so-called polytenic or polynemic chromosomes result (→ polytene chromosomes).

endonuclease — any phosphodiesterase that does not require a terminus for hydrolytic activity. The enzymes make internal cuts in nucleic acids. Their specificity may be confined to substrate conformation, nucleotide sequence (→ restriction e.), nucleic acid species (endodeoxyribonucleases, endoribonucleases), and the presence of modified nucleotides. Endodeoxyribonucleases are important components of DNA metabolism in both prokaryotes and eukaryotes. An endodeoxyribonuclease that specifically acts on damaged DNA resulting in correctional pathways in vivo (→ DNA repair) may be called a correndonuclease (→ nicking enzyme).

endophenotype (Lewis and John 1963) — → phenotype.

endoplasm (Pringsheim 1854) — the granular central portion of the → cytoplasm as opposed to → ectoplasm. Both e. and ectoplasm are defined by light microscopic studies (→ endoplasmic reticulum).

endoplasmic reticulum (Porter et al. 1945) — (ER); a complex set of membrane lamellae and tubules, which is the site at which major biosynthetic processes, such as protein secretion, membrane assembly, and organelle biosynthesis are initiated in eukaryotic cells. Sometimes the highly convoluted membranes enclose a discrete region of the cell called a cisterna.

In this sense, the membrane system has an inside and outside, and → ribosomes (rough ER) are bound only to the outside. ER free of ribosomes is called smooth ER. The ER can be rapidly dissociated into assembly-competent fragments.

Rough ER membranes play an important role in protein biosynthesis by providing sites for the binding of ribosomes involved in the synthesis of specific classes of polypeptides that are transported into the lumen of the ER or inserted into the membrane. ER membrane proteins also participate in ordered sequences of co- and post-translational modifications of polypeptides synthesized on bound → polysomes, including proteolytic processing, glycosylation, formation of disulfide bonds, and modifications of amino acid side chains (signal peptide).

endopolyploid — of cells whose chromosome number has been increased by → endomitosis and for which the degree of ploidy is proportional to the number of endomitoses which have taken place.

endoreduplication (Levan and Hauschka 1953) — → endomitosis.

endosome (Novikoff 1963) — in eukaryotic cells, any of a class of prelysosomal organelles that play a central role in → endocytosis (= receptosome). They serve as intermediates in plasma membrane recycling and in the transport of endocytosed ligands, fluid-phase components and membrane components to → lysosomes. They are involved in dissociation of ligand-receptor complexes and in the molecular sorting, processing, storage, and regulation of receptor expression.

Morphologically, endosomes are complex organelles composed of a large vacuolar component and a number of associated tubules.

endosperm — a tissue specialized in nourishing the developing embryo of flowering plants. The e. arises from double → fertilization by the second pollen (male) nucleus of two of the eight nuclei of the embryo sac (primary e. nucleus) and is typically triploid. Since it also originates from the fusion of one female nucleus with one or more others, or with a male nucleus, or with both, it may be diploid, triploid, tetraploid, pentaploid or even higher ploid depending on the species in question.

endosymbiont theory (Margulis 1970) — a theory according to which the → mitochondria of eukaryotes, and the → chloroplasts of plants and green flagellates, originated as free-living prokaryotes which invaded primitive eukaryotic cells and became established as permanent symbionts in the cytoplasm.

end point mutation (Demerec 1946) — any experimentally induced → gene mutation in bacteria which due to delayed manifestation (→ cell division lag; mutational lag) becomes recognizable only after a series of cell divisions (→ zero point mutation).

end product inhibition — a biological control mechanism in sequential → enzyme systems in which the accumulation of the final product of a sequence of metabolic reactions causes the inhibition of its own formation. End product inhibition is an example of "negative feedback control mechanisms", which may be grouped into competitive and noncompetitive ones. (1) **Competitive mechanisms:** The inhibition of an early enzyme in a sequence by the end product of the enzyme sequence is competitive with the substrate of an early enzyme. There is structural similarity between end product and substrate. (2) **Noncompetitive mechanism:** The end product of the enzyme sequence inhibits an earlier enzyme in the sequence in some manner other than by competition with the substrate for the catalytic site. The end product, which is inhibitory, becomes bound to an independent site of the enzyme. This causes a conformational change in the enzyme, altering its catalytic activity. The second site is specific for binding the end product of the multienzyme sequence. Enzymes of this types are called → allosteric.

enforced heterozygosity (Muller 1917, 1918) — → heterozygous.

enforced outbreeding — deliberate avoidance of mating between relatives (→ inbreeding).

enhancement — increased yield of viruses or increased cytopathic effects, or both, after mixed infection of cells by unrelated animal viruses, at least one of which is noncytocidal.

enhancer — → modifier gene.

enhancer sequence (Khoury and Gruss 1983) — any of a class of cis-acting short DNA sequences that increase transcriptional activity (transcriptional enhancer) of eukaryotic genes. Transacting cellular factors are required for activity of at least some e. sequences. Many RNA polymerase II transcription enhancer sequences contain a conserved octamer sequence motif ATTTGCAT with which two transcription factors interact.

Activation of transcription of a gene linked in cis to the e.s. occurs in a relatively orientation- and location-independent fashion. E.sequences display tissue and species specificity and activate transcription over large distances (→ silencer sequence).

enneaploid — of a → polyploid with nine chromosome sets in the somatic cells.

entry exclusion (Harada et al. 1961) — the phenomenon (= surface e.) whereby a number of conjugal → plasmids create a cell surface barrier to conjugal entry of genetically distinguishable derivatives of themselves. E.e. is quite distinct from → plasmid incompatibility, which operates at post-DNA penetration stages.

entry site — the ribosome site available for the initial binding of → transfer RNA during → genetic translation. In the process of initiation (→ initiation complex), → initiator tRNA binds first at this so-called A site and is then translocated to a second aminoacyl tRNA molecule which could now bind to the e.s.

enucleate — of a → cell lacking a → nucleus. The removal of the nucleus with very little cytoplasm is called enucleation and represents a classical method for investigating interactions between nucleus and cytoplasm (→ nuclear transplantation).

environment — the combination of all conditions external to the → genome that influence its expression and structure, and thereby the organism. Normally, an organism is adapted to the various different factors of its e. But the heterogeneity of the e. can have profound effects on evolutionary processes and can influence directly the genetic structure of → populations (→ environmental mutagenesis).

environmental mutagenesis — a field of chemical → mutagenesis studying potential mutagenic effects of, e.g., food additives, pesticides, drugs, cosmetics, and industrial compounds (= genetic toxicology). Many of these agents require "activation" following ingestion before any mutagenesis is manifest. The genetic changes resulting from environmental mutagens are of three kinds: (1) → Gene mutations; (2) → chromosome mutations; (3) changes in the number of chromosomes (→ aneuploid; polyploid).

To protect the genetic endowment of future generations, e.m. should be minimized or, whenever possible, avoided (→ mutagenicity testing).

environmental resistance — the sum of the physical and biological factors which prevent a → species from reproducing at its maximum rate (→ biotic potential).

environmental variance — (V_E); the portion of → phenotypic variance of a character which is caused by differences in the environment to which the individuals in a population have been exposed; studied by individuals which have the same genotype (inbred lines; monozygotic twins).

enzyme (Kühne 1878) — a protein which is the product of → genetic transcription and → genetic translation of the transcript of an enzyme-coding → gene. Enzymes speed up, enable, or control chemical reactions in living systems without being used up in the reactions. Enzymes are typically composed of a protein part (the apoenzyme) which confers specificity and a nonprotein part (the coenzyme) necessary for activity. Enzymes bind to substrates thereby affecting their breakdown or union with other substances.

Most enzymes contain more than one peptide chain. Normally, these polypeptide subunits show no catalytic activity by themselves. It is essential that the subunits be fitted together exactly in a specific, three-dimensional conformation to preserve enzyme activity.

Two fundamental processes are characteristic for e. action: (1) A **"recognition reaction"** during which an e. selectively binds a substrate and exhibits discrimination between closely related molecular structures (e.g., d- and l-forms); (2) a **"catalytic reaction"** in which covalent bonds are made or broken to form products.

A particular region of the e. possesses or can assume a conformation complementary to its substrate. This provides for binding which presents the substrate in an optimal configuration to the catalytic site of the e. or coenzyme. The catalytic site is that portion of the e. where the substrate is bound and where it is transformed into its reaction products. The catalytic sites contain specific charged groups, as well as hydrophobic groups, contributed by specific amino acid residues in the peptide chain of the enzyme. Different specific binding points of the active site may arise from widely separated points on the peptide chain, since the latter is ordinarily in a folded conformation in the intact enzyme structure. Changes in its secondary or tertiary conformation may greatly alter the enzyme's catalytic activity (→ allosteric).

During of after binding of specific substrate molecules, enzymes undergo a conformational change called the "induced fit" of the e. to its substrate. **Catalytic power** designates the ability of an e. to accelerate a chemical reaction. **Catalytic factor** is the term used to define the quantitative ratio of the enzymatic and nonenzymatic rates under some conditions. **Catalytic residues** are the amino acid residues directly involved in the covalent bond changes during enzyme action. **Specificity residues** designate those residues involved both in the binding of substrate and in subsequent processes of the chemical reaction which are not directly involved in covalent changes. **Contact residues** designate those residues which contain at least one atom within a bond distance (about 0.2 nm) of the ligand under observation and may include both catalytic and specificity residues. **Kinetic specificity** refers to the specificities in steps following the binding step. **Binding specificity** is the specificity of the binding process and **active site** designates that general region in the neighborhood of the catalytic residues identified with the binding and reaction of substrates.

Two or more enzymes in a linear sequence so that the product of the first is the substrate of the second, and so on, make up the simplest enyzme systems in a single homogeneous aqueous phase. In the cell, such systems may consist of as many as 12 or more separate enzymes acting linearly. Sometimes there is "branching" of enzyme systems allowing metabolic channeling. Some systems have been shown to be cyclic or closed.

Enzymes that are synthesized in fixed amounts, irrespective of the growth conditions and in absence of an → inducer are referred to as **"constitutive enzymes".** Those enzymes which are synthesized only in the presence of their substrates are called **„inducible enzymes".** In this case, the substrate acts as an inducer of enzyme synthesis ("enzyme induction"). In its absence the gene determining the enyzme in question is inactive (→ genetic transcription; genetic translation). According to the → operon model, the inducible enzyme synthesis is started when the inducer (→ effector) interacts with definite sites of the → repressor molecule inactivating it and thus activating the gene responsible for enzyme synthesis. Enzyme induction (as well as enzyme → repression) involves not a single e., but a sequence of enzymes acting in successive metabolic steps. Enzymes whose rates of production are observed to vary together show "coordinated enzyme synthesis". Enzyme synthesis is called "repressible" if it is inhibited by the product of the reaction which the enzyme catalyzes (→ end product inhibition; repression).

Enzymes appearing in multiple forms are classified into two classes: (1) → Isozymes, and (2) multiple molecular forms. Such enzymes catalyze the same reaction.

Five categories of reactions encompass virtually all the chemical reactions catalyzed by enzymes: (1) Hydrolysis, i.e., breaking a bond with the addition of the elements of water, and separation into two molecules; (2) group transfer, i.e., transfer of an integrated group of atoms from one molecule (donor) to another molecule (acceptor); (3) oxidation and reduction, i.e., transfer of one or more electrons, or hydrogen atoms (electron + proton), from one molecule (the one undergoing oxidation) to another molecule (the one undergoing reduction); (4) isomerization, i.e., rearrangement of the position of some atom, or group of atoms, within a molecule; (5) condensation, i.e., linking, by covalent bonds, of two molecules (like or unlike) to form a new molecule.

enzyme cluster (Gaertner and Cole 1977) — a term used to denote all forms of physically associated enzymes, including enzyme complexes, membrane-bound enzyme system, as well als multifunctional enzymes.

enzyme induction (Monod et al. 1951) — an increase, by → derepression of the rate of synthesis, of an enzyme as determined by the synthesis of a specific → messenger RNA. E.i. is a rather rapid monophasic response which is dependent upon the presence of an → effector, usually a metabolite. The opposite process is called **enzyme repression,** i.e., the disappearance or nonsynthesis of an enzyme as determined by the absence of a specific mRNA synthesis.

enzyme repression — → enzyme induction.

eobiogenesis (Pirie 1937) — a term used to designate the first instance of life (→ biopoesis; neobiogenesis).

eobiont (Pirie 1937) — → biopoesis.

eosome (Mc Carthy et al. 1962) — → ribosome.

epidermal growth factor — (EGF); a potent → mitogen that stimulates the proliferation of a wide variety of eukaryotic cell types. It acts through its cell surface receptor to induce a cascade of cellular events, including tyrosine phosphorylation, receptor clustering and internalization, eventually leading to → DNA replication and → cell division.

epigenetic (Waddington 1940) — of any change of gene activity during development of the organism from the fertilized egg to the adult. E. switches turn particular genes on or off during the developmental process, producing either transient changes in gene activity or a permanent pattern of activities. Owing to their stability, e. changes that arise in somatic cells are difficult to distinguish from rare, random mutations without direct genetic analysis.

The dynamic process with continuous feedback and modulation that assures that development proceeds towards specific targets or end states is called epigenesis.

epigenetic pleiotropy — the phenomenon wherein a given developmental event affects widely disparate parts of the overall phenotype.

epigenetics (Waddington 1942) — that branch of biology which deals with the causal analysis of development.

epigenotype (Waddington 1939) — the total developmental system consisting of a series of interrelated developmental pathways through which the adult form of an organism is realized. It comprises the totality of interactions among genes and between genes and the nongenetic environment resulting in the phenotype (→ epiphenotype). The e. of a cell is a stable, heritable (during at least many cell generations) character whose mode of impression is over and above or in addition to the classical genotype, i.e., the base sequence in DNA.

epiphenotype (Cahn 1969) — the aggregate of synthetic capabilities and other properties resulting from the expression of recognizable → cytodifferentiation (→ epigenotype).

episite — → lysogenization.

episome (Thompson 1931; Jacob and Wollmann 1958) — any bacterial → plasmid which can exist in two states within the cell, either independently in the cytoplasm or, following insertion, as an integral part of the host's chromosomes.

epistasis (Bateson 1907) — a form of → gene interaction whereby one gene interferes with the phenotypic expression of another nonallelic gene (or genes), so that the phenotype is determined effectively by the former and not by the latter when both genes occur together in the genotype. Genes whose expression is altered by nonallelic genes are said to be "hypostatic", or exhibit "hypostasis". Epistatic action may depend upon the presence of a dominant allele A, the recessive allele a, or the heterozygous genotype Aa. The hypostatic reaction of nonalleles may depend upon the presence of B, b, or Bb. (1) *E. of dominant genes:* if gene A is dominant over a and gene B is dominant over b, then A is recognized as the epistatic gene if the expression of B and b is suppressed and the genotypes. A.B. and A.bb appear the same. In this case, the normal F_2 dihybrid → segregation ratio (9:3:3:1) is modified to 12 (9A.B. + 3A.bb):3aaB.: 1aabb. (2) *E. of recessive genes:* in this case recessive gene a suppresses the manifestation of B and b. Under this condition the F_2 segregation ratio of 9A.B.: 3A.bb 4 (3aaB. + 1aabb) results.

In population genetics and quantitative genetics, the term epistasis is sometimes used to refer to *all* nonallelic gene interactions (→ interaction deviation).

epistatic deviation — = → interaction deviation.

epistatic disequilibrium (Moran 1967) — = → linkage disequilibrium.

epithelial — in tissue culture, of cells whose morphology is compact and resembles that of squamous epithelium as opposed to spindleshaped, fibroblast-like cells.

epivirus (Brinton 1972) — any → plasmid whose infectious form is obligatory cell attached. The infectious plasmid system is virus-like, i.e., it has a double-stranded → replicative form, a single-stranded infectious form, and an extracellular (but not cell-free) infectious particle; in contrast to virion viruses, epiviruses can introduce DNA into a cell that is not limited by the size of the virion "package".

equational division (Weismann 1887) — (1) = mitosis; (2) → reductional division.

equational exception — in the case of autopolyploid forms, of XXY flies of *Drosophila*, of crossing over in → attached-X females of *Drosophila*, or of crossing over and primary → nondisjunction, a gamete with chromosomes which contain sister chromatid segments (→ chromatid segregation).

equational separation — the separation of sister chromatids or sister chromatid segments and their distribution to different cell poles (= equation) (→ reductional separation).

equational split — the "split" (perpendicular to the reductional split) in homologous chromosomes, paired during → meiosis I (→ chromosome pairing), which separates the two chromatids of a reduplicated chromosome.

equator — the periphery of the → spindle on a plane perpendicular to the spindle axis and equidistant from the poles in a dividing cell; the periphery of the → metaphase plate.

equatorial plane — the plane midway between the two daughter nuclei of a dividing cell.

equatorial plate (van Beneden 1875) — = → metaphase plate.
equilibrium population (Crow 1948) — a population in which the → gene frequencies have reached an equilibrium between mutation pressure and selection pressure. The → genotype frequencies correspond to those expected for random mating and free combination between the loci.
ergastic (Meyer 1906) — of cell materials such as accumulations of nutritive materials and products of metabolic activity.
ergastoplasm (Garnier 1897) — the ribosome carrying components of the → cytoplasm which represent basophilic areas (storing basic dyestuffs) of the groundplasm of the cell. In the electron microscope the e. shows up as an accumulation of → ribosomes situated either on the parallel lamellae of the → endoplasmic reticulum or accumulated free in the groundplasm. The basophilic behavior of the e. is due to the → ribosomal RNA.
ergosome (Wettstein et al. 1963) — = → polysome or polyribosome.
error correction nuclease (Englund 1971) — an error-correcting 3'→ 5' exonuclease associated with prokaryotic → DNA polymerase.
error-prone DNA repair — (ERP); a type of → DNA repair (→ SOS repair) that processes damaged DNA in such a way that mutations result (= mutagenic repair), as opposed to error-proof DNA repair.

Organisms lacking ERP or suffering its mutational inactivation are usually immutable by diverse reactions and chemicals. ERP is probably initiated at → DNA damage which blocks the progress of → DNA replication (absence of a usable template at a damaged site; sterical impossibility of further elongation; efficient 3'-exonuclease action at a template which allows insertion). DNA polymerases (or a → DNA primase or an → RNA polymerase) must operate aberrantly in all these cases to conduct mutagenic DNA (or RNA) synthesis.

escape synthesis (Buttin 1960) — the derepression (constitutive expression) of a bacterial → operon as mediated by at least two separate mechanism: (1) Enzyme synthesis is dependent upon phage DNA replication (classic e. s.) and may be due to depletion of free repressor by phage-induced replication of the operon's → operator. (2) Extension of → genetic transcription into the bacterial operon from a promoter site of a → prophage during prophage induction. This type of e.s. is independent of phage replication, but is affected by mutations involving phage DNA transcription.
E site (Rheinberger et al. 1981) — in addition to the → A and P sites, a third tRNA binding site in (*E. coli*) → ribosomes which specifically binds deacylated tRNA (exit site). The deacylated tRNA leaving the P site during → genetic translation appears to be bound to this site before it is released from the ribosome.
ethnic group — a group of individuals or a population racially or historically related and having common cultural characteristics.
etiology — the demonstrated cause of a → trait or of a disease.
etioplast — a → plastid of dark-grown plants (angiosperms) containing crystalline prolamellar bodies, i.e., an arrangement of the membrane material which has been observed only under conditions that lead to protochlorophyll(ide) accumulation.
euapogamy(Farmer and Digby 1907) — a form of → apomixis in which the sporophyte develops from a gametophyte without fertilization and zygote formation (= diploid apogamy or unreduced apogamy).
eucell — the → eukaryotic cell.
eucentric (Darlington 1936) — of structural changes of chromosomes of the interchange type (→ chromosome mutation) in the course of which the transposed segments, as regards their linear gene order, do not change their positions relative to the centromere.
euchromatic (Heitz 1928) — of those chromosomes or regions of chromosomes which show the normal cycle of → chromosome coiling, normal staining properties and do not become → heteropycnotic, as opposed to → heterochromatic ones.

The euchromatic portions of the → chromosome appear to be almost free of rapidly reassociating, → repetitious DNA. They show the same DNA-histone ratio as → heterochromatin but three to four times more acidic chromosome proteins (→ hertone).

E. is the genetically active, dispersed part of → chromatin in eukaryotic → interphase nuclei.

euchromatization — the replacement of heterochromatin by euchromatin.

euchromocenter (Gregoire 1932) — in interphase nuclei, a → chromocenter which consists of the → heterochromatic segments located on both sides of the → centromere of one or more (in case of fusion of chromocenters) chromosomes.

euchromosome (McClung 1902) — = → autosome.

eugenic (Galton 1883) — of measures or trends which improve the genetic endowment of a human population, as opposed to → dysgenic.

eugenics(Galton 1883) — the study of agencies under social control that may improve or impair the hereditary qualities of future generations of man, either physically or mentally (Stern 1960).

E. may be subdivided then into negative (preventive) and positive (progressive) eugenics. The aim of negative e. is to prevent the increase or the presence of alleles which produce undesirable phenotypes. Positive e. is concerned with furthering the increase of alleles which cause desirable phenotypes or, at least, guarding against the decrease of such alleles (→ euphenics).

euhaploid (Katayama 1935) — → haploid.

euhomologous (Dvorak and McGuire 1981) — of homologous chromosomes within a homozygous inbred line.

eukaryon (Dougherty 1957) — the → nucleus of → eukaryotic organisms (→ prokaryon).

eukaryotic (Chatton 1925) — of plants and animals (eukaryotes) whose cells have, in contrast to prokaryotes, typical cell → nuclei with → nuclear envelopes, more than one multireplicon → chromosome, as well as nuclear divisions in form of → mitosis and → meiosis during which the chromosomes are visible by light microscopy in species-specific number and form. The e. cell has localized many of its metabolic processes into specific membrane-limited compartments or organelles. E. cells contain several sites of protein synthesis: The cytoplasm and the → mitochondria and, in the case of green plants, → chloroplasts as well (→ endosymbiont theory).

E. microorganisms and cells vary greatly in size but are usually much larger than those of prokaryotes.

eumeiosis (Battaglia 1945) — → meiosis.

eumitosis (Sagan 1967) — → premitosis, mitosis.

euphenics (Koltzov 1929; Lederberg 1963) — that part of → euthenics concerned particularly with the engineering of human development. E. means the improvement of phenotypes based on defective genetic constitutions and can compensate for, or redeem certain genetic defects.

euploid (Täckholm 1922) — of cells, tissues, and individuals with one complete → chromosome set ("monoploidy") or with whole multiples ("diploidy", "polyploidy") of the basic, monoploid number of chromosomes characteristic of the species. Each chromosome is represented once in each chromosome set (→ aneuploid).

eupsychics (Lerner 1968) — eductional and psychological engineering as a means of managing human biological resources.

eupycnotic — of chromosomes or chromosome segments which are normally coiled (→ chromosome coiling) and stained as opposed to → heteropycnotic.

eurygenic (Sallei and Zuckerkandl 1975) — of a state of → differentiation defined by the accessibility to transcriptional activation of a complex of tissue-specific genes (an euryon), i.e., of genes characterized by the order of magnitude of their rates of transcription (→ stenogenic).

euselectivity (Sedlmayer 1956) — true selective → fertilization as opposed to → paraselectivity.

eusexual — of organisms that show regular alternation of → karyogamy.

eutelegenesis — a means of positive → eugenics which proposes the use of preserved sperm from selected donors for voluntary artificial insemination, thereby increasing the frequency of desirable genes in the human → gene pool.

euthenics — improvement in the human environment as a means of managing human biological resources and an indispensable partner of → eugenics (→ euphenics; eupsychics).

eversporting (Bateson 1906) — of organisms that exhibit variations of a particular character from generation to generation, resulting (usually) from unstable, mutable genes (→ gene mutation).

The organisms never breed true, but produce in every succeeding generation the same segregations.

eversporting displacements (Muller 1930) — chromosomal structural changes which give rise as a consequence of position effect to → variegation or mosaicism in somatic tissues.

eviction — gene eviction.

evocation (Needham et al. 1934) — the → induction of → cytodifferentiation in embryonic cells (→ individuation).

evocator — a chemical substance which acts as an internal stimulus inducing developmental processes (→ differentiation; embryonic development). The agent responsible for e. action originates from an → organizer of the embryo and initiates the → determination of neighboring embryonic regions that are in the competent phase.

evolution — the transformation of the form and mode of existence of an organism in such a way that the descendants differ from their predecessors (Zimmermann 1953). E. is often divided into two basic processes: (1) Anagenesis, or change within a phyletic line, and (2) cladogenesis, or the multiplication of phyletic lines.

E. is a change in the genetic composition of a → population, the starting point of which is the formation of individuals with different genotypes.

The unit process in e. is → gene substitution and the elementary quantity is the → gene frequency which is the measure of genetic change in a population.

Evolutionary changes are brought about by the primary evolutionary forces which produce and sort out genetic variations and operate in a field of space and time.

The principal causes of biological e. are: (1) → **Mutations** supplying the raw materials, (2) → **selection** shaping these raw materials into the biologically fit genotypes of races and species, (3) random → **genetic drift** which can produce rapid changes in gene frequency in small populations, (4) differential → **migration** and → **gene flow** which can shift gene frequencies via exchange of individuals and genetic information between populations, (5) → **isolation** and → **annidation** which may act in a way similar to selection as a directive force in evolution and prevent the mixing of differentiations arisen in the population.

The direction of evolutionary change can be viewed either as a progressive movement of one population from one state to another, or as a divergence between two or more populations.

The indispensible basis for biological e. is the correlated change of genotype and phenotype within a population. Changes of this type are brought about by → gene mutations, → chromosome mutations and → genome mutations. If these genetic changes are not accompanied by changes in phenotype, they do not represent immediate evolutionary material but initially contribute merely to "cryptic genetic variability". This refers to the reservoir of variability of the population (= "potential genetic variability") which has not yet become manifest phenotypically. This part of genetic variability must first be converted to "free variability" by → segregation before selection can be effective as an evolutionary force.

In a relatively large population, the elementary evolutionary process of change is the change of the gene frequency. Three different types of change may be distinguished according to the degree of determinability of the changes for which they are responsible (Wright 1949): (1) **Systematic changes** in the population, which in principle may be formulated mathematically, thus allowing the evolutionary trend to be predicted with some certainty. To these belong (a) the pressure of recurrent mutations, (b) the pressure of imigrations and hybridizations, and (c) the pressure connected with the selection occurring within the evolving group of individuals. (2) **Random fluctuations** in the gene frequency. To these belong (a) fluctuations which are a consequence of sampling accidents, and (b) fluctuations in the systematic evolutionary pressures mentioned under (1). (3) **Nonrecurrent changes** in the population as (a) nonrecurrent mutations, (b) nonrecurrent hybridizations, (c) nonrecurrent selection processes, and (d) nonrecurrent extreme reductions in the number of individuals of the populations in question (→ population waves).

Evolutionary processes within a species or those leading to the formation of a new species (→ speciation) are termed "infraspecific evolution" (Rensch 1947) or "microevolution"; those processes leading to new genera, families, orders, etc. are called "transpecific evolution" or "macroevolution".

Microevolutionary events are usually viewed over a relatively short period of time, macroevolutionary events through the perspective of geological time.

The rate of e. is the time span needed for particular evolutionary changes in an organism or a population. Factors affecting the rate of e. may be the amount of → genetic variability in populations, population structure, the mode of → adaptation, intrinsic mutation rates, and others. Except for mutation rate, all these factors are functions of the → breeding system. Rates of e. are very diverse, not only between different groups of organisms living in different environments or even in the same environment but also between the same line at different periods in its evolutionary history and between different parts of the same organism (Stebbins 1950).

Rates of e. falling within the distribution most commonly found when a frequency distribution of such rates is plotted are called **"horotelic"** (Simpson 1944). E. at a much slower or faster rate than horotelic e. is referred to as **"bradytelic"** and **"tachytelic"**, respectively.

A contemporaneous group of individuals producing common offspring in the future is a "unit of evolution" (Thoday 1953). Depending on the → mating system it may be smaller than, correspond to, or comprise more than an individual species. Changes of the external and internal environmental conditions can lead to considerable changes in the size and complexity of the units of e.

The following **modes of evolution** (→ phyletic gradualism; punctuated equilibrium) may be distinguished: (1) **Phyletic e.** (Simpson 1944): Any evolutionary change occurring sequentially in a single line of descent without branching of lines. Form the point of view of paleontology, this mode of e. is characteristic for the more or less linear e. of phyla when measured against time. (2) **Quantum e.** (Simpson 1944): According to this evolutionary mode, a population which is not in adaptive equilibrium is displaced rapidly towards a new equilibrium by strong selection pressure. Quantum e. leads from one adaptive zone to another and no transitional forms occur between them since there is no adaptive equilibrium between the zones. (3) **Speciation** or **splitting** (Simpson 1944): A mode of e., in the course of which species evolve after splitting off and spatial isolation of parts of an initial population. (4) **Explosive e.** is evolution by means of "spontaneous variation", i.e., by natural selection.

Neo-Darwinian e. is the result of the re-examination of Darwin's theory of e. in the light of discoveries about the physical basis of inheritance. The counterpart of "spontaneous variation" is mutation. A difficulty is that the observed rates of molecular e. and the amount of variability in protein structure in natural populations are apparently too great to be accounted for on the theory of natural selection. This led to the theory of non-Darwinian e.

According to the theory of **non-Darwinian e.** (= neutral mutation-random drift theory of molecular evolution), the majority of evolutionary mutant substitutions in the species are caused by random fixation of selectively neutral (i.e., selectively equivalent) mutants rather than by positive Darwinian selection. Favorable mutations are viewed to be so rare as to be neglected in calculating rates of molecular evolution (→ coincidental evolution; horizontal evolution).

In all probability both neo- and non-Darwinian e. act in nature but their relative importances remain to be determined.

evolutional load (Kimura 1960) — = substitution load (→ genetic load; cost of natural selection).

evolutionary clock hypothesis — a hypothesis which assumes that neutral changes accumulate in a given protein or nucleic acid sequence at a constant rate. This rate can be expressed in UEP (unit evolutionary period), the time in millions of years (MY) required for the fixation of 1% changes in the amino acid or nucleotide sequence between two lines.

evolutionary distance — the number of base substitutions per homologous site that have occurred since the divergence of two DNA sequences.

evolutionary divergence — the degree of divergence, at the intra- and interspecific levels, of two (or more) populations, which presumably have evolved from a common ancestor. A direct test of the degree of evolutionary divergence of two compatible populations is the relative variability of their F_2 generation.

evolutionary force — any of those factors which bring about changes in → gene frequencies or chromosome frequencies in a population and are thus capable of causing evolutionary change. The major e. factors are: (1) Mutation; (2) gene flow; (3) natural selection; (4) genetic drift; and (5) meiotic drive. In addition there is the → founder effect, a compound of migration and drift.

evolutionary plasticity — the genetic adaptibility of populations or lines of descent. It depends on the presence in the population of → genetic variance in adaptive traits that are available for natural and artifical → selection to work on.

evolution pressure — the result of the combined action of → mutation pressure, → immigration and hybridization pressure, and of → selection pressure. It gives rise to systematic changes in the → gene frequency of a population. These changes, which are the basis for all evolutionary processes, may in principle be expressed mathematically.

evolutionary progress — the acquisition of new macromolecular and metabolic processes (adaptations) by which competitive superiority is achieved.

evolutionary rate — the amount of evolutionary change per unit of time. A number of factors affect e.f. (Grant 1985): (1) Stability of the environment; (2) population structure; (3) position of the organism in the hierarchy of nutrition; (4) its position in the hierarchy of reproduction; (5) the supply of genetic variability; (6) the length of generations.

E. rates are highly dissimilar at the species level.

evolutionary tree — (1) a diagram ("tree") that depicts the evolutionary relationship of protein (or nucleic acid) sequences. The topology, or branching order, and the branch length, which ought to be proportional to the true → evolutionary distances, are important features of an e.t. The construction of an e.t. based on sequence data is done by clustering the sequences according to their similarities. (2) A diagram that portrays a lineage or taxonomic group. The vertical axis represents time, the horizontal axis, diversity at any time (usually measured as number of coexisting taxa).

evolutionary trend — any trend in the evolution of phyletic lines. E. trends are the consequence of the unity of the genotype (→ genotypic cohesion) which greatly restricts the evolutionary potential.

evolvon (Edström 1968) — the operational unit in → evolution. It is assumed to consist of a DNA master sequence with a series of redundant sequences which constitute a repository of genetic information. The morphological correlate of the e. is assumed to be the → chromomere.

exaggeration (Mohr 1923) — the expression of a hypomorphic → allele placed opposite a deficiency or → deletion. In case of e., the recessive allele gives rise in single dose (hemizygous) to a more extreme mutant phentoype than in double dose (homozygous).

exchange — (1) transposition of chromatid or chromosome segments within one ("intrachange") or between two or more chromosomes ("interchange") giving rise to chromosome structural changes (→ chromosome mutation). (2) Reciprocal exchange of chromatid segments between chromosomes due to meiotic or mitotic → crossing over giving rise to → genetic recombination.

exchange pairing (Grell 1962) — → chromosome pairing; distributive pairing.

excision — the enzymatic removal of a nucleotide or polynucleotide fragment from a nucleic acid (DNA) polymer (→ excision repair).

excision linker (Mormeneo et al. 1987) — a bidirectionally cleaving blunt-ended DNA linker that can be used to generate defined nucleotide sequence modifications. The oligonucleotides containing two back-to-back recognition sites for class IIS endonucleases can be inserted into host DNA; digestion with the cognate class IIS enzyme results in a cleavage upstream and downstream from the adjoining enzyme recognition sites and thus allows directed deletion of a predetermined number of nucleotides within a gene.

excision postreplication repair (Sedliakova et al. 1975) — postreplication repair.

excision repair (Setlow and Carrier 1964; Boyce and Howard-Flanders 1964) — a pathway of → DNA repair which results in removal of a defective or unnatural base by cleaving the base-sugar bond (**base excision r.**) or, more commonly (except for uracil), by first making an endonucleolytic incision (nick on the 5' side) of the → DNA lesion, then excising the nucleotide by exonuclease cleavage, and finally patching the region (gap) by repair replication and ligation (**nucleotide excision repair**).

Base excision and nucleotide e.r. consist of the following operations: (1) **Base excision r.:** A damaged base is first removed as a free residue by base-specific → DNA glycosylases; the resulting → apurinic/apyrimidinic site is then removed by an AP endonuclease in concert with an exonuclease, followed by repair synthesis, and strand ligation. (2) **Nucleotide excision r.:** Several endonucleases detect DNA lesions and excise an oligonucleotide containing the damaged base by two sequential single-strand breaks (the first catalyzed by a specific damage-recognizing endonuclease, the second by an exonuclease); a DNA polymerase then inserts new nucleotides into the gap (repair synthesis), using the undamaged opposite strand for base-pairing information; subsequently the newly synthesized and preexisting strand segments are joined by DNA ligase.

E.r. operates on a wide spectrum of base modifications, such as mono- and diadducts, bulky adducts and DNA-protein cross-links, and uses a battery of different endonucleases and DNA glycosylases. Generally, base e.r. acts on lesions causing minor alternations in base pairing and base stacking, makes short repair patches (short patch r.) containing several nucleotides and goes to completion within about one hour after damage induction ("ionizing radiation-like" or alkylation lesions are the main substrate). Nucleotide e.r. operates on lesions causing major helix distortions, such as pyrimidine dimers („far UV light-like" lesions and strand crosslinks), forms long repair patches (long patch repair) of hundred or more nucleotides, and takes more than 24 h for completion.

Single-strand breaks are probably repaired by an abbreviated form of e.r. involving limited exonucleolytic digestion to "clean" the ends, repair synthesis, and strand ligation.

E.r. is a nearly error-free process; the initial recognition by site-specific nucleases which act on the primary chemical or photochemical damage is the rate limiting step in the excision process.

Repair-defective mutants may be defective in one or more of the steps of e.r. (→ germinative e.r.).

excison (Sakano et al. 1979) — a DNA sequence which is recombinationally excised in the

V-J joining of → immunoglobulin light chain genes. The e. carries at its ends characteristic inverted repeats about 30 bases long that could form a stem-loop structure somewhat similar to those of prokaryotic → insertion elements.

exclusion — → surface exclusion.

exclusion map — a → genetic map based on the phenomenon that all markers which are demonstrably heterozygous in an individual with a particular → deletion cannot be coded for by genes located on the deleted segment, since heterozygosity requires the combined expression of functional alleles.

exconjugant — → conjugation.

excretion vector — a → cloning vector which has the coding sequence for a signal peptide next to a restriction enzyme site into which foreign DNA can be cloned. If the foreign DNA fragment contains the coding sequence of a gene in phase with the signal peptide, the protein encoded by the foreign gene will be excreted from the host cell.

exocytosis (de Duve 1963) — the processes by which eukaryotic cells release material to the external extracellular environment. Lipids and proteins are secreted, post-translationally modified, and transported to the plasma membrane, the → lysosomes, and the organelles of the vacuolar system. Exocytotic vesicles bud off from the → Golgi apparatus and fuse with the plasma membrane, releasing their contents at the time of secretion. In some cell types, the vesicle membrane is incorporated into the plasma membrane (→ endocytosis).

exogamy (McLennan 1865) — all those systems of sexual → reproduction (→ mating system) in which there is a greater frequency of mating between unrelated or only distantly related partners than would be the case for purely random mating (panmixia) within the population (cross-fertilization or outbreeding). The genetic consequence of e. is an increased incidence of heterozygosity in the population (→ endogamy).

exogenote (Morse et al. 1956) — → endogenote.

exomutation (Imai 1936) — → plastom mutation.

exon (Gilbert 1978) — any of those parts of eukaryotic split or mosaic → genes that are expressed in the final protein or RNA product of these genes (→ RNA splicing). The exons of split genes are interrupted by → introns (or intervening sequences). Each of the exons of a protein coding gene may represent a structural or functional domain of the encoded protein. E. modification, duplication, and recombination may be a general mechanism for rapid evolution of eukaryotic → structural genes (→ exon shuffling).

exon binding site — → self-splicing.

exon/intron junction — → splice-junction sequence.

exon shuffling — in the case of eukaryotic split genes, the recombination of different coding regions (→ exons) through → crossing over. Since exons code for discrete, stable regions (domains) of proteins, e.s. might assemble new proteins through the novel reassortment of stable substructures and thus allows acceleration of evolution of new functions.

exonuclease — any of a family of enzymes (phosphodiesterases) which degrade nucleic acid molecules and act step-wise either from the 3' end (3'→ 5') or the 5' end (5'→ 3'). They are specific with regard to single-stranded or double-stranded DNA structure (→ DNA repair).

exophenotype (Lewis and John 1963) — → phenotype.

exoplasmosis (Zucker-Franklin and Hirsch 1964) — the release of enzymes from cells or into phagocytic vacuoles.

exorepressor system — a gene system (operon) characterized by the positioning of the → regulator gene outside the relevant operon (→ autorepressor system).

exosome (Fox et al. 1971) — a sequence of exogeneous DNA which after uptake by eukaryotic cells results in → genetic correction without integration of the exogeneous information into the linear structure of the chromosome (→ genetic transformation; transgenosis). The e. is assumed to replicate in step with the chromosome and can be lost or transmitted with it at the time of cell division. Either the exosomal gene or the corresponding chromosomal gene is transcribed, leading to phenotypic mosaicism.

explant — an excised fragment of a tissue or an organ used to initiate an in vitro culture (→ cell culture).

exploitation competition (Ricklefs 1973) — competition which occurs when one individual or species utilizes a limiting resource and thereby denies its use to another individual or species (→ interference competition).

exponential growth — a phase of growth of a cell culture or population in which the time of doubling in size (the population growth rate) is constant. Under e.g. (= logarithmic or geometric g.) the logarithm of the number of individuals increases linearly with time (→ logistic growth).

expression vector — a → cloning vector efficiently expressing inserted foreign genes. Ideally, such a vector should contain all the elements needed for high-level inducible snythesis of proteins: (1) A very strong → promoter (e.g., from a phage); (2) a → repressor gene for the promoter; (3) a strong translation initiation site.

The e.v. should be propagated in its host cell as a single copy genomic insert to enhance its stability and to facilitate repression of foreign genetic information. It should also respond to induction with rapid increase in copy number and high-level transcription of the foreign DNA.

expression extinction — in stable mammalian hybrid cell lines, the shutoff of expression of genes that are active in only one of the parent cell lines.

expressivity (Vogt 1926) — the phenotypic expression (kind or degree) of a penetrant gene or genotype which may be slight, intermediate or severe and may be described in either qualitative or quantitative terms (→ specificity). E. like → penetrance depends both on the genotype and on the external environment. It may be constant or variable; it may be similar or different in both sexes.

expressivity modifier — any factor which modifies the phenotypic expression of a major gene; plus modifiers enhance expression, minus modifiers suppress it.

expressor (Metzenberg 1972) — a chemically undefined positive regulator of → gene expression in eukaryotes.

extension gene — → modifier gene.

external (Darlington 1937) — of → chromosome mutations involving two nonhomologous chromosomes (→ fraternal; internal).

extinction — (1) in evolutionary biology, the failure of groups of organisms of varying size and inclusiveness to leave surviving descendants. (2) The reduction of differentiated, tissue-specific functions of their parents in somatic → cell hybrids (McMorris et al. 1973).

extracellular matrix (Gospodarowicz et al. 1979) — (ECM); an extracellular filamentous structure attached to the cell surface that provides cells with anchorage, traction for movement, and positional recognition.

extrachromosomal — of structures or processes that are not part of the chromosomes or proceed outside the chromosomes, respectively (→ inheritance; plasmid).

extrachromosomal DNA — DNA located outside the → chromosomes, such as → plasmid DNA (in prokaryotes and eukaryotes), → organelle DNA, amplified genes, and → small polydisperse circular and linear DNA's (in eukaryotes). The latter are intermediates and/or byproducts of → DNA transposition and → DNA amplification.

extrachromosomal inheritance — → inheritance.

extranuclear — of structures or processes which are located, proceed or are active outside the → nucleus in contrast to intranuclear structures or processes.

F

F_1, F_2, etc. (Bateson and Saunders 1902) — → filial generation.

factor (Mendel 1866) — in genetics, a determinant of a → character (nowadays a → gene) with respect to which an organism may show a Mendelian difference.

F-agent (Lederberg et al. 1952) — = → F-plasmid.

Fahrenholz rule — a rule postulating that parasites and their hosts speciate in synchrony (→ speciation).

family — (1) a group of individuals bearing a certain genetic relationship to each other by virtue of common descent (Lerner 1958); (2) a group of closely related → species of plants or animals.

family selection — a breeding method of selecting a pair on the basis of the average performance of their progeny.

fate map — a map of an embryo showing what will happen to each region of the embryo during normal development.

FB element (Potter et al. 1980) — acronym for → foldback element.

F duction (Jacob et al. 1960) — transfer of bacterial genetic markers (fragments of the bacterial chromosome) by the F-prime (F) plasmid (→ F plasmid), derived from Hfr strains (= sex duction). The markers became part of the F elements by → genetic recombination and stem from those chromosome regions with which the F element was previously associated. The F′-strains, like F^+-strains, transfer their F element to F^--cells with high efficiency.

The bacterial marker or group of closely linked markers transferred with the F′-element into F^--cells persists and replicates in the recipient cell without necessarily recombining with its chromosomal structure. The nucleotide sequences involved are under these circumstances present in duplicate, one set as part of the recipient cell's chromosome, and the other in the recipient cell's F′-element. Cells of this type are → heterogenotes and can serve in → complementation tests as efficiently as a transductional heterogenote.

fecúndity — potential fertility, i.e., the ability to produce functional gametes as measured by the quantity of gametes, particularly eggs, produced.

feedback (Warburton 1955) — a system in which a deviation of any size whatsoever from a certain value calls forth control factors which counteract (negative feedback) or increase (positive feedback) the deviation.

feedback control — a control system in which the accumulation of the product of a reaction leads to a decrease in its rate of production, or a deficiency of the product leads to an increase in its rate of production (→ feedback inhibition).

feedback inhibition (Umbarger 1956; Yates and Pardee 1956) — the inhibition of an early enzyme in a biosynthetic pathway by the end product of that pathway. In f.i. substrate and inhibitor may bind at two different sites on the enyzme (→ allosteric) and the inhibitor binding appears to produce a change in the tertiary or quaternary structure of the → protein. The inhibitor-specific site and the substrate-binding site may be on different polypeptide chains of a multichain enzyme. The obligatory requirement for excesses of two or more end metabolites to achieve inhibition is called "concerned" (Datta and Gest 1964) or "multivalent feedback inhibition" (Paulus and Gray 1964).

F.i. is a means by which cells adapt to their environment. In contrast to → induction and → repression of enzyme synthesis, f.i. acts at the enzyme level and thus is a rather direct and rapid means of regulating metabolic activity in response to an environmental change. By induction-repression, on the other hand, the expression of a gene is apparently altered, with the result that specific enzyme production may or may not occur, depending upon the particular environmental condition (Helsinski and Yanofsky 1966).

feeder effect (Earle et al. 1951) — density-dependent stimulation of cells growing in a culture as well as improved cloning efficiency of single cells when co-cultured with nongrowing feeder cells and stimulatory effect of conditioned medium upon sparse cultures.

F element (Dawid et al. 1981) — any of a class of → transposable genetic elements in *Drosophila* that resemble processed → pseudogenes in that they have polyadenylation signals. While F elements share with other transposable sequences the property of inducing target site duplications, they differ from these in that the length of the duplications varies from 8 to 13 bp. The F elements are highly homologous at their right termini; the left termini are subject to deletions and sequence heterogeneity. Whether or not their transposition is by way of an RNA intermediate and → reverse transcription is unknown.

F episome — = → F plasmid.

fertility — the reproductive potential of an individual or population as measured by the ability to produce viable offspring. "Normal" f. results when viable → gametes are produced in normal numbers, when union of gametes is normal, and when normal viability of → zygotes, embryos, and fetuses is followed by the production of viable offspring. When this chain of events is interrupted, or when deviations from average performance are great enough to be evident, the result is infertility or reduced f. Degrees of reduced f. vary all the way from that which is barely detectable down to complete → sterility (Hutt 1969).

"Effective f." (or probability of reproduction) is the term used in human genetics for the average number of children from patients with a particular hereditary disease in relation to the average number of children from comparable individuals not subject to the disease. The effective f. is a measure of the selective disadvantage associated with a hereditary illness. Effective f. may be reduced on biological, psychological, and sociological grounds in the case of a hereditary illness.

fertility inhibition (Egawa and Hirota 1962) — inhibition of fertility by other → plasmids (→ fi^+ plasmids) in bacterial (*E. coli* K 12) strains carrying the → F plasmid. The plasmids inhibiting fertility are called $f.i.^+$ (fi^+).

fertilization — the fusion of two → gametes of opposite sex (♂ and ♀, + and −) to form a zygote (= syngamy, zygosis, or cytogamy). Karyogamy or the fusion of nuclei of gametes is the culmination of fertilization. By this process the homologous chromosomes that were separated in the course of → meiosis, which is in the sexual cycle necessarily correlated with the process of f., are brought together again. F. is carried on in different organisms by different mechanisms: (1) **Self-fertilization** (selfing): Fusion, of male and female gametes from the same haploid, diploid, or polyploid organism. (2) **Cross-fertilization** (crossing): Fusion of male and female gametes from different haploid, diploid, or polyploid individuals. (3) **Selective or preferential fertilization:** Fusion of germ cells of different genotypes from one or both sexes in combinations having nonrandom frequencies. S.f. is a consequence of a direct reaction between egg cells and sperms or between embryo sacs and pollen tubes. This is called **"euselectivity"** by Owen (1945) in contrast to **"paraselectivity".** In the case of paraselectivity of fertilization, a selective fertilization process is imitated by decreased functioning or complete sterility of particular gamete types. (4) **Double fertilization:** In angiosperms, the fusion of one male gamete with the egg nucleus at the same time as a second male gamete fuses with a second female gamete to produce nutritive tissue (endosperm) (Darlington and Mather 1949).

fertilization-stasis (Boyes and Walker 1954) — an all inclusive concept for different abnormalities which occur following (species-) hybridization at the time of → fertilization and immediately thereafter. They are the result of disharmonies between the → genomes of the parent forms.

F factor — = → F plasmid.

fibrillar center — an area of low electron density in electron micrographs of the → nucleolus.

Functioning interphase nucleoli are surrounded by the dense fibrillar component which includes; (1) rDNA; (2) RNA polymerase I (nucleolar polymerase); (3) Ag-nucleolar organizing region (NOR) proteins.

fibroblast — a type of cell found in connective tissue of mammals and characterized by an irregular branching morphology. Fibroblasts are usually elongated, spindle-shaped cells which synthesize collagen.

fibronectin (Mynes 1976; Kuusela et al. 1976) — any of a class of large glycoproteins present on the surface of cells, in connective tissue matrices, and in extracellular fluids. They are involved in → cell-cell adhesion, cell-substratum adhesion, control of cell shape, and induction of cell migration.

fibrous lamina (Gray and Guillery 1963) — a distinct structural layer directly apposed to the inner membrane of the → nuclear envelope. It appears as a layer of fine granules and fibers typically 20 —80 nm thick (= zonula nucleum limitans).

field (Weiss 1925) — in developmental biology, a dynamic and patterned organizational whole, generated by the collective activity of a group of cells or a group of subunits of a single cell. The pattern is manifested in the form of tissues and organs as each individual cell of the group in turn responds to the f. by differentiating according to its position within it.

filament — a universal component of the cytoplasmic matrix which is 4 —5 nm in diameter and of indefinite length. Filaments vary greatly in abundance in different cells.

filial generation — any → generation (offspring) following the parental generation. The generation resulting from a cross of the parental individuals is called the first filial generation or F_1. From selfing or crossing of F_1 individuals the second filial generation or F_2 arises. The progeny of F_2 individuals makes up the third filial generation (F_3), etc.

filipodium (Trelstad et al. 1967) — any of the slender membrane-limited cell protrusions (0.2 μm in diameter and of varying length between 2 and 30 μm) which emerge from the cell surface of embryonic cells. The skeleton of the f. is the core of laterally associated actin filaments (= fibrous projection; finger-like projection; microspike).

filter hybridization — incubation on a nitrocellulose filter of a denatured DNA preparation with a solution of radioactively labeled RNA or DNA.

fimbrium (Duguid et al. 1955) — = → pilus.

F incompatibility — incompatibility which operates after an F-prime factor (→ F plasmid) has been successfully transferred to a recipient bacterial cell already carrying an F plasmid. If the resident F plasmid is in the autonomous state, either this or the incoming F-prime element is inherited unilaterally at cell division, ultimately giving a population of cells carrying one F plasmid or the other. If the resident F plasmid is integrated into the bacterial chromosome, it is always the incoming F-prime element that is inherited unilaterally. Both F.i. and → F surface exclusion are determined, at least in part, by the F plasmid itself.

finger motif (Miller et al. 1985) — → zinc finger motif.

fingerprint — (1) the pattern of the ridged skin of the distal phalanx of a finger; (2) a method for combining electrophoresis and chromatography to separate the components of a protein.

fi^+ plasmid (Watanabe et al. 1964) — any of a class of → resistance factors (R plasmids) able to repress the bacterial → F plasmid or sex factor. Those plasmids which do not repress the fertility of F cells are called fi^- (→ fertility inhibition). fi^+ plasmids produce a cytoplasmic → repressor which switches off synthesis of → sex pili.

firone (Varshavsky 1981) — any agent of intra- or extracellular origin which is able to induce → replicon misfiring in eukaryotes.

first division — the first of the two meiotic divisions (→ meiosis), formerly called the heterotypic or reductional division (→ second division).

first division segregation — the separation (→ segregation) of heterozygous → allele pairs during first division of meiosis (→ prereduction for the locus).

fission — (1) **binary f.:** The division of a one-celled organism by cell cleavage into two parts (daughter cells) in asexual reproduction immediately after → mitosis (or → endomitosis). The two daughter cells arising by binary f. are genetically like each other and like the original. (2) **Multiple f.:** Division of the → nucleus several times within the cell body. The cytoplasm fragments terminally into as many parts as there are nuclei present. Multiple f. may take the form of **"schizogony"** in which a number of new active individuals are produced, or of **"sporogony"** which results in a host of usually dormant spores.

fitness — in → population genetics, a quantitative measure of reproductive success of a given genotype, i.e., the average number of progeny left by this genotype as compared to the average number of progeny of other, competing genotypes (= adaptive value, selective value). **Inclusive f.** is the sum of an individual's f., quantified as the reproductive success of the individual and its relatives, with the relatives devalued in proportion to their → genetic distance.

Overdominance of f. it the situation in which the f. of a heterozygote exceeds the f. of both homozygotes. It plays a role in the maintenance of genetic variability in populations.

five-prime extension — (5′ extension); a single-stranded → cohesive end left at the 5′ terminus of a → restriction fragment.

fixation index (Wright 1922) — a measure of the correlation between two genes that results from the joint effects of → inbreeding and → selection.

fixed allele — an → allele which is homozygous in all members of a population under study; no other alleles for this locus exist in the population.

flagellin — aggregates of protein molecules of which bacterial flagella consist.

flagellum — any of the solid bacterial cylindrical structures (a few μm in length and 12 to 30 nm in diameter) responsible for bacterial motility. They are helical structures comprising helical arrangements of subunits each consisting of a single → flagellin molecule. Rotary locomotion is generated at the sites of insertion (with the basal body) of the organelles into the cell by rotation of the flagellar filament. Flagella have with rare exceptions a characteristic 9 + 2 arrangement of fibrils (a central pair of singlet tubules surrounded by nine parallel doublet tubules). (→ basal body; cilium).

flanking DNA sequence — any DNA sequence flanking close to the 5′ (5′f. sequence) end or 3′ end (3′f. sequence) of a gene. A prominent feature of the extragenic regions of most eukaryotic genes is that 300-nucleotid long repeats are interspersed with longer single copy sequences. 3′ flanking sequences may by responsible for insuring correct termination (frequently TTTT clusters). AT-rich regions have been observed in the 5′ flanking sequences; these bear a striking resemblance to that of the → Pribnow box sequences in prokaryotes.

flexibility (Thoday 1953) — (1) **Genetic f.:** The ability of a genotype to vary and adapt itself to changing conditions and thereby to survive. This faculty stems from the cryptic or potential → genetic variability (→ flow of variability). (2) **Phenotypic f.:** The ability of an individual to remain functionable in a whole series of different environmental conditions. Phenotypic f. may be subdividided into "developmental f." and "behavioral f.". **Developmental f.** (Thoday 1953) is the capacity of an individual to adapt reversibly to local conditions. The genotype of a developmentally flexible organism may either permit different phentoypes to be formed under different environmental conditions, such that each phenotype is best adapted to the surroundings in which it forms, or to be so balanced that development is extensively buffered against changes of environment. In the latter case, the same adaptive phenotype is produced in a series of different environments.

Waddington (1957) describes the first form of developmental f. as "adaptive flexibility", the second as "developmental canalization".

Behavioral f. (Thoday 1953) is the faculty of an individual to adapt itself reversibly to local or time-associated factors, or for a mobile individual, the ability to seek out that environment to which it is best adapted.

Salisbury (1940) used the term "plasticity" in the same sense as phenotypic f. In a relatively uniform and stable environment, selection towards genetic and phenotypic f. is slight; instead selection is directed towards adaptation and stability. These relationships are reversed under unstable environmental conditions which result in a strong selection towards genetic f.; the weaker this selection is, the stronger is the selection towards phenotypic f.

F-like sex factor — any of a group of conjugative → plasmids in bacteria which code for synthesis of → sex pili. The pili allow the adsorption of and infection by the F-specific bacteriophages.

floating (Darlington and Gairdner 1937) — of a chromosomal structural or gene change for which a mating group is not uniform (→ genetic polymorphism).

flow cytometry (Van Dilla et al. 1974) — a technique used for optical analysis and electrohydrodynamical separation of cells and metaphase chromosomes based on their light scatter and fluorescence signals. After labeling with fluorescent dyes, → chromatin can be analyzed quantitatively for DNA content, base composition, and structural proteins. An extension is the detection and quantification of specific DNA sequences by fluorescent DNA hybridization techniques.

F.c. also enables individual chromosomes to be purified for, e.g., → gene mapping.

flow karyotype (Gray et al. 1975) — a → karyotype based upon measurements by → flow cytometry which provides a precise description of the average chromosome complement of a cell population.

flow of variability (Fisher 1930) — the movement of → genetic variability within a population. It is the result of hybridization and → segregation. In the course of the movement, cryptic (potential) variability is transformed into free variability. The flow of variability can be influenced by selection if the allele frequencies, and with them the phenotypes, undergo a quantitative change.

FLP recombinase (Broach et al. 1982) — an enzyme in *Saccharomyces cerevisiae* that binds to specifc DNA sequences (FLP sites) and catalyzes the cleavage and rejoining of these sequences to mediate a site-specific inversion of the 2μm circle. FLP r. belongs to the integrase family of → recombinases.

flush end — = → blunt end (→ restriction endonuclease).

F-mediated transduction — = → F-duction or sex duction.

focal adhesion plaque (Abercrombie et al. 1971) — any of the local attachment points which form when cells attach to substrate or, during → phagocytosis, to particles.

foldback DNA (Britten and Smith 1970) — in eukaryotes, a structure formed when a double-stranded (nuclear) DNA molecule containing an → inverted repeat sequence (IR DNA) is denatured and then allowed to reanneal at low DNA concentrations. The repeated region allows the formation of a double-stranded region within each of the separated strands of the original molecule (= hairpin DNA).

foldback element (Potter 1982) — (FB element); any of a class of → transposable genetic elements characterized by long inverted repeats at their ends. FB's vary greatly in their inverted repeat sequences and the DNA separating these. The repeats are made up largely of tandemly repeated copies of a 155 bp repeat unit containing smaller subrepeats. Inverted repeats of different FB's vary in length from a few hundred to a few thousend bp. Transposase is probably coded by the central region of a f.e.

footprinting — → DNA footprinting.

forced cloning — a procedure of → molecular cloning in which → cloning vector and insert DNA are cut with the same pair of → restriction endonucleases. The linear molecule generated is free of complementary ends and self-ligation is prevented. Only recombinant molecules are the products of circularization.

Formenkreis — an aggregate of geographically representative (allopatric) species or subspecies.

formylmethionine tRNA — a → transfer RNA molecule ($tRNA^{Met}_f$) that participates in the initiation of bacterial protein synthesis (→ genetic translation; initiator tRNA).

forward mutation — a → gene mutation from wild type to the mutant condition, as opposed to → back mutation. The terms f.m. and back mutation are purely descriptive.

F. mutations may occur at any one of the large number of sites (→ muton) within a → gene or a → gene cluster. Secondary mutations (→ reversions or back mutations) may restore whatever activity was lost due to the f.m. Such mutations may occur at the originally mutated site (true reversion of the f.m.) or at some other site (false reversion, pseudoreversion, or partial reversion). When the reversion definitely occurs at a site separable by recombination from the site of the f.m., it is said to occur by → suppression, or by a → suppressor mutation.

F.m. rates (→ mutation rate) are the sum of variety of types of molecular events; these include → base pair substitutions, base pair additions, and deletions (→ frameshift mutation), and at a higher level chromosome rearrangements of various types.

A f.m. system has to fulfil three main criteria before it can be used for mutational studies (in continuous cultures of microorganisms): (1) The mutants obtained must be amenable to effective screening; (2) the spontaneous mutation frequency must be high enough to permit the recovery of mutants at the cell concentrations which can be maintained in the chemostat; (3) the mutants must not have a selective advantage or disadvantage over the nonmutants (McAthey and Kilbey 1976).

forward selection system — any selective system that allows the required mutant to survive.

founder effect (Mayr 1942) — the establishment of a new population by a few original individuals (founders) which carry only a small fraction of the total genetic variation of the parental population. The resulting new population may under these circumstances instantly become genetically different from the parental population (→ genetic drift). The f.e. is assumed to be important in → speciation.

The f.e. leads to two major changes in the genetic structure of a → founder population: (1) Random changes in → allele frequency, including loss and fixation (→ fixed allele), which lead to genetic divergence from the parental population and a reduction in genetic variability; (2) an accumulation and associated increases in levels of homozygosity.

In population genetics, the effect of (1) is measured by the "variance effective size". The effect of (2) is the accumulation of inbreeding as measured by the „inbreeding effective size".

founder population — a small number of individuals (founders) geographically isolated from their parental populations (→ founder effect).

FP factor — any → sex factor promoting bacterial host chromosome transfer.

F-pilus — (Brinton et al. 1964) — any of a class of bacterial filaments (hollow, cylindrical structures 8 nm in diameter with a central axial hole of 2 nm) extending typically 1 to 2 μm from the cell surface. Their average number per cell ranges from 0.5 to 1.5. F-pili are involved in bacterial → conjugation, in the transfer of genetic material between bacteria (sex pili), and they are the adsorption sites for certain → bacteriophages. They appear to contain only one protein subunit (F pilin) and are observed in cells containing the → F plasmid.

F plasmid (Lederberg et al. 1952) — a conjugative circular → plasmid (= F factor; F sex factor) mediating the *E. coli* mating process. The plasmid (94.5 kb) encodes information leading to: (1) The formation of F-specific → sex pili and conjugationally proficient cell surface properties; (2) conjugal DNA metabolism and DNA transfer; (3) prevention of self-mating (→ surface exclusion).

The plasmid denotes fertility since it can not only transfer itself from one bacterial strain to another but can mobilize portions of the bacterial chromosome. Bacteria harboring the F p. (F^+ or male donor cells) will transfer the F p. to F^- (female or recipient cells). Normal F p. transfer is initiated at a specific locus on the plasmid (ori T) and one of the DNA strands is transferred to the recipient (5′ terminus leading). DNA complementary to the transferred strand is synthesized in the recipient. Transfer ends with the circularization and establishment of the DNA in the recipient (these processes are called repliconation).

The F p. can integrate (via → insertion sequences) into the host chromosome converting the host cell from F^+ to Hfr (high frequency recombination). In such cells, the plasmid mediates not only its own transfer but also that of adjacent *(E. coli)* genes; these are then available for recombination with the recipient chromosome. The F p. is then referred to as an F prime (F′) plasmid. F′ formation occurs by an illegitimate recombinational event, analogous to the formation of specialized transducing bacteriophages (→ transduction).

Most of the conjugal functions of the F p. are related to a series of transfer (tra) genes included in a single operon (tra Y-Z) on the plasmid. The genes code for the production of sex pili, for surface exclusion, and for the metabolism of DNA during transfer. Two other regions of the F p. genome are necessary for recombination and control of replication.

F-prime (F′) plasmid — → F plasmid.

fragile chromosome site (Sutherland 1979) — (FS); any of the break-prone sites in (human) chromosomes that fail to compact for mitosis and thus show up as nonstaining gaps on one or both chromatids of metaphase chromosomes. Two major categories of FS's represent the rare or heritable ones and the common or constitutional fragile sites.

Three mechanisms for FS expression have been proposed: (1) A FS may be a DNA sequence which is particularly prone to misincorporation of uracil, thus eliciting a → DNA repair process that cannot be completed prior to mitosis owing to thymidylate deprivation; (2) the presence of uracil bases in FS DNA may itself disrupt chromatin conformation and thus results in cytogenetic expression; (3) a FS may be a DNA sequence that shows the progress of the replication fork and results in even more marked slowing and incomplete replication when the supply of DNA precursors is limited.

fragment — (1) a new centric or acentric product arising as a result of a chromosomal structural change of the deletion type (→ chromosome mutation); (2) a small → B-chromosome.

fragment mapping — = → cleavage mapping.

frameshift mutagen — any mutagenic agent (→ mutagen) with the property of causing a shift in the reading frame of the → messenger RNA synthesized from the altered DNA template (→ frameshift mutation). F. mutagens show strong preference for particular sites in DNA which are probably either long monotonous sequences or tandemly repeated base sequences. This means that particular genes might be strongly mutagenized by a particular f.m. while others would be insensitive.

Common to any f.m. is a heterocyclic ring structure and the ability to bind to DNA by intercalating into the base pair stack of DNA. There are indications that f. mutagens may exert mutagenic effects (in bacteria) in indirect ways, i.e., by producing a situation for error-prone → repair synthesis. This action may be superimposed on more direct effects.

frameshift mutation (Brenner et al. 1961; Crick et al. 1961) — a class of mutations (= reading frame mutations, sign mutations) that arise from the insertion (+) or deletion (−) of a nucleotide, or any number of nucleotides other than three or multiples of three into or from DNA (→ frameshift mutagen). F. mutations displace the starting point of → genetic transcription of the → genetic code and the resulting mRNA is "misread" by the translation process (out-of-phase translation) from the point of nucleotide addition or deletion. Thus once a f.m. is introduced into a gene the reading frame is shifted so that all codons distal to the mutation are read out of phase and reading frame, unless the particular genotype also carries a → f. supressor mutation (frameshift suppression).

A closely positioned f.m. of opposite sign results in a double f.m. (+ −) and restores the normal reading frame in mRNA except for the segment between the two frameshifts.

The codons which result after f.m. fall into three categories: (1) Sense codons, which are read or are translated the same as before f.m.; (2) missense codons which code for a different amino acid; (3) nonsense codons which code for no amino acid.

F.m. has been correlated with DNA sequences comprised of repeated bases mediating additions or deletions by local misalignments of the complementary DNA strands. Such aberrant DNA intermediates, if formed during → DNA replication, → genetic recombination, or → DNA repair, could then be the precursors of f. mutations.

frameshift suppression (Yourno et al. 1969) — the reversion of the effects of a frameshift mutation by a second mutation in the same gene (internal suppression) restoring the reading frame or by external suppressors (→ frameshift suppressor).

frameshift suppressor — any of a class of external → suppressor mutations specific for → frameshift mutations, i.e., capable of suppressing such mutations by changing (correcting) the phase of translation. Almost all suppressible frameshift mutations are of the addition type, most likely including an extra GC pair in the DNA containing repeated GC pairs. F. suppressors are genetic alternations outside the mutated cistron (external suppressor) and may be dominant or recessive to their corresponding wild-type allele. Dominant f. suppressors give rise to altered transfer RNA molecules that recognize quadruplets of bases in mRNA. Their anticodon contains a base quadruplet instead of a base triplet found in the wild-type tRNA. This allows the → suppressor tRNA to read nontriplet codons and to correct for single base pair additions in the message. Base deletions are probably not suppressible. The suppressors are specific for frameshift mutations, exhibit allele specificity and can be classified according to their patterns of cross-suppression. The efficiency of suppression is between 1 and 10%.

If the anticodon of frameshift suppressor tRNA is able to base pair equally well in two or three different reading frames, four general types of f. suppressors are expected: those which suppress mutations in sequences of repeated G, C, A, or U (in the mRNA). The codons GGG, CCC, AAA and UUU specify glycine, proline, lysine, and phenylalanine, respectively (→ genetic code).

Mutations of ribosomal proteins may also lead to the suppression of f. mutations. In this case, the structure of the → ribosome may promote or restrict the frequency of translation of quadruplets of bases by wild-type tRNA's.

frameshift suppressor tRNA — → frameshift suppressor; suppressor tRNA.

freemartinism — a modification of bovine gonadogenesis resulting in sterility of the genetic female co-twin of a male and accompanied by blood admixture between the twins in utero.

frequency-dependent fitness — → fitness differences the intensity of which changes with alterations in the relative frequency of genotypes in the population. If fitness is not determined by interaction with other individuals of the same species the phenomenon is called f.-independent f.

frequency-dependent selection — → selection in which the → fitness of genotypes or given alleles is dependent on their frequency in a → gene pool (population), i.e., when a certain allele is rare, it may confer a reproductive advantage on the individual carrying it; if it becomes common it may prove to disadvantageous.

If the fitness of genotypes is independent of their relative frequency in the population the selection is called frequency-independent s.

F surface exclusion — F-specific → exclusion (→ F plasmid) of the formation by bacterial → conjugation of an F-prime derivative of a cell that already carries an F plasmid and is, therefore, a poor recipient in conjugational crosses.

full mutant — → half-mutant.

full sibs — individuals that have both parents in common (→ half sibs). Full sib mating is the most extreme form of → inbreeding that can occur in bisexual diploid organisms.

functional suppressor — any → suppressor which is allele-unspecific in activity and opens up a metabolic bypass that may circumvent more than one step in a metabolic pathway and therefore leads to suppression of mutations in more than one gene.

fundamental number (Matthey 1945) — the number of chromosome arms (symbol FN) of a chromosome set (karyotype). Change of f.n. may occur by chromosome mutations, e.g., → centric fusions or → dissociations.

furrowing — → cleavage and → cytokinesis.

fusion gene — a hybrid gene consisting of coding sequences and a selected promoter which assures efficient transcription and translation. F. genes are produced by → recombinant DNA technology. The proteins encoded by f. genes are called fusion proteins and consist of all or part of the amino acid sequences of two or more proteins.

fusion nucleus — in flowering plants, a nucleus produced in the → embryo sac by the union of a nucleus from the chalazal and with a nucleus from the micropylar end. Usually the f.n. later unites with one sperm nucleus to give a triploid nucleus from which the → endosperm is derived.

fusion translocation — = → centric fusion.

fusogen (Pontecorvo et al. 1977) — any agent promoting in vitro → cell fusion, such as polyethylene glycol or Sendai virus (→ electrofusion).

fusome (Hirschler 1948) — any of the intercellular bridges in various tissues which undergo a common, synchronous development. Unlike the other intercellular complexes such as → desmosomes and → plasmodesmata, the fusomes merely consist of single, large, annular openings in the cell wall. Electron microscope observations support the theory that these bridges allow the free passage of cytoplasmic materials, including the mitochondria. The fusomes are probably the result of incomplete cell division where the annular furrow of the spindle residual-body does not divide completely during → cytokinesis. The fusomes show a strongly osmophilic annular thickening around the opening.

G

gametangiogamy (Kniep 1928) — the fusion (copulation) of whole → gametangia in certain zygo- and ascomycetes. The pairwise union of multinuclear cells is followed after a more or less long time (→ dikaryophase) by a pairwise fusion (karyogamy) of sexually different nuclei. In this case, karyogamy takes place simultaneously between the nuclei of many pairs of nuclei, not as in → gametogamy between two gametic nuclei ("polyfertilization" according to Hartmann).

gametangium — → germ cell.

gamete (Strasburger 1877) — a mature reproductive cell (in some cases only a nucleus), capable of fusing with a cell of similar origin but of opposite sex to give a → zygote. The gametes are produced from gametocytes during → gametogenesis. Normally, the chromosome number of the gametes is half that of the somatic chromosome number of the organism from which they derive; the reduction of the chromosome number is brought about by → meiosis. The genetic composition of gametes is determined by the genetic composition of the parent organisms in which the gametes are formed and by the behavior of the chromosomes in meiosis. Gametes possessing the somatic chromosome number are known as "unreduced gametes". They normally occur in individuals which undergo diploid → parthenogenesis, and occasionally in sexually reproducing organisms, owing to irregularities in meiosis. The fusion of two unreduced gametes yields a polyploid zygote (individual). In many one-celled organisms the gametes scarcely differ from the vegetative adult organism in size and form. This situation is termed "hologamy". In "merogamy", the gametes are smaller than the parent cells and are frequently motile and different in shape. They are usually produced from the vegetative parent cell by multiple division. The following distinctions are made on the basis of a difference in size and form between the male and female (+ and −) gametes: (1) **Isogametes:** male and female (+ and −) gametes are strikingly similar to each other. The mutual fertilization process is called "isogamy". (2) **Anisogametes:** male and female (+ and −) gametes are slightly different in size and structure. The fertilization process is referred to as "anisogamy". (3) **Heterogametes:** male and female gametes are strikingly different in size and form; the male gamete (sperm, spermatozoon) is usually very small and motile, the female (egg, ovum) relatively large and immobile. The fertilization process is referred to as "heterogamy".

Gametes are spoken of as balanced and unbalanced, respectively, according to whether they contain a complete chromosome complement or a complement in which there are → deletions or → duplications of genetic information.

gametic excess — = → linkage disequilibrium.

gametic disequilibrium — the phenomenon that alleles at different loci may not be randomly associated with each other in a population (often = linkage disequilibrium). G.d. may be a consequence of selection, genetic drift, migration, coancestry, mutation, and genetic hitchhiking. The rate of decay of g.d. is influenced by linkage, inbreeding, and the reproductive system (→ segregation distortion).

gametic mortality — → zygotic mortality.

gametoblast — archesporium tissue which is not yet differentiated (→ archesporium).

gametocyte — → germ cell.

gametogamy (Kniep 1928) — the sexual union → (copulation; fertilization) of two single-celled gametes of different sex and the fusion of the gamete nuclei to give the zygotic nucleus.

gametogenesis — the formation of female and male → gametes, that is sex cells (→ oogenesis; spermatogenesis).

gametogonium — → germ cell.

gametophyte (Hofmeister 1851) — the haploid, sexual generation which produces the → gametes in → diplo-haplontic plants characterized by having an → alternation of generation. In diplo-haplonts, the products of meiosis are → meiospores (→ megasporogenesis; microsporogenesis) which undergo mitotic divisions and develop into haploid individuals, → the gametophytes (→ sporophyte). Micro- (bearing sperms) and macrogametophytes (bearing an egg cell) may be distinguished.

gametotoky — = amphitoky (→ parthenogenesis).

gamobium — the sexual generation in those organisms which exhibit an → alternation of generations.

gamogenesis — = sexual → reproduction or gamogony.

gamogony (Hartmann 1904) — = sexual → reproduction.

gamone — any of a group of biological agents which initiate the processes leading to → fertilization. They are subdivided into: (1) **Gynogamones:** The female fertilization substances liberated from the egg cells. The gynogamone-I-complex effects the chemotactic attraction of the sperm, the activation of sperm movement, the antagonism towards androgamone I. Gynogamone II (probably identical with "fertilicine") causes sperm agglutination. (2) **Androgamones:** Male fertilization agents. Three complexes are distinguished. Little is yet known of the activity of androgamone I. Androgamone II acts to release the egg jelly, to precipitate the substance of the egg membrane, and to neutralize the agglutinating action of gynogamone II. Androgamone III probably exerts a liquefying action on the egg surfaces.

gamont (Hartmann 1904) — → germ cell.

gamontogamy — = → hologamy.

gamophase (Winkler 1920) — = → haplophase.

gap — (1) a region in a double-stranded DNA molecule where at least one nucleotide in one of the strands is missing (gapped region); (2) a totally achromatic region (lesion) in a chromatid (a chromatid g.) or both chromatids (a chromosome g. or an isochromatid g.) of a metaphase chromosome; (3) a discontinuity in variation.

gap junction (Robertson 1963) — any of the specialized areas (= nexuses) in the cell membranes of two apposed vertebrate or invertebrate cells in which fusion occurs, in contrast to occluding → tight junctions which act to seal intercellular spaces. G. junctions participate in electronic and metabolic coupling and are considered as low resistance pathways for intercellular communication.

G. junctions are composed of units, called connexons, which are embedded in the apposed (3.5 nm space) membranes, in register, and linked to each other. The connexon is often organized on a regular hexagonal lattice and is thought to be constructed from six protein oligomers (8.5 nm subunits) arranged in the form of an annulus and delineating a narrow channel through the membrane.

gap repair — the filling of gaps in one chain of a DNA duplex. The gaps arise during → excision repair.

gastrulation — → embryonic development.

GC box (Kadonaga et al. 1986) — in → promoters and → transcriptional enhancers of certain eukaryotic genes, a DNA sequence (GGGGCGGGGC and closely related sequences) which are recognized by sequence-specific eukaryotic → transcription factors (such as SP 1)

G + C ratio — the proportion of guanine (G) + cytosine (C) in DNA, expressed as a molar percentage of total bases (→ base pair ratio).

geitonogamy — → idiogamy.

gene (Johannsen 1909) — the basic unit of Mendelian → inheritance which represents a contiguous region of DNA (or RNA in some viruses) corresponding to usually one (less often two or more) transcripton unit or → transcripton. The particular sequence of nucleotides along the nucleic acid molecule represents a functional unit of inheritance (defined operationally by the → cis-trans test), a → cistron (→ split gene; pseudogene; orphon; retroposon). Genes can be divided into those which code for polypeptides (structural genes), those which are transcribed into RNA (→ transfer RNA, ribosomal RNA) but not translated (→ genetic transcription, genetic translation) into proteins, and possibly those whose functional significance does not demand that they are transcribed at all (DNA sequences which may govern the punctuation or regulation of genetic transcription).

A g. consists of a linear array of potentially mutable units (mutational sites or → mutons) between which intragenic → genetic recombination (→ recon) can occur and permits the → genetic mapping of mutational sites.

In **prokaryotes**, genes are often in tight clusters, resulting in overlapping → transcription units and RNAs (overlapping genes may also occur in eukaryotes and overlapping transcripts derived from opposite DNA strands have been reported). Several mechanisms exist in prokaryotes that allow one gene to specify more than one polypeptide: (1) Translation of a gene in more than one → reading frame; (2) internal initiation of translation in the same reading frame; (3) suppression of a → termination codon.

Many **eukaryotic genes** are discontinuous, with coding sequences (→ exons) alternating with essentially noncoding intervening regions (→ introns). Both exons and introns are subject to transcription, but introns are subsequently removed from the initial RNA transcript (→ RNA splicing). Additional modules of eukaryotic genes are leader and trailor regions bearing or near control elements (→ promoter). Most of the known DNA sequences that promote accurate and efficient transcription (by RNA polymerase II) are located in the flanking regions: the → TATA box is located about 30 nucleotides upstream (5′) from the → cap site and has the canonical sequence ${}^{T}_{C}$ATA${}^{T}_{A}$AAG (required for specific transcription); the → CAAT box is located about 80 nucleotides upstream from the site of initiation and has the canonical sequence GG${}^{T}_{C}$CAATCT (required for efficient transcription), as well as further upstream elements. All these are clustered within about 110 bp preceding the initiation site of → genetic transcription. Other regulatory sequences, such as → enhancer sequences, → promoters, and others, can be found upstream, downstream, or within the genes they activate.

Many eukaryotic genes display structural features suggesting that they have evolved through the amplification of intragenic sequences (internal repetition). During evolution eukaryotic genes have often evolved into → multigene families. Many functioning genes of eukaryotes are flanked by copies showing varying degrees of degeneracy (→ pseudogene).

Transcript heterogeneity observed for several eukaryotic genes may find its structural basis in the presence of two (or more) transcription start points and several → polyadenylation sites.

Several different eukaryotic genes, including those coding for tRNA, rRNA, and histones, occur in the form of tandemly iterated DNA sequences. Tandem arrays of amplified genes may also arise as a result of differentiation processes or in response to drug exposure (→ gene amplification).

On the basis of their repetition in the eukaryotic nuclear → genome several categories of genes may be distinguished (→ repetitious DNA): (1) genes present in a single copy; (2) clustered multicopy genes with copy numbers ranging from 2 to 10; (3) dispersed multicopy genes with the copies dispersed around the genome; (4) tandemly repeated, high-copy number genes.

Two types of sequence repetition are known to occur within protein-encoding eukaryotic genes: (1) A short nucleotide sequence is tandemly duplicated within the coding portion of a g. and is reflected by a repeating amino acid sequence of functional significance; (2) in members of multigene families, the existence of short intergenic sequence homologies between → domains of certain functionally related but nonidentical genes.

As a result of → gene mutation, alternative forms of a particular g. are produced which are referred to as → alleles. The existence of a gene is extrapolated from these alleles which generally influence the same phenotypic → character or trait(being the product of → gene expression). In the absence of information concerning the primary gene product (the polypeptide coded for by a cistron) and its mutational alterations, the operational definition of a g. is usually derived from cis-trans- or complementation tests.

Genes in → diploid organisms occur as pairs of alleles. The members of a pair segregate (→ segregation) during → meiosis so that any offspring receive only one member of a pair from each parent. The phenotypic expression of genes depends upon allelic and nonallelic interactions (→ gene interaction).

Genes are the basis of both continuous (qualitative) and discontinuous (quantitative) characters and usually produce effects on a wide variety of biochemical and morphological characters. So-called **"pleiotropic" "effects"** are the result of a variety of effects originating from a single primary action, i.e., coding for a definite polypeptide ("one cistron-one polypeptide model").

The collection of genes contained within one chromosome in linear sequence (→ genetic map) and in definite positions constitutes a → linkage group. There is evidence, in some microorganisms at least, that genes which control a series of related biochemical reactions are often adjacent to each other in the linkage structure (→ gene cluster). The same is true for the case of the mutational sites within one gene. Sites with similar properties show a tendency to cluster and are not distributed randomly within the nucleotide sequence defined functionally as a gene.

With respect to its specific location in higher organisms, a gene may be contained in a → sex chromosome ("sex-linkage") or in one of the autosomes. According to their relative location two nonallelic genes may be members of the same or different linkage groups.

In relation to allelic genes the manifestation of a gene may be → dominant, → recessive, semi-dominant, or → combinant; in relation to nonalleles → epistatic, hypostatic, → complementary, or indifferent to interaction processes. The → penetrance of a gene may be complete or incomplete, equal or different in both sexes, or confined to one sex only. Likewise, the → expressivity of a gene may be stable or variable, equal or different in both sexes.

Nucleated cells of eukaryotes may contain, besides the nuclear genes, **extrachromosomal or extranuclear "genes"** showing non-Mendelian inheritance. In some cases these extranuclear hereditary determinants seem to be foreign organisms (e.g., viruses, kappa particles in → killer strains of **Paramecium**); in others, they appear to be associated with normal constituents of the cell (plastids, mitochondria, kinetosomes, and kinetoplasts). Chromosomal and extrachromosomal genes are interrelated in two ways; the former are able to mutate the latter, and both may interact to produce particular phenotypes.

gene action — the expression of → genes by → genetic transcription into complementary RNA sequences and, in the case of structural genes, the subsequent translation of messenger RNA into polypeptide chains which form the ultimate protein products of g.a. (primary g.a.). In contrast to prokaryotes, where the transcription and translation of genetic information take place in one compartment (→ cell compartment), in eukaryotes the two processes are separated topologically by the → nuclear envelope. In addition, in these cells genetic information is also present and utilized within certain cytoplasmic organelles (→ organelle DNA), notably → mitochondria, and → chloroplasts (in plants). The operation of the nuclear and extranuclear → genomes is coordinated.

Gene action is called **"cell-specific"** if the expression of a particular gene is limited to a definite cell system, and **"phase-specific"** if the manifestation is limited to a definite developmental period. Both phenomena are consequences of differential → gene activation.

The action of a particular gene is **"sex-limited"** if the corresponding phenotypic character is realized in only one of the sexes. **"Sex-controlled"** g.a. is a type of action characterized by the different expression of a gene in male and female organisms. Genes showing either sex-limited or sex-controlled manifestation may be localized either in sex chromosomes or autosomes (→ gene expression).

gene activation — the process of activating a → gene resulting in a state of transcriptional readiness rather than transcription itself (→ gene expression). In eukaryotes, g.a. is usually accompanied by an altered structure of → chromatin that results in an increased susceptibility to nucleases (→ nuclease hypersensitive site).

gene alignment — a method based on the preferential loss of (human) chromosomes from (rodent-human) hybrid cells. Identification of a specific (human) gene product in the hybrid cells allows g.a. to be made by correlating the expression of the specific gene product with the presence of a particular (human) chromosome.

genealogy — a set of relationships among taxa which are based upon their evolutionary branching sequence.

gene amplification (Brown and David 1968) — an increase in the copy number of a → gene (→ DNA sequence amplification). G.a. must involve at least two general steps: (1) Several successive initiations of → DNA replication at one replication origin during one cell cycle (re-replication); (2) resolution, in which the multiple copies of sequences give rise (by recombination) to stable intrachromosomal (→ homogeneously staining region) or unstable extrachromosomal structures (→ double minute).

G.a. is thought to be frequently related to a developmental need for accumulation of a gene product in massive amounts in a relatively short period of time (ribosomal gene amplification) and also occurs in response to selective agents. The eukaryotic DNA sequences that become amplified appear to include the coding (exon) and noncoding (intron) regions as well as large amounts of flanking DNA.

gene assignment (McKusick and Ruddle 1977) — the localization of specific genes to individual chromosomes (linkage groups) by any of the following procedures (Eicher and Washburn 1978): (1) Hybridization of purified DNA sequences to chromosomes; (2) segregation of translocated chromosomes in somatic cell hybrids (→ gene alignment); (3) → deletion mapping; (4) → genetic recombination relative to → translocation breakpoints; (5) duplication-deficiency analysis which allows the assignment of specific genes to one side of a translocation breakpoint by comparing the phenotype of individuals with a defined chromosomal imbalance to that of chromosomally balanced individuals.

gene bank — a collection of bacterial clones containing hybrid → cloning vectors (plasmids or phages) with overlapping inserts of cloned DNA which together comprise the whole → genome of an organism (= gene library; colony bank). The most commonly used method for establishing a g.b. consists of the partial digestion of genomic DNA, its fractionation, and its ligation with purified phage DNA.

A cloning vector carrying an easily identifiable gene or an appropriate RNA or DNA probe can be used to locate other vectors carrying neighboring portions of the genome. Since a g.b. is initially constructed from randomly sheared DNA which provides the necessary random ends, vectors containing adjacent regions of the genome can be identified.

gene cloning — the isolation and amplification of specific genes by → molecular cloning.

gene cluster (Demerec and Hartmann 1959) — any group of two or more closely linked genes coding for the same or similar products (→ multigene family). The genes contained in a g.c. are co-ordinately expressed. G. clusters may have evolved as functional developmental units with sequential and differential expression of the genes within these units (→ operon; regulon).

genecology (Turesson 1923) — the study of the hereditary structures that adjust natural races to the ecology of their envionment.

gene complex (Brink 1932) — = → gene cluster (→ Renner complex).

gene conversion (Winkler 1930; Lindegren 1949) — the nonreciprocal recombinational transfer of genetic information between homologous DNA sequences (allelic or homologous nonallelic genes) without an accompanying information exchange.

By g.c. gene A interacts with A′ in such a way that the nucleotide sequence of part or all of gene A becomes identical to that of A′, yet each gene retains its integrity and physical location.

The c. process is said to be biased if for a pair of alleles the second allele is converted into the first more often than the first is converted into the second. With such a bias, g.c. acts as a repair mechanism if the second allele is a mutant while the first is of wild type.

G.c. between different alleles at the same locus (interallelic or intragenic c.) is in roughly 50% associated with reciprocal recombination of flanking marker genes; g.c. between alleles at different loci (interlocus or intergenic g.c.) is apparently not associated with reciprocal recombination events. Interchromosomal g.c. involves alleles on two homologous or nonhomologous chromosomes; intrachromosomal g.c. is possible when multiple copies of a gene are present on the same chromosome, and may be either intra-chromatid or may involve two sister chromatids. Both types of g.c. lead to sequence homogeneity (→ molecular drive; gene correction) and g.c. is, genetically, a mechanism yielding genes of identical sequence where the gene were nonidentical. In contrast to → unequal crossing over, the homogenizing influence of g.c. does not depend upon tandem-repeating sequence arrangement. G.c. occurs in mitotic and meiotic cells.

G.c. is assumed to involve → heteroduplex formation between one mutant and one wild-type allele and → mismatch repair on heteroduplex DNA (Holliday model). According to another model (Meselson-Radding model) asymmetric DNA synthesis is the source of g.c. The double-strand break repair model proposes transfer of duplex information between homologous duplex DNA. At least two types of conversions are known: (1) Long-patch c. in which several thousand bases are excised and DNA rapidly becomes homoduplex by a repair process specific for mismatched bases; (2) localized c. in which fewer than 27 bp are involved.

gene correction (Hood et al. 1975) — any mechanism compensating gene damage, such as → nonsense mutations and small deletions, and ensuring that members of particular eukaryotic repeated DNA families (→ multigene family) show greater homology within a species than between species (→ concerted evolution; molecular drive). In the main, two mechanisms have been proposed to function in the protection of conserved regions in multigene families from drift: (1) → Unequal crossing over; (2) → gene conversion. In addition, various → DNA repair mechanisms function in the production of properly paired nucleotide pairs from a sequence of heteroduplex DNA containing an illegitimate base pair.

gene diversity (Nei and Roychoudhuri 1974) — = → heterogeneity index.

gene dosage — the number of → alleles in a particular genotypes (→ gene amplification); this depends inter alia on the degree of → ploidy. In haploids, each allele is represented once, in diploids (with two chromosome sets) twice, in polyploids or polysomics more than twice. The completely → recessive genotype for a particular locus in the case of autopolyploidy or polysomy is referred to as **"nulliplex"** (aaa, aaaa, etc). **"Duplex"**, **"triplex"**, **"quadriplex"**, etc. refer to genotypes in which the corresponding → dominant alleles are present once (e.g., Aaa, Aaaa), twice (AAa, AAaa), three times (AAA, AAAa), and four times (AAAA in the case of "tetraploid").

The alleles of sex-linked genes located in the → X-chromosomes of organisms with an XX-XY- or XX-XO-mechanism of → sexdetermination are present twice in the homogametic (XX) sex, and only once in the heterogametic (XY or XO) sex, that is, show a dosage difference connected with the sex of the organism (→ dosage compensation).

gene duplication — the process occurring in various organisms by which new genes arise, i.e., the duplication of a single gene. G.d. and subsequent differentiation play an important role in the evolution of genetic systems. Since the original function can be maintained by one of the two copies, the duplicate is able to escape the pressure of natural selection and can accumulate the mutations necessary to develop a new function. G.d. may result in → multigene families, → supergenes, or → pseudogenes.

Two kinds of g.d. may be distinguished: (1) Contiguous, i.e., the duplicated sequence coexists within the boundaries set by the start and stop signals for protein synthesis of the original gene; (2) discrete, the duplicated sequence is outside these boundaries. In the first case, the result is a larger transcription product and a protein produced at the expense of the pre-existing protein. In the second case, two independent genes and gene products result.

Four categories of events may lead to g.d.: (1) Random nonhomologous chromosomal breakage and reunion; (2) unequal but homologous crossing over between two copies of some repeated DNA sequence that by chance flanks a single copy gene; (3) RNA-mediated → transposition (→ reverse transcription); (4) → gene amplifikation.

gene eviction — the retrieval of a chromosomal copy of any gene which has previously been cloned (→ molecular cloning). The method includes: (1) Transformation of a host with a recombinant → integrative plasmid; (2) by homologous recombination a duplication of the gene in question is created; (3) subsequent treatment of the transformant's DNA with → restriction endonuclease generates a linear molecule which contains bacterial sequences necessary for replication and selection in *E. coli* together with a hybrid region containing host and plasmid DNA from the chromosome segment originally cloned (together with some flanking sequences from the donor chromosome); (4) the fragment may be circularized by → DNA ligase and transformed back into *E. coli* where it will replicate.

Since DNA sequences adjacent to the gene of interest are retrieved by g.e. the method can be used for → chromosome walking.

gene evolution — the evolutionary transformation of a single gene into another one with new functions. G.e. is based on two fundamental processes: (1) Changes in the structure of the gene; (2) changes in its regulation. → Gene duplication generates the material upon which evolution will work.

gene expression — the phenotypic manifestation of genes; during g.e. information is transferred from its storage site (usually a double-stranded DNA region) to its site of action (usually a protein). The expression of (nuclear) genes as a cytoplasmic gene product entails several distinct events: (1) → Genetic transcription, (2) → RNA processing or maturation, (3) transportation of the → transcript to the cytoplasm (eukaryotes), and (in the case of → structural genes) → genetic translation of the transcript.

DNA-dependent → RNA polymerase is the key enzyme responsible for the readout of genetic information. Since only a small fraction of genes in prokaryotes and eukaryotes are expressed at any one time, it is crucial that in the first step of g.e. RNA polymerase recognizes proper sites on the DNA template where specific initiation of RNA chains can take place (→ promoter).

In **prokaryotes**, g.e. is regulated at the level of initiation of transcription (→ genetic regulation) and by termination signals located at the end or within the → transcriptional units (→ attenuator). The level of expression of a gene is determined by three elements: (1) The rate of transcription; (2) the stability of the RNA transcript; (3) the efficiency of translation.

Regulation of g.e. may occur by a negative element, a → repressor, or positive elements (activators) affecting the initiation of transcription, and antiterminators affecting the termination of RNA transcripts before they can be extended into the regulated structural genes (→ operon).

The reprogramming of g.e. leads to → cytodifferentiation in prokaryotic and eukaryotic systems.

In **eukaryotes**, g.e. may be controlled by two mechanisms: (1) Those in which g.e. is modulated by transcriptional, posttranscriptional, as well as translational and post-translational control; (2) those by which the genes are altered (DNA diminution, DNA amplification, DNA rearrangements).

Post-transcriptional and post-translational controls are involved in the fine tuning of g.e.; primary control occurs at the transcriptional level (→ DNA methylation). The faithful expression of a eukaryotic genes requires the presence of DNA sequences flanking the gene and units which promote the expression of genes at a specific developmental stage. During e. of eukaryotic split or mosaic genes, coding and intervening sequences (→ exon; intron) are copied faithfully into an initial RNA transcript. Excision of the introns and subsequent splicing (→ RNA splicing) of the exons occur in the → nucleus. The mature mRNA which after proper processing has a chemical group (5′ end) known as a cap and a so-called poly(A) tail (added at the 3′end) leaves the nucleus and translation occurs in the cytoplasm.

DNA sequences upstream of genes often play a crucial role in regulation of eukaryotic g.e. These include polymerase II recognition and transcription start sites (→ CAT box; TATA box), → transcriptional enhancers, and very short repeats which may be important in coordinate induction of genes. Site-specifc demethylation of DNA regions near the coding sequences may be involved in eukaryotic selective g.e.

One of the critical factors of eukaryotic g.e. appears to be → chromatin structure; transcriptionally active genes are largely confined to chromatin domains in which the DNA structure is in a relatively extended (open) configuration characterized by increased sensitivity to nuclease digestion (→ nuclease hypersensitive site), DNA hypomethylation, and possibly by differential binding of → nonhistone chromosomal proteins (→ high mobility group protein).

Most developmentally regulated genes exhibit tissue-specific expression conferred by interaction of DNA sequences in cis (within or near the genes) with trans-acting regulatory molecules in specialized cells.

During the developmental program of organogenesis, there is a serial appearance of phenotypically distinct cell types that exhibit selective patterns of g.e. (→ homeobox).

In some cases, expression of a single gene may result in the production of multiple forms of mature RNA transcripts. This can be achieved by alternative → RNA splicing of an initial (primary) RNA, utilization of different promoters or → polyadenylation sites, or a combination of these factors.

The expression of genes in foreign organisms as studied by means of → recombinant DNA technology is called **heterospecific g.e.** Expression of cloned genes can be experimentally elevated by several manipulations: (1) Increasing of cellular concentration of these genes (gene dosage) by cloning multiple copies of the gene or by use of a high-copy-number → cloning vector; (2) increasing the frequency of initiation of the gene's transcription by using cloning vectors containing strong promoters (inducible or constitutive) upstream of the cloning sites; (3) increasing the frequency of initiation of translation of the gene's mRNA by employing cloning vectors containing strong translation signals just upstream of the structural gene sequences.

gene family — = → multigene family.

gene farming — the expression of foreign genes and the multiplication of their products after their transfer (→ gene transfer) into an appropriate host (= genetic farming).

gene fixation time — the average time required for the fixation of a new gene (mutant allele) variant in a population (→ genetic load).

gene flow — (Birdsell 1950) — the movement or dispersal of genes within or among populations. G.f. occurs by movement of gametes, immature, and adult individuals prior to reproduction. While most animals move freely, plants are generally sedentary and g.f. takes place more or less passively by pollen and seeds.

G.f. is an important factor in determing the structure and cohesiveness of populations and → species. Spatial restriction of g.f. can lead to nonrandom mating and to subdivision of a population into genetic neighborhoods. Extensive g.f. over large geographical areas can overshadow the influence of → selection and may lead to genetic similarity among populations and increased uniformity within a species.

Actual g.f. can be measured directly by examining the spread of distinctive alleles in a population.

gene frequency — the proportion of one particular type of → allele to the total of all alleles at this genetic locus in a breeding population, or the probability of finding the specific gene under consideration when a gene is randomly chosen from the population (= allele frequency). The genetic composition (the → gene pool) of a population is described by the array of gene frequencies, i.e., by specifying the kinds and numbers or proportions of different alleles at each locus. Among a group of individuals, gene frequencies at a particular locus may be estimated form the → genotype frequencies. Both gene and genotype frequencies are constant from generation to generation, in the absence of migration (gene flow), mutation, and selection (→ Hardy-Weinberg law). Analyses of gene frequencies are an important tool of population genetics and to some extent permit a genetic reconstruction of the genetic history of populations. Changes in the g.f. depend mainly on three types of processes: (1) **Systematic processes,** e.g., recurrent mutation, recurrent selection, and recurrent immigration. (2) **Random processes** such as various fluctuations in g.f., in particular those due to accidents of sampling in selection as well as migration. (3) **Unique processes** such as nonrecurrent mutation, selection, migration, and accidental fixation of an allele.

Yet another category consist of changes in the system of coefficients determining the magnitude of the processes mentioned under 1 —3. These changes occur because the gene pool of the population has been undergoing transformation or the environment has changed.

Systematic processes tend to bring the g.f. to stable equilibria at particular values. Dispersive processes tend to distribute the g.f. away from equilibrium value. If not held in check by the systematic processes this would in the end lead to all genes being either fixed or lost in populations not infinite in size. In fact, the opposing tendencies of both types of process reach a point of balance at which the dispersion of g.f. is held in check by the systematic processes.

The gene frequencies of a certain locus may be inferred from the genotypic frequencies in the following manner: suppose that two alleles (A_1 and A_2) are present and that 100 genotypes have been classified, then an example situation is:

		A_1A_1	A_1A_2	A_2A_2	Total	
Number of genotypes		30	60	10	100	
Number of genes	A_1	60	60	0	120	200
	A_2	0	60	20	80	

Thus the frequency of A_1 is 60 or 0.6, that of A_2 40 or 0.4 in general:

Genes		Genotypes		
A_1	A_2	A_1A_1	A_1A_2	A_2A_2
Frequency p	q	P	H	Q

$p + q = 1$ and $P + H + Q = 1$. Since each diploid individual carries 2 genes of each locus, the frequency of A_1 alleles is $\frac{1}{2}(2P + H)$, and the relation between g.f. and genotype frequency in the individuals, $p = P + \frac{1}{2}H$, $q = Q + \frac{1}{2}H$.

gene fusion — the combination (fusion), either genetically or through use of → recombinant DNA technologies, of information from two separate DNA regions (structural genes) resulting in the formation of new protein coding sequences (cistron fusion), or the fusion of a regulatory region of a gene of interest to a structural gene coding for an easily assayed protein. G.f. can be used, e.g., to obtain eukaryotic gene products in bacteria and to understand the regulation of → gene expression.

gene identity — the probability that two randomly chosen alleles are identical by descent, i.e., are copies of an ancestral allele.

gene imprinting — the phenomenon that genetic activity of a locus depends on parental origin as well as allelic constitution (→ chromosome imprinting).

gene insert (Breathnach et al. 1977) — any DNA sequence that is part of a eukaryotic gene but not represented in mature mRNA (→ insertion sequence; intervening sequence).

gene interaction — interaction between allelic or nonallelic genes of the same genotype in the production of particular phenotypic → characters. Dominance is the major form of interactions between → alleles. On the basis of interaction with their wild-type alleles, mutant alleles are designated amorphs, hypomorphs, hypermorphs, antimorphs or neomorphs.

Nonallelic interactions involving two allele pairs result in variations of the normal dihybrid → segregation scheme. Conventional modifications include the following segregation ratios together with the groupings from 9A.B.:3A.bb:3aaB.:1aabb with which they can be equated: (1) 9A.B.:3A.bb:4(3aaB. + 1aabb) in the case of recessive epistasis. (2) 12(9A.B. + 3A.bb):3aaB.:1aabb with modifier gene interaction (dominant suppressor or dominant → epistasis. (3) 9A.B.:7(3A.bb + 3aaB. + 1aabb) with complementary gene interaction. (4) 13(9A.B. + 3aaB. + 1aabb):3A.bb with modifier gene interaction (recessive → suppressor). (5) 9A.B.:6(3A.bb + 3aaB.):1aabb in the case of additive effects of the dominant alleles. (6) 15(9A.B. + 3A.bb + 3aaB.):1aabb in the case of duplicate loci with dominance of both.

gene isolation — the identification and isolation of desired genes from the whole genome. G.i. often involves the use of specific hybridization probes (derived from natural sources such as mRNA or prepared from cDNA) or synthetic oligonucleotides (predicted from the amino acid sequence of a protein). Once isolated, a structural analysis and, in appropriate systems, a functional analysis can be conducted using the cloned DNA (→ molecular cloning).

gene library — = → gene bank.

gene location — the determination (by various methods) of the distances between linked genes of the same chromosome (→ chromosome map, genetic map), or their degree of → linkage.

gene locus (Morgan et al. 1915) — the position occupied by a → gene in the chromosome or → chromosome map. A g.l. may be established by using cytological (cytologic mapping) and genetic methods (→ genetic mapping). By means of intragenic recombination (→ genetic recombination) it may be subdivided into linearly arranged units called "sites" (→ gene map). Allelic gene loci are termed homologous if they occur in homologous chromosomes or chromosome segments (→ allele).

Actions and processes are "locus-specific" (Hadorn 1955) if they can be associated exclusively with a certain place on the chromosome.

gene loss — gene elimination (via chromosome elimination) from the cells during differentiation (in some protozoans, insects, and crustaceans). Only cells destined to produce germ cells maintain the complete DNA complement.

gene magnification (Ritossa 1968; Ritossa and Scala 1969) — the heritable reparative process within the genome (e.g., of *Drosophila melanogaster* males homozygous for the bobbed mutation) whereby organisms deficient in ribosomal DNA (→ rDNA) are capable of regenerating new nucleolus organizer material. G.m. is a phenomenon related to → gene amplification and probably involves a mechanism for the interstitial lengthening of chromosomal regions.

gene map — by analogy with the → chromosome map, the graphical linear arrangement of small chromosome segments; the → mutational sites belonging to the gene concerned ("cistron") are plotted to correspond with their relative distance from one another. As in the preparation of chromosome maps, the frequency of → genetic recombination (representing an intragenic recombination) between these sites, serves a as measure of the distance which separates them (→ genetic mapping).

gene migration — the migration of genes due to the active dispersal of individuals or biological species (→ gene flow).

gene modification — artifical change of the DNA sequence of a → gene by two general methods: (1) Oligonucleotide-directed in vitro mutagenesis (→ site-directed mutagenesis) for making small insertions, deletions, and base substitutions at specific sites in cloned genes (→ molecular cloning); (2) de novo construction of a completely synthetic gene (→ gene synthesis) using chemically synthesized oligomers that are ligated together.

gene module — a gene part such as an → exon, → intron, → leader sequence and trailor.

gene mutation — any heritable change within the limits of a single → gene ("intragenic mutation" or "point mutation"), as opposed to changes of chromosome structure (→ "chromosome mutation") or chromosome number (→ "genome mutation"). Discrimination between gene mutations and chromosomal structural changes (especially minute ones) is, in practice, frequently impossible.

As a result of g.m. alternative genes (called → alleles) arise. Gene mutations may arise spontaneously (in the absence of any definable cause) or may be induced experimentally by a variety of physical and chemical → mutagens.

Spontaneous g. mutations undoubtedly arise from a multiplicity of events and contain a variety of mutational lesions (base pair substitutions being the most frequent). The types of spontaneous g. mutations and their characteristic frequency depend upon the organism, the gene assayed, and to a certain extent on not readily defined extrinsic and intrinsic factors. They arise by several different mechanisms, such as depurination, deamination, and other forms of → DNA damage and → DNA repair, as well as error in → DNA replication and insertion or excision of → transposable genetic elements (insertion mutations).

A distinction can be made between g. mutations originating as untargeted errors (due to mispairing on intact templates during semi-conservative or repair replication) and mutations that originate as targeted errors (prompted by nonpairing → DNA lesions in the template during error-prone repair of DNA damage). Spontaneous g. mutations may thus be a reflection of the error probability of the appropriate enzyme systems involved in → DNA replication, → DNA repair, or → genetic recombination.

Several factors may cause a general increase in the frequency of g. mutation: (1) A defective DNA repair system; (2) a defective → DNA polymerase; (3) A defective DNA replication enzyme; (4) a defective → DNA ligase; (5) the production of aberrant DNA precursors.

A cell in which a primary mutational event has taken place may give rise either to mutant cells, or to both mutant and wild type cells. In the former case, the mutation is said to be complete; in the latter case it is referred to as a mosaic mutation.

Since the → genetic information carried by a gene is encoded in a specific nucleotide (base) sequence of DNA (RNA takes the role of DNA in some viruses) and is replicated by forming complementary nucleotide chains (→ deoxyribonucleic acid), intragenic mutation may result from any change in the normal sequence of entire nucleotides or their component bases. The types of g.m. are as follows:

A. **Forward mutations:**

(1) Single nucleotide pair (base pair) substitutions

(1.1) **transitions**, i.e., purine replaced by a different purine, or pyrimidine replaced by a different pyrimidine (AT → GC, GC → AT, CG → TA, TA → CG)

(1.2) **transversions**, i.e., purin replaced by a pyrimidine, or pyrimidine replaced by a purine (AT → CG, AT → TA, GC → TA, GC → CG, TA → GC, TA → AT, CG → AT, CG → GC)

(2) At the protein level

(2.1) Neutral mutations resulting in a triplet coding for the same amino acid or for a different but chemically equivalent amino acid.

(2.2) → Missense mutations to a triplet coding for a different and nonfunctional amino acid.

(2.3) → Nonsense mutations to a → chain terminating triplet.

(3) Single nucleotide pair addition or deletion, i.e., → frameshift mutation.

(4) Intragenic deletion of several nucleotide pairs.

B. → Reverse mutations:
(1) Exact reversions restoring the original nucleotide sequence (wild type).
(2) Equivalent reversions resulting in a pseudo wild type.
C. → Suppressor mutations:
(1) Frameshift of opposite sign at a second site.
(2) Second site missense mutation that restores a more or less wild type protein after primary distortion.
(3) Extragenic suppressors
(3.1) → Nonsense suppressors, i.e., a gene undergoes a mutation in its anticodon region that allows for the recognition of and alignment with a mutant nonsense codon resulting in the insertion of an amino acid thereby permitting the completion of translation.
(3.2) Missense suppressors.
(3.3) → Frameshift suppressors.
(3.4) Physiological suppressors, i.e., a defect in one chemical pathway is circumvented by another mutation which opens up another chemical pathway with the same result as in the wild type.

Induced mutations are produced via one or more intermediate steps that are initially reversible: Initially reparable DNA lesions ("mutation initiation") pass through a "premutational phase" and then reach a final phase ("mutation fixation") which begins with a new round of DNA replication. DNA replication and DNA repair can lead from premutational changes to permanent changes (fixed mutations) in the nucleotide sequence of DNA (→ reversion analysis).

The smallest alterable fragment of genetic material, and hence the basic unit of mutations is called the **"muton"** (Benzer 1957); it constitutes a mutational or mutable site within a gene or cistron. The number of **mutational sites** (Pontecorvo 1957) at which → heteroallelic mutations can occur is 500 —1500 per average gene. Each site can exist in several alternative forms. These sites are the deoxynucleotides from which DNA regions making up the genes are constructed. Just as with the order of genes in a chromosome, the rearrangement of mutational sites within a gene is strictly linear.

Not all intragenic mutations are deletions or substitutions of a single nucleotide or base. A commonly detected form is the **"multisite mutation"** involving several or many adjacent sites in the genetic region in question. Multisite mutations fail to recombine with two or more mutations that recombine with each other and their extent is revealed by recombination analysis. They do not revert to the original unmutated (wild type) condition (→ reversion, back mutation) either spontaneously or after treatment with mutagens and are probably always due to the → deletion of the sequence of mutational sites with which they fail to recombine.

The average frequency (chance) with which a particular mutational event takes place per cell (or individual) per generation is referred to as the **"mutation rate"** and may be estimated by various methods from the number of **"mutants".** A mutant is a phenotypically changed cell or organism resulting from a mutation (the standard of reference is the wild-type condition). **"Mutation frequency"** is the frequency of certain mutants in a population; it does not necessarily reflect the rate at which such mutants arose.

A number of factors are responsible for difficulties (especially in bacteria) in obtaining accurate mutation rates: (1) Inhibitory effects (produced by interference of the parent cells with the multiplication of mutant cells) and delayed increase in the number of descendants of a mutated cell (**"cell division lag"**); (2) delay in the phenotypic expression of mutational changes (→ end point mutations) due to the time required for changes to be manifest in the affected biochemical reactions (**"phenomic lag"**); (3) delay of phenotypic expression of a recessive mutation occurring in one nucleus of a multinucleate cell or in only part of the chromosomal subunits; this is due to the time required for the mutant nucleus or mutant strands to segregate from the nonmutated ones (**"segregation lag"**); (4) selective effects causing differential growth of mutant and parental cells.

Application of different mutagens usually results in different **"mutation spectra"** within one gene. The term mutation spectrum denotes the distribution of mutant sites within the limits of a gene after application of a number of mutagens, as determined by mapping a large number of mutations induced by each agent. Generally, the spectrum of spontaneous intragenic mutations is significantly different from that of any of an induced series both in respect to distribution of so-called **"hotspots"**, where many mutations are produced at a particular site, and in the virtual absence of coincidence of spontaneous and induced sites. This is shown by the fact that crosses between them always yield recombinants. The mutation spectra of various mutagens are in fact characterized by overlaps (many of the sites of each series coincide), but usually each differs from the other, especially in the distribution of "hotspots". „Hotspots" and mutagen-specific mutation spectra clearly show that mutational sites within one gene react differently towards different mutagens **("intragenic electivity")**. Furthermore, different genes may react to the same mutagenic treatment with strikingly different mutation rates **("intergenic electivity")**.

The **primary effect** of a g.m. is to cause an alteration in the amino acid sequence of the polypeptide molecule, the primary product of the nonmutated structural gene. Since the polypeptide is synthesized via → messenger RNA as the specific template, the sequence of events is: One mutated gene, one mutated messenger RNA, one mutated polypeptide chain. A single-site mutation causes the substitution of one amino acid for another in the polypeptide. The biological effect of such amino acid substitutions is dependent both on their type an their particular location within the polypeptide. Mutational changes in the polynucleotide sequense of DNA (missense mutation) and the amino acid sequence of the polypeptide are colinear (→ co-linearity).

The immediate effect of altering a protein with enzymatic properties is a change in the efficiency of a particular step in a metabolic sequence (→ gene action) causing a complete or incomplete → genetic block. Gene mutations may affect enzymes both quantitatively and qualitatively. Quantitative changes of constitutive as well as inducible → enzymes may result in a complete inhibition of synthesis or in the production of a decreased amount of the particular enzyme. Qualitative changes give rise to structurally altered enzymes often accompanied by changes in enzyme function. In extreme cases, the catalytic activity of an enzyme may be completely lost by gene mutation (→ cross-reactive material; nonsense mutation). The seemingly simple primary effect of a g.m. is followed by a multitude of secondary events of which the mutant (mostly → recessive) phenotype is the ultimate result. In the large majority of mutant phenotypes that have been studied, it is only the secondary effects which have been analyzed.

Gene mutations are the ultimate source of → genetic variability and thus one of the primary evolutionary forces. They supply the raw material for → evolution which is then sorted out by → selection, → genetic drift, and by recombination. In connection with the → adaptive value which is conveyed by a g.m. to its carrier, three types of gene mutations may be distinguished: (1) The adaptive value of the mutant is less than that of the normal form under all environmental conditions; the g.m. is ultimately eliminated. (2) Under the environmental conditions operative at the time of occurrence of the g.m., the adaptive value of the mutant is higher; the mutant type gradually supplants the original type. (3) Under certain environmental conditions the adaptive value of the mutant is higher than that of the original type, under other conditions lower; neither the mutant nor the original type are eliminated; natural selection establishes a state of equilibrium between the two types.

The majority of gene mutations are of negative selective value and of more or less unfavorable action. Their effect may be lethal or lethal only under certain conditions (→ lethal factor); their presence may be detrimental to the adaptive value of the bearer to a variable degree. In a particular genotypic milieu, they may have a neutral action and in rare cases they may even benefit the bearer organism.

gene overlap — the phenomenon that one sequence of DNA can code for more than one protein by use of different reading frames. Overlapping genes have been observed in various viruses, bacteria, and also higher organisms. The initiation signals of one gene overlap with the termination signals of an adjacent gene.

gene pool (Dobshansky 1951) — the total → genetic information encoded in the sum total of the genes in a breeding → population existing at a given time. The gametes of all breeding individuals in a population furnish a pool of genes from which the genes of the next generation are chosen.

A g.p. which preserves a constant composition throughout a series of generations is denoted as existing in equilibrium. Such a g.p. possesses a balanced structure and its components are "coadapted", i.e., changes of the allele frequencies of one genetic locus may involve changes of the allele frequencies of other loci until either the original balance is restored or a new one established.

The g.p. and the → genotype frequencies may remain unchanged only in successive generations if (1) the population size is large enough to exclude → genetic drift, (2) no mutations are produced preferentially in any given direction, (3) there is no differential selection with regard to certain genotypes, and (4) the genotypes of immigrating individuals correspond to the native genotypes.

If one of these conditions is not fulfilled under the given environment, changes of the gene and genotype frequencies result and these continue until a new equilibrium is established. Biological → evolution is based on changes in the gene pools of populations.

generation — (1) in the case of life cycles with an → alternation of generations, that phase which extends from one to the immediately successive reproduction. (2) In genetical research, a g. corresponds to a whole reproductive cycle and comprises those individuals of a population who are equally removed from a common ancestor.

generation time — the average time between two successive → generations; the time required by a cell population to double (→ cell cycle).

generative — → somatic.

generative nucleus — a haploid nucleus in the → pollen grain which produces two sperm nuclei by mitosis (→ microsporogenesis).

gene reactivation — the reactivation by → DNA rearrangements of silent genes deprived of their → promoters. Three types of g.r. are known (Charlier et al. 1983): (1) Insertion of a foreign DNA sequence (e.g., an → insertion element) in front of a gene which turns it on either because a promoter is created or because the inserted element carried a promoter; (2) a → deletion or a → translocation fused the gene to a sequence allowing efficient transcription; (3) an → unequal crossing over resulted in a tandem → duplication placing the gene in 3′ position with respect to a foreign promoter or creating a promoter at the novel juncture of the rearrangement.

gene rectification — in certain eukaryotic DNA sequence families (→ multigene family), a process of homogenization that results in a greater intra-specific homogenity of the repeated sequences as compared with inter-species homogeneity (= horizontal evolution). The mechanisms possibly involved in g.r. include → gene conversion and unequal recombination between sister chromatids or homologous chromosomes.

gene redundancy — the presence of genes in multiple forms in the DNA of eukaryotes. G.r. can be brought about by (1) a high degree of ploidy of the entire genome; (2) the presence of → polytene chromosomes with lateral multiplicity of genes; (3) extra copies of parts of the genome (→ gene amplification); (4) linear multiplicity of genes in the chromosomes (→ gene reiteration).

gene regulation — = → genetic regulation.

gene regulatory system — → genetic regulation.

gene reiteration — the amplification (→ DNA amplification, gene amplification) of specific genes (DNA sequences) as observed in → mitochondrial, viral, and eukaryotic genomes.

gene replacement (Morse et al. 1956; Orr-Weaver et al. 1981) — the site-specific substitution by homologous recombination of a cloned gene, or part of a gene, for the wild-type copy of the gene in the host chromosome (= marker exchange). The gene used for replacement may be mutated in vitro (→ transplacement; homogenotization). The basic procedure involves two steps: (1) Transforming DNA is integrated at the desired chromosomal location to produce a → heterogenote structure in which transforming and host alleles are separated by vector sequences; (2) the transforming DNA is excised from the genome and two classes of segregants are produced; one of these is indistinguishable from the original strain, in the other the original allele is replaced by the transforming allele.

Incorporation of the mutant gene is usually selected, e.g., by way of its containing a → transposon-encoded antibiotic resistance.

gene rescue (Perucho et al. 1980) — a method allowing the reisolation of a transferred gene from stably transformed (mammalian) cell lines by way of its activity in *E. coli.* The recovery of plasmids is called → plasmid rescue.

gene repression — → genetic regulation.

gene scanning (McKnigh and Kingsbury 1982) — a method in which the insertion of mutations at specific sites allows the determination of those DNA sequences needed for the activity of a gene. By construction of a series of mutants in which the position of the block is systematically moved, a length of DNA can be scanned rapidly for regions of importance for the protein product of a gene, or for regions involved in regulating → gene expression.

gene segment rearrangement — any form of positional change of genes or parts of these which may have important effects on → gene expression. G. rearrangements may be undirected or directed and occur by → DNA rearrangement. Regulatory g. s. rearrangements are of two types: (1) Those in which a structural gene and a promoter are put into contact or separated (usually not correlated with a developmental timetable and without predictable chromosomal locations); (2) those in which different segments of transcribed DNA are brought together in a programmed way to change the base sequence of the coding region (e.g., in the case of the structural genes encoding → immunoglobulins of the immune system).

G.s.r. expands the coding capacity of a locus, enabling a large repertoire of gene products to be generated from a relatively small number of components.

gene sharing (Piatigorsky and Wistow 1989) the phenomenon that a gene may acquire and maintain a second function without duplication and without loss of the primary function, i.e., the gene encodes two entirely different functions.

gene silencing (Ohno 1970) — degeneration of a structural or regulatory gene in a DNA sequence that cannot be expressed (→ gene expression). G.s. usually results from a → frameshift mutation or a chain-terminating → nonsense mutation.

genes in common — genes inherited by two individuals from a common ancestral source.

gene spacer — → spacer sequence

gene-specific probe — a DNA probe (recombinant DNA molecule) that in hybridization to chromosomal DNA only reacts with the gene sequence which it specifies. The approach is valid for the genes coding for rRNA or very abundant mRNA's.

gene splicing — the joining, by cutting and rearranging, of one DNA molecule to another (→ recombinant DNA technology) as, e.g., in the insertion of a desired gene into a → cloning vector (→ RNA splicing).

gene substitution — the replacement in a population of one allele by a mutationally arisen new allele (→ gene mutation). G.s. plays a key role in the evolution of species. It confers a load on the population which is called substitutional load or "cost of natural selection" (→ genetic load). The total variance loss for selection in all the generations involved in a g.s. is equal to the relative increase in → fitness brought about by that substitution (Crow 1970).

gene symbols — → genetic nomenclature.

gene synthesis — the assembly of synthetic genes from oligonucleotides. The technique of g.s. involves annealing of short overlapping complementary polydeoxyribonucleotides to form duplex DNA with single-strand breaks wich stepwise are ligated enzymatically (→ DNA ligase). Subsequently the gene is cloned (→ molecular cloning). During g.s. both DNA strands are constructed in response to a predetermined sequence.

G.s. offers the following advances over other possible ways in which a (eukaryotic) gene can be obtained and expressed in prokaryotes: (1) The exact desired sequence can be directly obtained; the coding and noncoding sequences can be designed at will for prokaryotic expression; restriction sites can be removed or built in, introns can be deleted; (2) g.s. bypasses the often difficult step of isolating the relevant genomic DNA or mRNA; (3) it simplifies the modification of the gene and its product by lengthening or shortening the coding region and allows changing of specific codons and the corresponding amino acids.

gene tagged — of chromosomes or → chromosome mutations marked genetically by marker genes (→ genetic marker).

gene targeting (Thomas and Capecchi 1986) — the intregration into the → genome of exogenous DNA at sites where its expression will be suitably controlled. The exogenous DNA can be in vitro manipulated; integration occurs by homologous recombination (→ gene transfer).

gene technology — = → recombinant DNA technology.

gene therapy — the insertion of a normal gene into an organism (patient) which may correct a genetic defect. Prerequisites of g.t. include isolation of the gene in question (→ gene isolation), its integration into the host cell at an appropriate chromosomal site, and correct regulation of its transcriptional and translational activity. Current approaches to g.t. focus mainly on the → retrovirus-mediated transfer of genes into pluripotent bone marrow stem cells which should allow continued expression of the gene in blood cells.

genetic — ref. to the structure, mutation, replication, and transmission of the genetic material (→ epigenetic; paragenetic).

genetically heterogeneous — of traits or characters where mutations at different loci can independently produce the same trait.

genetic analysis — the resolution of the genetic material into its component parts.

genetic antipolarity (Ito and Crawford 1965) — production of reduced amounts of an enzyme specified by the nearest unmutated gene on the operator side of a → polar and deletion mutant in an → operon (→ genetic polarity). The extent of g.a. is a function of the map location of the → nonsense codon. Nonsense mutations on the operator-proximal side of the gene have the most extreme polar and antipolar effects, i.e., a gradient of degree of antipolarity exists (its orientation runs in the same direction as the gradient of polarity). Antipolar effects may be partially relieved by introduction of a → nonsense suppressor gene.

genetic assimilation (Waddington 1942) — the "genetic fixation" (via selective processes) of "acquired" characters (phenotypes). Prior to fixation they seem to appear solely as a response to specifc environmental stimuli, i.e., they are → modifications whose range of variability is genetically controlled; after fixation (assimilation) they are also produced in the absence of the particular environmental condition ("pseudoexogeneous adaptations").

The phenotypes assumed to be somatically "acquired", stabilized, and genetically assimilated may in fact be a consequence of "threshold effects" (Stern 1958). In the case of "threshold selection", the special environmental conditions may reveal which individuals (from among a number) already carry polygenes and modifiers of the phenotype in question. The character differences may be subthreshold and thus not discernable in the original and above threshold in the new environment where they are endowed with positive selective value. Under these circumstances genotypes whose action is above threshold in both the original and the new environment may arise by crossing.

genetic background — = residual → genotype.

genetic balance (Bridges 1922) — the coordination and balance between coadapted genes and gene systems of a particular genotype. The internal integration of the genotype is prerequisite if the organism is to develop and function as an integrated whole.

genetic block — the reduction in → enzyme activity effected by a → gene mutation. A **complete g.b.** results in the complete absence of a particular enzyme, an **incomplete or leaky g.b.** in the formation of an altered enzyme that has a limited activity. In the case of genetic blocks, a particular product is not produced or is produced in insufficient quantitiy for normal cellular metabolism (→ gene action). As a consequence of such a block, precursor products of chemical reactions in a biosynthetic sequence preceding the blocked one may be accumulated.

Cells (or individuals) which are inviable in consequence of one or more genetic blocks are called "auxotrophic", in contrast to "prototrophic" ones. Their viability may be restored by identifying and supplying the lacking growth factors. Such cells are "monoauxotrophic" if the defect may be compensated by one factor and "polyauxotrophic" if more than one growth factor is necessary. Polyauxotrophy may be the result of one or several genetic blocks. A single block may cause polyauxotrophy (1) if it is located immediate by before a branching point of a biosynthetic sequence (inhibiting all reaction chains starting at that point in the absence of the block), or (2) if separate reaction chains are blocked simultaneously by the inhibition of the biosynthesis of one particular enzmye.

genetic carrier — in human genetics, an individual heterozygous for a wild-type allele and an allele that predisposes for a hereditary disease. Such a person may be clinically normal, but may participate in matings that produce offspring with the homozygous disease condition (→ genetic counseling).

genetic circularity — the presence of linkage structures (maps) without ends in bacteriophages, some animal viruses, bacteria, plasmids, mitochondria and chloroplasts. G.c. may be of various types: (1) The chromosomes are physically circular, i.e., closed molecules of double-stranded DNA. This DNA may have various structural forms including closed (supercoiled) and open circles that are either monomeric or multiple length, and also interlocked cyclic molecules. Both strands of the DNA duplex are covalently closed and free of single breaks in closed circular DNA; the open circular form of DNA is also circular but one or both strands have breaks (physically circular DNA is characteristic of some phages and viruses, bacteria, plasmids, mitochondria, and chloroplasts). (2) The chromosomes are linear but have gene sequences which are circular permutations of each other (e.g., bacteriophage T4 and possibly *Streptomyces*). (3) The chromosomes are linear with identical gene sequences, but they always engage in an even number of → crossovers (not reported up to now).

genetic coadaptation — the mutual adjustment of allele frequencies at two or more loci.

genetic code — the regularitis (coordination principles) according to which the information specifying the amino acid sequences of proteins and the start and stop signals for protein synthesis is stored in the heritable nucleic acid sequences of an organism or cell organelle. The g.c. is determined by both the specificity of the triplet → anticodon of → transfer RNA's for → codons in → messenger RNA's and the specificity with which tRNA's are charged with amino acids. Readout of the g.c. in DNA is by → genetic transcription into RNA (→ messenger RNA, → transfer RNA, → ribosomal RNA) and → genetic translation into polypeptides.

The g.c. distributes 64 triplet → dodons (each consisting of three contiguous nucleotides) to 20 amino acids, including initiation signals (→ initiator codon), and three termination signals (→ terminator codon) for the construction of protein molecules with specific sequences of amino acids (→ gene; gene expression; anticodon).

The g.c. is nonoverlapping, i.e., adjacent codons in DNA or RNA do not overlap, commaless, i.e., there are no special signals that give advice how to select the right codons for reading, and degenerate, i.e., most amino acids can be specified by more than one codon (Table 6).

Table 6. The nucleotide sequences of RNA codons

First Base	Second Base: U		C		A		G		Third Base
U	UUU	Phe	UCU	Ser	UAU	Tyr	UGU	Cys	U
	UUC		UCC		UAC		UGC		C
	UUA	Leu	UCA		UAA	Stop (ochre)	UGA	Stop (opal)	A
	UUG	Leu[a] F-Met	UCG		UAG	Stop (amber)	CGG	Trp	G
C	CUU	Leu	CCU	Pro	CAU	His	CGU	Arg	U
	CUC		CCC		CAC		CGC		C
	CUA		CCA		CAA	Gln	CGA		A
	CUG		CCG		CAG		CGG		G
A	AUU	Ileu	ACU	Thr	AAU	Asn	AGU	Ser	U
	AUC		ACC		AAC		AGC		C
	AUA		ACA		AAA	Lys	AGA	Arg	A
	AUG	Met[a] F-Met	ACG		AAG		AGG		G
G	GUU	Val	GCU	Ala	GAU	Asp	GGU	Gly	U
	GUC		GCC		GAC		GGC		C
	GUA		GCA		GAA	Glu	GGA		A
	GUG	Val[a] F-Met	GCG		GAG		GGG		G

[a] Codons for the initiation of polypeptide chains by f-met-transfer RNA when present at the 5'-hydroxyl terminus of a polynucleotide.

The 64 possible codons may be arranged into 16 families, each defined by the first two bases of its codons. In the standard code, eight of the 16 families are unmixed, i.e., all four codons in one family code for the same amino acid. In the mixed families, several patterns of assignment exist. In six of them the pyrimidine codons (U or C in the third or wobble position) signal one amino acid and the purine codons (A or G in the third position) another or, in one family, act as termination signals. Almost all of the degeneracy is confined to the third (III) codon position. With very few exceptions, the degeneracies are $U_{III} = C_{III}$ and $A_{III} = G_{III}$, with all four bases being equivalent in the third codon position in a number of cases (→ wobble). Thus most amino acids possess either two or four codons.

The transcription of the instructions encoded in DNA takes place when one of the polynucleotide chains of duplex DNA is used as a template in the synthesis of a single-strand of

RNA (genetic transcription). In this process, T, C, G, and A in DNA direct the respective incorporation of A, G, C, and U (uracil) into the RNA strand.

Deviations from the standard ("universal") g.c. are known for some organisms (and organelles) as diverse as enteric bacteria and vertebrates. A wide variety of variant codon assignments (Tab. 7), affecting both nonsense (stop) codons and sense codons, has been observed in the → mitochondrial genome (→ two-out-of-three reading hypothesis). The chloroplast g.c. is, however, of the standard type.

Notwithstanding these deviations, the g.c. is universal from a structural point of view, i.e., it is always made up of three-letter codons using the same four-letter nucleotide alphabet.

Table 7. Deviations from the universal code in mitochondria

Codon	Sense In mitochondrial code		In standard code
	Mammals	Yeast	
UGA	trp	trp	stop
AUA	met	met	ileu
CUA	leu	thr	leu
AGA, AGG	stop[a]	arg	arg

[a] Further stop codons in the mitochondrial code: UAA, UAG.

In bacteria, either AUG or GUG can act as initiator codons. AUG specifies the initiating amino acid n-formylmethionine as well as methionine at internal positions within a polypeptide chain. In eukaryotes, AUG is the only initiator and is translated as methionine. UAA, UAG, and UGA do not specify amino acids and act to terminate protein synthesis; they are terminator codons.

Although the standard g.c. is nearly universal, organisms have profound biases in codon use (→ codon bias; codon context effect) in degenerate codon families.

Ambiguity of the g.c. refers to the fact that one codon codes for more than one amino acid and suggests that a given codon can be recognized by more than one species of tRNA. Three types of ambiguity have been identified: (1) A codon may be misread due to environmental conditions around the tRNA-mRNA-ribosome complex which interfere with the codon recognition process (→ misreading in translation). (2) A codon may be misread because of a specific change in the primary, secondary, or tertiary structure of either the tRNA or the ribosome. (3) An → aminoacyl-tRNA synthetase may misrecognize either an amino acid or a tRNA molecule, leading to the formation of the wrong aminoacyl-tRNA.

Models proposed for the **evolution of the g.c.** may be classified into two categories: (1) Deterministic models based on the "stereochemical hypothesis" assume that codon assignment is due to preferential interactions between amino acids and their codons; (2) stochastic models (coevolution theory) suggest a random assignment of amino acids to codons at an early stage of evolution and subsequent fixation of this assignment ("frozen accident").

genetic coherence (Clausen and Hiesey 1958) — a tendency (= character adherence) for parental character combinations to remain associated in segregation of hybrid progenies.

genetic cohesion — the phenomenon whereby large central populations show continued inbreeding and phenotypic similarity. → Gene flow and → genetic homeostasis are supposed to confer g.c.

genetic colonization (Schell et al. 1979) — a case of successful natural → genetic engineering as examplified by parasitism practiced by **Agrobacterium** with plants, i.e., introduction of genetic information (→ Ti plasmid; T-DNA) into the DNA of host (tumor) cells which induces synthesis of products that only the inducer can use.

genetic compensation (Lewis and John 1970) — = intercistronic → genetic complementation.

genetic complementation (Fincham 1966) — the complementary action (cooperation) of homologous sets of genetic material (genomes) involving the interaction of mutant genes, or their products, in double mutants. Those combinations that result in marked improvement in the function under study or in the development of a character which cannot be realized by the individual action of single mutants are said to complement each other.

A prerequiste for complementation studies is the collation within the same cell of complete or incomplete homologous genomes in doubly → heterozygous, → heterokaryotic, or → heterogenotic condition. This is achieved in different genetic systems by different means: (1) **In eukaryotes** (higher organisms) with membrane-bounded → nuclei, different homologous genomes may be combined in the heterozygous condition in one nucleus after meiosis and fertilization, or in fungi characterized by a → dikaryophase they may be combined in haploid heterokaryons (in the heterokaryotic condition), that is, in separate nuclei which are present in a common cytoplasm. In both cases complete genomes coexist. (2) **In prokaryotic bacteria,** a fragment of one genome may be transferred from a donor cell to a recipient cell by the processes of → conjugation, → transduction, or → F-duction (sex-duction). (3) **In bacteriophages,** the combination of two complete virus genomes is achieved by mixed infection of bacterial host cells with two genetically different types of phage.

In the case of diploids, heterokaryons, and bacterial heterogenotes, the two genetic homologs can replicate and remain together for an indefinite number of cell divisions; in the case of bacterial merozygotes and abortive transduction, the complementation state is confined to the cell in which it arose.

G.c. is of fundamental importance in the definition of the → gene and in the analysis of gene function (gene-protein relationship). Complementation studies are based on the intracellular activities of two mutants in → cis-configuration (i.e., introduced by the same genome) or trans-configuration (introduced with different genomes). The cell with a double mutant in cis will contain one complete functional set of wild-type genes and thus be functionally active. In trans, either the two mutants will complement one another and the cell will be functionally active or they will not, and the cell will be deficient in a critical function. This → **cis-trans** or **complementation test** is one criterion for distinguishing allelic from nonallelic genes and for identifying physiologically distinct genes or → cistrons. Although it was originally supposed that allelic genes would not complement, while nonalleles would, there are numerous exceptions. A proportion of alleles show some degree of g.c., called intracistronic or interallelic complementation, in contrast to intercistronic complementation.

(1) **Intercistronic complementation** (Schlesinger and Levinthal 1963): Complementation between pairs of heterozygous nonallelic mutations belonging to different functional regions of the → genetic map (different cistrons).

(2) **Intracistronic complementation** (Schlesinger and Levinthal 1963): Complementation between heterozygous pairs of mutations which are on the basis of other criteria mutations within the same cistron thereby affecting the same polypeptide. Mutants, or classes of mutants, are said to overlap if they do not complement each other; they do not overlap in those cases where they complement each other.

Intercistronic complementation is explained by assuming that each cistron determines the structure of a separate polypeptide chain. The cell or organism functions normally if at least one genome (chromosome) produces each polypeptide in the normal configuration. Complete complementation is to be expected if the different polypeptides are part of one enzyme (protein) or if different enzymes determine the same biosynthetic sequence. Intracistronic complementation may then be the result of interpolypeptide complementation between aggregates of identical polypeptide chains, coded for by one cistron and which in aggregate (as dimers, trimers, etc.) make up one (hybrid)enzyme molecule. Interaction between two or more individually defective polypeptide chains may lead to one functional enzyme molecule, if mutual correction of faulty conformation occurs in the double mutant.

Generally, the amount of active enzyme produced by g.c. is dependent on the pair of mutations under consideration. In most cases, it is much higher with intercistronic than with intracistronic complementation.

The complementation pattern of a series of mutants belonging to a short genetic segment can be represented diagrammatically in form of a **complementation map** in which mutually complementing mutants are represented by nonoverlapping, and noncomplementing mutant pairs by overlapping, continuous lines. Such maps are often, but not invariably, linear (in some case circular). In many cases, there is a high degree of correspondence between the positions of mutants on genetic maps and on complementation maps; this is believed to result from → colinearity of the cistron and the polypeptide chain for which it codes.

genetic conservation — the collection and systematic preservation, in gene resource centers, of as wide a representation as practicable of the spectrum of genetic variation within plant and animal species, including samples of endangered wild progenitors and relatives.

genetic correction — in higher organisms, correction by exogeneous DNA of genetic defects caused by mutation (→ transgenosis). G.c. may be due to integration of exogeneous DNA into the chromosome or association with the homologous chromosome locus (→ exosome).

genetic correlation — the degree of association between two or more variables due to genetic (not environmental) causes, such as → pleiotropy, → linkage, → linkage disequilibrium, and change in → gene frequencies upon selection. G.c. is measured by the correlation coefficient and correlation (r) ranges from -1.0 to $+1.0$, with negative or positive signs indicating a negative or positive relationship between two variables. An r of 0.0 indicates no correlation, i.e., complete independence between the two variables. Since each parent provides half of its genes to each offspring, the g.c. between parent and immediate offspring is 0.5 (between identical twins 1.0; between full-sibs 0.5; between half-sibs 0.25; between first cousins 0.125, etc.).

genetic counseling — a service which provides people with information concerning the risk of having genetically abnormal progeny (→ recurrence risk). Prediction of recurrence may be based on Mendelian expectations for simple recessive and dominant conditions, or on empirically determined risks for diseases whose mode of inheritance is less well understood (→ amniocentesis). G.c. requires careful evaluation of the total clinical picture of the → genetic disease or malformation and assessment of the family history or kindred.

genetic covariance — the correlation among age-specific character values which describes quantitatively the genetic link between expressions of the same character at different points of → ontogeny.

genetic death (Muller 1950) — preferential elimination (by → selection) of genotypes which are carriers of mutationally arisen → alleles (→ genetic load) which reduce their → adaptive value or fitness (= genetic extinction). G.d. removes mutant genes from the gene pool of the population and may be the consequence of a reduction in physical vigor, sexual drive and/or fertility, or a high mortality rate in embryonic or juvenile stages of individuals possessing the critical genotype. G.d. is described as **"overlapping"**, if (1) an individual is eliminated as a consequence of the presence of one particular gene, but would also have suffered g.d. by one or more other independent genes in its genotype **("independent overlapping")**, or if (2) g.d. results from the synergistic effect of two or more mutant genes **("dependent overlapping")**.

The number of g. deaths which may be involved in the complete substitution of one allele by another varies and may be many times the number of individuals present in any one generation. If a mutant gene (together with all other genes of the genotype) is eliminated from the population in the same generation in which it arose by mutation, it is said to have a **"persistence"** of only one generation. On the average a dominant detrimental mutant with a selection coefficient of 0.2 and an adaptive values of 0.8 (as compared to normal), will persist for five generations before suffering genetic death. Thus, in each generation there is a 20% chance that the mutant individual will not transmit the gene to its progeny. The rate of g.d. of a particular mutant depends on the effect of the mutation. Its persistence is the shorter the more detrimental its effect. The average persistence of recessive mutations in man is approximately 40 generations or 1200 years.

genetic disease — any genetically conditioned disorder ultimately caused by a particular gene-determined biochemical defect ["inborn error of metabolism" according to Garrod (1908) or "molecular disease" according to Pauling (1964)]. In fact, all diseases have genetic and nongenetic (environmental) components whose importance may vary widely. Genetic disorders in the strict sense are only slightly influenced by environmental factors, as opposed to "common" disorders which are largely the result of environmental factors. The relationship between heredity and disease is the subject of "medical genetics".

A genetic d. is not necessarily incurable. Therapy may proceed along various lines depending upon the nature of disease under study. Direct detection (→ restriction fragment length polymorphism) and analysis of various genetic defects at the DNA level are now possible via cloned genes or oligonucleotide probes (→ gene therapy).

genetic disoperation (Bhalla and Sokal 1964) — → genetic facilitation.

genetic displacement — a change in → allele frequency which occurs when two species coexist; g.d. is a consequence of adaptation to the habitat by interspecific competition in the zone of sympatry and may result in character divergence or character convergence.

genetic dissection — the use of recombination (recombination dissection) and/or mutation (mutational dissection) to analyze the components of a given, genetically determined biological function.

genetic distance — (1) in population genetics, a quantitative measure of genetic relationship between two individuals or populations from which an evolutionary tree may be constructed.Measurementisintermsoftheprobabilityofthecommonpossessionofagivengeneor-character (coefficient of g.d.). The cumulative genetic difference between two populations is a summary of their evolutionary history and is proportional to the time of separation and inversely related to the intermigration between them (→ genotypic distance). (2) The distance between two gene loci as defined by the average number of cross over points per chromatid.

genetic diversity — an informative parameter of the genetic structure within populations which is a function of the number and frequency of alleles per gene in a certain → gene pool. G.d. can be used to reflect the integration of diversity at individual loci into multilocus genotypes, a process mediated primarily by the mode of reproduction.

genetic drift (Wright 1921) — (1) any change, either directed **("steady drift")** or undirected **("random drift")** in → gene frequency in a population (Wright 1955). (2) Irregular (random) fluctuations in gene frequency in a population from generation to generation due to finite population size (in "effectively" small populations whose → effective breeding size either remains small or periodically becomes small) or randomly fluctuating selection intensities (g.d. in the sense of random drift only). These fluctuations in gene frequency may lead to the fixation (→ random fixation) of one allele and the extinction of another without regard to their adaptive value. Random drift as a potential evolutionary factor (→ evolution) is referred to as the "Sewall Wright effect".

According to Waddington (1957), persistent and intermittent drift should be distinguished. In **"persistent drift"**, a population is small enough for statistical fluctuations to exert an important influence on the frequencies of the genes for a number of generations. **"Intermittent drift"** characterizes an occasional, sudden reduction in the number of effective breeding individuals, lasting for only one or two generations. The potential occurrence of random drift in a population depends on the interrelationship between the following factors (*Grant* 1963): (1) The number of breeding individuals (N), (2) the selective value of the allele(s), (3) mutation pressure (u), and (4) gene flow (m).

Before a change in gene frequency is attributable to random drift, it must be established that: (1) The effective population size is small enough, (2) the sub-populations are well isolated, and (3) the genes concerned are subject to very little selection.

According to the "drift theory" of non-Darwinian evolution (Kimura 1968; King and Jukes 1969) most gene substitutions are selectively neutral and caused by g.d. Such a type of → evolution has little substitutional load (→ genetic load) as opposed to the neo-Darwinian theory of evolution.

genetic endpoint test — a bioassay for detection of gene mutations or structural and numerical chromosome aberrations (→ mutagenicity testing).

genetic engineering — the use of in vitro techniques (→ recombinant DNA technology) for the deliberate manipulation (= genetic manipulation) of genes (within or between species) for the purpose of genetic analysis and product improvement. G.e. includes the production of manageable DNA fragments with → restriction endonucleases and the enzymatic splicing of the resulting fragments into an appropriate → cloning vector which can then be introduced (→ gene transfer) into the required recipient cell, whether bacterial, yeast, mammalian, or plant cell. Isolation of target DNA may be followed by site-specific deletion, insertion, or substitution of nucleotides; DNA obtained from → gene synthesis may also be used.

genetic equilibrium (Hardy 1908) — in a large randomly mating population, the situation in which both the → gene frequencies and the → genotype frequencies (in the absence of mutation, selection, and migration) are in accordance with the predictions of the → Hardy-Weinberg law ("Hardy-Weinberg equilibrium") and remain constant through successive generations.

Since any change fluctuations in gene frequency are perpetuated and there is no tendency for the above-specified system to revert to its original state, such an equilibrium is referred to as neutral rather than stable. Owing to systematic processes (mutation, selection, and migration) and dispersive processes (random sampling errors leading to differentiation between sub-populations, reduction of genetic variation in small populations, and increase in the frequency of homozygotes at the expense of heterozygotes) which tend to change the gene frequency, the genetic equilibrium of neutral populations is "labile". It is the tendency of the systematic processes to bring the gene frequencies to stable equilibria at particular values, whereas the dispersive processes tend to scatter gene frequencies away from these equilibrium values. These opposing trends reach a balance which is stable only for the interval during which the conditions remain unchanged.

genetic extinction — = → genetic death.

genetic facilitation (Lewontin 1955) — increased fitness for one or more populations (as measured by survival, length of developmental period, body weight, and other characteristics) in the nonoverlapping region of different species or strains competing for limited resources which may diminish the utilization of certain regions of their environmental hypervolume. The contrary response, i.e., reduced fitness of one or both competitors in response to increased competition, is called **genetic disoperation.**

genetic fitness — → fitness.

genetic flexibility — the capacity of a population to produce offspring different from their parents. G.f. is promoted by high → mutation rates, → heterozygosity, and → outbreeding.

genetic fusion — → gene fusion.

genetic hazard — any possibility that a primary change occurs in the quality or quantity of the DNA in any of the cells of an individual.

genetic heterogeneity — of apparently similar characters for which the genetic evidence indicates that different genes or different genetic mechanisms are involved in different kindreds.

genetic homeostasis (Lerner 1954) — the property of a population to regulate or stabilize its → gene pool constituents, i.e., to return to its original state following perturbations.

genetic homology — identity (complete g.h.) or near identity (partial g.h.) of nucleotide sequences in DNA (→ DNA-DNA hybridization).

genetic inertia (Darlington and Mather 1949) — → homeostasis.

genetic information — the information contained in a particular sequence of nucleotides in a DNA or RNA molecule. **Storage of primary g.i.** in permanent form is normally in → deoxyribonucleic acid (in RNA viruses, → ribonucleic acid takes the place of DNA and thus acts as the carrier of primary g.i.). All the g.i. of a genetic system may be contained in one linkage structure (→ chromosome), as in viruses, bacteria, and plasmids, or in several such structures, as in higher organisms. In eukaryotes, g.i. is encoded in the nuclear chromosomes, in → mitochondria and, in plants, in → chloroplasts.

Four main processes are engaged in the **transfer of genetic information** (Spiegelman and Hayashi 1963): (1) The replication (duplication, reproduction) of chromosomal DNA, i.e., the genetic material of all cells and some viruses. (2) The replication of RNA in those viruses whose genetic material is RNA and not DNA. (3) The → genetic transcription of g.i. from DNA to RNA [transformation of g.i. from a permanent (DNA) to a transient form (RNA) which is extant for a limited time interval only]. (4) The → genetic translation of g.i. from the transient RNA code or nucleotide sequence into the protein "text", or amino acid sequence.

genetic instability — various types of genetic mechanisms which give rise to phenotypic → mosaicism or → variegation (→ controlling element).

genetic interaction — the influence of more than one gene upon the presence or specificity of a particular → character (→ gene action; gene interaction).

genetic isolate — a relatively small breeding population which has little or no exchange of genes with other populations.

genetic load (Muller 1950) — the relative amount by which the average → fitness (or any other measurable character) of a population is lowered (due to the presence of genes that decrease survival relative to the maximum possible) compared to the genotype with maximum or optimal fitness.

The principal components of g.l. are the following: (1) **Mutational loads** due to deleterious mutations at loci which are homozygous for nondeleterious alleles (load maintained by recurrent mutation). (2) **Segregational loads** due to genes segregating from favored heterozygotes and giving rise to unfit homozygotes (load maintained by segregation from advantageous heterozygotes). Minor components of g.l. include **"input loads"**, i.e., the load of inferior alleles in a gene pool caused by mutation and immigration (gene flow); **"substitutional loads"**, i.e., the cost to a population of replacing a particular allele by another in the course of evolutionary change; **"selection loads"** due to selection for an intermediate optimum of a quantitative character, and **"interaction loads"** due to the interaction of deleterious genes. A g.l. is termed **"balanced load"** where decrease in the overall fitness of a population due to the segregation of inferior genotypes occur and whose component genes are maintained in the population because they improve fitness in different combinations (e.g., as heterozygotes).

genetic manipulation — = → genetic engineering.

genetic map — the representation of the genetic (relative) distance separating nonallelic gene loci in a linkage structure (the → genes of one → linkage group or chromosome), or the arrangement of the mutational sites (alleles) of a particular gene by utilizing the frequency of → genetic recombination via → crossing over between genes (chromosome map) and within genes (→ gene map; fine structure map).

Genetic maps are linear, unbranched, structures reflecting the sensitivity of genetic recombination tests in most genetic systems. In some bacteriophages and in bacteria, genetic maps taks the form of a circle (→ genetic circularity). The units of distance on the map are expressed as probabilities of recombination, or percentages. In many cases, genetic maps and "complementation maps" are parallel (→ genetic complementation). This is assumed to be the result of → co-linearity of the gene and the polypeptide chain whose amino acid sequence it determines.

The assignment of genes to specific linkage groups and the determination of their relative distance to other known genes is called **genetic mapping.**

The classical g.m. is based on the determination of meiotic recombination. Besides this, there is genetic mapping based on mitotic recombination and recombination in (parasexual) bacteria and viruses. To map the genes on the basis of intergenic crossing over or recombination data, it is necessary to determine all cross over or recombination frequencies for a minimum of three linked loci, since two points are insufficient to determine the specific linear arrangement. Since cross over and recombination frequencies fluctuate, because of variations in sample size and factors that affect either the crossing over process itself (temperature, nutrition, genotype, age), or exert their influence after crossing over occurs (e.g., differential viability), standard genetic maps are made under standard conditions optimal for crossing over and hence recombination.

The existence of linear linkage maps suggests that the frequency of recombinations for a given pair of gene loci is positively correlated with their physical distance on the chromosome. Each genetic marker (gene) is represented on the g.m. by a point in space. Distance between many pairs of markers is a function of their respective frequencies of recombination. Mapping of genetic markers is then based on the following facts (Esser and Kuenen 1965): (1) Frequency of recombination between any two markers is constant under standard experimental conditions. (2) With three linked markers (a, b, and c) the largest of the three recombination values (r_{a-b}, r_{a-c}, and r_{b-c}) which result from intergenic crossing over is always either equal to ("additive theorem" of recombination frequencies) or smaller than the sum of both the smaller values (e.g., $r_{a-c} + {}_{b-c} \geqq r_{a-c}$ if exchange between a—c is more frequent than between the other two marker pairs). Generally, $r_{a-b} + r_{b-c}$ is larger than r_{a-c}, and only in cases of close linkage or complete positive chromosome → interference may $r_{a-b} + r_{b-c}$ equal r_{a-c}.

On the basis of this relationship and with three markers, it is possible to order each of them in a unique sequence; the third marker is localized between those two others showing the largest recombination (exchange) frequency. In case of $r_{a-b} + r_{b-c} = r_{a-c}$, the marker sequence would then be either a —b—c or c—b—a. (3) Multiple crossing over is the most frequent source of nonadditivity. If at least one cross over takes place simultaneously in each of two adjacent regions of the chromosome (e.g., a—b and b—c), multiple crossing over in regions a—c may give rise to one recombination between adjacent markers and concurrently to a parental combination for the two external markers (a and c). (4) Crossing over frequencies are the values needed for exact mapping. In contrast to recombination values, they are always strictly additive; the sum of crossing over frequencies for two adjacent regions (e.g., a—b and b—c) is equal to the frequency of crossing over in the whole region (a—c). The crossing over frequency can be obtained from the observed recombination frequency by correcting for the effects of multiple (double) crossing over and → interference. Because of interference the correction is mathematically rather complex. The unit measure of chromosome length and distance between any two markers is the → **"map unit"** (= "crossing over unit" or Morgan unit) which is equal to the corrected recombination frequency, expressed as a percentage. One map unit is defined as the corrected recombination frequency of 1 percent or one centimorgan. The relative distance between two markers calculated from uncorrected recombination frequencies (the recombination value) is referred to as their **"apparent distance"**, as opposed to their **"actual distance"**, i.e., the distance calculated from the recombination frequency corrected for multiple crossing over (the "exchange value"). An estimate of actual distances between genes follows from the construction of cytological maps (→ chromosome map) or their equivalents. These are made independently of recombination data.

Mapping according to these principles represents the genes as seemingly dimensionless points on the g.m. The number of offspring tested for recombination is of the order of 10^2—10^3. Under these conditions, it is unlikely that rare recombinations will be discovered. In systems allowing the study of 10^6—10^7 progeny, unique types of (rare) recombinants are recovered as a result of crossing over within the limits of single genes (intragenic crossing over). In such systems, the genetic resolving power **("the genetic resolution power")** is greatly increased. The locus of a gene ceases to appear as a dimensionless point; by "genetic fine structure analysis" via recombination tests of high resolving power (→ deletion mapping), the gene is shown to be composed of a linear sequence of recombinable mutational sites (→ gene mutation) whose positions can be mapped (→ allele; genetic code).

In systems where for one reason or another recombination-type mapping cannot be effectively employed, some alternative approaches may be used. In prokaryotes, these approaches do not directly depend on any gene transfer mechanism and are based on the estimation of marker frequencies by → genetic transformation, analysis of density-labeled transducing phage, nucleic acid hybridization, studies based on the rates of induced enzyme synthesis, and mutation induction (→ replicational mapping; transcriptional mapping; heteroduplex mapping; temporal mapping).

genetic mapping — any method used in the measurement of positions and relative distances between genes of a → linkage group or sites within a gene (fine scale mapping). (→ genetic map).

genetic marker — any → allele used as an experimental probe to mark a nucleus, chromosome, or a gene.

In microbial genetics, a more or less arbitrary distinction is made between "selected" and "unselected" markers (Lederberg 1954). In recombination analysis, **selected markers** are those which prevent the growth of parental genotypes, manifesting a particular phenotype (e.g., growth factor auxotrophy), in particular environments (media), whereas recombinants for these markers are able to grow freely in the environment, i.e., have a selective advantage over the parental types, thus facilitating their quick identification.

Unselected markers are without selective advantage for the recombinants as compared to the parental genotypes. Here the identification of recombinants is determined primarily by the internal recombination mechanism and less by the technique applied. Under these conditions segregation is determined as a second step procedure.

In principle, the same markers may be used as selected or unselected ones dependent on the experimental conditions used.

genetic material — the carrier of primary → genetic information: single- or double-stranded → deoxyribonucleic acid (single in some, double in most bacteriophages, in bacteria and higher organisms) or → ribonucleic acid(in RNA-viruses). G.m. must fulfil at least two fundamental functions: (1) Serve as a template for its own → replication ("autocatalytic function"), (2) provide a template for the synthesis of other classes of macromolecules (specifically proteins), i.e., supply the structural and regulatory information it contains to the protein-synthesizing machinery of the cell ("heterocatalytic function").

The evolution of g.m. may be divided into at least three major phases: (1) Genomes of nucleic acid-like molecules; (2) genomes of RNA; (3) finally double-stranded DNA genomes.

genetic message — a → messenger RNA molecule coding for a single, functional polypeptide. A → polycistronic mRNA carries several messages.

genetic milieu (Chetverikov 1926) — the sum of genetic factors which act from inside on the manifestation (→ gene expression) of any gene.

genetic mimic — = → genocopy.

genetic mobility (Darlington 1958) – the ability of biological species to change their habitat or (in plants) the range of distribution of pollen and seeds. It is expressed differently in different individuals. The g. m. largely determines the manner whereby the mechanism of geographic → isolation becomes effective.

genetic mosaic — any individual (called → chimera or mosaic) which is composed of genetically different tissues and displays distinct characters in different sectors of the individual. Mosaic areas may correspond to the expression of different alleles contained in the particular genotype, or may arise by → gene mutation, → chromosome mutation, somatic crossing over and somatic segregation, aneuploidy, or double → fertilization. Genetic mosaics are called → gynandromorphs or gynanders if part of the individual is male, part is female.

genetic nomenclature — the designation of genes by symbols. Mendel (1866) was the first to use capitals (e. g. A) to represent dominant and small letters (e. g. a) to represent recessive characters without using letters specifically derived from the particular type of expression of the character pair under observation.

Today, genes (and characters) are designated by the letters of the Roman alphabet or by abbreviations of the designations (in Latin or English) given by the person who discovered the phenotypic character pair controlled by the gene pair in question. Generally, the recessive allele of a gene pair is marked by a small letter, the corresponding dominant allele by the same letter or letters, but beginning with a capital. Another notation for alternative alleles employs the same letter or letters with a (+) for the alleles of the standard or wild-type strain or variety. (+) for wild-type alleles is then used irrespective of their being dominant or recessive. Those genes not derived from a standard or wild type should be designated by letter-symbols which are in accordance with the particular phenotype controlled by them. Letter-symbols should contain no more than two to four letters; important genes may be symbolized by one letter only.

In bacteria and viruses, the prototroph is designated by letters and the auxotroph by the same letters with a superscript (−), e. g., lac, lac^-, try, try^-.

To designate pairs of allelic genes, different systems are used. Most often two alleles are placed in sequence (e. g., Aa or + a, etc.), sometimes they are indicated by "fractions" (e. g., $\frac{A}{a}$ or A/a). Linkage relationships may also be indicated in the form of fractions (e. g., AB/ab or $\frac{AB}{ab}$ and Ab/aB or $\frac{Ab}{aB}$ for two pairs of alleles linked in → cis- and trans-configurations, respectively).

genetic polarity (Jacob and Monod 1961; Franklin and Luria 1961) — the phenomenon wherein certain → gene mutations (→ polar mutation) in bacterial or viral → operons, in addition to inactivating the product (protein) of the gene in which they are located, also reduce the expression of all the operator-distal genes of the operon. G. p. seems to be a general characteristic of all operons transcribed into → polycistronic mRNA. Mutational changes which produce polar effects (in most cases → nonsense mutations, in some cases → frameshift, insertion, and deletion mutations) appear to have as a common alteration the introduction, either directly or indirectly, of a polypeptide chain terminating codon (→ terminator codon) within the gene.

This results in a decreased amount of mRNA for the operator distal genes. Much of the operon mRNA is shorter and corresponds to genes operator-proximal to the mutation site. The loss of mRNA from beyond the mutation site might result either from premature termination of transcription at or near the mutation or from an accelerated degradation of the messenger corresponding to the operator-distal genes. Missense mutations do not exert g. p.

The polar effect of a nonsense (or frameshift) mutation depends on its location in the gene, i. e., is a function of the distance of its site from the nearest operator-distal cistron boundary. The gradient of g. p. is such that the farther the mutation site from the end of the gene in which it is located, the greater is the reduction of proteins coded for by the subsequent genes. There are indications that the degree of g. p. depends, in part, on the distance from the site of the critical mutation to the next translational initiation site.

The mechanisms of g.p. may be as follows: (1) Nonsense codons by stopping the movement of ribosomes translating nascent RNA's prevent further transcription because transcription and translation are coupled (in bacteria). (2) Nascent mRNA chains continue to be elongated after a nonsense codon at some early position has prevented ribosomes from translating the messenger. Since the newly made stretches of mRNA are devoid of ribosomes they are rendered susceptible to attack by nucleases and are degraded (→ genetic antipolarity).

genetic polymorphism (Ford 1940) — The regular and simultaneous occurrence in the same population of two or more alleles (or genotypes) at the same locus at frequencies that cannot be accounted for by recurrent mutation alone (→ chromosome polymorphism).

In general, polymorphism in organisms is selected in a variety of situations, such as: (1) Protection for individuals (mimetic p., camouflage p.; apostatic p.); (2) permission of separation of reproductive or ecological roles; (3) permission of physiological flexibility.

(1) **Balanced polymorphism** (Ford 1940): A g.p. usually maintained by a selective superiority of the heterozygote over either type of homozygote and characterized by a stable and optimal relationship between genetically different morphs produced by an equilibrium of selection effects on the different allelic combinations. (2) **Transient polymorphism** (Ford 1940): A g.p. limited to that period of time in which one allele undergoes replacement by a superior one. (3) **Neutral polymorphism** (Ford 1940): A g.p. dependent on a gene action which is roughly neutral in its effect on the survival rate of the genotype in which it is contained. (4) **Regional or geographic polymorphism:** A polymorphism characterized by two or more genetically different morphs inhabiting different regions of the habitat of the population. (5) **Unisexual polymorphism:** A g.p. confined to one sex only owing to sex-limited manifestation of genes which can recombine in both sexes. (6) **Cryptic polymorphism:** A g.p. in which the genetically different morphs are not identifiable by their phenotype.

According to Huxley (1942), there are three possibilities which may help to install a balance of the selective advantages between different morphs: (1) Heterozygotes may be more efficient than homozygotes; (2) one or more of the morphs may show a selective advantage only if the number of individuals in the population falls below a certain threshold value; (3) each of the different morphs is most efficient under specific environmental conditions which are bound to different places at different times.

Cases (1) and (2) are associated with changes in the respective selective values.

G.p. in any form has been considered adaptive by providing an increased diversity of genotypes thus leading to reduced intraspecific competition and increased buffering against environmental changes.

genetic promiscuity (Ellis 1982) — the phenomenon whereby DNA sequences (promiscuous DNA) can be transposed between the eukaryotic cell nucleus and cell organelles: between the nucleus and the → mitochondria, from → chloroplasts to mitochondria, and from chloroplasts to the nucleus.

genetic recombination — the reassortment of series of nucleotides along nucleic acid molecules. These reassortments can occur within one DNA molecule (intrachromosomal r.) to produce → deletions, → inversions, → transpositions, or → duplications, or between two separate parental molecules (linkage groups, chromosomes) to produce a molecule (or two of them) derived in part from each parent structure (interchromosomal r.).

The process of new → linkage formation by interchromosomal r. occurs largely at random (→ chi element) along regions where the two homologous parental DNAs (chromosomes) are brought into register and is called **generalized g.r.** It includes a number of varieties: (1) Whether or not one of the parental molecules is single-stranded; (2) whether the product is one intact DNA sequence or two, and if two, whether or not they are reciprocal (→ gene conversion). Generalized r. involves breakage and reunion of DNA via the formation of → heteroduplex joints followed by lateral migration of the bridge and cleavage to release helices with various arrays of flanking markers.

In contrast to generalized g.r., r. events can also occur between DNA molecules that show little homology. When the r. event is a highly preferred molecular position (special sites) on one or both of the parental molecules, the phenomenon is called **site-specific g.r.** Two types are known in prokaryotes: (1) Replicative r., i.e., the DNA lying between a pair of special sites is duplicated and the replica is transferred to a new position (→ transposition); (2) conservative r., i.e., essentially breakage at a specific site and rejoining of broken ends in new permutations. Site-specific r. plays crucial roles in → DNA transposition, partitioning of extrachromosomal elements, → genetic regulation, and generation of genetic diversity.

The joining of DNA segments at DNA sites that lack extensive genetic homology is called **illegitimate r.** It is independent of normal recombination functions – such as recA in *E.coli* – and is of importance in the evolution of → plasmid, → bacteriophage, and bacterial genomes. **Nonhomologous r.** is the random insertion of foreign DNA into target DNA (→ recombinant DNA technology; transposable genetic elements).

The products of g.r. (the recombinants) may be detected after meiosis (meiotic recombination), after mitosis (in the case of mitotic recombination) or after equivalent processes occurring in bacteria and during multiplication of viruses.

G.r. requires the intimate association of all or part of the genetic material from two (or more) different genotypes. This is achieved by different mechanisms in different genetic systems: (1) **In eukaryotic organisms,** the intimate association is achieved by nuclear fusion and meiosis. Recombination may occur by three different processes: (a) → Chromosome assortment via the spindle mechanism during meiosis which leads to "interchromosomal recombination" of unlinked markers, (b) "intrachromosomal recombination" of linked markers by → crossing over, and (c) → gene conversion of one allele in the presence of the other resulting in nonreciprocal recombinants as opposed to the other two mechanisms which give rise to reciprocal recombination products (→ recombination system). (2) **In bacteria,** the association of different genetic material may occur, (a) by → genetic transformation, (b) by → conjugation, (c) by → transduction, and (d) by → F-duction, and may be followed by intrachromosomal recombination or gene conversion. (3) **In viruses** (bacteriophages), the same result may be achieved by mixed infection with two (or more) viruses differing in genotype.

"Interchromosomal recombination": This type of recombination is limited to eukaryotic organisms whose genetic information is contained in more than one linkage structure (chromosome) and represents the assortment of whole chromosomes in meiosis, leading to the independent → segregation of unlinked genes (allele pairs localized in nonhomologous chromosomes). Reduction of the chromosome number to one-half generally occurs during the first meiotic division and chromosome assortment is due to the random orientation of pairing configurations at metaphase in diploid organisms. The polarized distribution of particular chromosomes to one pole is of rare occurrence and leads to nonrandom assortment and nonrandom segregation of markers at loci on different chromosomes (→ affinity; meiotic drive; sex-ratio).

Apart from meiotic recombination, interchromosomal recombination may also occur in somatic cells during mitosis via the production of aneuploid nuclei which may ultimately become regulated down stepwise to yield balanced (euploid) descendants (→ haploidization).

"Intrachromosomal recombination of the intergenic type": Recombination of allelic markers at different loci on the same linkage structure in both eu-(higher organisms) and prokaryotic (bacteria, viruses) systems. Recombination of this type is by → crossing over, i.e., a process involving breakage and reunion of chromosome parts or segments, and gives rise to reciprocal recombination products. The frequency of recombination between any two markers ("the degree of → linkage") is constant under constant experimental conditions. The intergenic type of g.r. may take place during meiosis as well as mitosis (somatic or mitotic recombination of linked markers). Mitotic recombination occurs by a mechanism, which like the meiotic one gives rise to reciprocal recombinants and is thus called "mitotic crossing over". It is possible to map genes belonging to one linkage group (→ genetic map) by both meiotic and mitotic recombination.

"Intrachromosomal recombination of the intragenic type": Recombination between the subunits (mutational sites) of a gene; this may take place at many sites (→ recon). On the basis of the recombination frequencies, the mutational sites can be arranged in a linear order as a → gene map which is one-dimensional, like a linkage map. Intragenic recombinations are rare events with a tendency to show negative → interference. They are either reciprocal or nonreciprocal. Nonreciprocal recombination may be associated with numerically equal or unequal post-meiotic segregation in eukaryotes. Negative interference in intragenic r. is highly localized and polarity may be observed in respect to outside marker recombination. Intragenic recombination often appears to be impeded by heterozygosity and is frequently correlated with reciprocal r. (crossing over) in its immediate neighborhood.

Unequal g.r., as opposed to equal g.r., is intermolecular r. of homologous segments (chromosomes) in nonaligned molecules and generates tandem → duplications, triplications, etc., and corresponding → deletions.

Enhancement of homologous g.r. by special sites (→ chi element; recombination nodule) appears to be a common feature in diverse organisms. These sites are probably nucleotide sequences recognized by recombination-promoting proteins that act at or near the sites at a higher rate than elsewhere. (→ map expansion; marker effect; repair recombination).

G.r. plays a central role in evolution of gene diversification as well as in gene homogenization.

genetic rectification — a mechanism supposed to make redundant or reiterated genes (→ gene reiteration) evolve in parallel and behave as a single Mendelian factor (→ concerted evolution).

genetic regulation — the overall regulation of the type (specificity) and rate of cellular processes by regulation of the activity of specific genes controlling individual biochemical reactions, or of a block of functionally related genes which control a particular reaction pathway (→ operon).

Since the major steps in the production by → structural genes of → enzymes (proteins) that control the biosynthetic, metabolic, and catabolic processes are → genetic transcription (DNA → RNA), → genetic translation (RNA → polypeptide) and assembly (polypeptide → protein), gene activity as well as the resultant enzyme activity may be controlled at any of these stages. G.r. as described from experimental work with bacteria may involve: (1) Control of → gene expression by stimulation **(positive control)**, which involves → sigma factors and other positive activators (such as the → CRP) available only in response to certain external conditions, or repression **(negative control)** of → genetic transcription, which involves → repressor molecules that inactivate genes by interacting with their → promoters to prevent → RNA polymerase from initiating gene expression. Both modes represent **transcriptional regulation.** (2) Control of enzyme synthesis at the translation of → cistrons, i.e., the polypeptide formation from messenger RNA. (3) Control of enzyme activity after synthesis.

Without doubt, the main mechanism of g.r. in prokaryotes is transcriptional control.

G.r. in eukaryotes has retained elements of the prokaryotic system but may also be carried out at a higher DNA level (deoxyribonucleoprotein-structure). This type of structural regulation has been termed → transconformational control (→ chromatin). Since different genes are available for activation in different tissues at different stages in developing and adult organisms, two distinct levels of gene control, one related to the differentiation state (both temporal and type) of the particular cell and the other related to the response of particular genes to ambient signals must be distinguished.

Genetic regulation of gene expression could then occur during → genetic transcription, → RNA processing and transport, via mRNA stability, and via posttranslational events.

One mode of g.r. in eukaryotes operates directly at the level of the genome, including the loss, amplification, transposition and rearrangement, and the degree of interphase coiling (→ chromosome compaction) of (→ heterochromatic) chromosome regions and whole chromosomes (→ dosage compensation).

genetic relationship — the extent to which the parents of a forthcoming generation are genetically related to each other. G.r. is a variable quantity measured by the → coefficient of relationship.

genetic replication — the negative-positive image formation, by the master method through hydrogen bonding between pairs of purine and pyrimidine bases in DNA (→ deoxyribonucleic acid).

genetic revolution (Mayr 1954) – a model of → speciation which assumes that a small and isolated population of an originally widespread and highly variable species will undergo major genetic changes because of chance loss of alleles in the small founding sample and disruption of interacting gene complexes. The population may increase enormously in size ("population flush") with a consequent relaxation of selection against deviant individuals. This may lead quickly to the evolution of reproductive incompatibility with the parental population. Thus isolation is assumed to cause speciation, rather than allowing it to take place as an incidental result of slow adaptive evolution.

genetic risk — in → genetic toxicology, the product of the inherent or intrinsic mutagenic potency of a compound being evaluated and the ability of the compound to express any inherent activity on the DNA of target cells under normal exposure conditions. If either factor is zero, the risk product will also be zero, regardless of the value of the other.

The quantitative analysis of the expression of inherent genotoxic activity of a suspect compound under defined test conditions and the extrapolation of the results to human exposure is called genetic risk assessment.

genetics (Bateson 1905) — the science of heredity and variation. It uses viruses ("virus genetics"), plants ("plant genetics"), animals ("animal genetics"), and man ("human genetics") as objects of study. The subject matter is the phenomenology and physiology of heredity ("classical genetics") as well as the nature of the genetic material and the storage of → genetic information, its replication, mutation, transmission, recombination, and translation into systems by which the genetic material mediates its control over metabolism and development and determines the reappearance of parental characters among progeny **("molecular genetics"). "Population genetics"**, as distinguished from studies of inheritance at the familial level, describes in mathematical terms the consequences of inheritance on the population level and attempts to predict the behavior of future generations. It deals with the frequencies and interactions of genes in interbreeding populations and studies the agencies (e.g., mutation, natural and artificial selection, gene flow, migration, and chance factors) which tend to alter gene frequencies and thus to cause evolutionary changes. (→ reversed genetics).

genetic screening — the systematic search (= genetic monitoring) in a human population (generally for medical reasons) for individuals of certain genotypes and ascertaining potential genetically determined handicaps in them or in their progeny. Three types of g.s. may be distinguished: (1) newborn s. for identification of genetic diseases at birth; (2) fetal s. and prenatal diagnosis permitting selective termination of pregnancy; (3) carrier s. for identification of individuals heterozygous for a gene for a serious recessive disease who may be at risk for affected progeny.

Retrospective g.s. is the testing of relatives when an individual is diagnosed as having a → genetic disease. **Prospective g.s.** is the screening of individuals without known affected relatives.

G.s. may also be done for research purposes unrelated to disease or genetically determined susceptibilities to environmental agents. (→ mutagenicity testing).

genetic segregation (Bateson and Saunders 1902) – the separation of → alleles from one another and their distribution to different cells, usually at → meiosis ("meiotic g.s.") and sometimes (as a consequence of mitotic → crossing over) at → mitosis ("mitotic g.s."). G.s. can be observed only in genotypes with → heterozygous pairs of alleles.

Meiotic g.s. is the substance of Mendel's **second law** (→ inheritance): The paired (diploid) condition of chromosomes and genes becomes unpaired (haploid) as a result of random chromosome → assortment during meiosis. Meiosis specifies that one member of each pair of alleles is transmitted to each offspring whose diploidy is restored because a haploid → genotype is contributed by each of the two meiotically produced gametes taking part in fertilization. G.s. of a particular heterozygous pair of alleles may occur at the first ("first division g.s.") or second (→ "second division g.s.") meiotic division owing to either "pre-" or "postreductional" separation of the alleles under observation (→ meiosis).

A genotype heterozygous for one pair of alleles (Aa) gives rise by meiotic segregation to two types of gametes which are genotypically either A or a. By random union of these gametes during fertilization three kinds of "segregants" are produced by crosses between two such heterozygotes. Their genotypes will be either AA, Aa, or aa in a ratio of 1 AA:2 Aa:1 aa. This ratio is referred to as the **"genotypic segregation ratio".**

If allele A is dominant over a, the expected **"phenotypic segregation ratio"** (the proportion of phenotypic characters controlled by the pair of alleles in question in the progeny of the heterozygote) approximates 3A (the homozygous and heterozygous dominant genotypes AA and Aa): 1a (the homozygous recessive genotype aa). Three phenotypes are produced in those cases where no dominance between the alleles occurs (e.g., a^1 and a^2 giving rise to the genotypes a^1a^1, a^1a^2 and a^2a^2 in the approximate ratio 1:2:1 and all manifest distinct phenotypes).

Heterozygosity for more than one independent pair of alleles forms the substance of Mendel's **third law** of → inheritance. The segregation ratios of such polyhybrids result from the combinations of the independent monohybrid segregations for each of the heterozygous allele pairs (→ linkage). A hybrid with two pairs of alleles (e.g., Aa, Bb) with dominance of A over a and B over b forms combinations as follows: (3/4 A + 1/4 a) × (3/4 B + 1/4 b) = 9/16 A.B. + 3/16 A.bb + 3/16 aaB. + 1/16 aabb .

Modifications of this dihybrid phenotypic segregation ratio result when dominance is absent in one or both allele pairs, when phenotypes result from → gene interaction between nonalleles, when different combinations of alleles or nonalleles have different viabilities, or when the → gene loci under observation show → linkage, i.e., are located in the same chromosome (in the same linkage group), and, therefore, need not show independent segregation.

If two pairs of alleles (Aa and Bb) are linked in the cis-configuration (AB/ab), the frequency of the gamete genotypes is 1/2(1 − p) AB: 1/2p Ab: 1/2p aB: 1/2(1 − p) ab, and with linkage in trans-configuration (Ab/aB) their frequency is 1/2p AB: 1/2(1 − p) Ab: 1/2(1 − p) aB: 1/2p ab in which p symbolizes the recombination percentage.

Effective g.s. (Darlington 1931) is that which gives rise to viable gametic and zygotic combinations.

Secondary g.s. (Darlington 1928) is the s. (in an → allopolyploid) of differences between the ultimate diploid parents due to → chromosome pairing between chromosomes introduced by the same gamete (autosyndesis), as opposed to primary s., which is due to "allosyndesis" between chromosomes introduced by different gametes.

Preferential g.s. (Rhoades 1942) is due to nonrandom → assortment of a particular chromosome or chromosome segment (and the genes located therein) with respect to the four cells produced at meiosis (→ affinity; meiotic drive; segregation distortion). Preferential g.s. may be inferred from the cytological behavior of chromosomes or from the distortion of genetic segregation ratios, or from both. True preferential g.s. occurs, e.g., when during → oogenesis a certain chromosome or chromosome segment is included preferentially in the egg cell, while its homologue goes to the other nuclei (polar nuclei, or the equivalent nuclei in plant; → megasporogenesis) which do not participate in zygote formation. Cases of preferential g.s. occur during male gametogenesis when meiosis does not produce four spermatids, or when the four cells arising are not equally functional.

Somatic (or mitotic) g.s. is the mitotic formation of cells differing from one another in respect to their genetic make-up, by mutation, mitotic crossing over, or unequal assortment of extrachromosomal → hereditary determinants.

Contrary to nuclear genes, **organelle genes** contained in → mitochondrial DNA or → chloroplast DNA regularly segregate during mitotic cell division subsequent to a cross between organisms carrying different alleles for a gene in the organelle genomes (provided that the gametes of both sexes or mating types transfer the organelles and heteroplasmic zygotes originate). After a number of vegetative reproductions, the progeny of each zygote will be homoplasmic for the organelle genes. The adult organism or cell colony produced by this vegetative reproduction will be a genetic mosaic.

genetic stability — the capacity of individuals to produce adaptive offspring. G.s. is promoted by asexual reproduction, → inbreeding, → homozygosity, and low → mutation rates (Thoday 1975).

genetic step — in human genetics, a term used to define the relationship between one parent and one offspring (= one genetic step). By this single g.s. the genetic resemblence is reduced to 0.5 through the intervention of meiosis: a child receives only one half of its chromosomes and autosomal genes from one parent. Thus, two steps separate grandfather and grandson and consequently they share an average of 0.25 genotypic identity. In the case of full sibs, a special g.s. exists. Their similarity through the father involves two genetic steps and their total resemblance for paternal genes is thus 0.25. Two first cousins are separated by three genetic steps.

genetic suppression — in mutant organisms, the restoration of a pseudo-wild type phenotype by a secondary mutation. The primary mutation is maintained (→ suppressor).

genetic switch — a change (switch) in prokaryotic and eukaryotic → gene expression due to → DNA rearrangement. In prokaryotes, a system of DNA rearrangements which allows for the escape from the immune system (→ antigenic variation). In eukaryotes, g. switches by DNA rearrangement allow, e.g., activation of silent genes due to their transposition into the yeast MAT locus (→ mating type switch) or the generation of → antibodies against a large number of infections.

genetic system (Darlington 1939) — any of the species-specific ways of organization and transmission of the genetic material in pro- and eukaryotic organisms, which determine the balance between coherence and recombination of genes and control the amount and type of gene combinations. Evolution of genetic systems means the evolution of those mechanisms effecting and affecting → genetic variability. Factors which characterize a g.s. include the mode of → reproduction, the type of population dynamics (breeding size, sex ratio, degree of panmixia), the mode of chromosome organization (genetic information all in one linkage group or distributed to several such groups), the chromosome cycle (normal meiosis in both sexes, or abnormal in one in the case of eukaryotes), the → recombination index, and the presence or absence of genetic and → chromosome polymorphism.

The genetic system and its components determine the capability of a population to undergo evolutionary changes. Any g.s. is under genetic control.

genetic toxicology — a field of research (= environmental mutagenesis) which identifies and analyzes the action of agents with toxicity directed towards the genetic material of living organisms, i.e., agents able to cause → DNA damage, → gene mutations, or chromosome breakage and rearrangement (→ chromosome mutation; mutagenicity testing).

genetic transcription — the process by which → DNA-dependent RNA polymerases faithfully synthesize selected units (→ gene; operon) of genetic information stored in DNA (or RNA in some viruses). G.t. is the first step in → gene expression and involves initiation, elongation, and termination of RNA transcripts. Most of these are genetic messages that are translated (→ genetic translation) by → ribosomes to produce proteins (→ messenger RNA); some other transcripts (→ transfer RNA; ribosomal RNA) are not translated but directly incorporated into the cell's biochemical machinery. G.t. is controlled by specific base pair sequences encoded in DNA (→ genetic code) and by gene products that interact with DNA sequences (→ transcription factor) and the RNA polymerases (→ sigma factor; psi factor; rho factor).

The essential general chemical characteristics of g.t. are the following: (1) The precursors in the synthesis of RNA are the four ribonucleoside-5′-triphosphates (rNTP) ATP, GTP, CTP, and UPD. Two OH-groups (one each on the 2′- and 3′-carbon atoms) are the ribose moiety of each NTP. (2) During the polymerization reaction, a 3′-OH group of one nucleotide reacts with the 5′-triphosphate of a second nucleotide and a pyrophosphate is removed; a phosphodiester bond results. (3) The base sequence of an RNA molecule is determined by the base sequence of the DNA; each base added to the growing end of the RNA chain is chosen on the basis of its ability to base-pair with the DNA strand used as a template. (4) The DNA molecule being transcribed is double-stranded; in any particular region usually only one strand serves as a template. (5) The RNA chains grow in 5′ → 3′ direction (same direction as in → DNA replication); the RNA strand and the DNA template strand are antiparallel to one another. (6) RNA polymerases need no primer, i.e., are able to initiate chain growth. (7) Only ribonucleoside-5′-triphosphates participate in RNA synthesis; the first base to be laid down in the initiation events is a triphosphate whose 3′-OH group is the point of attachment of the subsequent nucleotide. (8) The overall polymerization reaction is:

$$\mathrm{nNTP} \rightarrow \mathrm{XTP} \xrightarrow[\mathrm{Mg^{2+}}]{\mathrm{DNA,\ RNA\ pol.}} (\mathrm{NMP})_n - \mathrm{XTP} + \mathrm{nPP_i}$$

(XTP = first nucleotide of the 5′-terminus of the RNA chain; NMP = a mononucleotide in the RNA chain; RNA pol = RNA polymerase; PP_i = pyrophosphate released each time a nucleotide is added to the growing chain). (9) Termination of RNA synthesis involves at least three distinct reactions, i.e., cessation of elongation, release of the nascent RNA chain, and release of RNA polymerase.

With respect to the g.t. of particular species of → RNA transcripts in a given cell at a given time, there are essentially three types of regulatory event possible: (1) initiation of g.t.; (2) rate (increase or decrease) of g.t.; and (3) termination of g.t.

The g.t. of double-stranded DNA may be asymmetric in the sense that only one of the two DNA strands in a given region serves as a template for DNA, or symmetric, i.e., both strands can be transcribed. Two modes of asymmetric g.t. are known: (1) T. occurs from only one (the codogenic) DNA strand; (2) there is a switch from one strand to its complement. In → mitochrondria both strands are transcribed (symmetric g.t.) but one of the two transcripts rapidly disappears.

In **prokaryotes**, g.t. involves interaction between a single → RNA polymerase and the DNA template. The polymerase holoenzyme core can recognize the → promoter of each transcribed gene. Initiation of g.t. is essentially a three-step process: (1) RNA polymerase holoenzyme (core + sigma factor) binds at promoter sites to form "closed" complexes; (2) "closed" or I complexes are converted into RS (rapid starting), or open, complexes perhaps by specific denaturation of 5–10 nucleotide pairs (thereby making single-stranded template DNA available to the enzyme; (3) RS complexes rapidly initiate the formation of the phosphodiester bonds in newly synthesized RNA chains.

Termination of g.t. also involves a number of discrete steps: (1) Pause or arrest of RNA polymerase at specific sites (terminators) on the DNA template; (2) release of RNA; (3) dissociation of RNA polymerase from the DNA. At independent termination sites these reactions occur without added transcrition factors. At rho-dependent termination sites (→ rho factor), rho activates one or more of these steps, possibly step (3). In the case of rho-independent terminators, termination occurs distal to a GC-rich region within a run of uridine residues. All these terminators display a region of dyad symmetry in the DNA molecule proximal to the termination point, i.e., the RNA molecule has the potential to form a stem-loop conformation close to the 3′-terminus.

The immediate product of g.t. of prokaryotic DNA sequences coding for proteins is → polycistronic messenger RNA. Transcription and translation systems are closely coupled in prokaryotes, i.e., g.t. can proceed only if the already transcribed part of the message is being translated. The basic features of **g.t. in eukaryotes** are similar to those in prokaryotes, but there are five essential differences: (1) Three different DNA-dependent RNA polymerases, each with a unique transcriptional function, are needed for transcription of the whole complement of genes; unique auxiliary proteins are required by each polymerase for promoter recognition. (2) Many mRNA molecules are very long-lived. (3) Both the 5′ and 3′ termini of mRNA are modified; a complex cap structure is found at the 5′ end and a long (up to 200 nucleotides) sequence of polyadenylic acid [poly (A)] is found at the 3′ end. (4) The mRNA that is used for genetic translation (the mature transcript) is usually only about one-tenth the size of the primary transcript since during → RNA processing introns are excised and the fragments are joined. (5) All eukaryotic → messenger RNA's are monocistronic.

RNA polymerase I (α) is nucleolar (→ nucleolus) in origin and responsible for the synthesis of → ribosomal RNA; RNA polymerase II (β) transcribes → heterogeneous nuclear RNA and mRNA; RNA polymerase III (γ) is present only in small quantities and is responsible for the synthesis of 5S rRNA, tRNA, and possibly other low molecular weight RNA species.

The g.t. of eukaryotic structural genes takes place at a nucleoprotein level of organization. This is true for both the templates for transcription (→ chromatin) and the products of g.t. which are rapidly assembled into ribonucleoprotein particles (hn RNP).

The transcriptional state of a (eukaryotic) gene is reflected in its chromatin structure, with actively expressed (or potentially active) genes typically exhibiting an overall sensitivity to nucleases as well as specific sites that are hypersensitive to digestion.

Convergent t. is the term used when two promoters are directing the g.t. of a region of DNA in opposing directions, i.e., where both strands of the DNA template would be transcribed simultaneously (→ transcription complex; transcriptional unit; spacer unit).

A special mode of g.t. is → **reverse transcription** by viral RNA-dependent → DNA polymerase (reverse transcriptase). It takes place in two steps: (1) Synthesis of a single DNA strand complementary to the viral DNA genome (→ antimessenger DNA) to yield a RNA-DNA hybrid. (2) Synthesis of a double-stranded DNA using this hybrid as a template.

genetic transduction — → transduction

genetic transformation (Griffith 1928; Avery et al. 1944) — the unidirectional transfer (→ gene transfer) and incorporation of foreign DNA by prokaryotic or eukaryotic cells and the subsequent recombination of part or all of that DNA into the cell's genome. **G.t. of bacteria** occurs according to the following sequence of events: (1) Binding of the DNA by competent cells; (2) uptake of the bound molecules which renders the transforming DNA single-stranded; (3) noncovalent incorporation of single-stranded segments (gram-positive bacteria) of transforming DNA into the recipient chromosome; (4) covalent binding of the single-stranded segment to the recipient chromosome, resulting in → heteroduplex DNA; (5) resolution of the heteroduplex DNA and integration of pieces of the single-stranded DNA into the recipient DNA; (6) expression of the newly aquired information.

Gram-negative bacteria preferentially interact with and transfer double-stranded DNA.

The state of **competence** for g.t. is generally related both to the conditions of growth and to the circumstances under which the cells and the DNA are combined.

As compared to homospecific bacterial t., heterospecific t. is of very low efficiency and this may be due to three major barriers: (1) Differential uptake of homologous and heterologous DNA; (2) enzymatic restriction of heterologous DNA inside the cell; (3) DNA sequence nonhomology leading to reduced recognition and chromosomal integration of heterologous DNA.

G.t. also occurs in eukaryotic cells (→ recombinant DNA technology). Integrated donor sequences are often localized within concatemers formed between donor and carrier DNA's, which are randomly inserted into the genome of the recipient cell. G.t. can also give rise to site-specific or transposition-like nonspecific integration and then results in the addition of genetic elements to the resident genome (→ DNA transposition).

genetic transfusion (Ferenczy et al. 1976) — the transfer of cell components (organelles, viruses) via → protoplast fusion.

genetic transilience (Templeton 1980) — a rapid shift (= genetic revolution) of a population to a new adaptive peak which is most likely to occur when a founder event (→ founder population) causes a rapid accumulation of inbreeding without a severe reduction in genetic variability.

genetic translation — the second step in the "read out" of → genetic information, whereby the → genetic code contained in specific nucleotide sequences of → messenger RNA molecules (synthesized during → genetic transcription) is translated into specific amino acid sequences, i.e., directs the order of the amino acids incorporated into poylpeptides during protein synthesis. During g.t. of the genetic code, the mRNA is read sequentially in sets of three nucleotides. Each nucleotide triplet, or → codon, specifies the amino acid which will be added to the growing polypeptide chain. The reading frame (→ frameshift mutation) of the codons is established at the initiation of polypeptide synthesis, and is maintained by a mechanism which moves the → ribosome along the message three nucleotides a time (→ mistranslation).

A variety of enzymes and → translation factors (→ initiation, transfer, and termination factor) is required for g.t. and for the unique events which serve in its accurate initiation and termination.

G.t. starts from the amino-terminal end of the polypeptide. The sequence into which the constituent amino acids are built is determined by the mRNA which is translated from its 5' end to its 3' end. The two reacting species in peptide bond formation are → aminoacyl tRNA and the growing polypeptide itself esterified to the terminal adenosine residue of a tRNA molecule. The usual recognition between the → anticodon of aminoacyl tRNA and mRNA is between the first two bases of the codon (from the 5' end) and the last two bases of the anticodon. Between the last base in the condon and the remaining base in the anticodon the pairing appears to be more flexible (→ wobble).

The unit of g.t. is a polycistronic (prokaryotes) or monocistronic (eukaryotes) product of genetic transcription.

In outline, g.t. consists of several stages: (1) Formation of an → initiation complex in which the ribosomal P site (→ ribosome) is filled with an → initiator tRNA molecule (bearing methionine) and in which the mRNA is correctly positioned so that a start → codon matches up with the initiator tRNA. (2) The ribosomal A site is occupied by an amino acid-tRNA complex (called → aminoacyl-tRNA); the particular aminoacyl-tRNA is determined by the codon on the mRNA that is adjacent to the start codon. (3) The first amino acid (the one on the → initiator tRNA) is transferred to and covalently joined to a second amino acid (the one attached to the tRNA in the A site) while both are still attached to tRNA. (4) The first tRNA molecule is ejected from the P site. (5) The aminoacyl-tRNA molecule in the A site advances to the P site. (6) The aminoacyl-tRNA corresponding to the third codon occupies the A site. (7) The process continues until a → terminator (stop) codon in the mRNA occupies the A site (→ E site). Then the completed polypeptide is released from the ribosome. (8) In eukaryotes, all mRNA molecules are monocistronic.

G.t. is remarkably accurate with an error frequency of 10^{-4} to 10^{-5} per amino acid incorporation. This is due to several mechanisms developed during evolution that increase the overall accuracy of translation (→ mistranslation). Fidelity of g.t. depends on the interaction of at least four macromolecular elements: (1) Activating enzymes; (2) tRNA's; (3) ribosomes; (4) the peptide termination system.

The process of polypeptide chain elongation is rather rapid. It occurs as many as 20 times a second in a bacterial cell and about once every second in a mammalian cell.

During evolution, two widely distributed and distinct translation systems have arisen, one in prokaryotes and one in eukaryotic cytoplasm. Eukaryotic cells also contain organelles that have their own transcription and translation systems. G.t. in → mitochondria and → chloroplasts shows some prokaryotic features; however, it also shows some eukaryotic and some unique features.

The main differences between g.t. in eukaryotic and prokaryotic cells are: (1) In eukaryotes, the initiating amino acid is methionine and not formyl-methionine as in bacteria. The initiating tRNA, which responds to the AUG codon (never to GUG), is designated $tRNA^{Met}_{init}$ or $tRNA^{Met}_{f}$ to distinguish it from the $tRNA^{Met}$ used in translation of internal AUG codons. (2) At least seven → initiation factors plus ATP und GTP are used for binding of $tRNA^{Met}_{f}$ to the → preinitiation complex. Two of these are similar to the prokaryotic factors that prevent binding of the large ribosomal subunit to the small subunit. One of these factors (eIF3) contains nine protein subunits. The initiating tRNA hydrolase, which removes the initial tRNA molecule after the first peptide bond forms, is also present in the complex. In prokaryotes, the corresponding enzyme (tRNA deacylase) is a ribosomal component. (3) Binding of $tRNA^{Met}_{f}$ must occur before mRNA can bind; in prokaryotes, the mRNA can bind either before or after binding of initiator tRNA. In eukaryotes, binding initially occurs at or near the 5' cap of mRNA and is mediated by the cap-binding factor (unnecessary for uncapped prokaryotic mRNA). The mRNA moves with respect to the pre-initiation complex until the → anticodon of the prior bound $tRNA^{Met}_{f}$ pairs with the first AUG encountered. No → Shine-Dalgarno box sequence is needed. It makes a significant difference between eukaryotes and prokaryotes that the AUG codon nearest the 5' end is always the initiating codon and plays an important role in the regulation of protein synthesis in animal viruses (→ ribosomal scanning model). (4) Eukaryotes need more factors for binding of the 60S ribosomal subunit than bacteria need for binding of their 50S subunit. (5) Eukaryotes need at least four → elongation factors which probably differ in structure and size in different tissues. (6) Termination of g.t. in eukaryotes is not well known.

The eukaryotic ribosome consists of the same general array of large and small subunit ribosomal proteins, rRNA's, and initiation, elongation, and termination factors, but the numbers, sizes, and shapes of these molecules differ somewhat. In eukaryotes, g.t. cannot be initiated as soon as a portion of a cistron has been transcribed (as it can in prokaryotes) because the mRNA must first be transported from the nucleus to the ribosomes in the cytoplasm.

The sequence (A/G)-C-C-A-U-G-G is the sequence at which eukaryotic t. initiation most probably begins. An AUG flanked by another (unfavored) sequence will be ignored, thereby allowing initiation at a downstream AUG.

In prokaryotes, g.t. can, in contrast to eukaryotic g.t., start either with an AUG or a GUG codon. In addition a purine-rich sequence called the Shine-Dalgarno sequence (the prototype is AGGAGGU), preceding the initiator codon usually 6 to 8 nucleotides, is implicated in binding the 30S ribosomal subunit to the beginning of the gene.

The termination of g.t. (→ release factor) occurs when the ribosome encounters one of three stop codons (UAA, UAG, UGA) in the correct reading frame. UAA use predominates over the other two for most mRNA's, but examples of the use of all three are found in both prokaryotes and eukaryotes and also in viral coding sequences. In mitochondria, UGA functions as a tryptophan codon with the mitochondrial $tRNA^{trp}$ reading UGA as well as the normal tryptophan codon UGG.

Translational errors (→ mistranslation) may occur during g.t. due to codon-anticodon mispairing and represent transient errors which are principally of two kinds: (1) Misreading of a codon in mRNA, i.e., the substitution of one amino acid for another; (2) errors that involve the maintenance of the correct reading frame. Translational errors are a function of the individual bases and not of the overall composition of the codon where the base occurs while the error pattern, i.e., what bases are mistaken for others, is more or less independent of the position in the codon. The order of decreasing error-proneness as a function of position in the codon is III > I > II. In each position the error pattern is UmC (i.e., U mistaken for C), > UmA, while UmG is least frequent or even undetectable. A similar pattern exists for C in that CmA ≫ CmG, while CmU is a rare mistake.

genetic transposition — the movement of genetic material within the genome mediated by → DNA transposition.

genetic unit — an operationally definable portion of the → genome, such as a → codon or → gene. Any g.u. is defined by the set of mutations that fall within it. A mutation used to define and map some g.u. is a marker. The position assigned to a g.u. on the → genetic map is its locus which covers a finite number of → mutational sites.

genetic variability — the formation of individuals differing in → genotype, or the presence of genotypically different individuals, in contrast to environmentally induced differences which, as a rule, cause only temporary, nonheritable changes of the phenotype. **Genetic variance** is the term used to denote that portion of the phenotypic variance which is caused by variation in the genetic constitution of the individuals in a population. G.v. is a universal feature of breeding populations and a necessary preliminary condition for evolutionary change. The g.v. of a population is usually measured by the average heterozygosity per locus. The magnitude of g.v. in populations is assumed to be determined by the action and interaction of numerous factors, such as mutation rates, the degree of → dominance of mutant genes, the mode of interaction (→ gene interaction) among loci, the effects of the environment, the nature of selection, stochastic processes (in the case of small population size), breeding systems, and population structure.

The chief sources of g.v. are the following processes: (1) → **Mutation** as the ultimate source of all genetic variability; it gives rise to changes in the molecular structure of the genes (→ gene mutation), in the linear sequence of chromosome segments (→ chromosome mutation) or in the number of chromosomes (→ genome mutation); (2) → **gene flow,** i.e., migration of individuals or gametes carrying new alleles from one population to another; (3) → **hybridization** between individuals having different genotypes followed by the production of offspring with new combinations via → genetic recombination.

Processes (1) and (2) may directly alter the → gene frequency in a population and are thus the basis of evolutionary change (→ evolution). Genetic recombination assembles an existing array of alleles of different genes into a variety of combinations without altering their frequencies. Although genetic recombination is, therefore, not an evolutionary force, it is by far the most effective source of individual g.v. in sexually reproducing organisms. Process (3) generates most of the intrapopulation genotypic differences which then become available for → selection and → genetic drift.

Regulation of g.v. in a population is primarily determined by its chromosome behavior and system of reproduction. Equilibrium conditions providing for the retention of g.v. can be established by the interaction of mutation and selection or through selection for heterozygotes.

According to Fisher (1930) and Mather (1943) two types of g.v. are to be distinguished: (1) **Free genetic variability:** that part of the total g.v. which is manifested by the phenotypes in a population and which is, therefore, exposed to the action of → selection. (2) **Potential (= "concealed" or cryptic") genetic variability:** That part of the g.v. not yet manifested in the phenotypes, and hence not exposed to selection in any given generation. Potential g.v. can become phenotypically manifest in later generations on passing from the potential to the free state. Both the store and the rate at which this type of g.v. becomes transformed into free g.v. are controlled by natural selection and are dependent on the breeding system of the population under observation. Asexual reproduction limits, or even prohibits, the realization of potential g.v.; in endogamic populations the store of potential g.v. is always small.

Potential g.v. is present in heterozygous genotypes; the potential g.v. of homozygotes is nil and thus their flexibility slight. The type and quantity of potential g.v. may become modified by particular cytogenetic mechanisms.

The underlying factors determining the store of concealed g.v. are: (a) Unrealized mutations and gene combinations; (b) the separate presence of genes with complementary action; (c) genes whose activity is reduced or prevented by modifier genes or epistatic interaction; (d) recessive alleles not manifest in the heterozygous state; (e) limitations by linkage, and, in polygenic systems, limitation in the production of recombinants, and (f) genes and gene systems giving rise to similar phenotypes under the prevailing environmental and developmental conditions.

Release of concealed g.v. is brought about by different mechanisms: in (a) and (b) by genetic recombination, in (c) by inactivation or elimination of the modifiers and epistatic genes, increased penetrance and recombination, in (d) by segregation or increased penetrance, in (e) by recombination via crossing over, and in (f) by changes of environmental conditions or phenotypic differentiation by placing the genes in question into different residual genotypes.

A store of concealed g.v. will contain variants which, when set free, may prove unsuccessful under all conditions, useful under a definite set of circumstances, neutral or even harmful at the time of realization, but useful later on. The significance of concealed g.v. lies in the provision of new material for selection when transformed into free g.v.

Decay of g.v. means the reduction of heterozygosity as a consequence of → genetic drift, leading to loss and fixation of alleles at various loci. Its rate is intimately tied to the population size.

genetic variance — that portion of the phenotypic variance which is caused by differences in the genetic constitution of individuals in a population (→ genetic variability). The g.v. may be broken down into (1) a contribution that is due to differences between homozygotes (additive v.), (2) a contribution that is due to specific effects of various alleles in heterozygotes (dominance v.), and (3) a contribution that is due to combinations of nonhomologous loci (epistatic v. or nonadditive v.).

gene transcription — → genetic transcription.

gene transfer — the introduction of functional (usually cloned) genes into cells and organisms. A variety of techniques can be used for g.t.: (1) Cell hybridization; (2) → microcell-mediated g.t.; (3) → chromosome-mediated g.t.; (4) → DNA-mediated g.t.

G.t. results in genetically transformed cells and invididuals (→ genetic transformation) and is a step in → recombinant DNA technology when cloned genes are being used for transfer.

gene transferent (Degnen et al. 1976) — a recipient cell that has taken up and expresses a gene incorporated by → gene transfer.

gene transplacement — the exchange of a DNA sequence on a recombinant → plasmid for the equivalent sequence in one of the host chromosomes (replacement of a wild-type copy of a gene by a mutant copy or vice versa).

The host organism is transformed (→ genetic transformation) with an integrative plasmid, the plasmid integrated by homologous recombination and thereby a duplication of the sequences of interest is created. Subsequently, the integrated plasmid and the copy of the gene originally contained in the chromosome loop out and undergo intrachromosomal recombination. The copy of the gene originally contained in the recombinant plasmid is left behind and is said to have transplaced.

genic instability — instability of a gene usually caused by the incorporation and excision of → transposable genetic elements.

genocline — → cline.

genocopy (Nachtsheim 1957) — the production of the same → phenotype by different genes ("mimetic genes").

genome (Winkler 1920) — in eukaryotes, the basic (monoploid) → chromosome set, consisting of a species-specific number of → linkage groups and the genes contained therein. Additional genomes are contained in → mitochondria (mitochondrial g.) and, in plant cells, in → chloroplasts (chloroplast genome). In prokaryotes, the complement of all hereditary factors (genes) included in one linkage group.

The smallest possible unit of the g. in mutation (→ gene mutation) and recombination (→ genetic recombination) is the individual nucleotide pair of → deoxyribonucleic acid (in some viruses of → ribonucleic acid), and is referred to as a muton and a recon, respectively. These nucleotides are arranged in a definite linear order into larger functional units of the g.: (1) → **Cistrons** (usually consisting of several hundred nucleotide pairs) are units (genes) which specify by → genetic transcription the nucleotide sequence of various species of → ribonucleic acids, and by → genetic translation, the amino acid sequence in specific polypeptides; (2) → **operons** are the units of transcription of DNA into → messenger RNA and include several functionally related cistrons; (3) → **polarons** are units in terms of the points where crossing over can be initiated and probably coincide with operons; (4) → **replicons** are units of replication and possibly consist of a series of operons; (5) whole linkage structures (chromosomes) may behave as units of → genetic regulation in eukaryotes under particular circumstances.

Two genomes are said to be exactly homologous if both contain the same genetic loci in identical sequence in their linkage structures, and homeologous, if only a part of the segments is identical in localization, due to structural rearrangements.

The diplophase of the life-cycle of diploid eukaryotes is characterized by the presence of two, the haplophase by the presence of only one genome. This → alternation of nuclear phases is brought about in the case of sexual reproduction by → fertilization and → meiosis. In genetic systems with parasexual modes of reproduction, the bringing together of two genomes or one genome and one genome fragment into the same cell is achieved by a variety of other means.

Organisms having more than two genomes in their cells are called → polyploid.

The basic organizational pattern of the nuclear genome of almost all higher organisms consists of an ordered arrangement of interspersed repetitive (→ repetitious DNA) and single copy (unique) DNA sequences (→ DNA interspersion). In addition to repeated DNA sequences, the eukaryotic g. contains repetitive inverted sequences capable of forming intra-strand base pairs (hairpins).

The nuclear g. of higher eukaryotes is in a continuous state of flux. The processes involved include → mutations and wide-range events such as → gene duplications, → gene amplifications, and → DNA transpositions. A relatively small proportion of the nuclear genomic DNA of eukaryotes is transcribed and translated.

The amount of DNA per g. (expressed as picograms or nucleotide pairs) ranges over eight orders of magnitude, from certain viruses to some algae, amphibians, and higher plants. Within taxa such as classes or orders, there is generally a 5- to 25-fold range in the amount of DNA per genome.

genome allopolyploid (Stebbins 1947) — → alloploid.

genome analysis (Kihara 1930) — a method for determination of the diploid ancestors of → allopolyploid (plant) species. It consists of the analysis of meiotic → chromosome pairing in hybrids between polyploids and diploids. If the diploid has at one time or another contributed to the formation of the polyploid, chromosome pairing should occur between two sets of homologous chromosomes in the hybrid. Final proof of the deductions from g.a.

is the resynthesis of the allopolyploid; when synthetic and natural alloploids resemble each other closely and have fertile F_1 progeny, it may be assumed that the results obtained by g.a. are correct.

genome blotting — a method (Southern blot technique) of transferring DNA fragments, separated by gel electrophoresis (agarose), to a nitrocellulose filter such that the relative positions of the fragments are maintained. Visualization of the DNA is usually by hybridization with a ^{32}P-labeled DNA or RNA probe.

genome elimination — the elimination during cell divisions of → genomes or parental sets of chromosomes. G.e. occurs in some cases as a regular developmental feature at meiosis **(Sciara)** or mitosis **(Metaseiulus)**, in others it occurs in hybrids **(Hordeum, Pocciliopsis, Rana)** and not always at defined developmental stages.

genome imprinting (Surani et al. 1984) — the phenomenon wherein maternal and paternal contributions to the mammalian zygote nucleus do not function equivalently during subsequent development, i.e., gene expression is differential during male and female gametogenesis. The underlying mechanism must fulfil at least three criteria: (1) the imprint must be physically linked to the pronucleus; (2) the imprint must persist through → DNA replication and cell division; (3) the mechanism must be capable of switching the identity of the imprint from one sex to the other in successive generations.

Possibly, differential → DNA methylation during gametogenesis is involved in g.i.

genome library — a → gene bank consisting of a set of independent clones that, statistically, contains the entire → genome of an organism among the → recombinant DNA molecules. Construction and screening of a complete library of clones for any genome is a starting point for isolating specific genes (→ cDNA library). The g.l. of eukaryotes includes both single copy and → repetitious DNA sequences.

genome-linked virus protein — any protein covalently linked to the → genome of certain single-stranded RNA and double-stranded DNA viruses. The proteins are bound via a phosphodiester bridge to the 5'-terminal base of the nucleotide chain and play a role in genome replication, possibly at the level of initiation.

genome mapping — the production of an ordered set of completely overlapping clones that covers an entire genome (→ genome library).

genome mutation — any spontaneous or experimentally induced change in the number of complete chromosomes giving rise to heteroploid cells or individuals (which are either → polyploid or → aneuploid) as compared with the number of → chromosome sets or individual chromosomes in the species under observation.

genome plasticity — the concept, based on → transposable genetic elements, that genes can be gained, lost, or shifted to new positions in the → genome by processes that do not require extensive nucleotide sequence homology (→ DNA transposition).

genome rearrangement — in prokaryotes and eukaryotes, a mechanism mediated by site-specific → genetic recombination that is involved in the control of → gene expression. In prokaryotes, the following three sets of factors are required: (1) The cis-specific DNA acting as the crossing point for g.r.; (2) site-specific trans-active factors coded for by genes mapping near the cis-specific DNA; (3) "host factors" encoded by genes which reside on the bacterial chromosome distant from the site of g.r. (→ DNA rearrangement).

genome segregation (Bauer 1943) — the separation of complete genomes (sets of chromosomes) during mitosis in eukaryotes; this is supposed to bring about "somatic → reduction" of the chromosome number in polyploids. Euploid segregation of → genomes is due to the occurrence of multipolar mitoses after polyploidization or cell hybridization.

genome transconformation — any of the environmentally induced changes of the higher-order structure of the → genome. Such changes are reversible and noninheritable but can modify basic genome functions (such as replication) and provide selective advantages or disadvantages in the absence of changes in the primary structure of DNA.

genome turnover (Dover and Tautz 1986) — the result of mechanisms that include → DNA transposition, unequal → crossing over, and → gene conversion. While transposition often involves site-specific → genetic recombination, unequal exchange and gene conversion are caused by homologous recombination between repeated DNA sequences.

genome walking — the study of extensive regions of a eukaryotic → genome by serial isolation of a number of overlapping clones (→ genome library).

genomic exclusion (Allen 1963; Nanney 1963) — in crosses of certain inbred strains (A) of **Tetrahymena pyriformis** with clones of other strains (B), the recovery in the progeny of genes from one parent (A) only although both exconjugants are present. The genes from the other parent strain (B) are excluded.

G.e. is characterized by two consecutive conjugations. Cells of the strains (B) whose genes are excluded have a defective micronucleus and in crosses with A cells, all B meiotic products disintegrate. During the first (abnormal) conjugation, the A conjugant undergoes meiosis normally generating sister haploid nuclei (the male and female pronuclei); the male pronucleus migrates to the B conjugant and then each pronucleus becomes diploid (probably by endoreduplication). In each conjugant the diploid synkaryon divides twice by mitosis and gives rise to two new macronuclei and two new micronuclei, and the old macronucleus starts to resorb. Once the exconjugants separate, the old macronucleus is retained and the new ones are resorbed. The progeny of the first conjugation are, therefore, heterokaryons with a micronucleus homozygous in genotype, express the macronuclear genes of the old macronucleus (distinct genotypically from the micronucleus), and are sexually mature. The second round of conjugation is normal and gives rise to a functional macronucleus similar in genotype to the micronucleus. The synkarya are derived from two meiotic products, one from each conjugant. If the two exconjugants from the same round — pair are remated, the (immature) progeny are, therefore, genetically homozygous at all known loci for the genes of the normal parent (B).

By g.e. the population becomes cleaned of defective cells and the development of new micronucleate lines is prevented. Thus, g.e. may be understood as a means for the reacquisition of a normal micronucleus (Allen 1967).

genomic flux — the phenomenon wherein eukaryotic → genomes are very dynamic structures due to the continued dispersal of members of interspersed repetitive sequence families (→ DNA interspersion) and genetic exchange processes that are involved in the reduction of sequence diversity in such families.

genomic footprinting (Becker and Wang 1984) — a technique used to investigate areas of protein/DNA interaction: Cells in suspension are treated with dimethylsulfate (DMS) or ultraviolet light to modify specific DNA bases, which undergo a quantitative cleavage reaction (β-elimination). The modified sites are then identified by genomic sequencing. Proteins bound to DNA protect these areas from modification by DMS and subsequent chemical cleavage, leaving their "footprint" on the DNA. In contrast, UV light cleaves at sites of protein-DNA interaction.

genomic imprinting — an epigenetic marking process which results in functional differences between homologous alleles depending on whether they come from the mother or the father (→ chromosome imprinting).

genomic instability — a phenomenon typical of neoplasia, that is manifested as a constellation of chromosomal aberrations including aneuploidy, translocations, deletions, duplications, gene mutations, and amplification of localized discrete chromosomal segments. Certain rearrangements are characteristically associated with particular forms of cancer and are referred to as nonrandom chromosomal changes.

genomic sequencing (Church and Gilbert 1984) — a method which provides the most accurate view of a DNA sequence in the chromosome and can be used, e.g., to study → DNA methylation patterns directly in uncloned DNA and to analyze DNA sequences associated with protein (protecting the DNA against chemical or enzymatic attach) in → chromatin. The protein association at specific sites results in characteristic"holes" in the sequence.

G.s. combines the chemical → DNA sequencing procedure of Maxam and Gilbert with the detection of DNA sequences by → Southern blotting: Fragmented DNA is electrophoresed, transferred to an inert support, and indirectly → end-labeled by hybridization to appropriate short radioactive DNA probes.

genomic stress (McClintock 1978) — any of those influences which may disrupt the stability of the → genome, such as environmental factors, altered genetic background (e.g., by production of hybrids), exposition of cells to cell culture, etc. G.s. may activate silent → transposable genetic elements and result in their transposition. Reactions on g.s. are viewed as a part of the survival strategy of a cell.

genophore (Ris 1961) — the physical equivalent of a linkage group (the linkage structure of chromosomes) in prokaryotes, bacterial → plasmids, → mitochondria, and → chloroplasts (= genomes) which is represented by a "naked" nucleic acid molecule. The term g. was proposed to emphasize the differences in structure of → chromosomes in prokaryotes and eukaryotes.

genospecies (Ravin 1963) — a genetically defined → species, i.e., a group of individuals potentially able to contribute to, or share in, a common gene pool.

genotoxic (Brookes et al. 1973; Ehrenberg et al. 1973) — of agents (genotoxicants) inducing toxic, lethal, or heritable effects to nuclear and extranuclear genetic material in germinal and somatic cells. G. agents may be activation-independent or activation-dependent. Usually they have chemical or physical properties that facilitate their interaction with nucleic acids (→ clastogenic; mutagenic; mutagenicity testing).

At subtoxic exposure level they produce genetic alterations. In dependence on the g. agent, g. effects can be induced by different mechanisms. Direct mutagenesis results from miscopying or miscoding → DNA lesions. Indirect mutagenesis results from an error-prone replication activity after induction of noncoding DNA alterations.

genotroph (Durrant 1962) — a plant (in some flax varieties) showing stable heritable changes induced when it is grown in certain environments. The origination of such heritable changes is hypothetically explained by variation in the number and arrangements of particular DNA sequences.

genotropic (Waddington 1962) — embryonic → induction.

genotype (Johannsen 1909) — (1) the sum total of the → genetic information (genes) contained in the linkage structures (chromosomes) of the pro- and eukaryotes, as distinguished from their → phenotype (→ idiotype). The g. determines not a unique phenotype, but a range of phenotypic capacities referred to as an individual's "norm of reaction" to the environment. The share of one particular g. in the → gene pool of the next generation is determined by the viability and reproductive success of its carriers. (2) The genetic constitution in respect to the → alleles at one or a few → pairs of genetic loci under observation. If specific loci primarily responsible for the appearance of a particular → character are studied, the rest of the genotype is referred to as the "residual genotype or background genotype" (→ epigenotype).

genotype-environment interaction — that part of → phenotypic variation which is the result of interaction between genotype and environment.

genotype frequency — the proportion or frequency of any particular genotype among the individuals of a population. Genotype frequencies are a function of → gene frequencies. Changes in frequencies of both are brought about by systematic and/or dispersive processes.

genotypic — ref. to any phenomenon or process connected with or controlled by the → genotype.

genotypic cohesion — the phenomenon wherein balanced and superior gene combinations (co-adapted gene complexes or → supergenes) are held together in the face of the centrifugal forces of → genetic recombination, thus reducing the frequency of deleterious recombinants, and with it the → genetic load.

genotypic distance — (1) for two individuals, A and B, and a given locus, the probability that the genotype of A is not the same as that of B; the distance between them is zero when they are of the same genotype at the particular locus, it is 1 when this is not so (g.d. between two individuals); (2) two populations have a g.d. of zero if the distribution of numbers of individuals among the various genotypes is the same in both, i.e., if they have the same genotypic structure or genic structure (g.d. between two populations).

genotypic mixing (Mc Bride 1962) — in animal viruses, after mixed infection with genetically different types of a particular virus, the enclosure within the virus envelope of two or more different genomes. Such genotypically mixed virions are different from → heterozygous bacteriophages in which the heterologous DNA is part of a single viral DNA molecule.

genotypic structure — (S); in population genetics, the set of the genotype frequencies of a population (homozygous and heterozygous genotypes). From the g.s. of a population the genic structure(s) can be derived when the exact genotype frequencies are known. If full information is not available, the g. probability s. can be calculated, i.e., the set of probabilities assigned, in accordance with the information available, to the various genotypes that an individual taken at random from the population could have. The genic probability structure is then the set of probabilities attached to the various genes which could be carried at a particular locus by a given chromosome taken at random from an individual member of the population (Jacquard 1974).

genotypic value — the value of an individual or character associated with genotype (→ phenotypic value).

genotypic variance — (V_G); a measure of the effect of genes on a character; it may be subdivided into additive, dominance, and dominance-additive (interaction) effects ($G = A + D + I$).

genus — a taxonomic unit consisting of a group of → species which are of common phylogenetic origin and appear to be clearly differentiated from other such groups.

geographic isolate — → isolate.

GERL (Novikoff 1973) — a hydrolase-rich region (structure) of the → endoplasmic reticulum (ER) situated at the trans phase of the → Golgi apparatus from which the → lysosomes appear to arise. GERL is considered to be a separate organelle from the Golgi apparatus because of its distinct cytochemical activity and ultrastructural connections with the ER. It is involved in the formation of autophagic vacuoles, secretory granules, the transport of lipids, and in receptor bound ligands.

germ cell (Engler and Prantl 1897) — any of the "reproductive" (generative) cells in multicellular organisms, as distinct from somatic cells. Germ cells are ultimately either → gametes or → agametes and arise from primordial germ cells or gonocytes. The term g.c. covers both the ancestral forms and their ultimate differentiation product. Germ cells may be classified as follows.

(1) **Agametes (= spores)** are noncopulating germ cells (in sporozoa and plants), usually asexual (in some cases sexually differentiated), which may arise by meiosis (**"meio-agametes"**, "gonospores", "meiospores", "tetra-spores" or "gonia") or by mitosis (**"mito-agametes"** or "gonidia"). Organs producing agametes are called agametangia (= sporangia, gonidangia); in cases of heterospory, microsporangia (= androsporangia) and megasporangia (= macrosporangia, gynosporangia) are distinguished (→ megasporogenesis, microsporogenesis).

(2) **Gametes** are sexually differentiated copulating germ cells which may arise by meiosis (**"meio-gametes"**, "gonogametes" or "gones") or by mitosis (**"mito-gametes"**). The cells giving rise to → gametes are referred to as "gametocytes" (= "gamonts", "gametangia", "gametogonia"). In metazoa, gametocytes are undifferentiated precursors ("primordial germ cells") of egg cells (ova) and sperm which are produced in the gonads (ovaries in females, testes in males). The process of egg cell production during gametogenesis proceeds via oogonia, primary oocytes, secondary oocytes, and ootids (→ oogenesis), the process of

sperm production via spermatogonia, primary and secondary spermatocytes, and spermatids (→ spermatogenesis).

Gametocytes in lower plants are in most cases sex organs and are called gametangia (isogametangia in thallophyta producing isogametes; anisogametangia in thallophyta producing anisogametes; oogonia and antheridia in thallophyta producing egg cells and spermatozoids; archegonia and antheridia in mosses and ferns; karpogonia and spermatangia in red algae; ascogonia and antheridia in fungi).

germinal spot (Purkinje 1825) — the nucleolus of the egg cell.

germinal vesicle (Purkinje 1825) — the → nucleus of the → oocyte during the growth phase which is arrested at first meiotic division in animal → oogenesis. At this stage, the oocyte synthesizes and accumulates large quantities of → ribosomes and → messenger RNA, considered to be a store of protein-synthesizing capacity, conserved during maturation for utilization in the early embryo after → fertilization.

germ layer — any one of the (two or three) fundamental cell layers referred to as ecto-, endo-, and mesoderm in an early embryo of a multicellular animal from which tissues and organs of the adult are formed (→ embryonic development).

germ line (Weismann 1885) — a lineage of "generative" cells (= germ track) ancestral to the gametes (sperm and egg cells proper) which, during the development of an organism (particularly in animals), are set aside as potential gamete-forming tissues. These ancestral cells, together with the gametes, are called → germ cells, as opposed to somatic cells. Location, nature, and time of formation of potential gamete-forming tissues are species-specific and may vary greatly from one species to another.

Generally, the separation of germ cells and somatic cells takes place in early ontogenesis. The germ line cells of some animals are characterized by so-called "germ line bodies" in their cytoplasm.

germ plasm (Weismann 1883, 1885) — the → genetic material which forms the physical basis of inherited qualities and is transmitted from generation to generation by the → germ cells (= Keimplasma).

germ track — = → germ line.

G factor (Watson 1964; Nishizuka and Lipman 1966) — a protein (tRNA translocase) involved in the → translocation (during bacterial → genetic translation) of the new peptidyl tRNA from the A to the P site of the → ribosome (→ tRNA releasing factor). It is involved in the hydrolysis of GTP during polypeptide elongation which is thought to assist the rearrangement of tRNA, mRNA, and ribosomal subunits. These rearrangements must occur between successive rounds of peptidyl transfer. Probably G enters and leaves the ribosome during each cycle of amino acid addition.

ghost layer (Tsai and Green 1973) — a monolayer of in vitro cultivated animal cells treated for a short time with a nonionic detergent in order to disrupt the membranes and to eliminate the problems connected with pool changes and uptake of the radioactive precursors. G. layers allow a rapid and sensitive assay for → genetic transcription utilizing a relatively small number of cells.

giant chromosome — → polytene chromosome.

global regulatory system — = → regulon.

glucocorticoid responsive element — (GRE); any of those specific DNA sequences with which hormone receptor complexes interact during transcriptional activation of → gene expression by glucocorticoid hormones. Deletion of GRE's abolishes glucocorticoid induction of → genetic transcription.

glucose effect (Epps and Gale 1942) — the → repression of the synthesis of certain (bacterial and yeast) enzymes by addition of glucose to derepressed cells (= carbon catabolite repression).

glycocalyx (Bennet 1963) — a component of the → cell surface located outside the → cell membrane and, in many instances, attached wholly or in part to its outer surface (= cell coat). The g. always contains sugars of some sort, frequently in company of, and often covalently bound to, other species of substances (such as amino acids, lipids, lignins, proteins, or other polymers). In contrast to the cell membrane, the g. components are readily pervaded by water, ions, and small molecules and show low electrical resistance. The structural and chemical characteristics of the g. vary widely.

One of the variants of the g. is the → cell wall of plant cells.

glycosylase — → DNA glycosylase; DNA repair.

glyoxysome (Greidenbach and Beevers 1967) — any of a specialized form of → peroxisomes found in plant tissues. Glyoxysomes are roughly spherical organelles (about 1 µm in diameter) comprising a protein matrix that sometimes contains a crystalline core, and are bounded by a single membrane. They occur in the → endosperm or cotyledonary cells of fat-storing seeds and contain a broad spectrum of enzymes including those of the glyoxylate cycle and the β-oxidation cycle in addition to catalase and the oxidase characteristic of peroxisomes.

Goldberg-Hogness box — = → TATA box.

Golgi apparatus (Golgi 1898; Ramon y Cajal 1908) — (GA); a single copy organelle in eukaryotic cells (= G. complex) that plays a key role in the processing and transport of secretory and lysosomal proteins as well as in the biogenesis and traffic of membranes. The G.a. begins to fragment at the onset of → mitosis and then looses its pericentriolar location. During telophase, reassembly of the G.a. occurs in the centrosomal region of each daughter cell.

The G.a. consists of → dictyosomes (a number of cisternal-flattened vesicles stacked together) plus the surrounding fine vesicular elements and the surrounding zone of organelle-free cytoplasm. The G.a. is an asymmetric structure, i.e., it consists of cis (or entry) and trans (or exit) faces that differ in morphology and histochemical properties. Vesicles are thought to enter the system by budding off specialized regions of the → endoplasmic reticulum (the transitional elements), and then to fuse with the cis compartment of the → G. stack. After a succession of sequential budding and fusion events that transport proteins from one cisterna to the next, the proteins exit by budding off the trans cisterna or cisternae.

The G.a. has the exclusive ability to modify secretory products by terminal glycosylation, sulfation, and proteolytic processing of preproteins.

Golgi stack (Whaley 1975) — the central structure (= dictyosome) of the → Golgi apparatus which is made up of flattened membrane-bounded cisternae associated with various vesicles.

gonad — in animals, a reproductive organ (ovary in females, testis in males, ovotestis in hermaphrodites) in which gametes (ova and sperm) are produced.

gone (Lotsy 1904) — any → germ cell arising by meiosis.

gonia — primordial sex cells such as oogonia and spermatogonia (→ oogenesis, spermatogenesis).

gonidia (Renner 1916) — → germ cell.

gonochoristic — of individuals having functional gonads of one sex only, either male or female (= geonochoric or dioecious).

gonocyte (Renner 1916) — any unripe → germ cell.

gonogenesis (Levitzki 1925) — the formation of → gones during meiosis.

gonomery (Haecker 1895) — the separate grouping of maternal and paternal chromosomes during → cleavage stages of some organisms.

gonospore (Renner 1916) — → germ cell.

gonotokont (Lotsy 1904) — any cell (= meiocyte) producing → gones by → meiosis.

gonozoospore (Oehlkers 1956) — a → zoospore produced in the course of meiosis.

G period (Howard and Pelc 1953) — a period of the interphase nucleus in eukaryotes.

G_1 is the period during which the chromosomes are released from their condensed mitotic state and protein and RNA synthesis resume at a rapid rate. Particular enzymes have been shown to be specifically synthesized in G_1, which ends with the beginning of the → S period of interphase.

G_2 is the interphase period which bridges the interval between the end of DNA synthesis (S period) and the initiation of mitosis. Condensation of chromosomes presumably begins during G_2 by supercoiling involving complexing → histones with DNA and subsequent folding up of the fibers.

grade (Huxley 1958) — a unit of biological improvement from an evolutionary point of view comprising a group of individuals similar in their level of organization.

gradient — levels or degrees in polarized systems (developmental potency, character formation, metabolism, etc.).

gradualism — the concept that historical change in general and → evolution in particular is slow, steady, and on the whole ameliorative (→ punctuated evolution).

graft-hybrid — → grafting.

grafting — the production of a → chimera by vegetative fusion of parts from different individuals. **Graft-hybrids** are plants in which tissues or cells of different → genotypes exist side by side, as some cell layers come from the scion and some from the stock.

granular component — a component of the → nucleolus which contains the 15 nm pre-ribosomal particles (→ dense fibrillar component).

granum (Meyer 1883) — a structure within the → chloroplast which is recognized as a green granule by the light microscope and as a column of flattened discs piled one on the top of the other by the electron microscope. Each disc consists of a pair of membranes appressed back to back (thickness 6 nm or more). Often grana are interconnected by double-membraned structures (which also extend into the stroma: stroma lamellae). The light-absorbing pigments are contained in the grana.

graphical genotype (Young and Tanksley 1989) — the portrait of → restriction fragment length polymorphism data in a graphical form for description of an entire genome in single graphic image. The g.g. describes the parental origin and allelic composition throughout the entire genome.

gratuitious inducer — a molecule resembling authentic → inducers of → genetic transcription but not being a substrate for the induced enzymes.

Gray (Gy); — 1 Gy = 100 → rad of ionizing radiation; 1 Gy represents the energy absorption of 1 Joule per kg.

g-region (Lee and Thomas 1973) — any of the relatively short DNA regions along the chromatids of eukaryotes which are characterized by clustering of → repetitive DNA sequences. The average length of such regions is about 5 nm to 10 nm (15,000 to 30,000 nucleotide pairs), or about one-half the average length of DNA per **(Drosophila)** chromomere. The g. regions are assumed to be composed of approximately regularly spaced, intermittently repetitious sequences of tandemly repeating sequences in which some portion (less than half) is nonrepetitious.

groundplasm — a highly complex, polyphasic system in which all the resolvable elements of the cytoplasm are suspended, including the larger organelles, such as the → ribosomes, the → mitochondria, the → plastids (in plant cells), → lipid droplets, and → vacuoles. It contains (in unknown degrees of organization) most of the enzymes of intermediary metabolism of the cell (including those involved in preliminary or alternative pathways of ATP synthesis), accommodates the metabolic reserves of the cell, and its aqueous phase contains the main cellular pool of soluble precursors. The g. also contains a population of contractile protein molecules responsible for most cell movements.

group selection (Wynne-Edwards 1962) — a subclass of natural → selection in which the group is the unit of selection, i.e., those cases in which changes in → gene frequency are brought about by the differential extinction and proliferation of the group (local population). By g.s. characters can be selected for that benefit the group rather than the individual (→ kin selection).

growth factor — any of the diverse families of protein factors required for proliferation and differentiation of many types of animal cells. They initially interact with receptors of the plasma membrane to deliver the mitogenetic signals for cell proliferation.

Grünstein-Hogness assay — a colony hybridization method for identification of → plasmid clones; colonies are transferred to a filter and hybridized with an appropriate probe.

GT-AG rule — a rule describing the presence of these dinucleotides at the first two and last two positions of → introns of eukaryotic nuclear genes.

guanine-rich sequence (Sen and Gilbert 1988) — any of the single-stranded DNA sequences which contain guanine-rich motifs that may self-associate to make four-stranded structures. In these the strands run in parallel fashion. G.-r. sequences occur in → immunoglobulin switch regions, in gene → promoters, and in chromosomal → telomeres. Self-recognition of g.-r. motifs of DNA may serve to bring together, and to zipper up in register, the four homologous chromatids during → meiosis.

guanosine 5', 3'-polyphosphate — a pleiotropic effector (ppGpp) that regulates various metabolic pathways as well as the transcription of certain → operons during nutritional stress in bacterial cells.

guide RNA (Church et al. 1979) — an RNA sequence the function of which is to provide a template for the alignment of → splice junctions so that the excision → ligation mechanisms (→ RNA splicing) may act. G. RNA sequences may be either external (EGS) or internal (IGS) to the → messenger RNA being processed and may hybridize to either → intron or → exon sequences close to the splice junction (→ small nuclear RNA).

gymnocyte — a cell without a → cell wall, as distinct from a lepocyte surrounded by a cell wall.

gymnoplast (Küster 1935) — in angiosperms, a cell without → cell wall (= gymnocyte), in contrast to dermatoplasts, i.e. plant cells with a cell wall.

gynaeceum — the female reproductive organ of higher plants; it is composed of pistil, style, and stigma.

gynander — = → gynandromorph.

gynandrism — a type of → genetic mosaicism; the individuals showing g. (= gynanders or gynandromorphs) are mosaics of XX and X cells (→ X chromosome).

gynandroid (Whiting et al. 1934) → gynandromorph.

gynandromorph (Goldschmidt 1915) — an individual of a dioecious (unisexual) species which is a sexual mosaic (a chimera), typically male in certain portions (sectors) of the body and typically female in others. Gynandromorphs result from the presence of chromosomes of both sexes in different parts of the body.

Gynandromorphs have been observed in animals (not yet in plants) where sex differentiation is apparently intracellular and not hormonal and may arise in several ways: (1) By loss of an → X-chromosome during early cleavage in cases of an XX-XY-mechanism of sex-determination, provided XO-cells are male in character (as in **Drosophila**); (2) by suppression of polar body formation giving rise to a binucleate egg cell. In cases of female heterogamety (as in silkworms), one of these nuclei contains an X-chromosome, the other a Y-chromosome. Fertilization of each of the nuclei by a sperm containing one X-chromosome may result in a g.; (3) by fertilization of only one of the two nuclei present in a binucleate egg cell in cases where the diploid nucleus gives rise to female and the haploid nucleus to male tissue (as in honey bees); (4) by loss of a whole chromosome set during the first cleavage division.

Both the size and position of the male and female sectors are dependent on the time and position of the onset of these processes. In animals with sex hormones, the production of clearcut gynandromorphs is prevented since these hormones give rise to an intersexual development irrespective of the different constitution of the tissues.

An exceptional kind of g. in **Habrobracon** is called **"gynandroid"** (Whiting et al. 1954). In **Habrobracon**, sex is determined by a system of multiple sex alleles (xa, xb, xc, etc.). Females arise in association with heterozygosity for the sex alleles, males in association with hemi- or homozygosity. Gynandroids are haploid and arise from unfertilized, binucleate eggs in which the two nuclei contain different sex alleles. In the region of the genitalia at the juncture between tissues of different genotypes for x, small female reproductive appendages arise, possibly by diffusion of gene products peculiar to one of the x tissues and interaction with locally produced substances of the other tissue; this may give rise to female differentiation.

gynephoric (Waardenburg 1932) — ref. to → sex-linked modes of inheritance whereby phenotypically normal females act as heterozygous → carriers of the genes in question.

gynoautosome (Yamamoto 1938) — an → autosome carrying → sex realizers for femaleness (→ androautosome).

gynodioecious (Darwin 1877) — of a plant species (angiosperms) which consists of separate → hermaphroditic and male-sterile (female) individuals, i.e., some plants are unable to transfer their genes to the next generation via pollen, while all plants can do so via ovules.

Male-sterility may be controlled by nuclear genes, cytoplasmically, or by interaction of cytoplasm and nuclear genes. In cases of cytoplasmic-genic male sterility, plants with male sterile cytoplasm are phenotypically hermaphrodite if they have an active fertility-restoring allele at the nuclear gene locus, and are pistillate if they do not.

With either genic or cytoplasmic-genic inheritance of male sterility, hermaphrodites have descendants through both pollen and ovules; male steriles contribute descendants only through ovules.

gynoecious (v. Uexküll-Syllenband 1901) — of plants having only female flowers (→ androecious).

gynogenesis (Wilson 1925) — female → parthenogenesis. After fertilization of the ovum, the male nucleus is eliminated and the haploid individual (described as "gynogenetic") so produced possesses the maternal chromosome set only (= pseudogamy; merospermy) (→ androgenesis).

gynomerogony — the development of an egg fragment containing only the female nucleus (→ andromerogony), i.e., with exclusively maternal chromosomes.

gynomonoecious (Darwin 1877) — → monoecious.

gynospore (Battaglia 1955) — = → megaspore.

gynosporogenesis (Battaglia 1955) — = → megasporogenesis.

gypsy element (Marlor et al. 1986) — in **Drosophila** a retrovirus-like → transposable genetic element composed of (1) two 482-bp-long terminal repeats (LTR's) containing transcription initiation and termination signals and (2) a central region with three different open reading frames that encode products homologous to retroviral proteins.

gyrase (Gellert et al. 1976) — = → DNA gyrase.

H

habitat — that place within a → biotope inhabited by a particular population, species, or species group.

"Habitat selection" is the selection by a dispersed individual of an appropriate species-specific h. and the avoidance of unsuitable ones. It is one of the most important factors leading to the localization of populations and to the restriction of species to their specific ecological niches (Mayr 1963).

"Habitat exclusion" is the mutual exclusion of the habitats of two related species.

habituation (Binns and Meins 1973) — a type of heritable cellular change in tissue cultures of various plant species characterized by a newly acquired capacity of cells to produce growth regulatory substances. Induced reversal of h. has been reported indicating that h. is not due to a permanent genetic change. H. may result from heritable alterations in the pattern of gene expression.

hairpin loop — any double-stranded region in single-stranded RNA and DNA caused by base pairing between complementary base sequences. Such regions might play a role in the stability of the molecule and their formation is due to the ability of the nucleotide chain to turn back on itself and form a particular secondary structure (→ palindrome).

Many segments of DNA contain sequences of twofold symmetry (e.g., 5'-CGGTAAT TACCG-3'); these include → operators, → promoters, and the → replication origin. This feature may allow the formation of hairpins in these regions after strand separation and, possibly, the mapping of these secondary structures since the h.l. can be visualized directly by the electron microscope.

In RNA, h. structures are also functionally important secondary structures (e.g., anticodon loop in → transfer RNA).

Haldane's rule — when among the offspring of crosses between species and varieties, one sex is absent, rare, or sterile, it is the → heterogametic sex.

half-heterogamy (Renner 1918) — → heterogamy.

half-mutant (de Vries 1917) — a "mutant" in → complex-heterozygous species of **Oenothera (Oe. lamarckiana)** which carries a modified nonlethal genome plus an unmodified (lamarckiana) genome in both sperm and egg. In each generation such mutants give rise to so-called "full mutants" containing the modified nonlethal genome in double dose (Cleland 1962).

According to Darlington (1931), half-mutants are the product of rare crossing over between homologous segments in otherwise nonhomologous chromosomes, thus giving rise to effective exchange between two "complexes" (retranslocation).

half-sibs — of individuals having one parent in common; the genetic correlation between h.s. is 0.25 (→ full sibs).

half-spindle — any of the two → spindles, each extending from the → centromeres of one group of → chromosomes to the nearer cell pole, plus the interzone between the separating centromeres. At anaphase (→ mitosis) each spindle is divided into two half-spindles.

half tetrad — two of the four strands (chromatids of a → bivalent) from a single meiotic event (→ tetrad). Recombination studies in **Drosophila** involving → attached X-chromosomes permit the analysis of half-tetrads. The same is true for mitotic recombination where the recovery of two of the four chromatids that may be engaged in mitotic → crossing over is possible.

half-translocation (Herskowitz and Muller 1953) — half of a reciprocal → translocation. It may result from: (1) The absence of one of the reciprocal fragment fusions giving rise to an incomplete translocation (loss or behavior of the unjoined fragments frequently causes descendant cells to die or to be abnormal); (2) segregation in heterozygotes for a symmetrical translocation whereby only one of the two reciprocal exchanges is included in a gamete.

hammock structure (Imai et al. 1988) — = → suspension arch structure.

haplochromosome (Darlington 1965) — any → chromosome arising in those cases where chromosome reduplication and chromosome splitting are inhibited. Each such chromosome (→ hemichromosome) consists of one chromatid of normal size. Haplochromosomes may be induced experimentally by irradiation and are characteristic of discordant or precocious → polymitosis.

haplo-diplontic — → diplo-haplontic.

haploid (Strasburger 1905) — of cells or individuals with a single genome or → chromosome set provided more than one linkage structure (as in eukaryotes), distinct in homology and function, is present.

Viruses and bacteria whose genome consists of only one linkage structure are haploid, and so are parts of the life cycle (referred to as "haplophase") of eukaryotes. In diploid species of eukaryotes, individuals may occasionally occur whose **somatic** cells are haploid. In most cases they result from spontaneous or induced parthenogenetic development of unfertilized eggs (→ parthenogenesis). Such individuals are referred to as **"monohaploids"**. They are to be distinguished from so-called "polyhaploids", i.e., haploid individuals arising from → polyploid species.

Exceptional haploids may be classified as follows, based on the special cytological features (Kimber and Riley 1963):

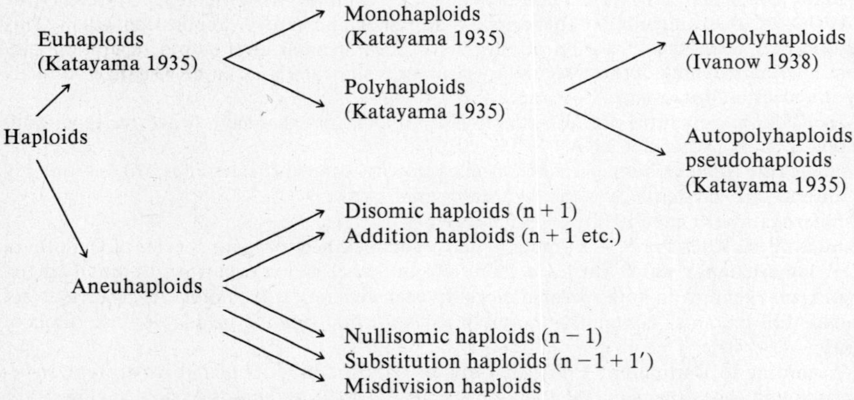

The cytogenetic constitution of **polyhaploids** varies and depends on the type of polyploid from which they arose. The chromosome numbers of **aneuhaploids** are not exact multiples of one of the basic numbers of the group, as compared with euhaploids. Aneuhaploids may have extra chromosomes, or may be deficient in chromosomes, relative to the euhaploids. Furthermore, the extra chromosomes may be members of the chromosome set of the species or they may be alien (e.g., in addition h.). Haploids deficient in a chromosome of the normal set (subhaploids or nullihaploids) have been rarely found. **Substitution haploids** have one or more alien chromosomes substituted for the species-specific one. **Misdivision haploids** are those with telocentric chromosomes and misdivision (→ centromere misdivision) derivatives of some chromosome(s). In accordance with their chromosomal constitution, the meiotic behavior (→ chromosome pairing) of these various types of haploids may vary considerably.

haploidization — the transformation of → diploid into → haploid cells (or strains) in certain fungi. H. is a process of the → parasexual cycle and apparently proceeds by the progressive loss of chromosomes, one chromosome at a time, via → nondisjunction until a haploid (euploid) nucleus is formed from the → aneuploid intermediates. During h., the chromosomes assort independently of one another and thus a variety of different haploids are obtained.

haplontic — of organism ("haplonts") with a life cycle in which meiosis occurs in the zygote giving rise to four haploid cells, e.g., most primitive unicellular or filamentous algae, and protozoa (→ diplohaplontic; diplontic). Only the zygote of haplonts is diploid. In some species, this cell becomes a resistant spore tiding the organism over hard times.

haplophase — the haploid phase or generation (= gamophase according to Winkler 1920) of the life cycle, lasting from meiosis to fertilization (zygote formation), as distinct from → diplophase.

haplosis — the meiotic reduction (→ meiosis) of the somatic chromosome number.

haplosomic — = → monosomic.

haplospory (Battaglia 1947) — apomictic development (→ apomixis) of a female gametophyte from a mother cell produced in the course of a normal → meiosis.

haplo-sufficient (Lindsley et al. 1972) — of → genes that allow the production of viable adults even when one copy of a gene in diploids is deleted from one of the homologous chromosomes, as opposed to haplo-insufficient genes.

haplo-triplo-disomic — of cells in which the number of chromosomes is unmodified (species-specific) but one normal chromosome is replaced by one of its → isochromosomes. Such variants are → monosomic for one, and → trisomic of the other arm of the same chromosome.

hapten — a small molecule which acts as an → antigen when conjugated to a protein.

Hardy-Weinberg law (Hardy 1908; Weinberg 1908) — a statement (= "H.-W. formula" or "H.-W. equilibrium principle") in mathematical terms concerning the theoretical relationship between → gene frequencies and → genotype frequencies within populations. It describes the genotypic equilibrium produced by a static gene pool. In a large random mating → population, both the gene frequencies and the genotype frequencies are constant from generation to generation in the absence of mutation, migration, and selection (the population is in H.-W. equilibrium). If the frequencies of allele A and a (of equal selective value and occurring together in a population) are q and $1-q$ and mating is at random, then male gametes $q\,A + (1-q)\,a$ fertilize eggs of $q\,A + (1-q)\,a$, producing offspring (genotypes) according to the following binomial law:

$$[qA + (1-q)\,a] \times [qA + (1-q)\,a]$$
$$= q^2AA + 2q\,(1-q)\,Aa + (1-q)^2\,aa.$$

harlequin chromosome — a pair of → sister chromatids that stain differently, so that one appears dark and the other light. H. chromosomes are observed after differential staining (harlequin staining) by fluorescent dyes of sister chromatids of metaphase chromosomes of eukaryotic cells grown for two replicative cycles in BrdUrd. The staining pattern is dependent upon whether one or both strands of DNA within each chromatid has incorporated bromodeoxyuridine (BrdUrd). → Sister chromatid exchanges (SCE's) can be demonstrated by h. staining in normal cells and elevated rates of SCEs in cells treated with various mutagens.

H-DNA (Hentschel 1982) — in the genomes of vertebrates and invertebrates an unusual (hinged) DNA structure which can be assumed by $(dT\text{-}dC)_n \cdot (dA\text{-}dG)_n$ repeats when exposed to negative supercoiling or low pH. H-DNA probably contains both single-stranded and triple-stranded regions, i.e., half of the Watson-Crick duplex is disrupted and the released polypyrimidine strand is folded back, enabling it to progress down the major groove of the other half of the repeat. This third strand associates with purines in the helical half of the repeat by Hoogsteen base pairs, forming $dT\text{-}dA \cdot dT$ and $dC\text{-}dG \cdot dC^+$ base triplets. The structure introduces a flexible kink into DNA, which serves as a hinge between the flanking duplexes.

headful hypothesis (Streisinger et al. 1967; Thomas et al. 1968) — a hypothesis explaining replication of bacteriophages with circularly permuted and terminally redundant DNA (→ genetic circularity; terminal redundancy). According to this hypothesis, long → concatemers of a viral genome are converted into mature virion chromosomes through the action of a genetic site-independent and → capsid volume-dependent (headful size) chromosome breakage system.

headful mechanism — a mechanism by which bacteriophage DNA is packaged into capsids: (1) Recognition and cleavage of a unique site (pac) on a concatemer of phage DNA; (2) unidirectional packaging from pac terminated, after the head has been filled, by a second cleavage reaction that has little or no sequence specificity (the headful cut); (3) packaging is progressive in that the nonpackaged end generated by the headful cut is used to initiate the next round of packaging.

health risk assessment — a major component of decisions, derived from the results of → mutagenicity testing, on the regulation of → mutagens, → carcinogens, and other → genotoxic agents. H.r.a. can be derived from the results of a two-step process: (1) Qualitative evaluation of toxicological data to determine the potential hazard of a chemical substance to humans (most frequently based on results obtained from short-term tests); (2) estimation of the magnitude of risk associated with human exposure to the agent (involving analysis of in vitro and in vivo data and extrapolation across species to human exposure situations).

heat shock response (Ritossa 1964) — in prokaryotic and eukaryotic organisms, a cellular response, triggered by elevated temperatures, that consists in an arrest or slow-down of normal protein synthesis (genetic transcription) and induction of a set of new polypeptides (heat shock proteins). Many other stresses are known to induce the same response.

The h.s.r. provides protection from thermal killing and is due to a dramatic change in the pattern of → gene expression, i.e., a small set of previously inactive genes (the heat shock genes) is activated. The extreme conservation of the h.s. genes and of the h.s. proteins suggests that these proteins carry out essential functions. Accompanying this gene activation are changes in eukaryotic chromosome structure that are evident from both the formation of → puffs in → polytene chromosomes and the increased sensitivity of these genes to DNase I in nuclei of heat-shocked cells.

In all genes activated during the h.s. response, a consensus sequence, the h.s. element (CTNGAANNTTCNAG) is found in multiple copies upstream of the genes. In response to increases in temperature, regulatory proteins bind to the HSE and activate nearby h.s. genes.

heavy shoulder DNA (Comings 1972) — a distinct family of DNA molecules in some eukaryotic species which manifests itself by a non-Gaussian skewing to the heavy (GC-rich) side of the → main band DNA (in analytical ultracentrifugation). H.s. DNA does not separate as a distinct band, does not contain a disproportionate amount of rapidly renaturing → repetitious DNA, and in some cases appears to be localized to a specific part of the nucleus or to specific chromosomes.

helicase — any of a class of ATP-dependent enzymes capable of unwinding the DNA double helix at a replication fork so that a → DNA polymerase may advance during → DNA replication.

helical repeat — the number of base pairs per helical turn of the DNA double helix (→ deoxyribonucleic acid).

helix — the natural conformation of many biological polymers (e.g., nucleic acid of the double-stranded type, proteins) characterized by a spiral structure with a repeating pattern described by rotation and translation.

helix-destabilizing protein (Alberts et al. 1968) — (HDP); any of a group of proteins which bind preferentially to single-stranded DNA and, therefore, destabilize double-stranded DNA. HDP's are involved in → DNA replication, → genetic recombination, and → DNA repair (= DNA-unwinding p.; DNA melting p.).

helix-turn-helix motif (Steitz et al. 1982) — in (bacterial) gene regulatory proteins, a sequence motif binding to specific DNA sequences. The h.-t.-h.m. is characterized by two successive helices juxtaposed at approximately 90° by a turn of four amino acids. Dimerization arranges one α-helix of each monomer with its analog so that they fit into successive major grooves of DNA (→ zinc finger motif; leucine zipper).

helper virus — a virus supplying one or more functions that a so-called "defective virus" is unable to perform. By infection of a cell with both a defective and helper virus the former is unable to multiply.

hemi-alloploid (Löve and Löve 1949) — = segmental allopolyploid (→ alloploid).

hemi-autoploid (Löve and Löve 1949) — of → polyploids which either started as autopolyploids and later became differentiated in the direction of allopolyploidy, or arose from more or less sterile intervarietal or intersubspecies hybrids.

hemicompatible — → di-mon mating.

hemihaploid — of cells, cell complexes, or individuals with only half the normal gametic ("haploid") chromosome number. Hemihaploids may arise from polyploids.

hemimethylated — of duplex DNA molecules produced during semiconservative DNA replication in which only the parental strand is methylated (→ DNA methylation). H. DNA is the preferred substrate for eukaryotic DNA (cytosine-5)-methyltransferase.

A maintenance methyltransferase is thought to methylate the daughter strand of newly replicated DNA, thus retaining the pattern of the parent strand through successive divisions.

hemiploid (F. v. Wettstein 1932) — of cells or individuals with half the normal somatic chromosome number (→ haploid).

hemizygous — in eukaryotes, of diploid cells or individuals (hemizygotes) in which one or more genes are present in only one, instead of two, copies; as in → differential segments of → sex chromosomes, as a result of → aneuploidy, or loss of chromosome segments.

hepatocyte DNA repair assay (Dunkel and Williamy 1981) — one of the short-term tests for evaluation of potential mutagens and carcinogens (→ mutagenicity testing). The test is an → unscheduled DNA synthesis (USD) assay.

hereditary determinant — a genetically functional unit which is replicated with conservation of specificity. A h.d. may be nuclear or extranuclear and its presence in an organism's → idiotype is essential for the appearance of a particular → character (→ inheritance).

hereditary disease — = → genetic disease.

hereditary factor (Mendel 1866) — = → hereditary determinant.

heredity (Spencer 1863) — the storage, transfer, and expression of genetic information. → Genetics is the science of heredity.

heritability (Lush 1941) — (h^2); (1) the ratio of genetically caused variability to total variability of a character in a population, i.e., the fraction of the → phenotypic variation that would remain if all genetic variation were removed. h^2 can range from one (no environmentally caused variation present) to zero (all variation is environmentally caused). Essentially, h^2 is a measure or correlation between genotype and phenotype.

(2) A measure of genetic response to selection (→ selection differential).

(3) The degree of genetic control of the development of a character.

heritable change — a nonlethal genetic change that is passed on to the descendants.

heritable translocation assay (Flamm 1974; Generoso et al. 1978) — in → mutagenicity testing, an in vivo mammalian (mouse) test which is based on the scoring of heritable → translocations induced in treated male germ cells and transmitted to viable first generation offspring. Mutant F_1 individuals are detected on the basis of either cytological evaluation of chromosome aberrations, or deviation from normal fertility (semi-sterility or full sterility).

hermaphroditic — of individuals ("hermaphrodites") bearing recognizable male and female tissues and producing mature male and female gametes either at the same ("simultaneous" or "synchroneous hermaphroditism") or at different periods ("consecutive hermaphroditism") of the life cycle. Protogynous hermaphrodites function first as females then transform into males, protandrous hermaphrodites transform males into females. The gametes of h. animals may be produced in sparate male and female gonads (testes and ovaries) or in the same gonad (ovotestis). Plant gametophytes are called h. when the two sexes are represented in the same flower (or prothallus in the case of cryptogams). They are generally referred to as monoecious when male and female gametes mature in different flowers on the same plant.

A population which is composed exclusively of h. individuals may be called "balanced hermaphroditic". A population composed of a majority of hermaphrodites with varying degrees of maleness or femaleness and a minority of pure male and female individuals may be designated "unbalanced hermaphroditic" (Bacci 1950).

hertone (Cole 1972) — = → nonhistone chromosomal protein.

He-T DNA (Traverse and Pardue 1989) — in **Drosophila**, a complex set of repeated DNA sequences located only in the pericentric regions and in the → telomeres of each chromosome arm.

heteroallelic (Roman 1956) — of genes ("heteroalleles") having mutations at different mutational sites, as distinct from homoallelic ones with mutations at the same site (= nonidentical and identical → alleles, respectively). Pairs of heteroallelic genes yield recombinants by intragenic recombination viâ → crossing over, pairs of homoallelic genes do not.

heterobrachial (Sorokin 1929) — → chromosome arm.

heterocapsidic — of viruses with → segmented genomes, the components of which are packaged in distinct → virions. If all fragments of the genome are present in a single virion, the virus is called isocapsidic.

heterocentric (Sears and Camara 1952) — of → dicentric chromosomes or chromatids (= "heterodicentric") whose → centromeres are of unequal strength. Heterodicentrics frequently behave like monocentrics, i.e., they do not give rise to a → chromosome or chromatid bridge during anaphase.

heterochromatic (Heitz 1928, 1929) — of eukaryotic chromosome segments or chromosomes that remain condensed (positive heteropycnosis) during interphase and stain differently from the mass of nuclear → euchromatin (→ chromosome banding). Heterochromatin may be associated with early or late → DNA replication, elimination processes (→ chromatin elimination), cyclical inactivation (→ dosage compensation), heteromorphism, genetic instability, imprinting (→ chromosome impriting), and → position effects. Heterochromatin is usually attached to the inside of the → nuclear envelope, while diffuse euchromatin is associated with intranuclear matrix fibrils (→ nuclear matrix).

Heterochromatin is generally recognized by the following features: (1) A condensed state throughout → interphase (positive heteropycnosis giving rise to → chromocenters); (2) out of phase replication of DNA (late replication); (3) transcription inactivity in the compacted state; (4) under-replication in → polytene chromosomes; (5) susceptibility to chromosome structural changes.

Heterochromatin is subclassified into two types (Brown 1966): (1) **Constitutive heterochromatin** is present in all cells at identical positions on both the homologous chromosomes and forms a permanent structural characteristic of a given chromosome pair. Generally, the DNA of constitutive heterochromatin consists of → repetitious DNA, nonrepetitive, AT-rich → mainband DNA, and nonrepetitive → heavy shoulder DNA (in some species). → Satellite DNA tends to be localized to centromeric heterochromatin (→ paracentromeric heterochromatin). (2) **Facultative heterochromatin** varies in its state in different cell types, at different developmental stages, or even from one homologous chromosome to another, i.e., is heteropycnotic in special cell types or at special stages. It represents inactivated and condensed segments of euchromatin (→ dosage compensation).

A multitude of functions have been ascribed to heterochromatin which may act (Cooper 1959): (1) On genes, to control → mutation, to modify specific → gene action, → penetrance, and → specificity, and to affect quantitatively the action of some or all of the genes. (2) Within chromosomes, to stabilize → centromeres and → telomeres, to affect the processes connected with the production of → chromosome mutations, to bring about or prevent meiotic → chromosome pairing, to regulate → crossing over and → chiasma localization, to produce specific, nonspecific, and reversed pairing and conjunctive properties, and to cause → variegation and chromosomal → sticky effects. (3) Trans-chromosomally, by controlling the dimensions of all the chromosomes of a nucleus, by regulating crossing over, pairing, and disjunction of other chromosomes, and by controlling variegation brought about by genes located on other chromosomes. (4) Metabolically, by performing or mediating special syntheses of nucleic acids, proteins, nucleolar material, and energy-rich substances, and by controlling the transfer of substances across the nuclear envelope. (5) On the cell, by playing a role in the control of cell size and by governing mitosis. (6) On development, by regulating rates of growth and differentiation, by regulating the very nature of differentiation itself, and perhaps also by playing a role in → sex determination. (7) In speciation, by providing neutral anchorage and supplementary chromosomal parts for chromosome mutation, increases in the number of chromosomal arms and, in addition, duplicate genes that may acquire new functions. (8) In theory, as the special seat of the unorthodox in genetic systems.

Some of the functions suggested above are well documented others not.

heterochromatin fusion — → chromosome pairing.

heterochromatin heteromorphism (Craig-Holmes and Shaw 1971) — any of the quantitative and/or qualitative variations of constitutive heterochromatin which result in widespread variability without overt phenotypic effect. H.h. is possibly due to unequal → crossing over between repetitive DNA sequences.

heterochromatinized — of euchromatic chromosome regions which, when juxtaposed to → heterochromatic ones, become transformed ("heterochromatinization") in the direction of or to a heterochromatic state (→ position effect) (Cooper 1959). Heterochromatinization is an epigenetic phenomenon due to → chromosome imprinting in which a section of coding DNA is temporarily and reversibly repressed (→ dosage compensation).

heterochromosome (Montgomery 1904) — originally, any eukaryotic chromosome differing from the → autosomes (or "euchromosomes") in size, shape or behavior. Today the term is generally used as a synonym for → sex chromosome (= allosome).

heterochronic — of genes (mutations) that cause particular cells in various cell lineages and tissues to adopt fates during postembryonic development that are normally associated with cells at earlier or later stages of development.

heterochrony — a change in order of appearance of a character during development between an ancestor and its evolutionary descendant.

heterocytonic (Catcheside 1958) — of a cell or mycelium (in fungi) consisting of genetically different cytoplasms, characterized by different extranuclear hereditary determinants (→ plasmon), as distinct from "homocytonic". The terms hetero- and homocyton are parallel to hetero- and homokaryon which designate equivalent nuclear states (→ heterokaryotic).

heterodichogamous — → dichogamous.

heteroduplex DNA — a double-stranded DNA molecule or hybrid DNA region in which the two strands do not have completely complementary base sequences. H. DNA can arise from mutations, recombination, or by annealing (→ DNA annealing) single strands in vitro: (1) Misincorporation during → DNA replication can lead to mispairs in DNA and → DNA mismatch repair can increase fidelity of replication; (2) deamination of DNA bases can form mispairs, and such lesions can be repaired by → DNA glycosylases; (3) mispaired bases can be formed during → genetic recombination and correction of mispairs, or failure of correction, can explain → gene conversion, → post-meiotic segregation, → localized negative interference, and → map expansion.

Formation of h. DNA is a critical intermediate step in homologous genetic recombination since it mediates the union of the two duplexes undergoing strand exchange. Mismatched bases and unpaired single strands can form in heteroduplexes between nonidentical parental DNAs (→ heteroduplex repair).

H. molecules (as opposed to homoduplex molecules with completely complementary base sequences) do not persist as such. During replication each strand serves as a template for the synthesis of two genetically distinct homoduplex molecules, provided h. regions are not eliminated prior to DNA replication by a correction mechanism (→ polaron hybrid DNA hypothesis; heteroduplex repair). Molecules corrected in this way would yield only one type of homoduplex molecule.

heteroduplex-induced mutagenesis (Thomas and Capecchi 1986) — induction of mutations due to incorrect repair (→ heteroduplex repair) of a heteroduplex DNA formed between an introduced and a homologous chromosomal → DNA rearrangement (→ gene targeting).

heteroduplex joint — a molecular splice containing paired strands from each of two DNA molecules; the h.j. is an intermediate in homologous → genetic recombination and its formation and processing is largely responsible for meiotic → gene conversion and aberrant meiotic segregation, i.e., phenomena associated with the reciprocal → crossing over.

heteroduplex mapping (Davis and Hyman 1971) — an electron microscopic procedure used for the detection and mapping of regions of sequence homology and nonhomology between related nucleic acid molecules. The appearance of single-stranded loops in regions of insufficient homology allows determination of the position and extent of heterologous DNA segments. The correlation of these results with genic differences between the sources of the nuclei acids permits construction of a partial or complete physical → gene map (of bacterial viruses and → plasmids).

heteroduplex repair — correction of heteroduplex regions in double-stranded DNA assumed to be involved in genetic recombination and various phenomena connected with it (e.g., → gene conversion, → high negative interference, → map expansion). The sequence of events involved in h.r. may be the following: (1) Single strands of recombining double-strands of DNA are cut; (2) base pairing occurs between different parental complementary strands; if these carry different genetic information, a heteroduplex DNA region may form in the course of the first step of genetic recombination; (3) conversion of the heteroduplex region by a repair-like mechanism (→ DNA repair) to a homoduplex structure; during this procedure genetic information from one strand could be transferred to its complementary strand.

Two modes of h.r. have been described in bacteria and mammalian cells: (1) One sensitive to the modification status of parental strands that may function in the correction of mismatches arising during → DNA replication; (2) another one that is methylation independent (→ DNA methylation) and probably involved in correction of mismatches arising during → genetic recombination.

heteroduplex-transformation assay (Chang et al. 1984) — a method which allows the generation of new → recombinant DNA molecules in vivo without solely relying on the presence of convenient restriction sites for manipulation of DNA fragments in vitro. In the h.-t.a. a pair of parental → plasmids sharing partial sequence homology is used; one or both of these carry a desirable sequence or mutant allele to be included in a new recombinant progeny plasmid: (1) The two parental plasmids are each linearized at a unique but different restriction site; (2) the termini of the molecules are then made blunt (→ blunt end) by S1 endonuclease with specificity for single-stranded DNA; (3) these two linearized parental plasmids are melted and annealed together; linear homoduplex and circularized heteroduplex molecules are generated. Mispaired sequences due to allelic differences can also be incorporated into the progeny plasmids.

heterodynamic (Waddington 1953) — → homodynamic.

heteroecious (Correns 1928) — (1) producing male and female gametes on different individuals (= dioecious): (a) diploheteroecy, i.e., dioecy in the diplophase, and (b) haploheteroecy, i.e., dioecy in the haplophase (Hartmann 1939); (2) of fungi spending different stages of their life cycle in different hosts.

heterofertilization (Sprague 1932) — in plants, the → fertilization of the endosperm- and embryo-forming nuclei by gametes of different genetic constitution (Darlington and Mather 1949).

heterogameon (Camp and Gilly 1942) — a → species comprised of varieties (races) which give rise to morphologically stable populations by selfing and several types of viable and fertile progeny after crossing.

heterogametic (Wilson 1910) — that one of the two sexes which during → meiosis gives rise to two types of gametes (male- und female-determining), in contrast to the "homogametic" sex. In the case of the XX-XY or XX-XO-types of → sex determination, the h. sex is that with both an X- and a Y-chromosome or with only one X-chromosome (= digametic). In most cases, the male sex is h., the female homogametic with respect to the → sex-chromosome constitution.

heterogamous (de Vries 1911) — of → complex heterozygous individuals whose male gametes transmit genes or gene-complexes different from those of the female gametes, as distinct from iso- or homogamous hybrids.

heterogamy — (1) = → anisogamy; (2) alternation of two sexual generations, one being truly sexual, the other parthenogenetic (→ alternation of generations; parthenogenesis); (3) the mating of unlike individuals, i.e., the preference of an individual for mating with another of unlike phenotype or genotype, as distinct from "homogamy" (→ mating system).

heterogeneity index (Nei and Roychoudhuri 1974) — a measure of genic variance of a population (= gene diversity) equal to the probability of nonidentity of randomly chosen genes.

heterogeneous closed circular DNA (Hotta and Bassel 1965) — in a wide variety of eukaryotic organisms, a class of small circular DNA molecules which are predominantly nuclear and exhibit a buoyant density in neutral CsCl indistinguishable from that of the main band nuclear DNA. The molecules (in **Drosophila**) range in size from approximately 300 to >7500 nucleotide pairs (average size of 3300 nucleotide pairs). The function of h.c.c. DNA is unknown but it may be involved in genetic regulatory processes via integration and excision from chromosomal DNA. The molecules show a high degree of homology with middle → repetitious chromosomal DNA.

heterogeneous nuclear ribonucleoprotein particle — (hn RNP); a complex (50 to >200S) between heterogeneous nuclear RNA(hn RNA) and specific nuclear proteins; such RNA particles (20% DNA, 80% protein) assemble on nascent chains of → pre-messenger RNA and contain a set of six proteins (mw 32,000 to 44,000) known as the hn RNP "core" proteins. In the electron microscope, these complexes probably correspond to the beaded relatively uniform particles (diameter 20–30 nm) spaced along the RNA chain (= informofere).

Packaging of hn RNA involves a condensation of nascent RNA with the specific proteins which are delivered from the cytoplasm to the nuclear interior. Probably this packaging allows a localized accumulation of RNA until transcription is complete, serves to condense and stabilize the nascent RNA and to sequester it from the transcriptional template. The RNA contained in hn RNP is accessible for immediate modification.

heterogeneous nuclear RNA (Scherrer and Darnell 1962) — (hn RNA); any of a class of eukaryotic → RNA polymerase II transcripts and their processed intermediates which are complexed with proteins (→ heterogeneous ribonucleoprotein particle). Post-transcriptional processing of hn RNA is operative in the generation of specific → messenger RNA molecules and stage- and tissue-specific mRNA populations. The usual order of the post-transcriptional events is: 1) hn RNA-ribonucleoprotein particle assembly; (2) → polyadenylation; (3) → cap formation; (4) → RNA splicing.

heterogeneric tRNA (White et al. 1973) — → transfer RNA.
heterogenesis — = → alternation of generations.
heterogenetic (Waddington 1939) — ref. to meiotic → chromosome pairing in hybrids (alloploids) when the pairing partners are derived from different original ancestors, as distinct from "homogenetic" pairing where the association is between chromosomes derived from one of the original ancestors. The frequency of h. pairing as compared with homogenetic depends on the relative degree of affinity between the completely homologous, as compared to the partly homologous ("homeologous"), chromosomes, i.e., on their → differential affinity.

If the parental genomes of a segmental allopolyploid are symbolized as $A_1A_1A_2A_2$, and the individual chromosomes as $1A_11A_11A_21A_2$, $2A_12A_12A_22A_2 \ldots nA_1nA_1nA_2nA_2$ (n being the basic haploid number), then pairing in meiosis between $1A_1$ and $1A_2$ chromosomes is h., and pairing between two $1A_1$ is homogenetic. The parallel (but not synonymous) terms → allo- and autosyndesis refer to the pairing between chromosomes derived from different parental gametes (allosyndesis) or from the same gamete (autosyndesis) (Stebbins 1950).
heterogenic (Lewis 1947) — (1) of a population, or (2) a gamete containing more than one → allele of a particular gene or genes, as distinct from a gamete which is "homogeneic" in this respect.
heterogenotic (Morse et al. 1956) — of bacterial "syngenotes", i.e., partially diploid bacteria **("heterogenotes")**, arising by → transduction, → F-duction, or → conjugation, which contain a → genome fragment (called an **exogenote**) from a donor cell in addition to their own (recipient) genome (called **endogenote**) and which are → heterozygous for pairs of marker genes localized in the segments concerned, as distinct from "homogenotic" cells ("homogenotes") which are homozygous in this respect. The heterozygous condition set up by the introduction of an exogenote into the recipient may be referred to as **cis-** $(++/m^1m^2)$ or **trans-heterogenotic** $(+m^2/m^1+)$ depending on whether two different mutational sites (either belonging to different → cistrons or to the same cistron) were introduced together (exogenote m^1m^2, endogenote $++$) or not (exogenote either $+m^2$ or m^1+, endogenote m^1+ and $+m^2$, respectively) by the genome fragment (exogenote) from the donor cell.
heterogony — (1) → parthenogenesis; (2) differential growth rates for two organs or parts of an individual, the rates showing a constant relationship to one another (Huxley 1932).
heterograft — a tissue graft from a donor of one species to a host of a different species (= xerograft). As a rule, heterografts are rejected faster than are → allografts.
heterohomologous (Dvořák and McGuire 1981) — of → homologous chromosomes from different inbred lines potentially differing in nucleotide sequences or gross chromosome structure (→ euhomologous).
heteroimmune — of phages that have a → repressor which recognizes its own operator but not the operator of another phage, and an operator sensitive to its own repressor but not to the repressor of another phage. Phages that share the same repressor-operator specificity are homoimmune or coimmune.
heterokaryon — a bi- or multinucleate cell, or a tissue composed of such cells, whose nuclei are of more than one genetic type, as distinct from a homokaryon, in which the nuclei are genetically identical. The property of being hetero- or homokaryotic is referred to as heterokaryosis and homokaryosis, respectively. In certain fungi (Asco- or Basidiomycetes), heterokaryosis constitutes the principal vegetative phase of the life cycle.

In contrast to → heterozygous diploids, the different alleles of a particular gene in binucleate heterokaryons are separated by → nuclear envelopes.
heterokaryon test (Jinks 1958) — a test for cytoplasmic mutations (in fungi) which is based on new associations of phenotypes in cells derived from specially marked heterokaryons. The simplest form of the h.t. involves uniting two homokaryons (each marked by a single contrasting chromosomal gene) to form a heterokaryon which is then resolved into its origi-

nal components. If extrachromosomal genes are present they will not necessarily reassort with the parental, homokaryotic marker genes and so the phenotypes may differ from those of the original parents.

heterokaryosis (Hansen and Smith 1932) — the association of genetically different nuclei in a common cytoplasm. H. is widespread in fungi and may initiate → plasmogamy leading to sexual as well as to → parasexual recombination.

In fungi, h. represents an alternative to diploidy. Two types of heterokaryons may be distinguished and may occur in the same species (Raper and San Antonio 1954): (1) **Indefinite heterokaryons** in which the ratio of component nuclear types is indefinite and subject to change. Heterokaryons of this type are basically vegetative. (2) **Dikaryons** represent genetically balanced heterokaryons, each containing two haploid genomes in a ratio of 1:1. Heterokaryons of this type constitute an integral phase of the sexual cycle of all higher fungi, and in the Basidiomycetes are capable of indefinite vegetative propagation.

heterokaryotype — a → karyotype which is heterozygous for a → chromosome mutation (symbolized HTK), as opposed to a structurally homozygous "homokaryotype" (symbolized by HOK).

heterolabeling — → sister chromatid exchange.

heteromeric — of genes which in combination control the manifestation of a particular character, each gene having a definite, but individually different share in the production of that character, in contrast to homomeric genes, whose individual share is quantitatively similar (→ gene interaction).

heteromixis (Burnett 1956) — mating systems in (heteromictic) fungi; sexual reproduction results from the fusion of genetically different nuclei derived from different thalli, in contrast to → "homomixis", H. includes: (1) **Dimixis:** Two types of complementary nuclei control mating. The nuclear types are determined by two alleles at a single locus (= heterothallic; haplodioecious; two allele morphological and physiological → incompatibility; bithallic). (2) **Diaphoromixis:** Several types of complementary nuclei control mating. The nuclear types are determined by multiple alleles at one or two loci, the bipolar and tetrapolar conditions, respectively (= multipolar sexuality; bipolar sexuality; multiple allele physiological heterothallism; bithallic; incompatibility factors at one or two loci). (3) **Homo-heteromixis:** Sexual reproduction results only from the fusion of genetically different nuclei normally derived from the same thallus; a reproductive mode derived from (1) or (2) and hence "homodimictic" or "homo-diaphoromictic" (= secondarily homothallic; amphithallic; pseudomonothallic).

heteromorphic (Carothers 1917) — of "homologous" chromosomes differing in size or form (→ bivalent).

heteromorphosis — = → homeosis.

heteronuclear (Krooth 1965) — of → cell cultures showing a marked variation (numerical and structural) from cell to cell in their chromosome complement. If h. cultures are cloned, each colony will usually regenerate the same spectrum of chromosome complements as was present in the parental mass culture. Homonuclear cultures show almost no variation from cell to cell in the chromosome complement, except when such cultures are developed from chromosomal mosaics.

heterophenogamy (Strandskov 1950) — → isophenogamy.

heteroplasmic — = → heteroplasmonic.

heteroplasmonic — of cells containing two or more types of cytoplasmic hereditary determinants (= heteroplasmic), in contrast to homoplasmonic. The term h. is equivalent to → heterozygous, → heterogenotic, and → heterokaryotic in the case of extranuclear hereditary determinants. A "heteroplasmon" (→ plasmon) may arise by mutation of some but not all the homologs of an extranuclear determinant which are present in a cell or by fusion of cells containing alternative forms of such a determinant. A heteroplasmon may give rise to → somatic segregation.

heteroplastidic (Michaelis 1957) — of cells containing → plastids of different shape, in contrast to homoplastidic ones.

heteroploid (Winkler 1916) — of all chromosome numbers deviating from the normal chromosome number of the diplophase (in organisms with a predominating diplophase), or of the haplophase (in organisms with a predominating haplophase) of typical representatives of a given species (Levan and Müntzing 1963). H. chromosome numbers may be either → euploid or → aneuploid.

heteropycnotic (Gutherz 1907) — of chromosomes or chromosome regions which, in comparison with "isopycnotic" ones (Östergren 1950), are out of phase (→ allocycly) in respect to their coiling cycle (→ chromosome coiling) and staining properties (→ heterochromatic). (1) **Positively h.:** Chromosomes or chromosome segments are tightly coiled by prophase and remain so as others (isopycnotic) are undergoing decondensation at anaphase. During interphase they show up as the → chromocenters. Positive heteropycnosis is characteristic of heterochromatin. (2) **Negatively h.:** Chromosomes or chromosome segments condense more slowly or to a lesser extent during prophase than do the positively h. ones and undergo decondensation more rapidly at anaphase.

Chromosomes and chromosome regions which are positively h. at one stage of the cell cycle, or in one kind of cell or tissue, or in one of the sexes, may be negatively h. at another stage, in another cell, or in the other sex.

hetero-R state (Hashimoto and Hirota 1966) — → resistance transfer factor.

heteroselection (Carlson 1959) — natural → selection which predominates in central populations. It is assumed to favor polymorphism, restricted recombination, and a general adaptedness to a multiplicity of conditions. In contrast, marginal populations are dominated by homoselection favoring a relative homozygosity, random genetic drift, free recombination, and adaptive specialization.

heterosequential (Carson et al. 1967) — → band.

heterosis (Shull 1911) — the superiority of heterozygous genotypes with respect to one or more characters in comparison with the corresponding homozygotes (= hybrid vigor). H. is the phenotypic result of → gene interaction in heterozygotes and is thus confined (at least in its maximal amount) to that state. It can be disrupted by → inbreeding and restored by interbreeding of the inbred lines.

heterosomal — of chromosomal structural changes (→ chromosome mutation) involving two or more nonhomologous chromosomes (→ allelosomal; homosomal).

heterosome — = → sex-chromosome, as opposed to → autosome.

heterospory (Ernst-Schwarzenbach 1939) — the production of micro- and megaspores (→ spore), as opposed to homospory.

heterostylous (Darwin 1865) — of plants in which physiological → incompatibility is genetically linked with morphological differences between pistils and stamens. Heterostyly promotes → outbreeding and means the presence in a species of two (distyly) or three (tristyly) different types of individuals, distinguished by the relative positions of stigma and anthers in the flower, as distinct from homostyly. Usually, pollination and fertilization is between flowers differing in the lengths of their styles (heteromorphic incompatibility).

heterosynkaryon (Harris 1970) — → synkaryon.

heterothallic (Blakeslee 1904) — → heteromixis; homothallic.

heterotropic (Wilson 1906) — of a → sex chromosome which does not find an exactly homologous partner in the heterogametic sex (e.g., in XX-XY and XX-XO systems of sex determination).

heterotypic (Flemming 1887) — ref. to the first meiotic division (→ meiosis).

heterozygote advantage — an advantage (better survival or reproduction) over the homozygotes of individuals (genotypes) being → heterozygous for one or more → allele pairs.

heterozygous (Bateson and Saunders 1902) — (1) in eukaryotes, of → diploid or → polyploid individuals ("heterozygotes") having different → alleles for one or more genes in homologous chromosome segments, in contrast to homozygotes with identical alleles at these loci.

Heterozygosity provides a store of genetic reserves and potential plasticity and permits a large proportion of the individuals to exhibit combinations of phenotypic properties near the optimum. Underlying both is the greater → homeostasis of heterozygotes.
Structural heterozygotes (structural hybrids) are individuals heterozygous for → chromosome mutations, i.e., → "heterokaryotypes". Their parental gametes differed in the arrangements of genes in corresponding linkage structures, in contrast to "homokaryotypes".

Permanent (enforced) heterozygotes are individuals in which the heterozygous state of specific parts of the genome is fixed or made obligatory by factors contributing to the maintenance of heterozygosity. Fixation of heterozygosity may be a function of balanced or partially balanced → lethal factors, heterozygosity for → translocations (→ complex heterozygous) or → inversions, or → chiasma localization. → Parthenogenesis is an important ingredient in many cases of permanent heterozygosity.

(2) In bacteriophages and bacteria, of partially diploid particles and cells, respectively, which carry two copies (containing different alleles of one or more genes) of at least some of their genetic information.

Heterozygous bacteriophages are called HET's. Single factor crosses (mixed infections) produce two distinct types of h. progeny. One of these probably corresponds to a "terminal redundancy" of the phage linkage structure (Streisinger et al. 1964) resulting from the incorporation of two double-stranded segments of a homologous sequence of genetic information into a single genome. The segments in question are localized at opposite ends of the linear genome ("overlap HET's"). When this redundant information stems from genetically different parents the result is a region of heterozygosity corresponding to the length of redundancy:

DNA molecule $\dfrac{a^{+} \qquad\qquad a}{a^{+} \qquad\qquad a}$ (duplicated information terminally represented by different alleles). The other type is a → heteroduplex structure (Levinthal 1959) formed by a breakage-reunion recombination (also called "internal heterozygote" or "recombinational heterozygote"). The information on each strand of the double-strand DNA molecule is then genetically distinct within the h. region and should be lost by semiconservative replication:

DNA molecule $\dfrac{a^{+}}{a}$ (different information in the two polynucleotide chains).

Heterozygosity in phages concerns only a very limited region of the whole genome.

The heteroduplex HET's cannot persist throughout a cycle of DNA replication, but the overlap HET's may persist for several replication cycles.

Heterozygous bacterial cells may arise by → transduction, → F-duction, and → conjugation. They generally contain an additional genome fragment (an → "exogenote") and are h. (→ heterogenotic") for one or more marker genes where the donor and recipient cells carried different alleles. In contrast to the HET's, they are quite comparable to the heterozygotes in higher organisms which arise by the union of gametes differing in genotype.

HET's (Hershey and Chase 1952) — a shorthand for partially → heterozygous phages.

hexaploid — of a → polyploid (allo-or autoploid) with six chromosome sets in the somatic cells.

hexasomic — of an otherwise diploid cell or individual having one chromosome represented six times instead of twice ($2n+4$) (→ polysomic).

Hfr cell — a cell harboring the → F plasmid integrated into the bacterial chromosome which may or may not (in case of a defective F plasmid) confer on the cell the phenotype of high frequency of recombination. Each Hfr strain transfers the bacterial chromosome with a definite polarity to F^{-}-recipient cells during bacterial → conjugation. Hfr strains originate by opening up the circular chromosome of the bacterial (F^+) cell at one of many possible sites and insertion of the F plasmid at this place.

high mobility group nonhistone protein (Johns 1964; Goodwin et al. 1973) — (HMG); in eukaryotes, any of a family of nonhistone chromosomal proteins that bind to both → histones and DNA in → nucleosomes and can induce changes in DNA helical structure. They show high electrophoretic mobility at low pH (mw ranges from 8,000 to 26,000) and have very unusual amino acid compositions. Their presence is linked to transcriptional activity of → chromatin and coupled with enhanced DNase I sensitivity of transcribing DNA sequences. Characteristics of HMG proteins are: (1) Their high concentration (10^5 to 10^6 molecules per cell); (2) their location in both the nucleus and cytoplasm; (3) their existence as integral components of chromatin; (4) their nonrandom stoichiometric distribution on the nucleosome core particle; (5) their ability to alter the T_m of DNA; (6) their association with specific transcriptionally active chromatin fractions; (7) their capacity to modulate deacetylase activity; (8) their potential to be postsynthetically modified.

The four major HMG proteins (HMG1, HMG2, HMG14, and HMG17) are ubiquitous in occurrence and are conserved in various tissues and cell types. Modifications of HMG's after synthesis involve ADP-ribosylation and acetylation of HMG1 und HMG2, and phosphorylation of HMG14 and HMG17.

high negative interference (Chase and Doerman 1948) — (HNI); the appearance of multiple recombinants (exchanges) at a frequency much higher than would be expected as the product of independent single cross over frequences. NHI is observed (in bacteriophages and fungi) in crosses involving tightly linked markers whose exchanges are not statistically independent of one another. HNI may be due to → heteroduplex repair.

histocompatibility — the capacity to accept a tissue or organ graft (→ major histocompatibility complex).

histocompatibility antigen — any of the cell surface glycoproteins which are important in the rejection of → allografts and function in the discrimination of altered self-antigens, e.g., in distinguishing normal cells from virally infected or transformed cells. (→ major histocompatibility complex).

histocompatibility gene — in mammals, any of the genes (H genes) of the → major histocompatibility complex which controls cell surface determinants, immune response differences, some components of the component system and perhaps other related functions connected in general with cell-cell recognition.

histogenesis — the formation and development of tissues (= histogeny).

histone (Kossel 1884) — any of a set of five basic chromosomal proteins found in all eukaryotic organisms. They are complexed with DNA and responsible for the packaging of eukaryotic DNA into → nucleosomes and for the transcriptional properties of → chromatin. Posttranslational modifications of histones are involved in the modulation of nucleosomal structure during → genetic transcription and → DNA replication. Six types are known: (1) Acetylation; (2) phosphorylation; (3) methylation; (4) ADP-ribosylation; (5) conjugation of a nonhistone peptide (ubiquitin) to an inner h. (H2A) of the nucleosome to give a branched protein; (6) proteolytic cleavage.

The five major species of histones (H1, H2A, H2B, H3, and H4) are present in chromatin in a molar ratio of 0.5:1:1:1:1. H. synthesis by transcription of the genes coding for the histones in cycling cells is divided into two types: (1) S-phase synthesis; (2) basal h. synthesis (accounting for about 10% of the total h. synthesis) continues throughout the cell cycle at a similar rate. Certain h. variants participate only in basal synthesis and other variants only in S-phase synthesis.

The core histones (H2A, H2B, H3, and H4) are responsible for the primary level of DNA folding with the formation of nucleosomes. The linker histones (H1; largely replaced by H5 in chicken reticulocytes) are associated with the linker regions between the nucleosomes and thought to direct the next highest level of DNA folding. The synthesis of the core histones is usually temporally coupled with DNA replication. Histones show a remarkable degree of sequence conservation. Different h. subtypes (differing in primary structure) are expressed in a tissue-specific manner. Each h. molecule consists of two kinds of domain: (1) A highly basic portion interacts primarily with DNA, and (2) a hydrophobic region which is the site of h.-h. interaction (four of the five histones are associated pairwise with one another in the nucleosomes).

The genes coding for histones possess some unusual features: (1) Absence of → introns; (2) absence of mRNA polyadenylation; (3) presence of a 14 bp G + C-rich palindromic structure followed by four bases (A-C-C-A) near the 3' ends of their mRNA's which is essential for correct processing of the 3' termini. The h. genes are a family of moderately repetitive DNA sequences with about 20–100 copies of each h. gene. In all eukaryotes, the great majority of h. genes are grouped into clusters but must not be repeated in single tandem arrays observed in the sea urchin and *Drosophila* (repeating units in the order H4, H2B, H3, H2A, and H1). Regions at the beginning and end of the structural genes are called prelude and trailing sequences, respectively. Order and direction of transcription of the repeating units differ between organisms.

H. gene organization shows an increasing diversity during → evolution. For invertebrates and lower vertebrates, regular, tandemly repeated h. gene quintets are characteristic. Higher vertebrates show a divergence to a much more varied organization of h. gene clusters. Individual genes within a cluster are separated by spacer DNA sequences; the order, orientation, and identity of the genes in the clusters may differ.

hitch-hiking effect (Kojima and Schaffer 1967) — the increase in frequency of a selectively neutral or nearly neutral gene closely linked to a gene which is selectively favored. Changes in gene frequency, changes in the association of alleles at different loci, and gametic disequilibrium may be caused by the h.-h. e.

HMG protein (Johns 1964) — → **h**igh **m**obility **g**roup nonhistone protein.

HNI — acronym for → **h**igh **n**egative **i**nterference.

hn RNA — acronym for → **h**eterogeneous **n**uclear RNA.

hnRNP — acronym for → **h**eterogeneous **n**uclear **r**ibo**n**ucleo**p**rotein particle.

Hogness-Goldberg box — = → TATA box.

holandric (Enriques 1922) — of a type of → inheritance controlled by genes completely linked to the → Y-chromosome. The genes in question are transferred exclusively from father to sons in case of XX(♀)-XY(♂)-systems of → sex-determination, and thus their manifestation is limited to the male sex (→ hologynic).

Holliday junction (Mizuuchi et al. 1982) — = → Holliday structure.

Holliday structure (Sigal and Alberts 1972) — a four-stranded branched (crossed) DNA intermediate which arises in homologous and site-specific → genetic recombination by exchange of single strands between two homologous DNA duplexes. The branch site (a twofold symmetric complex whose four arms form two stacking domains) is called the Holliday junction. H structures are presumably cleaved at two points across the junction by resolving enzymes and are subsequently religated to generate either the original DNA duplexes or recombinational duplexes.

Three conformations for the H.s. have been proposed: (1) The branches are base-stacked and co-planar such that the recombining molecules have unperturbed helix axes; (2) it is a planar structure, but not base-stacked, with the branches at right angles to each other such that the H.s. possesses pseudo-fourfold symmetry; (3) an intermediary conformation wherein the branches of the junction are arranged in a tetrahedral fashion.

In prokaryotes, such H. structures are quickly lost by branch migration to the free ends of the constituent DNA molecules.

holocentric — of → chromosomes with diffuse kinetic activity due to the presence of nonlocalized (diffuse) instead of localized → centromeres (= holokinetic). A chromosome with a diffuse centromere has no localized organelle which elaborates chromosomal fibers, but the entire body of the chromosome forms a sheet of such fibers on the poleward surface and these converge to the pole. Sister chromatids are not associated at any point by centric connection and a complete spatial separation is possible which may make the chromatid an autonomous entity (→ reductional division; meiosis). Such chromosomes move broadside to the poles during anaphase and do not have the V- or J-shaped configuration of chromosomes with a localized centromere.

holoenzyme — an → enzyme composed of apo- and coenzyme.

hologamy — a mode of → reproduction characterized by the fusion (plasmogamy and karyogamy) of whole (unicellular) individuals (= gamontogamy, macrogamy).

hologynic (Enriques 1922) — of → sex-limited characters which are manifest only in the female sex (→ holandric).

holokinetic (Bauer 1952) — = → holocentric.

holozygote — → zygote.

homeoallelic (Washington 1971) — in → polyploids, of duplicated, triplicated, etc. genes similar to each other in function but located in → homeologous chromosomes with different evolutionary origin.

homeobox (Garber et al. 1983) — in the coding region of development-controlling (homeotic) regulatory genes, a highly conserved DNA sequence (about 180 bp) which codes for a protein domain (60 amino acids) that exhibits structural similarity to certain prokaryotic and eukaryotic → DNA-binding proteins. The homeo-domain proteins are probably involved in the control of → gene expression during morphogenesis and development. The transcripts are detected in a tissue- and region-specific manner in the vertebrate embryos, undergo qualitative and quantitative changes during embryogenesis, and are limited to specific tissues and cell types of the adult organism (mouse).

In *Drosophila*, mutation in the h.-containing genes results in exchange of segmentation identity and disruption of segmentation patterns.

Two classes of homeoboxes have been identified in vertebrate, sea urchin, and insect genes indicating that the diversification of function of these boxes began early in the evolution of metazoans (→ homeo-gene).

homeo-gene — any of the → homeobox-containing genes which play a central role in pattern formation during animal embryogenesis and participate in, e.g., the establishment of segmentation and in the specification of segment identity. Homeo-genes show tissue- and region-specific expression and specify positional identity along the anteroposterior axis.

homeologous (Huskins 1931) — of partially homologous chromosomes. "Homeology" designates the residual homology of originally completely → homologous chromosomes.

homeorhesis (Waddington 1957) — an equilibrium property characterizing the regulatory pathway of development which is the result of the activity of many genes controlling different synthetic processes. These processes interact in a way which assures that their combined activity delimits a pathway exhibiting homeostatic properties and is referred to as h. or canalization (→ homeostasis).

homeosis (Bateson 1894) — the transformation of one structure of the body into a homologous structure of another body segment (= homoeosis). Mutant genes giving rise to such alterations, by interference with primary processes in embryonic development, are called → homeotic mutant genes.

homeostasis (Cannon 1929) — the tendency of a system to be specified by context (individual, population, process, etc.) to maintain a dynamic equilibrium and in case of disturbance to restore the equilibrium by its own regulatory mechanisms.

(1) **Genetic or collective h.** (Lerner 1950; Lewontin 1955): The tendency of a population to equilibrate its → gene pool and to maintain a genetic composition leading to an optimum balance (= "genetic inertia" of Darlington and Mather 1949) in the environment in which the population lives. G.h. determines the extent to which a gene pool can respond to → selection. It is manifest by the tendency of phenotypes to return to the original condition

when selection is relaxed after a population has been exposed to severe selection pressure for a particular phenotypic character. Since phenotypes are the product of genotypes, selection for a new phenotype is coupled with the trend to abandon the previously integrated genotype, thus leading to lowered fitness. Decrease of → selection pressure for the new phenotype then allows at least partial return (by natural selection) to the original gene combination with maximal fitness accompanied by a (partial) restoration of the original phenotype.

(2) **Epigenetic h.** (Nanney 1958): The persistence of differences at the cellular level in the absence of genetic or environmental "differentials", i.e., their relative stability to disturbances. The term may also cover what has been called **"developmental h."** (Lerner 1954), "canalization" of "homeorhesis" (Waddington 1957), namely the capacity of the developmental pathways to produce a normal phenotype in spite of developmental or environmental disturbances.

(3) **Physiological h.** (Cannon 1932): A type of phenotypic reaction buffering the organism against fluctuating external conditions by internal self-regulatory mechanisms which enable the individual to adjust to variable environments not by conforming to changes in the external environment but by resisting those changes (either by varying phenotypically or by remaining constant in the face of environmental changes). Physiological h. is frequently associated with heterozygosity and then confers a higher average fitness in various environments to the carriers of such genotypes, as compared to their less homeostatic relatives (homozygotes) (Grant 1963).

homeotic (Goldschmidt 1945) — of genes that assign spatial identity to groups of cells with respect to their morphogenetic fates; they are required (by segmented organisms) for the proper morphogenesis of a distinct region and probably act by controlling the activities of other genes during development (→ homeobox).

Mutations that cause cells to switch from one developmental fate to another and to produce specific intersegmental transformations of identity by causing embryonic founder cells at one position to adopt the fates reserved normally for those at another position are called h. mutations.

homoduplex DNA — a DNA duplex in which the nitrogenous bases of the two strands are precisely complementary, as opposed to → heteroduplex DNA.

homogametic (Wilson 1911) — of that sex which produces only male or female determining gametes in sex digamic organisms, in contrast to the heterogametic sex.

homogamous (Sprengel 1793) — of hermaphroditic flowers and animals whose male and female sex organs function at the same time, in contrast to → dichogamous ones.

homogamy — the preference of a mating individual for another with similar pheno- or genotype, as opposed to heterogamy.

Mechanisms of h. are positive → assortative mating, selective → fertilization, and premating reproductive → isolation. They characterize closely related → sympatric, biparental species.

homogeneric tRNA (White et al. 1973) — → transfer RNA.

homogenetic (Waddington 1939) — → heterogenetic.

homogenic (Fisher 1928) — → heterogenic.

homogeneously staining region — (HSR); in human chromosomes, a more or less extended alternatively banding chromosome segment found predominantly in cancer cells. HSR's represent sites of → gene amplification (→ double minute).

homogenotic (Morse et al. 1956) — → heterogenotic.

homogenotization — = → gene replacement.

homograft — a tissue graft between two members of the same species. Normally homografts are rejected ("homograft reaction"), but accepted if the donor and host are isogenic or nearly so (identical twins, highly inbred animal strains).

homo-heteromixis (Burnett 1956) — → heteromixis.

homoimmune — → heteroimmune.
homokaryon — → heterokaryon.
homokaryotype — → heterokaryotype.
homologous — (1) of chromosomes or chromosome segments ("homologs") which are identical with respect to their constituent → genetic loci (the same loci in the same sequence) and their visible structure, as distinct from nonhomologous and → homeologous chromosomes or parts of chromosomes; (2) of structures similar in the same or different organisms due to similar hereditary and developmental origin (Owen 1840); (3) of similar variations ("law of homologous genes" (Vavilov 1922)).
homomeric — → heteromeric.
homomixis (Burnett 1956) — sexual reproduction in homomictic fungi resulting from the fusion of genetically similar nuclei normally derived from the same thallus (= homothallism, monothallism). (→ heteromixis; amixis).
homonuclear (Krooth 1965) — → heteronuclear.
homoplasmic — = homoplasmonic (→ heteroplasmonic).
homoplasmonic — → heteroplasmonic.
homoplastidic (Michaelis 1957) — → heteroplastidic.
homopolymer tail — a sequence of only one sort of nucleotides (a homopolymer) at the end of a DNA or RNA molecule (most eukaryotic mRNAs have, e.g., a h.t. of pA at the 3' end).
homopurine-homopyrimidine sequence — in various eukaryotic genes, any of the common sequences consisting of alternating $d(C-T)_n - d(G-A)_n$.
homoselection (Carson 1959) — → heteroselection.
homosequential (Carson et al. 1967) — → band.
homosteric (McElroy et al. 1967) — of enzymes which combine at the catalytic site with their normal substrate or a structurally similar compound, resulting in a modified reaction of the bound intermediate. Homosteric enzymes are to be distinguished from → allosteric ones where the modifier molecule combines at a site other than the catalytic site.
homothallic (Blakeslee 1904) — → heteromixis.
homozygote disadvantage load (Dobzhansky 1965) — = segregational load (→ genetic load).
homozygous (Bateson and Saunders 1902) — → heterozygous.
horizontal evolution (Brown et al. 1972) — → gene rectification.
horizontal templating (Alberts et al. 1977) — in eukaryotes, the propagation of the active structure of expressed genes down the → transcription unit, as opposed to vertical t., i.e., the propagation of the structure from parental to daughter chromosomes at the time the chromosome is replicated. During h.t. the basic → nucleosome structure is modified in active genes to allow → RNA polymerase to read the DNA and transcribe the gene (→ gene expression).
hormone (Starling 1905) — any type of chemical messenger responsible for the timing and regulation of growth and development and the homeostatic maintenance of the fully developed organism. Many hormones achieve their effects by (indirectly) altering gene activity.

A distinction can be drawn between rapidly acting **primary hormones** (messengers) and **secondary hormones** which generally act over a longer time period and characteristically provide trophic stimulation to their target cells. The primary hormones use cyclic nucleotides as second messenger.

In the target cells, h.-receptor complexes interact with an effector structure according to two major patterns: (1) In the case of polypeptide hormones and neurotransmitters, interaction involves the plasma membrane level influencing membrane enzymes and/or cation transfer systems; (2) in the case of steroid and thyroid hormones, h.-receptor complexes interact in the cell nucleus with elements of the → gene expression machinery and modify → genetic transcription.
hormone-responsive element — (HRE); in most hormonally regulated genes, specific regula-

tory nucleotide sequences which are located in close proximity to the → promoter. They appear to mediate the action of steroid hormones after their formation of complexes with specific receptors.

horotelic (Simpson 1944) — ref. to standard rates of → evolution (→ bradytelic; tachytelic).

host cell — a cell whose metabolism is used for the growth and reproduction of a virus.

host-cell reactivation (Garen and Zinder 1955) — enzymatic → dark repair (Symbol HCR) of lethal and nonlethal lesions induced by UV and some chemical agents (damage by ionizing radiation is only slightly repaired) in the DNA of bacteriophages after infection of host cells (→ reactivation). The same processes also repair lesions in bacterial DNA. Bacteria performing h.-c.r. are designated HCR(+), mutants which do not so (they are highly sensitive to UV) HCR(−). The repair process overlaps with photoenzymatic repair (→ photoreactivation). HCR is the result of excision of damaged portions from DNA and their replacement by material newly synthesized in conformation with the information contained in the undamaged complementary DNA strand. With single-stranded DNA and RNA, HCR fails to occur (Rupert and Harm 1966).

host-controlled DNA modification — → DNA modification-restriction.

host-controlled DNA restriction — → DNA modification-restriction.

host-controlled variation (Luria and Human 1952; Bertani and Weigle 1953) — a general phenomenon in which DNA may be modified when it is synthesized in one cytoplasm and then undergoes → restriction upon entering another cytoplasm (= host-induced modification, host-controlled modification). H.-c.v. has been observed in viruses (extension or restriction of the host range), in restriction of both → F plasmids and col-I in bacterial strains carrying a certain → prophage, in decreased → zygotic induction, and in the formation of recombinants when crosses between specific strains of bacteria were performed.

host-killing efficiency (Luria and Delbrück 1942) — the capacity of "inactivated" bacteriophages to kill their host cells.

host-mediated assay (Gabridge and Legator 1969) — an indirect method for evaluating potential mutagenic agents (→ mutagenicity testing) introduced to bridge the gap between in vitro microbial studies and tests in mammals. The mammal (usually the mouse) is injected intraperitoneally with a microorganism (bacterium, yeast) as an indicator for induced → gene mutations and then treated with a potential mutagen. The comparison between the action of the compound on the microorganism directly and in the h.-m.a. indicates whether the host can detoxify the compound or mutagenic products can be formed as a result of host metabolism. To obtain an even more precise evaluation of persistance, organ distribution, retention, and metabolism of a potential mutagen, some modifications, of the initial procedure are being used, e.g., incubation of microorganisms in extracts of organs and diffusion bags, in urine, and in blood and liver.

host range — the spectrum of strains of a given bacterial species (or the number of different species) in which a given → plasmid, → bacteriophage or other virus will replicate.

host range mutant (Luria 1945) — any mutant of bacteriophages able to overcome the resistance of phage-resistant mutant bacteria. Host range mutants differ from wild-type phages in the structure of their adsorption organs. Such a structural change permits the stereo-specific fit with the phage receptor in the envelope of the phage-resistent bacteria in spite of the modification of the cell surface which prevents attachment of the wild-type phage. Most of these mutations extend the host range beyond that of wild type (Stent 1963).

hot spot (Benzer 1955) — a site of a → gene or → chromosome at which the frequency of → mutation (mutation h.s.; aberration h.s.) or → genetic recombination is very much increased under the action of a particular mutagen or recombinagen.

housekeeping gene — any of the genes that are active in all cell types; they produce rare to moderately abundant messenger RNA's, as opposed to tissue-specifically expressed genes. H. genes are regulated primarily at the level of mRNA maturation and/or stability.

Hpa II tiny fragment — = → HTF island.

HTF island (Bird et al. 1985) — in vertebrate DNA, any of about 30,000 single-copy, unmethylated loci containing a high density of → restriction sites for the → restriction endonuclease Hpa II (= Hpa tiny fragments) in which CpG is abundant. HTF islands are associated with the transcriptional → promoter region of "housekeeping" genes and are usually absent from tissue-specific genes. The HTF islands are large (between 1500 and 2000 bp) and cover an extended region of the 5' end, often including the first few → exons of the genes in question as well as upstream sequences. HTF-associated genes are probably inactivated if the islands become methylated (→ DNA methylation).

HT-transducer (Schmieger 1972) — phage mutants with increased ability to transduce bacterial markers (→ transduction).

human genetics — the study of the genetically controlled similarities and differences (physical and mental, normal and abnormal) in human beings, in respect to their causes and modes of transmission (inheritance) from generation to generation (→ medical genetics, pharmacogenetics).

hyaloplasm (Pfeffer 1877) — the ground substance of the cell (= cell sap; cytoplasmic matrix; ground plasm; soluble phase of the cell) which lies between the canals of the → endoplasmic reticulum and contains the cell organelles (mitochondria, plastids in plants, Golgi apparatus, etc.). The h. is structureless under the electron microscope except for the presence of scattered → ribosomes and very fine → microfilaments and → microtubules. It contains some soluble RNA and the enzymes concerned with anaerobic metabolism.

hyaluronic acid — a significant component (a linear polydisaccharide of β-D-glucaronic acid and 2-acetamido-2-deoxy-β-D-glucose) of the extracellular material of certain streptococcal strains as well as a wide variety of eukaryotes.

hybrid — (1) as adjective = → heterozygous. (2) Any offspring of a → hybridization or cross between two homozygous, genetically unlike individuals.

Genetic h. (Darlington 1937): The result of the fusion of two gametes differing in their genetic constitution (in the → alleles they carry) or produced by mutation in a homozygote.

Structural h. (Darlington 1929): The result of fusion of two gametes which differ in the arrangement of their genes owing to chromosomal structural changes (heterozygosity for → translocations, inversions, deletions, etc.). **Cryptic structural hybrids** (Stebbins 1945) are those whose structural differences are so small that they do not generally lead to disturbances in meiotic → chromosome pairing.

Numerical h. (Darlington 1931): The result of fusion of two gametes which differ in respect to the number of their chromosomes.

Permanent h.: A stable hybrid maintained as a heterozygote of balanced lethals by the loss of homozygous lethal genotypes (→ complex heterozygous).

hybrid advantage — → hybrid zone.

hybrid-arrested translation (Izant and Weintraub 1985) — a technique used to identify the proteins coded for by a cloned DNA sequence or the → cDNA corresponding to a → messenger RNA by relying on the ability to base pair with RNA in vitro. The method involves DNA-RNA hybridization and is as follows: (1) A crude cellular mRNA preparation, composed of many types of mRNA, is hybridized with a cloned, denatured DNA; (2) mRNA homologous to the cloned DNA will anneal to it; (3) the rest of the mRNA molecules are put into an → in vitro translation system and the protein products are compared with the proteins obtained by use of the whole mRNA preparation.

Hybridization of mRNA with → antisense DNA (or antisense RNA) blocks translation or processing of specific mRNA's by hybridization.

hybrid breakdown — → isolation

hybrid cell — a cell formed by the fusion of two different cells (→ cell fusion).

hybrid complex — any group of plant species in which the morphological differences of the basic diploids have been masked by hybridization. They may be classified as follows (Grant 1953): (1) **Homogamic c.:** Usually the hybrids are diploid and exhibit normal meiosis. Structural heterozygosity (if present in the F_1) is eliminated by natural selection; (2) **clone complex:** the hybrids produce mainly or exclusively by asexual means; (3) **heterogamic c.:** the hybrids are permanent structural heterozygotes, as, e.g., in *Oenothera* (→ complex heterozygous); (4) **polyploid c.:** the hybrids are polyploids and undergo reproduction; (5) **agamic c.:** the hybrids or their derivatives reproduce mainly or exclusively via unfertilized seeds or bulbils.

hybrid disadvantage — → hybrid zone.

hybrid DNA — a double-stranded DNA molecule with → heteroduplex regions.

hybrid DNA technology — = → recombinant DNA technology.

hybrid dysgenesis (Sved 1976; Kidwell and Kidwell 1976) — in certain *Drosophila melanogaster* interstrain crosses and groups, a set of germline abnormalities (including temperature-dependent sterility, high rates of mutation, male recombination, chromosome breakage, and several other traits) caused by → transposable genetic elements (the P and I elements) that are present in multiple copies in the genomes of some strains and absent from others. Hybrids from crosses show dysgenic traits that are reduced or absent in the reciprocal hybrids.

Two independent systems associated with two families of transposable elements are known: (1) in the P-M system, h.d. results from interaction between chromosomally linked transposable elements (2.9 kb P elements present in P strains) and cytotypes (M or P), known to be determined by absence (M) or presence (P) of the chromosomal factors, and confer susceptibility (M) or resistance (P) to the P elements and h.d. (2) In the I-R system, I (inducer) strains are analogous to P strains and, likewise, R (reactive) strains to M strains. In this system, h.d. is limited to females, in the P-M system, dysgenic traits occur in hybrids of both sexes.

The transposable genetic elements do not normally cause dysgenic traits within homogeneous P- or I-strains because of immunity conferred by the cellular condition known as the P or I cytotype.

hybrid gene — any → gene, usually constructed by use of → recombinant DNA technology, containing varying segments from parental genes. The chimeric genes are composed of, e.g., a fragment of the gene of interest ligated to some other heterologous DNA sequence which permits defining regions responsive to regulatory signals.

Three primary requirements of h.g. expression are: (1) A cloned gene; (2) a suitable expression system; (3) the ability to distinguish the inserted h.g. from its natural counterpart (→ reporter gene; minigene).

hybrid incapacitation — a collective term comprising → hybrid sterility and → hybrid inviability.

hybrid inviability — the lowering of somatic vigor due to aberrant developmental processes in hybrids. H.i. is due to failure of the maternal and paternal portions of the hybrid genomes to cooperate and to direct harmonious growth and development. The lethal crisis may occur at any time during development.

hybridity — the state of being → heterozygous or hybrid. **"H. equilibrium"** (Darlington and Mather 1949) is the term proposed to designate the equilibrium state of hybridity in a stable breeding population which is dependent on: (1) The rate of recovery of homozygosity (itself depending on the type and amount of → inbreeding in the population), (2) the frequency of lapses from inbreeding, and (3) the amount of heterozygosity produced by lapses (this in turn depends on the amount of genetic variation in the breeding population). The **"h. optimum"** (Darlington and Mather 1949), i.e., the optimal amount of h. in a particular population, is thus normally related to the mode of breeding. It is low in inbreeders, and high in outbreeders.

hybridization — (1) the formation of a → hybrid by performing a cross (→ cell hybridization). The individuals being crossed may belong to two unlike natural populations that have come into contact secondarily. Five kinds of h. may be distinguished when h. is viewed as a population phenomenon (Mayr 1963):

Occasional h. of sympatric species, usually resulting in hybrid individuals that are either inviable or sterile (no back crossing possible with the parental species);

Occacional or recurrent h. resulting in more or less fertile hybrids between sympatric species (some hybrids may backcross to one or both of the parental species);

Partial interbreeding between two formerly isolated populations in a secondary zone of contact (→ intergradation) if the populations failed to aquire complete reproductive isolation during the period of geographic isolation;

Complete local breakdown of reproductive isolation beetween two sympatric species, resulting in the production of → hybrid swarms that may include the whole range of variability of the parental species;

Production of a new specific entity in plants due to h. and subsequent doubling of chromosomes, giving rise to allopolyploidy (→ alloploid).

(2) The annealing of nuclei acid strands from different sources (→ DNA hybridization).

(3) → In situ hybridization.

hybridization probe — a small (about 1000 bp) labeled DNA or RNA molecule which is used to detect complementary nucleic acid sequences by → hybridization.

hybridoma (Milstein and Köhler 1975) — any somatic cell hybrid between myeloma cells and → antibody secreting cells that produces large amounts of pure (or monoclonal) antibodies of desired specificity and homogeneous binding characteristics.

Hybridomes can be classified into three major categories: (1) Those mediating specific immune responses, such as induction of plaque-forming cells and specific tumor cell lysis; (2) those producing specific T cell factors, including hapten- or carrier-specific helper factors, and suppressor factors; (3) those secreting nonspecific lymphokines, e.g., macrophage migration inhibitory factors (MIF), macrophage activating factors (MAF), T cell-replicating factor (TRF), T cell growth factor (TCGF), and granulocyte-macrophage colony-stimulating factor (CSF).

hybrid optimum (Darlington 1963) — a concept according to which each → species is adapted to a certain optimal degree of heterozygosity.

hybrid panel analysis — a method used for assigning (human) → linkage groups to specific chromosomes. A DNA probe from a linkage group is hybridized against DNA from various hybrid cell lines, each of which contains an incomplete set of chromosomes.

hybrid plasmid — any → plasmid chimera containing inserted DNA sequences and originating by use of the → recombinant DNA technology. Frequently newly constructed h. plasmids are not stable. Instability is of two types (Primrose and Ehrlich 1981): (1) **Segregational instability**, i.e., loss of the entire plasmid from the cell; (2) **structural instability**, i.e., rearrangements and loss of plasmid most frequently via → deletions.

hybrid release (Woodruff et al. 1979) — the breakdown in suppression of mutator activity (in *Drosophila*) upon crossing a wild strain and a laboratory stock. H.r. results in an explosive increase in genetic variation.

hybrid released translation — a method used for detection of the protein coded for by cloned DNA: (1) The cloned DNA is bound to a nitrocellulose filter and a crude preparation of mRNA is hybridized to the filter-bound DNA; (2) only mRNA sequences homologous to the cloned DNA will be retained on the filter; (3) these mRNA's can be removed by high temperature or by formamide; (4) the purified mRNA is plated in an → in vitro translation system and the protein encoded can be analyzed (electrophoresis through a polyacrylamide gel).

hybrid replicon — two → replicons covalently joined by use of → recombinant DNA technology (→ hybrid plasmid).

hybrid sterility — the impairment (partial sterility) or suppression (complete sterility) of the reproductive capacity of F_1 or of later generation hybrids between genetically different populations (usually belonging to different species). H.s. is an isolating mechanism (→ isolation) which may partially or completely prevent the → gene flow between populations. The basis of h.s. may be genic, chromosomal, or cytoplasmic (Dobzhansky 1937; Ehrman 1962). (1) **Genic h.s.:** The result of abnormal developmental processes caused by specific gene complexes in the hybrid. A special type of this category is h.s. due to intersexuality where the interference occurs with those developmental processes leading to the normal differentiation of male and female sex. (2) **Chromosomal h.s.:** The result of structural differences between parental chromosomes which interfere with their pairing (→ chromosome pairing) and disjunction at meiosis. (3) **Cytoplasmic h.s.:** The result of discordance between the chromosomal genes and the cytoplasm. The cytoplasmic component of the system may be due either to autonomous hereditary determinants in the cytoplasm or to formative influences exerted by chromosomal genes which were present in the egg cell before meiosis (→ predetermination)

Another classification distinguishes between h.s. due to disturbances of the haploid or the diploid phase of the life cycle, i.e., between **haplontic** and **diplontic sterility**.

hybrid swarm — a collection of hybrids produced by complete or local breakdown of isolating barriers between two → sympatric species (= hybrid flock). The h.s. replaces the parental species and serves as a continuous bridge between the two parental extremes (Mayr 1963).

hybrid vigor — → heterosis.

hybrid zone — a geographic zone in which two populations, formerly separated by geographic → isolation, hybridize after the breakdown of the geographic barrier provided there is no reproductive isolation between them. A hybrid population or zone of secondary → intergradation is produced by hybridization. The hybrid populations are highly variable; some individuals resemble one or the other subspecies but most are hybrids more or less intermediate between the two original populations.

H. zones may be of two kinds: (1) Those in which hybrids are less fit than either parental type (hybrid disadvantage); (2) those in which hybrids are more fit than either parental type (hybrid advantage).

hyparchic — of genes in → mosaics, prevented from phenotypic manifestation by genes from genetically different neighboring tissues, as opposed to "autarchic" genes.

hyperchimera — a plant → chimera with both sectorial and periclinal distribution of genetically different tissues.

hyperchromicity — The increase in optical density that occurs when DNA is denatured.

hypermorph (Muller 1932) — → allele.

hyperplastoid (Martin and Sprague 1973) — of cell lines from mammalian tissues which have finite replicative life-spans and (in the majority of cases) display the → karyotype of the donor (→ neoplastoid). Cessation of replication of h. cell lines and clones may be due to the accumulation of errors in macromolecule synthesis or due to differentiation to death (terminal differentiation).

hyperploid (Bělăr 1928) — of → cells or individuals with one or more added chromosomes or chromosome segments in their complements, as opposed to **hypoploid** cells and individuals which are deficient for one or more chromosomes or chromosome segments. Depending on the degree of ploidy in the original cell or individual, hyper- and hypohaploids, hyper- and hypodiploids, as well as hyper- und hypopolyploids may be distinguished (Winkler 1916).

hypomorph (Muller 1932) — → allele.

hypoploid (Bělăr 1928) — → hyperploid.

hypostasis (Bateson 1907) — → epistasis.

I

IAP element (Lueders and Kuff 1977) — any of an intracisternal type of particles representing endogenous → retrovirus-like but noninfectious structures observed in rodent cells. IAP's contain a polyadenylated RNS molecule, a → reverse transcriptase, and a 73-kDA protein.

identical by descent (Malecot 1948) — of two genes if they are both derived from the same gene in some ancestor.

identifier sequence (Sutcliffe et al. 1982) — (ID sequence); in some tissue-specific genes, an "identification sequence" of 82 nucleotides located (often in multiple copies) within the → introns that interrupt protein-coding segments of eukaryotic genes. ID's are transcribed into RNA and may, like → transposable genetic elements, be able to jump around the genome. Probably, the ID must be transcribed before the gene containing it can be switched on. ID's show structural analogy with some short repeated DNA sequences and terminate, like → Alu sequences, in oligo (dA).

identity disequilibrium (Weir and Cockerham 1969) — (ID); the difference between the joint probability of → identity by descent and the product of the separate probabilities of identity by descent for two loci. It occurs in finite random mating populations, even for unlinked loci when it is a measure of the variation in → inbreeding coefficients of members of the population with varying pedigrees, and is further enhanced by linkage.

idiochromosome (Henking 1891) — = → sex chromosome.

idiogamy — = → self-fertilization.

idiogram (Navashin 1912) — a diagrammatic representation of chromosome morphology used diagnostically for comparison of the → karyotypes of different species and varieties. The construction of idiograms is based on measurements (in several or many cells) of total chromosome length and arm length ratio as well as on → centromere position, → nucleolus organizer position, position of → heteropycnotic segments, etc., for any one chromosome during metaphase of → mitosis. Centromere position is usually calculated from the total length, c, and the lengths of the long and short arms, l. and s., respectively, of the chromosome in question and expressed either as a difference $d = l/s$, as a ratio $r = l/s$ or as a centromere index $i = 100\ s/c$.

Chromosomes in an i. may be classified according to their centromere position as follows (Levan et al. 1964):

Centromere position	Arm ratio	Chromosome	designation
Median sensu strictu	1.0	M	(Metacentric)
Median region	1.7	m	
Submedian	3.0	sm	(Submetacentric)
Subterminal	7.0	st	(Subtelocentric)
Terminal region		t	(Acrocentric)
Terminal sensu strictu	∞	T	(Telocentric)

idioplasm (Naegeli 1884) — the sum total of → hereditary determinants (→ idiotype).

idiotype (Naegeli 1884) — (1) the sum total of all hereditary determinants (nuclear and extranuclear) of an organism (Naegeli 1884). (2) The structural features of the variable regions of the → antibody that itself provokes the formation of other antibodies. Operationally, an i. is defined by the class of other antibodies with which it reacts uniquely (immunoglobulin).

I element — in *Drosophila,* a → transposable genetic element (5.4 kb) involved in the I-R systems of → hybrid dysgenesis.

I-Line — = → inbred line.

illegitimate recombination → genetic recombination between nonhomologous DNA sequences present on separate DNA molecules or located in various regions of simple-sequence DNA molecules. I.r. may produce → deletion, → duplication, and many other rearrangements.

illegitimate transcription — low transcription of a tissue-specific gene in nonspecific cells.

immigration — → gene flow from one population into another („genetic input") as mediated by immigrating individuals capable of interbreeding with members of that population. The percentage of individuals in a population who become members by immigration is referred to as the **"immigration coefficient"**. **"Immigration pressure"** is the rate of change of → gene frequency in a population subject to i. It depends on the i. rate and on the difference in gene frequency between immigrants and natives. **"Immigration load"** is that part of the → genetic load which is due to the incorporation of those alien genes into the → gene pool that reduce fitness in their new environment, irrespective of whether these genes had been advantageous or → deleterious in their former environment. The load of inferior alleles in a gene pool caused by → gene mutation and i. is referred to as its "input load" (Mayr 1963).

immune system — in higher organisms (vertebrates), a system that has evolved the capacity to react specifically with a very large number of foreign molecules (antigens) with which it had no previous contact, while avoiding reactivity for autologous molecules, i.e., a system which discriminates between self and nonself so that normally immune responses are only mounted against nonself components.

The i.s. is composed of two distinct cell types (B lymphocytes and T lymphocytes) each of which is independently capable of recognizing a large number of distinct entities through their antigen receptors: Surface → immunoglobulin in the case of the former, and T cell receptor in the case of the latter. In both cell types, the genes encoding the antigen receptor consist of multiple gene segments which recombine (gene segment rearrangement) during maturation to produce many possible peptides. It is B cells that are stimulated to antibody synthesis during the immune response and therefore the binding of antigen to helper T cells must result in the delivery of a signal to B cells.

immunity substance (Jacob and Wollman 1956) — in → lysogenic bacteria, a temperature-sensitive substance (repressor) which prevents the transcription of → prophage genes whose transcripts are needed for vegetative replication of the → prophage. The same substance is responsible for the immunity of lysogenic cells to infection by phage of the same type as their prophage.

immunogenetics — a branch of genetics encompassing all the areas of overlap between immunology and genetics. I. includes the application of methods of immunology to the recognition of individual differences for genetic studies (e.g., blood groups and protein allotypes), and the application of genetic methods to the study of immunological phenomena (e.g., the mechanisms determining antibody induction, structure, and specificity).

immunoglobulin — any of the symmetrical, globular protein molecules (Y-shaped) composed of two heavy (H) and two light (L) chains; both types of chain contribute to the structure of the → antibody combining site (→ immune system) through interactions of their NH_2-terminal variable (V_H and V_L) domains. The variability of the sequence of the V regions of antibodies is the feature that gives rise to the diversity of the antibody (immune) response.

Genes coding for i. require gene rearrangement or somatic recombination for the creation of a complete, active transcription unit. For light chains, a variable (V) gene segment recombines via a joining (J) segment to the 5' flanking sequence of the constant (C) region producing a functional VJ-C-gene. Heavy chain gene rearrangement involves multiple steps: The joining of a diverse (D) segment to a J segment prior to VDJ-C joining and switch recombination which results in the expression of the same VDJ-sequences with different subclasses of heavy chain constant genes.

Three separate → multigene families (lambda, kappa, and heavy) encode the i. light and heavy chains. Each polypeptide chain is encoded by several distinct gene segments: Variable (V_L), joining (J_L), and constant (C_L) for light chains, and V_H, diversity (D), J_H, and C_H for heavy chains. The gene segments are separated by → intervening sequences in the germ line and are rearranged with respect to each other during the differentiation of antibody producing cells. → RNA processing removes the intervening sequences from the primary transcript to generate the mature mRNA for the variable and constant regions. The joined V gene is separated from the C gene segment by intervening sequences. In the heavy chain family, transcripts can be processed in one of two ways to give rise to alternate forms of mRNA which encode membrane-bound or secreted polypeptides.

Both multiplicity and diversity of i. variable regions are essential features for keeping the → immune system in a state of maximum efficiency. Additionally to the presence of a large number of V genes in the germ line, diversity is partly increased by somatic processes such as V-D-J recombination, flexibility in the joining site, and somatic mutations.

imprinting — the phenomenon that very early exposure to a given stimulus imposes a certain behavior pattern on individuals (→ chromosome imprinting).

inactivating DNA alteration (Freese et al. 1969) — any → DNA damage which blocks → DNA replication. If not eliminated by → DNA repair processes, such damage is usually lethal to the cell. Occasionally i. DNA alterations may give rise to → gene mutations and frequently (by misrepair) to chromosome structural changes (→ chromosome mutation).

inborn error of metabolism (Garrod 1902) — a genetically determined biochemical disorder in which a specific enzyme defect produces a metabolic block. The block may have pathological consequences (= molecular disease).

inbred — the progeny of matings between relatives. An inbred individual is connected through both the female and male parent to the same recent ancestor and can thus receive two copies of a gene that was carried by this ancestor (these copies are called "identical by descent").

inbred line — any more or less → homozygous line derived from an outbreeding population by repeated → inbreeding (self-fertilization, sib mating, etc.) of an individual and its progeny. A population in which inbreeding is inevitable, because the population size (→ breeding size) is limited thereby forcing relatives to mate to each other, is equivalent to an i.l. The rate of loss of → genetic variation is then a function of the size of the population.

inbreeding — a → mating system (= endogamy) which, in contrast to → out-or crossbreeding (= exogamy), involves the breeding together of individuals more closely related than mates chosen at random from a population. The degree of i. in any particular population or mating is a function of the relationship between the mating partners (→ inbreeding coefficient). The closest form of i. is self-fertilization; other forms are sib matings and half-sib matings.

All forms of i. result in an increase of homozygosity, i.e., lead to genetic fixation. When i. is imposed on a normally outbreeding population, the population is broken up into smaller groups accompanied by a decrease in total heterozygosity as fixation occurs in the subgroups. If no selection for particular homozygotes occurs, → genetic variance as a whole is increased in the population (transformation of potential into free → genetic variation). I. in normally outbreeding populations generally leads to a loss of fitness referred to as → inbreeding depression.

In plants, morphological and physiological mechanism favoring i. are common, e.g., failure of flowers to open (cleistogamy) or opening of flowers only after the pollen has been shed. The relative restriction of outbreeding mediated by these and other mechanisms ranges from facultative self-pollination to obligate self-fertilization.

inbreeding coefficient (Wright 1922) — a measure of the rate of → inbreeding or the degree to which an individual is inbred. (1) (F); the probability that the two alleles at a locus of an individual are derived from a common ancestral gene, or alternatively, a measure of the proportional decline of the average expected number of heterozygous loci in an individual, compared to the average number in its ancestors (i.c. of an individual). (2) (α); the probability that the two genes at a locus of a member of a population taken at random are → identical by descent, or the average decline in population heterozygosity at all loci (i.c. of a population).

The i.c. may range from zero, in a large random-mating population, to 1.0 when complete homozygosity at all loci is attained.

inbreeding depression — a decrease in → fitness and vigor, resulting from → inbreeding imposed on individuals that are normally outbreeding (= inbreeding degeneration). The extent of i.d. is not the same in all lines produced by inbreeding. That point after several generations of inbreeding at which no further decrease of fitness and vigor occurs is referred to as the "inbreeding minimum". I.d. is the consequence of increased homozygosity for → deleterious recessive genes (→ genetic load) and the break-up of balanced polygenic systems (Mather 1941). The crossing of inbred lines which have reached their inbreeding minimum frequently results in → heterosis.

inbreeding load (Morton et al. 1956) — the extent to which an equilibrium population under random mating is impaired by inbreeding which ordinarily increases → genetic load.

incest — the mating between parents and their offspring or between siblings. The cultural regulation of sexuality in human society through rules which prevent inbreeding with close relatives is called incest taboo and forces outmarriage.

inchondriosis (La Bella and Krass 1968) — ingestion by cells of particulate material with or without extracellular medium (= phagocytosis; endocytosis; heterophagy).

inchylocytosis (La Bella and Krass 1968) — ingestion by cells of fluid material by convolution of the cell membrane and vesiculation of engulfed bulk fluid.

inchylosis (La Bella and Krass 1968) — penetration of water and solutions across an apparently intact cell membrane (→ inchylocytosis).

incidence — in human genetics, the frequency of new occurrence of a genetically caused medical condition among individuals of a specific population, within a certain period of time (e.g., incidence of → genetic disease). From the i. (symbolized by x) of a character and its mode of inheritance the frequency, q, of the "abnormal" allele may be calculated. If each of the affected individuals is → heterozygous, then the frequency of the allele in question is equal to one half the i. of the character: $q = \frac{1}{2}x$. (→ prevalence).

incipient species — populations that are in the process of diverging to the point of → speciation but which still have the potential to interbreed; usually they are prevented from interbreeding by specific barriers (→ super species).

incision endonuclease — any of a class of → deoxyribonucleases that recognize mismatched bases (→ mismatch repair) and other defects of the DNA duplex and make a single phosphodiester bond cleavage that leaves a 5'-phosphate group on the defective chain (→ excision repair).

I. endonucleases recognize different types of → DNA lesions: (1) A base modified by alkylation or by other addition or transformation (monoadduct damage); (2) a mismatched base; (3) more than one base modified such as pyrimidine dimers (diadduct damage); (4) absence of a base due to the action of, e.g., → DNA glycosylase, heat, or acid; (5) cross linking of the DNA strands.

These lesions or the DNA helix disformations caused by them are recognized by various such enzymes (→ DNA repair).

incompatibility (Stout 1918) — the selective restriction of mating competence by extrasexual "incompatibility genes" which limit or prevent zygote formation (fertilization) by certain classes of gametes and thus in an otherwise freely interbreeding group limits either self-fertilization ("self-incompatibility") or cross-fertilization ("cross-incompatibility") between certain self-incompatible or self-compatible individuals if these carry genes for self-incompatibility. Members of the same i. class are typically strictly self-fertile and the various i. systems which occur in the plant kingdom usually promote outbreeding. Mating between individuals of compatible classes is usually reciprocal.

In general, homo- and heterogenic i. systems may be distinguished (Raper and Esser 1964): (1) **Homogenic i.:** Zygote formation is prevented if the mating partners carry identical i. genes. It may be controlled by one or two gene loci. In the first case, there are at least two mating types (bipolar i.), in the second, at least four (tetrapolar i.). The number of mating types equals the number of alternative alleles in bipolar systems. Inbreeding is reduced to 50% and 25%, respectively, while outbreeding is practically unrestricted. Homogenic i. may be based either upon two alles of one series, multiple alleles in one series, or multiple alleles in each of two series. (2) **Heterogenic i.:** This system consists of more than two gene loci. Mating partners homozygous at all loci interact freely, but those heterozygous for the loci in different combinations either interact unilaterally or do not interact at all. It is a unique system which, in contrast to homogenic i., limits the sexual compatibility of heterogenic partners and thus promotes inbreeding, while homogenic i. restricts the mating between genetically similar organisms and thus favors outbreeding. Heterogenic i. is known only in fungi.

A more detailed classification of i. systems in angiosperms is as follows (Crowe 1964):

(1) **Heteromorphic i.** characterized as follows: (a) Each mating type is morphologically distinct and recognizable without breeding tests (e.g., heterostyly); (b) the mating type of the pollen is determined sporophytically, i.e., the mating behavior of the pollen is controlled by the sporophyte of the pollen producer irrespective of the pollen's incompatibility genes; (c) dominance between the alleles of the incompatibility gene(s) is expressed both in the pollen and the style; (d) the style inhibits the growth of incompatible pollen tubes. This type of system may be controlled by one gene with two alleles (dimorphy) or by two genes each with two alleles.

(2) **Homomorphic i.** characterized as follows: (a) There are no morphological differences between mating types, thus they can be identified only by breeding tests; (b) there are numerous mating types; (c) there are numerous alleles of the incompatibility gene(s).

The control of the mating type of the pollen in homomorphic systems of i. may be either gametophytic or sporophytic. In the case of gametophytic i., alleles of the i. gene(s) act individually in the style and incompatible pollen tubes are inhibited there. The control may be by alleles at one or two loci. In the case of **sporophytic i.,** alleles of the i. gene can exhibit either dominance of individual action in both male and female organs. Inhibition of incompatible pollen tubes may be on the style or i. may be expressed between gametes after fertilization (post-fertilization i.).

A special system of i. is **cytoplasmic i.** (Jones 1951) or **protoplasmic i.** (Garnjobst and Wilson 1956), which consists in deleterious interactions of the genome or cytoplasm of one species, hybrid, or sex with the cytoplasm of another.

Unilateral i. is the phenomenon whereby the pollen of a population is inhibited from functioning on the plants of another population while in the reciprocal cross no inhibition occurs. It occurs between populations of self-incompatible plants as well as populations of self-compatible plants, both in the cross of self-incompatible × self-compatible and self-compatible × self-incompatible. Unilateral i. is based on an incompatibility gene system.

The failure of fertilization of related plants due to nuclear or cytoplasmic factors is called **prezygotic i.** When fertilized eggs cease to develop at some early stage after fertilization, the phenomenon is called **postzygotic i.**

incompatibility group (Datta et al. 1971) — → plasmid incompatibility.

incorporation error — → gene mutation.

incubation medium recovery — → liquid-holding recovery.

independent assortment — → assortment; genetic segregation.

index case — → proband.

indirect end labeling — a method used for examining the organization of DNA by making a cut at a specific site and isolating all fragments containing the sequence adjacent to one side of the cut, i.e.l. reveals the distance from the cut to the next breaks(s) in DNA.

individual — from a genetic point of view, a unit of life representing either a single cell (unicellular organisms) or comprising all the cellular derivates of one fertilization or all the derivates of the four spores produced by meiosis up to the succeeding fertilization.

individual seletion — → selection.

individuation (Waddington and Schmidt 1933) — in embryonic → development, the process whereby a spatially organized tissue, organ, or whole embryo comes into being as a result of → induction.

individuation field (Waddington and Schmidt 1933) — → embryonic field.

inducer (Jacob and Monod 1961) — an anti-inhibitor (effector) that triggers transcription of specific negatively regulated (→ genetic regulation) genes by binding to a regulator protein (→ operon). Anti-inducers stabilize the → repressor-operator complex.

inducible — of a gene or gene product (→ gene expression) which can be expressed or increased in amount, respectively, when the cell is exposed to an → effector (→ repressible).

inductest — a method for the detection of → mutagenic and potentially → carcinogenic agents by use of → prophage induction.

induction — (1) **enzyme induction** (Monod et al. 1951): The induced synthesis of so-called → inducible enzymes (proteins) following the addition of a metabolite referred to as → inducer. In contrast to → constitutive enzymes, inducible as well as → repressible enzymes fluctuate widely in their amount as a function of changes in the cellular environment. Enzyme i. is probably a common phenomenon found in all types of organisms.

Enzyme induction is said to be "sequential i." if cells that can respond with induced synthesis of a particular enzyme, in the presence of its specific substrate, produce intracellular intermediates that in turn serve as substrates for the i. of certain other inducible enzymes. In this case, the availability of only one substrate may initiate a sort of chain reaction resulting in the successive induction of a series of specific enzymes each of which reacts with only one of a number of related substrates.

(2) **Prophage induction** (Lwoff 1953): The breakdown of the functional integration of the → prophage into the genetic system of a → lysogenic bacterium followed by the vegetative replication and maturation of the phage progeny and the ultimate lysis of the bacterial cell. Prophage i. occurs spontaneously with a low probability, $10^{-3} - 10^{-6}$, per cell division, and can be experimentally increased in practically all cells of a lysogenic population by a variety of physical and chemical agents, or by transfer of prophage DNA to a nonlysogenic cell ("zygotic" or "transfer i."). Prophage i. requires two events: (a) Release of repression (accomplished by the inducing effect), and (b) phage-directed excision of the prophage DNA from the bacterial chromosome.

In more general terms, prophage i. means the initiation of any event which depends on the presence of the prophage in a lysogenic cell (Lieb 1966).

Cross-induction (Borek and Ryan 1958) is the i. of phage development in lysogenic bacteria by the transfer of radiation-sensitive metabolic products from UV irradiated to nonirradiated cells during → conjugation. This transfer is unidirectional from F^+ to F^- cells.

(3) **Zygotic induction** (Jacob and Wollmann 1956): The conversion of a prophage to the vegetative state and the eventual lysis of the "zygote" after transfer of a prophage to a nonlysogenic recipient bacterium. Zygotic i. takes place after → conjugation between a lysogenic (ly^+) → Hfr cell and a nonlysogenic (ly^-) F^- cell (→ F plasmid). Owing to the absence in the F^- cell of the → immunity substance which permits the maturation of lytic phage following the transfer of the chromosome-associated prophage into the F^- cell, this recipient cell is lysed.

Z.i. is recognizable by the appearance of → plaques or lytic centers on plates containing the "zygote" together with a multitude of nonlysogenic cells. Z.i. is limited to crosses of the type Hfr (ly^+) × F^- (ly^-) and does not occur in Hfr (Ly^+) × F^- (Ly^+) or Hfr (ly^-) × F^- (ly^+) crosses (Braun 1965).

(4) **Embryonic induction:** The determination of the developmental fate of one cell or tissue by another, the latter designated as the "inductor". Embryonic i. is the result of an interaction, in which one group of cells (inductor or evocator) evokes and controls the differentiation of other groups of cells. The kind and extent of embryonic i. depend on the embryonic region from which the inductor stems as well as on the region to which it is transferred in the case of transplantation. The inductive stimulus (probably a chemical one) exerted by the inductor seems to result in the activation ("derepression") of the → genetic material of the reacting system and may either impinge directly on the → structural genes ("genotropic action") or on the → regulator gene products ("plasmotropic action").

The propensity of a group of cells or of a tissue to form a particular organ in reaction to embryonic i. is referred to as **"competence"**. During → embryonic development there is a sequential appearance and loss of competence, i.e., time-regulated changing patterns of competence occur. The appearance of competence may be understood as a priming of some genes of the reacting system. Embryonic i. may proceed over a long time span with cumulative effects in the competent tissue, or it may start from a high level of → "predetermination", i.e., very early determinative events. In the latter, the time course is shorter than in the former.

inert — of chromosome segments (usually → heterochromatic) which are supposed to be genetically inactive as judged by the absence of → gene mutations or effects on genetic balance.

infectious unit — the smallest number of virus particles which will cause a lytic infection in a susceptible cell in which the virus can multiply.

infertility — → fertility.

informatin (Krichevskaya and Georgiev 1969) — a protein which is assumed to associate with nascent nuclear RNA and to be involved in its processing (→ heterogeneous nuclear RNA; informofere).

information — → genetic information.

informational macromolecule (Vogel et al. 1963) — any macromolecule (DNA, RNA, protein) with a unique role in determining the properties of living matter (= semantide) by affecting either short-term biochemical reactions, medium-term ontogenetic events, or long-term evolutionary events.

informational suppressor — → suppressor.

informaton (Samarina et al. 1968) — → informofere.

informofere (Samarina et al. 1967) — = → heterogenous nuclear ribonucleoprotein particle.

informosome (Spirin et al. 1964) — = → heterogeneous nuclear ribonucleoprotein particle.

inheritance — the transmission of → genetic information from parents and ancestors to offspring.

(1) **Nuclear or Mendelian inheritance (heredity):** I. as controlled by the replication, mutation, recombination, transmission, and readout of the genetic information stored in form of a → genetic code in particulate, chromosomal → genes whose sum total makes up the genotype (Johannsen 1909). Nuclear i. is based on principles first discovered experimentally by Gregor Mendel (1866) and rediscovered independently by Correns, Tschermak, and De Vries (1900). **Mendel's first law:** When purebred (→ homozygous) strains differing in a particular → character and the corresponding → allele pair (AA and aa, respectively) are crossed, the F_1 individuals (Aa) are uniform (independent of the direction of the cross: either AA × aa or aa × AA) and usually express one of the two characters (that controlled by the → dominant allele A) to the exclusion of the other (the → recessive one) **("principle of uniformity in F_1")**.

Mendel's second law: Recessive characters which are masked in the → heterozygous F_1 arising from a cross between two pure-bred strains, reappear in a specific proportion of the F_2, that is, the members of an allele pair (Aa) separate from each other when an individual forms → germ cells **("principle of → segregation")**.

Mendel's third law: Members of different allele pairs (e.g., Aa and Bb) assort independently of each other when germ cells are randomly formed provided the genes in question are unlinked (→ linkage) **("principle of independent assortment)**.

The basic deductions of "Mendelism" which have been correlated with the behavior of → chromosomes ("chromosome theory of i.") may be summarized as follows: (a) Hereditarily determined characters are controlled by hereditary determinants, or genes, which occupy definite → genetic loci in the chromosomes and are transmitted from parent to progeny (→ genetic information). (b) In → diploids, these genes exist as pairs of alleles in somatic cells. Each germ cell produced by → meiosis receives only one of the two alleles. (c) Fusion of the germ cells at → fertilization is usually random, i.e., germ cells carrying a particular allele do not preferentially unite with others containing particular alleles. (d) Allele pairs controlling different characters are transmitted independently. During germ cell formation, particular genes do not remain preferentially associated, provided the genes concerned are not linked, i.e., are not localized in the same → linkage group or chromosome (→ genetic recombination). (e) When an organism receives during fertilization different alleles for a particular character, only one of these is usually (→ codominant) expressed (the → "dominant") and the other (the "recessive"), though maintained and later transmitted, is not manifest phenotypically.

"Monogenic" or **"oligogenic" i.** is manifest by characters which are mainly controlled by the alleles at one genetic locus (gene). → Variation is discontinuous.

"Multigenic", "polygenic", or **"quantitative" i.** is manifest by characters or components of the → phenotype which are controlled and affected by several interacting genes (→ gene interaction). Variation is continuous.

Unilateral i. (Winge 1927) is associated with → linkage in the → Y-chromosome of the genes in question and implies the resemblance of offspring and parent of the same sex (→ holandric). **Criss-cross** inheritance (Bridges 1913) is associated with linkage in the X-chromosome (sex linkage) of the genes concerned and implies the resemblance in particular characters of sons to their maternal and of daughters to their paternal parent.

Delayed i. (Boycott and Diver 1923) is associated with → predetermination, i.e., the determination of particular characters prior to fertilization by the genotype of the female parent which are manifested in the following generation. The phenotype of a hybrid is thus "matroclinous" in respect to these characters and corresponds to the genotype of the premeiotic oögonium (= "maternal effect").

(2) **Extranuclear, non-Mendelian, or cytoplasmic inheritance (heredity):** I. is controlled by extranuclear, cytoplasmic heredity determinants whose sum total makes up the "plasmotype" (Imai 1936; Jollos 1939), in contrast to the genotype or all hereditary determinants contained in the chromosomes.

Extranuclear i. may be due to → plasmids or the genetic information contained in → mitochondrial and → chloroplast genomes. These genetic determinants in their totality represent the plasmon or cytoplasmon. Extranuclear genetic information is usually transmitted maternally in higher plants. In animals, paternal transmission is relatively rare (uniparental i.). Maternal transmission may be achieved by various mechanisms in different organisms: Input bias due either to differences in organelle content between the gametes or, in isogamous forms, to preferential destruction of organelles or their genomes, with random drift or random sampling to complete the process.

inhomologous — = → nonhomologous.

initial spindle — → spindle.

initiation complex (Nomura and Cowry 1967) — an obligatory intermediate of protein synthesis (→ genetic translation) in pro- and eukaryotes which sonsists of → messenger RNA, programmed with → initiator codons at the 5'-OH end of the smaller ribosomal subunit (→ ribosome), → initiation factors, and → initiator tRNA.

(1) In **prokaryotes**, a complex containing the 30S ribosomal subunit, transfer RNA Met, messenger RNA, and → initiation factors. The i.c. is apparently stabilized by base pairing

between the 3'-terminal sequence of 16S ribosomal RNA (...CCUCCUUA), and a purine-rich nucleotide stretch located in mRNA on the 5'side of the AUG initiator codon. Subsequent to its transformation to a 70S i.c. by addition of a 50S ribosomal subunit, the i.c. is competent to bind the aminoacyl tRNA specified by the second → codon.

(2) In **eukaryotes**, the i.c. is a ternary complex containing the initiator tRNA (Met-tRNA) which reacts with GTP and an → initiation factor (eIF-2); the ternary complex interacts with the 40S ribosomal subunit to form the 40S preinitiation complex. Subsequently, mRNA binds to the preinitiation complex and the 80S i.c. is formed by joining the 60S ribosomal subunit to the 40S intermediary complex in which the $Met-tRNA_f$ is at the P site. This series of reactions serves to decode the initiator codon in mRNA. Chain elongation components then react with the 80S i.c. to translate the internal codons in mRNA in a cyclic series of reactions. Formation of the 80S complex is accompanied by GTP hydrolysis and eIF-2 release (→ ribosome binding site).

initiation factor — (IF); any of the protein factors that associate with the small subunit of the → ribosome at the stage of initiation of → genetic translation (protein synthesis). IF's are requisite for the formation of the → initiation complex (→ elongation factor; termination factor; interference factor). In the presence of IF ribosomes bind to specific → ribosome binding sites on mRNA.

In **prokaryotes**, three IFs have been recognized: (1) the formylmethionine-tRNA binding factor IF-2; (2) the mRNA binding factor IF-3; (3) IF-1 which is involved in the recycling of IF-2.

IF-3 and ribosomal protein S1 bind to the 30S ribosomal subunit at sites close to the 3' terminus of 16S rRNA and S1 interacts specifically with 16S rRNA at a pyrimidine-rich sequence (ACCUCCU).

Table 8. Nomenclature for initiation factors. (After Schreiber et al. 1977)

Nomenclature	Function in natural initiation	*E. coli* equivalent
eIF-1	Stimulation of ($met-tRNA_f·40S$) and ($mRNA·Met-tRNA·40S$)	IF-I
eIF-2	Ternary complex ($Met-tRNA_f·GTP·eIF-2$) ↓ ($Met-tRNA_f·40S$)	IF-2
eIF-3 (Stabilization → $Met-tRNA_f·40S$) →	Ribosome dissociation	
eIF-3, eIF-4A, eIF-4B, eIF-4C →	mRNA binding ($mRNA·Met-tRNA_f·40S$)	IF-3
eIF-5	60S joining → 80S complex GTP hydrolysis, EIF release	IF-2
eIF-2A		
eIF-4D	? elongation? (Mg^{2+} shift)	

In **eukaryotes**, up to now nine IF's have been observed (Tabelle 8) The eukaryotic i.f. eIF-3 is the largest and structurally most complex one; it binds to the small ribosomal subunit and participates in the following functions: (1) Prevention of ribosomal subunit association; (2) stabilization of binding of the ternary initiation complex to the small ribosomal subunit; (3) mRNA binding to the 40S preinitiation complex.

At least five eIF's are involved in peptide chain initiation.

initiation mass (Donachi 1968; Helmstetter et al. 1968) — the cell mass per chromosome origin (initiator) at which bacterial chromosome replication is initiated. The i.m. is constant over a broad range of growth rates. This ensures that the frequency of new rounds of replication and the frequency with which the culture mass doubles will be identical. In a formal sense, this relationship constitutes a mechanism for the control of DNA synthesis.

initiation protein — any of a class of proteins that enable priming of the leading strand of a DNA replication fork at the origin of → DNA replication.

initiator (Jacob et al. 1964) — (1) according to the → replicon model of bacterial DNA replication, a gene product that interacts specifically with a particular site of the replicon, the replicator. The result of this interaction is the initiation at the replicator site of a new round of DNA replication. (2) = a → promoter, i.e., the starting point for the transcription and/or translation process (Margolin and Bauerle 1966).

initiator codon — any → codon (→ genetic code) in → messenger RNA (and its DNA template) that directs initiation of → genetic translation by stimulating the binding of initiator tRNA and by phasing the translation of mRNA (= initiating codon; chain-initiating codon).

In **prokaryotes** (bacteria), either AUG or GUG can act as i. codons. AUG specifies the initiating amino acid N-formylmethionine as well as methionine at internal positions within a polypeptide chain. In **eukaryotes**, AUG is the only initiator and is translated as methionine.

I. codons in prokaryotic mRNA are preceded by a conserved purine-rich sequence (4–8 nucleotides) which plays a role in directing the → ribosome to the correct AUG codon. Comparison of the 5'noncoding regions of eukaryotic mRNA's failed to reveal any sequence homology or conservation, i.e., there is probably no sequence functionally analogous to the bacterial → ribosome binding site (→ scanning hypothesis).

A not formylated methionine-specific tRNA (Met-$tRNA_f$) serves the purpose of bacterial transfer RNA_f^{Met} as initiator tRNA in eukaryotes (→ translational reinitiation; Shine-Dalgarno sequence).

initiator region — = → ribosome binding site.

initiator RNA (Reichard et al. 1974) — (iRNA); an oligoribonucleotide (9–10 nucleotides) of near random sequence which acts a primer (= primer RNA) for the initiation of → Okazaki fragments during → DNA replication. iRNA is covalently linked to the 5' termini of nascent DNA chains and initiates discontinuous synthesis of DNA via Okazaki fragments. iRNA is characterized by the following features: (1) Transient linkage to nascent DNA chains; (2) purines at the 5' terminus; (3) random sequences at the RNA-DNA junction; (4) unique size.

iRNA is synthesized by copying a particular base sequence from one DNA strand and differs from a typical RNA molecule in that after its synthesis the primer remains hydrogen-bonded to the DNA template. In bacteria, two different enzymes synthesize iRNA molecules, a specific RNA polymerase and RNA primase.

The first step in iRNA synthesis is the formation of a complex called preprimosome which (in *E. coli*) contains five proteins (n, n′, n″, DnaB, and DnaC). Protein n′ binds to single-stranded DNA and acquires a bound ATP molecule; then primase joins the preprimosome forming a unit called primosome. The energy of ATP is used by n′ to move the primosome along the DNA until a priming site is found and synthesis of iRNA (5′ → 3′) is initiated. The primer is removed before ligation of Okazaki fragments.

initiator transfer RNA — a unique species of methionine-transfer RNA used for initiation of protein synthesis (→ genetic translation) in prokaryotes, eukaryotes, and in cell organelles. The i.tRNA species of prokaryotes and organelles are, after aminoacylation, formylated and the fMet-tRNA is then used for initiation. The eukaryotic cytoplasmic i.tRNA's are used as Met-tRNA's without formylation.

Both prokaryotic and eukaryotic i.tRNA's have unique sequence and structural features that are absent from → elongator tRNA. Both contain a sequence (in the anticodon stem) of three guanines and three cytosines forming three consecutive G-C pairs. Prokaryotic i.tRNA's share two additional features: (1) Absence of a Watson-Crick base pair at the end of the acceptor stem; (2) presence of a purine (11): pyrimidine (24) base pair instead of a pyrimidine (11): purine (24) bp in the dihydrouridine loop.

The ability of i.tRNA's to interact with the translation → initiation complex is due their unique primary sequence and secondary structure. As opposed to its prokaryotic counterpart, eukaryotic $tRNA_i^{Met}$ becomes associated with the 40S preinitiation complex prior to binding the messenger RNA.

Other properties of i.tRNA's besides their binding to → initiation factors are: (1) i.tRNA's bind directly to the P site on the ribosome, → elongator tRNA's initially bind to the A site; (2) since i.tRNA's do not bind to the A site, they are prevented from inserting methionine into internal peptide linkages.

injection chimera — in animals, a → chimera produced experimentally by injection of embryonic cells of contrasting genetic constitution into the preimplantation blastocyst (→ aggregation chimera).

input load — → genetic load; immigration.

insert — (1) a piece of foreign DNA introduced into a → cloning vector molecule (= insert DNA); (2) any region of noncoding DNA of varying lengths which interrupts the coding region of a eukaryotic mosaic gene at one or several points (= → intron; intervening region).

insertase — an enzyme which can insert a correct base into an → AP site.

insertion — → translocation; DNA transposition.

insertional inactivation — a technique in which foreign DNA is cloned into a restriction site which occupies a position within the coding sequence of a gene in the → cloning vector molecule. Insertion interrupts the gene's sequence such that its original function is no longer expressed.

insertional restoration test (Cronan et al. 1988) — a test which markedly enhances the proper assembly of synthetic genes and can be used to detect → plasmids that contain synthetic DNA sequences of the proper length, termini, and reading frame. The i.r.t. is the reverse of the β-galactosidase α-complementation (blue-to-white) test used to detect the insertion of DNA segments into the polylinker sequence of phage M13 mp, plasmid pUC, and related → cloning vectors. A vector defective (due to frameshift or DNA insertion) in α-complementation is used for insertion (between the cohesive sites generated by digestion of two polylinker restriction sites) of the synthetic DNA segment to restore α-complementation. Formation of blue bacterial colonies requires insertion of a DNA segment of appropriate length and termini to reconstruct the lac Z′ open reading frame and this is more selective than the → insertional inactivation strategy.

insertion element (Hirsch et al. 1972) — (IS element); any of a class of distinct prokaryotic segments of DNA (generally shorter than 2 kb) which are able to transpose to numerous sites on bacterial → plasmids, chromosomes, and bacteriophages. They contain no known genes unrelated to their insertion function (transposase) and exert their effects in cis position only. IS elements are able to modulate gene expression and cause rearrangements of genomes or replicons. Recombination between pairs of identical i. elements can produce insertions, additions, or deletions of large DNA segments. Additionally, i. elements are active units of transposition in larger composite bacterial → transposons and some can mediate cointegration of a replicon carrying an i.e. with a second replicon into structures with two copies of the i.e. at the junction between the two replicons.

IS elements have inverted repeats (10–40 bp) at both ends which are essential for transposition and cointegration. They contain one or more genes encoding proteins required for the recombination events mediated by the element. Upon transposition or cointegration, IS elements generate a duplication of a few base pairs at the target site. Insertion of an IS e. may interfere with local transcription patterns in host chromosomes by causing → genetic polarity (→ insertion mutation).

insertion heteroduplex — → heteroduplex DNA resulting from insertion of a segment of one strand of DNA of one parent into a duplex DNA molecule of the other parent in place of its corresponding segment.

insertion hot spot (Kopecko et al. 1976) — a region of a genome at which → transposable genetic elements preferentially insert. Such sites are usually characterized by the presence of specific → consensus sequences that play a major role in determining insertion site selection.

insertion mutation (Jordan et al. 1968; Shapiro 1969) — a → mutation caused by insertion of a → transposable genetic element in or near a host gene. Insertion of such an element into the coding region of a gene may disrupt the reading frame (→ frameshift mutation) and give rise to an aberrant product (→ genetic polarity); insertion can also affect initiation or termination of transcription, or → RNA splicing when the element inserts into sequences that control these processes.

The phenotype of the i.m. depends upon the particular transposable genetic element, the position (with respect to the gene) at which it has inserted and the genetic background; it can be affected by transacting genes unlinked to the insertion.

I mutations may be unstable and revert by changes promoted by the element. Precise excision will restore the original phenotype; imprecise excision occurs more frequently and is often accompanied by → inversion or → deletion of chromosomal segments within or adjacent to the element. Many of the deletions leave small segments of the element at the site of insertion.

insertion mutagenesis — → mutagenesis due to insertion of a → transposable element in vivo or of synthetic DNA linkers in vitro. A DNA fragment to be used for i.m. should have three properties: (1) Be selectable; (2) abolish the expression of the genetic unit into which it has been inserted; (3) be easy to localize.

insertion sequence — any of the prokaryotic → insertion elements; many are integral parts of a class of → transposable genetic elements known as compound → transposons, i.e., elements composed of two i. sequences that bracket a gene or genes that usually encode one or more proteins conferring antibiotic resistances.

insertion site — (1) a unique → restriction site in a → cloning vector DNA molecule into which foreign DNA can be inserted (= cloning site); (2) the position of integration of a → transposable genetic element.

insertosome (Malamy et al. 1972) — = → insertion element.

in situ nucleic acid hybridization (Gall and Pardue 1969; John et al. 1969) — the annealing (hybridization) of radioactive single-stranded DNA or RNA probes to denatured cellular DNA on microscopic slides and their detection by autoradiography. In situ n.a.h. (= cytological h.) permits the localization of DNA sequences (genes) complementary to the probes in morphologically intact chromosomes. Biotin-labeled probes may also be used for i.s.n.a.h. (higher resolution of hybridization sites and lower background interference than with conventional hybridization).

in situ transcription (Tecott et al. 1988) — a technique achieved within fixed tissues that allows to generate → complementary DNA (cDNA). In situ t. is initiated by → in situ hybridization of an oligonucleotide complementary to a specific messenger RNA, providing a primer required for the polymerase activity of → reverse transcriptase. The priming step allows to select the individual mRNA to be transcribed. Subsequently, the enzyme extends the primer with the mRNA serving as a template. Radiolabeled deoxynucleotides may be incorporated into transcripts during the polymerase reaction, resulting in high specific activity cDNA.

instability (Mather 1953) — a type of variation which appears to be random in direction. It is not genetic in origin and has no identifiable environmental cause. I. is sometimes explained as being due to developmental errors ("developmental noise") arising from sudden changes during development, unconnected with particular environmental influences. "Stability" refers to the opposite condition, i.e., nonoccurrence of variations of the type described above, or more generally, to any condition characterized by the absence of → plasticity (Bradshaw 1965).

integrase (Zissler 1967) — a viral enzyme that catalyzes site-specific recombination involved in the integration of the genome of a → temperate bacteriophage (a prophage) into the bacterial host chromosome (→ terminase).

integration — insertion of a viral or another DNA sequence (→ transposable genetic element) into a host genome as a region covalently linked on either side of the host sequences. I. involves a single → cross over when the molecules are circular, two cross overs when they are linear (→ integrase).

In the case of infection of bacteria with → temperate phages (lysogenic response), the i. system is an → integrase-mediated mechanism for → prophage i. at a specific site in the host DNA. The i. system includes a region required for recognition of the proper prophage site in the host DNA and one or more genes encoding proteins that catalyze the i. process (→ terminase).

integration vector — any → cloning vector that lacks sequences which would enable it to replicate autonomously. The propagation of an i.v. depends on its integration in one of the host chromosomes (→ CEN-ARS vector).

integrative compatibility (Watson and Scaife 1980) — the phenomenon wherein the incompatibility function of some → plasmids is phenotypically switched off when they integrate into the bacterial chromosome. The excised plasmid regains its incompatibility (→ plasmid incompatibility).

integrative recombination — a mode of → genetic recombination which results in the → integration of bacteriophage genomes or → transposable genetic elements into chromosomal DNA.

integrative suppression (Nishimura et al. 1971) — the phenotypic suppression of a bacterial temperature-sensitive mutation by integration of a → plasmid at various sites on the chromosome. Initiation of replication then takes place at the plasmid initiation site, i.e., the entire chromosome becomes part of the plasmid → replicon.

integrator gene (Britten and Davidson 1969) — a (hypothetical) positive acting → regulatory gene which mediates the induction and concomitant expression of two or more noncontiguous → structural genes of eukaryotes (→ sensor gene).

integrin (Hynes 1987) — any member of a protein superfamily consisting of heterodimers of noncovalently associated transmembrane glycoproteins (cell adhesion receptors). Integrins play a crucial role in → cell adhesion.

intensifier — → modifier gene.

interaction deviation — → variation.

interaction of genes — → gene interaction.

interallelic complementation — → genetic complementation.

interarm fiber — any of the DNA-containing fibers as seen in electron micrographs of whole mount mitotic chromosomes which cross at the → centromere and at the ends of arms in chromosomes and in some cases all along the length of the sister chromatids.

interarm pairing — → chromosome pairing between the two identical, completely homologous arms of one → isochromosome (= internal pairing).

interband (Painter 1939) — → band.
interbreeding — of individuals capable of actual or potential gene exchange by hybridization. I. is the process that holds individuals together in → populations, and populations together in subspecies (races) and species.
intercalary — ref. to chromosome segments not terminally localized and distinguished from other segments terminal in position.
intercalating agent — any planar molecule which is able to insert itself between adjacent base pairs in a double-stranded DNA (or RNA) molecule. Such agents may inhibit → DNA replication and → genetic transcription, promote the curing of plasmids from their hosts, and reduce the buoyant density of DNA in solution.
intercellular adhesion — the occurrence of cellular recognition and adhesion by a multistep process. I.a. plays an important role in morphogenesis, intercellular communication, regulation of growth and cell motility, and in malignancy.
intercellular invasion — the active migration of cells of one type into the interiors of tissue consisting of dissimilar cell types. A necessary condition for a cell type to be invasive to a given host tissue is that it lacks → contact paralysis of locomotion during collision with cells of that host tissue.
intercellular junction — any of the specialized regions of contact between the membranes of adjacent eukaryotic cells. They are of different types (→ gap junction; tight j.; synaptic j.; desmosome; zonula adherens) and serve to integrate and coordinate the various kinds of cellular interactions. There are three functional categories of i. junctions: (1) **Impermeable junctions** that enable an organism or organ to maintain an internal environment which is chemically distinct from its surroundings; (2) **adhering junctions** that promote adhesion between the cells of a tissue, reinforcing its physical integrity; (3) **communicating junctions** that enable cells to exchange nutrients and signal molecules and thereby coordinate their activities.
intercentric (Sears and Camara 1950) — ref. to the intercentromeric segment localized between the two → centromeres of a → dicentric chromosome or chromatid.
intercept (Franke and Scheer 1978) — the region of a → chromatin strand between any two given morphologically identified points (usually between two transcriptional complexes).
interchange (Belling 1925) — = reciprocal → translocation.
interchange trisomic (Belling 1927) — → trisomic.
interchange quadrivalent — in heterozygotes (heterokaryotypes) for a reciprocal → translocation (an interchange) involving two nonhomologous chromosomes, an association of four chromosomes at first meiotic division. I.qu. formation depends on the position of breakpoints and the location of → chiasmata.
interchromomere — any of the regions (more or less uncoiled fibrils) connecting adjacent → chromomeres in the chromosomes. In → polytene chromosomes of Diptera, interchromomeres are called "interbands".
interchromosomal — ref. to effects, phenomena, or processes in which two or more chromosomes participate such as, e.g., a chromosomal structural change in one chromosome influencing the gene action (→ position effect) or crossing over frequency in another chromosome, as opposed to intrachromosomal.
intercistronic divide (Rechler and Martin 1969) — = → intercistronic region.
intercistronic region — (1) the DNA sequence separating two genes; the i.r. may contain regulatory signals. (2) The nucleotide sequence between the → terminator codon of one gene and the → initiator codon of the next gene in a polycistronic transcription unit (messenger RNA). Usually, such regions consist of only a few base pairs; in cases where the initiator codon of one gene overlaps the terminator codon of the previous gene, the i.r. is absent. In other cases, the i.r. may be up to several hundred base pairs long. I. regions are usually not translated. (3) Any untranslated region of the genome of single-stranded DNA phages.

These regions, or their transcripts, have the potential to form hairpin loop structures. (= Intercistronic divide; i. stretch; i. space).

intercistronic stretch (Streitz 1969) — = → intercistronic region.

intercompartmental gene transfer (Börner 1984) — → gene transfer between different compartments (nucleo-cytoplasmic, mitochondrial, plastidal) of the eukaryotic cell (→ promiscuous DNA). Two mechanisms of DNA transfer are possible: (1) Some kind of vector, such as a → transposable genetic element; (2) fusion of the compartments and sequence transfer.

interdeme selection (Wright 1956) — → group selection.

interference (Muller 1916) — (1) **Chromosome interference** (Muller 1916): In eukaryotes, the effect of one → crossing over event in either decreasing **(positive chromosome i.)** or increasing **(negative chromosome i.)** the probability that a second cross over will take place in its vicinity. In both positive and negative chromosome i., the frequency of multiple (double) crossovers is not in accordance with a random distribution of subsequent cross overs. Chromosome i. is also called "chiasma i." (Mather 1933), "chiasma position i." (Carter and Robertson 1952), "cross over i.", "cross over position i.", or "position i." (Whitehouse 1965).

Strengths of chromosome i. are commonly measured as coefficients of → coincidence (Muller 1916), i.e., as the quotient of the number of observed double cross overs and the number expected from the Poisson distribution:

$$K = \frac{\text{actual frequencies of doubles}}{\text{expected frequencies of doubles}}.$$

Coincidence varies inversely with interference, and coincidence values usually range from o (complete positive i.) to 1 (no i. whatsoever). In those cases where the occurrence of one crossover increases the probability of a second nearby (localized negative i.), the number of double crossovers is in excess of the random expectation and the coincidence values are greater than 1. It is improbable that this kind of i. has anything to do with chiasma formation (as is assumed in positive i., hence "chiasma i."). One explanation for localized negative i. is the assumption that pairing of linkage structures, which actually gives rise to recombination (so-called "effective pairing"), is usually restricted to short regions which pair discontinuously and within which the probability of recombination is high. According to this hypothesis, recombination in two intervals of the linkage structures is greater than expected on the basis of random probabilities, provided these intervals are so close together that they frequently belong to the same effectively paired region.

(2) **Chromatid interference** (Mather 1933): The nonrandom participation of the four chromatids of two homologous chromosomes in successive crossovers. This type of i. results in a deviation in the frequencies of so-called two-strand, three-strand, and four-strand double crossovers (→ crossing over) from the 1:2:1 ratio expected if crossing over is random with respect to the strands (chromatids) involved in two successive crossovers. Positive chromatid i. increases four-strand double crossovers, whereas negative chromatid i. increases two-strand doubles. Positive i. of this type has not yet been found; negative chromatid i. has been reported in some fungi. Chromosome i. influences the crossover frequency, chromatid i. the mode of crossing over. Both types of i. are usually connected with changes in the recombination frequency as compared with random expectation.

(3) **Interchromosomal interference:** The compensation of a decreased crossover frequency (positive chromosome i.) in particular paired chromosomes (which may result from, e.g., heterozygosity of chromosomal structural changes) by an increased crossover frequency (negative chromosome i.) in others.

(4) In population biology, a type of competition, distinct from → exploitation, which occurs when the competitors reduce one another's → fitness in nonexploitative manners (cannibalism and other forms of killing and interactions that reduce survival or reproduction qualify as i.).

interference distance (Mather 1936) — the average distance within which successive cross overs and chiasmata are formed after the first has been produced in the → differential distance from the centromere. The i.d. is genome-specific and controlled by chromosome → interference.

interference factor (Groner et al. 1972; Lee-Huang and Ochoa 1972) — any of a class of protein factors (translational repressors) which change the cistron specificity of → initiation factor IF-3 during → genetic translation in prokaryotes. I. factors affect the recognition of mRNA initiation sites by the ribosome and stimulate the translation of certain cistrons while inhibiting that of others (cistron specificity of action).

interference range — the minimal genetic distance sufficiently large for two crossing over events not to interfere with each other, i.e., where their coincidence value becomes 1. The i.r. seems to be different in different linkage groups and in different regions of the same linkage group.

interfering protein (Revel et al. 1973) — = → interference factor.

interferon (Isaacs and Lindemann 1975) — any of a family of species-specific glycoproteins (cytokinins) produced by almost all vertebrates when induced by viral infection or other inducers such as double-stranded RNA and certain nonviral microorganisms. The proteins exert antiviral, antiproliferative, and immunomodulatory activities in target cells and mediate their diverse biological responses by inducing a limited set of genes in the responsive cells and promotion of specific mRNA's which are then translated into so-called antiviral proteins. Different interferons are distinguished according to their cellular origin and to the inducing agent.

intergenic (Muller 1941) — of → mutations (chromosomal structural changes or → chromosome mutations) involving more than one gene, in contrast to intragenic mutations (→ gene mutation).

intergenic conversion (Slightom et al. 1980) — localized information transfer between nonallelic genes of related nucleotide sequence by a mechanism analogous to → gene conversion. It occurs between and within chromosomes in mitosis and meiosis and provides a mechanism of → gene correction (→ concerted evolution). Information transfer seems to be initiated by a single sequence hot spot of variable length which may be located in a noncoding → intron.

intergenic DNA — any of the DNA sequences (= i. sequence; i. region; i. spacer; intercistronic region) flanking the eukaryotic genes 3' and/or 5'. The function of i.DNA is probably due to structural features such as base composition, pyrimidine runs, or sequence length. I.DNA may be an untranslated region or may encode functions that regulate → gene expression at the level of either transcription or post-transcriptional processing. Long noncoding i.DNA sequences tend to make co-transcription of genes inefficient.

intergenic spacer (Coen and Dover 1983) — = → intergenic DNA.

intergenomic coordination — the functional coordination of the mitocondrial and nucleocytoplasmic or the chloroplast and nucleocytoplasmic genetic systems in eukaryotes.

intergenote (Bernheimer et al. 1967) — in hybrid bacterial strains, a chromosome with integrated DNA of heterologous origin (→ heterogenote).

intergradation (Mayr 1942) — a descriptive term for the coming into contact and hybridization of phenotypically different populations or groups of populations along a zone of primary or secondary contact, thus giving rise to character gradients (clines).

Primary intergradation: Takes place while all the populations involved are in continuous contact. The populations are interconnected by gradients of phenotypic characters and → gene frequencies. The gradients developed gradually.

Secondary intergradation (→ introgression): The populations now connected by a character gradient were originally completely separated and came into contact anew. Since their reproductive isolation is incomplete, on becoming sympatric, gene exchange occurs. In this case, the character gradient usually has a steep slope. Zones of secondary i. are hybrid belts or → hybrid zones between populations which had become differentiated during a preceding period of isolation. They are usually characterized by a high degree of individual variation (Mayr 1963).

intergradation index (Ginsburg 1954) — a measure of the range of variability in all populations of a species with respect to a particular character. The total variability of the character under observation is referred to as the "divergency index". The indices may be calculated mathematically and are used in species diagnosis.

intergroup selection — an evolutionary process due to differences in growth rate or survival of different competing or noncompeting groups of individuals (→ group selection).

interkinesis (Gregoire 1905) — a short "resting stage" which may occur between the first and second meiotic division (→ meiosis). In contrast to the mitotic interphase, there is no chromosome reduplication during i. Each chromosome consists of two chromatids, which do not lie parallel but diverge due to a lapse of attraction. The regions where the sister chromatids are in association are localized to both sides of the → centromere.

interlocking (Mather 1932) — (1) **True i. of meiotic chromosomes:** The intertwisting of different (nonhomologous) chromosomes which ensues during meiotic pairing (→ chromosome pairing) if a foreign chromosome passes through a loop between associations of two homologous chromosomes. True i. takes place during zygotene-pachytene of the first meiotic division and chiasma formation is a prerequisite for its maintenance up to metaphase I.

According to Darlington (1937), simple i. in a species with → chiasma terminalization may be classified at metaphase into three types according to whether the ciasmata formed by the interlocked chromosomes are moving away from the point of i. in both bivalents, in one, or neither. (a) **Proximal i.:** The chiasmata move away from the point of i. in both pairs of chromosomes, since the i. is within the centromere loop. (b) **Distal i.:** The chiasmata on either side of the i. move in the same direction, i.e., towards one end, in both of the interlocked pairs, owing to the i. being distal to the most proximal chiasma. (c) **Proximal-distal i.:** The interlocking is proximal with respect to one pair of chromosomes and distal with respect to the other. These different types of i. [in practice types (b) and (c) are not usually distinguishable and both may be described as distal] may occur between pairs of chromosomes (bivalents) and between separate chromosomes of multiple rings, in cases of heterozygosity for reciprocal → translocations or in polyploids.

Another type of i. is known as "double" i., i.e., two successive loops of a bivalent have chromosomes interlocked with them. Under these circumstances each loop separates the original parental chromosomes.

(2) **Spurious i. of meiotic chromosomes:** I. characterized by a whole bivalent passing through a loop between the pairing partners of another bivalent.

(3) **I. in dicentric chromosomes:** I. during mitosis in → dicentric chromosomes with long intercentromeric segments resulting in single or double anaphase bridges (→ chromosome bridge). This type of i. is due to single ("criss-cross separation") or multiple torsion ("interlocked separation") in the intercentromeric region.

intermediary — of characters controlled by a → heterozygous pair of alleles (aa′) which are intermediate in phenotype in comparison with the phenotypes controlled by the corresponding homozygous pairs of alleles (aa and a′a′). Intermediacy is a consequence of incomplete dominance (→ semi-dominant).

intermediate filament (Ishikawa et al. 1968) — any of a class of compositionally different but structurally similar, long, unbranched cytoplasmic protein filaments (diameter 7–12 nm) that have been implicated in various intracellular roles such as nuclear anchoring, organelle interactions, and gene expression. I. filaments are distinct from the → microfilaments and the → microtubules; all three are components of the → cytoskeleton. At least five subclasses of i. filaments (generally associated with particular cell types or origins) may be distinguished in various cell types of higher vertebrates: (1) Tonofilaments in epithelial cells composed of several different polypeptides (cytokeratins) (2) filaments in fibroblasts (and a variety of other cell types) consisting of a protein called vimentin; (3) filaments in muscle cells containing desmin; (4) glial filaments in astrocytes containing glial fibrillary acidic protein (GFAP); (5) neurofilaments in peripheral and central neurons consisting of the proteins collectively called the neurofilament triplet.

A common feature of all i. filaments is that they are phosphorylated. More than one class of filaments can coexist within the same cell; the same i.f. may contain more than one subunit protein.

Unlike other cytoskeletal structures, which are largely dissolved when cells undergo mitosis, i. filaments are retained but appear to be reorganized: They form a cagelike structure around the mitotic → spindle in early (animal) mitosis.

intermitotic — ref. to the interphase of the mitotic → cell cycle.

internal (Darlington 1937) — (1) ref. to an intrachromosomal structural change (e.g., → inversion) in contrast to → external structural changes, in which two or more nonhomologous chromosomes are involved; (2) of meiotic pairing (→ chromosome pairing) between the two homologous arms of an → isochromosome; (3) of a coil (→ chromosome coiling) within the single chromatid between mitotic pro- and anaphase, or at meiosis jointly for two sister chromatids, as distinguished from the "external coil" (including relic and rational coils) which is recognizable when the internal coiling is hidden (Darlington 1935).

internal balance — → coadaptation.

internal control region (Kressmann et al. 1979) — (ICR); in all eukaryotic genes transcribed by → RNA polymerase III, a small DNA sequence which directs upstream initiation by the enzyme (= intragenic c.r.). ICR's occur within the DNA encoding the mature RNA. Specific → transcription factors appear to bind initially to the 3' region of the ICR. Transcription of eukaryotic tRNA genes is controlled by two i.c. regions, the D-control r. (which in the tRNA codes for the D-stem and -loop) and the T-control region (which in the tRNA codes for the t_4T loop).

internal guide — → self-splicing.

internal guide sequence (Davies et al. 1982) — (IGS); in group I → introns, a sequence pairing with exon sequences adjacent to both the 5' and 3' → splice sites. The IGS usually begins with 20 nucleotides of the 5' splice site. IGS · 5' exon pairing is important in both steps of → RNA splicing.

internal nucleolar matrix — in many animal cell types, a fibrogranular network which extends inwards from the lamina and includes compact residual nucleolar masses (the nucleolar matrices or cores).

internal promoter — in eukaryotic 5S rRNA and tRNA genes (which are transcribed by → RNA polymerase III), a → promoter comprised of two centrally located, but physically separated, boxes (A and B). These internal control regions are preferentially exposed in → chromatin and appear to have an altered DNA conformation.

internucleosomal spacer — a linker DNA sequence (40–80 bp) connecting the → nucleosome cores. The i.s. is complexed with → histone H1 (H5). The length of the i.s. may vary between species and cell types (in different functional chromatin regions).

interorganellar gene transfer — transfer of DNA sequences (or RNA intermediates followed by local reverse transcription) between nuclear, chloroplast, and mitochondrial genomes (→ promiscuous DNA) that result in the presence of these sequences in more than one cellular genome.

interphase (Lundegardh 1912) — the part of the eukaryotic → cell cycle during which the → chromosomes are more or less decompacted and enclosed in the → nuclear envelope (i. nucleus). On the basis of chromosome replication, the i. is conventionally subdivided into the → G1, S, and G2 subphases. I. chromosomes are attached to the nuclear envelope via specific chromosome segments such as → centromeres, → telomeres, and transcriptionally active genes (→ genetic transcription). Ordered arrangement of chromosomes in i. cells is apparently one way in which genetic activity is coordinated.

The transition from i. to → mitosis is marked by the reorganization of the i. → microtubule network and eventually results in the formation of the mitotic (meiotic) → spindle (→ microtubule organizing center).

interphase nucleus — → interphase and → nucleus.

interrupted mating — a technique used to map (→ genetic mapping) bacterial genes by determining the sequence in which donor genes enter recipient cells during bacterial → conjugation (→ F plasmid).

intersex (Goldschmidt 1915) — any individual of a unisexual (dioecious) species whose reproductive organs and/or secondary sex characters are partly of one and partly of the other sex without showing genetically different parts (→ gynandromorph), i.e., sex intergrades who are neither clearly male nor clearly female but represent a mixture of male, female, and intermediate characters. Intersexes are generally infertile or at most produce mature gametes of one sex only. Hence it is always possible to distinguish them from → hermaphroditic individuals, which produce mature gametes of both sexes. Intersexes may start their development in conformity with the genetically determined sex and then switch over to the other sex. The degree of intersexuality is determined by the timing of the switchover in differentiation.

Intersexes may arise (in mammals) by: (1) Aberrations of genetic and chromosomal origin; (2) aberrations of gonadogenesis; (3) reversal of the sex of the accessory genital organs; (4) gene mutations.

interspecific competition — competition between → sympatric species having similar ecological requirements and utilizing some particular limited resource. I.c. is either consumptive competition, interference c., or a combination of both. I.c. leads to → interspecific selection and may result in species replacement.

interspecies selection — = → species selection.

interspersed repetitious DNA — in eukaryotes, a class of → repetitious DNA sequences interspersed throughout the nuclear → genome. Two groups of i.r. DNA sequences may be distinguished: (1) SINE's, i.e., repetitive sequences usually less than 500 bp in length and occurring on the order of 10^5 times within the genome; (2) LINE's, i.e., long interspersed repetitive sequences several kilobases in length. Some may be → retroposons.

Eukaryotes contain representatives of both types of DNA in their genomes in different relative proportions. SINEs and LINEs may consist, at least in part, of → transposable genetic elements; these are transcribed (by → RNA polymerase II) into abundant and heterogeneous species of RNA. Possibly LINE's are involved in the coordinate regulation of → gene expression.

interspersion — a more or less regular pattern of an alternating arrangement of → repetitious DNA and nonrepetitious DNA sequences in the nuclear → genome of eukaryotes (→ DNA interspersion; interspersed repetitious DNA).

intersterility — cross-sterility confined to specific genetically determined groups of individuals ("intersterility groups"). Under these circumstances, a group of individuals (A) may be incapable of mating with another group (B), but may successfully mate with group C, which in turn may be unable to mate with group D. A and B on the one hand, and C and D on the other are then said to represent intersterility groups.

interstitial distance (Mather 1936) — the distance between successive chiasmata in those cases where more than one is present in a bivalent (→ differential distance).

interstitial segment — in the case of heterozygosity for a reciprocal → translocation between two nonhomologous chromosomes, any of the four chromosome regions between the → centromeres and the loci of exchange within the characteristic cross-configuration, consisting of four partially homologous chromosomes and produced by strictly homologous meiotic pairing (→ chromosome pairing), as distinguished from the so-called "pairing segments", i.e., the chromosome arms not involved in the translocation, together with the regions distal to the points of exchange in those arms which are involved (→ differential segments)

intertrabecular space — → microtrabeculae.

intervening sequence (Jeffreys and Flavell 1977) — = → intron.

interzonal connection (Schrader 1932) — any connection between the chromatids during their separation in mitotic anaphase (= connecting fiber or interzonal fiber).

intrabreeding — mating only between individuals of the same population (= endogamy).

intracellular promiscuity (Ellis 1982) — → intercompartmental gene transfer; promiscuous DNA.

intracellular protein topogenesis (Blobel 1980) — those intracellular processes which result (concomitant with or shortly after protein synthesis) in the unidirectional transfer of proteins across membranes or in their asymmetrical integration into translocation-competent membranes (→ signal peptide).

intrachange — any → exchange of segments within one chromosome (either within the same or between chromosome arms) resulting in chromosomal structural changes (→ chromosome mutation), as distinguished from the interchange of segments between two (or more) chromosomes. The units of structural change in case of i. and interchange may be both chromatids of the chromosome or only one chromatid.

intrachromosomal — → interchromosomal.

intracistronic complementation (Jackson and Yanofsky 1974) — a type of → complementation between mutant variants of certain genes (cistrons) encoding catalytic enzymes, repressor type proteins, or membrane proteins. The functional form of the proteins often consists of two or more homologous polypeptide chains (homopolymers).

Complementation is the result of the combination of mutant polypeptides (a hybrid protein) in which each mutant chain provides or corrects the conformation and the site(s) of activity of the protein.

intracodon recombination — → genetic (intragenic) recombination between adjacent nucleotides within a → coding unit (a nucleotide triplet).

intragenic (Muller 1941) — → intergenic.

intragenic promoter — a → promoter located within the coding region of a gene. Such promoters are characteristic of genes coding for eukaryotic tRNA, 5S RNA, and adenovirus-associated RNA, as well as some families of short dispersed repetitious nucleotide sequences (all transcribed by → RNA polymerase III).

intragenic spacer — = → intron.

intragroup selection — → selection which acts on the individuals within a group of organisms (→ intergroup s.).

intrahaploid pairing (Darlington 1937) — meiotic → chromosome pairing between chromosomes of the basic (haploid) chromosome set. I.p. may be homologous pairing between duplicated chromosome segments or (less often) nonhomologous torsion pairing of genetically and structurally nonhomologous chromosomes or chromosome segments. I.p. of the homologous type may give rise to chromosomal structural changes by → crossing over between duplications.

intranuclear — → extranuclear.

intranuclear virus — any of those DNA or RNA virions the major components of which are formed or assembled inside the cell → nucleus, giving rise to the accumulation of either complete or incomplete virus particles in this organelle.

intranucleolar chromatin — → chromatin within the → nucleolus and connected with the → perinucleolar chromatin by septalike structures. The states in which i.c. has been observed are: (1) Chromatin filaments (diameter 2 nm) which are dispersed or present in chromatin clusters; (2) fibrils of various diameter which can be dispersed or organized in clusters that represent condensed chromatin.

introgression (Anderson and Hubricht 1938) — the incorporation of genes of one species into the → gene pool of another species by hybridization ("introgressive hybridization") and backcrossing. According to Anderson (1953), sympatric i. (in which old and new genotypes produced by i. coexist in the same populations) and allopatric i. (in which new genotypes produced by i. exist separately from the old genotypes) may be distinguished.

intron (Gilbert 1978) — any of those sequences (= intervening sequence) in eukaryotic (and archebacterial) → mosaic (or split) genes that interrupt the coding sequence (exons) and are transcribed as part of a precursor RNA (pre-messenger RNA, pre-ribosomal RNA or pre-transfer RNA). Excision of introns and ligation of → exons is accomplished by an RNA processing reaction called → RNA splicing.

Based on conserved nucleotide sequences and/or RNA structures within or adjacent to introns, introns may be categorized into four major categories (Cech 1988): (1) Group I introns; (2) group II introns; (3) nuclear mRNA introns; (4) nuclear tRNA introns. Each of these categories of introns has its own RNA splicing mechanism (see Table 13 p. 438.). Introns have been observed within genes encoding messenger RNA, ribosomal RNA, and transfer RNA (nuclear, mitochondrial, chloroplast, and bacteriophage); they are essentially absent from eubacterial genes and rare in lower eukaryotes.

Three alternative models have been proposed concerning the time of i. appearance during evolution: (1) Introns existed in the progenitor genes but were lost from prokaryotes; (2) introns existed in the progenitor genes, but i. insertion or deletion may occur later during evolution; (3) prokaryotic genes resemble the progenitor genes, and the addition of introns occurred at or after the emergence of eukaryotes.

According to present evidence, introns probably existed before the divergence of prokaryotes and eukaryotes (models 1 and 2). They may have been created by reversal of the RNA splicing reaction, i.e., insertion of a self-splicing i. into an mRNA, which could then have integrated in DNA after → reverse transcription. A second view on the origin of introns suggests that they arose by insertion of → transposable genetic elements into preexisting genes. The length of introns varies between 50 and 10,000 bp (in most cases the upper limit of i. length is about 600 bp) and they frequently divide protein-coding genes into regions (exons) that encode functional domains or subdomains in proteins.

The great majority of introns appears to be incapable of coding for proteins; a few encode proteins of their own, which in some cases are involved in precise excision of the i. from the precursor RNA or may be able to code for reverse transcription (→ intronic or nested genes).

Removal of introns by splicing sometimes allows for multiple protein products from a single gene by alternative processing of precursor RNA (alternative splicing); in some cases, the existence of introns also appears to have allowed more rapid evolution of proteins by mechanisms such as → exon shuffling.

Nearly all mRNA introns are bounded at the 5' end by a conserved sequence of nine nucleotides called a 5' splice site and begin with the dinucleotide GU (5'-exon'AAG/GU-GAGU-intron-3'). At the 3' end they are bounded by a 3' splice site that is a pyrimidine-rich region of about 11 nucleotides followed by N-Y-A-G (N is any nucleotide, Y is a pyrimidine).

Mitochondrial introns do not obey the GU...AG rule and thus belong to a special class of introns, together with all chloroplast introns and those nuclear-encoded introns that interrupt structural RNA genes. Fungal mitochondrial and other introns have been classified into group I and group II introns on the basis of conserved sequence elements and potential secondary structures: (1) In group I introns, a 16-nucleotide-long consensus sequence
CC GA
(GUUCAACGACUACANG), which is always far away from the splice site in the primary se-
U
quence, is probably needed for RNA splicing; (2) group II introns include consensus sequences GUGCG and $\overset{\text{UAUU}}{\text{CGCCUAC---AU}}$ at the 5' and 3' ends, respectively, and a characteristic 14 bp potential hairpin containing a bulged guanosine on its 3' side and located within the last 100 residues on the 3' terminus of the i.

Both groups of introns contain open reading frames (ORF's), which are in most cases in frame with the upstream exons.

Yeast introns are distinguished from mammalian introns principally by the possession of a conserved sequence termed UACUAAC box which specifies the branch point for the lariat (→ RNA lariat) and to which U2 (→ small nuclear RNA) binds by complementary base pairing.

intron binding site — → RNA self-splicing.

intron-exon junction — a DNA sequence at the frontiers of → introns and the exons (the 5' and 3' splice sites) of mosaic genes, i.e., in those regions at which introns interrupt the coding portions of the genes. These regions constitute the basis for specifying the site of → RNA splicing. Sequence homologies exist at the i.-e. junctions of protein-coding genes from diverse organisms; the tRNA i.-e. boundaries differ from these, indicating that different splicing activities for the processing of an intron exist.

intron-exon junctional sliding — a hypothesis assuming that translation of → intron-exon junctions in mosaic genes along the sequence may be a mechanism to account for peptide sequence length variation within protein families.

intron splicing — the precise splicing out of the initial transcript of → introns before the RNA is translated (→ RNA splicing).

intron transfer (Dujon 1980) — the movement by site-specific recombination of certain mobile group I and group II → introns from intron$^+$ to intron$^-$ alleles. I.t. is correlated with a double-strand break in the recipient DNA and is dependent upon the product of the intron open reading frame (ORF) which codes for a site-specific DNA endonuclease.

intronic gene — a gene located (nested) inside an → intron and surrounded by noncoding regions.

inversduplication — → duplication.

inverse polymerase chain reaction (*Ochman* et al. 1988) — (IPCR); → polymerase chain reaction.

inversion (Sturtevant 1926) — (1) The alteration of a DNA molecule by removing a fragment, reversing its orientation, and putting it back into its place (→ DNA rearrangement). Since the backbones of the two complementary DNA strands have opposite polarity, an inverted piece of DNA can be fitted into its original double helix only after having rotated round its own axis.

Site-specific i. regulates → gene expression in bacteria (expression of genes coding for different flagellar antigens) and some bacteriophages (regulation of host range) and occurs between two crossover sites located within the inverted repeat sequences that flank the invertible DNA segment. (2) In eukaryotic chromosomes, a → chromosome mutation of the "intrachange" type (→ chromosome mutation), characterized by the reversal of a chromosome (or chromatid) segment, and the gene sequence contained therein, relative to the standard arrangement of the linkage group in question. There is no experimental evidence for the existence of terminal inversions. According to the "breakage-reunion model", intercalary inversions are the result of two breaks (referred to as inversion points) in one chromosome, 180° rotation of the segment localized between them and reunion at the sites of breakage.

The following i.-types may be distinguished, according to the number of inverted segments within a chromosome and the location of the i. points with respect to each other:

(1) **Single inversions:** Only one segment of the chromosome is inverted. (a) **Paracentric inversions:** Both i. points are located in the same chromosome arm (= acentric, dyscentric, parakinetic, asymmetrical inversions). (b) **Pericentric inversions:** The two i. points are located in different chromosome arms and the inverted segment includes the centromere (= transcentric, transkinetic, symmetrical inversions).

(2) **Complex inversions:** Several segments of the chromosome are inverted. (a) **Independent inversions:** The inverted segments are separated from one another by an uninverted segment (a b c d e f g h → a/cb/de/gf/h). (b) **Direct tandem inversions:** Two inverted segments are

mutually adjacent (a b c d e f g h → a/cb/ed/fgh). (c) **Reversed tandem inversions:** Two inverted segments are adjacent but mutually interchanged (a b c d e f g h → a/ed/cb/fgh). (d) **Included inversions:** A segment within an inverted segment is inverted again (a b c d e f g h → a/gfedcb/h → a/gf/de/cb/h). (e) **Overlapping inversions:** Part of an i. is inverted a second time together with a segment which was not previously inverted (abcdefgh → a/edcb/fgh → a/ed/gf/bc/h).

Complications in meiosis arise in conjunction with heterozygous inversions, i.e., if a structurally normal homologous chromosome is present in addition to the i. chromosome. The meiotic pairing behavior of the i. chromosome and its structurally normal partner depends on the length of the i. and the longitudinal relationship of the inverted and uninverted chromosome segments. If the i. is long, chromosome pairing involves the formation of a retrograde loop in the homologous, structurally unchanged chromosome; the i. chromosome affixes itself to its partner in such a way that homologous loci pair with one another. (Similarly, loop formation occurs in i. heterozygosity in the somatic pairing of → polytene chromosomes of Diptera).

If the inverted segment is so short that loop formation is not possible, either the i. segment remains unpaired or pairing of nonhomologous segments may take place. If the i. segment is very long, it may pair without loop formation and the uninverted terminal segments then remain unpaired.

Crossing over and chiasma formation within (i. chiasma) and outside the inverted segment give rise to secondary structural changes (duplications and deletions), depending on (1) the i.-type, (2) the number of chiasmata, and (3) the localization of the chiasmata. Such structural changes lead in turn to meiotic products (gones) with unbalanced sets of chromosomes.

Inversions may be detected in the case of i. homo- and heterozygosity in various ways, using genetic and/or cytological methods: (1) By detecting changed → linkage relations with the aid of genetic linkage studies; (2) by the detection of eventual changes in chromosome morphology during the metaphase of mitosis; (3) by observing changes in the band pattern of → polytene chromosomes.

The following features are diagnostic of inversions in the case of i. heterozygosity: (1) I. loops during the prophase of meiosis; (2) formation of bridges, fragments, and loop univalents in the anaphase of meiosis; (3) aberrant meiotic products which may be uncovered by means of → tetrad analysis; (4) decreased fertility resulting from the production of genetically unbalanced meiotic products via → crossing over.

Inversions often affect the crossing over frequency. In this context a distinction is made between: (1) Effects within the i. bivalents (according to experiments of Sturtevant and Beadle 1936 on *Drosophila*). (a) Heterozygous inversions lead to a localized crossing over reduction in the i. bivalent which may be only partly explained in terms of the elimination of crossover chromatids with duplications and deletions. (b) In the case of heterozygosity, the reduction in crossing over is greater for short inversions than for long ones, i.e., the reduction of the crossing over frequency is inversely proportional to the length of the inverted segments. (c) Outside the inverted segment, reduced crossing over occurs only in the neighborhood of the i. points. (d) The crossing over reduction occurs equally with heterozygous paracentric inversions and pericentric inversions.

(2) Interchromosomal effects of inversions:

The presence of an i. in the chromosome complement and the associated reduction of crossing over may result in an increased crossing over frequency in the remaining heterologous bivalents. An increased crossing over in the region of the centromere and at the chromosome ends is characteristic of the so-called "Schultz Redfield effect". The resultant crossing over frequency within the chromosome complement is usually unchanged. As a rule, the reduced crossing over in the i. bivalent is compensated by an increase (conditioned by the Schultz-Redfield effect) in the remaining bivalents.

Inversions represent a factor in → karyotype evolution. Their occurrence may result in → chromosome polymorphism within a population especially among insect species (e.g. *Drosophila*), less often in plants. The result is that in addition to a heterozygous i. type (heterokaryotype) two homozygous forms (homokaryotypes) occur: one containing the standard pair of chromosomes, the other the inversion pair.

The general evolutionary significance of inversions is as follows: Because of the extensive crossing over reduction within heterozygous i., balanced gene complexes are built up ("super-genes"). These frequently confer a greater biological → fitness upon the heterokaryotypes. In comparison with the homokaryotypes, the heterokaryotypes often prove to be heterotic (→ heterosis) and homeostatic (→ homeostasis) with respect to characters which favor their → adaptation. If the karyotypes with inversions prove to be adaptable to certain environmental conditions they can be the starting point for the production of new local populations.

inverted repeat (Davidson et al. 1973) — (1) (i.r. DNA); any of a class of short DNA sequences (100 to 130 bp) which have the same nucleotide sequence but in opposite orientation to one another (such as ABCDE E'D'C'B'A'). Adjacent i. repeats may form double hairpin or double stem-and-loop structures (palindromes). Regions of i.r. DNA's are present in the chromosomal DNA of many prokaryotes and eukaryotes and involve the → promoter regions, recognition sites of → restriction endonucleases, and → repressor binding sites. Possible functions of inverted repeats are in the initiation of → DNA replication, transcription termination, and → transposition. (2) Inverted r. sequences also occur in RNA, where they contribute to secondary structures that are important in regulating → gene expression (→ attenuator) and in processing primary transcripts (→ RNA processing).

inverted terminal repeat — any of the short related or identical sequences present in reverse orientation at the ends of some → transposable genetic elements.

invertible DNA (Iino and Kutsubabe 1983) — any DNA segment with → inverted repeat sequences at both ends. At least some of such prokaryotic regions are involved in regulation of → gene expression (→ inversion).

invisibles (Goldschmidt 1938) — → gene mutations or chromosomal structural changes (→ chromosome mutation) without visible effects.

in vitro — of studies under artificially controlled conditions outside the organism as opposed to in vivo studies carried out in the intact organism.

in vitro DNA recombination — → recombinant DNA technology.

in vitro genetic assay — any method of → mutagenicity testing which is based on qualitative plate tests. Procedures of screening of genetic activity of chemical agents have been described for the detection of lambda phage → induction, reverse mutation in bacteria, and mitotic → gene conversion and reverse mutation induction in fungi. Some technical modifications of plate tests involve the utilization of hepatic drug-metabolizing enzymes for the potentiation of certain chemicals, construction of highly sensitive indicator strains through alterations in their capacity to repair genetic damage or to limit permeability of specific molecules, and the development of nonrestrictive assays such as mitotic gene conversion or forward mutation (→ host-mediated assay).

in vitro mutagenesis — → mutagenesis of cloned genes. It uses → recombinant DNA technology and is a tool that allows for the analysis and specific change of a particular sequence of nucleotides by chemical and/or enzymatic manipulation in a segment of DNA after its removal from its normal site in an intact genome. Three broad categories of i.v.m. are known: (1) Methods that restructure segments of DNA; (2) localized random mutagenesis; (3) oligonucleotide (site)-directed mutagenesis.

All methods of i.v.m. have some in vivo component, often → molecular cloning or expression of the altered DNA sequence (→ reverse genetics).

in vitro packaging — a method of introducing → recombinant DNA into bacteria *(E. coli)*. A functional viral particle is formed around phage lambda or cosmid DNA in vitro. Semi-complete, preformed phage head and tail components are made from two lambda → lysogens; these are mixed in vitro with any concatemeric DNA which contains two cos sites separated by about 49 kb of DNA and the DNA will then be packaged into an infective viral particle.

in vitro transcription — the accurate → cell-free transcription of purified DNA into RNA.

in vitro translation — the → cell-free translation of protein from purified mRNA molecules.

in vitro unscheduled DNA synthesis assay — (UDS assay); a method for the direct and economical observation in primary rat hepatocytes of damage to DNA; resynthesis of a section of DNA is detected through incorporation of radioactive thymidine. Compounds which induce damage resulting in long-patch repair can be detected and quantitated autoradiographically (→ excision repair).

in vivo — → in vitro.

I pilus — → sex pilus.

IS-associated deletion (Reif and Saedler 1975) — any of a class of → deletions generated by → insertion elements (IS) in adjacent DNA sequences at higher frequencies than → transposition. Such deletions have a fixed end point located at one end of the IS; the other end point exhibits the same specificity as the target sites used in transposition.

IS element (Hirsch et al. 1972) — = → insertion element.

isoacceptor tRNA (Doctor et al. 1961) — any of different families of two or more → transfer RNA molecules with different coding response but specifying the same amino acid during → genetic translation (= isoaccepting tRNA). Their multiplicity and widely varying distribution indicate that i. tRNA's play a regulatory role in cell function. Isoacceptor tRNA's may differ in their primary sequences, in post-transcriptional modifications, and in structural configuration, but are recognized by the same → aminoacyl-tRNA synthetase. Structurally different isoacceptors may or may not function identically in protein synthesis. Differences in the ribosome recognition areas of tRNA may result in different affinities for the ribosomes and affect the relative rates at which the isoacceptors can donate their amino acids to the peptide chain. Even though an isoacceptor recognizes two or more → codons, it may not do so with equal efficiencies.

I. tRNA's are involved in the production of → suppressor tRNA species.

isoallele (Stern and Schaeffer 1943) — any → allele so similar in its phenotypic effect to that of other independently occurring alleles as to require special techniques to distinguish among them.

iso-anisosyndetic (Lilienfeld 1951) — → isosyndetic.

isobrachial (Sorokin 1929) — → chromosome arm.

isocapsidic — → heterocapsidic.

isochore (Bernardi and Bernardi 1985) — in vertebrate genomes, any of the very long DNA segments (larger than 200–300 kb) that are compositionally fairly homogeneous (at least sizes above 3 kb) and belong to a small number of classes characterized by different GC levels. Isochores correspond to the DNA segments of individual or contiguous chromatin domains and represent selection units in the vertebrate genome.

isochromatid break (Sax 1941) — a discontinuity in both chromatids of a chromosome at apparently identical sites, as identified by genetic methods or by cytological observation in mitotic and meiotic meta- or anaphase (= isochromatid deletion, isolocus break, or isolocus deletion).

isochromocentric (Dangeard 1947) — of nuclei showing as many → chromocenters as they contain chromosomes (→ oligosomal).

isochromosome (Darlington 1940) — a monocentric or dicentric (→ isodicentric) chromosome with homologous (equal and genetically identical) arms which are mirror images of one another (ABC·CBA). Isochromosomes may arise in three different ways: (1) Through → centric fusion between two homologous → acrocentric chromosomes; (2) through replacement, by two successive → translocations, of the arms of a heterochromatic chromosome by homologous acrocentrics; (3) by sister strand reunion, following breakage through or immediately adjacent to the centromere (centromere misdivision).

Meiotic → chromosome pairing of isochromosomes may be either internal, fraternal, or internal-fraternal (Sen 1952): (a) **Internal pairing:** The two arms of an i. pair with each other to produce a ring → univalent. (b) **Fraternal pairing:** One of the arms of an i. pairs with a homologous arm of another chromosome. (c) **Internal-fraternal pairing:** Pairing of the i. is partly internal, partly fraternal.

isocoding — of genes coding for one and the same polypeptide, such as, e.g., those coding for → histones (→ gene reiteration; repetitious DNA).

isodicentric (Darlington and Wylie 1953) — of a → dicentric chromosome or chromatid in which the free arms on both sides of the two centromeres are of equal length and homologous (→ isochromosome).

isoenzyme — = → isozyme.

isogamete — → gamete.

isogamous (de Vries 1911) — (1) of → complex heterozygous species whose male and female gametes transfer the same genes or gene complexes, in contrast to → heterogamous hybrids; (2) of a population derived from parents chosen either at random or because of their coancestry and in which all pairs have the same fecundity (Malecot 1948).

isogamy — → anisogamy, gamete.

isogenic (Johannsen 1926) — (1) of any group of individuals which possess the same genotype, irrespective of their being homo- or heterozygous; (2) of two alleles or chromosome segments that are identical by descent (→ allogenic).

isograft — a tissue graft between two individuals of identical genotypes (= isogeneic graft, syngeneic graft), e.g., monozygotic twins, members of inbred strains, and the F_1 of crosses between two inbred strains (→ allograft).

isolabeling — in the case of tritium autoradiography of chromosomes, the occasional finding of both chromatids (sister chromatids) apparently labeled in a situation where single-chromatid labeling should prevail (→ sister-chromatid exchange).

isolate (Wahlund 1928) — a group of individuals (a population or group of populations) who, because of their isolation, choose their mating partners from members of the same group. One i. is completely separated from other such isolates. In human genetics, geographic and social isolates are distinguished. Social isolates may be the consequence of economic, religious, or traditional preferences in marriage.

isolation — the restriction, or limitation, of the extent of → gene flow (exchange of → genetic information) between discrete populations owing to barriers to interbreeding. I. is a factor in → evolution and the various mechanisms of i. may be classified as follows (Grant 1963): (1) **Spatial i.:** Gene flow between populations is reduced or prevented by the distance between them; (a) **geographical i.:** The populations concerned occupy different territories, i.e., are allopatric; (2) **environmental i.:** Gene flow between populations is limited by the availability of habitats suitable for the growth and subsistance of their hybrid progeny (selection against the hybrids and their derivatives); (a) **ecological or habitat i.:** The populations occupy different habitats in the same territory, i.e., are sympatric; (3) **reproductive i.:** Gene flow is prevented or restricted by genotypically controlled differences in the reproductive behavior and fertility relationships of the individuals which are members of different populations; (a) **external reproductive i.:** The barrier of gene flow lies outside the individuals and is effective prior to fertilization; (a_1) **mechanical i.:** Male and female reproductive organs are insufficiently coadapted structurally; (a_2) **ethnological i.:** Psychological or behavioral factors prevent the mating of male and female individuals which are members of different populations; (a_3) **seasonal i.:** The period of mating activity is different in the different populations; (a_4) **gametic i.:** In cases of external fertilization, the free-living male and female gametes produced by different populations fail to attract one another; (b) **internal reproductive i.:** The barrier to gene flow operates inside the individuals and results from the unsuccessful interaction between gametophytes, gametes, chromosomes, or genes of the different populations; (b_1) **incompatibility:** In the case of internal fertilization, cross-insemination or pollination takes place between individuals of different populations but does not result in hybrid progeny; (b_2) **hybrid inviability:** Hybrid progeny are produced but are of reduced viability; (b_3) **hybrid sterility:** F_1 hybrids grow to maturity but either their sex organs or their gametes are abortive and nonfunctional; (b_4) **hybrid breakdown:** Hybrids produce F_2 or backcross progeny consisting partly or entirely of inviable or sterile individuals.

I. between species does not generally operate by a single isolating mechanism but by a combination of different mechanisms. Isolating mechanisms function to prevent the breakdown of highly integrated genetic systems (e.g., species) which are adapted to a particular environmental niche. The evolutionary significance of i. for the processes of speciation is its effect on limiting random mating and on population size. I. is a directive force in evolution and prevents or restricts the mixing of genetically controlled differences which have arisen in a population (→ annidation). Breakdown of i. barriers between different breeding populations allows hybridization to occur between members of the two groups. The results of such hybridizations depend among others on the genetic similarity of the populations concerned. If they are genetically rather similar, hybridization facilitates gene flow between contiguous populations and has a tendency to reduce or even to eliminate the genetic differences between them (prevention of local differentiation of populations within a species). Hybridization between species or subspecies may lead to what is called → introgressive hybridization, i.e., the introduction of genes from one population into the other (increase of genetic variation).

Mechanisms of reproductive isolation may originate in two ways (Grant 1966): (1) As byproducts of evolutionary divergence; (2) as products of selection for reproductive i. per se, i.e., by an increase within each species in the frequency of genetic factors which determine a block or aversion to hybridization between two sympatric species over the course of generations. Such selective processes (which may be called the "Wallace effect" as proposed by Grant 1966) may reinforce reproductive i. which developed as a byproduct of divergence but did not necessarily result in a complete barrier to hybridization.

isolation estimate (Merrell 1950) — a measure of genetically controlled sexual → isolation between populations (A and B) based upon the relationship between homo- and heterogamic matings that take place. There are four possible kinds of mating between individuals of two populations composed of A^0, A_0 and B^0, B_0 respectively; two homogamic mating types $A^0 \times A_0$ and $B^0 \times B_0$, and two heterogamic types $A^0 \times B_0$ or $B^0 \times A_0$. The i.e. is calculated as the quotient

$$i = \frac{(A \times B) + (B \times A)}{(A \times A) + (B \times B)}.$$

It may be used if equal numbers of the types to be compared are at hand. If there is complete sexual isolation (no mating between populations A and B) i. is zero; with no sexual isolation i. is 1.

isolation gene (Brieger 1958) — any gene causing a reduction in viability or fertility when present in the heterozygous form. Under these circumstances the homozygotes are kept partially or completely isolated by the elimination of the heterozygotes.

isolation index — a measure of genetically controlled sexual isolation between two populations, based on matings of equal numbers of females of each population with males of one population. After an interval large enough to allow approximately 50% insemination, the distribution of all females between the populations is recorded. The percentage of intrapopulational copulation minus the percentage of interpopulational copulations divided by the sum of both, is referred to as i.i. The index is +1 if all copulations were intrapopulational, −1 if all were interpopulational, and o if equal numbers of females of both populations were inseminated.

isolecithal — of egg cells in which the yolk is scattered evenly throughout the cytoplasm (→ centrolecithal, telolecithal).
isolocus break (Thoday 1953) — = → isochromatid break.
isomeric (Sirks 1933) — ref. to equivalent genes which can each produce the same phenotype (→ anisomeric). If several i. genes are present in one genotype they may either be cumulative ("cumulative isomery" or true "polymery") or noncumulative ("noncumulative isomery") in their action.
isomorph (Serra 1949) — = hypomorph (→ allele).
isonymous substitution — any DNA base pair substitution (→ gene mutation) that does not change the amino acid coded for.
isophene — a line on a map connecting points of equal expression of a clinally (→ cline) varying character (Mayr 1963).
isophenic — of genes with similar phenotypic effects (= → isomeric).
isophenogamy (Strandskov 1950) — a → mating system in which mating between individuals of identical or very similar phenotype is more frequent than expected on the basis of random mating, in contrast to "heterophenogamy", in which mating occurs more frequently between individuals of dissimilar phenotype as compared with random mating. If the phenotypes preferentially mating in both i. and heterophenogamy are genetically controlled, then the first system may lead to an increase in homozygotes, the second to an increase in heterozygotes for the alleles determining the phenotypes in question.
isophenous — being of the same → phenotype.
isoplith (Hall and Sinsheimer 1963) — = → isostich.
isopycnotic (Östergren 1950) — of chromosomes or chromosome regions which do not differ from the majority of chromosomes and chromosome regions of the cell in their appearance, in contrast to → heteropycnotic chromosomes and chromosome regions.
isoschizomeric — of a pair of → restriction endonucleases which have the same target sequence.
isostaining (Luchnik and Porjadkova 1977) — the phenomenon wherein segments of sister chromatids stain with equal intensity after use of certain → sister-chromatid exchange staining techniques (→ isolabeling).
isosteric — of a regulatory mechanism by which enzyme activity is inhibited when the concentration of the end product approaches a certain value. The inhibitor structurally resembles one of the substrates and binds to the same site on the enzyme surface as the substrate. If the binding of the inhibitor occurs at a site other than the substrate site, the mechanism is called → allosteric inhibition.
isostich (Shapiro and Chargaff 1964) — any cluster of → oligonucleotides whose lengths are equivalent. Thus dinucleotides pTpT and pCpC are isostichs but not isomers, whereas pTpC are both isostichs and isomers.
isosyndetic (Lilienfeld 1951) — of allopolyploids (→ alloploid) in which meiotic → chromosome pairing is restricted to the chromosomes of those chromosome sets derived from one species, as opposed to isoanisosyndetic allopolyploids, in which chromosomes derived from different species are partially homologous and, therefore, able to pair to a limited extent. I. allopolyploids are called "genome allopolyploids" by Stebbins (1945); isoanisosyndetic allopolyploids are referred to as "segmental allopolyploids".
isotelocompensating trisomic (Kimber and Sears 1968) — a compensating → trisomic (= monotelo-monoisotrisomic) in which the missing chromosome is compensated for by one telocentric chromosome and one tertiary chromosome (→ diiso-compensating trisomic).
isotertiary compensating trisomic (Kimber and Sears 1968) — a compensating → trisomic in which the missing chromosome is compensated for by one isochromosome and one tertiary chromosome (→ diiso-compensating trisomic).
isotrisomic (Renner 1941) — → trisomic.

isozygotic — of a genotype homozygous in respect to all genetic loci.

isozyme (Markert and Moller 1959) — any of a class of multiple, separable forms of enzymes (= isoenzymes) occurring within the same organism and having similar or identical catalytic activities. Isozymes may have different kinetic properties. The relative concentrations of i. often differ from one tissue to another and may change during the course of development. The molecular multiplicity plays a role in the differentiation of cells during embryonic development and also increases the evolutionary adaptibility of organisms. Sets of i. can be generated by both genetic and epigenetic mechanisms.

Three major classes of isozymes may be distinguished: (1) Primary isozymes (occurrence of multiple gene loci coding for structurally distinct polypeptide chains of the enzyme); (2) allelozymes (occurrence of multiple allelism at a single locus determining structurally distinct versions of a particular polypeptide chain); (3) secondary isozymes (i. formation due to post-translational modifications of a given enzyme).

J

joint DNA molecule — any DNA molecule containing material contributions from two different DNA duplexes as a result of breakage and reunion recombination (in vivo) or due to → recombinant DNA technology (in vitro).

juggernaut polymerase (Adhya et al. 1974) — a transcriptase (DNA-dependent RNA polymerase) that ignores all signals for termination of transcription due to the → rho factor. J.p. is assumed to originate by attachment of other proteins conferring anti-termination ability.

jumping gene — any gene associated with → transposable genetic elements.

junctional complex (Farquhar and Palade 1963) — any specialized area of adhesive contact between cells. Junctional complexes consist of thickened plasma membranes and variable amounts of associated intra- and intercellular materials (→ desmosome). In some cases, the junctional complexes are in the form of rings encircling each cell apex.

Besides linking cells together, two more functions are attributed to the j.c.: (1) They establish a barrier between the apical and basal sides of cells reducing diffusion between the two compartments and allowing active transport. (2) They are responsible for the electrical couplings found in many tissues.

junctional membrane — any membrane region where the cells are in contact. Such regions are much more permeable to small ions than the membrane regions facing the exterior (nonjunctional membrane). In the areas of membrane contact organized regions are localized known as → tight junctions, → gap junctions, and → septate junctions.

junction sequence — any of the two terminal regions of the → intron in RNA precursors (→ small nuclear RNA).

junk DNA — in eukaryotes, DNA sequences which appear to have no function (→ selfish DNA).

K

Kappa particle (Sonneborn 1943) — a bacterial endosymbiont *(Caedobacter taenospiralis)* found in certain stocks of *Paramecium* which is responsible for the → killer phenotype.

karyoclastic (Dustin 1934) — of agents (referred to as "amitotic", → "mitotic poisons", or "mitotic inhibitors"), which typically are able to reversibly suppress mitosis, without being lethal to the cell.

karyodesma (Scott 1950) — = → nucleodesma.

karyogamy — the fusion of nuclei. K. is one process of the → parasexual cycle in fungi and the ultimate event of → fertilization. By means of k. homologous chromosomes that were separated in the course of → meiosis are brought together again. The result of k. during sexual reproduction is a fertilized egg cell or → zygote (→ plasmogamy).

karyogram (Chiarugi 1933) — the sum total of all chromosomes of a cell described in terms of their morphology (→ idiogram).

karyoid — = → nucleoid.

karyokinesis (Schleicher 1878) — the division of the cell nucleus (= → mitosis), as distinguished from cytoplasmic division or → cytokinesis. K. represents an orderly system by which the → genetic information contained in the chromosomes of eukaryotes is distributed to daughter nuclei which are genetically identical with the mother nucleus provided no disturbances occur in the chain of events which make up the karyokinetic process. The basic mechanisms of k. are the same in most plant and animal cells. Differences involve the formation of the → spindle and the mode of cytokinesis which normally follows k.

karyology (Trow 1895) — = → nuclear cytology.

karyolymph (Haeckel 1891) — → nucleoplasm.

karyolysis (Auerbach 1874) — (1) the disappearance of the interphase nucleus at the beginning of → karyokinesis (Auerbach); (2) in a pathological sense, the dissolution of the nucleus either with or without previous nuclear fragmentation (karyorrhexis).

karyomere (Fol 1896) — any of a series of small nuclei („micronuclei") formed in cells in which the chromosomes diverge at anaphase, i.e., are widely spaced on the → spindle. This phenomenon is common in cleavage divisions in some animals and can be induced experimentally by various chemical means.

karyon (Haeckel 1891) — = cell → nucleus.

karyonidal inheritance — in ciliates, differentiations to alternative states occurring during the development of the "macronuclear anlage" (→ nuclear dimorphism).

karyoplasm (Flemming 1882) — = → nucleoplasm.

karyoplast (Strasburger 1905) — (1) = cell → nucleus. (2) A nucleus-containing cell fragment (= → minicell) consisting of a nucleus and a small portion of perinuclear cytoplasm enclosed by an intact plasma membrane (→ cytoplast).

karyorrhexis — a form of nuclear degeneration (→ karyolysis) consisting of the fragmentation of the nucleus.

karyoskeleton — = → nuclear matrix or nuclear scaffold.

karyosphere — the small fraction of the total nuclear volume to which the chromosomes are confined in nuclei extremely rich in → karyolymph (e.g., egg cell nuclei of amphibia).

karyotheca — = → nuclear envelope.

karyotype (Levitsky 1924) — the particular chromosome complement of an individual or of a related group of individuals, as defined both by the number and morphology of the chromosomes, usually in mitotic metaphase. A diagrammatic representation of a k. based on the morphological features of the chromosomes (the "endophenotype") is called an → "idiogram". The absence, duplication, or positional rearrangement of certain morphological features of the chromosomes, as compared with their typical topography (with the standard k.), is visible evidence of → chromosome mutation.

Diploid individuals possessing a k. in which two chromosomes of a pair have different gene arrangements as a result of chromosomal structural changes, are referred to as **"heterokaryotypic"** (or are "heterokaryotypes"), whereas those in which the two chromosomes of a pair are similar in gene arrangement are **"homokaryotypic"** ("homokaryotypes").

Karyotypes may differ in respect to their basic chromosome number, the shape and relative size of the chromosomes, the number and size of secondary → constrictions, the absolute size of the chromosomes, and the distribution and size of → hetero- and euchromatic chromosome segments.

Karyotypes which consist of chromosomes essentially similar in size (all with median or submedian centromeres), and those with chromosomes differing greatly in size and/or (partly) with subterminal or terminal centromeres, may be called **"symmetric"** and **"asymmetric" karyotypes,** respectively (Stebbins 1950). In comparison with plants, the karyotypes of animals are usually far more asymmetric. **Karyotype evolution** as a consequence of chromosomal structural changes (inversions, translocations, centric fusions, etc.) is usually associated with a progressive reduction of the basic chromosome number and an increased asymmetry.

Fundamental k. (Battaglia 1952) is the term which may be used to refer to the original k. of a species, genus, etc. (symbol k^f). A **derived k.** is any k. evolved from the fundamental k. by the processes of karyotype evolution (symbol k_1, k_2, etc.). **Semikaryotype** (symbol k/2) and **haplokaryotype** are terms proposed to denote the karyotype of the gametes and the haplophase, respectively.

karyotypic orthoselection (White 1954) — in some evolutionary lineages, the tendency for similar chromosome structural changes (→ chromosome mutation) to establish themselves in one chromosome member of the karyotype after another. This may result in a situation where all chromosomes are structurally alike.

karyotyping — a variety of methods used for characterization and identification of each → chromosome of the chromosome complement in question (→ chromosome banding).

k DNA — acronym for → kinetoplast DNA.

killer plasmid (Markower and Bevan 1963) — in yeast *(Saccharomyces cerevisiae),* any of two major species of capsid-associated, linear, double-stranded RNA's (dsRNA), denoted KIL-K_1 and KIL-K_2, which specify secreted, heterodimeric protein toxins and resistance functions. The toxin is lethal to strains not carrying the plasmids.

killer strain (Sonneborn 1938) — any of a number of *Paramecium* strains containing a bacterial endosymbiont (*Caedobacter taenospiralis;* Kappa) which is responsible for the killer phenotype (killer). Kappa is of two forms in killer strains: Nonbrights and brights. The former are the reproductive and infective forms of Kappa and can, when induced, give rise to brights. Brights are responsible for killing, and contain refractile bodies (R bodies) representing coiled ribbons of protein. On the inside end of the coiled R body, there are either spherical or helical phage-like structures composed of equal amounts of DNA and protein (→ mate killer).

When released into the medium, the toxin is able to kill sensitive strains. The killer character is dependent upon two components: The genotype (presence of genes K, S_1 und S_2) of the *Paramecium* cell and the presence of Kappa particles.

kilobase — (kb); 1000 base pairs (bp) in duplex DNA or 1000 bases in single-stranded RNA.

kindred — → pedigree.

kinetic complexity — the complexity of a DNA component as measured by the kinetics of DNA reassociation.

kinetic proofreading (Hopfield 1974) — a model according to which errors during nucleic acid or protein synthesis are reduced by the introduction of a time delay between the formation of an activated complex and the formation of product which allows removal of incorrect nucleotides or amino acids.

kinetid — any of the locomotory structures in eucells called → cilia and → flagella. The → basal bodies of kinetids are termed kinetosomes. All of the kinetids in a cell constitute the kinetome and, if they are arranged in orderly rows, each row is a kinety.

A k. has a nonad (9 + 2) symmetry, i.e., nine pairs of → microtubules are arranged in form of a ring with two more in the center. In the kinetosomes the outer doublets generally become triplets and the central two disappear.

kinetochore (Schrader 1936) — (1) = → centromere. (2) In centromeres, a specialized electron-dense structure in the primary constriction which is the site of attachment of spindle → microtubules during mitosis and meiosis. The k. is (in mammalian chromosomes) a morphologically defined trilaminar organelle consisting of three closely apposed disc-shaped plates positioned along the surface of the chromosome. The k. plates and the associated centromeres are fibrous structures. The fibers in the outer plate of the k. interact with the spindle microtubules during chromosome movement in mitosis and meiosis. The structures found in interphase cells are termed prekinetochores. The k. is an integral part of the → chromosome scaffold.

In the case of → holocentric chromosomes, kinetochores are elongated trilaminar plates with attached microtubules running the entire length of mitotic (or meiotic) chromosomes.

kinetochore fiber — in early anaphase of mitosis and meiosis, any of the fibers (→ microtubules and associated proteins) which connect and anchor the → kinetochore to the polar region of the → spindle. Their shortening results in movement of the chromosomes towards the spindle poles (pole-to-pole movement).

kinetochore organizer (Holmquist and Dancis 1979) — by analogy to the → nucleolus organizer, a DNA region located in the → centromere which segregates with the chromatids and is activated at a specific stage of the → cell cycle.

kinetochore plate (Jokelainen 1967) — the site of spindle fiber attachment and → microtubule polymerization in the → centromere. It probably also induces the ancillary attributes of an active centromere, including primary constriction at metaphase, pericentric regions of tight sister-chromatid association, and C banding (→ chromosome banding).

kinetome — → kinetid.

kinetonucleus (Woodcock 1906) — = → kinetoplast.

kinetoplast (Alexeeff 1917; Lwoff 1949) — in flagellate protozoans of the Trypanosomatidae family, a distinct part of the branched → mitochondrion in which the → kinetoplast DNA is located (= blepharoplast; parabasal body).

kinetoplast DNA (Simpson 1972) — (k DNA); the → mitochondrial DNA (→ kinetoplast) of trypanosomes and related parasitic protozoa, which consists of a catenated network of rings of two classes: (1) A large number of small DNA duplex rings (minicircles) around 2000 bp in length; (2) several dozen much larger rings (maxicircles) of about 37 kb which are functionally similar to the mitochondrial DNA of other eukaryotes. The biological function of minicircles is obscure.

During replication of k DNA, minicircles are released from the network as individual covalently closed circles which contain long → open reading frames; they replicate as free molecules. The minicircle L strand replicates continuously and unidirectionally starting at a unique position; the H strand replicates discontinuously. Size and sequences of minicircles are dependent on the species, and even within a single k DNA network there are different classes of minicircles with different sequences.

The doubled k DNA network splits into two progeny networks which aggregate into daughter cells during cell division.

kinetosome (Allen 1912) — the 9 + 0 → basal body from which all → undulipodia develop. The k. contains RNA and protein (→ centriole).

kinety — an orderly arranged row of → kinetids.

kin founding (Hedrick and Levin 1984) — the instance in which a group founding a population is composed of related individuals.

kink (Crick and Klug 1975) — an abrupt and significant bend (→ DNA bend) between two helical segments of DNA achieved by unstacking one base pair and twisting (kinking) the polynucleotide backbone.

kin selection (Hamilton 1963) — selection for characteristics which favor the survival of close relatives of a given individual, as in the case of selection for altruistic behavior between related individuals (= kinship selection).

Kleinschmidt spread — a technique which permits visualization of single- and double-stranded RNA or DNA molecules in the electron microscope. Minute quantities of nucleic acids are coated with a basic protein and spread on a denatured protein monolayer at an air-water interface and then shadowed with a heavy metal prior to being viewed. The thickness of the coated molecules is proportional to their original diameter and therefore single- and double-stranded molecules can be discriminated.

Klenow fragment — the large proteolytic fragment of bacterial → DNA polymerase I obtained after cleavage with a protease (such as subtilisin). The k.f. retains 5'→3' polymerase and 3'→5' exonuclease activities and can be used in the Sanger method of → DNA sequencing.

knob (Longley 1937) — a large, heavily staining → chromomere or chromomere group found at or near the ends of meiotic → pachytene chromosomes or mitotic prophase chromosomes (less frequently interstitially) of certain organisms (e.g., maize). Knobs may be used as cytological markers for the identification of particular chromosomes (→ idiogram).

Kozak sequence — in the 5' untranslated mRNA region (UTR), a base sequence (CCRCCATGG; R = purine) which appears to be required for the most efficient recognition of the correct → initiator codon by the ribosomes in eukaryotic cells.

K selection (MacArthur and Wilson 1967) — a type of → selection which operates where a resource is constantly available but in limited amounts, so that rapid population growth is unimportant and the population is maintained at or near the carrying capacity (K) of the habitat. K.s. produces superior competitive ability to survive a harsh environment, including competitors and predators (→ r selection; A selection).

A population or species with slow development, relatively high competitive ability, late reproduction, large body size, and interoparity is called K-selected.

L

labeling index — the proportion of cells which are labeled by a radioactive isotope (corrected for background in case of autoradiography).

lagging — of chromosomes or pairing configurations (referred to as "laggards") which show retarded movement, or no movement at all, in comparison with the bulk of chromosomes during mitotic or meiotic meta- and anaphase. A frequent consequence of lagging is the failure of the chromosomes concerned to be included in one of the two daughter nuclei. Thus lagging may give rise to "aneuploidy".

lagging strand — the discontinuously (in form of → Okazaki fragments) synthesized DNA strand (5'→3') whose synthesis is initiated only periodically (→ leading strand).

lag load (Maynard-Smith 1976) — a type of → genetic load which measures the extent to which a population is lagging behind the contemporary environment. The concept behind the l.l. is that, as the environment changes, the population tracks it genetically. With increasing rate of environmental change, the population will lag further behind, i.e., the load will increase; if the environmental change is too rapid, the load will become excessive and the population will become extinct.

Lamarckism — the erroneous concept, advocated by Lamarck, that biological → evolution is brought about by environmental induction of adaptive changes in living organisms. Such changes were thought to be inherited by the progeny ("inheritance of acquired characters") (→ evolution).

lamella — a double membrane structure (= vesicle), formed by two → unit membranes lying more or less parallel to each other and in close apposition.

lamellipodium — = → ruffled edge.

lamin — any of the karyoskeletal proteins that form an insoluble fibrillar meshwork (the nuclear lamina) intercalated between the → chromatin and the inner membrane of the → nuclear envelope.

lampbrush chromosome (Rückert 1892) — (LBC); any of a special type of chromosomes characteristic of the diplotene stage of meiosis in oocytes of many animal species. LBC's are highly extended chromosomes showing characteristic paired loops which are active in RNA synthesis (genetic transcription). The newly synthesized RNA molecules associate in a ribonucleoprotein (RNP) matrix. Many of the loops show a gross morphology related to the shape and density of their RNP matrix (granular, globular, or dense) and thus permit recognition and mapping of the lampbrush bivalents. The basic structure of the RNP matrix of all kinds of loops is a 30 nm RNP particle. The extent of aggregation of these particles may be responsible for the morphological differences between loops. Part of the transcripts may represent maternal mRNA's required for subsequent development of the zygote. The RNA fibrils are attached to the loop axis via → RNA polymerase-containing granules; the lengths of the fibrils form continuous gradients from the starting point of transcription. Each loop may contain one or more → transcription units.

At specific loops, DNA rings (copies of the loops) are synthesized and extruded into the cytoplasm and replicate extrachromosomally. They form additional nucleoli in the nucleoplasm (→ gene amplification; rDNA).

landmark — any distinct morphological or other feature (→ chromosome banding; restriction fragment polymorphism) that can be used in identifying a chromosome. Characteristic bands and position of, e.g., the → centromere, the → nucleolus organizer (secondary constrictions), and → satellites are such features. A chromosome region is the area between the midlines of two adjacent landmarks.

land race — any of the geographically distinct (plant) populations which evolved under cultivation and are conspicuously diverse in their genetic composition both within and between populations. They differ from varieties developed by modern breeding in that they have not been deliberately selected to a predetermined level of genetic heterogeneity and performance.

large-patch repair (Cooper and Hanawalt 1972) — → excision repair.

lariat replicative intermediate — in → mitochondrial DNA replication *(Paramecium)*, a circle-plus-tail intermediate formed during unidirectional replication which terminates in a dimer length molecule.

lariat RNA (Padgett et al. 1984) — an RNA intermediate generated during splicing (→ RNA splicing) of mRNA precursors. The lariats have a circular component with an extending tail and are formed by a branch where an adenosine residue is linked (by a 2′→5′ phosphodiester bond) to the 5′ end of an → intron and (by a 3′→5′ phosphodiester bond close to the 3′ splice site) to the remainder of the intron. The site of branch formation is typically 20–50 nucleotides upstream of the 3′ splice site. The excised intron is released intact in form of the lariat RNA and retains the branch site; it terminates in a guanosine residue with a 3′ hydroxyl group. The phosphate group at the splice site junction connecting the two exons originates form the 3′ splice junction.

leader peptide — → leader region.

leader region (Bronson et al. 1973) — (1) a DNA sequence between the transcription start site (the → promoter) of certain → operons and the first structural gene. The l.r. codes for a short peptide called the leader peptide. (2) The nontranslated sequence at the 5′ end (before the coding region) of a → messenger RNA molecule which (in prokaryotes) may contain a regulatory sequence (the → attenuator) determining the rate of → genetic transcription. The l.r. is involved in selection of the AUG → initiator codon by eukaryotic ribosomes to initiate → genetic translation. (3) In a variety of secretory proteins synthesized on membrane-bound ribosomes, NH_2-terminal extensions of 15–30 predominantly hydrophobic amino acid residues (= leader sequence; leader peptide). The l.r. peptide initiates the secretory pathway by directing vectorial discharge of the protein from the ribosomes through the membrane of the → endoplasmatic reticulum. The peptides are cleaved from the nascent chains during translation and thus are not found in the secreted forms of proteins.

leader transcript (Zurawski et al. 1978) – a short RNA molecule produced by transcription termination at an → attenuator. L. transcripts of bacterial amino acid biosynthetic operons code for short peptides (leader peptides) and are involved in translational control of gene expression.

leading strand — in → DNA replication, the continuously synthesized strand (5′→3′ direction) as opposed to the → lagging strand.

leaky — of (1) a → genetic block which is incomplete ("leaky block"); (2) a protein coded for by a mutant gene which shows some residual activity ("leaky protein"); (3) a → gene mutation (often a → missense mutation) which fails to completely prevent the action of a gene and permits some residual expression of the function in question ("leaky mutation"). Although completely changing the sequence of codons, → frameshift mutations in the promoter-distal part of the affected gene may also give rise to a very low activity of the mutant gene product.

lecanoid meiosis — a special type of → meiosis in which the first division is equational and the second division segregational; it separates the paternal heterochromatic chromosome set from the euchromatic set.

lectin — any of the proteins which are able to bind to specific cell surface oligosaccharides. Most are multivalent and can agglutinate cells; some are mitogenic for specific cell types, some are cytotoxic.

left splicing junction — the boundary between the right end of an → exon and the left end of an → intron (→ RNA splicing; lariat RNA).

lepocyte — a cell surrounded by a → cell wall, as opposed to a → gymnocyte.
leptonema (Gregoire 1907) — = → leptotene.
leptotene (v. Winiwarter 1900) — → meiosis.
leptotene bouquet — → bouquet stage; attachment plate.
lesion — a damaged site in a gene (a mutant site), a chromosome, or a protein. (→ DNA lesion).
LET — acronym for → linear energy transfer.
lethal — of a gene or genotype which, when expressed, is fatal to its carrier (→ lethal factor).
lethal DNA alteration — any irreparable change in DNA which blocks → DNA replication as opposed to mutagenic DNA alterations (→ DNA damage).
lethal equivalent value (Morton et al. 1956) — the mean number of harmful (detrimental) genes carried by each member of a population multiplied by the mean probability that each gene will cause → genetic death when homozygous.
lethal factor — a gene mutation or chromosome structural change (deletion or duplication, as well as inversion and translocation connected with a so-called → position effect) whose presence in the genotype blocks the normal development of the carrier (most often by failure to produce one or more essential gene products) and is fatal to the individual before sexual maturity.

Lethals may be classified in various ways (Hadorn 1949):

(1) **According to their degree of → penetrance:** (a) Lethal factors sensu strictu causing the death of all carriers which contain the particular lethal factor in active dose in their genotype (lethality at least 90% as compared to standard). (b) Semi-lethals causing the death of more than 50% of the carriers (about 50 to 90% lethality of the carriers). (c) Subvitals (Hadorn 1948) ("detrimental mutations" according to Muller 1948) causing the death of less than 50% of the carriers (about 10 to 50% lethality). (d) Conditional lethals resulting in the loss of some vital function, but only under environmental conditions over which the experimenter may exert some control. (e) Quasi-normals, causing less than 10% lethality.

(2) **According to their phase of activity** ("phase specificity of lethals"): (a) Gametic (Mohr 1926), gonic (Renner 1924), or haplophasic lethals (Hadorn 1949) which cause the death of gones or gametes and prevent zygote formation, and (b) zygotic lethals (Mohr 1926) which cause the death of zygotes at some stage of zygote development. Such lethals may be further subdivided into embryonic, larval, pupal, and early imaginal lethals in insects and into embryonic, post-embryonic, and juvenile lethals in birds and mammals.

(3) **According to the influences of the external and internal environment** on the action of the lethals: (a) Unconditional lethals; their penetrance and expressivity may not be influenced nor compensated by experimental circumstances, and (b) conditional lethals; their action is dependent on the growth or environmental conditions (e.g., temperature-sensitive lethals, or sex limited lethals, the action of which is expressed in only one of the sexes).

(4) **According to their localization:** (a) Autosomal lethals localized in the → autosomes, and (b) sex-linked lethals localized in the → sex chromosomes.

(5) **According to their being → Dominant or recessive:** (a) dominant lethals, which are lethal in single dose, (b) dominant Mendelian factors with recessive lethal effect, which are lethal only when present in the homozygous form, and (c) recessive lethals without dominance effect, which are ineffective in heterozygous individuals due to the complete dominance of the wild-type allele.

Synthetic lethals (Dobzhansky 1946) are genes that in some combinations cause little or no apparent disadvantage to their carriers, but in other combinations are lethal. Synthetic l. are a class of genes having a direct effect on viability, but where the average effect of a particular gene substitution is less than unity.

Lethals are often maintained at high frequencies in natural populations by various genetic mechanisms. **"Balanced lethality"** (Muller 1917) is such a mechanism. It represents a special case of linkage which assures the maintenance of recessive lethals in a system of permanent heterozygosity. The simplest of such systems consists of two linked, recessive lethal factors (l_1 and l_2) in the heterozygous condition (L_1l_2/l_1L_2). The genotype is phenotypically normal, but heterozygotes may segregate lethal homozygotes when crossed. In the case of loose linkage of the two loci or in the absence of factors (heterozygous inversions or translocations) reducing or preventing crossing over between them, the lethal segregants from a cross of two such heterozygotes approach rather closely the following ratio: 0.25 l_1L_2/l_1L_2 (lethal): 0.50 l_1L_2/L_1l_2 (viable): 0.25 L_1l_2/L_1l_2 (lethal). Systems in which recombination between the two heterozygous loci concerned is not completely prevented are referred to as "partially balanced" and give rise to recombination gametes of the types l_1l_2 and L_1L_2 in addition to the nonrecombinant gametes of type L_1l_2 and l_1L_2. The lethals in systems of balanced lethality may act zygotically or gametically, or partly zygotically and partly gametically in one or the other sex. In all such cases only the heterozygous genotypes l_1L_2/L_1l_2 are viable. If both lethals are zygotic lethals, the fertility is reduced to 50%; if one is zygotic, the other gametic in action, the fecundity (number of zygotes) is 50% but the fertility (number of the individuals which can reproduce) is 25%.

Polygenic lethality (Hadorn 1955) is lethality realized by the combined action of at least two mutated loci.

Effective lethal phase (Hadorn 1955) is the term proposed for that phase in ontogeny beyond which a lethal mutant is normally unable to survive. On the basis of their effective lethal phases, lethals may be classified as follows (Hadorn 1955): a) Monophasic lethals, i.e., those with only one effective lethal phase. (b) Pluriphasic lethals, i.e., those characterized by two or more effective lethal phases separated by developmental periods during which no impairment of development takes place. (c) Aphasic lethals, i.e., those which may cause a breakdown of the developmental system of the carrier at any time during the whole of ontogeny.

The causes underlying phase specificity of lethals may be different: (1) A general biochemical disturbance may block or change indispensable physiological functions, resulting in death during early embryogeny or later, when the system is confronted with specific stresses. (2) Loss or disturbances of a particular function in definite cell systems resulting in a lethal crisis when the specific function is needed during development. (3) Gross disturbances of a morphological nature (e.g., absence of an organ) which are initiated as embryonic disturbances but lead to death much later.

lethal lysogenization (Ohashi and Dove 1976) – a response of mutant lambda phage insensitive to epistatic replication inhibition by repressor; such phages lysogenize sensitive host cells and establish defective complexes which neither lyse nor multiply as lysogens. L.l. is prevented by mutations that inactivate either of the viral replication genes (O or P).

lethal sectoring (Haefner 1965; James and Werner 1966) — in yeast, bacteria, algae, and mammalian cells in tissue culture, the sporadic appearance of clusters (sectors) of dead cells in the post-irradiation vegetative generations of surviving cells. L.s. tends to diminish with advancing generations of an irradiated population.

lethal zygosis (Alföldi et al. 1957) — a decrease in the number of viable bacterial F^- cells (→ F plasmid) and an associated decrease in recombinant cells when recipient (F^-) cells (of *E. coli* K-12) are mated with an excess of donor (Hfr) cells. In analogous $F^+ \times F^-$ matings, F^- bacteria are usually not killed (immunity to l.z.).

leucine zipper (Landschulz et al. 1988) — in → DNA-binding proteins, an amino acid sequence (30 residues with leucine at every seventh amino acid in the region of similarity) that binds specifically to the → CAAT box and → transcriptional enhancers. The leucine repeat regions in the protein are capable of forming regular α-helix, with the leucine side chains projecting out from the helix at regular intervals. Leucine side chains of one molecule may interdigitate with those from a second protein, thus forming a zipper that correctly holds the molecules together. Such dimers may be the entities that interact with DNA and regulate gene expression (→ helix-turn-helix motif; zinc finger motif).

leucoplast (Schimper 1885) — a → plastid containing no visible pigments (→ elaioplast, aleuroplast, amyloplast).

Lewis effect (Pontecorvo 1955) — → position effect.

L form (Gilpin and Patterson 1976) — in bacteria, a cell differing from the normal situation in that the cell wall is missing or reduced to a template moiety. L forms arise under the action of penicillin with so-called large bodies ("multinucleated" bodies of irregular shape) as an intermediate state.

Stable L forms are bacteria which have permanently lost their cell wall but still multiply and produce small colonies on agar medium. Most stable L-forms require the addition of osmotic stabilizers to the culture medium. The term stable protoplast (spheroplast) is often used synonymously with L f.

life cycle — (1) the sequence of events from the individual's origin as a → zygote to its death. (2) All the species-specific stages through which an organism passes between the production of gametes by one generation and the next. Each life cycle is based upon and controlled by a particular → genetic system. In many plants and some animals, the l.c. involves a regular → alternation of generations. Four basic types of l. cycles are immediately distinguishable: (1) **The haploid l.c.:** The entire life history is passed in the haploid state except for the diploid zygote nucleus. (2) **The diploid l.c.:** The entire life history is passed in the diploid state except for the immediate products of meiosis. (3) **The haplo-diploid l.c.:** Alternation of haploid and diploid vegetative phases. (4) **The haplo-dikaryotic l.c.:** The stable association (in higher fungi) of paired haploid nuclei that is propagated vegetatively (the dikaryophase is the substitute for a vegetative diplophase). This type of l.c. proceeds through haploid and dikaryotic phases of varying proportions prior to a single nuclear generation in the diploid state and thence to meiosis.

ligand — a small molecule (including activators, substrates, and inhibitors of enzyme activity) bound to a protein by noncovalent forces. Interactions between like and unlike ligands may be called homotropic and heterotropic, respectively.

ligase — = → DNA ligase; sealase; joining enzyme.

ligation — the process of joining (via a phosphodiester bond) of two adjacent bases separated by a → nick in one strand of a DNA duplex or of two linear nucleic acid molecules. L. can be used for → blunt end joining and also for joining of RNA molecules.

light-regulated gene expression — the organ-specific expression of genes in higher plants (light-regulated genes) which is correlated with the presence of → chloroplasts in a cell. → Enhancer-like elements possess the necessary DNA sequences for expression in the correct cell types.

light repair — = → photoreactivation (→ dark repair).

lignin — the most abundant biopolymer after cellulose which is an integral → cell wall constituent of all vascular plants. It imparts mechanical rigidity to plant tissues specialized in solute conductance or structural support.

limited chromosome — any chromosome that may be either eliminated or diminished during cleavage producing clones of different functional karyotypes (→ nuclear differentiation).

line — a group of identical pure-breeding diploid or polyploid organisms, distinguished from other individuals of the same species by some unique genotype or phenotype.

LINE (Singer 1982) — acronym for a → **l**ong **i**nterspersed **n**ucleotide **e**lement in the chromosomal DNA of eukaryotes (→ interspersed repetitious DNA).

lineage — a line of common descent (→ cell lineage).

linear energy transfer — (LET); a measure of energy deposited by a charged particle within a unit distance, i.e., the average energy lost per μm path for the primary ionizing particle.

linearize — the introduction of a single double-strand break in a covalently closed circular DNA molecule resulting in its conversion into a linear molecule.

linkage (Morgan 1910) — the association in inheritance of certain genes (and, in turn, of the phenotypic characters they control) due to their being localized in the same chromosome or nucleic acid molecule. Genes borne in → homologous chromosomes belong to the same → linkage group. The number of linkage groups is identical with the number of chromosomes of one → chromosome set.

Linkage of genes is only rarely complete. It is broken (→ genetic recombination) when homologous linkage structures exchange corresponding parts, by a process referred to as → crossing over ("partial l."). Crossing over in eukaryotes may occur during meiosis (meiotic crossing over) or mitosis (mitotic crossing over). In prokaryotes (such as bacteria and viruses) it can occur without meiosis or mitosis by functionally analogous processes at the level of nucleic acid molecules.

Linkage (rather than independent → genetic segregation) of two → genetic markers (e.g., allelic pairs A/a and B/b) is evident if the parental gene combinations (e.g., AB and ab) used in a crossing experiment show a tendency to remain together, i.e., the proportion of so-called recombinants (Ab and aB) is less than the proportion of parental combinations (AB and ab). If two marker genes are borne on nonhomologous linkage structures, independent segregation occurs and the parental and recombination types occur in equal frequency.

In the case of linkage between two markers (A and B), a third marker C may be either linked or unlinked with these two. Lack of l. between two markers is indicated in crossing experiments by their random distribution and with three markers by the impossibility of arranging them in a linear order when the recombination frequencies are used as the basis for their ordering. These relationships allow the unequivocal localization of genes in particular linkage groups. The orderly arrangement of genes in a chromosome is represented by a → chromosome map which may be either genetic (→ genetic map) or cytological. Strength of l. between two markers is dependent on the distance separating them in the linkage group. Linked genes at loci relatively close together tend to be preferentially transmitted together from generation to generation, while those relatively far apart may frequently become separated by crossing over and may then segregate nearly as independently as do unlinked genes.

Sex linkage (Morgan 1914) is the special type of l. of those genes localized in the → sex chromosomes of eukaryotes. According to their position in either nonhomologous (differential) or homologous segments of the X and Y chromosomes, such genes may either be completely (localization in "differential segments" in which no crossing over takes place) or partially sex-linked. The latter genes, localized in homologous segments of X- and Y-chromosomes, may be exchanged between the sex chromosomes by crossing over.

Interchromosomal linkage (Stern 1933) or "false linkage" (Longley 1945) is a type of spurious linkage between genes belonging to different linkage groups and results from nonrandom (polarized) → assortment of certain chromosomes (and the linkage groups they contain) during meiosis (→ affinity; meiotic drive) or from random assortment, if only particular chromosome distribution patterns give rise to functional gametes or, subsequently, functional zygotes (→ complex heterozygous; translocation).

linkage disequilibrium (Kimura 1956) — the nonrandom association of alleles at different loci at frequencies higher than expected from the random combinations of their frequencies in the population (= gametic excess; gametic phase unbalance; epistatic equilibrium). L.d. is often regarded as evidence of selection, i.e., selection may favor particular combinations of alleles at different loci (epistasis between loci). Besides epistatic selection, l.d. can also come about by: (1) Random genetic drift due to small population size; (2) population mixing or migration; (3) genetic → hitch-hiking.

In a large, random-mating population (without selection) l.d. should be nearly zero. If l.d. is introduced, it would rapidly decrease towards zero with each successive generation. (→ gametic disequilibrium).

linkage group (Morgan 1911) — a group of coupled genes located in a chromosome (→ linkage); specific l. groups correspond to specific chromosomes. The number of linkage groups is limited, the maximum number equalling the number of chromosomes per → genome (the haploid number of chromosomes in diploid organisms).

linkage map — a chromosome map, i.e., an abstract map of the genes contained in a → linkage group associated with a particular chromosome. Crossover and recombination values are utilized to demonstrate a definite serial order of the genes in a chromosome and provide an index of the relative distance between any two genes (→ genetic map).

linkage number (Crick 1976) — = → linking number.

linkat (Demarly 1979) — a set of genes (a multitranscriptional unit) which aggregated in the same chromosome region during species differentiation. The clusters usually encode coadapted functions.

linked — two genes whose loci occupy different sites on the same chromosomes are linked; in genetic analysis two genes showing less than 50% recombination, which is the normal amount for genes of different → linkage groups showing independent → genetic segregation. Linked genes tend to be transmitted together (synthenic).

linker — a synthetic oligonucleotide which contains a → restriction site and can be ligated (by → blunt end ligation) onto the ends of DNA fragments. Linkers can be used for insertion of an exogenous piece of DNA and subsequent → molecular cloning of the fragment into a → cloning vector.

linker DNA — the extra-core DNA (about 40—60 bp) connecting adjacent → nucleosomes in chromatin. Linker DNA and the nucleosome core particle make up the nucleosome. The lysine-rich → histones H1 and H5 are associated with the l. DNA and induce higher-order folding in the polynucleosome fibers. Linker histone H1 protects about 20 bp of l. DNA.

linker insertion mutagenesis (Heffron et al. 1978) — an → in vitro mutagenesis technique used to generate a large library of mutants with specific alternations scattered throughout a cloned DNA and to screen these mutants for a desired phenotype. L.i.m. allows the definition of the boundaries and functional regions of genetic elements and involves the → blunt end insertion (at the position of a restriction site) of a short synthetic duplex after random linearization of a circular target DNA. A unique restriction site in the linker facilitates mapping of the mutations, sequence determination, and construction of secondary mutants.

linking number (Fuller 1971) — a topological parameter which characterizes closed circular duplex DNA and specifies the number of times the two complementary strands are intertwined in the DNA circle when the DNA is conceptually constrained to lie on a plane. The l.n. can only be changed by breaking and resealing the DNA (→ DNA topoisomerase).

lipid — a class of diverse chemical compounds that includes fats, waxes, phospholipids, carotenoids, and sterols. Lipids serve as food reserves and as parts of structures of the cell. All lipids are soluble in fat solvents.

lipochondrium (Baker 1951) — a spherical or subspherical lipid body of variable size present in the cytoplasm. In many types of cells, the distribution of lipochondria is restricted to a zone in the vicinity of the nucleus; this zone represents the → Golgi apparatus.

liposome (Bangham et al 1965) — any of the lipid vesicles (spherules) consisting of a lipid bilayer enclosing a single aqueous compartment (unilamellar vesicles) or a number of concentric bilayers enclosing an equal number of aqueous spaces (multilamellar vesicles). Liposomes can be used to introduce a variety of biological molecules and molecular complexes (such as chromosomes) into cells (→ liposome-mediated gene transfer). Introduction of the l. contents into the cytoplasm of cells may occur by three mechanisms: (1) Endocytosis followed by intracellular processing of the foreign material; (2) adsorption of the 1. to the cell surface, rather than intracellular incorporation, and entrance of the contents through a permeable region of the cell membrane; (3) fusion of the lipid bilayer via the plasma membrane of a cell.

liposome-mediated gene transfer — → gene transfer by means of synthetic → liposomes into eukaryotic cells in culture or into protoplasts resulting in → genetic transformation. Transformation by naked DNA is DNase-sensitive, that by liposomes is not.

liquid-holding recovery (Roberts and Aldous 1949) — in bacteria (yeast and mammalian cells in culture), recovery from → DNA damage in UV-irradiated cells which takes place during liquid holding, i.e., under conditions in whih the cells are not allowed to actively metabolize or divide. L.-h. r. is a recombination-dependent pathway of → DNA repair.

localized mutagenesis (Hong and Ames 1971) — the creation, by various methods, of mutations (base substitutions) which are localized to a specific region of the DNA and cause a series of random changes within this region (= located m.). (→ site-directed mutagenesis; cassette mutagenesis).

local population — a breeding unit in sexual organisms; the members of a l.p. share a common → gene pool. A certain amount of interbreeding occurs between l. populations due to migration and gene flow. L. populations are linked by their breeding relationships into → races and → species.

locus (Morgan et al. 1915) — any genetically defined site, i.e., a gene, a part of a gene, or a DNA sequence with a regulatory role.

A compound l. is one composed of two or more genes (in the sense of → cistrons).

locus-specific (Hadorn 1955) — ref. to phenomena or activities correlated exclusively with one particular chromosomal → locus; e.g., l.—s. control of the formation of a particular character, l.—s. → puffing in → polytene chromosomes, l.—s. loop formation in → lampbrush chromosomes, etc.

logarithmic growth — = → exponential growth.

logistic growth — a pattern of population growth in which the growth rate decreases with increasing number of individuals until it becomes zero when the population reaches a maximum (saturation size). When the number of individuals is small, the population increases almost exponentially (exponential growth), but as the growth rate decreases, the growth curve (population versus time) gradually levels out toward a constant value at the saturation size, thus forming a characteristic "S-shaped"curve.

lomasome — (Moore and McAlear 1961) — any of the cells' boundary formations (border bodies) which protrude into the interior of the cell (evaginations delimited by the cell membrane) and enclose cytoplasmic and multivesicular structures somewhat variable in form. Lomasomes are located between the → cell membrane and the → cell wall and have been found in fungi, algae, and higher plants. They may play a role in secretion, in increasing the surface area for the diffusion of substances involved in cell wall formation or wall breakdown, in membrane proliferation, and in endocytosis.

long interspersed nucleotide element (Singer 1982) — (LINE); any of the long (6—7 kb), repetitious (about 10^4times), interspersed DNA sequence elements in mammalian nuclear genomes which represent, at least in their majority, → retroposons. Most LINE's terminate at the 3′ end with an A-rich tract, sometimes preceded by a polyadenylation signal, and some are flanked by short direct repeats that are target site duplications. LINE's contain one or more → open reading frames (ORF's), they are mobile, and make target site duplications at the site of insertion. Most of them are truncated and lack varying amounts of sequence at the end conventionally designated the 5′ end. LINE elements are known to cause mutations in man by insertion into specific genes. (→ short interspersed nucleotide element; interspersed repetitious DNA).

long patch repair — → excision repair.

long-range restriction site mapping — the construction of long-range maps of genomic DNA by use of → restriction endonucleases that cut the DNA infrequently.

long terminal repeat — (LTR); in the provial DNA of → retroviruses, bacterial and eukaryotic → transposable genetic elements and in bacteriophage Mu two long (usually a few hundred bp) direct repeats at the termini (frequently flanked by short inverted repeats) which carry → promoter sequences capable of initiating transcription (by RNA polymerase II) of genes located outside the LTR elements.

loop chromatid — → inversion.

loop joint — in homologous → genetic recombination, an overlap region observed at several stages of the recombination process.

low molecular weight nuclear RNA — = → small nuclear RNA.

Ludwig theorem — a theory according to which new genotypes can be added to a population if they are able to utilize new components of the environment (i.e., occupy a new subniche), even if these are inferior to the ancestral niche (Mayr 1963). This phenomenon is also called "annidation".

Luria-Delbrück fluctuation test — a method used to estimate → mutation rates in cell populations; originally devised for bacteria and now widely applied also in somatic cell genetics and cancer biology.

luxuriance (Renner 1929) — → heterosis.

lyonization (Lyon 1972) — a mode of → dosage compensation in mammals; the inactivation of the paternal or maternal → X chromosome (at random) early in the development of the female (XX) embryo. Once established, the inactivation of one of the two X chromosomes is maintained in the cell and all its descendants.

lysis — the bursting of a cell by the destruction of the cell membrane.

lysogen — a → lysogenic bacterial cell.

lysogenic (Lwoff and Gutman 1949) — of bacteria (lysogens) containing a repressed → prophage as part of their genome. If some phage genes are absent and the page is thus unable to complete a → lytic cycle, the cell is a defective lysogen. Two important properties of l. bacteria are : (1) A lysogen cannot be reinfected by a phage of the type that first lysogenized the cell (resistance to superinfection or immunity); (2) even after many cell generations, a lysogen can initiate a lytic cycle (the process is called prophage induction and the phage genes are excised as a single segment of DNA).

The protein responsible for preventing a prophage from reentering the lytic cycle is called a lysogenic repressor.

lysogenization (Ball 1925; Bordet 1925) — the establishment of → lysogeny by infection of a nonlysogenic bacterial cell with a → temperate bacteriophage and the attachment of the phage to the bacterial chromosome at a specific region, called the attachment region or attachment site. The infecting phage DNA cyclizes and is then linearly inserted into the host chromosome by reciprocal recombination between what is called the bacterial chromosite and the homologous segment of the phage genome, the episite. For the site-specific recombination involved in integration a function specified by a phage gene is needed. Phage mutants unable to perform this function are called integration-defective. Integration of the phage genome into the bacterial chromosome and its passive replication with it is called **stable lysogenization.** If the temperate phage fails to attach to the bacterial chromosome and is inherited unilaterally among the progeny of the infected cell (and eventually diluted out of the population) this is termed **abortive lysogenization.**

lysogeny — the potentiality of bacterial cells for production and release of bacteriophages as a stable, heritable character. It is widespread among strains of different bacterial species. Such strains are referred to as → lysogenic or lysogens.

lysosome (de Duve et al. 1955) — any of a class of morphologically heterogeneous cytoplasmic particles (ranging in size from 0.25 to 0.8μm) in animal and plant tissues characterized by their content of hydrolytic enzymes (acid hydrolases, including various phosphatases, nucleases, glycosidases, peptidases, sulfatases, and lipases), and the structure-linked latency of these enzymes. The intracellular functions of lysosomes depend on their lytic potential. The single → unit membrane of the l. acts as a barrier between the enzymes enclosed in the l. and the external substrate. The activity of the enzymes contained in lysosomes is limited or nil unless the vesicle in which they are enclosed is ruptured. Such rupture is supposed to be under metabolic (hormonal) control.

Lysosomes start functioning when the cell takes up substances either by phagocytosis or pinocytosis. Insoluble remnants of digestion within the lysosomes usually remain and the particles become packed with different inclusions. The results of the digestive process are normally amino acids, sugars, and simple organic compounds which may be used as synthetic building blocks by the cell. Under certain circumstances lysosomes start to digest the cell in which they occur by opening their membrane and setting free their digestive enzymes.

The size and density of lysosomes lies between that of → mitochondria and → microsomes (fragments of the → endoplasmic reticulum.) Their fine structure varies from cell to cell and within the same cell type during the life of the lysosomes.

The lysosomes appear to originate through cooperative action of the → endoplasmic reticulum and the → Golgi apparatus (→ GERL). Two main groups of lysosomes may be distinguished (Novikoff et al. 1964; de Duve and Wattiaux 1966): (1) **Primary lysosomes,** i.e., those which may be formed from small Golgi vesicles or larger vacuoles cut off from the tips of Golgi saccules, and have not yet been involved in digestive processes. (2) **Secondary lysosomes,** i.e., those digestive vacuoles which are or have been site of digestive activity. Lysosomes belonging to this group may be classified into two separate types, heterophagic and autophagic, based on the exo- or endogeneous origin of the material undergoing digestion. Each of these types may be further subdivided into prelysosomal, lysosomal, and postlysosomal forms.

lytic — of a phage life cycle (→ lysogenic) during which the bacterial host cell lyses and releases many phages (usual outcome) or during which phage are continually released without killing the host (in the case of some filamentous phages). Phages with a l. life cycle are called virulent phages, as opposed to → temperate phages.

M

M I, M II — symbols for the metaphase of the first (M I) and second meiotic division (M II). Similarly, A I, A II, T I, and T II for anaphase and telophase (→ meiosis).

macroevolution (Philiptschenko 1927) — transspecific → evolution which, within the span of geological time, leads to the origin of higher taxonomic categories such as new genera, families, and orders with new characters and a new adaptive mode. M., also called "quantum evolution" (Simpson 1944), implies a more or less rapid shift of a small population to a new equilibrium state (a new adaptive mode) distinctly unlike the ancestral one.

As compared to evolution at the species level, referred to as "microevolution" or → "speciation", m. is neither a fundamentally distinct phenomenon, nor does ist employ a different mechanism, although this has frequently been stated by some authors.

macrogametophyte — = → embryo sac.

macromolecule — a large molecule with colloidal properties; the molecular weight is > 1000.

macromutation (Goldschmidt 1940) — any genetic change leading to a striking change of the phenotype. Such mutations are likely to be ill-adapted until compensating changes at other loci have occurred.

macronucleus (Bütschli 1876) — in ciliated protozoans, the large nucleus which contains a large amount of DNA; the DNA molecules are gene-sized (2000—3000 bp in length) and result from sequence-specific breakdown of DNA during macronuclear development from a → micronucleus. Micronuclei have normal chromosomes and are without metabolic function in vegetative growth (→ nuclear dimorphism).

macrophylogenesis (Zimmermann 1948) — = → macroevolution.

macrospore — = → megaspore.

macrotubule (Tyson and Bulger 1973) — any of a class of tubules that occur during → meiosis (some fishes) and during organization of the → spindle. They are larger (30—50 nm in diameter) than → microtubules and are associated with the polar region of the meiotic apparatus. They may arise by the polymerization of microtubule proteins.

macula adherens (Farquhar and Palade 1963) — → junctional complex; → cell junction.

magnification (Ritossa 1968) — → ribosomal DNA.

main band DNA — a peak of eukaryotic genomic DNA on a density gradient, excluding any visible → satellite DNA's which form separate bands.

maintenance methylation — → DNA methylation.

major gene (Mather 1941) — any gene (→ oligogene) individually associated with pronounced phenotypic effects. Major genes control the production of discontinuous or qualitative → characters, in contrast to "minor genes" or → polygenes with individually small effects. Major genes segregate clearly and are easily subject to Mendelian analysis. The classification of genes into major and minor determinants is more or less arbitrary, since both types are in all probability only end points in a more or less continuous series of gene actions and gene integrations (→ gene interaction).

major histocompatibility complex — (MHC); in vertebrates, a tightly linked multigene family consisting of numerous histocompatibility genes controlling cell surface determinants, immune responce differences, some components of the complement system, and perhaps other related functions connected in general with cell-cell recognition. The MHC can be subdivided by recombinational studies into a number of regions coding for classes of gene products (→ immunoglobulin).

major spiral (Sax 1935) — → chromosome coiling.

male gametocide — any of a group of chemical agents that disrupt → microsporogenesis in plants; seed set on the treated plants is under these conditions not self-pollinated.

male multiple choice — → pair mating.

male recombination element (Hiraizumi 1979) — (MRE); in *Drosophila,* any of those → transposable genetic elements that are characterized by the following features: (1) Induction of male recombination; (2) induction of gene and chromosome mutations; (3) induction of female and male sterility; (4) induction of → nondisjunction; (5) induction of transposition of → P-elements.

male-specific bacteriophage — any → bacteriophage specific for bacteria with F-like → sex pili. Such phages are of two types: Isomeric RNA phages attaching along the length of the sex pilus and filamentous DNA phages attaching to the tip of the pilus.

malignant — of cells which exhibit various constant behavioral features and show what is called the transformed phenotype: Anchorage-independent growth, decreased serum requirements, rounded cell morphology, loss of microfilaments, increased hexose uptake, increased plasminogen activator secretion, and decreased cell surface fibronectin. Cells may be transformed by introduction of one of a number of → oncogenes or by activation of → protooncogenes within a cell genome. The ability of m. cells to grow progressively, to invade, and to kill the host is called malignancy.

mammalian spot test (Fahrig 1977) — in → mutagenicity testing, a rapid test system for gene mutations in mammals in which the consequences of coat color gene mutations induced in embryonic mice are detected as patches on the fur of the animals a few weeks after birth.

map — (1) a diagram showing the relative positions of genes (or restriction sites) and the distances between them (→ linkage map); (2) the procedure to determine the relative positions of genes (or restriction sites) on a DNA molecule (→ genetic mapping).

map contraction (Esposito 1968) — → map expansion.

map distance (Bridges 1932) — the distance, in terms of percentage of → crossing over obtained from genetic experiments, between particular linked genes (→ genetic map). The m.d. is measured as → centi-Morgans (= percent recombination). Since m.d. is based on the total frequency of crossing over in a chromosome region, it is directly translatable into units of m.d. or → map units. One unit of m.d. between linked genes is that interval within which 1% crossing over takes place.

The relationship between "recombination frequency" and m. d. is not one-to-one, since recombination frequency measures only those crossovers that lead to reciprocal exchanges of → genetic markers. The farther apart the genes, the more striking the discrepancies become because double crossovers are scored as nonrecombinants. Expression that relate m.d. to recombination frequencies and correct for the probable double crossovers which have occurred in the interval under consideration are referred to as "mapping functions" (→ marker effect; map expansion).

map expansion (Holliday 1964) — the tendency, observed in many linear fine-structure maps (intragenic → genetic recombination), for recombination frequencies between more widely spaced mutants within a gene to exceed estimates of the same intervals obtained by adding the recombination frequencies given by more closely spaced mutants. M.e. is a general → marker effect and the opposite of what is expected from negative → interference.

M.e. has been explained on the basis of an independent correction of base-pair mismatches (→ heteroduplex repair) included in the same hybrid DNA segment. A possible additional source of m.e. may be due to the contribution of reciprocal crossing over to the total frequency of recombination. This contribution will increase with the square of the distance between the mutant sites of a two-point cross if the pairs of single-strand breaks serving for hybrid DNA formation between the mutant sites of a two-point cross are distributed at random within a given chromosome region.

The opposite effect is called **map contraction,** i.e., the recombination value observed is less than the value expected by summing adjacent intervals.

map length — → Morgan unit.

mapping function (Haldane 1919) — a formula expressing the quantitative relation between distance in a → linkage map (crossing over) and recombination frequency:

$$y = \frac{1}{2}(1 - e^{-2x})$$

(Y is the recombination frequency between two marker genes and x is the number of crossovers).

The main reason behind the interest in this relation is the fact that, whereas the number of crossovers are additive, recombination frequencies are not. By use of the m.f. a nonadditive metric (recombination frequency) is transformed into an additive metric (number of crossovers).

mapping mark (Renwick 1971) — the beginning and the end of a chromosomal lesion that has led to a difference between homologous chromosomes and is used (in human genetics) for estimating the → map distance of a certain locus from one or more mapping marks.

map unit —a unit of distance in a → linkage map, i.e., the distance separating two linked genes for which, in a standard cross, recombinants appear in a frequency of 1% among the products of meiosis (→ Morgan unit). The m.u. equals the total crossover frequency between the genes concerned, corrected for the effects of double → crossing over and chromosome → interference. The corrected crossover frequency of 1% equals 1 m.u.

marginal population (Carson 1955) — a → population which exists on the edge of a species range.

marker chromosome — any morphologically distinguishable abnormal chromosome.

marker effect — the phenomenon wherein the frequencies and kinds of → genetic recombination events are affected by the nature of the mutations used to monitor them.

marker rescue — = → cross reactivation.

masked messenger RNA — = → maternal messenger RNA.

massule (Bessis et al. 1958) — a → centriole satellite.

master-slave hypothesis (Callan and Lloyd 1960) — a hypothesis proposed to explain the large differences in DNA content among various eukaryotes on the basis of a uninemic chromosome model. According to the m.s.h. each gene may be represented 100 to 10,000 times in a cluster: One copy is the master and the tandem duplications are slave genes. Although each slave gene could mutate independently, it is postulated that the slaves fold back upon the master, match their base sequences, and are corrected wherever incongruities happen to be present.

mastigoneme (Deflandre 1934) — any of the fine hairlike appendages (= flimmer) which extend from the shaft of the → flagellum in a variety of motile cells and serve to move water towards the flagellum base.

mate killer (Siegel 1953) — in *Paramecium,* individuals which contain mu particles and cause their sensitive partners to die after conjugation; unlike to the → killer strains they liberate no toxin into the medium. Cell-to-cell contact is necessary for mate-killing.

maternal effect — any specific effect of the maternal genotype or maternal environment (phenotype) on the immediate offspring (→ predetermination).

maternal inheritance — → inheritance controlled by extranuclear hereditary determinants (= non-Mendelian or cytoplasmic inheritance).

maternal messenger RNA — in oocytes of many animals, large quantities of mRNA stored during → oogenesis for translation in the course of maturation in early embryogenesis. Masking proteins are associated with m.mRNA and suppress translation (= masked messenger RNA). A large portion of this RNA is not assembled into polyribosomes until the completion of meiosis in the oocyte or fertilization in the egg.

mating — (1) bacterial m., i.e., mating mediated by → conjugation systems that are encoded by → plasmids (or → transposons). A number of different conjugation systems with distinct genetic, morphologic, and physiological properties are known. All plasmid transfer systems code for → sex pili and other components on the donor cell which interact with surface structures on the recipient during m.

Mating aggregate is the term used to denote both pairs (and larger aggregates) of m. bacteria.

(2) In eukaryotes, the pairwise union of unisexual individuals for the purpose of sexual reproduction (and by extension the "self-mating" of bisexual organisms"), ultimately resulting in the formation of → zygotes.

Among sexually reproducing species mating systems do not normally fall into discrete classes, since natural populations may show several of the arbitrarily delimited types of m. in an infinite variety of variations and combinations. Furthermore, the type of mating system may be affected by the selective advantage of particular characters as well as by population size. Nonetheless, in sexually reproducing organisms three main mating systems are distinguished: (1) **Random mating or panmixia:** Any individual of one sex has an equal probability of m. with any individual of the opposite sex. (2) **Genotypic assortative mating:** Mating involves pairing of males and females on the basis of relationship. M. of individuals more closely related than the average is **"positive genetic assortative m."** or → inbreeding which increases the changes of mating by individuals with like genotypes. M. of individuals less closely related than the average is **"negative genetic assortative m."** or → outbreeding. (3) **Phenotypic assortative mating:** Mating involves pairing of males and females on the basis of their phenotypes and may be either **"positive phenotypic assortative m."**, when like is preferentially mated to like, or **"negative phenotypic assortative m."**, when phenotypically unlike individuals are more likely to mate than would be expected on the basis of change alone.

M. structure in plants is in part controlled by the vectors utilized in transferring pollen from anther to stigma. These vectors may be simply wind or gravity, or they may be insects or other animals manipulated into performing pollination by the floral characteristics.

mating aggregate (Achtman 1975) — pairs (mating pairs) or larger aggregates of conjugating bacteria.

mating continuum (Darlington and Mather 1949) — a group of interbreeding individuals, systematically exchanging chromosomes and genes among one another. Any member of a m.c. may have received genetic material from individuals located anywhere in the area occupied by the group. A m.c. may be divided spatially into various subgroups.

mating frequency — the number of pairs of individuals which mated in a given time period. The limit of m.f. as the time interval study approaches zero is called the m. rate.

mating group — a group of individuals, haploid or diploid, characterized by genetic or environmental conditions which favor → mating within the group at the expense of mating externally (Darlington and Mather 1949).

mating pair (Anderson 1958) — → mating aggregate.

mating propensity — the probability of a given individual for mating.

mating system — the pattern of matings between individuals of a population, including such factors as extent of → inbreeding, pair-bonding, and number of simultaneous mates. The m.s. is of major importance in determining both the genetic structure and evolutionary potential of natural populations. Three main types (→ mating): (1) Assortative mating; (2) disassortative mating; (3) random mating.

mating type (Sonneborn 1937) — the genetically controlled mating capacity of an individual. Individuals of the same m.t. do not usually mate amongst themselves but with individuals of a complementary mating type.

mating type interconversion (Hicks and Herskowitz 1977) — in *Saccharomyces cerevisiae,* a switch (= mating type switching) of the two mating types a and α (encoded by a single locus called MAT) by unidirectional transposition of genetic information. During m.t.i. the yeast cell switches mating type from α to a, or vice versa.

According to the cassette model, the cassette at MAT is removed and replaced by a cassette originating from the loci HML (α) or MMR (a) which are located far from MAT. These loci are assumed to contain silent copies of the mating type information (silent cassettes) and mating type i. occurs by duplication and subsequent transposition of replicas of such silent cassettes to MAT where they replace the MAT allele and are transcribed ("played"). The particular mating type gene activated in a cell determines the mating type of that cell. M.t. i is due to → DNA rearrangement.

mating type switching — = → mating type interconversion.

mat mRNA — acronym for → maternal messenger RNA.

matrilinear — ref. to a mode of → inheritance mediated by nuclear hereditary determinants which are transmitted only in the female line.

matrix (Sharp 1929) — → nuclear matrix.

matrix association region (Gasser and Laemmli 1986) — (MAR or SAR: scaffold association region); in eukaryotic interphase nuclei, any of the specific sites to which chromatin loop domains are anchored. MAR's play a role in the functional organization of genes and → DNA methylation may determine or reflect the DNA-nuclear matrix interactions.

matrix bridge — a → chromosome bridge which is the result of so-called "chromatic agglutination", i.e., → sticky effects that may impair the separation of chromatids in anaphase of mitosis or of chromosomes in anaphase of the first meiotic division. Matrix bridges are also called "pseudobridges", as opposed to "true bridges" which arise as a consequence of chromosomal structural changes or of crossing over in structural hybrids (→ inversion).

matrix unit (Franke and Scheer 1978) — an → intercept of the chromatin fiber covered with a series of lateral fibrils that increase in length from the starting point of transcription (sometimes used synonymously to → transcription unit).

matroclinal (Kerner 1881) — ref. to modes of → inheritance where the offspring resemble the mother more closely than the father ("matrocliny"), as opposed to patroclinal.

maturase (Lazowska et al. 1980; Cech 1985) — any of the proteins encoded by self-splicing introns that are, together with nuclear encoded proteins, needed for splicing, i.e., fold the RNA into the catalytically active conformation. All m. function only in splicing the intron in which they are encoded (or closely related introns).

maturation — in virology, the overall process that leads to the incorporation of viral genomes into → capsids and complete → virions.

maturation divisions — = → meiosis.

Maxam and Gilbert method — → DNA sequencing.

maximum parsimony hypothesis — the hypothesis that the optimum reconstruction of ancestral character states is the one which requires the fewest mutations in the → phylogenetic tree to account for contemporary character states.

mdg element — → copia-like element; retroposon.

mean lethal dose — the dose or concentration of a mutagenic agent required to reduce the surviving cell fraction to $1/e$ (e = the base of the natural logarithm). The sensitivity of the material to a mutagenic agent is expressed by the value of the m.l.d. (D_o).

medical genetics — a field of → human genetics concerned with the relationship between heredity and disease.

medusoid fibril body (Diaz de le Espina et al. 1982) — in the nucleolar periphery and elsewhere in the cytoplasm of oocytes and other cells, any of a class of distinct spheroidal units (0.5 to 3 μm in diameter) which display a characteristic tumbleweed-like or medusoid organization and contain a type of protein filaments with periodical arrays of beaded structures (diameter 9 to 14 nm).

megaevolution (Simpson 1944) — the → evolution of higher systematic categories, such as families, orders, classes, and phyla (→ macroevolution).

megagamete — the larger of two anisogametes; usually the female gamete (= macrogamete).

megaheterochromatic (White 1943) — of populations with a chromosome complement containing a large number of → heterochromatic segments, as distinct from "microheterochromatic" which refers to populations with a chromosome complement containing a small number of heterochromatic segments.

megameric (White 1951) — of → bivalents (found in many grasshoppers) containing large → heterochromatic segments; the chromosomes (autosomes) condense precociously, beginning at zygotene, and open out into a diplotene stage well in advance of the other members of the chromosome complement. Chiasmata in such bivalents form in small euchromatic segments.

megaphenic (Morton et al. 1977) — of genetic and/or environmental factors which are individually of large effect relative to the phenotypic standard deviation (→ microphenic).

megasporangium — a sporangium (i.e., a spore sac, capsule, or cell in which spores develop) producing → mega- or macrospores (= macrosporangium).

megaspore — the larger of the two kinds of haploid spores produced by heterosporous vascular plants (= macrospore). On division, it develops into the embryo sac, the gametophytic generation (→ gametophyte; megasporogenesis) of seed plants, as opposed to a → microspore.

megaspore competition (Renner 1921) — in the genus *Oenothera*, competition among the four meiotically produced and genetically different megaspores of a linear megaspore quartet which determines which spore develops and forms the embryo sac. M. c. is also called "Renner effect" (Darlington 1932) and represents a case of early postmeiotic selection of the functional megaspore.

megasporocyte — the embryo sac mother cell (→ megasporogenesis).

megasporogenesis — in angiosperms, the formation of → megaspores and the production of the embryo sac, i.e., the female gametophyte (megagametophyte), from one or occasionally several cells (megasporocytes, gonotoconts, or embryo sac mother cells) of the subepidermal cell layer within the ovule of a closed ovary (megasporangium). The megasporocyte is normally diploid.

Generally but with numerous species-specific variations, four haploid cells (megaspores) are produced from the megasporocyte in the course of → meiosis. The megaspores are usually arranged in a linear order (e.g., in maize). Three megaspores ultimately degenerate; the fourth enlarges and forms the embryo sac (the megagametophyte), which initially contains a single haploid nucleus. Before fertilization this nucleus undergoes three mitoses resulting in eight haploid nuclei lying free in the embryo sac cytoplasm. After completion of the first mitosis, one of the two nuclei migrates to the chalazal pole, the other to the micropylar pole (the micropyle being to the opening between the integuments through which the pollen tube will enter). Subsequently, two mitoses result in four nuclei at each pole.

Of these, three are enclosed by cell walls. Of the group of cells at the micropylar end, one cell (in most cases the central one) becomes the egg cell, the other two are the synergids without apparent function.

The three cells at the chalazal end of the embryo sac are called antipodals. In most cases they degenerate, at the latest, after fertilization. The two remaining free nuclei either stay adjacent to each other in the middle of the embryo sac or eventually fuse to produce the diploid "secondary endosperm nucleus". This is the condition of the embryo sac at the time of fertilization (→ microsporogenesis).

meiocyte — any cell in which the nucleus divides by → meiosis. In most animals, the primary oocytes and primary spermatocytes (→ oogenesis; spermatogenesis) are the meiocytes; in most higher plants, the meiocytes are sporocytes giving rise to the embryo sac (female sex) and to pollen grains (male sex) (→ megasporogenesis; microsporogenesis; germ cell).

meiosis (Farmer and Moore 1905) — a fundamental process in all sexually reproducing eukaryotes that, from a morphological point of view involves four major and independent events: (1) Pairing (→ chromosome pairing) of homologous chromosomes or chromosome segments; (2) the exchange by → crossing over of segments between nonsister chromatids leading to the formation of → chiasmata and homologous → genetic recombination; (3) the coorientation of homologous → centromere pairs within each paired entity (→ bivalent); (4) the segregation of homologous chromsomes.

M. occurs by two successive divisions (meiosis I and m. II) of the meiocyte → nucleus preceding the formation of → gametes or → meiospores. During m., chromosomes replicate once only (as compared to two replications in the case of two successive → mitoses), and undergo assortment so that each of the four meiotic products resulting from one m. receives one representative of each chromosome set. Thus, the zygotic (diploid) chromosome number is meiotically reduced to the gametic (haploid) number characteristic for the gametes or haploid phase of the life cycle. The point of insertion of m. into the life cycle of → haplontic (zygotic meiosis). → diplontic (gametic meiosis), or → diplohaplontic (intermediary meiosis) species differs. A special type of m. in some protozoa is referred to as "one step meiosis" (Cleveland 1947).

M. is a genetically controlled process, definitely ordered in both space and time, with a number of variations. These variations, all under some form of genetic control, concern the amount of chromosome pairing, the production, frequency, and localization of → chiasmata, the orientation and distribution (→ chromosome orientation) of the pairing partners in the → spindle, the → chromosome movements, and the modes of cytokinesis following the nuclear divisions. Such variations may be either species-specific properties which do not impair the basic functions of m. in any way, or may represent more or less pronounced abnormalities, limiting or preventing the cytological and genetical functions of meiosis (John and Lewis 1965).

A normal m., the basis of sexuality of all higher plants and animals, usually serves the following main functions (Lewis and John 1964): (1) It provides for the random → assortment of paired chromosomes (→ meiotic drive), the halving (reduction) of the somatic chromosome number and marks the transition from the diplophase to the haplophase of the life cycle (→ alternation of nuclear phases). In the absence of meiosis, fertilization as an integral part of the sexual cycle would lead to a geometric increase in chromosome number. With the intervention of m. the species-specific chromosome number established at zygote formation is maintained. (2) It determines the accurate → segregation of allelic differences (→ allele) as well as the random recombination of unlinked genes (interchromosomal → genetic recombination) and the nonrandom recombination of linked genes (intrachromosomal genetic recombination).

The stages into which a normal m. is typically divided are as follows:

(1) **First meiotic division:** (a) **Leptotene:** The chromosomes appear as long, widely despiralized (→ chromosome coiling) and optically single threads (no differentiation into → chromatids yet recognizable). (b) **Zygotene:** Homologous chromosomes begin to pair (→ chromosome pairing) in a highly specific manner and become twisted around one another ("relational coiling"). Pairing is initiated at one or more so-called "contact points" (→ zygomere) and proceeds from these along the chromosomes (→ synaptonemal complex). (c) **Pachytene:** The homologs are completely paired, i.e., are in intimate contact along their entire length. In the case of diploid organisms, the number of pairing configurations ("bivalents") is equal to half their somatic chromosome number. In → autopolyploids, the pairing configurations may consist of more than two chromosomes (→ "multivalents"). At any one point within these configurations, pairing affinity is evidently saturated by the intimate apposition of two pairing segments. In comparison to leptotene and zygotene, the chromosomes in pachytene are shorter as a result of internal coiling. During pachytene it may be apparent in favorable material that each of the paired chromosomes consists of two chromatids and that there are reciprocal exchanges of segments between homologous nonsister chromatids. Such exchanges are recognizable as cross configurations called chiasmata. Each chiasma is understood to represent the physical result of → crossing over which, in turn, is the process giving rise to intrachromosomal genetic recombination. (d) **Diplotene:**

Chromosome contraction by coiling continues. The homologous chromosomes repel one another and tend to move apart giving a picture of an opening out between homologs. Chiasmata hold the homologs together until their separation in anaphase. At this stage chiasmata are usually clearly observable and scoreable. (e) **Diakinesis:** Chromosome contraction is near maximum and the pairing configurations are well spread throughout the cell with a tendency to lie at the periphery of the nucleus near the nuclear envelope. Successive loops produced by repulsion between pairing partners on both sides of the chiasmata come to lie in planes which are at right angles to each other. If there is only one chiasma in a bivalent this appears cross-like in one plane. Chromosomes at this stage show two types of internal coiling which are referred to as major and minor coils or spirals. During diakinesis the nucleolus normally disappears and the nuclear envelope is disrupted. (f) **Prometaphase I:** Simultaneous with the disruption of the nuclear envelope the → spindle is organized. The pairing configurations become attached to the spindle and their centromeres begin to coorientate (→ centromere orientation) by movements which are mediated by the spindle (→ prometaphase stretch). (g) **Metaphase I:** The pairing configurations line up on the equatorial plate by movements referred to as "congression". Coorientation of centromeres is finished and homologous centromeres come to lie in the longitudinal axis of the spindle in such a way that these are ultimately equidistant from the equator and their respective poles. (h) **Anaphase I:** Homologous centromeres move toward opposite cell poles pulling two chromatids (being sisters in some and homologous chromatids in other regions, depending on where crossing over has occurred) behind them. (i) **Telophase I:** Regrouping of the chromosomes at the cell poles. Each pole receives one-half the original chromosome number of the meiocyte, i.e., one set of chromosomes in the case of diploidy. This stage as well as the following interkinesis may be absent in some organisms. (k) **Interkinesis:** The interphase stage between the first and second meiotic divisions, which may be of variable duration if present at all. Whether or not a cell membrane is laid down between the two nuclei at this stage depends on the organism in question. During interkinesis the chromosomes are partly uncoiled. Homologous chromatids diverge considerably but remain associated in the segments situated on both sides of the centromere.

(2) **Second meiotic divisions:** Meiosis II is mechanically similar to a mitosis and proceeds via prophase II, metaphase II, anaphase II, and telophase II. (a) **Prophase II:** This stage is omitted in organisms which omit interkinesis. It is a stage of chromosome contraction by coiling. (b) **Metaphase II:** Each of the two cells of a dyad or each area of cytoplasm organizes a spindle which may (but need not) have a consistent pattern of arrangement with respect to the other. The centromeres line up on the equator of the second division spindles. (c) **Anaphase II:** Sister centromeres, together with the component chromatids of each chromosome, separate to the poles. (d) **Telophase II:** Interphase nuclei are reconstituted and cell membranes (or cell walls) are laid down between the four nuclei (each with half the somatic chromosome number of the organism) giving rise to four cells, whose ultimate fate differs in the two sexes of the organism. In males, all four usually develop into gametes or spores (→ spermatogenesis; microsporogenesis). In female animals, three of the four usually abort as "polar nuclei" and only one develops into a female gamete, whereas in flowering plants two or all four take part in the production of the → embryo sac (→ oogenesis; megasporogenesis).

A special variant of this "chiasmate" course of m. is the so-called "**achiasmate meiosis**" which proceeds regularly without crossing over and chiasma formation. This type of meiosis is generally restricted to only one of the two sexes (e.g., the male of *Drosophila*). Presumably, small heterochromatic regions (→ collochore) fulfil the role of chiasmata by causing chromosome association and mediating the orderly assortment of homologous chromosomes during anaphase I. In cases of achiasmate meiosis the unit of recombination is the whole chromosome. Such meioses provide a means of reducing extent of genetic recombination and hence variation in the species.

A normal chiasmate m. in organisms with localized centromeres is generally "reductional" in its first division as far as the halving of chromosome number is concerned. But it may be "reductional" or "equational" in respect to the disjunction of the alleles at particular heterozygous loci. With random assortment of chromosomes, whether or not for two loci the first division of meiosis is "reductional" or "equational" (prereduction vs. postreduction) depends upon the location of crossing over. Prereductional separation for a locus represented by the alleles A and a means that the two chromatids with allele A move to one, the two with the allele a to the opposite pole. However, in the case of postreductional separation or disjunction for the same locus, the two chromatids which move together to each of the two poles have one A and one a allele, i.e., are identical in respect to their genetic constitution for that particular locus.

In the case of prereduction, anaphase II will be equational for the locus A, in postreduction it will be reductional for that locus. If there is no crossing over between the centromere and the locus A/a, anaphase I is always reductional, anaphase II always equational.

The situation is different in organisms with nonlocalized centromeres. In this case, the chromatids are active in the spindle throughout their entire length and are wholly autonomous and independent structures. Under such circumstances, two completely separated chromatids move to each pole at anaphase I. In this case, the chromatid is considered to be a chromosome, and at telophase I each pole has an unreduced number of chromosomes. During interphase in such a system, a reassociation of two homologous chromatids occurs and in metaphase II the chromatids are "paired". The actual reduction of chromosome number takes place at anaphase II. This type of meiotic behavior, limited to organisms with nonlocalized centromeres, is termed **"postreductional meiosis"** with respect to the halving of somatic chromosome number, and has been observed in some insects (coccids, aphids and Lepidoptera), chilopods, and in plants in the Juncaceae (*Luzula*).

Another variant of the normal meiotic pattern is the socalled **"one-step meiosis"** (Cleveland 1947) observed in some protozoa. This type is characterized by a single nuclear division. Chromosomes which are apparently single (unreplicated) pair loosely or not at all. In the latter case, the centromeres of homologous chromosomes become arranged parallel to the division axis at a certain distance from the spindle equator and are then regularly distributed to the cell poles. Each pole receives one-half of the somatic chromosome number. Crossing over and chiasma formation are absent.

Variations of the normal meiotic sequence may be classified according to their kind and the situation in which they occur (John and Lewis 1965). They may occur as adaptive changes, characteristic for different species, between the two sexes of one species, between individuals of the same sex and species, between cells within individuals, between nuclei within cells, between chromosomes within nuclei, as well as between parts of chromosomes.

A major period of chromosomal DNA synthesis occurs during the interphase preceding chiasmata m.; unlike the premitotic interphase, premeiotic interphase does not include complete chromosomal DNA replication. About 0.3 to 2% of the total chromosomal DNA fails to replicate by the termination of the premeiotic S phase (→ zygotene DNA). Its replication is delayed to the leptotene, zygotene, and pachytene. Delayed semi-conservative replication of DNA occurs when chromosome pairing takes place; delayed, nonsemi-conservative repair type DNA synthesis is coupled with the processes of meiotic genetic recombination (recombinational repair) during zygotene/pachytene. This type of DNA synthesis is distinct from the premeiotic S phase synthesis in GC value.

meiosome (Höfler 1957) — = → spherosome.

meiospore — a haploid cell which results from → meiosis and undergoes mitotic divisions to produce a gamete-bearing, multicellular state, the → gametophyte (= meio-agamete; tetraspore; gonium). (→ germ cell).

meiotic — ref. to or produced by → meiosis. M. mutants (Sandler et al. 1968) are due to mutations affecting one or more of the processes which together comprise meiosis.

meiotic drive (Sandler and Novitski 1957) — any meiotic deviation from Mendelian segregation ratio in heterozygotes (→ genetic segregation) resulting exclusively from a disturbance in the meiotic mechanism, i.e., an exception to the rule of 1:1 recovery of segregating alleles (= segregation distortion). M.d. may be an evolutionary force since a driven allele or chromosome may increase in frequency even when deleterious (inadaptive evolution). Factors which either enhance or suppress m.d. are known so that in some cases m.d. appears to be the result of the action of complex genetic systems rather than a single gene (→ meiotic drive locus).

meiotic drive locus (Sandler and Novitski 1957) — a locus (or allele) that drastically modifies segregation ratio. M.d. loci (as, e.g., the segregation distorter alleles in *Drosophila* or the t alleles in mice) are often found in more than 90% of the gametes produced by heterozygotes. M.d. alleles may severely reduce → fitness.

meiotic effect (Magni 1964) — in fungi, the phenomenon that the frequency of spontaneous mutants recovered is much higher after → meiosis than after mitosis. It is assumed that the m.e. is caused by some inaccuracy of → genetic recombination.

meiotic inversion (Brown and Cleveland 1968) — in organisms with → holocentric (or holokinetic) chromosomes, an equational divisions of the chromatids at the first meiotic division and reductional separation of two chromatids (one from each homologue) at the second division. Without m.i., pairs of chromatids separate from each other reductionally in the first meiotic division and the two members of each pair separate equationally at the second meiotic division. Both normal and inverted meiosis may occur in species with holocentric chromosomes. Among the Hemiptera, m.i. is restricted to the advanced Homoptera (all coccids and aphids so far analyzed).

meiotic mutation — any mutation affecting the fidelity of → meiosis; m. mutations may be sex-specific in their effect on meiosis and may affect one or the other, or both meiotic divisions.

meiotic synapsis — → chromosome pairing.

meiotrophic (Engelsberg and Ingraham 1957) — of mutants which arise from naturally deficient strains of bacteria and can dispense with one of the requirements of these strains, i.e., require fewer growth factors than these. The term permits a distinction between auxotrophs (requiring additional growth factors) and autotrophs (requiring fewer growth factors).

melting temperature — the temperature at which 50% of the total hyperchromicity of double-stranded DNA is achieved. Tm is the symbol used to denote the midpoint of the 260 nm transition profile of DNA melted out by gradual increase in temperature (→ denaturation).

membrane — in biology, any of a class of dynamic structures that represent bilayers of lipids (with polar groups at the surfaces) providing a two-dimensional fluid matrix in which proteins are embedded and can move laterally. Biologic membranes are essential elements of all cells and act as highly selective permeability barriers. They are functionally characterized by a number of features: (1) They impart on cells and organelles (mitochondria, chloroplasts, cell nuclei, lysomes, etc). their individuality by forming boundaries around them and compartmentalizing specialized cellular environments; (2) they play a central role in biological communication as they contain receptors and respond to external stimuli; (3) photosynthesis in chloroplasts and oxidative phosphorylation in mitochondria are m.-dependent processes. Proteins (integral and peripheral) confer on membranes their functionality and mediate information transfer, transport molecules, and perform various metabolic activites.

membrane carrier — any of those proteins which function to facilitate the movement of small molecules across the cell membranes. Without such m. carriers many extracellular substances are inaccessible to intracellular enzymes, and as such they remain functionally cryptic.

membrane fusion — the fusion of biologic → membranes involved in and mediating events such as secretion, endocytosis, fertilization, muscle development, cell hybridization, and intracellular transport. The sequence of processes during m.f. may be divided into four stages: (1) Close approximation and contact; (2) induction of m. f., i.e., establishment of an adequate physical-chemical milieu at the site of fusion; (3) fusion proper, i.e., establishment of intermembrane linkages; (4) stabilization, i.e., reestablishment of a physical-chemical equilibrium within the region of fusion.

Intercellular m.f. is a process vital for the maintenance of tissue integrity (→ cell junction). **Intracellular m.f.** serves to link discrete, discontinuous cellular compartments leading to the mixing, export, or intake of exogenous and biosynthesized macromolecules.

membrane skeleton (Kirkpatrick 1979) — a layer of protein lying just beneath the →membrane bilayer (on the cytoplasmic side of the membrane) and interacting with it (= m. cytoskeleton). (→ cell cortex).

membrane system — → endomembrane system.

Mendelian character —a differentiating → character, which is formed under the control of chromosomal genes according to the rules of "Mendelian" → inheritance.

Mendelian inheritance (Castle 1906) — the mode of → inheritance of nuclear genes, in contrast to "non-Mendelian" or "extra-nuclear" inheritance which is controlled by cytoplasmic heredity determinants.

Mendelian mutation (Stubbe 1940) — a → mutation fulfilling "monohybrid" Mendelian segregation. This group of mutations comprises true → gene mutations as well as chromosomal structural changes (→ chromosome mutation) which may be either microscopically visible or submicroscopic.

Mendelian population (Dobzhansky 1935) — an interbreeding group of individuals sharing a common → gene pool (→ population). The m.p. is the basic unit of study in population genetics. Usually, there is no clearcut boundary between two adjacent Mendelian populations owing to → gene flow between them. If gene flow is limited, such populations may show considerable differences with respect of their gene pools. → Gene frequency is the most important single index characterizing a m. p.

Mendelian ratio — a → segregation ratio which is in accordance with the expectation on the basis of Mendel's laws of → inheritance.

Mendelism (Punnett 1905) — particulate → inheritance of nuclear genes according to the chromosome theory of heredity.

Mendelize — of genes undergoing Mendelian → segregation (→ inheritance).

Mendel's laws of inheritance —→ inheritance.

mentor pollen (Pandey 1975) — compatible, irradiated → pollen which produces a growth tube but fails to fertilize the ovule. In certain genotypic combinations substances are released by the m.p. which allow normally incompatible pollen to produce a growth tube able to fertilize the ovule.

mericlinal (Jørgensen and Crane 1927) — of a → chimera with inner tissue of a particular genetic constitution only partly surrounded by outer tissue of a different genetic constitution.

meristem — an area of rapidly dividing plant cells. The m. may be a single cell (as in ferns), or it may include many cells.

meristic — of characters (traits) affected by genes at many loci; the variation can be counted (→ metric).

merodiploid — → merozygote.

merogamy (Meisenheimer 1921) — a mode of fertilization in Protozoa brought about by the fusion of so-called "merogametes", i.e., gametes produced by fission or multiple division of vegetative cells and differing from these in size and structure, in contrast to → "hologamy" or syngamy between whole individuals. M. may take the form of iso-, aniso-, or oogamy.

merogenote (Clark and Adelberg 1962) — a fragment of a bacterial → chromosome, irrespective of its length or of the mechanism by which it has arisen in the bacterial donor cell.

merogony (Delage 1899) — experimentally induced development of an egg fragment containing the sperm nucleus but lacking the egg nucleus.
meromixis (Wollman et al. 1956) — a process in bacteria during which only a fraction of the → genome of a donor cell is transferred to a recipient cell which contributes both its cytoplasm and its entire genome. The result of m. is a → merozygote.
merospermy — fusion of an egg cell with a sperm which has lost its nucleus and does not take part in → karyogamy with the egg nucleus (= → gynogenesis). M. is a normal event in some nematodes and may be induced experimentally.
merostathmokinesis (Gavaudan 1943) — a → mitosis (= partial → C-mitosis) which gives rise to a → multipolar anaphase due to a partial inactivation of the → spindle. M. results from the action of C mitotic agents applied in doses too low for complete spindle inhibition.
merotelic — of → centromeres possessing → microtubules extending toward both spindle poles.
merozygote (Wollman et al. 1956) — in bacteria, an incomplete → zygote which is diploid only for part of its genetic material and haploid for the remainder. The process of m. formation is called → meromixis. A m. contains the whole → genome of the recipient cell and a → merogenote derived from a donor cell. Merozygotes are usually very unstable and may be → hetero- or homogenotic with respect to their genetic constitution. The potential for → genetic recombination in a m. is limited to the diploid part of its genome (→ genetic transformation; transduction; bacterial → conjugation).
mesokaryotic (Dodge 1964) — of organisms which share the characteristics of both → prokaryotic and → eukaryotic organisms. Dinoflagellates are representative of this group. They are cytoplasmically eukaryotic in possessing → mitochondria and → chloroplasts as well as ribosomes, endoplasmic reticulum and Golgi membranes, while their genetic system is a combination of eukaryotic and prokaryotic features. The genome is contained in a nucleus without nuclear pores and which does not undergo dissolution during cell division (→ dinomitosis). The chromosome-associated proteins constitute a much smaller proportion of the chromatin mass by weight than is typical of the eukaryotes. The DNA has usually a higher proportion of methyluracil replacing thymidine and is not associated with histones.
mesoplasm (Plowe 1931) — that part of the → cytoplasm which is surrounded by the cell membrane on the one hand, and by the → tonoplast of vacuoles on the other.
mesosome (Fritz-James 1960) — any of the intracytoplasmic membraneous structures of bacteria (= plasmalemmasome), which appear to originate from the cell membrane by an invagination (infolding) and "pinching off" process. In sections, mesosomes appear as clusters of vesicles and tubules, as membraneous whorls, or as a combination of these types of ultrastructure. The list of proposed m. functions includes their participation in energy production, DNA replication and → nucleoid separation, cell division, photosynthesis, nitrogen fixation, nitrification, and endosporulation (→ bacterial sporulation).
message — → genetic message.
messenger DNA —single-stranded DNA (e.g., DNA isolated from certain bacteriophages) that can act as a messenger in an in vitro system of protein synthesis. It binds → initiator tRNA (fMet-tRNA) to → ribosomes in the presence of → initiation factors and GTP. The ability to translate (→ genetic translation) a DNA strand depends critically on the Mg^{2+} concentration.
messenger-like RNA (Scherrer and Marcaud 1968) — = → pre-messenger RNA.
messenger recognition factor (Heywood 1970) — any specific factor involved in binding → messenger RNA to ribosomes during the formation of the → initiation complex. M.r. factors represent binding factors in both pro- and eukaryotes; they may exert a positive control on protein synthesis and may play an important role in → cytodifferentiation and in the stability of the differentiated state in eukaryotes.
messenger ribonucleoprotein particle (Spirin et al. 1964) — (mRNP); in eukaryotic cells, the

association product of cytoplasmic → messenger RNA's with various proteins. It is of two classes; (1) Free mRNP not bound to → ribosomes which contains the mRNA in a repressed form; (2) mRNP present in polyribosomes (polysomal RNP).

Free mRNP can serve as a direct precursor to polysomal RNP. Additional functional roles of free mRNP may be: (1) Sequestration of different frequency classes of mRNA; (2) storage of particular mRNA's in inactive form; (3) efficient packaging of accessory factors required for protein synthesis.

messenger RNA (Brenner et al. 1961; Jacob and Monod 1961) — (mRNA); a population of large → ribonucleic acid molecules of heterogeneous size (= informational RNA; template RNA; unstable RNA, or complementary RNA) synthesized during the → genetic transcription of → deoxyribonucleic acid and serving as a template for protein synthesis (stimulating amino acid incorporation reaction in a cell-free polypeptide synthesizing system). The mRNA from a given organism may vary in size over a range of ca. 8S to 45S. Eukaryotic mRNA is produced by DNA-dependent RNA synthesis (one strand of the DNA duplex is transcribed into a strand of mRNA by the action of DNA-dependent → RNA polymerase II, the substrates being 5′-nucleoside triphosphates. The polarity of mRNA synthesis is 5′ → 3′). It has a nucleotide sequence complementary to that of the transcribed DNA strand (specific hybridization with an appreciable fraction of homologous DNA: → DNA-RNA hybrid), and the information content contained in its particular sequence of nucleotides (→ genetic code; genetic information) is translated (→ genetic translation) sequentially to produce specific polypeptide chains.

Generally, the lifetime of bacterial mRNA is short (usually less than 5 min). The enzyme machinery for degrading bacterial mRNA's attacks different classes of messengers at different rates and must be able to distinguish one messenger from another. Overall degradation is due to a 5′ → 3′ exonuclease activity following immediately behind the last ribosome, i.e., degradation is intimately coupled to genetic translation.

The chain growth rate of bacterial mRNA is 55 nucleotides/second. Although only a few (< 3%) bacterial mRNA species have poly(A) at the 3′ terminus, they lack the capped and methylated 5′ termini that are found on most eukaryotic mRNA's.

In **eukaryotes**, mRNA's are usually functionally monocistronic and in virtually all instances the 5′-proximal AUG (which lies closest to the 5′ terminal cap) functions as the → initiator codon (prokaryotic mRNA's have multiple translation initiation sites). A 5′-terminal nucleotide sequence preceding the initiator codon is termed → leader sequence.

The primary transcript of eukaryotic genes (pre-mRNA) is subjected to an extensive series of post-transcriptional processing steps (→ RNA processing): (1) 5′ terminal capping, i.e., formation of the 5′-5′-pyrophosphate linkage of a methylated guanosine residue (7-methylguanosine: 7-MeG), to either 2′-O-methylguanosine (2′-oMeG) or 2′-O-methyladenosine (2′-oMeA); (2) in most, but not all, cases 3′ terminal polyadenylation, i.e., addition of a poly(A) tract by poly(A) polymerase (20-200 nucleotides); (3) internal base modification; (4) in the case of mRNA's transcribed from → mosaic genes, the excision of the → introns by → RNA splicing.

The half-life of eukaryotic mRNA's, which represent differentiated products of gene transcription, is measured in hours to days (→ messenger ribonucleoprotein particle). The complex multistep process of eukaryotic mRNA biosynthesis is regulated both at the levels of transcription initiation and posttranscriptional RNA processing.

Besides polyadenylated poly(A)$^+$ mRNA's, poly(A)$^-$ mRNA's with a base composition similar to that of poly(A)$^+$ mRNA have been observed (bimorphic mRNA's).

The cap structure protects eukaryotic mRNA from exonucleolytic degradation and plays an important role in the initiation of protein synthesis.

In a wide variety of eukaryotic cells, the mRNA is distributed among multiple frequency or abundance classes: (1) The socalled **low complexity class** consists of a few species each present in many copies per cell; (2) the so-called **high complexity class** consists of several thousand species each present in only a few copies per cell; (3) one or more **intermediate classes** of moderate complexity and frequency.

In addition to serving as templates for genetic translation, mRNA's are involved in several other cellular activities including transport from the nucleus, attachment to the cytoskeleton, and localization within the cytoplasm.

Several mechanisms for the production of multiple RNA's from a single gene are known: (1) Association of genetic sequences previously far apart; (2) initiation of several primary transcripts at alternative → promoters; (3) differential termination or 3′ post transcriptional processing (but not splicing); (4) alternative splicing of the same primary transcript.

Selective translation of mRNA's is one of the mechanisms that eukaryotic cells use to modulate levels of → gene expression. The sites required for such regulation reside within the 5′ untranslated region (UTR) of these mRNA's. 3′ untranslated regions (between terminator codon and poly(A) tail) are also present in eukaryotic mRNA. Both 5′ and 3′ UTR's may differ in size and are probably involved in post-transcriptional controls of gene expression either by governing mRNA translation efficiency or stability (→ Kozak sequence).

The eukaryotic mRNA structure participates in at least two RNA-RNA interactions during positioning of the → ribosome to form the → initiation complex: (1) Specific portions of the message base-pair with the → initiator tRNA and (2) with the 3′ end of the 16S → ribosomal RNA. During the ribosome-binding process, mRNA secondary structure functions as a controlling element, covering unwanted internal initiators and regulating the availability of authentic initiation sites (→ scanning model).

Besides containing the information for specifying a particular amino acid sequence, mRNA also determines the efficiency of translation. This is determined both by the frequency of initiation and by the proper choice of codons. → Codon usage is markedly different in highly expressed genes compared with genes coding for rare proteins (like → repressors).

The turnover rate of mRNA ultimately determines how long it can function. Different mRNA's possess different half-lives.

In prokaryotes (bacteria), mRNA's generally are primary transcripts and exact replicas of DNA sequences in the genome. The are usually polycistronic, i.e., metabolically related genes from one regulatory unit (an → operon) which is transcribed into a polycistronic mRNA coding for several proteins.

Eukaryotic mRNAs contain about 2.2 methyl groups per 1000 nucleotides, i.e., about 6 or 7 per average cellular messenger of 3000 nucleotides. Methylation is probably of functional significance in the processing of heterogeneous nuclear RNA.

The RNA of RNA viruses has a dual function: it represents the genome of the viruses (formally analogous to DNA) and may, in part, serve itself as messenger (→ antimessenger DNA).

messenger RNA binding factor — = → messenger recognition factor.

messenger RNA capping (Rottman et al. 1974) — the post-transcriptional addition to most eukaryotic → messenger RNA's (and most mRNA's of animal and plant viruses) of a 5′ terminal methylated cap structure: $m^7G(5')ppp(5')X$(cap 0); $m^7G(5')ppp(5')X^m$(cap 1); or $m^7G(5')ppp(5')X^mY^m$(cap 2). Caps play an important role in the initiation of protein synthesis. mRNA's deprived of the cap moiety are translated at lower efficiency; the cap structure seems to be involved in mRNA binding to 80S → ribosomes.

A multifunctional enzyme complex is responsible for mRNA capping. The about 30 bp long → TATA box is located upstream from the genomic location of the cap site.

messenger RNA editing — alteration of → messenger RNA sequences by subtraction (or addition) of information present (or not present) in heritable DNA.

messenger RNA interruption (Huang et al. 1988) — a 50-nucleotide untranslated region present within the coding sequence of the mRNA transcribed from bacteriophage T4 gene 60, which encodes one of the subunits for its type II → DNA topoisomerase. The mRNA i. is bracketed by a direct repeat of 5 bp. Its folding may bring together codons separated by the interruption so that the elongating ribosome may skip the 50 nucleotides during translation.

messenger RNA leader sequence — a 5' sequence joined to the coding sequences of → messenger RNA. Its probable function is facilitation of translation of the protein coding sequence of the mRNA.

messenger RNA maturation — in eukaryotes, all the steps occurring after transcription of the primary mRNA: Addition of poly(A) to the 3' end, capping and methylation of the 5' end, processing of larger precursors to shorter mature mRNA by → RNA splicing.

messenger RNA nuclear export — the transport of the fully processed (→ RNA processing) mRNA from the eukaryotic → nucleus to the cytoplasm.

messenger RNA polyadenylation — the posttranscriptional covalent addition of 50—250 adenylate residues (A) to most eukaryotic pre-mRNA's; mRNA p. is catalyzed by the nuclear enzyme poly(A) polymerase. Most genes coding for polyadenylated mRNAs contain the sequence AATAAA (about 10—30 nucleotides upstream from their polyadenylation site). The hexanucleotide (AAUAAA in mRNA) is an important signal for efficient processing of the 3' terminus of the primary RNA transcript. The poly(A) tail is associated with proteins and undergoes size reduction in the cytoplasm in an age-related process. The proteins associated with the 3'-poly(A) tail may serve to inhibit the action of specific endogenous nucleases and so influence mRNA degradation.

messenger RNA splicing — → RNA splicing.

messenger RNA stability — the differential life-time of individual → messenger RNA species as inferred from their rate of decay within the same cell. In bacteria, the half-lives of transcripts range from seconds to as long as 20 min. In eukaryotic cells, the rate of degradation of individual mRNA's can change to various signals and the half-life may be hours or even days. Regulation of mRNA s. plays an important role in the control of → gene expression. Most mRNA's are degraded exonucleolytically from the 3' end; specialized internal sequences can greatly destabilize the RNA, possibly by acting as a site for an endonuclease.

messenger RNP particle — (mRNP); = → messenger ribonucleoprotein particle.

messenger RNA turnover — a process of central importance in determining levels of protein synthesis in prokaryotes and eukaryotes. The rate of synthesis of a protein (→ genetic translation) usually depends upon the concentration of the mRNA encoding it (transcription rate), which is determined by a balance between the rates of mRNA synthesis and decay. Disparate stabilities of mRNA's (due to differences in sequence and structure) are a consequence of their differential stabilities to degradation by cellular endoribonucleases and 3'-exoribonucleases and can contribute to differential expression of genes.

messenger RNA translation — → genetic translation.

metabolic cooperation (Subak-Sharpe et al. 1966) — the exchange of molecules between cells through permeable → gap junctions formed at the sites of cell contact. M.c. is usually detected as the phenotypic modification of mutant cells when in contact with wild-type cells (= cross feeding).

metabolic DNA (Pelc 1968; Anker et al. 1971) — nonmitochondrial DNA in the cytoplasm of differentiated animal cells with a metabolic turnover related to cell aging (= cytoplasmic DNA).

metabolic nucleus — = → interphase nucleus.

metabolite — a product of metabolism.

metacentric — of → chromosomes whose → centromere is localized in a roughly median position (lengthwise). This gives them a V- or J-shaped appearance at metaphase (→ idiogram).

metafemale (Stern 1959) — an individual which has according to the balance theory of → sex determination a greater dose of female determiners than normal females (e.g., *Drosophila* individuals with 3X-chromosomes and two sets of autosomes A). The term "superfemale", used synonymously with m., should be replaced by m. since the expression of femaleness reaches its upper limit in normal 2X 2A individuals. Metafemale Drosophilae show hypoplastic ovaries and both malformed wings and eyes; the majority die as larvae or pupae. Human beings with two sets of autosomes and three X-chromosomes are underdeveloped with respect to their primary and secondary sex characters.

metagamic — (1) of reproductive cycles that alternate between sexual and asexual phases. (2) → Sex determination in which the sex of the offspring is not fixed genetically but is largely due to external environmental influences. (3) Of nuclear divisions immediately following → fertilization.

metagenesis (Haeckel 1866) — = → alternation of generations.

metagon (Gibson and Beale 1961) — a stable messenger RNA with viruslike properties postulated to exist in mate-killing *Paramecium aurelia* (→ killer) and responsible for the maintenance of mu particles in a stock that contains at least one of two macronuclear genes denoted M 1 and M 2. The m. is able to replicate in another organism *(Didinium).*

metagynic (Loew 1895) — → monoecious.

metahermaphroditic (Bacci 1965) — of a population which is composed of half hermaphroditic and half gonochoric individuals (→ hermaphroditic).

metakinesis (Flemming 1879) — (1) the separation of the two chromatids of each chromosome and their movement to opposite spindle poles during anaphase of → mitosis (Flemming 1879);

(2) = → chromosome congression to the spindle equator (Wassermann 1926).

metandric (Loew 1895) — → monoecious.

metaphase (Strasburger 1884) — → mitosis, meiosis.

metaphase arrest — the arresting of cell division at metaphase of → mitosis or meiosis by the application of "antimitotic" agents (→ mitotic poisons; mitotic inhibition). Such a block is usually associated with structural abnormalities or the apparent dissolution of the cell's → mitotic apparatus (→ C-meiosis; C-mitosis). In some organisms, the egg nucleus is normally blocked at first or second metaphase (or anaphase) of meiosis until fertilization occurs.

metaphase inhibitor (Dustin 1947) — → mitotic poison.

metaphase pairing index (Togby 1943) — that fraction of the sum total of all cells observed in which two particular chromosomes are found in the paired condition (→ chromosome pairing) at metaphase I of → meiosis.

metaphase plate — the equilibrium position in the equatorial region of the → spindle (= equatorial plate or nuclear plate) reached by the chromosomes via their congressional movement (→ chromosome congression; centromere orientation) at metaphase of → mitosis and meiosis.

metaphase spindle — → spindle.

metaplasm (v. Hanstein 1868) — a collective term for the ergastic materials of the → cell (→ paraplasm).

metapopulation — a group of conspecific populations coexisting in time but not in space.

metareduplication (Hsu and Moorhead 1956) — → interreduplication.

metaxenia — → xenia.

methionyl-transfer RNA binding factor (Levin et al. 1973) — in eukaryotes, a protein factor which forms a ternary complex with the → initiator tRNA and GTP (→ initiation complex). Formation of this complex is followed by joining of Met-tRNA to the 40S ribosomal subunit. This latter reaction requires additional → initation factors but does not require added → messenger RNA.

methionyl-transfer RNA transformylase — an enzyme catalyzing the N-transformylation of → initiator transfer RNA (methionyl-$tRNA_f^{Met}$).

methylation-instructed mismatch repair — a bacterial → DNA repair system assumed to follow the replication fork, scrutinizing the newly replicated DNA for mismatches (→ DNA mismatch) resulting from errors of DNA replication. The mismatches are then corrected using the undermethylation (at GATC sites) of the newly synthesized strand to distinguish the correct from the incorrect half of the mismatch (→ DNA methylation). The correction process involves a repair synthesis event of considerable length (up to 1000 nucleotides) which probably stretches between the mismatch site and the nearest (dam) methylation site. Both these sites have to be present for the repair synthesis to occur.

metric — of → characters that are measurable on a continuous scale (→ meristic). Most m. characters are polygenic, i.e., controlled by many genes.

metroclinal — = → matroclinal.

M factor (Davison et al. 1969) — a protein factor (in *E. coli*) which stimulates in vitro transcription by purified → RNA polymerase (20- to 30-fold) from several templates. The M f. is distinct from the → rho factor and the → sigma factor but has several properties in common with sigma. Both are thermolabile proteins which sediment at approximately 4 to 5S and both stimulate RNA synthesis from a variety of templates. M might be a continuation factor preventing abortive termination of initiated RNA chains or might confer additional specificity on sigma-directed initiation (→ psi factor). Stimulation is probably due to interaction of the M f. with polymerase.

mic RNA (Mizuno et al. 1983) — messenger RNA interfering complementary RNA, i.e., RNA molecules able to bind to the → transcripts of particular genes and consequently preventing their translation (→ antisense RNA).

microbody (Rhodin 1954) — any of a distinct class of cytoplasmic organelles in microbial, prokaryotic and eukaryotic cells which display some or all of the following features: (1) A round to elongate structure (0.3 μm to 1 μm in diameter) delimited by a single membrane; (2) close association with the → endoplasmic reticulum and/or lipid bodies; (3) a dense granular matrix of moderate electron density in which opaque inclusions and/or crystalloid bodies may be included.

Microbodies usually contain enzymes (such as oxidases and catalases); at least three distinct subclasses are recognized on the basis of enzymatic components, metabolic roles, and morphology: (1) → Peroxisomes; (2) → glyoxysomes; (3) → microperoxisomes.

microcell (Ege and Ringertz 1974) — a → micronucleus surrounded by a thin layer of cytoplasm and an intact membrane; subdiploid microcells may contain anything from single chromosomes or chromosome fragments to almost complete chromosome complements. They can be used as vectors to transfer single chromosomes or partial genomes to recipient cells by cell fusion (m.-mediated chromosome transfer).

microcell-mediated chromosome transfer (Fournier and Ruddle 1977) — the transfer of intact chromosomes from → microcells into intact cells via (Sendai virus-mediated) cell fusion and production of hybrid cells (heterokaryons) which can be isolated by use of appropriate selection systems. The generated cell lines (microcell hybrids) are powerful → gene mapping tools (chromosome assignment of genes).

microevolution (Philipschenko 1927) — intraspecific → evolution or speciation, i.e., all processes of species formation and species differentiation brought about by the combined action of the various evolutionary factors.

microfilament — in eukaryotic cells, any of a class of long (5—8 nm in diameter) intracellular fibers that comprise polymerized F actin. Microfilaments together with myosin and associated structural and regulatory proteins provide the machinery for much of the contractile activity of cells. Additionally, the microfilaments are involved in establishment of cell shape and motility and are probably responsible for forming a contractile ring which accomplishes the final separation of the two daughter cells following mitosis in animal cells.

In different types of cells, microfilaments exist in quite different states of organization and their regions of attachment to membranes appear quite different structurally (→ cytoskeleton).

microheterochromatic (White 1943) — → megaheterochromatic.

micronucleolus (Poessnerova and Smetana 1966) — a very small → nucleolus containing a small fibrillar center, little dense material, and some granular material in a segregated form.

micronucleus (Howell 1891; Jolly 1905) — (1) the generative nucleus of some foraminifera and most ciliates (→ nuclear dimorphism; macronucleus). (2) A small → nucleus, separate from the main nucleus of eukaryotic cells. Such micronuclei originate from → acentric chromosome fragments or from whole chromosomes/chromatids that are not incorporated into daughter nuclei at the time of cell division (→ micronucleus test).

micronucleus test (Boller and Schmid 1970) — in → mutagenicity testing, an assay based upon the occurrence of → micronuclei (e.g., in bone-marrow erythroblasts of animals exposed to toxic chemicals) and used to detect in vivo both → clastogens and agents that affect the regular mitotic distribution of chromosomes.

microparapatric — of populations that coexist in the same locality and in adjacent but not overlapping microhabitats so that → gene flow between them is possible.

microperoxisome (Novikoff and Novikoff 1973) — any of a subclass of → microbodies in a large number of mammalian tissues and characterized by their small size (0.15—0.25 µm in diameter) as well as by their catalase and oxidase activities. They possess numerous slender connections to the smooth → endoplasmic reticulum and lack crystalline cores.

microphenic (Morton et al. 1977) — of genetic and/or environmental factors which are numerous but individually of small effect relative to the phenotypic standard deviation (→ megaphenic).

microphylogenesis (Zimmermann 1943) — = → microevolution.

micropia (Henning et al. 1983) — any of a class of *Drosophila* → retroposons which share several similarities to DNA viruses and retroviruses, retrotransposons and nonviral elements. The gene organization of m. is represented by genes coding for core proteins, a protease, reverse transcriptase, RNase and integrase. In addition, m. contains a region coding for class I major histocompatibility complex antigens of mammals, one open reading frame (~ 4000 bp long), a nonprotein coding region (~ 500 bp) between the 3'end of the ORF and the 5'start of the 3'long terminal repeat (including 32 bp tandem repeats).

micropinocytosis (Yamada 1955) — → pinocytosis.

micropinocytotic vesicle — = → caveola intracellulare, surface vesicle or surface pit.

microsomal mutagenicity assay — an in vitro technique which allows tissues to activate promutagens in the presence of a test organism in which induced mutation frequencies are simply determined. Basically, test organisms such as *Salmonella thyphimurium* histidine auxotrophs are incubated at body temperature with mammalian tissue (usually liver microsomes), the promutagen to be tested, and whatever cofactors are necessary for biotransformation of the promutagen to mutagenic metabolites.

The m.m.a. is useful as a screening technique only where a specific type of metabolic activation to a mutagenic product is highly suspected (→ mutagenicity testing).

microsome fraction (Hanstein 1880; Palade and Siekevitz 1955) — a subcellular fraction composed of small particles (size 50 to 150 nm) isolated by differential centrifugation and composed of three characteristic structural features: Rough vesicles, smooth vesicles, and ribosomes. The microsomal fraction may also contain elements of the → Golgi complex, fragments of the endomembrane system of the cell, mitochondrial components, endocytic vacuoles, and other subcellular entities in addition to vesicles from the → endoplasmic reticulum. Numerous enzymatic activites are associated with the microsomal fraction.

microspecies — a small, morphologically uniform population reproducing mainly if not exclusively by uniparental methods and occupying a definite geographical or microgeographical area. M. are differentiated morphologically from related species and frequently possess a hybrid constitution (Grant 1971). Different kinds of m. can be recognized according to the mode of uniparental reproduction (→ jordanon).

microspike (Taylor 1966) — any of the narrow (0.2 µm) cytoplasmic projections which can reach 20 µm in length. They may extend and retract from the → cell surface and probably play a sensory role.

microsporangium — a sporangium producing → microspores (a pollen sac or anther lobe of phanerogams) (→ microsporogenesis).

microspore — the smaller of two kinds of → microspores produced by heterosporous (vascular) plants; the male spore (→ macrospore). The production of microspores is termed → microsporogenesis, i.e., the production of the male gametophyte.

microsporocyte — the pollen mother cell (→ microsporogenesis).

microsporogenesis — in angiosperms, the formation of → microspores in the microsporangia (anthers) and the production of the male gametophyte (microgametophyte or pollen), normally from the diploid microsporocytes (pollen mother cells). In the archespore of an anther, a large number of microsporocytes are produced. These undergo meiosis, which results in the formation of four haploid microspores per microsporocyte. In most cases, the microspores form a tetrad and, after they separate from one another, they form pollen. After some time a further division (pollen grain mitosis) takes place in the pollen grain. This results in the production of a smaller "generative" cell (or nucleus) and a larger "vegetative" cell (or nucleus). The generative nucleus is generally very dense, somewhat elongated, and stains darkly, while the vegetative or "tube" nucleus is rounded and takes a lighter stain. The generative cell divides producing two male → gametes, in some cases before the anther has opened, in others at a later stage during the growth of the pollen tube.

microsubspecies (Huxley 1940) — any of the small local populations that comprise a → subspecies.

microtrabecular lattice (Wolosewick and Porter 1976) — (MTL); a dynamic three-dimensional lattice-like network of slender strands (3—9 nm in diameter), called microtrabeculae, which interconnect the three major filament systems of the cytoskeleton (the → microtubuli, the → microfilaments, and the → intermediate filaments) and other cell organelles. The MTL is coextensive with the outer → cell membrane and the components in some manner combine functionally to control cell shape and motility.

microtubule (Slautterback 1963) — (MT); any of a class of tubular structures in eukaryotic cells which are hollow cylinders (average diameter 25 nm) and composed primarily of a heterodimer (6S), termed → tubulin, and various associated high molecular weight proteins (MAP's). Microtubules can organize stable structures (such as → centrioles and → axonemes) and transitory structures that appear during cell division (such as the → spindle). The transitory structures disappear by depolymerization of microtubuli at the end of cell division (spindle).

The microtubules are a component of the → cytoskeleton and intimately involved in generating and maintaining cell polarity.

The hollow core of the m. is 15 nm in diameter and the wall of the tubule is 5 nm thick. Most MT's display 13 globular subunits in their walls; they also compose the 9 + 2 central core or axoneme of cilia and flagella and the 9 + 0 organization of centrioles and basal bodies.

Nucleotides have a central role in the assembly of tubulin into microtubules (→ microtubule organizing center). The tubulin dimer binds two moles of guanine nucleotide: One (at the E site) which can exchange with exogenous nucleotide, and one (at the N site) which does not exchange. E site GTP is hydrolyzed during the assembly process.

microtubule organizing center (Pickett-Heaps 1969) — (MTOC); any of the cellular sites for initiation, assembly, anchorage, and stabilization of → microtubules (MT) and MT-associated proteins. MTOC's (= nucleating sites) are discrete foci around which MT's are organized and are assumed to be composed of at least two essential components: (1) An enzyme (tubulin polymerase) or other membrane-bound factor for formation of the active → tubulin dimer; (2) nucleating centers required for polymerization of tubulin.

The most easily recognizable MTOC's are the → centromeres and the spindle poles (where large numbers of MT's converge into a zone called the → centriole) and vesicles of the → Golgi apparatus and the → endoplasmic reticulum as well as the axoneme of cilia and flagella.

microvillus — in animal cells, any of the tubular projections (from 0.2 to several μm long and about 0.1 μm in external diameter) from the cell surface which have various functions connected with an increase in surface area. They can be present as stable, differentiated structures or as labile, dynamic structures. Stable microvilli function to increase membrane transport, labile microvilli may act as a site of membrane reserve for cell movement, division, and membrane transport.

Microvilli are invested by a membrane continuous with the cell membrane and comprising part of it.

mictic (Storch 1924) — of females producing female and male offspring by → apomixis or by → amphigony (= amphogenous).

mictohaplontic (Kniep 1928) — of a haploid organism with genotypically different nuclei.

micton (Camp and Gilly 1942) — a widely distributed population (produced by → hybridization between two or more species) which is fully fertile in crosses with the parent species.

midbody — a shaft of → microtubules originating during cell cleavage (→ cytokinesis) and consisting of interzone spindle fibers (= spindle bridge or Flemming's body). The m. persists as an isthmus between the daughter cells for varying periods up to a complete cell cycle. Speculations as to the function ot the m. include stabilization of the cleavage furrow, retardation of cell separation, and definition of direction of the first separation movements of the daughter cells.

middle lamella — the outermost layer of the plant → cell wall connecting two adjacent cells. It is composed primarily of pectic substances.

midparent value — the mean of the values of a quantitative phenotype for two specific parents.

mid-spindle elongation — the elongation of the → spindle region which lies between the two separating groups of daughter chromosomes in mitotic anaphase (→ mitosis). M.-s.e. takes place in most animal and some plant cells and aids in the movement of chromatids to the poles. This movement ceases when the centromeres have moved about two-thirds the distance between the spindle equator and the poles. Further movement takes place by m.-s.e.

migration — in genetics, the transfer of → genetic information among populations by the movement of individuals or groups of individuals from one population ("emigration") into another ("immigration"). M. may result in changes of → gene frequencies in the populations concerned (→ gene flow) and is thus a factor in → evolution.

The effect of m. will be determined in large measure by the relative sizes of the migrant and nonmigrant populations. Its genetic consequences will be functionally related to the migrant's sex, age, and biological relationship to other individuals in the population. The level of m. may itself be under genetic control (**migrational selection** of Fisher 1930).

If the probability of m. depends on individual phenotype, this is called **selective m.** (Fisher 1930; Thoday 1972).

mimetic — = → mimic.

mimic — of nonallelic genes (= mimics) with similar phenotypic effects (= mimetic) (→ genocopy; isoallele).

miniature spindle — → spindle.

minicell (Adler et al. 1967) — (1) a small, quasi-spherical bacterial cell without chromosomal DNA produced by special mutants and abnormal cell division of rod-shaped bacteria. Minicells can synthesize any protein encoded by exogenous DNA and are useful for studies of the proteins defined by a → cloning vector. (2) = → microcell.

minichromosome (Griffith 1975; Yasuda and Hirota 1977) — (1) viral DNA complexed with → histone in a → chromatin-like structure in permissive cells infected with a virus. (2) A linear → plasmid constructed by joining together cloned DNA sequences (prokaryotic or eukaryotic), such as replication origins (AR's), → centromeres (CEN) and → telomeres. The centromere is necessary both for stable attachment to the → spindle apparatus and for regulation of the copy number of such chromosomes and their proper segregation during mitosis and meiosis. (3) A eukaryotic → chromosome reduced in size by deletion or extrusion which might exist as a component additional to the normal chromosome complement.

mini-F (Timmis et al. 1975) — a fragment of the bacterial → F plasmid that exhibits the replication and incompatibility characteristics of the parent F plasmid.

minigene — a shortened version of a natural gene constructed by deleting internal sequences with the result that the new gene (→ hybrid gene) produces a much shorter mRNA.

minimal recognition length — the shortest length of complementary nucleotide sequences forming a stable double-stranded DNA complex in → genetic recombination.

mini-muduction (Faelen et al. 1979) — a type of Rec-independent → transduction mediated by mini-Mu phage, i.e., a derivative of the phage Mu which carries a long internal deletion. After infection with a Mu/mini Mu mixed lysate, transduction occurs at frequencies about ten times lower than generalized transduction. The mini-muductants are diploid for the transduced marker which is recovered integrated at random in the bacterial host genome (sandwiched between two mini-Mu genomes in the same orientation).

M. allows construction of → transposons carrying any identifiable gene from a bacterium in which Mu can be introduced and grow.

minimum interaction hypothesis (Imai et al. 1986) — the m.i.h. predicts that karyotypes of eukaryotes evolve toward reducing the genetic risk resulting from reciprocal → translocations and that the modes of karyotype evolution are framed mostly by the balance between nuclear volume at → pachytene and genome size. A high ratio of genome size to nuclear volume will result in frequent interaction of chromosomes and an increase of the probability of occurrence of reciprocal translocations (→ suspension arch structure).

minimum mutation distance method (Jukes 1963) — a method which uses the → genetic code for estimation of the number of nucleotide point fixations separating two genes that encode two homologous proteins since the time of their separation from a common ancestral gene. Paired amino acids are examined (a position at a time) for the minimal number of nucleotide differences between their codons. The sum of the minimum nucleotide differences over all aligned positions is the minimum mutation distance for the pair of amino acid sequences.

miniphage (Griffith and Kornberg 1974) — any → bacteriophage of less than unit length as observed among filamentous single-stranded DNA phages. M.contains only a fraction of the viral genome including the origin of replication and requires a → helper phage for its propagation.

miniplasmid (Timmis et al. 1975) — any → plasmid reduced in size by means of → recombinant DNA technology. Such miniplasmids contain only an origin of replication and plasmid functions necessary for plasmid replication together with selective markers for identification purposes.

minisatellite sequence (Jeffreys et al. 1985) — in the vertebrate genome, any of a class of many dispersed arrays of short (10—50 bp) tandem direct repeat motifs that contain variants of a common core sequence (10—15 bp). They exhibit a high degree of length variability, probably due to changes in the copy number of tandem repeats.

minor gene — = → polygene and → modifier gene.

minor transfer RNA — the less-frequent forms of a group of chemically distinct → transfer RNA molecules which all have the same → anticodon. Their role is unknown, but mutant m.tRNA molecules sometimes serve as intergenic → suppressors.

minor spiral (Sax 1932) — = minor coil (→ chromosome coiling).

minus 10 sequence — → Pribnow box.

minus 35 sequence — a region of DNA upstream of prokaryotic → promoters which is centered about 35 nucleotides from the mRNA initiation site. The −35 s. is thought to be involved in the initial recognition between → RNA polymerase and the promoter site.

The conserved sequence in the −35 region of most prokaryotic promoters is:

5′	T	T	G	A	C	A	3′
3′	A	A	C	T	G	T	5′
	−36	−35	−34	−33	−32	−31	

Minute (Brehme 1939) — in *Drosophila*, any of a group of dominant mutants (probably → deletions of sites of → transfer RNA genes) that exhibit (in heterozygous condition) a fairly uniform phenotype. The mutants show a slow rate of development, decreased body size, smaller bristles, and residual fertility. Minutes are scattered throughout the genome and all are lethal in homozygous condition as expected from the absence of a tRNA species.

minute fragment (Muller 1940) — a chromosome fragment smaller than the diameter of one chromatid.

mis-aminoacylation — → aminoacylation.

mischarging mutation (Ozeki et al. 1980) — → transfer RNA mischarging.

miscoding — = → mistranslation.

miscopying (Beadle 1957) — → gene mutation.

misdivision (Darlington 1939) — → centromere misdivision.

misdivision haploid (Kimber and Riley 1963) — → haploid.

misincorporation frequency — the overall rate of base misinsertion during → DNA replication by a → DNA polymerase. M.f. is affected by two factors: The base misinsertion frequency, i.e., the frequency that the polymerase adds an incorrect nucleotide to the 3′ terminus, times the frequency that the incorrectly incorporated base escapes the proofreading step. In the absence of proofreading, the m.f. is equal to the misinsertion frequency. The accuracy of polymerases that do not proofread (most eukaryotic enzymes) is thus limited by the misinsertion frequency.

misinsertion frequency — → misincorporation frequency.

mismatch — → DNA mismatch.

mismatch correction — → methylation-instructed mismatch repair.

mismatch repair — → DNA mismatch repair.

mispairing — the presence in one polynucleotide chain of a DNA double helix of a nucleotide which is not the complement of that at the corresponding position in the other polynucleotide chain (→ gene mutation; heteroduplex).

misreading — = → mistranslation.

misrepair — → DNA repair which gives rise to gene mutations (→ SOS repair) or the production of chromosome structural changes (→ misrepair mutagenesis).

misrepair mutagenesis (Witkin 1969) — the production of → gene mutations and/or chromosome structural changes (→ chromosome mutation) due to → DNA lesions which have been misrepaired or not repaired at all (→ DNA repair). M.m. typically results in a wide spectrum of mutations, including both frameshift and base pair substitution mutations. Deletions may arise from the gaps (by joining together the ends of gaps) induced in replicating DNA by lesions that cannot act as templates (→ SOS repair).

misreplication — erroneous base pairing during → DNA replication directed by, e.g., base analogs and mutagen-modified bases. Agents resulting in such mispairing lesions are called "direct" mutagens as opposed to "indirect" mutagens which cause mutations via the action of error-prone DNA repair systems.

missense codon — → missense mutation.

missense mutation (Brenner et al. 1961) — a → base substitution (→ gene mutation) that results, during → genetic translation of the messenger RNA, in amino acid substitution at a specific position in the polypeptide chain (→ samesense mutation; nonsense mutation).

Whether a mutant phenotype is readily detectable depends on the particular amino acid substitution. Certain types of amino acid substitutions affect the assigned functions of a → cistron less drastically than others (conservative subsitution). Some change the net electric charge of the polypeptide chain, so that a mutant chain becomes distinguishable from the original wild-type chain by electrophoresis. A m.m. may escape detection in those cases where the insertion of the incorrect amino acid does not lead to an appreciable change in the conformation and subsequent function of the protein. A m.m. can be suppressed (→ missense suppressor) by secondary mutations affecting transfer RNA structure and function (→ suppressor tRNA).

M. mutations may have one or several of the following attributes: (1) Leakiness; (2) conditional expression; (3) formation of cross-reacting material; (4) intracistronic complementation.

missense suppressor (Yanofsky et al. 1961) — a → suppressor that acts on a → missense mutation and inserts the correct amino acid at the site of the missense codon. Missense suppressors represent mutationally altered tRNAs (→ suppressor tRNA) which accept the relevant amino acid and insert it into a polypeptide chain in response to an altered (missense) codon. Since a codon is missense at some mutated position and sense everywhere else, suppressor tRNA species cause generalized → mistranslation. They are, therefore, generally very inefficient and cells which carry them grow more slowly than normal (→ nonsense suppressor).

mistranslation — erroneous translation of the → genetic code by → codon misreading, i.e., by incorporation (misinsertion) into proteins of amino acids other than the one dictated by the standard codon-anticodon recognition. The basal-level m. of certain sense codons can equal or exceed 5×10^{-3} (→ context effects have been noted). A small amount of ambiguity (recognition of a given codon by more than one species of tRNA) is probably an intrinsic property of the translation process. Ambiguity in the → genetic code potentially resulting in m. may be due to: (1) A codon is misread owing to environmental conditions existing around the tRNA-mRNA-ribosome complex which interfere with the codon recognition process. (2) A codon is misread because of a specific change in the primary, secondary, or tertiary structure of either the tRNA or the ribosome such that it affects the codon-anticodon recognition. (3) A tRNA synthetase may misrecognize either an amino acid or a tRNA, leading to the formation of the wrong → aminoacyl tRNA.

→ Suppressor tRNA's, besides their major decoding specificity (including → wobble), possess some minor decoding properties which result in m. (→ missense suppressor). A mutation which affects the major decoding specificity of a tRNA species also entails a specific change in its m. specificity.

mitochondrial complementation — increased biochemical activity (as measured by the rate of oxidation and efficiency of phosphorylation) of a 1:1 in vitro mixture of → mitochondria isolated from two potential parents over the arithmetic mean of the two preparations when assayed separately. It has been claimed that there is a correlation between m.c. and → m. heterosis in F_1 hybrid plants.

mitochondrial crest — any of the complex infoldings of the inner membrane of the → mitochondrion which project into the matrix. Shape and disposition of m. crests vary in different cells.

mitochondrial DNA (Nass and Nass 1962; Luck and Reich 1964) — (mt DNA); the double-stranded DNA of eukaryotic → mitochondria (the mitochondrial genome) which codes for a single set of → ribosomal RNA's, a minimum set of 24 → transfer RNA's, and about 10 proteins (among them subunits of ATPase, cytochrome oxidase, and cytochrome b apoprotein). Plant mt DNA is characterized by a large and variable size, a low rate of nucleotide sequence divergence (in coding sequences), and a high extent and frequency of genome rearrangements by homologous recombination. Vertebrate animal mt RNA shows little size variation, diverges in sequence at a relatively rapid rate within coding sequences, and has maintained an invariant gene order, i.e., mt DNA evolution has taken different courses in plants and animals. Some of the fungal mt DNA's, but not animal mt DNA's, have been found to contain → introns in genes encoding the large rRNA and in certain protein encoding genes. mt DNA is associated with proteins and acquires a → chromatin-like condition.

mt DNA's from most animal cells are closed circular molecules about 15 kb in length; mt DNA from lower eukaryotes can be circular (yeast) or linear (*Paramecium*) and range in length from 15 kb up to 75 kb. Plant mt DNA is the largest, with a size of 330—2 400 kb, and is usually isolated as linear molecules of heterogeneous length.

There is evidence that → nuclei may contain sequences homologous to mitochondrial genes, i.e., mt DNA can travel within the cell (→ promiscuous DNA).

Among the features of mt DNA variability pertinent to the analysis of evolutionary processes are the following: (1) A rapid rate of evolution involving primarily base substitutions and very small additions/deletions; (2) remarkable stability of gene content and arrangement; (3) apparent mt DNA sequence homogeneity (homoplasmicity) in somatic cells within an individual organism; (4) extensive sequence polymorphism among conspecies; (5) maternal transmission (in the majority of cases) of mt DNA to progeny.

Recombination between different mitochondrial genomes is known to occur (yeast) and yields recombinant mitochondrial genotypes. There is evidence for symmetrial → genetic transcription of mt DNA, i.e., both strands of the double-stranded molecule are transcribed into RNA (tRNA). Animal mt DNA is replicated unidirectionally from its origin of replication (→ DNA replication). Most of the enzymes involved in replication, transcription, and translation of mt DNA in animals and fungi are coded in the nuclear genome, synthesized in the cytoplasm, and transported into the mitochondrion.

Unusual features of the → genetic code and of the codon recognition pattern are utilized in mitochondria from different organisms (Table 9). Most conspicuous, and found in all mitochondrial genomes, is the use of UGA as a tryptophan codon rather than as a stop codon. Additionally, there are genetic code differences between the individual mitochondrial genetic systems (even within the same phylogenetic group) and a different decoding mechanism that allows reading of all codons by a smaller number of tRNA's (24) than in the standard code (two-out-of-three base interaction between codon and anticodon in the four-codon family boxes).

Besides the large complex genome, the mitochondrion of many higher plants contains smaller, either circular or linear, DNA molecules which are assumed to replicate autonomously of the high molecular weight genome and are referred to as mitochondrial plasmids.

Table 9: Exceptional codons employed by mt DNA

Mammals	UGA	Trp
	AUA	Met
	AGA, AGG	termination
	AUA and (possibly) AUU	initiation
Aspergillus nidulans	UGA	Trp
Neurospora crassa	UGA	Trp
Saccharomyces cerevisiae	UGA	Trp
Drosophila	AUA	Met
	UGA	Trp
	AGA (possibly)	Ser
Zea mays	UGA	Trp
	CGC	Trp

mitochondrial genome — the circular or linear → genome of → mitochondria specified by → mitochondrial DNA. Recombination can occur between m. genomes of different species as well as within the same genome. Mitochondrial plasmids and autonomously replicating sequences from mt DNA have been described. Sequences homologous to plasmid-like mitochondrial DNAs have been found also in → chloroplast and → nuclear genomes. The mt genome differs from the nuclear genome by being outside the nucleus, existing in a multitude (thousands) of copies per cell, being inherited maternally, and evolving quickly.

mitochondrial heterosis (McDaniel and Sarkissian 1966) — the superior activity (due to → mitochondrial complementation) of mitochondria isolated from heterotic (→ heterosis) F_1 hybrids as compared with mitochondria from the two parents of the hybrid. M.h. is claimed to indicate the yield potential in a number of plants.

mitochondrial plasmid (Pring 1977) — in the → mitochondria of several higher plants and in some fungi, any of the small → plasmid-like DNA molecules (minicircles and minilinears) which can be classified into two groups: (1) Those which show some sequence homology with the major → mitochondrial DNA; (2) those showing no sequence homology with the major mt DNA.

Some of the known m. plasmids resemble linear DNA viruses in having proteins attached to their ends, terminal inverted repeats and long open reading frames *(Zea mays).*

mitochondrial ribosome — (1) the 72—74S mitochondrial ribosomes of the Ascomycetes; (2) the 55—60S mitochondrial ribosomes of mammalian and amphibian cells; (3) the 80S mitochondrial ribosomes of plants and protozoa.

mitochondrial RNA — RNA complementary to → mitochondrial DNA.

mitochondrial sheath — → spermatozoon.

mitochondrial transformation (Clark and Shay 1982) — → mitochondrion-mediated transfer of genetic information, i.e., the → genetic transformation of eukaryotic (mammalian) cells by endocytotic uptake of purified → mitochondria. The resultant cells are termed mitochondrial transformants.

mitochondrion (Benda 1897) — any of a class of semi-autonomous eukaryotic cell organelles (about 0.5 to 1.0 μm in diameter, and 5 to 10 μm in length) having an own genetic system (the mitochondrial genome) encoded in → mitochondrial DNA.

The primary function of mitochondria in the cell is to synthesize ATP from ADP and P_i and all mitochondria contain the enzymes, coenzymes, and other cofactors required for individual enzyme complexes which catalyze respiration and phosphorylation.

Morphologically, there are two types of membranes (each about 5 nm thick), the inner and the outer membranes. The former constitute a closed system consisting of two parts, the cristae mitochondriales (lamellar or tubular) and a peripheral part extending along the inner surface of the outer membrane. The cristae are folds of this peripheral membrane, and the inner membranes can be viewed as parts of a sac with a number of folds extending toward the center of the sac. The two parts of the inner membranes are chemically and functionally identified.

The biogenesis of the m. requires the function of two genetic systems. It entails the post-translational ATP-dependent import of the majority of the mitochondrial proteins (encoded by nuclear genes) and involves: (1) Specific interaction of precursors with the mitochondrial outer membrane; (2) energy-dependent translocation of polypeptides across one or both mitochondrial membranes; (3) processing of the precursors by a matrix-localized (the space enclosed by the inner membrane) protease which removes a cleavable presequence.

The m. contains distinctive ribosomes, ribosomal RNA (21S and 12S), tRNA species, aminoacyl-tRNA synthetases, initiation, elongation, and termination factors, i. e., the components of a minor system for synthesis of proteins coded for by mitochondrial DNA. Poly(A)-containing → messenger RNA has been found in animal mitochondria.

The endosymbiontic theory of the origin of mitochondria implies that the m. originated from trapped prokaryotes (bacteria).

mitodepression (Östergren et al. 1958) — a decrease in mitotic frequency which may be experimentally induced by a large number of → mitotic poisons or → antimitotic agents.

mitogenic — of any agent (mitogen) which stimulates quiescent (G_0) cells to enter the → cell cycle, i.e., stimulates → DNA replication, mitosis, and growth of cells (→ mitogen-releasable protein).

mitogen-releasable protein (Nilsen-Hamilton et al. 1980) — (MRP); any protein whose synthesis is specifically stimulated by → mitogenic agents. M.-r. proteins are released into the medium and represent a group of glycoproteins which have a medium, monomer molecular weight of 34.000 Da. MRP's are possibly involved in the growth-stimulatory response or are perhaps growth factors themselves.

mitoplast — (1) a → mitochondrion from which the outer membrane has been removed. (2) A → cytoplast of mitotic cells after extrusion of the chromosomes.

mitoribosome — a → mitochondrial ribosome.

mitosis (Flemming 1882) — in eukaryotes, a mode of nuclear division (= karyokinesis) which provides for the production of two daughter nuclei that contain identical chromosome sets and are genetically identical to one another and to the parent → nucleus from which they arose (→ amitosis; endomitosis).

Two main types: (1) In the "closed" type (intranuclear division), the chromosomes remain within the nucleus during the whole mitotic cycle; (2) in the "open" type, the → nuclear envelope disintegrates and disappears during prophase and reappears during telophase. Type (1) is present in lower eukaryotes of both plant and animal Kingdoms (→ dinomitosis).

M. marks the completion of one → cell cycle and the beginning of the next and involves the periodically assembled structure referred to as the → mitotic apparatus. It is a genetically controlled process coordinated with the division of the cell into two genetically equivalent daughter cells (→ cytokinesis; cell division) and represents a continuous sequence of events which may be divided into five main stages as follows: (1) **Prophase:** The chromosomes which reduplicated during → interphase of the cell cycle make their appearance in species-specific number. At the beginning of this stage they are long threads consisting of two → chromatids each. During prophase, which ends with the disruption of the nuclear envelope, the chromosomes become progressively shorter and more compact owing to → chromosome coiling which transforms them from the functional interphase form into the condensed "transport form". (2) **Prometaphase:** The → spindle is formed to which the → centromeres of each chromosome become attached. Subsequently the chromosomes undergo congression movement (→ chromosome movement) and finally come to lie at the spindle equator. (3) **Metaphase:** The chromosomes reach their metaphase arrangement on the metaphase plate, an equilibrium position with all centromeres aligned at the spindle equator (→ centromere orientation). (4) **Anaphase:** The two sister chromatids of each chromosome separate by their movement toward opposite poles of the spindle. (5) **Telophase:** The chromatids ("daughter chromosomes") of each chromosome arrive at the poles and reconstitution of a nuclear envelope around each of the two daughter nuclei takes place. Each of the two nuclei is identical in its chromosome complement to the other, to the original complement, and similar in the content of → genetic information coded in the chromosomal DNA.

The mitotic division cycle of the nucleus characterized by the regular alternation of interphase chromosome reduplication and accurate distribution of the reduplication products is typically followed by → cytokinesis, i.e., division of the cell's cytoplasm.

The mitotic pattern leading to the production of two separate nuclei may show species-specific variations which are classifiable as follows (Mazia 1956):

(1) According to nuclear behavior during prophase: (a) Disappearance of the nuclear envelope (most plants and animals); (b) persistence of the nuclear envelope (many protozoa).

(2) According to chromosome alignment at metaphase: (a) All chromosomes collected together in the metaphase plate (many large cells with small chromosomes); (b) centromeres only aligned on the equatorial plane; the arms of chromosomes dangle from the equatorial

plane (many small cells with large chromosomes); (c) equatorial alignment not evident (pollen tubes).

(3) According to the form of the spindle (→ mitotic apparatus): (a) Continuous spindle fibers from pole to pole intermingled with chromosomal fibers connecting chromosomes to poles (many plant and animal cells); (b) continuous "central spindle" distinct from "half spindles" connecting chromosomes to poles; chromosomes may be arrayed around the circumference of a central spindle (many plant and animal cells); (c) spindle functions entirely within the nuclear membrane (many protozoa; some plant and animal cells); (d) each chromosome moves within a distinct spindle element of its own (certain insects).

(4) According to → mitotic centers: (a) Distinct centers with asters ("astral" or "amphiastral mitosis") which may be represented by → centrioles lying in a larger → centrosome, or large centrospheres without centrioles (most animals, many lower plants); (b) not distinct centers at the poles ("anastral mitosis") (most higher plants and some invertebrates).

(5) According to chromosome-to-pole connections and chromosome movement: (a) Chromosomes move toward the poles as the chromosomal fibers shorten (most plant and animal cells); (b) chromosomes behave as though they are being dragged by fibers attached along their entire length (diffuse centromere activity) (some animals and plants); (c) chromosomes move with their arms ahead of the centromere ("neocentric activity") (some insects and plants).

(6) According to the relative rates of movement of individual chromosomes: (a) All chromosomes move at the same rate, independent of their size (most cases); (b) certain chromosomes move at a different rate in comparison with others (e.g., sex chromosomes).

(7) According to elongation of the spindle and shortening of chromosomal fibers: (a) Chromosome sets separate by simultaneous shortening of chromosomal fibers and elongation of the spindle; (b) shortening of chromosomal fibers precedes elongation of the spindle.

Further variants concern the process of → cytokinesis, i.e., the segregation of daughter nuclei in separate cells.

Chromosome reduplication and cell growth without formation of daughter nuclei or new cells may take place and is referred to as → "endomitosis" or "endoreduplication". It results in either endopolyploid cells or polynemic chromosomes if the products of endoreduplication do not become individualized (→ polytene chromosome).

Since every sexual cycle consists of two phases, a haploid between → meiosis and → fertilization and a diploid one between fertilization and meiosis, m. can occur during either or both of these phases. In → haplontic organism m. is confined to the haplophase, in → diplontic organisms it is confined to the diplophase, and in → diplo-haplontic ones m. occurs during both haplo- and diplophase.

mitospore — any haploid or diploid cell resulting from mitotic division that undergoes mitotic growth to reproduce the kind of organism from which it arose (→ germ cell).

mitostatic — → antimitotic.

mitotic — ref. to or produced by → mitosis.

mitotic apparatus (Mazia and Dan 1952) — a transient ensemble of structures which appears during → mitosis (and meiosis) and functions in the ordered distribution of chromosomes during mitosis (and meiosis). The major functional components of the m.a. are: (1) The → spindle poles (→ centriole); (2) → centromere regions of the chromosomes; (3) → microtubules.

The fully formed m. a. contains three classes of microtubules (MT's): (1) Kinetochore (or centromere) MT's with one end attached to the kinetochore and the other free or attached to a pole; (2) polar MT's having one end attached to the pole and the other free or at the other pole; (3) MT's with both ends free.

Almost all blocks of cell division are associated with either structural or functional abnormalities of the m.a. (→ C-mitosis; mitotic poison).

mitotic center — a structure that regulates → spindle assembly and defines a mitotic pole towards which the chromosomes move during → mitosis (and meiosis). In some organisms, the functions of the m. centers are associated with the → centrioles. After maturation, many → microtubules are associated with each m.c., forming a radial array.

mitotic cycle — the sequence of steps by which the nuclear genetic material is equally partitioned prior to division of eukaryotic cells into daughter cells (→ mitosis).

mitotic delay — delayed progress of the mitotic cycle characteristically observed after treatment with high energy radiations and chemical mutagens.

mitotic haploidization (Pontecorvo et al. 1954) — a process in the → parasexual cycle of fungi, during which whole chromosomes reassort at random during mitosis, eventually producing nuclei with half the diploid chromosome complement by repeated → nondisjunction.

mitotic index (Minot 1908) — that fraction of the total number of cells scored in a meristematic tissue which are actively engaged in → mitosis, irrespective of their stage (prophase to telophase).

The m.i. appears to be a function of the generation time for a given species under the conditions provided and the actual length of time a cell of this species remains in mitosis. The m.i. for a cell population in which the period of mitosis is brief will be lower than for a population with a comparable → cell cycle span in which mitosis is prolonged.

mitotic inhibition (Gavaudan 1943) — the inhibition of → mitosis (and by extension also → meiosis) accomplished naturally or artificially by (Mazia 1961): (1) Diverting cells from the preparations for division; (2) inhibiting any one of the preparations for division; (3) arresting cells at the transition stage between interphase and division; (4) disabling the mitotic apparatus (→ metaphase arrest). Theoretically, all such blocks are reversible and not fatal to the cell.

mitotic nonconformity (Bainbridge and Roper 1966) — mitotic instability resulting in parent and daughter cells that do not conform in genotype. In *Aspergillus* strains, m.n. is due to the presence of a chromosome segment in duplicate (one in normal position, one translocated to another linkage group), i.e., a consequence of chromosomal unbalance.

mitotic poison (Dustin 1934) — operationally defined as any of many different kinds of noxious agents which act to impair → mitosis (and also by extension → meiosis), irrespective of whether the impairment takes place during or following the treatment of actively dividing cells. The mitotic disturbances may result in morphological changes of the chromosomes and/or inhibition of mitosis as well as cell death.

Mitotic poisons may be classified according to various principles.

Bauch (1947) proposed the distinctions → spindle poisons, cell division poisons, and chromosome poisons. Ahlström (1951) recommended the terms spindle poisons, interphase poisons, karyoclastic poisons; Dustin (1952), spindle and metaphase poisons, perfect and partial → radiomimetic poisons; d'Amato (1954), inhibitors of cytokinesis, inhibitors of the spindle, preprophase and prophase poisons; Heilmeyer (1950), interphase poisons, mitotic poisons sensu strictu, spindle poisons, and poisons of cytokinesis; and Loveless and Revell (1949) distinguished mitotic poisons with and without radiometic effects and mitotic poisons of the spindle-poison type.

mixed oligonucleotide probe (Wallace et al. 1981) — a mixture of synthetic oligodeoxynucleotides representing all possible → codon combinations for a short peptide sequence. M.o. probes are used for the screening of libraries of cloned → cDNA (→ cDNA bank).

mixoploid (Nemec 1910) — of cell populations whose component cells differ in their chromosome numbers, irrespective of whether these numbers are → euploid or → aneuploid. Thus, the term mixoploidy covers all types of mosaics or → chimeras in which the heterogeneity between different elements involves differences in chromosome number. Mixoploidy may stem from a variety of mitotic irregularities, from cellular and nuclear fusions, or even from amitotic processes, if these give rise to viable products. A special case of mixoploidy is endopolyploidy arising by → endomitosis or endoreduplication of chromosomes (Levan and Müntzing 1963).

mixture disequilibrium (Chakraborsky and Weiss 1988) — a form of gametic association between two loci occurring at admixture between two populations with different allele frequencies at these loci, as distinguished from gametic association between closely linked loci.

mobile dispersed genetic element (Ilyin et al. 1980) — (mdg); = → transposable genetic element.

mobile intron (Osiewacz and Esser 1984) — an → intron which may be excised at the DNA level (DNA splicing) and thus become a → mitochondrial plasmid (causing senescence in *Padospora*).

mobilizable — of a → plasmid able to provide the gene products which mediate its own → mobilization.

mobilization (Novick et al. 1976) — promotion by a conjugative bacterial → plasmid of conjugal transfer of normally unlinked replicons (nonconjugative plasmids as well as the main bacterial chromosome). (→ F plasmid).

modal dominance — → dominance.

modal number (White 1945) — (1) the most common chromosome number of a particular taxonomic group (genus, family, class, order), there being no implication that it is the ancestral number of the group in question (→ type number).

(2) The most common chromosome number in a tumor cell population observed in a direct preparation or following short-term culture (incubation for a few days without → mitogenic agents).

mode — a peak in a frequency distribution. The modal value represents a trait value whose frequency is greater than those of adjacent higher or lower values. If the distribution is unimodal (has only one m.), the modal value corresponds to the trait value having the highest frequency in the distribution.

modificability — the property of a → genotype to react in response to varying developmental conditions by particular alterations in its → resultant phenotype (→ reaction norm) called → modifications. M. is genotypically controlled with respect to its extent and form (→ flexibility).

modification (Naegeli 1884) — any environmentally induced, nonheritable change in the → phenotype of an individual, in contrast to a → mutation (→ dauermodification). That period of time during which the mode of manifestation of a particular → character may be modified is referred to as the "period of modification" (→ modificability) of the character concerned. Many characters are modifiable during a large part of the individual's life cycle, others may be influenced only during one or a few short periods ("sensitive periods") of development preceding their definite determination.

modification methylase — any of a group of site-specific enzymes that protect bacterial DNA from endogenous type II → restriction endonucleases (→ DNA modification-restriction). The type II m. methylases require S-adenosylmethionine (SAM) but not Mg^{2+} for their activity. These methylases methylate within the sequence recognized by the corresponding endonuclease at either adenine or cytosine.

modification-restriction system — → DNA modification-restriction.

modified base — any of the nucleic acid bases except the usual four from which DNA (T, C, A, G) or RNA (U, C, A, G) are synthesized; m. bases result from post synthetic changes in the nucleic acid.

modifier gene (Bridges 1919) — any gene that by interaction (→ gene interaction) affects the phenotypic expression of genes at other loci. Many modifiers can be detected only by their effects on the expression of other nonallelic genes, but others may have phenotypic effects of their own. According to the type of modifying action such genes may be classified into: (1) **Enhancers** (= intensifiers, extension genes) which intensify the phenotypic effect or increase the → mutability ("mutator genes") of other genes. (2) **Reducers** (= restriction genes) which decrease the effect of other genes or ("inhibitors") completely prevent their manifestation. They are called → suppressor genes when they shift the manifestation of a mutant allele toward the → phenotype of the corresponding wild type or normal allele (→ reversion).

modulating codon (Ames and Hartman 1963) — any → codon responsible (directly or indirectly) for controlling the frequency with which cistrons are transcribed and, perhaps, the frequency with which messengers are translated (→ modulation).

modulation — (1) any of those reversible alterations which cells may undergo in reaction to different environmental conditions without changing their essential nature (e.g., any of the histological types which a given tissue may assume dependent on the nature of its environment) (Weiss 1939).

(2) The more frequent reading of certain sequences of polycistronic → messenger RNA during → genetic translation. The polycistronic messenger produced by → genetic transcription of an → operon shows a certain constant probability of being read in its entirety; a higher probability of its first half being read, and an even higher probability of only its initial quarter being read. This influence on the read-out process results in the disengagement of the ribosomes from the messenger or in destruction of the terminal section of the messenger RNA beyond the point where genetic translation was temporarily halted. The result of this phenomenon, which was termed modulation by Ames and Hartman (1963), is a greater number of polypeptide chains formed for those genes in the operon which are proximal rather than distal to the → operator end.

Modulation of this type is unterstood as a genetic adjustment in the relative rates of synthesis through coding for → transfer RNA species of different abundance. According to Stent (1964), the regulation of protein synthesis may be achieved by the control of the availability of so-called modulating transfer RNA species which correspond to specific "modulating condons" or "modulating triplets" in the messenger RNA. The "modulating codons" are presumed to concern degenerate representations of particular amino acids, to affect the reading of the messenger by the ribosomes and thus to determine the relative molar quantities of each polypeptide coded for by the operon in question. Translation of any messenger containing one or more of these codons is assumed to proceed only so long as the corresponding modulating transfer RNA species capable of base pairing with the modulating codons are at hand.

molecular biology (Astbury 1950) — a branch of modern biology in which biological phenomena and processes are studied not only from a phenomenological point of view but by physical-chemical and biochemical investigations at the molecular level. M.b. is based on developments in genetics, biochemistry, physical chemistry of macromolecules, and chemical physics, and represents an effort to account for biological events in terms of the established principles of physics and chemistry.

molecular chaperon (Laskey et al. 1978) — any of a class of cellular proteins whose function is to ensure that the folding of certain other polypeptide chains and their assembly into oligomeric structures occur correctly.

molecular clock hypothesis (Zuckerkandl and Pauling 1965) — a hypothesis postulating that the rate of molecular evolution is approximately constant over time among different evolutionary lineages and reflects the divergence time between taxa. The m.c.h. implies that gene mutations occur at a sufficiently regular interval to permit dating of phylogenetic dichotomies.

molecular cloning (Cohen et al. 1973) — the multiplication of DNA sequences usually involving the isolation of appropriate DNA fragments and their in vitro joining (insertion into a restriction site) to a → cloning vector capable of replication when introduced into an appropriate host. M.c. requires: (1) DNA of interest (foreign, passenger, or target DNA); (2) a cloning vector; (3) → restriction endonucleases; (4) → DNA ligase; (5) a prokaryotic or eukaryotic cell to serve as the biological host.

→ Transposable genetic elements can be used to transpose and clone genes in vivo by a process involving multiple transposition steps that results in two copies of the transposable element flanking the DNA fragment to be cloned. This structure is usually placed on a new replicative element such as a conjugative → plasmid because transposable genetic elements do not usually have replicons.

M.c. is a powerful tool for analyzing the structure and function of both prokaryotic and eukaryotic DNA's, and for amplifying specific gene products (→ recombinant DNA technology).

If a gene product to be amplified by m.c. is derived from a bacterial or phage gene, the procedures for protein production are straight forward. If the gene is from a eukaryote, special problems arise: (1) The eukaryotic promoter may not be recognized by a bacterial host RNA polymerase; (2) the mRNA transcribed from eukaryotic genes lacks the → Shine-Dalgarno sequence needed for binding the bacterial ribosomes; (3) the mRNA may contain → introns that must be excised; (4) the gene product often must be processed; (5) eukaryotic proteins are often recognized by bacterial proteases as foreign and are cleaved.

molecular disease (Pauling et al. 1949) — a genetically determined biochemical disorder in which a specific enzyme defect produces a metabolic block with pathological consequences (= inborn error of metabolism).

molecular drive (Dover 1982) — the hypothetical mechanism giving rise to → concerted evolution of → multigene families. M.d. is based on nonreciprocal exchanges of sequences between member genes (unequal crossing over, gene conversion, transposition) with the long-term effect of homogenizing a family throughout a population of individuals for newly arisen variant repeats. The relative rate of sequence homogenization processes within defined regions of the genome (intrachromosomal homogenization) versus the rate of homogenization throughout the genome (interchromosomal homogenization) is assumed to determine most of the actual divergence among the individual members of a repeat family. The mechanism underlying nonreciprocal exchanges, which may be random or biased in activity, provides the driving force of cohesive genetic change in a population. M.d. is seen as a mechanism which operates independently of traditional evolutionary forces, selection, and genetic drift, in shaping the extent of genetic divergence among populations.

molecular evolution — → evolution at the molecular level of DNA sequences and proteins. The neutral theory of m.e. (Kimura 1968) assumes that the selective advantages of most alleles are effectively zero, so that the difference of homologous DNA sequences or proteins of different species should be unterstood rather as a result of random fixation of one of the selectively neutral alleles than as a result of natural selection.

molecular genetics — a branch of → genetics which strives to account for genetical events in terms of the established principles of physics and chemistry. In the broadest sense, m.g. is the study of those aspects of genetic systems, or components thereof, which can be described at the molecular level. In a more restrictive sense, the study of the molecular aspects of genetic mechanisms and the control of metabolic processes by genetic material (Steiner 1965).

molecular hybridization — = annealing (→ DNA/DNA hybridization).

molecular linkage map — any → linkage map based largely on cloned DNA sequences; variation is detected directly at the DNA level, the alleles behave in a codominant manner, and normally have no visible effect on the phenotype (→ restriction fragment length polymorphism). Practical applications of m.l. maps include tagging and tracking of major genes.

molecular phylogenetic tree — an evolutionary tree, i.e., a phylogenetic topology, based on molecular data (nucleotide or amino acid sequence comparisons) derived from the presence and expression of a single structural gene. Step-by-step reconstruction of the evolution of these sequences is based upon the "maximum parsimony procedure" which minimizes the nucleotide changes or point mutations necessary to account for the descent of a contemporary sequence (→ molecular clock hypothesis; molecular evolution).

molecular spiral (Darlington 1935) — → chromosome coiling.

monad — an individual cell produced by a meiocyte in place of a → tetrad as a consequence of a meiotic abnormality.

monaster (Wilson 1901) — a unipolar → spindle which may arise, e.g., by the suppression of → centriole division, and which results in an irregular (unipolar) nuclear division giving rise to a → restitution nucleus.

monitor gene — = → reporter gene (→ gene fusion).

monoallelic (Atwood 1944) — of → polyploids with multiple allelic loci all of which have the same allele: $A_1A_1A_1A_1$ or $A_2A_2A_2A_2$ in a tetraploid. The other possible genotypes involving a multiple allelic locus in a tetraploid are:

diallelic	$A_1A_1A_1A_2$,	$A_1A_1A_2A_2$ etc.
triallelic	$A_1A_2A_3A_3$,	$A_1A_1A_2A_3$ etc.
tetraallelic	$A_1A_2A_3A_4$,	$A_2A_3A_4A_5$ etc.

monobrachial (Sorokin 1929) — → chromosome arm.

monocentric — (1) of chromosomes or chromatids with one localized → centromere (= monokinetic), in contrast to → di- and polycentric chromosomes; (2) of nuclear divisions with a unipolar → spindle (→ monaster).

monochlamydeous — = haplochlamydeous (→ periclinal chimera).

monocistronic — of → messenger RNA molecules that code for one polypeptide chain (→ polycistronic; transcription).

monoclinous — having male and female germ cells in one and the same flower (= synoecious or hermaphroditic).

monoecious (Darwin 1877) — of plants with male and female sex organs on the same individual and producing male and female → gametes (→ dioecious). Plants with male and hermaphroditic flowers or female and hermaphroditic flowers on the same individual are called andromonoecious and gynomonoecious, respectively. M. fungi include (Esser 1971): (1) All species which produce male and female sex organs on the same mycelium; (2) all species without sex organs in which each mycelium can serve both as nuclear donor and nuclear recipient.

monogametic organelle transmission (Birky et al. 1982) — in eukaryotes, the purely maternal inheritance of cell organelles due to failure of one of the parental gametes to transmit organelle genomes to the zygote, or due to destroyal of organelles or organelle DNA derived from one parent.

monogenesis — = → monogony.

monogenic — of character differences controlled by the → alleles for one particular locus, as opposed to di-, tri-, or polygenic control exerted by two, three, or many nonallelic genes.

monogeny — the production of only male ("arrhenogeny") or female ("thelygeny") offspring.

monogony (Haeckel) — asexual reproduction.

monohaploid (Katayama 1935) — → haploid.

monohybrid (de Vries 1900) — of individuals → heterozygous (e.g., Aa) with respect to one pair of → alleles and by extension also of crosses (e.g., AA × aa) involving parents that differ with respect to the alleles of one locus, as opposed to di-, tri-, and polyhybrid heterozygotes and crosses involving differences in two, three, or more pairs of alleles.

monoisodisomic (Kimber and Sears 1968) — of a cell or individual deficient in one chromosome but which has an → isochromosome for one of the arms of the missing chromosome (= haplo-triplodisomic: Khush and Rick 1971). (→ monotelodisomic; monotelosomic).

monoisosomic (Kimber and Sears 1968) — of a cell or individual which lacks one chromosome pair but has an → isochromosome for one arm of the missing pair (→ diisosomic).

monokaryon (Boveri 1907) — (1) a uninucleate cell, a spore, or a tissue consisting of such cells, as opposed to a → dikaryon (or multikaryon); (2) = → pronucleus (Boveri).

monolayer — a single layer of cells growing on a surface.
monolayering (Abercrombie and Heaysman 1953) — the tendency of cultured (animal) cells to occupy space evenly and to spread out in thin layers.
monomeric — = → monogenic.
mononeme — of chromosomes which contain only a single DNA double helix prior to replication, as opposed to → polytene chromosomes.
monophyletic — of individuals derived in the course of evolution from a single interbreeding population or phyletic stock, as opposed to polyphyletic.
monoploid (Langlet 1927) — of cells or individuals (monoploids) having one chromosome set with the → basic number of chromosomes, i.e., the lowest → haploid number of chromosomes in a polyploid series.
monoplontic — of a → monoploid individual (a monoplont) or phase of the life cycle (→ haplontic).
monosome (Haselkorn and Fried 1964) — (1) a single → ribosome bound to messenger RNA (→ polysome); (2) a chromosome in the chromosome complement lacking a homologous partner (→ monosomic).
monosomic (Blakeslee 1921) — of an → aneuploid (2n − 1) cell, tissue (in case of → chimeras), or an individual with one (or more than one: doubly monosomic, etc.) chromosome missing from an otherwise diploid or allopolyploid (and thus functionally diploid) chromosome complement. Monosomics characterized by the absence of one structurally normal chromosome are called **"primary monosomic".**

Monosomics may be obtained from several sources (Khush 1973): (1) Normal disomics, by spontaneous occurrence or by treatment with physical and chemical agents (→ nondisjunction); (2) asynaptic disomics and aneuploids; (3) polyploids, e.g., haploids and triploids; (4) intervarietal and interspecific crosses; (5) translocation heterozygotes; (6) trisomics; and (7) monosomics.

As a result of their unbalanced chromosome complement, monosomics which are diploid in respect to their other chromosomes are characterized by the presence of one unpaired → univalent in meiosis. The expected meiotic products are theoretically half monoploid and half (n − 1). Frequently, however, the univalents lag in chromosome movement and thus more (n − 1) gametes are produced than expected. Normally (n − 1) male gametophytes of diploid plants fail to function, but such female gametophytes of plants, as well as male and female gametes of animals, generally function.

A case of regular "monosomy" is represented by XX-XO-systems of → sex determination and the X-chromosome of the heterogametic sex is said to be a "monosome" (Montgomery 1904).

"Tertiary monosomic" (Rick 1943) is a term applicable to cells, tissues, or individuals that possess one product of a reciprocal → translocation but lack the reciprocal product and thus are composed of a segment of each of two different chromosomes of the chromosome complement.

"Pseudomonosomic" (Hiorth 1948) are cells or individuals which lack one chromosome (one centromere) but contain (due to → translocation) the essential segments of two nonhomologous chromosomes in one of their other chromosomes.
monotelic (Bauer et al. 1961) — → centromere orientation.
monotelodisomic (Kimber and Sears 1968) — of diploid → karyotypes with a chromosome pair represented by one normal and one → telocentric chromosome, the latter consisting of one of the two arms of the normal chromosome (→ trisomic; monoisodisomic; diisosomic; teloisosomic).
monotelomonoisosomic (Kimber and Sears 1968) — a cell or individual in which one chromosome pair is missing but a telocentric for one arm of the missing pair and an isochromosome for the other arm are present (→ teloisodisomic).
monotelotrisomic - → trisomic.

monozygotic — → dizygotic.

Morgan unit (Haldane 1919) — a unit of → genetic recombination by → crossing over (= 100 centimorgans), i.e., the distance separating two genes (loci) between which there is on average one crossing over per → meiosis (→ map unit). A centimorgan indicates one percent of recombination. For short intervals, the map length in M. units is equal to the → recombination fraction expressed as a decimal, since doublecrossovers will have a negligible frequency.

morph (Huxley 1955) — any of the genetic forms (individual variants) that account for → genetic polymorphism.

morphism (Huxley 1955) — = → genetic polymorphism.

morphogen (Turing 1952; Gierer 1977) — any factor with the characteristics of occurring naturally, exerting morphogenetic effects at low concentrations, and in a way that correlates with morphogenetic activities in time and space (→ morphogenesis).

morphogenesis — the ensemble of processes through which embryos and parts of embryos change their form and an adult organism arises. M. involves → cell division and → cytodifferentiation and results in the development of a three-dimensional structure which involves arrangements of cell populations in precise positioning and organized manner (→ morphogen). Cell-cell recognition and cell-cell interactions are important during m. to obtain the morphological and physiological relationships unique to the specific tissue or organ and the disposition of these components in the embryo. The basic morphogenetic plan (→ morphogenetic information) of multicellular organisms is established by cell interactions during embryonic development.

morphogenetic information (Soll and Waddell 1975) — information for the genesis of a particular morphology which cells accumulate as they progress toward that morphology (→ morphogenesis). Once accumulated, m.i. is retained throughout the following stages.

morphosis (Sachs 1894) — any nonadaptive and usually unstable variation (→ modification) in an individual's → morphogenesis associated with certain changes in its external environment. Morphoses which mimic known mutations in their effect are referred to as → phenocopies.

morph-ratio cline (Huxley 1955) — a gradual change in the proportion of two or more distinct varieties from one place to another.

mosaic evolution — the tendency of different characteristics of the organism to evolve at different rates in the same evolutionary lineage.

mosaic gene — in eukaryotes and their viruses, any → gene consisting of contiguous → exons and → introns (= chimeric gene).

mosaicism — the presence within one individual (called a "mosaic") of cells differing with respect to their → idiotype, chromosome structure, or chromosome number (→ mixoploid). The different sectors of a mosaic may result from the mutation of chromosomal and extrachromosomal → hereditary determinants, somatic → crossing over (→ twin spot), changes in chromosome number (→ genome mutation), chromosomal structural changes (→ chromosome mutation), or "V-type" → position effects.

A "phenotypic sex mosaic" is an individual whose "mosaic constitution" is due to a development initially directed toward the male and then toward the female phenotype or vice versa (→ gynandromorph).

mottling (Muller 1930) — (1) = mosaicism or variegation; (2) → chromosome mottling.

mouse spot test — in → mutagenicity testing, an in vivo somatic mutation test which detects expression (or uncovering) of recessive markers which can occur by any one of a number of mechanisms: (1) Gene mutation in the wild-type allele; (2) deletion of a chromosome segment involving the wild-type allele; (3) nondisjunctional loss of the chromosome carrying the wild-type allele; (4) somatic recombination.

The genetic alteration is induced in a mouse melanocyte precursor cell of a midgestation embryo that is heterozygous for several recessive coat color genes. Animals with a mutant clone are detected postnatally (after hair growth has occurred) by examiniation for mosaic patches.

The m.s.t. is a quick prescreen for the more extensive tests of heritable → DNA damage.

movable genetic element — = → transposable genetic element.

movement index (Slizynsky 1955) — → chiasma movement index.

M protein — a protein which stimulates bacterial DNA transcription by complete → RNA polymerase under a variety of conditions. Possibly, M associates with RNA polymerase and provides for an increased frequency of initiation of RNA chains. It might function at the binding or initiation stages of the RNA polymerase reaction.

mRNA — acronym for → messenger RNA.

mRNA interfering complementary RNA (Mizuno et al. 1984) — → mic RNA.

mRNP — acronym for → messenger ribonucleoprotein.

mt DNA — acronym for → mitochondrial DNA (mit DNA).

multicentric — of eukaryotic → chromosomes with more than one → centromere (→ dicentric).

multicompartmental genome — in many viruses (multicompartmental viruses), the division of the genetic information between separately encapsidated nucleic acid molecules.

multicopy inhibition — the suppression of the transposition-related functions of a single → transposon by multiple copies of its terminal repeat elsewhere in the cell.

multicopy plasmid — a → plasmid present in bacteria at amounts greater than one per bacterial chromosome.

multicopy single-stranded DNA (Yee et al. 1984) — (ms DNA); in some bacteria, extrachromosomal DNA molecules composed of a single strand of DNA; attached to the 5′ end is a branched RNA molecule. Attachment is via a unique 2′, 5′ phosphodiester bond. ms DNA is probably synthesized by → reverse transcriptase using a folded RNA precursor as a primer as well as a template.

multienzyme system — an aggregate of different, functionally related (consecutive) enzymes of a metabolic sequence. Three types: (1) M. complexes, i.e., noncovalently associated enzymes; (2) m. conjugates, i.e., covalently linked enzymes; (3) enzymes linked by, e.g., binding to some membrane.

multiforked — of a bacterial chromosome that has more than one replication fork, because a second initiation has occurred before the first cycle of → DNA replication has been completed.

multigene cluster — a group of closely linked genes representing a → multigene family which probably evolved by → gene duplication and subsequent divergence of pre-existing genes.

multigene family (Brown et al. 1971; Hood 1972) — in eukaryotes, any set of sequence-related genes (or nucleotide sequences) that apparently share a common ancestor. The members of a m.f., although related to each other, may code for identical, related, or even functionally different proteins or proteins expressed at different times during development (divergent m.f.). Two basic patterns of m. families have emerged: (1) Linked clusters of related genes in which individual genes are separated by long regions of noncoding DNA; (2) dispersed families in which individual genes are scattered over widely separated chromosome locations.

Unit structure and family organization are highly variable among species, but a high level of homogeneity is maintained within each; the different members of a family apparently evolve (in some cases at least) in unison (→ concerted evolution; molecular drive).

M. families may include truncated genes, → pseudogenes, or → processed pseudogenes, and could arise by a sequential series of duplications over a long time period or by an event giving rise to many gene copies in one generation. In sequential → gene duplication the first duplication is most likely a rare event, while subsequent duplications may originate more frequently by unequal → crossing over within the duplicated sequences. Quick amplification could be a consequence of aberrant DNA replication in germ line cells (more frequent replication of a DNA segment than the rest of the genome) giving rise to multiple gene copies which are initially in an extrachromosomal state but may eventually be integrated in one or more chromosomal loci as linear arrays (→ orphanon).

Tandemly repetitious m. families are represented by the → histone genes, the globin genes, rRNA genes and 5S RNA genes. The m. families may be distinguished from one another by two general properties: (1) The gene composition, and (2) the size of the family.

multigenic — controlled by several genes, as opposed to → monogenic.

multihybrid — → heterozygous for more than one gene (dihybrid, trihybrid, etc.).

multikaryon — a multinucleate cell (→ dikaryon) or a → coenocyte, as opposed to a → monokaryon.

multimer (Crick and Orgel 1964) — an active aggregate resulting from the interaction of polypeptides (monomers). It is probable that complementation of the intragenic type generally results from multimer production by differently defective monomers (→ genetic complementation).

multinucleate — of cells that contain more than one → nucleus. Heterophasic m. cells are the result of fusion between similar cells in different cell cycle phases (→ premature chromosome condensation); homophasic m. cells are formed by fusion of cells being in the same phase of the cell cycle (→ cell hybridization).

Synchrony is a rule in m. cells; asynchrony in relation to DNA synthesis and mitosis has been observed in various systems.

multiple allelism (Morgan 1914) — the existence of several known → alleles of a gene.

multiple codon recognition — the recognition of more than one → codon in → messenger RNA by a single → transfer RNA species due to codon-anticodon pairing involving nonstandard base pairs (→ wobble) or the use of multiple tRNA adaptors.

multiple complex (Moses 1968) — any of the laminate or tubular arrays of units, each of which is similar to the → synaptonemal complex (SC). Such complexes have been observed in the germinal cells of insects and other organisms. M. complexes are probably formed from SC material disjoined from the chromosomes.

multiple gene — (1) any of two or more nonallelic genes with similar or complementary cumulative effects on a single character. (2) Any member of a → multigene family.

multiple promoter — in a single gene, the presence of more than one → promoter. The use of m. promoters may confer qualitatively and quantitatively different expression to genes (alternative promoter utilization). M. promoters may originate from (Schibler and Sierra 1987): (1) De novo creation of an additional promoter sequence by a series of point mutations and/or small deletions and insertions upstream or downstream of the first promoter; (2) acquisition of a second promoter by → transposable genetic elements, → translocation, or → inversion; (3) promoter duplication followed by promoter diversification.

multiplicity of infection — the average number of → bacteriophage particles that infect a single bacterial cell in a specific experiment.

multiplicity reactivation (Luria 1947) — the interaction in a multiply infected cell of inactivated (by, e.g., UV light, X-rays, gamma rays, ^{32}p decay, nitrous acid) bacteriophages or animal viruses, inviable in single infection, so as to produce viable progeny. M.r. is probably due to → replication-dependent recombination repair that results in exchanges between damaged viral genomes present in the same cell and gives rise to intact recombinant virus genomes.

multipolar (Mayzel 1875; Strasburger 1880) — of a → spindle (and mitosis) with more than two spindle poles which may result from the presence of several nuclei in the cell (each with its centers), from polyspermy in eggs, or from supernumerary divisions of the cell centers (→ centriole). The type of mitosis resulting from the presence of a m. spindle is called a multipolar mitosis and is generally connected with an aberrant mode of chromosome distribution.

M. mitosis appears to be an important, perhaps the most important, mechanism whereby polyploid cells can reduce their level of ploidy.

multisite mutation — a mutation of one → genetic locus involving changes of two or more adjacent → mutational sites. Multisite mutations (→ gene mutation) occur spontaneously and are inducible experimentally by the application of → mutagens. They do not revert to wild type spontaneously or after mutagen treatment (→ reversion) and some in all probability are true → deletions (→ deletion mapping).

multivalent — in eukaryotes, any association of more than two completely or partially homologous chromosomes whose homologous regions are held together by meiotic → chromosome pairing and chiasma formation from prophase to metaphase of the first division → meiosis (as distinguished from → univalents and → bivalents). Multivalent formation is characteristic of → autopolyploid and some types of → allopolyploid individuals as well as some forms of structural hybrids (e.g., those heterozygous for reciprocal → translocations). Depending on whether three, four, five, six, seven, or eight chromosomes are meiotically paired, the multivalents are called trivalents, quadrivalents, pentavalents, hexavalents, heptavalents, and octovalents, respectively.

The configuration of multivalents in meiosis I depends upon the number of chromosomes participating in pairing, the number and location of chiasmata, and the degree of → chiasma terminalization. The arrangement of multivalents within the spindle at metaphase I of meiosis is determined by the particular type of → centromere coorientation, which in turn is dependent on the position of the chiasmata, the relationship in space between the centromeres of the m., and the space available in the spindle.

Oksala (1952) distinguishes between **"primary"** and **"definitive" multivalents.** The former are those held together by the pairing forces between homologous chromosomes or chromosome segments from zygotene up to diplotene, the latter those held together mechanically by chiasmata up to anaphase I of meiosis.

Pseudomultivalents are multivalents formed by nonhomologous chromosomes (e.g., by → sticky effects).

multivesicular body (Yamada 1955) — a round to oval structure (diameter from 0.1 to 0.6 μm) which functions as a digestive vacuole for exo- and endogenous materials. Multivesicular bodies (= vesicular conglomerates) are found in a variety of cells and are thought to be self-duplicating organelles (Sotelo and Porter 1959). They are surrounded by a smooth membrane and their lumina contain a rather large number of round vesicles (diameter from 10—50 nm).

Mu phage (Taylor 1963) — a temperate → bacteriophage with properties similar to those of → transposable genetic elements: It is able to insert randomly into the bacterial genome, to transpose, to inactivate, and to cause → DNA rearrangements. The phage genome is a linear, double-stranded DNA molecule (36—38 kb in length) and replicates by a process that involves → DNA transposition. Upon infection of a susceptible host, Mu DNA can integrate into host DNA and either establish → lysogeny or grow lytically as a phage with hundreds of transposition events. Transposition requires both phage- and host-encoded functions. Sequences required in cis for Mu transposition are located at both termini (within 163 bp on the left end and 52 bp of the right end). At each end Mu has many repeated sequences, including the terminal 2 bp that are inversely repeated and three nonidentical 30 bp sequences at each end for binding the Mu A transposase protein.

mutability — the property of any → gene and → genotype to undergo → mutation. M. provides the ultimate basis for the adjustment (→ adaptation) of → gene pools to changes in the environment. Generally it is advantageous for a species to maintain a constant low level of m., since most mutations are deleterious. This relative constancy of m. is achieved by manifold mechanisms which may consist of hereditary changes that primarily affect the nongenic parts of the cell (e.g., alterations of the pattern of resistance to the admission of → mutagens to the chromosomes, neutralization of the action of mutagens, etc.) or hereditary changes that primarily affect the genetic system per se (e.g., changes in the structure of the → genetic material resulting in changes in its susceptibility to mutagens, changes in mechanisms that repair lesions induced by mutagens in the genetic material; → DNA repair). Specific genes giving rise to significant increases in the m. of other genes are called → mutator genes.

mutable gene — any of a class of labile or unstable genes which in the course of development mutate spontaneously rather frequently among two or more possible states, in comparison with the majority of genes which as a rule are elements of considerable stability.

Although there is no sharp boundary between "stable" and "unstable" genes, a gene in multicellular organisms is usually referred to as mutable or unstable when its spontaneous → mutation rate in the individual is so high, that the gene and its mutant derivatives result in → variegation. A particular gene may be unstable during all stages of ontogenetic development or in only one or a few of them. The high mutation rate (→ hot spot) of a m.g. is either an inherent, autonomous property of the gene itself or is controlled by → transposable genetic elements (→ mutator gene; gene conversion) independent of the mutating gene itself.

Mutable genes may give rise with high frequency to a whole range of more or less stable alleles (quantitatively graded in their expression), or to other unstable alleles, differing from the original in the timing or frequency of mutation, or in the kind of stable derivatives most often produced. Frequently, the mutation rate of unstable genes is found to be drastically affected by changes of environmental conditions (e.g., temperature), or by genes elsewhere in the genotype which exert no evident effect on mutation in general (→ genomic stress).

mutable site — = → mutational site.

mutagen — any → mutagenic agent. The genetically significant concentration or dose of mutagens is the concentration or dose of the active form of a m. in the immediate neighborhood of the genome.

Most mutagens exert several chemical effects on DNA, some of which induce → mutagenic DNA alterations and others → inactivating DNA alterations.

mutagen burden — the amount of unavoidable exposure to various chemical → mutagens by analogy to exposure to → background radiation.

mutagenesis — the mutational process, i.e., the mechanisms by which → gene mutations or → chromosome mutations occur spontaneously or are induced by → mutagenic agents. M. is a multistep process which starts when a mutagen enters or is formed in the cell and interacts with DNA. It ends with the recovery of a → mutant individual. Various factors affect m., such as factors affecting proofreading during → DNA replication, error-prone → DNA repair and → mismatch correction.

M. by a number of mutagenic agents can be separated into two classes: (1) **Targeted m.** refers to a mutagenic change that is caused in the mutated gene and probably involves the mode in which a → DNA polymerase interacts with a mutagenic lesion in a DNA template; (2) **nontargeted m.** refers to mutagenic events that take place in undamaged regions of the DNA and probably results from a loss or limitation of the fidelity of DNA replication.

Site-specific (or local **m.**) is the preferential induction of mutations within a particular region of a given base sequence of DNA (or RNA) molecules (→ in vitro mutagenesis; transposon mutagenesis) or within a given genome.

mutagen dose — the amount of a → mutagenic agent that reacts with a selected target molecule as measured by the accumulation of a selected reaction product (→ mutagen exposure).

mutagen exposure — exposure of the whole organism to a → mutagenic agent as opposed to → mutagen dose. A multitude of physiological effects and chemical interactions can occur after exposure to a chemical mutagen (assimilation, circulation, transport across cell membranes, metabolic activity, enzymatic degradation, or promutagen activation).

mutagen hypersensitivity — in many prokaryotic and eukaryotic organisms, increased → mutagen sensitivity usually associated with abnormalities in → DNA repair processes.

mutagen specificity — the propensity of a given → mutagen to produce one type of mutation rather than another. M.s. may give rise to allele- or locus-specific induction of → gene mutations or → reversions by different mutagens due to specific reactions between the mutagens used and the nucleotide sequence of the tested chromosome region.

mutagenic — of any physical or chemical agent (mutagen) capable to covalently modify the DNA and inducing → gene mutations and/or → chromosome mutations, via → premutational lesions. → Transposable genetic elements and mutator phages (→ Mu phage) may be called "genetic mutagens" (→ insertion mutagenesis). In addition to mutations (induced as opposed to spontaneous mutations), m. agents have been found to induce a variety of other genetic alterations, such as mitotic → crossing over, → gene conversion, and → sister-chromatid exchanges.

Since the active forms of most known carcinogens are m., screening systems (→ mutagenicity testing) to detect mutagens can also serve to detect many potential → carcinogenic compounds.

Two classes of m. agents may be distinguished: (1) **Direct-acting mutagens:** They result in mispairing lesions (base modifications such as base analogs and 5-bromouracil, bisulfite- and hydroxylamine-modified bases, oxygen alkylated bases as well as tautomers and isomers of normal bases). The premutational lesion is converted into a mutation by misreplication involving an erroneous base-pairing directed by the lesion. (2) **Indirect-acting mutagens:** They induce nonpairing lesions (usually bulky modifications of DNA induced by, e.g., UV and ionizing radiations and many powerful m. carcinogens) which may cause arrest of DNA chain elongation and are converted into a mutation by inducible error-prone → DNA repair functions that process the premutational lesion into a mutation (→ misrepair mutagenesis; SOS repair). Because of their dependence upon the induction of inducible misrepair processes such agents are called indirect mutagens.

mutagenic DNA alteration (Freese et al. 1969) — any of those primary DNA alterations which do not prevent DNA replication and may give rise to → gene mutations. Such lesions are not responsible for the production of large chromosome structural changes, i.e., → chromosome mutations (→ inactivating DNA alteration). They may be due to mistakes in base pairing (induced either by base analogs or by minor chemical alterations of DNA bases), or may result from mistakes by the → DNA polymerase.

mutagenic effectiveness — the relation of the mutation frequency to the molar concentration of the mutagen applied.

mutagenic efficiency — the relation of the mutation frequency to some other biological effect of the mutagen such as killing, chromosome breakage, or → genetic recombination.

mutagenicity — any event resulting in alterations of the genetic material.

mutagenicity assay — any system for → mutagenicity testing.

mutagenicity screening — = → mutagenicity testing.

mutagenicity testing — testing, by use of various mutagenicity assays, for the detection of → mutagens or potential mutagens (environmental mutagens) and extrapolation of the data obtained to man. Most test systems assay damage to DNA by monitoring gene mutations, chromosome structural changes, DNA repair induction, or transformation of cultured mammalian cells. Since no single assay system can detect all mutagenic compounds, a battery of tests must be used (→ tier system). (→ inductest; protein mapping test; mouse spot test; specific locus test; heritable translocation test; numerical sex chromosome anomaly test; nick translation assay).

mutagenic repair (Hastings et al. 1976) — any pathway of (error-prone) → DNA repair which results in mutations.

mutagenic specificity — the preference for the type of the mutagenic event (e.g., the type of base-substitution) initiated by a certain → mutagen.

mutagen persistence — the stability of an environmental → mutagen in the environment and in the human body as defined by the half-life in the environment and the endogenous half-life, respectively. The efficiency of environmental mutagens strongly depends on the m.p.

mutant — any variant strain, individual, cell, or gene resulting from → mutation. Unstable mutants are characterized by frequent changes to other mutant states and/or high → reversion rates. The standard reference is the → wild-type condition. Any heritable variation from it due to mutation is a m.

mutant-distribution spectrum — the spectrum of the distribution of mutated sites along the DNA after treatment with → mutagenic agents (→ mutation-type spectrum).

mutant enrichment — a technique of increasing the proportion of → mutant cells, which cannot be selected for directly, in a culture: Under the chosen conditions the desired mutants will not grow; growing wild-type cells are then killed by some physical or chemical treatment which is without effect on nongrowing (mutant) cells.

mutant frequency — the ultimate proportion of mutants detectable in a population, regardless of how and when they arose (Zamenhof 1967).

mutant site — any damaged or altered area within a mutated gene.

mutant substitution — a change at the population level due to the spreading of a → mutation to the entire species. Each m.s. takes a long time and only a small fraction of mutations manage to lead to m.s. (→ genetic load).

mutate — to undergo or exhibit → mutation.

mutation (de Vries 1901) — any heritable alteration, whether a single base change in DNA (or RNA) or a large-scale rearrangement, in the genetic material of a living cell or virus. Such alterations can be grouped into two classes: (1) → Gene mutations and (2) gross chromosome structural changes, i.e., → chromosome mutations. Provided they are not lethal, mutations can be transmitted to somatic daughter cells and even to succeeding generations giving rise to mutant cells or mutant individuals. It is through the genetic variation provided by mutations that natural → selection and genetic → drift can alter the genetic composition of populations, i.e., mutations provide the ultimate driving force behind → evolution. A m. that does not result in a significant selective advantage or disadvantage is called a neutral mutation.

The mutational units of a → genotype are, in order of decreasing size, the genome, the chromosome (→ genome mutation), the chromosomal segment involving more than one gene (→ chromosome mutation), and the gene (→ gene mutation). Since a functional gene (cistron) consists of many → mutational sites, a gene mutation may involve the entire functional unit or only one or more of the mutational sites of which it consists.

Mutations may occur spontaneously (→ mutability) and can be induced experimentally by application of → mutagenic agents. There appears to be no qualitative difference between spontaneous and induced mutations, the former being the result of several unknown and ubiquitous mutagenic agents or effects. A considerable fraction of the spontaneous mutations seem to depend upon the intracellular production of mutagens and → antimutagens.

Since the products of mutation are in general deleterious, the process of m. is harmful for a proportion of the individuals in a population (→ genetic load). It is, therefore, advantageous for a population to maintain a relatively constant and low level of → mutability; this may be achieved by various means. When an individual dies or fails to reproduce as a consequence of the reduced → fitness of its mutated genotype, it is a victim of → "genetic death".

A special category of mutations concerns the hereditary determinants which are extranuclear and show non-Mendelian → inheritance. These are mutational changes not of the genotypic but of the plasmotypic component of the → idiotype (→ plasmon, plastom).

mutational delay (Newcombe 1953) — = → mutational lag.

mutational dissection — the study of the components of a biological function by investigating mutations which affect this function.

mutational equilibrium — the frequency of mutated genes per individual. It is equal to the frequency of mutations per individual per generation multiplied by the average number of generations during which a mutated gene is maintained before it is eliminated (→ genetic load).

mutational hot spot (Benzer 1955) — = → mutation hot spot.

mutational lag (Demerec and Cahn 1953) — in addition to → copy error lag, → cell division lag, → phenomic lag, and → segregation lag, a delay (= "mutational delay", Newcombe 1953) in the appearance of bacterial mutant clones by delayed → gene mutation.

mutational load (Muller 1950) — → genetic load.

mutational mosaic — an organism or (in the case of microbes) a clone composed of both mutant and nonmutant cells.

mutational site (Pontecorvo 1958) — any position (nucleotide) along a → gene (or chromosome) at which → mutations can occur.

Based on the presence or absence of intragenic recombinations (→ genetic recombination) between mutational sites, three types of mutationally produced → alleles of one gene are distinguished: (1) Those giving rise to the wild type by recombination. They are interpreted to be the result of mutation in different sites of the same genetic locus and are referred to as "nonidentical" alleles (Demerec 1956) or "heteroalleles" (Roman 1956); (2) those giving rise to the wild type by recombination with certain alleles of type 1., but which do not recombine with all of them. They are interpreted to be the result of changes in two or more adjacent sites, i.e., represent "multisite mutations"; (3) those not recombining with each other but with all mutant alleles of type 1. They are interpreted to be the result of recurrent but not necessarily identical mutation of the same site and are referred to as "identical alleles" (Demerec 1956) or "homoalleles" (Roman 1956).

mutational specificity — the specificity characteristics of each → mutagenic agent that arises from a "preference" both for certain types of mutations and for certain → mutational sites called hot spots.

mutation epidemiology — the study of the distribution and determinants of abrupt and sustained changes in the genetic material of human beings and of the disease processes aroused by such changes.

mutation fixation (Witkin 1956) — conversion of a premutational → DNA lesion to final mutation by errors in → DNA repair or in → DNA replication (most often).

mutation frequency— the frequency of certain mutants in a population. This does not necessarily reflect the rate (→ mutation rate) at which the mutations arose (→ mutation yield).

mutation frequency decline (Doudney and Haas 1958) — (MFD); the rapid loss (by a cellular repair process) of potential de novo nonsense → suppressor mutations of tRNA genes when excision-proficient bacteria *(E. coli)* are incubated in glucose-minimal medium after mutagenesis by UV or in complex medium but with protein synthesis inhibited. The mechanism of removal of potentially mutagenic lesions from → suppressor tRNA loci appears to be → excision repair.

mutation hot spot (Benzer 1955) — any site in the DNA sequence or chromosome preferentially involved in mutations or chromosome structural changes (aberration hot spot). M.h. spots may be sites in DNA with high affinity to → mutagenic agents or for sequence-dependent processing of premutational lesions. In the latter case, a m.h.s. can arise from the inability of error-free repair enzymes to excise a lesion at a given sequence.

mutation induction — the process during which a premutational → DNA lesion, induced by a → mutagenic agent, is converted by → mutation fixation to a final mutation.

mutation initiation — → gene mutation.

mutation isoallele (Lefevre 1955) — any one of a number of → alleles with similar functional activity but different rates of mutation (→ isoallele).

mutation load (Crow 1958) — the equilibrium reduction in mean → fitness caused by deleterious mutations; it is equal to the haploid mutation rate if the mutants are recessive, and twice this if they are dominant or partially so.

mutation map (Muller 1927; Muller and Painter 1932) — the frequency of mutation plotted as a measure of chromosome length (→ chromosome map).

mutation monitoring — the surveillance of background → mutation rates in populations.

mutation pressure (Wright 1921) — the continued recurrent production of an → allele by mutation tending to increase its frequency in the → gene pool of a population (→ gene frequency). If no other forces intervene, the mutation of allele A to a leads to changed proportions of these alleles which might ultimately result in the replacement of A by a (the entire population would then be aa in respect to the A locus). However, since → back mutations from a to A also occur (generally at different rates), the population never becomes homozygous for one type of allele by mutation alone. Instead of "fixation" of one allele, an equilibrium will be established at that point where the number of mutations from A to a equals the back mutation rate from a to A.

mutation rate (Muller 1928) — the frequency of → gene mutations per generation and biological entity (virus, cell, individual) at a given locus. The apparent m.r. depends upon a number of contingencies: (1) The primary error rate; (2) the probability of repair (→ DNA repair); (3) error avoidance by the apparatus of → DNA replication; (4) the probability of detection of mutations. Genetic control of apparent m.r. is evident from two phenomena: (1) Large differences in average m.r. per base pair in diverse organisms; (2) mutations in specific organisms which alter average m. rates (→ mutator gene; antimutator gene; meiotic effect).

mutation spectrum — the frequency distribution of induced or spontaneous mutations in different genes of a genome or in different sites within a gene. The frequencies at which different mutations are represented in a m.s. may be a function of both the distribution of → DNA damage and the competitive efficiencies of error-free and error-prone → DNA repair.

mutation-type spectrum — the types of DNA alterations observed in a DNA molecule after → mutagenic treatment (→ mutant-distribution spectrum).

mutation yield — the number of surviving mutants per total cells assayed.

mutator gene (Demerec 1937) — any mutant gene that increases the spontaneous mutation frequencies of other genes. Some of the proteins specified by m. genes are probably concerned with → DNA repair, → DNA replication, and precursor synthesis.

mutator phage (Taylor 1963) — → Mu phage.

muton (Benzer 1957) — the basic unit of → gene mutation, represented by a → mutational site (a nucleotide pair) within a gene or → cistron. The muton is the smallest alterable element within the one-dimensional array of mutational sites in the → genetic material which, when altered, may give rise to a → mutant organism (→ recon).

N

n, 2n (Blakeslee 1921) — shorthands for the → haploid (gametic) and → diploid (zygotic and somatic) chromosome numbers of eukaryotic organisms, respectively. Symbol x (the basic number of chromosomes in a → genome) should be used, instead of n to denote → polyploidy levels (3x = triploid, 4x = tetraploid, etc.).

nascent — of protein chains in the process of being formed on a → messenger RNA template in polyribosomes.

nascent DNA maturation (Schwartzman and van't Hof 1982) — the joining of nascent → replicons and replicon clusters to form chromosome-sized molecules in eukaryotes during → DNA replication. N. DNA m. occurs during late S (→ cell cycle) and requires replication of DNA sequences that are interspersed within replicon clusters that replicate early in the S phase.

Nebenkern (Bütschli 1871) — = → paranucleus.

negative complementation (Müller-Hill et al. 1968) — the phenomenon that interallelic complementation allows a mutant protein subunit to suppress the activity of a wild-type subunit in a mixed hybrid multimeric protein (as observed for repressor mutants defective in DNA binding) (→ negative dominant mutation).

negative control — in genetics, a type of → genetic regulation of the transcription of a gene where a → repressor binds to an → operator site upstream from the coding region and prevents transcription by → RNA polymerase (= negative gene regulation). An → inducer (an anti-inhibitor) is needed to turn on transcription (→ positive control).

negative dominant mutation — a type of mutation concerning genes encoding multisubunit proteins. Dominance of the mutant phenotype over the wild-type allele is explained by the formation of mixed oligomers between the wild-type and mutant subunits of the protein (= → negative complementation or anticomplementation).

negative liquid holding (Parry 1972) — the reduction in cell viability produced by post-UV holding treatment in saline. N.l.h. is probably the result of the reduced activity of an enzyme system involved in the → repair of **nonphotoreactivable** UV-induced lesions in some UV-sensitive yeast mutants (→ liquid holding recovery).

negative regulator — in genetics, any regulator that functions by switching off transcription (→ repressor) or translation (→ negative control).

negative supercoiling — → DNA supercoiling.

neighborhood (Wright 1931) — a subpopulation unit of a geographically structured population in those cases where the breeding system brings about significant departures from local panmixia, as opposed to the → panmictic unit. N. size is the number of breeding individuals in the n. area, i.e., the size of the space occupied by the individuals constituting the n.

neobiogenesis (Keosian 1965) — a term proposed to describe the presumed repeated origin of life in nature since life began (→ eobiogenesis).

neobiont (Keosian 1965) — an organism originating by → neobiogenesis. Neobionts may be commensals, symbionts, or parasites, as well as free-living.

neocentric activity (Rhoades and Kerr 1949) — an exceptional behavior of (plant) chromosomes where certain chromosome regions act as secondary attachment sites for spindle fibers (neocentromeres), usually at both meiotic divisions but not at mitosis.

neo-Darwinian evolution — a theory explaining biological → evolution by essentially two elements: (1) Gene (point) mutations within → structural genes are the source of genetic variability in organisms and evolutionary change results from a shift in the frequency of genes within populations; the origin of species and the development of trends in groupe of species is viewed as a consequence of the gradual accumulation of small genetic differences (gradualism). (2) The direction of evolutionary change is determined by natural → selection; the variants that survive are those that are best fitted to their environments (adaptationism) (→ neutral evolution).

neomorph (Muller 1932) — → allele.

neoplastic transformation — → cell transformation.

neoplastoid (Martin and Sprague 1973) — of → cell lines from mammalian tissues (e.g., HeLa) which are immortal and sometimes behave like neoplasms in vivo. They are typically → heteroploid, but may be euploid (→ hyperplastoid).

neoteny (Kollmann 1885) — (1) the retention of juvenile characters in the adult state; (2) the occurrence of adult characteristics in the juvenile state.

neo-XY system — → sex chromosome.

neo-Y-chromosome (White 1935) — → sex chromosome.

neutral mutation — any mutation which is without adaptive significance (selectively neutral genes) to the organism or without phenotypic effect. Such genes are unaffected by → selection but subject to → genetic drift, → migration, and possibly → segregation distortion. The "neutral mutation hypothesis" of → evolution states that the great majority of molecular changes in evolution are selectively neutral or nearly so and evolution occurs by accumulation of n. genes (mutations) through random drift and mutational pressure.

nexus (Dewey and Barr 1962) — any of the specialized regions of contact (= gap junction) between the apposed cell membranes of adjacent cells.

N-formyl-methionyl transfer RNA (Marcker and Sanger 1964) — the → initiator tRNA in bacteria, → chloroplasts, and → mitochondria. It arises by formylation of the α amino group of methionine, which takes place after the methionine has become attached to its → transfer RNA. Formylation is catalyzed by a transformylase. The specificity of N-formylmethionyl tRNA as a polypeptide initiator during → genetic translation resides in the structure of this tRNA and not in the formyl group.

niche (Grinnell 1904) — all the relationships of an organism to its physical and biotic environment, i.e., the genetically determined capacity (range of tolerance) and pattern of biological response of an individual, a species population, or the whole species to environmental conditions (→ ecological niche). The variety of resources a population exploits is called the niche width and may be narrow or wide.

nick — in double-stranded DNA, a site where the phosphate-sugar backbone of one strand is hydrolysed, but with all the complementary nucleotides still in place (→ cut), i.e., without removal of any nucleotides (= single-strand break) (→ nick translation). Nuclease action to sever the sugar-phosphate backbone in one DNA strand at one specific site is called nicking.

nicking-closing enzyme (Wang 1971) — = type I → DNA topoisomerase.

nicking enzyme (Hotta and Stern 1974) — in organisms with → chiasmate meiosis, a specific endonuclease believed to produce single-strand breaks in the DNA duplex during zygotene and pachytene (lacking in cases of achiasmate meiosis). The n.e. may act as an initiator of → crossing over.

nick translation (Kelly et al. 1970) — a technique for radioactively labeling DNA in vitro using → DNA polymerase I which initiates synthesis at single or multiple → nicks. N.t. allows the production of a DNA probe which may be used in some hybridization techniques, such as → Southern blotting.

N.t. in vivo occurs during bacterial → excision repair: After cleavage of the DNA helix at or near a lesion by a specific endonuclease, simultaneous excision of an oligonucleotide containing the lesion and reincorporation of nucleotides at the gap created by 5′ → 3′ exonuclease and polymerase activities of prokaryotic polymerase I occur. In organisms other than *E. coli*, the excision and → repair replication steps may be catalyzed by a separate exonuclease and polymerase.

nick translation assay (Nose and Okamoto 1983) — in → mutagenicity testing, an assay that monitors both DNA damage (appearing as strand breaks) and repair. DNA strand breaks with free 3′ OH termini serve as template sites for incorporation into DNA of radiolabeled deoxycytidine triphosphate (dCTP) by DNA polymerase I. Decrease of this incorporation with time indicates repair of the initial damage.

NIH guidelines — the guidelines issued by the National Institutes of Health (USA) which prescribe the use of biological and physical → containments to ensure safety in experiments in which propagation of → recombinant DNA outside the confines of the experiment might pose a real or hypothetical hazard (→ recombinant DNA technology).

n_0 coalescent (Kingman 1982) — → coalescent.

nodoc (Cantoni et al. 1963) — the anticodon of → transfer DNA

nodule (Schrantz 1970) — → recombination nodule; synaptonemal complex.

nomadic DNA family (Young 1979) — in eukaryotes, any family of dispersed repetitious DNA sequences which occupy no fixed chromosomal positions.

nombre fondamental (Matthey 1949) — in eukaryotes, the sum total of chromosome arms in the chromosome complement of a species, not including the small arms of acrocentric, rod-like chromosomes.

nonchiasmate — = → achiasmate.

noncoding DNA sequence — any DNA sequence apparently not coding for a transcription product.

noncoding DNA strand — by convention, the base sequence for a transcribed region on this strand is that of the messenger RNA transcribed from it (advantage: the amino acid encoding codons and the start and stop codons are in their standard forms and thus easily recognizable).

noncongression (Darlington 1937) — the failure of pairing configurations (in → meiosis I) or chromosomes (in → mitosis and meiosis II) to come into orderly arrangement on the spindle equator (metaphase or equatorial plate). Noncongressing chromosomes are usually not included in the nuclei produced by meiosis or mitosis (→ chromosome congression).

nonconjugative plasmid — any → plasmid that does not enable its bacterial host to conjugate. When another plasmid in the same host promotes → conjugation, n.-c. plasmids can be transmitted to recipient cells during mating.

nonconjunction (Belling 1925) — absence of meiotic → chromosome pairing (→ nondisjunction).

nonconservative — of a mode of DNA replication in which there is no net increase in the amount of DNA but only a replacement of parental DNA with newly synthesized DNA (→ repair replication).

non-Darwinian evolution (Harris et al. 1968; King and Jukes 1969) — → evolution.

nondisjunction (Bridges 1913) — (1) **cytological n.-d:** The failure of sister-chromatids (mitotic n.-d.) or homologous chromosomes (meiotic n.-d.) to disjoin at anaphase to the cell poles. N.-d. results in → aneuploid (complementary hyper- and hypoploid) products of mitosis and meiosis (gametes or meiospores), respectively. (2) **Genetic n.-d.:** All types of abnormal meiotic behavior resulting in n. as inferred by genetic methods. (a) **Chromatid n.:** (Sansome 1933): The distribution of segments of sister chromatids, distal to a → chiasma, to the same gamete or meiospore at meiosis. In → autopolyploids, this type of n. gives rise to → "chromatid segregation". (b) **Primary and secondary n.** (Bridges 1916): In XX-XY-systems of → sex determination, n. leading to the production (by an XX-individual) of egg cells with two X-chromosomes or none ("primary n.") or to the production of egg cells with two X-chromosomes, X + Y, one X, or one Y-chromosome, by an XXY-individual ("secondary n.").

Primary n. may also occur in male individuals (XY) during first (resulting in XY- and O-sperm) or second meiotic division (resulting in XX- and O-sperm or YY- and O-sperm). (3) **Directed nondisjunction:** The preferential distribution of particular chromosomes (e.g., → B-chromosomes) to one of the two cell poles in mitosis or meiosis.

nonhistone chromosomal protein (Mayfield and Ellison 1975) — (NHCP); in eukaryotes a very complex and heterogeneous subset of acidic proteins associated with → histones and DNA to yield the complex structure referred to as → chromatin. A group of NHCP's is called the → high mobility group (HMG) of chromosomal proteins. The NHCP's are tissue- and species-specific and include proteins implicated in → gene expression in general and in the control of transcription in particular. They undergo covalent modifications like phosphorylation and acetylation. Synthesis of NHCP's continues throughout the → cell cycle in continuously dividing cells, as well as in quiescent cells stimulated to proliferate, and is not dependent on → DNA replication. They exhibit large variations in binding strengths to DNA (→ chromosome scaffold).

nonhomologous — of chromosomes or chromosome segments which contain dissimilar genes and do not pair during meiosis (→ chromosome pairing), as opposed to → homologous or → homeologous ones. Rare meiotic pairing of such segments in haploids, polysomics, or structural hybrids is referred to as "nonhomologous pairing" and does not lead to → crossing over and chiasma formation.

non-isolabeling — → sister-chromatid exchange.

non-Mendelian — the mode of → inheritance shown by extranuclear hereditary determinants.

nonpermissive — of a cell which does not support viral replication, as opposed to permissive cells.

nonrandom assortment — → assortment.

nonreciprocal recombination — = → gene conversion.

nonrecurrent parent — → recurrent parent.

nonreduction (Belling 1925) — the failure of meiotic → reduction (halving) of the somatic chromosome number due to disturbed → chromosome pairing (→ asynaptic; desynaptic) resulting in the production of a → restitution nucleus with the somatic number of chromosomes.

nonrepetitive DNA — = → unique DNA or single-copy DNA.

nonsense codon (Brenner et al. 1961) — → nonsense mutation.

nonsense mutation (Brenner et al. 1961) – any mutation (base substitution or frameshift mutation) that converts a sense (amino acid specifying) → codon into a → chain terminating codon (UAG, UAA, or UGA in mRNA). N. mutations result in premature termination of polypeptide chain synthesis at the site where the nonsense codon occurs in the → messenger RNA.

N. mutations can lead to a pleiotropic polar effect (→ genetic polarity; polar mutation), reducing the level of expression of genes operator-distal to the n.m. in the bacterial → operon. To select for n. mutations in viral functions it is necessary to isolate the mutations in a genotype carrying a → nonsense suppressor (→ nonsense suppression).

nonsense suppression — suppression of the polypeptide chain termination effect of a nonsense codon (→ nonsense mutation) and partial or complete restoration of the normal phenotype. In the case of n.s., nonsense codons can act as sense (amino acid specifying) codons. This may be due to the presence of a → nonsense suppressor gene or a chemical suppressor (such as 5-fluoro-uracil and streptomycin). If, for instance, the nonsense triplet is UAA, then the substitution of 5-fluoro-uracil (5-FU) for U in the mRNA may result in the reading of FUAA as CAA, a sense codon. This will allow translation to continue, since the tRNA will mistake FU for C with a higher probability than if U were present.

The efficiency of n.s. depends on the balance between polypeptide chain termination and amino acid insertion and may be drastically affected by the composition of the nucleotides adjacent to the nonsense codon (reading context).

nonsense suppressor (Capecci and Gussin 1965) — a → suppressor gene that allows the insertion of an amino acid into the growing polypeptide chain (→ genetic translation) in response to a nonsense codon. In the absence of such a n.s. premature chain termination occurs. Most of the n. suppressors so far tested are dominant over their wild-type alleles. Typically, they arise via base changes in the tRNA anticodon which are complementary to the chain terminating codons (→ suppressor tRNA). A tRNA species can only be converted to a n.s. if the normal functions performed by that tRNA are dispensable (minor tRNA's), i.e., if the genome contains multiple gene copies for that tRNA. Some n. suppressors suppress a single class of nonsense mutations only (UAG, UAA, or UGA), other suppress UAG and UAA mutants. A few can suppress all three nonsense codons.

The polar effect of a → nonsense mutation is also corrected by n. suppressors. The degree of this polarity alleviation correlates in most cases with the level of phenotypic suppression.

nonsister chromatids — → sister chromatids.

nonsister label exchange (Craig et al. 1970) — → sister label exchange.

nonspecific pairing — → chromosome pairing.

nontranscribed spacer — any of the nontranscribed intergenic DNA sequences that separate transcriptional units.

nontriplet decoding — a ribosomal frameshift mechanism which results in the synthesis of a (minor) protein that is either elongated or truncated (depending on the distribution of terminator codons in the new reading frame), as compared to the major protein. N.d. may be used to control the expression of certain proteins that are only required in small amounts.

non-unit membrane — any membrane less than 6 nm wide, and appearing in section as a single, electron-dense layer (→ unit membrane).

norm of reaction (Woltereck 1909) — the range of potential phenotypes that a single genotype could develop if exposed to a specified range of environmental conditions.

Northern blotting — a technique (= Northern transfer), analogous to → Southern blotting, used for transferring RNA from an agarose gel to a nitrocellulose filter on which it can be hybridized to a complementary (radioactive) single-stranded DNA or RNA probe.

novel joint (Hershey 1971) — the point of fusion by illegitimate → genetic recombination of DNA segments that were not originally adjacent.

N-terminus — → genetic translation.

nu body (Olins and Olins 1973) — = → nucleosome.

nucellear embryony — a form of → apomixis in which the embryo arises directly from the nucellus, i.e., a parenchymatous tissue (localized between ovule or megaspore and its inner integument) which extends from the chalaza at the base to the micropyle at the apex on the embryo sac.

nuclear — ref. to phenomena, processes, and constituents associated with the → nucleus of the cell.

nuclear acidic protein — → nonhistone chromosomal protein.

nuclear ADP-ribosyl transferase — (ADPRT); an enzyme which transfers the ADP-ribose portion of NAD^+ to → chromatin proteins, thereby synthesizing mono-, oligo-, and poly-(ADP-ribose). ADPRT is dependent on DNA for activity and is activated by breaks in the DNA. ADP-ribosylation of chromatin is probably involved in → DNA repair and in → cytodifferentiation.

nuclear association (Buller 1941) — a transitory stage during fertilization. It follows plasmogamy and precedes karyogamy. It is characterized by the presence of one or more pairs of nuclei within the same cell (→ dikaryophase).

nuclear blebbing — → nuclear envelope bleb.

nuclear body (Weber and Froomes 1963) — (1) any of the nuclear organelles made of proteins and found in normal and pathological conditions (chiefly in tumor and virus-infected cells). They may be loaded with ribonucleoproteins of nucleolar origin. Their function is unknown but they may be involved in nucleo-cytoplasmic exchange of proteins and/or the regulation of nuclear activity by proteins synthesized in the cytoplasm. (2) The region in the bacterial cell to which the chromosomal DNA is confined (= nucleoid).

nuclear cage (Jackson et al. 1984) — = → nuclear matrix.

nuclear cap — in certain organisms, an unusual membrane-enclosed organelle surrounding the anterior part of the → nucleus.

nuclear cloning — surgical placement of nuclei from accessible tissues (such as skin, intestine) into an unfertilized, enucleated egg. By this procedure whole organisms may develop with precisely those features of the organism from which the nuclei where obtained.

nuclear cortex — a subnuclear fraction composed of a layer of dense → chromatin associated with the → nuclear matrix and/or dense lamina and nuclear pore complexes. The n.c. can be isolated as a whole.

nuclear differentiation — the production of two (or more) types of → nuclei which differ strikingly in morphology and/or nucleic acid content. N.d. within a single cytoplasmic compartment occurs in many animal and plant cell types (developing eggs, neuroblast cells, pollen grains, ciliates, and foraminiferans). (→ chromatin diminution; chromatin elimination).

nuclear dimorphism — in ciliated protozoa, the separation of genetic function into two distinct types of nuclei per cell. The transcriptionally inactive micronucleus contains long multigenic DNA molecules typical of eukaryotic → chromosomes. The transcriptionally active, somatic macromolecules, derived from the diploid germline micronucleus by a complex process involving the polytenization of the micronuclear genome and subsequent fragmentation and selective degradation of the polytene chromosomes, consists of a heterogeneous distribution of conserved DNA sequences (400 to 20.000 bp). These linear DNA molecules are extensively replicated to form the macronuclear genome and function as single-gene units.

nuclear division — in eukaryotes, the division of the cell → nucleus by → mitosis, → meiosis, or → amitosis.

nuclear DNA — in eukaryotes, the DNA contained in the cell → nucleus, as opposed to the → organelle DNA. Movement of DNA sequences between the intracellular genomes (mitochondrial, chloroplast, and nuclear) is known. Nuclear genes follow the pattern of Mendelian → inheritance, while organelle genes and DNA's are inherited in a non-Mendelian way (→ genetic segregation).

nuclear envelope (Anderson 1953) — (NE); in eukaryotes, the double membrane structure that forms the boundary of the → nucleus. The major structural components of the NE are the inner and outer nuclear membranes (7—8 nm in diameter) enclosing a lumen (the perinuclear space) and the nuclear lamina and → nuclear pore complexes. The outer membrane on the cytoplasmic side is continuous with the rough → endoplasmic reticulum and covered with → ribosomes on its outermost surface; the inner membrane is smooth. The nuclear pore complexes are located in those regions where the two membranes fuse. The nuclear lamina is a filamentous protein (lamin) meshwork (20—40 nm thick) lining the nucleoplasmic surface of the NE and probably provides an anchoring site at the nuclear periphery for interphase → chromosomes. The NE is a barrier responsible for segregating the cytoplasm from the nucleoplasm and thereby channeling all transport across the n. membranes through the nuclear pores. Since the NE is the site at which the first events of → chromosome condensation occur, it is probably involved in the maintenance of interphase chromosome organization. The NE breaks down during prophase-metaphase of → mitosis and reforms in anaphase-telophase.

nuclear envelope bleb (Gondos and Bhiraleus 1970) — in almost all mammalian zygotes, any of the small evaginations of the inner leaflet of the → nuclear envelope (= tertiary nucleoli; nuclear extrusions). Blebs represent a step in the transport of some material from the nucleus to the cytoplasm at the beginning of development.

nuclear fragmentation — a degenerative breakdown of the → nucleus into two or more parts of equal or unequal size.

nuclear framework (Berezney and Coffey 1975) — the nuclear protein matrix and → nuclear lamina which is believed to maintain the structure and organization of the → nucleus and to mediate the spatial ordering and processing of replicative and nonreplicative chromosomal DNA during replication and mitosis.

nuclear genome — in eukaryotes, the genetic information encoded in the nuclear DNA, as opposed to the organellar genomes (mitochondrial and chloroplast genomes).

nuclear ghost (Aaranson and Blobel 1974) — = → nuclear pore complex-lamina fraction.

nuclear inclusion — any structure in the interphase → nucleus which is not known to result from a redistribution of normal nuclear structures or from trapping of cytoplasmic material (→ nuclear body).

nuclear lamina (Harris and James 1952) — a fibrillar protein (lamin) meshwork that lines the inner (nucleoplasmic) surface of the → nuclear envelope (= fibrous lamina; dense lamella; zonula nucleum limitans). The n.l. may both regulate nuclear envelope structure and anchor interphase chromosomes at the nuclear periphery. The lamina are members of the → intermediate filament protein family.

nuclear location sequence — (NLS); in proteins actively transported across the → nuclear envelope, a sequence which mediates the nuclear import of the respective proteins.

nuclear matrix (Berezney and Coffey 1974) — a structural framework (= karyo-skeleton; nuclear scaffold; nuclear cage) remaining when isolated → nuclei are depleted of their membranes, soluble molecules, and chromatin by means of subsequent treatments with detergents. The n.m. is a specialized compartment of the skeletal structure of the nucleus involved in nuclear functions, such as → DNA replication, → genetic transcription and processing of transcripts, and hormone-receptor binding. It is composed of at least three structural components: (1) A residual → nucleolus (nucleolar matrix); (2) a residual peripheral (pore-complex) lamina; (3) an interconnecting fibrillar component reminiscent of a ribonucleoprotein network.

Newly replicated DNA is intimately associated with the n.m. which probably provides a structural support for DNA synthesis in eukaryotes analogous to the cell wall of prokaryotes. The sites of the n.m. to which topologically constrained loops of DNA (about 100 kb) are anchored are referred to as n.m. attachment sites. These sites are nonrandomly distributed and are generally found in nontranscribed A+T-rich DNA regions in close proximity to transcription units and to upstream → enhancer sequences. Nuclear actin is a major n.m. protein.

During metaphase the n.m. is termed → chromosome scaffold.

nuclear membrane (Hertwig 1893) — → nuclear envelope.

nuclear phenotype (Darlington 1932) — the genetically controlled →phenotype of the chromosomes, which is species-specific, varies considerably with respect to chromosome appearance and chromosome behavior in the course of the → cell cycle (→ mitosis; meiosis; endoreduplication; nuclear differentiation), and frequently to changes of genotype (by segregation, hybridization, or mutation) in much the same way as do characters belonging to the "exophenotype". Variation of the n.p. depends on genotypic (the chromosomes carrying the genes are subject to the action of these genes) and structural properties (represented by the order rather than the action of the genes in the chromosomes) of the → karyotype, the evolution of which results from the action of natural → selection on the heritable variability of chromosome shape and chromosome function and gives rise to the several different variants of the chromosome mechanism found in nature (→ endophenotype).

nuclear plaque — in yeast, any of the regions on the → nuclear envelope from which → spindle microtubules originate.

nuclear pore complex (Watson 1959; Callan and Tomlin 1950) — any of the nonrandomly distributed octagonal orifices (diameter about 60 nm) of the → nuclear envelope that form channels for nucleo-cytoplasmic transport of molecules and particles. The n.p.c. has a symmetrical structure framed by two widely separated, coaxial rings. Each ring is composed of eight globular subunits, and attaches to the nuclear membranes. Connected to these rings and extending radially inwards from them along a central plane are elongated structures called spokes. These appear to contact a large central spherical protein structure (the plug or annulus), i.e., n.p. complexes are not freely communicating openings.

nuclear pore complex-lamina fraction (Aaranson and Blobel 1975) — a stable supramolecular fraction of the → nuclear envelope consisting of the → nuclear lamina isolated together with morphologically associated → nuclear pore complexes. It underlies the inner nuclear membrane and connects → chromatin to the nuclear envelope (= nuclear ghost).

nuclear protein — any protein uptaken by the → nucleus; uptake is highly selective and such proteins must therefore contain within their final structure a signal that specifies selective accumulation in the nucleus.

nuclear protein kinase — any of a class of enzymes that transfer γ-phosphate to serine, threonine, lysine, or histidine in proteins. N.p. kinases with a preference for basic proteins are called → histone kinases, those preferring acidic proteins are casein kinases.

Both nuclear and cytoplasmic p. kinases appear to have a key regulatory role in the eukaryotic cell, particularly in metabolism, control of cell proliferation, and in → gene expression.

nuclear receptor — any of a family of proteins which coordinate complex events controlling morphogenesis and homeostasis in response to the binding of their cognate ligands. N. receptors act as → transcriptional factors that can regulate → gene expression positively or negatively by interaction with specific DNA sequences (ligand-responsive elements, transcriptional enhancers). Binding of the ligand is prerequisite for specific DNA binding and for transcriptional activation (→ steroid hormone receptor).

nuclear ribonucleoprotein — RNA-protein complexes in the → nucleus which contain the primary RNA transcripts (→ heterogeneous nuclear RNS; informofer). Electronmicroscopically, n.r. particles appear as a loose arrangement of fine fibrous material which itself is particulate. The protein which constitutes the main mass of the particules is arranged as 10—20 nm subparticles in bead-like fashion on the string or strings which make up the total particle.

nuclear RNA (Harris 1963) — any of the high molecular weight RNA molecules which are the first product of eukaryotic → gene expression. nRNA's undergo a series of processing steps (→ RNA processing), including the removal of introns, capping at the 5′ end, and polyadenylation at the 3′ end, to produce the mature → messenger RNA that is ultimately transported to the cytoplasm for translation.

nuclear run-on transcription (Groudine et al. 1981) — in vitro measurement of RNA elongation. N. r.-o. t. allows detection of transcription of single-copy genes and rests on small amounts of embryonic material.

nuclear sap — → nucleoplasm.

nuclear scaffold — = → nuclear matrix.

nuclear segregation — the segregation of genetically different nuclei contained in a → heterokaryotic multikaryon.

nuclear sieve complex (Virkki and Kimura 1978) — in certain organisms, any of the shallow, cup-like indentations of the → nuclear envelope into which the nuclear pores are clustered.

nuclear skeleton — = → nuclear matrix.

nuclear tetrad — the four haploid nuclei arising from the two meiotic divisions (→ meiosis; tetrad).

nuclear transplantation — the replacement, by artificial means (micromanipulation, cell fusion), of the → nucleus of one cell with that of another. Frequently, the recipient cell is deprived of its own nucleus by merotomy (section into two halves) prior to n.t.

nuclear vacuole — in sperm nuclei of mammals, any of the irregular, clear areas in the otherwise densely packed → chromatin mass of the nucleus. N. vacuoles, unlike true cytoplasmic → vacuoles, do not possess a limiting membrane.

nuclease — any enzyme hydrolyzing a phosphodiester bond in a nucleic acid. They are usually specific for DNA (DNase) or RNA (RNase) and for either single-stranded or double-stranded molecules. Exonucleases release nucleotides from the end of the molecule, endonucleases cleave the polynucleotide chain at an internal site.

nuclease hypersensitive site (Varshavsky et al. 1978; Scott and Wigmore 1978) — in eukaryotic chromatin, any of the nucleosome-free regions that are exceptionally sensitive to different nucleases and allow enhanced access of resident cis-acting DNA sequences to trans-acting protein factors. N.h. sites usually represent a minor (about 1 %) and highly selective fraction of the genome. They are usually associated with sequences that function in → genetic transcription, → DNA replication, → genetic recombination, or chromosome segregation. → Histones are absent from n.h. sites; several classes of nuclear proteins (such as → DNA topoisomerase I and II, → RNA polymerase II, and transcription factors) have been found to be associated with a subset of n.h. sites.

A common feature of n.h. sites must be an alteration of DNA structure. Possibilities include bent DNA, the unwound and A-form DNA associated with transcription, "alternate" (including unwound and supercoiled) B-form DNA structure, the left-handed → Z-DNA, and DNA modified by cytosine methylation.

Nuclease sensitivity of actively transcribed DNA sequences is dependent upon their association with two → high mobility group proteins (HMG14 and HMG17). N.h. sites have been mapped to homopurine sequences.

Within a given cell, n.h. sites fall into two major categories: (1) Constitutive sites often present in → promoter regions of genes; (2) inducible sites which often persist long after removal of the inducing agent.

N.h. sites that differ between cell types are called tissue-specific sites; those tissue-specific sites which appear transiently during embryogenesis are termed developmental n.h. sites.

nucleate — having a → nucleus.

nucleating site (Tucker 1972) — → microtubule.

nucleic acid (Altmann 1889) — → deoxyribonucleic acid; ribonucleic acid.

nucleic acid blotting — transfer of DNA and RNA from gels onto a solid carrier. Subsequent characterization of the transferred nucleic acid molecules is limited by hybridization to radioactive probes (→ Southern blotting).

nucleic acid hybridization — the formation of stable DNA-DNA or DNA-RNA hybrid molecules via Watson-Crick base pairing (annealing). The efficiency of n.a.h. is a test of homology of the sequences hybridized. N.a.h. typically requires that both the probe and target to be single-stranded, i.e., double-stranded nucleic acids need to be denatured (by heating or treatment at high or low pH) prior to h. (→ in situ nucleic acid hybridization; Cot value).

nucleic acid microinjection — the injection of DNA or RNA into intact eukaryotic cells, which provides a means for studying transcription and translation of specific genes.

nucleic acid probe — a gene, fragment of a gene, or its transcription product (RNA) which is used to assay a gene.

nucleocapsid — the → capsid-nucleic acid unit of a virus. If the virus is without an envelope, the n. itself is the virion.

nucleo-cytoplasmic interaction — interaction between nuclear genes and the heterogeneous cytoplasm which may be divided into (1) effects of the nucleus on the activity of the cytoplasm, and (2) effects of the cytoplasm on the activity of the nucleus. N.-c.i. involves the passage of molecules from one compartment to another of the eucell; it may result in a wide range of changes in chromosomal and gene activity (→ gene activation) and, thereby, in → cytodifferentiation.

nucleodesma (Scott 1950) — a fibrillar connection between cell nucleus and cytoplasm (= karyodesma).

nucleofilament (Finch and Klug 1976) — the linear array (loose coil) of → nucleosomes in the 20- to 30-nm-thick fibrils characteristic of the in situ organization of transcriptionally inactive, condensed → chromatin of interphase and metaphase chromosomes. Three different models of n. organization within the higher-order fibrils have been proposed: (1) The chromatin fibril represents a continuous solenoid of a helically wound n.; (2) The chromatin fibril is composed of closely packed, regularly spaced globular subunits of 20 to 25 nm diameter, each granule (or "superbead") being an aggregate of a defined number of nucleosomes within a n.; (3) the chromatin fibril is a continuous, uniformly thick (nonsolenoidal) array of nucleofilaments characterized by densely packed nucleosomes.

nucleohistone — a complex of → deoxyribonucleic acid and → histone with the histone bound to the large groove of DNA. Two models of n. structure have been proposed: (1) Nonbasic residues of the histone are looped away from the DNA, enabling the basic residues to neutralize successive phosphate groups of DNA. (2) One histone is stretched along the length of about 12 turns of the DNA helix, a segment in which about 75 % of the DNA phosphate groups would not be neutralized by histone basic residues and is, therefore, free for interactions with other cations.

Extraction of n. is always accompanied by extraction of DNA-bound → nonhistone chromosomal protein, suggesting that multimolecular structural complexes exist in native nucleoprotein. Different types of histones vary in their tendency to aggregate and in the closeness of their binding to DNA.

nucleoid (Piekarsky 1937) — (1) the region in the bacterial cell to which the negatively supercoiled DNA (about one supercoil per 200 bp) is confined; the bacterial DNA is associated with small, basic proteins. N. structure is of importance in → gene expression, → DNA replication, → DNA inversion, and → genetic recombination. (2) Any of the structures resembling → nuclei and released by lysing eukaryotic cells gently in solutions containing non-ionic detergents and high concentrations of salt. Such nucleoids are complex templates containing DNA, nuclear RNA, some protein and unknown quantities of lipids.

nucleolar chromosome — any chromosome with a → nucleolar organizer.

nucleolar constriction — a secondary → constriction associated with the formation of the → nucleolus.

nucleolar DNA — in eukaryotes, the DNA of the → nucleolus organizer associated with → nucleologenesis and coding for → ribosomal RNA.

nucleolar dominance (Navashin 1934) — in interspecific hybrids preferential transcription of → ribosomal DNA (and formation of nucleoli) of one of the parent species (= differential amphiplasty) and suppression of the transcription of rDNA in the other parent. N.d. also occurs in interspecific somatic cell hybrids and in some translocation lines with all nucleolus organizing regions combined in a single chromosome pair.

nucleolar interstices (Bernhard and Granboulan 1968) — light areas (nucleolar vacuoles) surrounding the nucleolonemas of the → nucleolus; n.i. vary in size and content and are characterized by a low concentration of structural elements.

nucleolar lacuna (Chouinard 1970) — any of the small electron-clear areas within the → nucleolus which contain fibrils of 6 to 10 nm in diameter.

nucleolar matrix — the residual structure of the → nucleolus left after certain extraction procedures; it consists of an amorphous background material in which the fibrils of the nucleolar dense fibrillar component and the granules of the granular component are embedded.

nucleolar organizer (McClintock 1934) — = → nucleolus organizer.

nucleolar organizer track — a meandering structure (about 1 µm in diameter) which marks the course of the → nucleolus organizer through the → nucleolus in plants.

nucleolar reorganization — the postmitotic reformation of a functional → nucleolus. N.r. involves the precise targeting of old nucleolar material to the → nucleolus organizer of the telophase chromosomes of daughter nuclei, the rapid assembly into the nucleolar body, and the resumption of pre-rRNA synthesis.

nucleolar ribonucleoprotein particle (Warner and Soeiro 1967) — → ribosomal precursor RNA.

nucleolar segregation (Bernhard et al. 1965) — the redistribution of morphologically discrete nucleolar components induced by various drugs (actinomycin D and agents acting by similar biochemical mechanisms on DNA-dependent RNA synthesis) and characterized by the sorting out into distinct zones of the fibrillar and granular components of the → nucleolus.

nucleolar vacuole (Chouinard 1970) — any of the distinct electron transparent areas within the granular zone of the → nucleolus. Such vacuoles contain scattered granules and fibrils and their presence, number and size are correlated with various physiological conditions of the cell. They are assumed to be involved in the release of nucleolar products (RNP precursors of → ribosomes).

nucleolar zone (Serra 1942) — any chromosomes region, irrespective of whether or not it is a secondary → constriction that is associated with the formation of the nucleolus during telophase.

nucleolin (Orrick et al. 1973) — the major (multifunctional) protein of the → nucleolus associated with intranucleolar chromatin and preribosomal particles. N. is thought to play a direct role in pre-rRNA transcription and → ribosome assembly. N. holds the whole transcription apparatus in an arrested state until it is phosphorylated and cleaved. It is the main silver-staining protein of nucleoli.

nucleolinus — a nucleolar fibrillar center. Nucleolini usually appear as small rounded bodies of different size which are surrounded by dense fibrillar RNP components. In contrast to → nucleolar interstices or light areas, they contain a large amount of fine filaments and dense fibrils. Their function is not clear.

nucleologenesis — the formation of the → nucleolus which takes place once per → cell cycle in postmitotic nuclei (late anaphase/telophase) when nucleolar material appears in the vicinity of and on the surfaces of the chromosomes, sometimes in the form of a perichromosomal sheath. This material then condenses or collects into prenucleolar bodies (PNB's) which subsequently fuse at the → nucleolus organizers (NOR's) in telophase and early interphase. N. depends on reinitiation of transcription of rDNA and the kinetics of the process is a linear function of the number of rRNA genes present in the nucleus.

nucleolonema (Estable and Sotelo 1950; Bernhard et al. 1952) — any of the threadlike ribonucleoprotein structures (about 0.1 μm in diameter or larger) contained in the → nucleolus of most cells. They are surrounded by light areas (nucleolar interstices). Nucleolonemas are composed of fibrillar and/or granular elements. The n. may contain chromatin structures in the form of fine filaments or strands.

nucleolus (Valentine 1836) — a distinct spheroidal body in interphase nuclei which is the site of synthesis and processing of → pre-ribosomal RNA and of assembly of the → ribosomal proteins and ribosomal RNA into pre-ribosomal particles (→ ribosome). → Small nuclear ribonucleoprotein particles (U3 and U8 in particular) are also accumulated in the n. and may function in rRNA processing (→ nucleolar matrix). Nucleoli contain, in addition to pre-rRNA transcription units (→ nucleolus organizing region) and preribosomal particles, abundant proteins that are thought to provide a structural support for the spatial arrangement of the rRNA genes and to play a role in the ordered translocation of preribosomal particles from their primary site of assembly to the nucleolar periphery.

Three nucleolar components can be distinguished: (1) The granular component (pars granularis; GC), consisting of distinct, nearly spherical granula (about 15 nm diameter); (2) the dense fibrillar component (DFC), consisting of tightly packed 3- to 5-nm-thick fibrils; (3) the fibrillar centers (FC) consisting of a loose network of fibrils with a little larger average diameter (4-8 nm) than in the dense fibrillar component and containing the genes encoding rRNA.

The n., like the → nucleus, contains a scaffold or skeletal matrix material and three types of nucleoli (reflecting different levels of functional activity) may be distinguished: (1) Ring-shaped nucleoli (low trancriptional activity) with one centrally located fibrillar center surrounded by a small zone of dense fibrillar component; (2) nucleoli with → nucleolonema (beginning of high transcriptional activity) consisting mainly of a three-dimensional network of the dense fibrillar component and with the granular component situated around the dense fibrillar strains in usually small amounts; (3) compact nucleoli (fully active and in equilibrium of transcription of rRNA precursors, processing, and transport) consisting of several fibrillar centers, each surrounded by a layer of the dense fibrillar component.

During mitotic prophase nucleoli disintegrate and their constituents are released and become differentially distributed within the dividing cell. During telophase these constituents rapidly reassemble (→ nucleologenesis) in an apparently coordinate fashion and accumulate in the nucleolus organizer regions of the chromosomes.

Primary steps of pre-ribosome formation occur in the DFC; subsequent maturation stages occur in the GC. Nucleoli contain exclusively → RNA polymerase I (located in the FC) which transcribes rDNA. Probably the nascent transcripts of the rRNA genes serve as specific nucleation sites for DFC assembly.

nucleolus-associated chromatin (Thorell 1944) — condensed → chromatin found adjacent to the → nucleolus (perinucleolar chromatin).

nucleolus organizer region (McClintock 1934) — (NOR); the chromosome region containing the tandem arrays of repeating units of genes (rDNA) coding for the 17-18S, 5,8S, and 25-28S → ribosomal RNA's and transcribed by → RNA polymerase I. The NOR is in many cases a secondary constriction in metaphase chromosomes and plays a crucial role in the topologically specific assembly of nucleolar material (prenucleolar bodies) during telophase. The number of NOR's may vary from species to species, but within a given → karyotype, the number and the locations of NOR's are more or less constant (= nucleolus organizing region; nucleolus organizer).

nucleolytic — → nuclease.

nucleomere (Kiryanov et al. 1976; Franke et al. 1977) — any of the supranucleosomal structures (superbeads) consisting of about ten closely packed → nucleosomes and forming a discontinuous periodic structure which appears as a higher-order chain of tightly adjoined globlular aggregates (the ~30 nm chromatin fibril or nucleomeric fibril).

nucleomixis (Diamelidis 1951) — the establishment of connections between the cell → nuclei during → meiosis (→ cytomixis).

nucleoplasm (Strasburger 1882) — the nonstaining or slightly chromophilic (liquid or semi-liquid) ground substance (the "achromatin") of the interphase nucleus which fills the nuclear space around the → chromosomes and the nucleoli.

nucleoplasmic index (Hertwig 1903) — the ratio of nuclear to cytoplasmic volume expressed numerically as

$$NP = \frac{V_n}{V_c - V_n}$$

(NP = nucleoplasmic index; V_n = nuclear volume; V_c = volume of the cell).

nucleoplasmin (Earnshaw et al. 1980) — a pentameric (anionic) protein which is located at a high concentration in the soluble phase of the (oocyte) → nucleus, but not bound to DNA or chromatin. It interacts with, and stabilizes the octamer of → histones found in → nucleosomes.

nucleoprotamine — in eukaryotes, the salt-like complex between basic → protamine and DNA found in certain cell types (e.g., developing and mature sperms of some organisms), where it is usually the only nucleoprotein present (→ histone).

nucleosid (Levene 1909) — → nucleotide.

nucleosome (Oudet et al. 1975) — any of the repeat units (flat cylinders 11 nm diameter, 9.5 nm high; 10.5—12S) of eukaryotic → chromatin consisting of DNA folded around a → histone core as a 1.8-turn left-handed superhelix. The midpoint of the bound DNA is termed the dyad, at which point each copy of the octamer of histones (two each of H2A, H2B, H3, and H4) can be related to its partner by an axis of twofold symmetry. The minor

DNA helix groove faces outwards from the protein core. The particle mass (~ 206 kDA) of the n. core is equally distributed between 146 (± 2) bp of DNA and the histone octamer. The fifth histone (H1) interacts with an additional 20 bp of DNA to complete two turns of DNA around the histone core.

The n. cores are then linked by DNA (linker DNA) less intimately associated with histones to form structures that appear as "beads on a string". The amount of linker DNA varies from organism to organism, cell type to cell type within an organism, and even within chromatin of a single nucleus. This results in an overall n. size of around 160 bp to about 240 bp (200 bp in most vertebrates).

The structural and mechanical properties of DNA change according to its base sequence, and therefore the ability of a DNA molecule to bend around the histone octamer is thought to be a major determinant of nucleosome positioning.

The principal function of nucleosomes is to compact DNA into the 30-nm filament; transcriptionally active DNA sequences possess special features that permit → RNA polymerase access and passage (→ nuclease hypersensitive site). Three levels of chemical heterogeneity may contribute to this transcriptionally active state: (1) Different primary structures of the core histones create the potential to generate (in mammals) 180 compositionally unique histone octamers; (2) the core histone subtypes are post-translationally differentially modified (by acetylation, methylation, phosphorylation, poly(ADP-ribosyl)ation, and ubiquitation); (3) histone H1 subtypes, → high mobility group proteins, and other accessory proteins interact with specific binding sites on nucleosomes.

Essentially all the nascent core histones are conservatively deposited and segregated in nucleosomes. They bind preferentially to newly synthesized DNA; parental histones segregate to one arm of the DNA replication fork (the leading side for → DNA replication). The conservative assembly and segregation of n. cores is a means whereby information controlling gene structure and expression can be stably incorporated into chromatin.

There are two extreme models to account for the folding of DNA in the n.: (1) A smooth, continuous bending of the DNA (2) localized distorsions at regular points ("kinks") with straight regions of multiples of 10 bp.

nucleosome core particle — any of the particles derived from → chromatin by digestion of linker DNA sequences and removal of histone H1. The n.c.p. consists of 146 (± 2) bp of DNA wrapped around the outside of an octamer of histones (→ nucleosome). It is a flat cylinder (5.5 nm high and 11 nm in diameter). A special type of n.c.p. is called → chromatosome. At least three types of interactions are involved in n.c. assembly: (1) Stereospecific bonding between histones to form octamers; (2) electrostatic repulsion between the charged histones; (3) electrostatic attraction of the histones for DNA (→ nucleoplasmin).

nucleosome filament — a linear array of connected → nucleosomes which abut each other (diameter of the n.f. 10 nm). The n.f. is the → chromatin fiber and can be coiled into a solenoid, thereby forming a 20—30 nm diameter filament or "thick fiber" which may be further folded into a radial loop with the loop bases organized into the central portion of the → chromatid.

nucleosome linker — → linker DNA.

nucleosome phasing (McGhee and Felsenfeld 1980) — the preferred positioning of the → nucleosome core histone octamer with respect to the sequence of the DNA wrapped around it. Three different situations of n.p. are known: (1) The distance of regularly spaced nucleosomes may be related to the repeat length of a reiterated DNA base sequence; (2) not regularly spaced nucleosomes may be specifically positioned; (3) regularly spaced nucleosomes may be associated in one or several distinct modes with a given DNA sequence.

nucleosome spacing — the disposition of → nucleosome cores one to another, i.e., the average length of → linker DNA which may vary from 20 to 100 bp depending on the organism or tissue under study.

nucleotide (Levene 1909) — any of the monomeric units which are the building blocks of the polynucleotides referred to as → nucleic acids (Table 10). A n. is a phosphate ester of the N-glycoside of a nitrogenous base and consists of a purine or pyrimidine base, a pentose (D-ribose in → ribonucleic acid, 2′-deoxy-D-ribose in → deoxyribonucleic acid) and a phosphate (PO_4) group. The combination of nitrogenous base and sugar is called a ribonucleoside (when the pentose is D-ribose) and a deoxyribonucleoside (when the pentose is 2′-deoxy-D-ribose). The nucleotides are formed by the addition of a phosphate group to the nucleosides and are either ribonucleotides (in RNA) or deoxyribonucleotides (in DNA).

Table 10. The main components of DNA and RNA

Nucleic acid	Main purine (PU) and pyrimidine bases (PY)	Pentose	Nucleoside	Nucleotide
		2'-deoxy-D-ribose	Deoxyriboside	Deoxyribotide
	Cytosine (PY)		Deoxycytidine	Deoxycytidylic acid
	Thymine (PY)		Deoxythymidine	Deoxythymidylic acid
DNA	Adenine (PU)		Deoxyadenosine	Deoxyadenylic acid
	Guanine (PU)		Deoxyguanosine	Deoxyguanylic acid
		D-ribose	Riboside	Ribotide
	Cytosine (PY)		Cytidine	5'Cytidylic acid
	Uracil (PY)		Uridine	5'Uridylic acid
RNA	Adenine (PU)		Adenosine	5'Adenylic acid
	Guanine (PU)		Guanosine	5'Guanylic acid

nucleotide-excision repair — a type of → DNA repair which removes damaged bases from DNA as an oligonucleotide (→ excision repair; base-excision repair). The resulting gap is patched by → repair synthesis. Mono- and diadducts are removed by n.-e.r. (interstrand croslinks are repaired by concerted action of n.-e.r. and → recombinational repair pathways).

nucleotide replacement site — any position in a → codon where, due to → gene mutation, a base has been substituted.

nucleotide sequence — a specifically ordered row of → nucleotides in the DNA or RNA of organisms. A n.s. which occurs many times in the DNA of one cell is called a repeated n.s. (→ repetitious DNA). A particular n.s. which occurs only once in a genome is called a unique n.s. (→ unique DNA).

nucleotide sequencing — → DNA sequencing.

nucleotide substitution — the presence of different nucleotides in equivalent positions of homologous DNA or RNA sequences.

nucleotide synthesis — in virtually all cells, synthesis of → nucleotides by two fundamentally different kinds of pathways: (1) The **de novo pathway** in which ribose phosphate, certain amino acids, CO_2 and NH_3 are combined in successive reactions to form the nucleotides (free bases are not used as intermediates); (2) the **salvage pathway,** in which free bases or nucleosides, produced by breakdown of nucleic acids, are converted back to the nucleotides.

nucleotidyl cap — → messenger RNA capping.

nucleotype (Bennet 1972) — the gross physical characteristics of the → nucleus, its mass, and particularly the amount of DNA contained in its constituent chromosomes.

nucleus (Brown 1831) — a membrane-enclosed cell organelle which represents one of the two main compartments of the eucell and contains the bulk of its → genetic information (nuclear DNA) in the form of → chromatin. Replication of chromosomal DNA, → genetic transcription of DNA and processing of DNA transcripts (→ heterogeneous nuclear RNA; messenger RNA, ribosomal RNA, transfer RNA) occur in the n.

Four main structural components can be visualized in nuclei: (1) The surrounding → nuclear envelope; (2) the → nucleoli; (3) dense chromatin patches (heterochromatin) beneath the nuclear envelope and surrounding the nucleoli (perinucleolar chromatin), as well as in other interior regions of the nuclei; (4) interchromatinic areas between condensed chromatin regions, which contain the diffuse chromatin (euchromatin) and various granular and fibrous components.

Interphase nuclei exhibit an enormous polymorphism. They vary in size, shape, and fine structure both in different cell types and in a given cell type depending on the → cell cycle, the physiological state, the circadian periodicity, etc.

The n. is filled with a network or lattice mainly composed of soluble proteins, → ribonucleoprotein particles (RNP's), and chromatin that extends throughout the interior to form a continuum from the nucleoli to the nuclear envelope and pore complexes (→ nucleoplasm; nuclear matrix).

The n. divides by → mitosis (in somatic tissues), by meiosis (in germinal tissues resulting in the production of → gametes or meiospores), or (rarely) by → amitosis. It may become endopolyploid by the process of → endomitosis (→ nuclear differentiation).

Typical membrane-bounded nuclei are present only in the cell during → interphase (**"interphase nuclei"**), prophase, and telophase; **"prophase nuclei"** are dissolved (breakdown of the nuclear envelope) at the end of mitotic prophase, and **„telophase nuclei"** are reformed at the end of mitosis and meiosis.

The **lifetime of interphase nuclei** which undergo further mitosis may be subdivided into subphases if DNA synthesis and chromosome reduplication (which occupy only a part of interphase) are used as the criteria for classification. Interphase of the mitotic cell cycle consists of three parts referred to as G_1 (gap 1), S (synthesis), and G_2 (gap 2). During G_1 the chromosomes are as yet unreduplicated; during S DNA synthesis and chromosome reduplication take place; and G_2 is the postsynthetic phase which ends with the beginning of mitotic prophase. A somewhat similar subdivision is possible for those interphases which precede meiosis, but in this case G_2 is usually short or entirely absent.

null allele — an → allele whose effect is either absence of a normal gene product at the molecular level or absence of a normal function at the phenotypic level (= amorph). N. alleles are due to null mutations arising by base substitution, → transposition events, or loss from the genome of the locus coding for the protein (deletion).

nulli-haploid (Riley and Chapman 1958) — → haploid.

nulliplex (Belling and Blakeslee 1923) — of a → polyploid in which all chromosomes of one homologous type carry the same recessive → allele of a particular → gene (e.g., aaa in case of triploidy, aaaa in case of tetraploidy, etc.), in contrast to simplex, duplex, triplex, quadruplex, etc., in which the dominant allele is represented one, two, three, four, etc., times (Darlington and Mather 1949).

nullisomic (Blakeslee 1921; Sears 1941) — of an → aneuploid cell, tissue (in the case of → chimeras), or individual with both members of one (or more than one: double nullisomic, etc.) particular pair of homologous chromosomes missing from the chromosome complement. Normally, nullisomics (or nullisomes) cannot survive in diploids, but they are recoverable in analogous forms, e.g., in → allopolyploids (which are functionally diploid) and → autopolyploids. Nullisomy may arise from selfing or intercrossing of → monosomics by fusion of two gametes lacking the same chromosome ("n − 1") of a particular chromosome pair.

numerical sex chromosome anomaly test — in → mutagenicity testing, a method based on the fact that some of the end results of two types of basic damage are detectable postnatally by the use of genetic markers present in affected → X and/or Y chromosomes. These types of damage are: (1) A chromosome breakage in germ cells which leads to chromosome loss and thus monosomy of the conceptus; (2) meiotic → nondisjunction, i.e., failure of homologous chromosomes to separate, which leads to monosomy and complementary trisomy.

O

occluding junction — = → tight junction.

ochre mutation (Brenner et al. 1965) — a → nonsense mutation giving rise to the terminator codon UAA in → messenger RNA.

octoploid — of → auto- or → alloploid cells, tissues, and organisms, respectively, with eight chromosome sets in their nuclei (symbol 8 x).

Okazaki fragment — any of the short (2000 nucleotides in bacteria, 200 nucleotides in eukaryotes) nascent DNA sequences (= O. pieces) that grow discontinuously on the lagging arm (5′ → 3′) of the → DNA replication fork. They are linked to a 5′-terminal oligoribonucleotide (8—11 nucleotides in length) called → initiator RNA. Following ligation of O. fragments at their 3′ end (iRNA distal), iRNA must be removed prior to the next ligation event at the 5′ end (iRNA proximal) since no intercalated RNA sequences are found in long nascent DNA chains. RNA primers for O. fragments are synthesized by a specific RNA polymerase called → primase.

oligo (dA-dT) cluster (Shenkin and Burdon 1974) — in nuclear DNA of all eukaryotes, any of the DNA sequences consisting predominantly of nonalternating oligo (dA-dT) and about 15 to 25 bp long. Their frequency varies from several hundred per genome (lower eukaryotes) to 10^4 or more (higher eukaryotes).

oligodeoxynucleotide-directed mutagenesis (Razin et al. 1978) — the use of synthetic oligodeoxynucleotides (= oligo-directed m.) complementary to the region surrounding the site of mutagenesis (target site) for induction of defined and random mutations within a precisely targeted segment of DNA (→ site-directed mutagenesis) by enzymatic misincorporation of noncomplementary nucleotides. Usually the target region, present in a ss-phage DNA or in a ss-plasmid DNA, is annealed to the oligo and converted into the ds form in an in vitro polymerization-ligation reaction. The mutants are selected after cell transformation.

oligogenic (Mather 1941) — of characters controlled by a few genes (oligogenes or major genes), as opposed to → polygenic characters.

oligolecithal — = → isolecithal.

oligomer (Changeux 1964) — a protein composed of identical subunits (polypeptide chains).

oligonucleotide probe — any of the very short probes (about 20 nucleotides) that can identify single base changes in DNA.

omega protein (Burgess et al. 1969) — a (bacterial) DNA untwisting enzyme (→ RNA polymerase).

omnipotent suppressor (Inge-Vechtomov and Andrianova 1970) — in yeast, any of a class of → suppressors that are codon-nonspecific and cause translational errors that lead to the misreading of all three → nonsense codons. Translational inaccuracy probably results from altered ribosomes.

oncogene (Huebner and Todaro 1969) — (onc gene); any of a family of cellular DNA sequences (cellular oncogenes) which possess the potential to become malignant through incorporation into the genome of a → retrovirus (viral onc genes) or by undergoing alterations within the cell (i.e., they are activated homologs of → protooncogenes). Viral and nonviral onc genes can be classified into four distinct groups according to their functional properties: (1) Protein kinases; (2) GTPases; (3) nuclear proteins; (4) growth factors. One class of viral onc genes that does not appear to have cellular analogs are the transforming genes of DNA tumor viruses.

Major mechanisms implicated in the induction of the transforming activity of cellular onc genes are: (1) Integration of a retroviral genome in the vicinity of a cellular o. and consequent increase in the level of expression; (2) generation of a point mutation in the o. coding region leading to the formation of an altered gene product; (3) → gene amplification by which the o. copy number increases; (4) o. → translocation to another chromosome.

The onc genes of retroviruses (v-onc) arose by transduction of cellular genes (c-onc).

Genes giving rise to tumors by loss of function are called recessive onc genes or cancer suppressor genes, since the presence of one or more normal alleles is sufficient to prevent expression of the cancer phenotype.

oncogenic — of viruses that are tumor-inducing in animals. Viruses having this property are found in almost all families of DNA viruses but among RNA viruses only the → retroviruses produce tumors (→ oncogene).

one gene — one enzyme hypothesis (Beadle and Tatum 1941) — a hypothesis which states that the information required for the synthesis of each protein is encoded in a → gene unique for that protein. The model has to be revised since the following variants are known: (1) Oligomeric proteins composed of two or n identical polypeptide chains (one gene → one polypeptide chain and two or n polypeptide chains → one enzyme); (2) oligomeric proteins composed of different polypeptide chains (two genes → two polypeptide chains and n (two) polypeptide chains → one enzyme); (3) one polypeptide chain with several enzyme activities (one gene → one polypeptide → two enzymes); (4) → antibodies (two genes → one polypeptide chain); (5) one gene → two polypeptide chains (→ overlapping genes).

one operon — one messenger hypothesis — → operon.

one step meiosis (Cleveland 1947) — → meiosis.

ontogenetic — ref. to the development (particularly → embryonic development) of an individual.

ontogeny (Haeckel 1886) — the course of development of an individual organism, an organ, or an organelle (→ recapitulation).

oocyte (Boveri 1891) — the egg mother cell ("primary o.") giving rise, in the first meiotic division, to the "secondary o." and the first polar body. At the second meiotic division, the secondary o. gives rise to the ovum and a second polar body (→ oogenesis).

oocyte assay of mRNA (Gurdon et al. 1971) — a sensitive technique for the assay of mRNA involving microinjections in nanogram amounts of polyribosomes or of other preparations containg mRNA into *Xenopus* oocytes for efficient and long-term translation. Translation of exogenous mRNAs in the oocyte has been shown with minimal estimates to be 100—1000 times more efficient than in the reticulocyte cell-free system. The oocyte lacks specificity with respect to the type of mRNA capable of being translated.

oogamy — the union during → fertilization of a nonmotile female → gamete (egg cell) and a motile male gamete.

oogenesis — the development of the female → germ cell (egg cell or ovum) of animals which takes place in the gonad. The gonad (ovary) contains the primordial germ cells (primary oogonia) which enter a phase of rapid multiplication giving rise to the secondary oogonia. At the end of the multiplication period, the secondary oogonia transform into oocytes which grow until the end of o. The period of oocyte development is divisible into three subphases. During the first (premeiotic) phase immediately following the formation of the oocyte from an oogonium, growth is very slow. At the end of that phase the nucleus starts to swell and the oocyte enters a second phase of slow growth (mainly due to synthesis of cytoplasm). During the third phase the growth rate increases rapidly and in a relatively short time the oocyte reaches its final size.

The so-called premeiotic phenomena normally begin immediately after the completion of the last oogonial division. During the first and second meiotic division, oogonia give rise, via primary and secondary oocytes, to one functional gamete (the ovum or egg cell) and three or (rarely) merely two abortive cells (the polar bodies or polocytes). Meiosis II takes place in the secondary oocytes and is generally arrested at meta- or anaphase II up to the time of fertilization. After plasmogamy the second meiotic division is completed and the female and male "pronuclei" unite at karyogamy (→ cleavage).

oogonium (Boveri 1891) — (1) a primordial → germ cell which gives rise to primary → oocytes by mitosis (→ oogenesis); (2) a female → gametangium of Thallophyta.

ooplasm — the → cytoplasm of an egg cell. Different regions of the egg cytoplasm, spoken of as "ooplasms", may have specific properties, so that a particular region can only develop (→ embryonic development) in one definite way. Such regions may be precisely localized (sharp boundaries between them), or they may merge into one another (Waddington 1956).

ooplasmic segregation — → determination.

oosome (Sylvestri 1914) — = → germ line body or germ cell determinant.

oospore — (1) a fertilized ovum (= zygote, oosperm, or ooblastema); (2) an encysted zygote in certain protozoa.

ootid — any of the four meiotic products arising during → oogenesis (= ovotid).

opal mutation (Brenner et al. 1967) — a → nonsense mutation giving rise to the → terminator codon UGA in → messenger RNA (→ amber mutation; ochre mutation).

open-promoter complex — an active intermediate in RNA chain initiation during → genetic transcription, i.e., a stable complex in which a local unwinding (melting) of the DNA duplex occurs starting about 10 bp from the left end of the → Pribnow box and extending to the position of the first transcribed base. Melting is necessary for pairing of the incoming ribonucleotides.

open reading frame — (ORF); a stretch of triplet → codons with an → initiator codon (in most cases ATG) at one end a → stop codon at the other as identifiable by → nucleotide sequences. The ORF is potentially capable of coding for an as yet unidentified polypeptide.

operator (Jacob and Monod 1959) — in prokaryotic → operons, any of the short specific DNA sequences to which proteins (activators or repressors) bind that either activate or prevent the transcription of the associated genes into mRNA. The operator is a symmetrical structure, as is the structure of the binding proteins themselves (dimers consisting of two identical polypeptide chains). The o. is located just after, or overlaps with, the → promoter of a gene or operon.

In negatively controlled operons, the o. occupies the positions between +20 and −50 (where +1 is the first nucleotide transcribed) covered by a promoter-bound → RNA polymerase or, as in the case of the P_1 promoter of the *E. coli* gal operon, within the cAMP-CAP (→ catabolite activator protein) binding sites. Binding of a → repressor to an o. sterically interferes with the binding of RNA polymerase or prevents the formation of an active complex between the polymerase and the promoter.

Operators bearing A·T or T·A bp at positions 6 to 9 bind the repressor more strongly than those bearing G·C or C·G bp at these positions. Base composition near the center of the bacterial o. affects the operator's affinity for repressor by altering the ease with which the o. can be overtwisted into the proper configuration for complex formation with repressor. Both DNA and repressor flexibility influence the strength of o.-repressor interaction. O. specific mutations (symbol O^x) result in an altered specificity of response to the repressor. O. constitutive mutations (symbol O^c) result in a loss of response to the repressor and interfere with repressor binding to the o. Thus all associated genes of the operon function so as to produce their enzyme products at relatively high unregulated (constitutive) levels. Such mutations are usually deletions of all or part of the o. in an inducible operon. **operator zero mutations** (symbol O^o) abolish the expression of a whole operon; some mutants have been found to reduce the maximum rate of expression of the operon without suppressing its expression totally. They then render the residual synthesis constitutively.

operon (Jacob et al. 1960) — in prokaryotes, any group of contiguous, functionally related genes organized into a single transcriptional unit, the expression of which is controlled by positive and/or negative elements acting on a single adjacent regulatory region (the → operator). Operons are operationally defined by the isolation of cis dominant regulatory mutations exhibiting pleiotropic effects on the expression of closely linked structural genes. An o. divergently transcribed from an internal control region is called a bipolar o.

In the case of **repressible operons** (catabolic system), the regulatory protein has no activity by itself (the aporepressor). It takes on the properties of a → repressor only when it binds the appropriate small molecule, called a corepressor. The small molecule in such systems is generally the end product of a biosynthetic pathway or a molecule closely related to the end product.

In the case of **inducible operons** (biosynthetic systems), the regulatory protein is, by itself, the active repressor. The inducer binds to it and thereby either renders it unable to bind to the operator (negative control) or imparts to it a new activity, i.e., that of an activator able to bind to the initiator (positive control). In general, the inducer is the substrate of a catabolic pathway or a molecule closely related to the substrate.

In eukaryotes, functionally related genes are typically scattered on different chromosomes; neither operons nor polycistronic mRNA's have been identified (→ regulon). (→ attenuation; genetic polarity).

operon fusion (Jacob et al. 1965) — the joining of unrelated → operons; usually of the regulatory elements of one operon to the structural genes of another.

operon overlap — the overlapping of neighboring operons as obseved in *E. coli.*

order — a systematic category (below the class) which is composed of one or more families.

ORF — acronym for → open reading frame.

organelle — any usually membrane-bound substructure of the eukaryotic cell performing specialized functions, such as the cell → nucleus, the → mitochondria, the → plastids, the → Golgi apparatus, the → lysosomes, etc. Some organelles contain DNA, called → organelle DNA as opposed to nuclear DNA.

organelle DNA — extranuclear DNA (→ chloroplast DNA, mitochondrial DNA) of eukaryotes localized in cell organelles other than the → nucleus and usually inherited in a uniparental (usually maternal) fashion in plants and animals. O. DNA carries a set of organelle genes and is present in many copies in every cell (organelle genomes). The genetics of cell organelles containing DNA as genetic material is called organelle genetics and comprises organelle gene transmission, non-Mendelian segregation, and recombination.

organelle intron — → intron.

organelle ribosome — any of the → ribosomes present in → mitochondria and → plastids. They are distinct from the 80S ribosomes free in the cytoplasm and synthesize a subset of organelle-specific proteins which are encoded in → organelle DNA.

organism — the unit of a continuous lineage with an individual evolutionary history (Luria and Darnell 1967).

organization effects (Levitan 1954) — any effect (e.g., differential → fitness of individuals which are → heterokaryotypic for two chromosomal structural changes present in either → cis- or trans-configuration) resulting from interaction among adjacent → genetic loci owing to some feature of chromosomal structural organization (→ position effect).

organizer (Spemann 1918) — a region ("organization center") of an embryo from which inducing grafts (embryonic → induction) can be obtained, indicating that the group of cells in question may act as an embryonic inductor (or evocator) which evokes and controls the → differentiation of other groups of cells, i.e., determines the fate of cell masses with which it comes into contact.

organogenesis — the specific assembly of cell populations (during development) creating a functional unit, → an organ. The process is governed by positional and differentiation-inducing signals.

origin — → replication origin.

orphon (Childs et al. 1981) — in eukaryotes, any of the dispersed solitary genes which may be derived from tandem → multigene families. Since these sequence elements are separated from their families by large physical distances in the DNA and appear to be unrelated in flanking sequence they probably originate by a transposition event involving excision of a segment of repeat unit and its reintegration (possibly via RNA or cDNA intermediates) at a remote genetic locus. Transposed orphons presumably become → pseudogenes after mutational divergence from their ancestors.

orthologous (Fitch 1977) — of two genes (or proteins) if they diverged at the same time that the species harboring them diverged (→ paralogous; duplicate genes).

orthoselection — natural → selection which acts in the same direction for a long period of time and gives rise to what has been called "orthogenesis" (→ karyotype o.).

osmiophilic halo — = → centrosome.

osmotic remediability (Hawthorne and Friis 1964; Metzenburg 1968) — a phenomenon demonstrated with a variety of auxotrophs in several organism which show phenotypic repair in growth media of high osmolarities. O.r. is probably the result, in some instances at least, of an effect which prevents the mutant protein from assuming an unstable configuration at the time of synthesis or changes the protein from an unstable to a stable form by some unknown cellular mechanism.

outbreeding — in contrast to → inbreeding, a mating system (= crossbreeding) in which → mating is between individuals less closely related than average pairs chosen at random from the population.

overdominant (Hull 1946) — of two → alleles if the heterozygote (Aa) is superior in comparison with the two homozygotes (AA and aa), i.e., lies outside the range of the homozygotes as to the genotypic value of the character under observation (= superdominant).

Real overdominance in the above sense is to be distinguished from apparent or **pseudo-overdominance** which is not a property of a single → genetic locus, but of blocks of loci or of chromosome segments. Evidence for real overdominance is obtainable only from a genetic locus which has mutated in a highly inbred line, or from a population in which → coupling and repulsion linkages are in equilibrium (Falconer 1960).

Overdominance might result if (1) one allele of the allele pair is neutral in effect and allele A in single dose (Aa) is nearer to optimal gene dosage than in double dose; (2) both alleles (A and A′) are active and complementary in respect to separate supplementary functions; (3) both alleles (A and A′) are active in respect to different primary functions and give rise to a larger effect if present together; (4) an alternative genetic pathway is available for the heterozygote, thus resulting in the formation of a product different from those manufactured by each allele in the absence of the other.

overlapping genes (Barrell et al. 1976) — a single nucleotide sequence encoding (from one or opposite strands) two functional proteins (in bacteriophages, plant and animal viruses, in some mitochondrial genomes and, to a minor extent, in the bacterial genome, as opposed to → split genes). The sequence in question contains two → open reading frames in different phases, both of which can be recognized by the translation machinery. Translation into unique gene products can occur by alternative → RNA splicing of one and the same pre-mRNA (eukaryotes), by alternative recognition of 5′ terminal and internal start sequences and start codons resulting in differently large gene products with identical 3′ terminal sequences (pro- and eukaryotes), by read-through of stop codons (some phages and tobacco mosaic virus), or by use of a second or an additional third reading frame (viruses and some phages).

Possibly o.g. have evolved to maximize the coding capacity at a minimal replication cost or to coordinate gene expression at the translational level in a way that is difficult to achieve in separated genes.

overlapping transcripts (Henikoff et al. 1986) — transcripts arising from opposite DNA strands in some eukaryotic genes (→ overlapping genes).

ovotid — = → ootid.

ovule — the megasporangium of a flowering plant that becomes the seed. It includes the nucellus and the integuments.

oxydosome (Deley 1960) — → mitochondrion.

oxygen enhancement ratio — (OER); the ratio of the radiation effect in fully oxygenated conditions to that in the absence of oxygen.

oxygen-mimic — of a variety of electron-affinic radiosensitizers capable of sensitizing hypoxic bacteria and mammalian cell populations to the lethal effects of radiation. Such compounds include various substituted forms of nitrofurans, nitroimidazoles, and nitrobenzenes.

P

P (Bateson and Saunders 1902) — a shorthand for the parental generations in crosses. Starting with any individual, P_1 denotes the generation of the parents (the immediate parents of a first filial generation or F_1), P_2 the grandparental, P_3 the great-grandparental generation, etc.

pachynema (Gregoire 1907) — = pachytene (→ meiosis).

pachytene(v. Winiwater 1900) — → meiosis.

pachytene DNA (Hotta and Stern 1971) — DNA synthesized during the pachytene stage of → meiosis as the product of → repair replication DNA synthesis. The synthesis is concomitant with → genetic recombination.

pachytene small nuclear DNA (Hotta and Stern 1971; 1984) — (Psn DNA); any of a unique class of moderately repeated DNA sequences which are the substrate for a meiosis-specific endonuclease that introduces nicks during zygotene (early pachytene) at both 5′ ends (→ pachytene DNA). Chromatin containing Psn DNA releases a unique class of RNA molecules (about 125 nucleotides long) called pachytene small nuclear RNA (Psn RNA). This RNA regulates the accessibility of Psn DNA to endonuclease nicking during meiosis and is synthesized during meiotic → chromosome pairing. Pachytene small nuclear protein is required for Psn DNA/Psn RNA hybridization.

packing factor (Darlington and Upcott 1939) — a measure for the degree of → chromosome coiling. It is determined from the diameter of the helix divided by the diameter of the chromosome thread.

packing ratio (Du Praw 1970) — the ratio of the length of a segment of DNA to the length of the → chromatin fiber it is packed in.

paedogamy (Guilliermont 1910) — a form of automixis (obligatory self-fertilization) characterized by the fusion of iso- or anisogametes produced (in the course of → meiosis) by one and the same individual. Either a parental cell (gamont, gametangium) produces directly copulating gametes, or the individual gives rise to male and female sexual organs (antheridia and oogonia) the products of which fuse. Other types of automixis are → autogamy and → parthenogamy.

paedogenesis (v. Baer 1886) — (1) reproduction by pathenogenetic development of egg cells in the larval or embryonic stages of the individual; (2) = → neoteny.

paedomorphosis — a mode of ontogenic change, i.e., the adult descendant resembles the juvenile stage of an ancestor. Two types may be distinguished: (1) Progenesis, i.e., any process bringing about early reproduction; (2) neoteny, i.e., a retardation of somatic growth without a comparable retardation of maturation.

pairing block (Darlington and Mather 1932) — a chromosome segment of definite length (consisting of a definite number of → chromosomes) which acts as a functional unit in meiotic → chromosome pairing. If there are more than two homologous chromosomes in the meiocyte, there may be changes of partners which may lead to → multivalent formation (→ zygomere).

pairing index (Pätau 1941) — → pairing number.

pairing number (Pätau 1941) — the number of chromosome arms, in diploids and polyploids, associated in twos by → chromosome pairing which resulted in a minimum of one → chiasma in any of the arm pairs concerned. The proportion between the observed and theoretically expected (i.e., the number of chromosome arms which might be paired in diploids and polyploids) p.n. is referred to as the "pairing index".

pairing segment (Darlington 1931) — → interstitial segment.

pair mating — a procedure used to determine the degree of sexual isolation between two groups (A and B) of individuals. Separate tests of mating success are made for the four possible mating combinations, $A^o \times A_o$, $B^o \times B_o$, (= homogenic matings), $A^o \times B_o$, and $B^o \times A_o$ (heterogenic matings). The mating system is "male multiple choice", if males of one group (A or B) are allowed to mate with females of both groups; it is "female multiple choice", if females of one group are allowed to mate with females from both groups.

palindromic (Wilson and Thomas 1974) — (PU or REP); of ubiquitous → inverted repeats in nucleic acid sequences (DNA and RNA) which are self-complementary and can exist in two alternative structures: (1) The regular DNA duplex with base pairing between strands; (2) a cruciform structure with intrastrand base pairing of the self-complementary sequence (bases at the symmetrical center of the palindrome are at the ends of hairpin or cruciform arms with respect to flanking non-p. sequences). Perfect palindromes (e.g., GAATTC) frequently occur in binding sites for → restriction endonucleases; less perfect palindromes (e.g., TACCTCTGCGTGATA) frequently occur in binding sites for other proteins, such as in operators. Interrupted palindromes (e.g., inverted repeats such as GGTTXXXAACC) afford the possibility in single-stranded nucleic acids for the loop-stem (hairpin) structures as in → transfer RNA's.

P. sequences can be rather long (300—1200 bp). PU sequences in prokaryotic genes (20—40 bp long) may correspond to stable stem and loop structures whose function is elusive. The stem is GC-rich; it has a lower part of 8 highly conserved bp and an upper part of 0 to 5 bp. The two parts are separated by a C-T mismatch. The loop is AT-rich and variable in sequence and length.

Besides in DNA, p. sequences also exist in transfer RNA, ribosomal DNA, heterogeneous nuclear RNA, and messenger RNA.

palingenesis (Haeckel 1874) — recapitulation of ancestral characters during embryonic development.

panmictic (Weismann 1895) — of a randomly interbreeding population, i.e., a p. unit (Wright 1943). P. units represent the smallest → Mendelian populations and are integrated into the system of populations composing the → species.

panmictic index — the complement (symbolized by P) of the → inbreeding coefficient, 1-F.

parabasal body — = → kinetoplast.

paracentric — of intrachromosomal structural changes (→ inversion) in eukaryotes that do not include the → centromere region of the chromosome involved in the "intrachange", as opposed to pericentric ones which include the centromere.

paracentromeric heterochromatin (Luyks 1974) — → heterochromatic regions on both sides of the centromere which contain, in many cases at least, highly repetitious → satellite DNA. In → meiosis the primary role of p.h. may be (Luyks 1974): (1) To hold sister chromatids together until the second meiotic division; (2) to determine a particular kind of arrangement of chromatin fibers underlying the centromere itself and determining the orientation and disposion of the sister centromeres on the surface of the chromosomes.

paragenetic (Brink 1962) — of changes within the eukaryotic chromosomes that affect the expression (exemplified by V-type → position effects, unstable loci in plants, mottled expression of sex-linked characters in the female of mammals in consequence of the → dosage compensation mechanism), but not the constitution of the genes concerned. Changes affecting gene constitution are referred to as "genetic".

parallelism — in evolutionary biology, the development of similar characters separately in two or more related lineages (convergence).

paraloci (Dunn 1954) — closely linked genetic loci with similar or identical activity (= pseudoalleles).

paralogous (Fitch 1977) — → duplicate genes.

paramutation (Brink 1958) — an interaction between alleles leading to the directed, heritable change of one of the alleles. The changing allele is called the paramutable allele, the other is paramutagenic, and the resultant individual a paramutant. P. represents a kind of gene instability due to heterochromatization at or near the locus at which metastable repression of gene activity occurs.

paranemic joint — a region in which two complementary sequences of DNA are associated side by side instead of being intertwined in a double helical structure.

paranucleic body — a cap made up of cytoplasmic RNA bound to proteins which surrounds the nuclei of gametes and zoospores of Allomyces. The p.b. dissolves at the onset of mitosis or meiosis.

parapatric (Smith 1955) — (1) of a distribution where two closely related taxa meet in a zone of extremely narrow overlap, in which hybridization occurs; (2) of a mode of → speciation from two or more populations that initially have some → gene flow among them, as differentiated from → allopatric speciation, which requires that there first be complete absence of gene flow, and → sympatric speciation, which requires that there initially be random mating (→ stasipatric).

paraplasm (Kupffer 1875) — a type of eukaryotic cell component (demonstrable by cytological or cytochemical methods) located as transient inclusions in the → cytoplasm of certain eukaryotic cell types or during certain periods of cell activity.

paraselectivity (Sedlmayr 1956) — → fertilization.

parasexual (Pontecorvo 1949) — of any mode of reproduction in addition to or in place of normal sexual reproduction and also of mechanisms providing for nonmeiotic recombination of the genetic material (as observed in viruses, bacteria, and in fungi).

parasitic DNA — = → selfish DNA.

parasterility (Brieger 1928) — sterility which results not from genetically or chromosomally defective gametes but from → incompatibility mechanisms which limit or prevent zygote formation by certain classes of gametes.

parental generation (Bateson and Saunders 1902) — → filial generation; → P.

parental investment (Trivers 1972) — any investment by the parent in an individual offspring that increases the offspring's chance of surviving (and hence reproductive success) at the cost of the parent's ability to invest in other offspring.

par locus (Meacock and Cohen 1980) — a cis-acting DNA element needed to promote the equipartition of replicating → plasmid molecules (in *E. coli*). The par sequences are functional in either orientation in the plasmid and par loci of different plasmids are interchangeable.

parsimony — the use of the shortest number of evolutionary steps as a criterion for constructing a → cladogram.

parthenapogamy — = diploid → parthenogenesis.

parthenocarpy (Noll 1902) — the development either of seedless fruits or of fruits in which the seeds lack embryos (→ apomixis). P. results from a failure of pollination (irrespective of the presence or absence of artificial stimulation), a failure in fertilization (→ incompatibility or gametic sterility), or a failure in embryo development (zygotic sterility).

parthenogamy (Guilliermond 1910) — (1) = → pathenogenesis (Guilliermond); (2) an extreme case of → automixis: two nuclei fuse within a single undivided cell which has the character of a female gamete or gametangium; the fusion results in the production of a new individual (→ paedogamy), (3) automictic → parthenogenesis, i.e. oogenesis involves a normal meiosis and reduction of chromosome number, but the zygotic chromosome number is restored by the fusion of two haploid nuclei, or by the formation of a → restitution nucleus, or by → endomitosis (= meiotic parthenogenesis).

parthenogenesis (Owen 1849) — the production of an embryo from a female gamete without the participation of a male gamete, with or without eventual development into an adult. The parthenogenetic organism may be called a "parthenote" of "parthenogenome" (Beatty 1957).

P. can be subdivided into several types which in turn may be grouped into several systems. Classification may be based on the mode of reproduction, the mechanisms of sex determination, and cytology (Suomaleinen 1950).

Classification according to the mode of reproduction:

I. Occasional or accidental p. (= tychoparthenogenesis), i.e., unfertilized eggs occasionally develop through p. II. Normal parthenogenesis.

(1) Obligatory p., i.e., the egg develops exclusively by parthenogenesis.

(a) Constant or complete p., where all generations are parthenogenetic (= acyclic p.). (b) Cyclic p., where one or more parthenogenetic generations alternate with bisexual (amphigonic) generations (= heterogony). (c) Paedogenesis, where the eggs of individuals at larval stages develop parthenogenetically.

(2) Facultative p., where an egg may either be fertilized or develop parthenogenetically.

Classification according to the mechanism of sex determination:

I. Arrhenotoky, where unfertilized eggs develop parthenogenetically into males and fertilized eggs give rise to females.

II. Thelytoky, where unfertilized eggs develop into females.

III. Deuterotoky (= amphitoky), where unfertilized eggs develop into either sex.

Classification based on cytological data:

I. Generative (or haploid) p., where parthenogenetically produced individuals develop from egg cells in which meiotic reduction of chromosome number took place.

II. Somatic p., where parthenogenetically produced individuals have a zygotic (diploid or polyploid) chromosome number.

(1) Automictic (= meiotic) p. or parthenogamy, where regular → chromosome pairing and reduction of chromosome number occur but the zygotic number of chromosomes is restored by fusion of two haploid nuclei, the formation of a → restitution nucleus, or → endomitosis. (2) Apomictic (somatic or ameiotic) p., where neither chromosome reduction nor fusion of nuclei nor any corresponding phenomenon takes place in the egg cell which develops parthenogenetically.

The genetic consequences of p. differ with the cytological features of the parthenogenetic system and may be classified as follows (Suomaleinen 1950):

(1) The progeny are genotypically similar to the mother (as in apomictic p.); (2) heterozygosity is obligatorily replaced by homozygosity of the progeny (as in automictic p. when the somatic (zygotic) chromosome number is restored by fusion of two cleavage nuclei with reduced chromosome number); (3) heterozygosity is replaced by homozygosity if the alleles of heterozygous loci separate prereductionally, i.e., at → meiosis I, but is maintained if they show postreductional disjunction, i.e. separate at meiosis II (this type of genetic behavior occurs in organisms with automictic p. in which the somatic (zygotic) chromosome number is restored by fusion of the second polar nucleus with the egg cell nucleus. It is also characteristic of those cases of automixis in which the zygotic number of chromosomes is reestablished by endomitosis in both anaphase-I groups); (4) heterozygosity is maintained if the alleles of heterozygous loci separate postreductionally. In the case of prereduction for the alleles in question, a mosaic of the two homozygous cell types results; (5) heterozygosity is maintained in the case of prereductional separation of heterozygous allele pairs. If they separate post-reductionally, a heterozygous individual results in 50 % of the cases, and a mosaic formed by the two homozygous cell types in 50 %.

The property of parthenogenesis conferred on amphimictic organisms results from the gradual development of a polygenic system of modifier genes. The advantages connected with p. depend on the system of p. present. Generally, p. is connected with an increase in the reproductive potential since only half as many parents are needed per individual. An additional advantage may be fixation of the genotype and the preservation of extensive heterozygosity as well as → heterosis.

parthenogenone (Beatty 1957) — the parthenogenetic organism (→ parthenogenesis).

parthenomixis (Winkler 1908) — the fusion of two female nuclei within a multinucleate oogonium. This may replace true fertilization.

parthenote — = → parthenogenone.

paternal — pertaining to, or derived from, the male parent (→ maternal).

patroclinal (Kerner 1881) — → matroclinal.

patrogenesis (Collins and Kempton 1916) — = → androgenesis.

pattern formation — in → embryogenesis, the process specifying the spatial pattern of → cytodifferentiation in an initially near-homogeneous structure. P.f. occurs in two stages: (1) A cell first acquires information about its position, and then (2) interprets this according to its own genetic information content and developmental history.

pattern gene — a gene whose action concerns a particular pattern established in the course of → cytodifferentiation.

pattern modifier (Waddington 1939) — any gene modifying the pattern of → differentiation in a quantitative rather than a qualitative way. Pattern modifiers may be grouped into: (1) "Disruption genes" which prevent the formation of a particular pattern of differentiation; (2) "reorganization genes" which may lead to disorganization of a pattern of differentiation when present in their normal genotypic milieu but may give rise to new nondeleterious effects in a genotype which has been selected for this activity; (3) genes which, when present, result in a new, functional pattern of differentiation.

pattern of damage (Hadorn 1945) — the pattern of action of a → lethal factor. It comprises the sum total of all the locus-specific, abnormal characters resulting in malformation and lethality connected with the presence of a lethal mutation in effective dosage (→ pattern of manifestation) in the genotype.

pattern of manifestation (Hadorn 1945) — the sum total of primary and secondary processes and characters which are characteristic products of the action of a particular gene (→ gene action). The p. of m. is defined by comparison of the wild type and mutant, whereby all statistically significant differences between the two are regarded as direct or indirect consequences of the particular mutation which changed, inactivated, or eliminated the normal allele (→ gene mutation). The p. of m. is thus a differential pattern denoting the differences in phenotype between wild type and mutant. A so-called "residual pattern" representing those activities which are identical for the normal and mutant allele may remain since it is not accessible to analysis. The total pattern of manifestation is, therefore, composed of the differential and the residual patterns of action. The more drastic the mutational alteration, the clearer the differential pattern becomes, and the smaller is the residual pattern. In the case of → lethal factors, the p. of m. is referred to as the → pattern of damage.

pause site — → transcription pausing.

pecalosome (Perucho et al. 1980) — a structure consisting of concatemerized exogenously acquired DNA, residing either extrachromosomally or integrated as a unit in chromosomal DNA of efficiently transformed cells (= → transgenome).

pedigree — a formal list or register of ancestry or genealogy of an individual or family group. Pedigree diagrams are frequently used in human genetics in the analysis of Mendelian inheritance. Females in such a p. are symbolized by either circles or by the symbol ♀, males by squares or the symbol ♂. Parent symbols are joined by a horizontal "marriage line", and symbols for their offspring are placed in a horizontal row below this line to which they are connected by verticals. The children of a parental pair represent a sibship, and the individual children are referred to as sibs or siblings, irrespective of their sex. They are listed and numbered in a p. from left to right in the order of birth. If an individual possesses the character whose inheritance forms the subject of the p., it is designated by a black symbol. An open symbol denotes absence of the character under observation. Twins are represented by two symbols connected by lines to the same point of the sibship line; a number enclosed in a large symbol shows the number of sibs who are not listed separately. The pattern of distribution of genetically controlled characters in kindreds, i.e., large groups of persons interconnected by descent and marriage, is called the "pedigree pattern". It provides information on the Mendelian principles of → inheritance. Observation of pedigree patterns is one of the chief methods of genetic study in man since critically informative matings made by design are impossible in human beings.

pegmatypic (Kalmus and Maynard Smith 1966) — of a → mating system in which imprinting operates during a sensitive period of the individual's youth. The term **pegmatype** is used for the imprinted characteristic of an individual, or (according to the context) for an individual or class of individuals bearing this acquired characteristic.

P element (Rubin et al. 1982) — in *Drosophila*, any of a class of moderately repetitive → transposable genetic elements (→ hybrid dysgenesis; MR element). A complete P e. is 2.9 kb long and contains short 31 bp terminal inverted repeats flanking four → open reading frames that encode a transposase function. P e. transcription is restricted to the germline cells due to a tissue-specific splicing event required for the production of transposase. P elements form a heterogeneous family of elements originating from the autonomous 2.9 kb P factors by internal deletion. Insertion sites of P elements are clustered in target genes (often at the 5′ end, either upstream of the transcribed region or in the transcribed, untranslated portion of the gene).

P elements are being used as vectors for germline transformation in the form of recombinants developed from a P e.

Pelham box — a conserved, short DNA sequence (14 bp) in the 5′ noncoding region, immediately upstream from the → TATA box, which serves as the → promoter for transcription of → heat shock genes.

penetrance (Vogt 1926) — the frequency (in percent) with which a (dominant or homozygous recessive) gene or gene combination manifests itself in the → phenotype of the carriers. P. (as well as → expressivity) depends both on the genotype and the environment.

P. is complete when all the homozygous recessives show one phenotype, when all of the homozygous dominants show another phenotype, and when all of the heterozygotes are alike. If less than 100 % of the carriers of a certain genotype manifest that phenotype characteristic for the class, p. is reduced or incomplete; p. of a gene may be identical or different in either sex of dioecious organisms, or in extreme cases, may be limited to one or the other sex ("sex-limited genes").

pentaploid — of → auto- or alloploid cells, tissues, and organisms, respectively, with five chromosome sets in their nuclei (symbol 5x).

pentasomic — of a → polysomic cell, tissue, or organism with one chromosome in the complement represented five times (e.g., a pentasomic diploid $2n+3$; a pentasomic tetraploid $4n+1$, etc.).

peptide chain elongation — in → genetic translation, the addition of amino acids, one at a time, to a growing polypeptide in a sequence dictated by → messenger RNA. P.c.e. occurs in three steps: (1) After formation of the → initiation complex or during growth of the polypeptide, the codon-directed binding of aminoacyl-tRNA to an occupied → P site of the ribosome; (2) peptide bond formation during which peptidyl transfer from fMet-$tRNA_f^{Met}$ or the peptidyl-tRNA in the P site to aminoacyl-tRNA in the A site occurs; (3) translocation of the peptidyl-tRNA back into the P site to permit another incoming aminoacyl-tRNA to enter the A site, with concomitant ejection of the discharged tRNA from the P site (this translocation involves relative movement of the ribosome and the mRNA).

peptide synthetase — = → peptidyl transferase.

peptidyl site — → genetic translation; ribosome.

peptidyl transferase — a ribosome-associated activity which catalyzes the reaction between → peptidyl-transfer RNA on the ribosomal A site and the newly bound aminoacyl-tRNA and results in the formation of a peptide bond (transfer of the peptidyl moiety to the amino group of the aminoacyl-tRNA). The newly formed peptidyl-tRNA at the A site of the ribosome is translocated to the P site (→ aminoacyl transferase).

peptidyl-transfer RNA — (pt RNA); the → transfer RNA molecule that dissociates from the → ribosome during protein synthesis. After dissociation, pt RNA is degraded; the peptide is removed from the pt RNA by peptidyl-tRNA-hydrolase, producing an aminoacylatable RNA (→ ribosome editor hypothesis).

peptidyl transferase — an enzyme (= peptide synthetase) which catalyzes the formation of peptide bonds (peptide transfer) during → genetic translation. This takes place by transfer of the growing peptidyl group from the CCA terminus of one molecule of → transfer RNA to an aminoacyl group attached to the CCA terminus of a second, incoming molecule of tRNA. The reaction possesses properties of a single group transfer reaction and the p.t. is an integral part of the larger subunit of the → ribosome (→ aminoacyl transferase; translocation).

peptidyl-transfer RNA hydrolase (Menniger et al. 1970) — an enzyme which catalyzes the removal by hydrolysis of peptides (or N-acyl-amino acids) from → peptidyl-transfer RNA (or N-acyl-aminoacyl-tRNA) that is not bound to ribosomes. Aminoacyl-tRNA and the prokaryote initiator tRNA (fMet-$tRNA_f^{Met}$) are no substrates for the prokaryotic hydrolase.

percent substitution — an index of the relative evolutionary rate at the nucleic acid level calculated by dividing the number of sites that show substitutions by the total number of sites and multiplying by 100. For calculation of the absolute rates, p.s. values must be corrected for multiple substitutions and divided by the time that elapsed since divergence of the pair of sequences studied.

performance index — a single-valued measure of the multiple-trait performance of a population (→ selection index).

pericentric — → paracentric.

pericentriolar cloud — a cloud (= p. material) of amorphous substance surrounding the pair of → centrioles (→ centrosome). The p.c. is the focus of → microtubule organization, i.e., it can nucleate and anchor microtubules.

pericentriolar satellite (Brinkley and Stubblefield 1970) — any of the small (about 55 nm in diameter), electron-opaque spheres scattered radially outward the → centriole.

pericentromeric region — the DNA sequences around the → centromeres of eukaryotic chromosomes; they are composed of diverse classes of → repetitious DNA's (which provide a source of variability).

perichromatin fibril (Monneron and Bernhard 1969) — any of the fibrillar ribonucleoprotein components surrounding areas of condensed → chromatin in the cell → nucleus. Quantitative and topological changes of p. fibrils occur after increased or decreased RNA synthesis and correspond to similar changes of → heterogeneous nuclear RNA.

perichromatin granule (Yamamoto et al. 1969) — any of the ribonucleoprotein (RNP) granules (30—45 nm in diameter) contained in the sites of → gene expression.

periclinal (Baur 1909) — of a → chimera in which the distinct, idiotypically different tissues are arranged in concentric layers.

For plants in which the vegetative point (the apical meristem) is composed of three tissue layers (L I, L II, and L III), four different types of p. chimeras are distinguishable: (1) "Haplochlamydeous" p. chimeras in which only the outermost (L I) of the three layers is idiotypically different from the others; (2) "diplochlamydeous" p. chimeras in which only the innermost (L III) of the three layers is idiotypically different from the others; (3) "mesochimeras" in which only L II (the middle layer) is idiotypically different from the other two layers; (4) "trichimeras" in which all three layers are idiotypically different from each other.

perinuclear cisterna (Watson 1955) — → nuclear envelope (= perinuclear space).

perinuclear DNA (Sparvoli et al. 1977) — DNA which replicates at the → nuclear envelope at the end of S phase. It correlates in part with → heterochromatin.

perinuclear space (Policard and Bessis 1956) — → nuclear envelope.
perinucleolar chromatin — = → nucleolus-associated chromatin.
perissoploid — → artioploid.
permanent heterozygosity — → heterozygous.
permissive — (1) of a cell that supports lytic infection by a specific virus. Non.-p. (or restrictive) cells do not; there are indications that the permissiveness of the cells to virus infection is regulated at the initiation of protein synthesis. (2) Of the situation when a conditional mutant shows the non-mutant phenotype.
peroxisome (de Duve 1965) — a nearly ubiquitous organelle in eukaryotic cells, spherical or dumbbell in shape (mean diameter 0.2 to 1.0 μm), delimited by a single unit membrane and possessing a granular matrix (sometimes a crystalline core). Peroxisomes perform many biochemical functions depending on cell type and organism (→ microbody; glyoxysomes) but share at least two biochemical capabilities: Respiration (based on H_2O_2 metabolism) and fatty acid oxidation. They grow by post-tranlational incorporation of new content and membrane proteins into preexisting peroxisomes, which then divide to form daughter peroxisomes.

From → mitochondria and → chloroplasts, the p.s. differ in two fundamental aspects: They have only one membrane, and they lack DNA. Their proteins are encoded exclusively by nuclear genes and are synthesized on free cytoplasmic polysomes (without precursor extensions).

Two major subclasses of peroxisomes are: (1) Those in which the oxidative processes predominate (found in most animal and plant tissues); (2) those involved in metabolic conversions of lipids to carbohydrates via the glyoxylate cycle (found in specialized plant tissues and some invertebrates).
perpetuation (Tsuchida et al. 1966) — the maintenance and multiplication of a virus genome in (bacterial) host cells without killing them due to a weakening of the virus virulence under appropriate conditions (e.g., incubation at low or high temperature).
persitence — → genetic death.
petit mutant — in yeast, cytoplasmic mutants resulting from large deletions of the wild-type → mitochondrial DNA compensated by numerous periodic repetitions of the nondeleted segments.
phage — = → bacteriophage.
phage-associated gene transfer — a → gene transfer system with similarities to generalized → transduction.
phage attachment site — any of the sites (bacterial polymers) at the bacterial cell surface to which bacteriophages attach.
phage conversion — any change of the bacterial cell's phenotype due to the presence of a bacteriophage in the culture. In those cases in which the modified phenotype is the direct or indirect result of the presence of a stable → prophage, either integrated or autonomous, the term lysogenic c. can be applied.
phage cross — multiple infection of a single bacterium by bacteriophages that differ at one or more genetic sites. This leads to the production of recombinant progeny phage.
phage exclusion (Delbrück 1945) — the gross reduction by an infecting bacteriophage of the number of progeny produced by a related, but distinct phage which happens to infect the same bacterium. Due to p.e. the number of recombinant progeny emerging from such a mixed infection is markedly reduced, despite the fact that extensive genetic exchange between the two parental genomes is otherwise possible.
phage ghost — the phage protein coat prepared of viable phage by osmotic shock. P. ghosts retain no detectable DNA with genetic functions. They may, however, cause cessation of bacterial synthetic processes.
phage heterozygosity — → heterozygous.
phagemid (Messing 1983) — = → phasmid.

phage restriction — the prevention of phage replication due to a → DNA modification-restriction system in the bacterial host cell.
phagocytosis (Metchnikoff 1883) — the ability of cells to ingest ("engulf") solid particles (→ pinocytosis).
phagolysosome — the fusion product of a → lysosome and a → phagosome (= lysophagosome or telolysosome).
phagosome (Strauss 1958) — any phagocytotic or pinocytotic vacuole or vesicle before the fusion with → lysosomes (endosome). Phagolysosome and lysophagosome are the terms used for the combined granules.
pharmacogenetics (Vogel 1958) — the study of the relation between an individual's genotype and his reaction to various pharmaceutical agents.
phase variation — the reversible switch between two relatively stable genotypes by, e.g., DNA inversion: Host-range variation in → Mu phage (and related phage P1), flagellar variation of *Salmonella*, and P inversion in *E. coli.*
phasmid (Kahn and Helinski 1978) — a chimeric plasmid-phage construct with replicative functions of both plasmid and phage (bifunctional construct) that can behave as a → plasmid, as a → bacteriophage, or as a → prophage, i.e., can replicate as a plasmid (nonlytically) or as a phage (lytically). Phasmids allow → molecular cloning, → oligonucleotide-directed mutagenesis, and expression of mutant proteins from the vector (= phagemid).
phene (Johannsen 1909) — a genetically controlled phenotypic → character.
phene therapy (Siniscalco 1972) — all methods applicable for correcting the consequences of expression of mutant genes and the manifestation of → genetic diseases (such as dietary measures, avoidance of toxic drugs, replacement of missing gene products or defective tissues, enzyme induction, etc.) (→ eugenics; euphenics).
phenetic distance — a measure of the difference in → phenotype between any two points on a → phylogenetic tree.
phenetics — classification involving grouping of taxa by means of their overall similarity.
phenocline (Huxley 1939) — → cline.
phenocopy (Goldschmidt 1935) — a nonhereditary, phenotypic → modification (caused by special environmental conditions) that mimics a similar phenotype caused by a → gene mutation (→ genocopy).

According to Landauer (1958) a true p. is characterized by: (1) An additive effect between the gene action in question and the particular environmental conditions resulting in a p.; (2) a larger effect of the environmental condition in heterozygotes than in homozygotes for the standard allele; (3) an increase of the environmental effect producing a p. in the presence of mutants with low → penetrance; (4) a similar effect of modifiers on both the penetrance and → expressivity of the gene whose effect is phenocopied and on the strength of activity of the phenocopying environmental condition.

phenocritic (Haecker 1918) — of that phase in ontogony when the developmental pathways of two different genotypes, e.g., wild type and a mutant genotype, begin to differ, either morphologically, physiologically, or biochemically.
phenodeviant (Lerner 1954) — a phenotypically abnormal individual that deviates from the populations mean (or norm). The abnormality is due to special gene combinations and normally fails to follow a simple → inheritance pattern.

Phenodeviants are characterized by the following features: (1) They arise sporadically; (2) they occur more frequently in → inbred lines; (3) their frequency and → expressivity may be increased by selection and inbreeding; (4) the genes causally responsible for the production of phenodeviants in different strains are nonalleles; (5) the genes responsible for the production of phenodeviants are difficult to localize; (6) in crosses, phenodeviants behave in a variable and unpredictable manner; (7) in strains, in which phenodeviants frequently arise, the reproductive efficiency is decreased.

Phenodeviants (whose origin is poorly understood) have been explained as an excessive proportion of homozygous loci (Lerner 1954), as the breakdown of balanced polygenic systems coadapted in heterozygotes (Dubinin 1948; Goldschmidt 1949), and as defective regulatory systems giving rise to phenodeviants by the erroneous interception by particular genes of information intended for others (Wallace 1963).

phenogenesis (Fischer 1939) — the development of genetically controlled phenotypic → characters or phenes in the course of ontogenesis. P. is the result of the interaction between → genotype and environmental factors and results in a definite → phenotype (→ phenogenetics).

phenogenetics (Haecker 1918) — a branch of genetics involved in the study of the relationships between the genotype and its manifestation (the phenotype). One aspect of p. is developmental genetics.

phenology — the study of the seasonal timing of any life cycle defined as a statistical distribution by parameters such as time of occurrence, duration, synchrony, and skewness.

phenomic lag (Davis 1949) — a delay in the phenotypic expression of a → gene mutation owing to the time required for the manifestation of changes in the affected biochemical reactions (→ segregational lag).

phenotype (Johannsen 1909) — the observable properties (structural and functional) of an organism, produced by the interaction between the organism's genetic potential (its → genotype) and the environment in which it finds itself. The genotype sets the "norm of reaction" on the environmental influences. The term p. can be applied either to the totality of expressions of the genotype or to only a part, i.e., to particular characters, traits, or phenes.

Phenotypic expression of → genetic information is mediated by → messenger RNA, which carries the transcript of DNA (→ genetic transcription) to the polyribosome, i.e., the ribosomal complex engaged in protein synthesis (→genetic translation).

Those (adaptational) components of the phenotype which are not directly related to the environment and the survival of the individual may be called **endophenotypic** (Lewis and John 1963), as opposed to the (adaptive) **exophenotypic** components which affect an individual's ability to compete and to survive. It is the exophenotype on which → selection acts directly. Both endo- and exophenotype are controlled by the genotype (→ epiphenotype).

Heredity and environment overlap extensively in the shaping of the phenotype. Phenotypic similarities between individuals may occur when they possess the same or different genotypes, and phenotypic differences between individuals may or may not be accompanied by genotypic differences. Environmental differences can cause the same genotype to produce a variety of phenotypes. This phenotypic range of reactions of a particular genotype is referred to as its "norm of reaction" or "reaction norm". The capacity of an organism to function in a range of environments, which may include plastic and stable responses, is called **"phenotypic flexibility"** (Thoday 1953). **"Phenotypic correlations"** between different → characters result from combinations of genetic and environmental correlations. Genetic correlations can arise from → pleiotropy, → linkage, or common introduction of the genes involved into a population. Environmental correlations may be produced by the development of two or more characters in the same individual and their exposure to the same environment.

phenotypic expression — the manifestation via → gene expression of a particular gene present in effective dosage in an individual's genotype. Gene expression results in the production of particular phenotypic characters.

phenotypic flexibility (Thoday 1953) — the extent to which an organism can grow and reproduce in a range of environments, either by varying its genotype (plasticity) or by maintaining a constant phenotype (homeostasis or stability).

phenotypic intergradation — the occurrence of variability and intermediate forms in areas of geographic overlap between populations which elsewhere are distinct and relatively homogeneous. P.i. often indicates interspecific or interracial hybridization.

phenotypic lag — the delayed phenotypic manifestation of a → gene mutation (e.g., a mutation from prototrophy to auxotrophy) in bacteria. The delay may last several generations and is due (in most cases) either to → phenomic lag or → segregational lag.

phenotypic masking (Gorini et al. 1967) — in bacteria, the masking of the true phenotype in drug (antibiotic)-dependent strains. P.m. is due to alterations of the ribosomes by antibiotics.

phenotypic mixing (Novick and Szilard 1951; Hershey et al. 1951) — the production in the same host cell of virus particles with discrepant genome and → virion specificities due to the incorporation into the same → capsid or envelope of proteins made under the direction of different viral genomes. The genomes in question produce different structural components of the protein "coats". The components of the coats are drawn, without regard to the genotype of the particle, from a pool of mutant and wild type protein molecules and, thus, mixed coats may be produced giving rise to p.m. In bacteriophages, phenotypic mixing has been observed for stability at high temperature, serological properties, the properties of particles to attach to different strains of bacteria, as well as other features depending on the protein moiety of mature phages.

phenotypic plasticity — the amount by which the expression of individual characteristics of a genotype are changed by different environments. P.p. is genetically controlled (→ developmental instability).

phenotypic reversion — the nongenetic restoration, partial or complete, of the functional activity of mutated genes (missense and nonsense mutations) without directly substituting the required function itself. P.r. may be due to miscoding or → mistranslation. A number of agents are able to provoke p.r. and result in nonheritable phenotypic changes.

phenotypic stability (Lewis 1954) — = developmental → homeostasis.

phenotypic suppression (Gorini 1974) — any case in which expression of a mutant phenotype (due to nonsense or missense mutation) can be prevented by, e.g., compensating errors of → genetic translation or palliative treatments (drugs that promote mistranslation in vivo; the mutant phenotype returns after removal of the drug).

phenotypic value — the value of a character as determined by the genotype (genotypic value) and the environment (environmental deviations), i.e., all nongenetic circumstances that influence the p.v.

phenotypic variation — the total variation among members of a population which consists of variation due to genetic differences and that due to environmental ones. Some of the p.v. may be due to interaction between genotype and environment. The genetic fraction of p.v. can be subdivided into fractions attributable to dominance, epistasis, and straightforward additive effects of genes. The p.v. due to environment can be subdivided into the effects produced by controlled (tangible) and those produced by uncontrolled (untangible) ones.

phosphodiesterase I — (PDase); an enzyme that removes, by hydrolysis, 5′ nucleotides from the 3′ hydroxy termini of oligonucleotides with 3′ ends (= 5′ exonuclease).

phospholipid vesicle (Huang 1969; Johnson et al. 1973) — a unilammellar liposome consisting of a single, closed spherical bilayer of phospholipid (outside diameter approximately 30 nm) which defines an internal aqueous space (about 20 nm in diameter). Water-soluble molecules can readily be incorporated into the vesicle's interior space, and lipophilic molecules can be incorporated into the phospholipid bilayer.

photodynamic action — a process whereby a photoactive dye absorbs incident light energy and is excited to a short-lived, high-energy singlet state which evolves spontaneously to a longer-lived, unstable triplet state. Some of the photoactive dyes intercalate between the base pairs of DNA where incitation of p.a. results in oxidation of adjacent guanine residues.

photoenzymatic repair (Harm 1968) — (PER); enzyme-mediated, light-dependent monomerization of pyrimidine dimers induced by UV in DNA (→ DNA repair). The photolyase contains a flavin-adenine dinucleotide chromophore and absorbs near-ultraviolet and visible light to gain the energy needed for splitting the cyclobutene ring.

photofootprinting (Selleck and Majors 1987) — a procedure which monitors the sensitivity of individual base pairs to UV light-induced covalent modification in intact cells. Since protein binding to DNA affects the modification pattern, p. affords the opportunity for looking, in vivo, at DNA-protein interactions within control regions of regulated → promoters.

photolyase — any of those enzymes that, during → photoenzymatic repair, specifically repair the unique class of cis-syn cyclobutyl pyrimidine dimers in DNA.

photomorphosis — the control which may be exerted by light over growth, development, and differentiation of a plant, independently of photosynthesis. Light may be regarded as an elective factor which influences the manner in which those genes present in the particular organ are expressed.

photoprotection (Latarjet and Gray 1954) — (PP); a decreased sensitivity of a biological system to ultraviolet light caused by pre-irradiation treatment with light. Generally, PP requires larger light doses than → photoreactivation and the PP action spectrum covers a narrower band of wave lengths than does that for photoreactivation. PP is practically temperature independent during illumination, whereas photoreactivation is usually strongly temperature-dependent. PP operates by providing more time for intracellular → dark repair to act before DNA replication leads to → "mutation fixation", since the photoprotecting wavelength (338 nm) induces delayed growth. → Liquid-holding recovery is absent in photoprotected cells (Rupert and Harm 1966).

photoreactivation (Kelner 1949) — (PR); the partial elimination of the lethal and mutagenic effects of ultraviolet light (290—310 nm) by subsequent exposure of irradiated cells to other nonionizing radiation (330—450 nm) leading to a decrease in the effective UV dose. PR has been found to proceed by at least three distinct mechanisms (Rupert and Harm 1966): (1) Direct → photoenzymatic repair of UV-damage to DNA by splitting the pyrimidine dimers; apparently this is the major mechanism of PR in bacteria and bacteriophages (Rupert 1960; 1961); (2) indirect processes (related to → photoprotection) by which lightinduced growth delay favors a light independent repair system (Jagger and Stafford 1965); (3) direct nonenzymatic photochemical repair as observed in transforming (→ genetic transformation) DNA but not in any organism (Setlow and Setlow 1963);

Not every UV induced lesion is photorestorable. The fraction of photoreactivable lesions is called the **"photoreactivable sector"**.

Except for marsupials, mammalian cells lack a photoreactivating enzyme. Cells from nonmammalian vertebrates possess a photoreactivating enzyme.

photoreversal — = → photoreactivation.

phragmoplast (Errera 1888) — a defined plasma body representing a modified → spindle region in the equatorial plane of dividing plant cells which is identifiable after → chromosome movement to the poles. The p. is involved in the formation of the → cell plate (→ phragmosome).

The p. consists of fibrillar elements (microtubules), which arise at the cell plate and radiate outward, and numerous droplets which are moved to and incorporated into the plate. The microtubules of the p. first appear in the midregion of the cell in early stages of plate formation and assist in establishing the zone along which the cell plate forms.

phragmosome (Porter and Caulfield 1958) — any of the circular to oval structures in dividing plant cells which (in addition to the → endoplasmic reticulum) are associated with the formation of the → cell plate. Such particles (0.2—0.5 μm in diameter) are surrounded by a membrane enclosing their contents which appear homogeneous or finely granular depending on the type of fixation.

phyletic evolution (Simpson 1944) — → evolution; phyletic gradualism.

phyletic gradualism (Simpson 1944) — a model of evolution (= phyletic evolution) according to which most of the evolutionary change along lineages is cumulative and gradual (= gradualism). The fact that morphological gaps appear between ancestral and descendant populations is taken as a function of the imperfect nature of the fossil record (→ punctuated equilibrium).

phyletic line — an unbroken ancestral-descendant sequence of populations that exists for a long period of time.

phyletic trend (Grant 1963) — a gradual unidirectional evolutionary change within a phyletic line (speciational trend).

phylogenetic — of the evolutionary relationships within and between groups.

phylogenetic denseness (Moore et al. 1976) — a concept for correcting the estimated number of nucleotide replacements separating two nucleotide sequences descendant from a common ancestral gene in a region of a phylogenetic tree represented by only a few contemporary species (the uncorrected estimates being usually too low). The method uses the information in those portions of the phylogenetic tree represented by many contemporary species. Regions with few and/or distantly related species are called "sparse" and those with many and/or closely related species are called "dense".

phylogenetic tree — a branching diagram in the form of a tree that is meant to express the nature of the evolutionary relationships between different lines of descent (= evolutionary tree; phylogram). (→ molecular phylogenetic tree).

phylogeny (Haeckel 1866) — the evolutionary history of an organism or taxonomic group (= phylogenesis).

physical containment — the use of physical barriers to prevent escape of genetically engineered organisms from the laboratory to the environment (maintenance of negative air pressure, fume hoods with filters).

physical genome map — the structure of the DNA of a → genome. The p.g.m. is constructed from genomic DNA by use of → restriction endonucleases, separation of fragments by pulsed field gel electrophoreses, and ordering of the fragments in the genome from genetic information and analysis of partial digest patterns. The resulting ordered set of fragments is a macrorestriction map.

phytohormone — any of the plant hormones to which plant tissues respond by expressing specific genes. Promoters respond specifically to the hormones and act to regulate the initiation of gene transcription.

pilus (Brinton et al. 1954) — any of the nonflagellar filamentous appendages (of varying diameter and length) on the cell surface of bacteria. Two basic groups may be distinguished: (1) **Conjugative (sex) pili** that are encoded by self-transmissible → plasmids and are involved in bacterial → mating; (2) **common pili** which are probably encoded by the bacterial chromosome.

Both types of pili (= fimbriae) can be the attachment sites for filamentous bacteriophages (that adsorb to the tip of the pili) and isometric bacteriophages (that adsorb along the side of pili). The protein filaments of pili consist of several hundred or more helically arranged and predominantly identical subunits.

pinocytosis (Lewis 1931) — the process of active intake of fluid by the cell. It proceeds by the elevation of thin undulating folds that impound sizeable droplets of fluids, by inpocketing of the surface to form shallow vesicles of submicroscopic size that detach and move into the cytoplasm, or by deeper invagination of the cell membrane to form channels that bud off small fluid-filled vacuoles at their end (Fawcett 1965).

The term "micropinocytosis" is used for the submicroscopic vesiculation of the plasma membrane by invagination of smooth surfaced vesicles or local specializations of both surfaces of the membrane prior to it invagination to form a vesicle (Odor 1956).

pistil (Clusius 1601; Tournefort 1700) — in flowers, the female seed-bearing organ (the gynoecium) consisting of the ovary, style, and stigma. Flowers bearing pistils are termed "pistillate" (→ stamen).

pit — in higher plants, a thin region (in the light microscope, a canal divided into two pit cavities by a pit membrane) of the secondary → cell wall, functioning in the exchange of material between neighboring cells. According to electron microscopie studies, the primary cell wall is penetrated in those places where pits will later arise by clusters of → plasmodesmata which remain visible in the pit membrane. Cellulose microfibrils are deposited around the future p. and plugs of cytoplasm on both sides of the pit area prevent the apposition of the secondary wall.

In cells which lose their living contents and act as capillaries for water conduction, the simple p. is replaced by so-called "bordered pits". These possess a large pit membrane which is circularly overarched by a border originally laid down as a slender rim of the primary wall. This border is separated from the pit membrane and covered on both the inner and outer sides with a tertiary layer; the space referred to as the pit cavity in simple pits represents a narrow entrance (porus) and a widened pit chamber. The membrane consists of a cushion-like thickening (torus) and a thin marginal zone (margo).

pitch — the number of base pairs per turn of the DNA double helix.

planogamete — a motile → gamete, as opposed to a nonmotile aplanogamete.

plaque — a round clear area in a confluent bacterial cell lawn which results from the killing or lysis of contiguous cells by several cycles of → bacteriophage growth.

plaque hybridization — an adaptation of the technique of → Southern hybridization (= Benton-Davis technique; phage lift, plaque lift): Bacteriophage plaques are transferred from an agar plate to a nitrocellulose filter (or other DNA-binding matrix) to which a small amount of phage and unpackaged phage DNA bind, leaving behind the plaques. Phage DNA is denatured on the filter and baked. The bound DNA can then be hybridized with a probe (^{32}P-labeled DNA or RNA) and any homology between the probe and the DNA on the filter will allow the two species to anneal and the hybrids formed can be detected by autoradiography.

plasmagene (Winkler 1920; Darlington 1939) — any extranuclear → hereditary determinant showing non-Mendelian → inheritance. The sum total of plasmagenes represents the → plasmotype which is one component of the → idiotype.

plasmagynogamous (Michaelis 1955) — of a → fertilization whereby the cytoplasm of the → zygote is exclusively derived from the female parent, in contrast to "plasmaisogamous" and "plasmaheterogamous" fertilization.

plasmaheterogamous (Michaelis 1955) — → plasmagynogamous.

plasmaisogamous (Michaelis 1955) — → plasmagynogamous.

plasmalemma (Seifriz 1928) — = → cell membrane or plasma membrane.

plasmalemmasome (Edwards and Stevens 1963) — = → mesosome.

plasma membrane — = → cell membrane.

plasmatic — pertaining to or mediated by the → cytoplasm (as, e.g., cytoplasmic → inheritance).

plasmatic inheritance — = → cytoplasmic inheritance.

plasmid (Lederberg 1952) — any of the autonomous, covalently closed circular (in rare cases linear), double-stranded and self-replicating DNA molecules (→ plasmid DNA) found in most bacterial species and in some species of eukaryotes. Under normal circumstances, a particular p. is dispensable for the host cell; in certain environments, p. genes may be essential (→ R plasmid; F plasmid; colicinogenic plasmid). Plasmids may, potentially at least, be integrated into the host genome, range in size from about 2 to 500 kb, and are present in host bacteria at cellular levels between 1 or 2 (stringent or low-copy-number plasmids) to 10 to 100 (relaxed or high-copy-number plasmids) copies per chromosome. Characteristic features of (bacterial) plasmids are (Kornberg 1974): (1) The control of their replication that keeps their number fixed from one generation to the next; (2) the capacity of some to integrate into the host chromosome (= episome); (3) the transfer of some to other cells by viral-mediated transduction, or by mating via sex pili (= conjugal transfer); (4) the promotion by F plasmids of conjugal transfer of the host chromosome; (5) the variety of p. gene products which result in distinctive traits of the cells in which they are established.

Most of the genetic information contained in plasmid DNA (coding capacity for about 3 to 500 genes) concerns p. replication, stability, and autonomous transferability. Ten to 20%

is concerned with the so-called somatic functions which under certain conditions secure p. survival.

P. replicons exhibit a variety of DNA replication modes. Some replicate unidirectionally from a unique origin, others possess two initiation sites and one termination site (asymmetrical bidirectional replication, i.e., replication begins at an origin and proceeds in the opposite direction from the same origin to the same termination site), still others replicate either unidirectionally or bidirectionally with equal probability from a unique origin. All plasmids are almost totally dependent on host replication functions, but different plasmids have different requirements.

In the replication of linear plasmids (found in, e.g., *Streptomyces*, yeast, and mitochondria of maize), unique structures such as cohesive ends, terminal repetitions, and terminal proteins (TP's), or hairpins are usually involved since DNA polymerases require a primer to initiate replication.

On the basis of these replicative behaviors the bacterial plasmids can be divided into two broad categories: (1) **Amplifiable plasmids** which continue to replicate in the absence of de novo protein synthesis and do not require any p.-specific replication protein; (2) **nonamplifiable plasmids** whose replication appears to be more or less tightly coupled to the synthesis of p.-encoded initiator protein(s).

A further distinction can be made between conjugative and nonjugative plasmids: (1) **Conjugative, or self-transmissible plasmids,** encode all the functions needed to effect transfer of the p. from the donor cell to the recipients; (2) **nonconjugative plasmids** are incapable of self-transfer, but some can be transferred at high efficiency if certain conjugative plasmids are simultaneously present in the donor cell (p. mobilization).

P. transfer consists of four stages: (1) Formation of specific donor-recipient cell pairs (effective contact); (2) preparation for DNA transfer (mobilization); (3) DNA transfer; (4) formation of a replicative, functional p. in the recipient cell (repliconation).

Maintenance of any p. consists of two main processes, i.e., replication and partitioning. If two distinct plasmids can co-exist and be stably co-inherited in individual dividing bacteria, they are termed **compatible**; if they cannot be stably co-inherited, they are termed **incompatible** (→ p. incompatibility).

Plasmids that lack any defined functions other than those required for replication and transfer are called cryptic plasmids. **Chimeric** or hybrid plasmids carry foreign DNA and are being used as → cloning vectors.

plasmid cointegrate (Guerry et al. 1974; Oliver et al. 1974) — a recombinant → plasmid consisting of two different plasmids cointegrated at specific sites by → insertion elements and → transposons giving rise to a direct duplication of the element that mediated the cointegration.

plasmid curing — the elimination of a plasmid from its host cells (bacterial or eukaryotic) by many agents which interfere with → DNA replication (usually by intercalation).

plasmid DNA — the DNA characteristic of a certain type of → plasmid. Typically p. DNA is a supercoiled circular, double-stranded DNA molecule (occasionally a linear molecule) whose molecular weight ranges from about 10^6 (for the smallest plasmids) to slightly more than 10^8 (for the largest ones).

There is an enormous variation in both enzymology and mechanics of p. DNA replication. Plasmids rely on the host replication apparatus for their replication, some seem to use host gene products exclusively, others require some plasmid gene products. All replicate semi-conservatively (either bidirectionally and/or unidirectionally) and maintain circularity throughout the replication cycle. Termination of bidirectional replication occurs either when the growing forks reach the same region or at a fixed termination site that is sometimes reached by one replication fork before the other fork reaches it. In most plasmids, replication occurs by the so-called butterfly mode, i.e., in a partially replicated molecule, the replicated parts are untwisted, but the unreplicated portion is supercoiled. One of the circles must be cleaved (possibly by → DNA gyrase) when the replication cycle is completed. The result after one round of replication is therefore one nicked and one supercoiled molecule. The nicked molecule is subsequently sealed and, somewhat later, supercoiled.

plasmid incompatibility (Scaife and Gross 1962; Echols 1963) — the inability of two different → plasmids (with extensive structural similarity) to be maintained stably together in the same (bacterial) cell in the absence of selection (incompatible plasmids). Distantly related plasmids that stably co-reside in the absence of selection are called compatible plasmids. Stable maintenance of a plasmid requires, in addition to a replication control mechanism, either random segregation or a specific system for the proper partitioning of plasmid molecules at cell division. P.i. is due to negative control factors that act either at the level of transcription of the p.-specific replication proteins or at the level of initiation of new rounds of DNA synthesis. This results in a lower copy number of each of the plasmids and this, coupled to partition of the plasmids between daughter cells, leads to their more frequent loss.

On the basis of whether or not a given pair of plasmids can coexist stably, plasmids are classified into **i. groups**. The assignment to an i. group is based on incompatibility between autonomous plasmids, rather than on incompatibility between an integrated and an autonomous plasmid. Two plasmids are said to belong to different incompatibility groups if they coexist stably in a cell (→ plasmid superinfection immunity).

plasmid incompatibility group (Datta and Hedges 1971) — → plasmid incompatibility.

plasmid instability — instability of maintenance of → plasmids in bacterial cells which may be caused by: (1) **Segregational instability**, i.e., the loss of plasmids during cell division because of inaccurate distribution of the plasmid into daughter cells; (2) **structural i.**, i.e., instability due to deletions, insertions, or rearrangements; (3) **instability on transfer** (conjugation or transformation).

Instability is a typical character of constructed plasmid → cloning vectors.

plasmid integration — the integration of a → plasmid into the cellular genome. P.i. is a sporadic event, involves preferred sites, and is usually not mutagenic (→ insertion mutation) but may sometimes suppress the replication defect of certain (bacterial) mutants.

plasmid-mediated DNA repair — any mechanism for repairing DNA damage mediated by plasmid-encoded → DNA repair systems.

plasmid mobilization (Hamer 1977) — the high efficiency passive transfer to an appropriate recipient cell of a nonconjugative → plasmid when the host cell also contains certain conjugative plasmids (conjugal mobilizing plasmids).

plasmid multimer — a → plasmid which due to recombination in recombination-proficient cells consists of dimers, trimers, and higher oligomers of identical plasmid molecules.

plasmid prophage — a → prophage (e.g., coliphages P1 and P7) which is a → plasmid present in one to two copies per bacterial chromosome, and whose replicas are distributed faithfully to each daughter cell. Such stringency of replication and segregation in the absence of integration within the host chromosome is a common feature of the large bacterial plasmids.

plasmid rescue transformation (Contente and Dubnau 1979) — a method involving the use of a genetically marked plasmid DNA to transform competent cells carrying a resident homologous plasmid. It allows gene isolation (based on the gene's ability to cause a phenotypic change in cells) as well as recovery of transferred genes from mammalian cells. It occurs by recombination and allows reproducible re-isolation of transfected genes for further analysis.

plasmid superinfection immunity (Meynell et al. 1968) — the reduction in recipient ability for one → plasmid caused by a second plasmid (= superinfection inhibition). P.s.i. involves at least two operationally distinguishable processes called → entry exclusion and → plasmid incompatibility (intracellular incompatibility).

plasmid transduction — → transduction by transducing particles formed by integration (via recombination between homologous sequences) of → plasmid DNA into the bacteriophage chromosome. Another recombination after the transducing DNA enters a recipient cell serves to excise the plasmid from the phage DNA. P.t. is a tool for moving plasmids among different bacterial strains.

plasmid vector — a → plasmid used as a → cloning vector.

plasmodesma (Tangl 1879; Strasburger 1882) — in multicellular plants, any of the small intercellular channels (30—50 nm width and roughly circular in cross-section) which frequently contain a central rod (plasmodesmoduct) and consist of two → unit membranes separated from each other by a thin layer of cytoplasm and surrounding a thin lumen. They are channels for symplastic transport and are located (in groups) within depressions of the cell wall called → pits. The perforations of the cell wall in which plasmodesmata are situated are called plasmodesmoducts or p. pores.

plasmodesmoduct (Mecuse 1957) — → plasmodesma.

plasmodium — = → coenocyte or syncytium.

plasmogamy — the fusion of the cytoplasm of two (or more) cells. P. precedes → karyogamy in the process of fertilization (→ dikaryophase).

plasmolytic vesicle — in plant → protoplasts, any of a class of single-membraned vesicles, several nm across and originating as gross infoldings of the plasmalemma into the cytoplasm as the protoplast contracts during plasmolysis (→ pinocytotic vesicle).

plasmon (v. Wettstein 1927) — the cytoplasmic genetic system of eukaryotes. The major constituents are → mitochondrial DNA and (in plants) → chloroplast DNA. Each of these molecules contains a complete genome. In the organelles, the molecules may be localized in discrete regions called nucleoids. A cell containing two different alleles of an organelle gene (in any ratio) is called heteroplasmic, as opposed to a homoplasmic cell.

plasmon mutation — any mutational change of extranuclear hereditary determinants which may give rise to genetically different cytoplasms (= "plasmon alteration" according to Marquardt 1952).

plasmon "segregation" — the "segregation" (by dissociation) of qualitatively different extranuclear hereditary determinants as manifested by the production of new plasmonic or plasmotypic combinations from a heterogeneous plasmon or plasmotype. P.s. taking place within one individual may give rise to chimeras.

plasmon-sensitive (Renner and Kupper 1921) — of genes whose action is confined to a particular cytoplasm (Goldschmidt 1955).

plasmotropic (Waddington 1962) — embryonic → induction.

plasmotype (Imai 1936) — = → plasmon.

plast (Bĕlăr 1928) — any of the differential, self-reduplicating cell organelles, localized in the cytoplasm, e.g., → plastids, → centrioles, → mitochondria, → kinetosomes.

plasticity — the ratio of plastic variance (i.e., the deviation of the mean phenotype of a genotype within an environment from the mean phenotype of that genotype across all environments) to total phenotypic variance (→ flexibility).

plastid — in eukaryotic plant cells, any of a class of specialized organelles which contain pigments and may synthesize and accumulate various substances: Leucoplasts are colorless plastids; amyloplasts produce starch; proteinoplasts accumulate proteins; elaioplasts produce fats and essential oils; chromoplasts are colored plastids that contain less chlorophyll than the → chloroplasts, but more carotenoid pigments.

All plastids contain comparable plastom (plastid) DNA (→ chloroplast DNA), irrespective of their state of differentiation in different cells. P. development is a complex process that involves the cooperative expression of both nuclear and p. genes. The developmental program of p. gene expression depends on the external factor light, which operates to a large extent through the phytochrome photoreceptor system. Specific mRNA's for p. proteins accumulate in response to light.

All types of plastids arise from so-called "proplastids" which are bounded by a double membrane and filled with a relatively dense ground substance. The inner of the two unit membranes frequently shows tubular invaginations. The proplastids may give rise to leucoplasts which are transformable into chloro- and chromoplasts, or directly to chloroplasts transformable into chromoplasts.

plastid DNA (pt DNA); the circular, double-stranded DNA of → plastids which is comparable irrespective of the states of differentiation of plastids in different cells (→ chloroplast DNA).

plastid inheritance — non-Mendelian → inheritance due to hereditary determinants located in the → plastids of plant cells and controlling (generally in cooperation with nuclear genes) plastid characters.

plastidome (Dangeard 1920) — collectively the → plastids of a cell.

plastidotype — that component of the → idiotype of plants which comprises the sum total of the hereditary determinants located in the → plastids (→ plastom).

plastom (Renner 1924, 1934) — the → plastid-specific genetic information encoded in plastid DNA (→ chloroplast DNA). Mutational changes of hereditary determinants encoded in the p. are p. mutations.

plating efficiency — the percentage of cells which form clones when a cell culture is started.

platysome (Finch et al. 1977) — the → nucleosome core particle.

PLD repair — repair of potentially lethal damage (PLD) as evident from the enhanced survival of mammalian cells maintained in a density-inhibited state for an appropriate time after treatment with DNA damaging agents. PLD r. is closely associated with the → excision repair pathway and may be due to prolonged intervention of excision repair during suppression of → DNA replication, thereby reducing the → DNA damage and increasing survival.

pleiotropic (Plate 1910) — of genes or mutations that result in the production of apparently unrelated multiple (manifold) effects at the phenotypic level (pleiotropy = polypheny). Pleiotropy is the consequence of the fact that the biochemical pathways starting from different genes intersect in many places, inhibiting, deflecting, and variously modifying each other.

According to Goldschmidt (1955) four types of pleiotropy may be distinguished: (1) **Syndromic p.** (= spurious p. of Grüneberg 1938): The effect of a particular mutant gene is a syndrome of pathological features which represent a kind of mechanical consequence of primary, early embryonic damage. (2) **Pattern p.:** The mutant gene produces a primary product which affects one process most conspicuously, but simultaneously influences related processes at other places, sometimes in a qualitatively different way. (3) **Dichotomic p.:** The primary activity of the mutant gene (usually an inhibitory one) is of a generalized type affecting the whole of the organism simultaneously according to the reactivity of its parts. (4) **Interference p.:** The nearest appromixation to the unlikely type of "genuine" p. whereby one and the same gene affects completely different processes. A genetically controlled process interferes with other independently determined actions with the result that they shift in a way which might have been brought about by independent mutation.

Another classification of p. has been proposed by Hadorn (1954): (1) **Mosaic p.:** Genuine p., due to the differential action of the same gene in different cytoplasms and resulting in a pattern of mutant gene manifestation which consists of a number of independent → autophenes. (2) **Relational p.:** A mutant gene acts directly on only one type of cells but exerts indirect influence on other cells; each of these reacts in its own fashion. The intercellular influence presumably occurs through diffusible substances such as hormones, amino acids, and the like. This is a form of "spurious" p. The pattern of manifestation of the pleiotropic gene consists of autophenes and secondarily arisen allophenes.

Both mosaic and relational p. are based on intercellular mechanisms of gene activity, as opposed to the following types controlled by intracellular mechanisms of gene action: (3) **Cell reactive p.:** A mutant gene exerts identical primary actions in different cell systems but owing to their differential reaction different characters arise. (4) **Induced gene active**

p.: The primary action of a mutant gene is not identical in different cell systems because the developmental factors are different in these systems and, therefore, different characters arise in each cell systems.

ploidy — a term referring to the number of chromosome sets per cell (→ haploid; diploid; polyploid).

point error (La Cour and Rutishauser 1953) — → pseudochiasma.

point mutation (Bridges 1923) — any mutation (→ gene mutation) arising from the substitution, addition, or deletion of one base (or a few bases) initially, which become stabilized as a base pair change upon replication past the altered site (→ transition; transversion; frameshift mutation).

point stickiness — localized sticky effects in chromatids assumed to give rise to a → pseudochiasma.

polar body (Robin 1862) — in → oogenesis, the smaller cells produced during meiosis that do not develop into functional egg cells.

polar cap — = → polar zone.

polar insertion (Malamy et al. 1972) — in bacteria and viruses, any of a class of extreme → polar mutations (insertion mutations) which result from the insertion (→ insertosome) of some "foreign" DNA into an → operon. The insertion process seems to include recognition of a specific sequence of the insertable DNA which serves as the integration site. The insertion may result in complete inactivation of the gene in which DNA insertion occurs and in most cases also of genes within an operon distal to the site of the insertion.

polarity (Franklin and Luria 1961; Jacob and Monod 1961) — (1) in polycistronic → operons, expression of → promoter distal genes at a lower level than the promoter proximal ones. This p. may occur in certain operons under the usual growth conditions (natural p.) or by specific metabolic alterations within the cell (metabolic p.). P. is probably due to the coupling between → genetic translation of polycistronic messengers and premature termination of their transcription, mediated by transcription termination factor rho (→ genetic polarity). (2) Position-dependent preferential → gene conversion of one of two alleles in an intragenic cross. This type of p. is accounted for by postulating a discontinuous distribution of recombination events (Lissouba et al. 1962).

polarity gradient — the quantitative effect a → polar mutation exerts on the expression of genes located more operator-distal in the same operon. The p.g. is a function of the distance between the polar mutation and the next polypeptide chain initiation signal. The greater the distance of the mutation from the end of the gene, the greater the reduction in synthesis of the proteins encoded by the subsequent genes (→ genetic polarity).

polarity suppressor (Beckwith 1963) — any → suppressor mutation which relieves → genetic polarity and may, in addition, reduce the frequency of transcription termination at known termination sites (→ attenuator). P. suppressors differ from other → nonsense suppressors in that they restore expression of distal genes of an → operon without allowing translation of nonsense codons. They are not codon-specific and presumably can relieve the polarity caused by → nonsense mutations in any operon.

P. suppressors act by stabilizing mRNA after premature termination of protein synthesis or are deficient in an endonuclease activity and thus mRNA distal to the → polar mutation does not become degraded. Their efficiency varies considerably.

polar mutation (Franklin and Luria 1961; Jacob and Monod 1961) — any mutation of the → nonsense, → frameshift, → deletion, or → insertion mutation type which results in → genetic polarity. In the case of a p.m. (= dual effect mutation of Englesberg 1961) in an operon (1) the production of the particular enzyme for which the nonmutated gene codes is prevented, and (2) the functioning of the operator-distal genes of the operon is relatively decreased (mRNA distal to the mutation remains free of ribosomes and is highly susceptible to endonucleolytic attack) (→ polarity gradient).

Certain p. mutations have been found to cause an "antipolar effect" (Ito and Crawford

1965), i.e., the enzyme specified by the gene of the operon immediately preceding the mutated gene is formed in reduced amounts (→ genetic antipolarity; polarity suppressor).

polar rays — fibrillar cytoplasmic differentiations arising from the → centrioles.

polar zone — the functional center for mitosis in cells of higher plants devoid of → centrioles. → Spindle fibers as a rule converge on these cytoplasmic poles (= polar caps), which lack a distinct structure and represent clear cytoplasmic areas.

polarization — (1) of chromosomes at mitotic telophase and the following interphase, if their proximal parts (adjacent to the → centromere) remain directed to the polar side of the nucleus; (2) of chromosomes at prophase of meiosis I, if their ends are directed towards a definite point on the nuclear surface (→ bouquet stage); (3) of a nonrandom "segregation" (polarized segregation) of the members of a pair of homologous chromosomes, or of a pair of daughter chromosomes (chromatids), with respect to the two poles of the spindle in meiosis (→ assortment); (4) of → genetic recombination via → conversion, if a continuous gradient in the conversion frequency of mutational sites occurs from one end to the other of a → polaron; (5) of replication (→ replicon) and transfer processes of the → genetic material as well as its → genetic transcription and → genetic translation (→ operon) which proceed with a definite polarity.

polaron effect (Lissouba and Rizet 1960) — a gradient in the frequency of → gene conversion from one and of the locus (polaron) to the other when → heteroalleles within a locus are considered.

pole — one of the two opposing ends of the → spindle towards which chromosomes move (→ chromosome movement; centriole) at mitotic or meiotic anaphase.

pollen grain — the angiosperm microspore which germinates to form the male → gametophyte. Any of the sporophyte cells which undergo meiosis (in the anther) during → microsporogenesis and give rise to four (haploid) p. grains is called a pollen mother cell or microsporophyte.

pollen grain mitosis — → microsporogenesis.

pollen mother cell — → pollen grain; microsporogenesis.

pollen tube competition — → certation.

pollination — the placing (naturally or artificially) of pollen on the receptive area (stigma in angiosperms, micropyle in gymnosperms) of a female flower.

pollution — an undesirable change in the physical, chemical, or biological characteristics of air, land, and water that may or will harmfully affect human life or that of relevant species.

polocyte (Waldeyer 1888) — = → polar body.

polyadenylation (Edmonds and Abrams 1960) — a nearly universal → messenger RNA processing event in eukaryotes which is characterized by the nontranscriptive addition of 50 to 250 adenosine residues to the 3′ terminus of the mRNA [poly(A) tail]. Shorter poly(A) segments are found in many bacterial mRNA's.

The eukaryotic poly(A) tail is associated with proteins and undergoes size reduction in the cytoplasm in an age-related process. Oligonucleotides in DNA, from four to six bases long (e.g., CATTG and AATAAA) serve as polyadenylation signals and their transcripts precede the poly(A) tract in mRNA.

A consensus sequence ($TT\genfrac{}{}{0pt}{}{G}{A}NNNTTTTTT$) is located 5 to 20 bp downstream from the poly(A) addition site and this element may be involved in the cleavage/polyadenylation reaction. The mRNA 3′ end is generated by endonucleolytic cleavage and subsequent p. of the mRNA precursors [→ poly(A) polymerase].

poly (ADP-ribose) (Chambon et al. 1963) — a homopolymeric chain synthesized from NAD^+ by nuclear (chromatin-bound) poly(A) synthetase and consisting of a linear sequence of repeating (1 to 30) ADP-ribose units post-translationally attached by DNA-associated proteins. Poly (ADP-ribosyl)ation is involved in the regulation of various nuclear functions, such as → DNA replication, → DNA repair, → gene expression, and → cytodifferentiation. Single-strand breaks in DNA act as stimuli for poly (ADP-ribose) synthetase.

poly(ADP-ribose) synthetase (Chambon et al. 1966) — = → poly(A) polymerase.

polyandry — one of the two basic forms (→ polygyny) of → polygamy. In polyandrous mating systems, a female forms pair-bonds simultaneously or sequentially with more than one male; males usually assume most or all of the parental duties.

poly(A) polymerase (David et al. 1971) — any of the terminal riboadenylate transferases found in eukaryotes (in nuclei, cytoplasm, and mitochondria) and prokaryotes which catalyze both the synthesis and degradation (in the absence of ATP) of poly(A) tails of → messenger RNA. P.(A) p. adds adenylate residues to a primer (usually a polyribonucleotide) sequentially to form poly(A).

poly(A)$^+$ RNA — any of a population of polyadenylated RNA molecules in prokaryotes and eukaryotes (→ messenger RNA).

poly(A) site selection — in eukaryotes, tissue-specific selection (in certain transcription units) of alternative poly(A) sites (→ polyadenylation) and production of multiple → messenger RNA's from one gene.

poly(A) tail — at the 3′ end of prokaryotic and eukaryotic RNA molecules, a sequence of adenine residues added posttranscriptionally (→ polyadenylation).

polybasic (Darlington et al. 1945) — of → alloploids containing chromosome complements derived from two, three, or more species which differ in respect to their → basic number of chromosomes (two different basic numbers = dibasic; three different basic numbers = tribasic, etc.).

polycentric (Darlington 1937) — of chromosomes or chromatids (= polykinetic) with more than one localized → centromere: → Dicentric (with two centromeres), tricentric (with three centromeres), etc. Polycentrics are in most cases the result of → chromosome mutations. In special cases (→ nuclear differentiation), compound polycentric chromosomes represent a normal feature of chromosome organization.

polycistronic — of prokaryotic → messenger RNA carrying the information for the synthesis of more than one protein. The single large protein is either processed by proteolysis into individual units, or the mRNA is sequentially cleaved into individual cistronic units, each being translated into a single protein (→ operon).

poly(dA)·poly(dT) tract — any region of DNA containing uninterrupted dA residues on one strand and dT residues on the complementary strand. Such tracts are present in a wide variety of lower and higher eukaryotes as abundant and dispersed species. They may be involved in a wide range of functions including transcription, recombination, and chromatin structure.

Poly(A) tracts (up to 32 bp long) may be linked to other DNA sequences including the 5′ regulatory regions upstream of → structural genes and sequences adjacent to the 3′ end of both → pseudogenes and → transposable genetic elements. In the latter cases, the tracts presumably arise by → reverse transcription of poly(A) RNA followed by insertion into the genome.

polydisperse circular DNA (Schindler and Rush 1985) — (spc DNA); in eukaryotes, any of a class of small extrachromosomal DNA molecules derived from chromosomal DNA by both precise (sequence-dependent) and imprecise (sequence-independent) mechanisms.

polyembryony (Braun 1851) — the production of more than one embryo from a single egg cell, or from other gametophytic or sporophytic cells (in plants). (1) **Simple p.:** Several egg cells develop from one → megaspore and each is fertilized by a separate sperm or spore, or their development is parthenogenetic. (2) **Cleavage p.:** More than one embryo results by division (during early cleavage) of the zygote into two or more units, each developing into an embryo. The resultant embryos are "monozygotic" in origin and genetically indentical. (3) **Sporophytic p.:** Adventitious embryos arise by sporophytic budding from the nucellus and from the integument in flowering plants. The embryos are pseudogamic and usually identical to each other and to the mother plant.

polygenic (Plate 1913; Mather 1941) — of traits or → characters (quantitative characters), whose expression is controlled by many genes (polygenes) which individually exert a slight effect on the → phenotype. The mode of inheritance of p. characters is called p. inheritance and its main properties may be summarized as follows (Lerner 1968): (1) Most metric and meristic characters are affected by a number of genetic loci. (2) The effects of allelic substitution at each of the segregating genes (→ genetic segregation) are usually relatively small and interchangeable in the sense that identical phenotypes may be displayed by a great variety of genotypes. (3) The phenotypic expression of p. characters is subject to considerable modification by environmental influences. (4) P. characters show a continuous rather than a discontinuous distribution. (5) Balanced systems of p. inheritance in a population contain a great deal of potential → genetic variability stored in the heterozygous condition and released by small increments through → genetic recombination between linked polygenes.

polygyny — one of two basic forms (→ polyandry) of → polygamy. In polygynous → mating systems, a male forms pair bonds simultaneously or sequentially with more than one female; females assume most or all of the parental duties.

polyhaploid (Katayama 1934) — → haploid.

polyhybrid (de Vries 1900) — → monohybrid.

polykaryotic — of cells with several → nuclei.

polykinetic — = → polycentric.

polylysogenic — of a bacterial strain → lysogenic for more than one → prophage.

polymerase — any of the enzymes (→ DNA p.; RNA p.) which catalyze the assembly of nucleotides or deoxynucleotides into RNA or DNA on a DNA or RNA (reverse transcription) template, processes known as → genetic transcription and DNA replication (→ deoxyribonucleic acid).

polymerase chain reaction (Saikai et al. 1985) — (PCR); a technique used for enzymatic in vitro amplification (PCR amplification) of specific DNA sequences without utilizing conventional procedures of → molecular cloning. It allows the amplification of a DNA region situated between two convergent primers and utilizes oligonucleotide primers that hybridize to opposite strands. Primer extension proceeds inward across the region between the two primers. The product of DNA synthesis of one primer serves as a template for the other primer; repeated cycles of DNA denaturation, annealing of primers, and extension result in an exponential increase in the number of copies of the region bounded by the primers. The sequence of the double-stranded DNA fragment can be identified indirectly by hybridization to allele-specific oligonucleotide probes representing the various alleles studied or whose sequence can be determined.

Inverse p.c.r. (IPCR) allows rapid in vitro amplification of DNA sequences that flank a region of known sequence. It uses PCR but has the primers oriented in the reverse direction of the usual orientation.

polymeric (Nilsson-Ehle 1908) — of genes (→ gene interaction) whose individual effects are equivalent, but whose actions intensify each other (→ cumulative genes). Interactions among p. genes which give rise to so-called **"polymery"** (the production of a particular character by cooperation among several p. genes) are of two types: Dominance interactions between alleles, and nonallelic interactions. These interactions may be expressed statistically as components of genetic variance (→ variation).

polymitosis (Beadle 1933) — the intercalation of extra → mitoses at a particular stage of the life cycle in abnormal individuals or varieties of a species (Darlington 1965).

polymorphism — → chromosome polymorphism; genetic polymorphism.

polynucleosome (Tata and Baker 1978) — an aggregate of 6 to 30 linked → nucleosomes involved in the higher orders of → chromatin folding.

polynucleotide — a linear sequence of → nucleotides in DNA or RNA in which the 3′ position of the sugar of one nucleotide is linked through a phosphate group to the 5′ position on the sugar of the adjacent nucleotide.

polynucleotide kinase (Richardson 1965) — an enzyme which catalyzes γ-phosphoryl transfer from ATP to 5′-OH termini of DNA, RNA, and 3′-mononucleotides. The enzyme also harbors a 3′-phosphatase function.

polynucleotide ligase (Weiss and Richardson 1967) — = → DNA ligase.

polynucleotide phosphorylase — (PNPase); an enzyme, present in prokaryotic and eukaryotic cells, which catalyzes the stepwise, primer-dependent, reversible polymerization of 5′-nucleoside diphosphates with the release of P_i.

polypeptide — a polymeric, covalently bonded arrangement of → amino acids joined together by peptide linkages, i.e., covalent linkage between two amino acids in which the alpha-amino group of one acid is linked to the alpha-carboxyl group of the other with the elimination of H_2O (→ protein; elongation factor).

polypeptide chain elongation — the translation (→ genetic translation) of internal → codons by ribonucleoprotein particles (ribosomes) carrying a → peptidyl-tRNA, an N-acylaminoacyl-tRNA, or an aminoacyl-tRNA. Amino acids from aminoacyl-tRNA are added one at a time through a cyclic series of reactions (→ translocation), resulting in the elongation of the polypeptide chain from the N-terminal residue toward the C-terminal amino acid residue (→ aminoacyl transferase; peptidyl transferase). P.c.e. requires the participation of several protein factors, called → elongation factors.

polypeptide chain initiation — the first step in the synthesis of a polypeptide during → genetic translation of → messenger RNA.

polypeptide chain termination — in → genetic translation the final step in the synthesis of a polypeptide. It is triggered by a → terminator codon in mRNA (UAG, UAA, or UGA) located in the A site of the → ribosome: (1) Hydrolysis of the peptidyl-tRNA ester linkage to release the completed polypeptide from the ribosome; (2) disassembly of the ribosome·mRNA·tRNA complex to allow recycling of those components for new rounds of protein synthesis.

polyphenic (Haecker 1925) — = → pleiotropic.

polyphyletic — → monophyletic.

polyploid (Strasburger 1910; Winkler 1916) — of somatic cells and tissues, as well as of individuals (polyploids) having three (triploid), four (tetraploid), five (pentaploid), or more complete → chromosome sets instead of two as in diploids. The state of being p. is referred to as **"polyploidy"** and may arise spontaneously or be induced experimentally by → mitotic poisons (→ C-mitosis; C-meiosis). A polyploid that originated from a zygote containing unlike chromosome sets (usually due to hybridization of two species) is called an → allopolyploid, as opposed to a polyploid containing more than two homologous chromosome sets in its somatic cells, which is called an → autopolyploid.

secondary polyploid (Darlington and Moffett 1930) is the term used to denote polyploids with a chromosome complement in which particular chromosomes in the basic set are represented more frequently than other (due to → nondisjunction).

Di-, tri-, or polybasic polyploids (Darlington and Janaki Ammal 1945) are allopolyploids derived from the combination of diploid chromosome complements with two, three, or more basic numbers.

In higher plants, polyploidy in the form of → allopolyploidy has been of major significance in → evolution. Among lower plants and animals, polyploidy is quite rare and is, in animals, almost confined to some hermaphroditic groups and to forms reproducing by → parthenogenesis.

"Intra-individual somatic polyploidy" (resulting from → endomitosis and → endoreduplication) is frequently associated with the processes of differentiation in higher plants and includes conspicuous parts of the plant soma especially those cells and tissues with the most varied degrees of differentiation and of function. Generally, the diploid condition is retained in embryonic cells, apical initials, procambium and some cambial regions, the pericycle, and those cell lines which assure genetic continuity of the species. Intra-individual somatic polyploidy via endomitosis and endoreduplication results when the exact coupling of chromosome duplication and mitosis, controlling diploidy, is relaxed by disturbances or upsets of the spindle function.

polyploidization — the conversion (spontaneous or experimentally induced) of cells or individuals from the haploid or diploid state to a → polyploid state (→ C-mitosis; C-meiosis; endomitosis). Agents capable of inducing polyploidy may be called "polyploidogenic" (→ mitotic poison).

polyploidizing mitosis (Brodsky and Uryvaeva 1977) — any → mitosis resulting in polyploidy.

polypurine/polypyrimidine segment (Birnboim et al. 1973)- (plps); in eukaryotes, any of a special subset of middle → repetitious DNA. Such conserved segments (25 to 250 bp in length) contain exclusively purine residues in one and pyrimidine residues in the other strand of the DNA duplex. They are distributed at fairly regular intervals and in close association with repetitious DNA.

polyribosome (Warner et al. 1962) — = → polysome.

polysomatic (Langlet 1927) — of tissues or individuals containing diploid and polyploid cells side by side (→ mixoploid).

polysome (Warner et al. 1962) — a multiribosomal structure (= ergosome, polyribosome) representing a linear array of → ribosomes held together by → messenger RNA (the single ribosome bound to mRNA is referred to as a monosome by Haselkorn and Fried 1964). Polysomes represent the "active complexes" in cellular protein synthesis and are able to incorporate amino acids into polypeptides both in vivo and in vitro.

The ribosomes contained in polysomes isolated from → chloroplasts, → mitochondria, and the surrounding cytoplasm of eucells can be distinguished from one another by sedimentation analysis. Chloroplast and mitochondrion ribosomes are closely related to bacterial ribosomes.

Cytoplasmic polysomes of eucells (metazoa) occur in at least two and possibly three forms, which are biochemically and morphologically distinguishable. One class of polysomes exists apparently free in the cytoplasm, while a significant fraction of polysomes is closely associated with the → endoplasmic reticulum (especially in cells synthesizing protein for secretion). The fraction of ribosomes occurring on membranes is dependent on the type and state of the cell.

polysomic (Blakeslee 1921) — of an otherwise diploid cell or individual with one or more chromosomes represented three (→ trisomic: $2n+1$; doubly trisomic: $2n+1+1$) or four (→ tetrasomic: $2n+2$) times, etc., instead of twice. **Polysomic inheritance** (Blakeslee et al. 1923) is the mode of inheritance which arises when any chromosome (in the case of polysomy or polyploidy) finds more than one partner for → chromosome pairing at meiosis.

polyspermy — the entry of more than one sperm into the egg cell, irrespective of whether the supernumerary sperm are effective or ineffective in fertilization. P. may be a normal or abnormal condition.

polyspore (Renner 1916) — = gonospore (→ germ cell).

polytene chromosome (Koltzoff 1934; Koller 1935) — in various cell types of protozoa, plants, and animals, a special type of → chromosome consisting of extended chromosome strands that have replicated repeatedly and show a more or less clear suborganization into bands and interbands. In larval dipteran p. chromosomes the DNA strands are tightly paired. The bands are made up of aligned folded domains (chromomeres).

The so-called "banding pattern" (→ band) which results from the intimate association (pairing) at the same level of identical → chromomeres of the constituent fibrils is highly chromosome-specific and may be used for the identification of individual chromosomes as well as for the construction of detailed chromosome maps. Structural modifications of the banding pattern, in the form of reversible "puffs" and "Balbiani rings" are associated with differential → gene activation (→ puffing).

polyteny (Koltzoff 1934; Darlington 1937) — a special nuclear differentiation in larval and adult Diptera, Collembola, Protista, and angiosperm ovular nuclei represented by the presence of → polytene chromosomes. P. arises during differentiation of specific cell types which represent end-products of the cell-lineage in question.

polytopic — of a taxonomic category composed of widely separated but phenotypically identical populations (Mayr 1963).

polytypic (Mayr 1942) — of a taxonomic category (e.g., a species) composed of two or more immediately subordinate categories (e.g., geographic races or subspecies).

population (Johannsen 1903) — a community of potentially interbreeding individuals (at a given locality) which share a common → gene pool. Such a → "Mendelian population" (Dobzhansky 1935) represents a (dynamic) panmictic unit (→ breeding size). The sum of all factors governing the pattern by which gametes from various individuals unite with each other during fertilization makes up the population structure.

The largest group of potentially interbreeding individuals and thus the largest Mendelian population is the → species. It in turn is composed of numerous local populations, each intercommunicating and intergrading with others.

A p. freely exposed to → gene flow and subject to the input of alien genes, due to immigration, is called an **"open p."**, as opposed to a **"closed p."** with no genetic input other than by → mutation.

Population systems in nature may be classified on the basis of their phenotypic variation, their relative geographical distribution, and their mode of isolation (Grant 1963) as follows: (1) The population systems intergrade continuously in the phenotypic characters and are, therefore, judged to interbreed freely. (2) The population systems intergrade discontinuously or partially and are, therefore, judged to interbreed on a restricted scale. (3) The population systems are separated by a discontinuity in the pattern of morphological and physiological variations and are evidently not interbreeding.

population density — the number of individuals of a population per unit of space in the habitat.

population equilibrium — = → genetic equilibrium.

population genetics — a branch of → genetics that describes (in mathematical terms) the consequences of Mendelian → inheritance on the population level. It deals with the frequencies and interactions of genes in so-called "Mendelian → populations" and studies the influences of agencies such as → mutation, natural and artificial → selection, → migration, mixing of races and chance factors which tend to alter → gene frequencies and thus to cause evolutionary changes (→ evolution).

population monitoring — surveying biochemical, cytogenetic, genetic, and phenotypic characteristics of populations in order to obtain information about, e.g., spontaneous mutation rates, natural variations in these rates in certain subgroups of the population, and possible increases in the mutation rate which might result from environmental mutagens (→ mutagenicity testing).

population size — = → breeding size.

population structure — the genetic structure of a population, i.e., the spatial relationship on a micro- or macrogeographic scale between genetically differentiated → demes. P.s. is a consequence of → gene flow rates and environmental heterogeneity. It influences the potential for adaptation to spatially varying → selection pressures and the level at which → selection acts.

population system — any of the aggregates of → populations which form breeding units of large extent and long duration (Grant 1963).

population wave — any of the irregular or rhythmic changes of the effective → breeding size of a → population, or any territorial shift ("territorial p.w.") within the area occupied by a population. Such changes may be accompanied by random and relatively rapid alterations of → gene frequency (→ genetic drift) and may thus be a force in evolutionary change (Ludwig 1954). Population waves either restrict panmixis or (in case of territorial population waves) confront the population's gene pool with new constellations of environmental influences and thus new conditions of → selection.

pore complex (Watson 1959) — → nuclear envelope.

pore orifice (Watson 1955) — → nuclear pore complex; nuclear envelope.

portable promoter (Backman et al. 1976) — a DNA fragment (generated by → restriction endonuclease cleavage) which bears a (bacterial) → promoter and is capable of directing efficient transcription in vivo of genes to which it has been fused in vitro (→ recombinant DNA technology; gene fusion).

position effect (Sturtevant 1925) — a change in the phenotypic effect of one or more genes due to a change in their position with respect to other genes in the → genotype; this may be the result of chromosomal structural changes (→ chromosome mutation) or of → crossing over. Differences in the phenotypic expression of a gene caused by a change in its spatial relation to other genes in the chromosomes may be brought about by: (1) Changes in the production of a given → messenger RNA (→ genetic transcription), (2) changes in the → genetic translation of the messenger RNA in question, or (3) changes in the interactions between the polypeptides resulting from the translation of messenger RNA.

According to Lewis (1950), two types of position effects may be distinguished: The "cis-trans type" (or "stable type") and the "variegated type".

position effect variegation (Muller 1930) — mosaic expression of a euchromatic gene relocated to a chromosomal region adjacent to a heterochromatin breakpoint. The translocated gene is expressed in only a subset of cells of a given tissue and it is repressed in others. The variegating phenotype can be affected by both genetic (enhancers and suppressors) and environmental modifying factors. In most cases, the inactivating effect of heterochromatin extends for a limited distance from the heterochromatic breakpoint into the adjacent euchromatic region and affects several genes (polarized spreading effect).

positional information (Wolpert 1969) — information directing the spatial specification of differentiation probably via morphogenetic gradients which direct the differentiation of each cell in an embryo. Depending on the location of the cell within the (chemical) gradient, it becomes "instructed" to differentiate along a particular pathway (→ cortex).

positive control — in genetics, a mechanism of gene regulation in which an effector molecule (a protein, a small molecule, or a molecular complex) activates a → promoter. An inhibitor is not overridden in positive regulation. The positive regulator binds to a nucleotide sequence upstream of the -35 sequence to which → RNA polymerase binds and allows transcription to start (→ genetic regulation). Some genes or → operons are both positively and negatively regulated and need two regulators.

positive selection vector — a → cloning vector that allows a one-step identification of colonies harboring recombinant plasmids by direct positive selection. Transformants carrying religated or uncut vectors are eliminated by genetic means. The vectors typically rely upon the inactivation of a lethal gene, inactivation of a dominant function conferring cell sensitivity to metabolites, removal of a lethal site, or the derepression of an antibiotic-resistance function.

positive supercoiling — the coiling of a DNA duplex in space in the same direction as the winding of the two strands of the double helix itself (→ DNA supercoiling).

postadaptation — any adaptive genetic change specifically induced by an environmental stress (→ SOS response).

postmeiotic segregation — (PMS); the segregation of two strands of a duplex DNA that bear different information when a subsequent replication allows the strands to separate. PMC of genetic variants occurs when a DNA heteroduplex formed during → meiosis goes undetected by repair enzymes and is transmitted unresolved to the meiotic products. It is an integral part of → genetic recombination and is usually indicated by aberrant segregation ratios (e.g., 5:3 instead of 4:4 in a single eight-spore ascus of fungi) in lower eukaryotes where all meiotic products can be examined and the segregation of phenotypes in early mitotic divisions can be readily detected.

postreduction (Korschelt and Heider 1903) — → meiosis.

postreplication mismatch repair (Wagner and Meselson 1976) — a mechanism which eliminates DNA biosynthetic errors on the newly synthesized strand; transient undermethylation of the newly synthesized DNA could provide the strand discrimination necessary for elimination of replication errors (→ DNA methylation).

postreplication repair (Rupp and Howard-Flanders 1968) (PR); a → DNA repair process (= daughter strand gap repair) which involves recombinational patching of gaps or discontinuities generated in newly synthesized DNA upon replication of the damaged template. In bacteria, PR takes place as follows: (1) Replication proceeds past the → DNA lesion in the parental strand, leaving gaps 500- to 1000-nucleotides wide in the daughter strand; (2) the gaps in the daughter strands are filled with material from the parental strands by single-strand exchanges between the sister duplexes; (3) the gaps (about 15000 nucleotides wide) formed in the parental strands as a result of the recombination process are filled by → repair replication.

In mammalian cells, gaps (about 100 nucleotides wide) are formed by a bypass mechanism similar to that in bacteria but, although recombinational exchanges are involved in gap filling, the gaps are less frequent and/or much shorter than those in bacterial DNA.

PR is an error-prone process; mutations may arise either during recombinational exchange or by enzymatic insertion of incorrect bases into gaps opposite DNA lesions.

postsplit aberration — a chromosomal structural change (→ chromosome mutation) which takes place (in the chromosome segments already duplicated) during or after chromosome duplication, as opposed to a "presplit aberration" affecting the unreduplicated chromosomes.

post-transcriptional control — a mechanism of → genetic regulation which can involve several processes: (1) Premature transcription termination; (2) RNA sense-antisense interactions (→ antisense RNA) and mRNA decay; (3) translation initiation, elongation, and termination. (→ transcriptional control; translational control).

post-transcriptional processing — in eukaryotes, the stepwise tailoring (trimming) of DNA transcripts (→ heterogeneous nuclear RNA; pre-messenger RNA; pre-ribosomal RNA; pre-transfer RNA) which results in a shortening and modification of the primary nucleotide sequence of the transcript.

post-translocation complex — in → genetic translation, the messenger-ribosome complex with peptidyl-tRNA in the ribosomal P site and a vacant A site (→ pretranslocation complex).

POU domain (Herr et al. 1988) — any of a class of bipartite DNA-binding protein structures in which the POU homeo (C-terminal) and the POU-specific (N-terminal) regions form two subdomains (coupled by a flexible linker) required for DNA binding. The POU d. is a highly charged 155 to 162 amino acids region of sequence similarity found in mammalian → transcription factors. The POU sequence is sufficient for sequence-specific DNA-binding by a helix-turn-helix region of the domain.

preadaptation (Davenport 1903) — the possession of (preadaptive) characters that enable an organism to immediately adapt to an environment to which it has not been previously exposed (→ adaptation). Such characters may be due to the presence of duplicate genes.

precursor RNA — in both prokaryotes and eukaryotes, any of the high molecular weight primary transcripts from which mature RNA molecules arise by → RNA processing (→ pre-messenger RNA; pre-ribosomal RNA; pre-transfer RNA). Maturation of p. RNA's involves their ordered and specific recognition by cleavage and modifying enzymes. Precursor-specific RNA elements participate in many functions: (1) They may manipulate the folding of the remainder of an RNA precursor; (2) they may function during transcription (during chain initiation and termination); (3) some p. RNA sequences may serve directly as regulatory elements (→ attenuation); (4) precursor segments may supply recognition elements (sequence or conformation) required by specific processing enzymes.

predetermination (Kühn 1927) — the determination of gene-controlled characters by the maternal genotype prior to the fertilization of the egg cell. Hybrid progeny arising from predetermined zygotes are matroclinous with respect to characters showing p.

preferential fertilization — = selective → fertilization.

preferential segregation — → genetic segregation.

preinitiation complex — → initiation complex.

preinitiation stretch (Schäfer et al. 1973) — an untranscribed stretch of nucleotides in the → promoter region of an → operon which is assumed to serve as a storage site for → RNA polymerase molecules.

premature chromosome condensation (Johnson and Rao 1970) — (PCC); after fusion between mitotic and interphase cells, a rapid breakdown of the interphase nuclear framework and reorganization of interphase chromatin into prematurely condensed chromosomes. PCC achieved by → cell fusion falls into three categories: (1) G1 PCC with single, extended chromatids; (2) elongated G2 PCC more closely resembling mitotic chromosomes; (3) S-phase PCC which usually appears fragmented.

Regions of S- and G2-PCC that have completed replication are highly condensed, closely resembling the double chromatid arms of prophase chromosomes.

premature transcription termination — in a number of eukaryotic genes, premature termination of transcription shortly downstream from the transcription initiation site. P.t.t. may be a means by which the rate of transcription is regulated in analogy to prokaryotic → attenuation.

pre-messenger RNA (Georgiev and Mantieva 1962; Scherrer and Darnell 1962) — (pre-mRNA); in eukaryotes, the primary transcript of most structural genes (transcribed by → RNA polymerase II) which is the precursor molecule of cytoplasmic and polyribosomal → messenger RNA and contains (in the case of → mosaic genes) the → introns that are removed by → RNA splicing. Pre-mRNA's are complexed with proteins during transcription to form RNP particles (assembled at specific regions of pre-mRNA).

Pre-mRNA is characterized by the following features (Scherrer 1973): (1) The coding sequence which represents mRNA is read from DNA as a small part of pre-mRNA molecules. (2) Oligo(A) and poly(A) regions. The former comprise 5 to 30 adenylic residues, the latter (added after the first cleavage during processing) comprise 50 to 250 adenylic residues. (3) Sequences (of about 30 nucleotides) consisting of up to 80% of uridylate (oligo(U) sequences) which make up about 0.1 to 0.2% of the total pre-mRNA molecule. (4) Double-stranded sequences (about 3% of the molecule) fall into two size classes (one of more than 200.000 mol weight, the other of a 20.000 to 50.000 mol weight range). These regions appear to occur ten times per nascent pre-mRNA molecule.

Components (2) to (4) may represent signals in the programming sequence of pre-mRNA, attachment sites for stabilizing proteins, or may expose specific RNA fractions to cleaving enzymes. Programming sequences are probably involved in the mechanisme of post-transcriptional processing and modification of pre-mRNA and may contain signals for transcriptional and/or post-transcriptional regulation, transport, and translation.

After the first cleavage of nascent pre-mRNA, intermediate-sized pre-mRNA molecules are adenylated post-transcriptionally. The resulting poly(A) sequences are eventually up to 250 nucleotides long; small pre-mRNA molecules originating after further processing contain poly(A) sequences of the same lenght.

Pre-mRNA molecules are much larger (up to 10^7 daltons) than messenger-sized molecules (10^5 to 10^6 mol. weight range).

premetaphase stretch (White 1941) — in some animal species, the directed orientation of paired centromeres (→ centromere orientation) towards opposite poles, resulting in a visible stretching of chromosome arms on the spindle. It takes place at prometaphase of first meiotic division. The p. s. does not persist and the centromeres soon approach each other again.

premitosis (Sagan 1967) — a type of → mitosis in certain protozoa characterized by an intranuclear → centriole. The → mitotic apparatus is formed within the → nuclear envelope. In flagellates and all higher organisms, the centrioles (or their substitutes) have become extranuclear and division is by eumitosis.

premutational lesion — any → DNA lesion whose conversion into a mutation leads to substitution, insertion, or deletion of one or more bases. Conversion into a mutation occurs either by misreplication involving an erroneous base-pairing directed by the lesion (direct-acting mutagens) or by error-prone → DNA repair if the lesion fails to base-pair and causes arrest of → DNA replication (indirect mutagens) (→ mutation fixation).

prenatal diagnosis — monitoring of pregnancy through → amniocentesis in the diagnosis of → genetic diseases of metabolism and of changes in normal chromosome number of man.

prepattern — a descriptive term for the general organization of a developing system before a specific and observable pattern of organization is reached.

prenucleolar body — (PNB); any of the dense fibrogranular structures observed during → nucleologenesis. They fuse at the NOR's.

preprophase band (Gunning 1982) — (PPB); in a variety of higher plant species, a bundle of → microtubules girdling the → cell cortex around the nucleus prior of prophase.

preprophase inhibitor (D'Amato 1954) — → mitotic poison.

preprimosome — → initiator RNA.

prereducation (Korschelt and Heider 1903) — → meiosis.

prereplicative repair — any → DNA repair system that can remove → DNA lesion before they can be converted to mutations by → DNA replication (→ mutation fixation).

preribosomal particle — any of the ribonucleoprotein particles representing precursors of the cytoplasmic → ribosomes. In eukaryotes, their maturation takes place within the → nucleolus.

preribosomal RNA — (pre-rRNA); the primary transcript of the genes coding for → ribosomal RNA.

In **bacteria**, no large molecular weight precursor containing the joint sequences of 5′ 16S-(tRNA)-23S-5 S-3′ (some tRNA's are coded in bacteria with the rRNA's) accumulates, i. e., the growing RNA chain is cleaved by RNase III to yield the immediate precursors of the mature rRNA molecules. The precursors are called p16, p23, and p5 and accumulate in the absence of protein synthesis, but during normal growth they are rapidly converted by cleavage events to the sizes of their mature counterparts (m16, m23, m5). 250 nucleotides are endonucleolytically removed from the 5′ and 3′ termini of the p16 and p23 molecules. The p5 molecule (in **E.coli**) is three nucleotides larger than m5 rRNA (the excess lenght is removed exonucleolytically).

Eukaryotic rRNA is transcribed as a long precursor molecule which is converted to mature 18S, 5.8S, and 28S rRNA by a series of endonucleolytic cleavages. Excision of 18S rRNA sequences is an early event during the maturation process, while 5.8S and 28S rRNA's remain covalently linked in a common intermediate precursor (32S in mammalian cells) until internal spacer RNA is removed by processing endonucleases. After this, 5.8S and 28S rRNA remain associated by hydrogen bonding throughout the ribosome cycle. The fourth rRNA species (5S RNA) is transcribed at extranucleolar sites; the others are transcribed from the → nucleolus organizer region (rDNA) inside the → nucleolus.

Synthesis and processing of eukaryotic pre-rRNA proceeds as follows: (1) Synthesis of a 35 to 45S precursor to rRNA (Table 11) by transcription of → ribosomal DNA located in the nucleolar organizer. In mammals, the precursor has a lifetime of several minutes before being cleaved to produce 41S RNA from 45S RNA.

Table 11. rRNA synthesis in various eukaryotic systems

Fungi (yeast)	
35S RNA	20S RNA → 18S rRNA 27S RNA → 25S rRNA
Insects *(Chironomus)*	
38S RNA	18S rRNA 30S RNA → 28S rRNA
Amphibians *(Xenopus)*	
40S RNA	18S rRNA 32S RNA → 28S rRNA
Mammals (rat, human)	
45S RNA	23S RNA → 18S rRNA 35S RNA → 28S rRNA

(2) Methylation (primarily 2′-0-methylation) of the 18S and 28S regions of the pre-rRNA occurring during its transcription. The nonconserved portions at the 3′-OH ends contain few, if any, methyl groups, and are characterized by a high G + C content. (3) Association of pre-rRNA with protein (including species found on mature ribosomes) to form a complex sedimentating at about 80S. (4) Processing of pre-rRNA in the nucleolus to yield 16—19S and 32S species by cleavage and loss of nonmethylated parts. (5) Cleavage of the 32S RNA to yield 28—30S rRNA and incorporation of 5S RNA into the large nascent ribosomal subunit (65S nucleolar particle). Concomitant change of the ribonucleoprotein particle from 65S to 60S. (6) Rapid transport of ribosomal subunits to the cytoplasm.

The processing of the large pre-rRNA is possibly dependent on the presence of helical base-paired segments at specific processing sites which may serve to facilitate specific recognition by specific processing enzymes (convertases). As much as half of the original precursor species may be lost during this processing.

pre-transfer RNA — (pre-tRNA); the primary product of transcription of genes (tDNA) coding for → transfer RNA. The pre-tRNA of prokaryotes and eukaryotes has extranucleotides at the 5′ and the 3′ extremities; in some eukaryotic pre-tRNA's → introns are also present. Maturation of tRNA precursors is a multistep enzymatic process consisting of nucleolytic size-reducing reactions and of nucleotide modifications (such as methylation, thiolation, and pseudouridylation).

Prokaryotic pre-tRNA is composed of about 120 nucleotides, eukaryotic pre-tRNA (mammals) of 300 to 400 nucleotides. The 5′-leader sequence of eukaryotic pre-tRNA is cleaved by an endonuclease activity at a position seven nucleotides upstream of the constant uridine residue (position 8 of mature tRNA). After removal of the 3′-flanking nucleotides, the C-C-A terminus of mature tRNA is added, and the intron (if present) is excised. Nucleotide modification occurs at various stages during maturation.

pretranslocation complex — in → genetic translation, the messenger-ribosome complex with peptidyl-tRNA in the ribosomal A site and discharged tRNA in the P site (→ post-translocation complex).

prevalence — in human genetics, the frequency with which a medical condition with a genetic basis is found at a particular time in the general population (as, e. g., p. of a → genetic disease) (→ incidence).

Pribnow box — in prokaryotic → promoters, a highly conserved sequence element (positions -12 to -8) located about 10 bp upstream from the transcriptional start site:

5′ TATAATG 3′
3′ ATATTAC 5′
-12 -6

Analogous sequences have been found in the 5′ flanking DNA regions of eukaryotic genes (→ TATA box; Hogness box).

primary transcript — the immediate RNA product of → genetic transcription corresponding to a transcription unit (→ pre-messenger RNA; pre-ribosomal RNA; pre-transfer RNA; RNA processing).

primase (Lark 1972; Rowen and Kornberg 1978) — any of a class of enzymes (either free or in association with → DNA polymerase) which prime the initiation of → DNA replication, i. e., initiate the synthesis of 5′-terminal ribooligonucleotides or mixed ribo-deoxyribooligonucleotides that can be extended by DNA polymerase. Eukaryotic p. (an oligoribonucleotide polymerase tightly bound to DNA polymerase α) synthesizes a stretch of 4 to 10 nucleotides transiently linked in the DNA of → Okazaki fragments (→ RNA primer; primosome).

primer — (1) a short RNA sequence that is paired with one strand of DNA and provides a free 3′-OH terminus at which a → DNA polymerase starts synthesis of a deoxyribonucleotide chain (→ primase); (2) a short sequence (about 10 bp) of DNA which is complementary to a sequence of → messenger RNA and allows → reverse transcriptase to start copying the adjacent sequences of mRNA (→ cDNA); (3) a cellular transfer RNA (primer tRNA) whose elongation initiates RNA-directed DNA synthesis by the DNA polymerase of → retroviruses; primer tRNA is tightly associated with the genome RNA in the viral 70S RNA complex.

primer-restriction-end adaptor (Coleclough and Erlitz 1985) — (PRE); any of a class of bifunctional synthetic oligonucleotides; one end serves as a → primer for polymerase reactions, the other end can be ligated to → restriction endonuclease digested DNA. PRE adapters can be used for insertion of single-stranded cDNA into → cloning vectors.

primosome (Low et al. 1981) — a mobile multienzyme DNA replication priming complex that assembles at a specific site on single-stranded *E.coli* DNA (called a p. assembly site) and is thought to migrate 5′ → 3′, occasionally synthesizing a → primer through the action of → primase. The p., in combination with *E. coli* DNA polymerase III holoenzyme, can form replication forks in vitro.

prion (Prusinger 1982) — any of a class of proteinaceous infectious particles (about 100 times smaller than the smallest virus) that resist inactivation by procedures that attack nucleic acids. They cause neurologic disorders of animals and humans (such as Scrapie and Creutzfeldt-Jacob disease) and possibly replicate via protein-directed protein synthesis.

proband — in human genetics, the „affected" person with whom the study of a particular → character in a family (→ pedigree) usually begins (= propositus if male, proposita if female, or index case). A method of comparing the proportion of progeny in families in which an „index case" or p. shows a specific character with the proportion expected if this character fulfils Mendelian inheritance is referred to as the „proband method".

probe — a defined (radioactively labeled) DNA or RNA sequence used to detect complementary sequences by hybridization techniques (→ Southern blotting; Nothern blotting; colony hybridization).

procentriole (Gall 1961) — a short new → centriole arising perpendicular to the old centriole and growing distally as it matures. The maturation and differentiation of a p. involves (Pitelka 1969): (1) The synthesis of required polypeptides; (2) the assembly of those into → microtubules, dense material, and „cartwheel" substance; (3) the elaboration from these of the polarized, nine-membered cylinder; (4) the synthesis and assembly of the protein of the fibrous appendages at specific times and places.

processed gene (Hollis et al. 1982) — → processed retropseudogene.

processed retropseudogene — in eukaryotes, any DNA sequence (retroposon) resembling a → cDNA copy of a fully processed → messenger RNA species (formerly p. gene). Such → pseudogenes include the 3′ terminal → poly(A) tail of the parental mRNA species, lack any → intron present in the parental gene, and often extend to the normal 5′ cap site. They are flanked by a direct repeat varying in length from 10 to 20 bp (both 5′ and 3′); truncations of processed mRNA's are not uncommon (→ Alu-like sequence; P. element).

The pathway to create a p. r. probably involves two steps: (1) Conversion of RNA sequences to DNA sequences by → reverse transcription; (2) integration of these sequences by a transposase-type mechanism that produces varying lengths of direct repeats.

processing endoribonuclease — any of a class of specific ribonucleases involving in a complex set of reactions which process RNA's (→ RNA processing).

processon (Weissman 1979) — a nucleic acid sequence which has to be removed from → precursor RNA (→ RNA processing).

progamic (Haecker 1902) — → sex determination.

progeny — the offspring of a particular → mating, of a particular mate, or of a particular individual in the case of apomictic reproduction (→ apomixis).

progeny selection — the evaluation of an individual's progeny as customarily employed in artifical → selection. Methods of p. s. vary with the types of → matings that are possible with a given animal or plant.

progeny test — a method for assessing the → genotype of an individual or of the performance of a parent by a study of its progeny under controlled conditions.

programmed gene rearrangement — any → DNA rearrangement whose outcome is predictable, as opposed to unprogrammed or mutational g. r. P. g. r. is usually carried out by specific recombination enzymes and is developmentally regulated. Three catagories: (1) Amplification or deletion of specific genes; (2) assembly of genes from gene segments; (3) DNA rearrangements which alter → gene expression.

progressive double crossing over (Weinstein 1936) — → crossing over.

prokaryon (Dougherty 1957) — the nuclear equivalent (→ nucleus; nucleoid) of → prokaryotic organisms (→ eukaryon) which is not bounded by a nuclear envelope.

prokaryotic (Chatton 1925) — of viruses and bacteria (prokaryotes) as opposed to eukaryotic organisms. They have no membrane-enclosed nuclei dividing by mitosis and meiosis, no spindle mechanisms, and no condensation cycles of the → chromosomes. Nuclear equivalents (nucleoids) replace true nuclei to which they are functionally homologous. There are two groups of p. bacteria, the archebacteria and the eubacteria, which are as different from each other as they are different from eukaryotes.

Most biochemical processes known in eukaryotes also occur in p. bacteria; in addition there is a wide variety of activities unknown in eukaryotes, e. g., anoxygenic photosynthesis, energy production by oxidation of inorganic compounds, and oxidative metabolism based on electron acceptors other than oxygen, nitrogen fixation.

prolamellar body (Hodge et al. 1956) — an accumulation of vesicles formed by the invagination of the proplastid membrane during etiolation (→ plastid).

prometaphase (Lawrence 1931) — → mitosis, → meiosis.

promiscuous DNA (Ellis 1982) — any DNA sequence found in more than one of the membrane-bound organellar genetic compartments of eukaryotes (nucleus, mitochondria, chloroplasts).

promoter (Jacob et al. 1964) — a DNA sequence that directs → RNA polymerase to bind and initiate specific transcription of → genes or → operons (→ genetic transcription). The interaction of RNA polymerase with the p. passes through a series of steps which involve transitions in the state of the polymerase-p. complex. Promoters differ in the rate of polymerase binding, the stability of the complex, the transition ratio from closed (recognition) to open (rapid start) complexes, and the efficiency with which the transcription is initiated.

Although the binding sites for RNA polymerase generally extend from about 45 bp upstream from the startpoint of transcription to about 20 bp downstream, p. recognition in **bacteria** *(E. coli)* is typically governed by two limited regions: The recognition or "−35" region and the → Pribnow box or "−10" region (centered at the positions 35 and 10 bp, respectively, preceding the transcription start site). The consensus −35 and −10 sequences are 5′-TTGACA-3′ and 5′-TATAAT-3′, respectively. Optimal spacing between these regions is believed to be 17 bp. In some cases, there are additional signals (upstream from the p.), that affect the level of transcription (cis-acting upstream signals) similary to the → transcriptional enhancers in eukaryotes. They contain AT tracts that bend when the DNA of these regions is isolated.

P. strength is primarily determined by: (1) The binding affinity to RNA polymerase; (2) the rate of isomerization from "closed" p. complexes to "open" p. complexes in which the DNA duplex is partially opened at the p. region. Strong promoters have high values for both the binding constant for RNA polymerase and the rate of open complex formation; weak promoters have low values.

In one class of bacterial promoters, the DNA sequence contains all information for binding of RNA polymerase and formation of a stable complex for productive initiation of transcription. In other promoters, the sequence information (the DNA structure) is insufficient by itself to allow such stable complexes to be formed in the absence of additional protein factors (cyclic AMP-dependent promoters fall in this class).

A typical *E. coli* p. can be represented by the following consensus sequence (N represents any of the four nucleotides; start is the transcription start point):

TTGACA — N_{15-19} — TATAAT — N_{5-7} — Start.

A class of positively controlled promoters lack the features associated with transcription initiation sites in prokaryotes. Instead of the consensus −35 and −10 elements. They contain conserved nucleotides around −24 and −12 (CTGGCAV and TTGCA, resp.), the −24 consensus contains the invariant dinucleotide GC and the same dinucleotide is found around −12 (the spacing of 10 bp between these dinucleotides is critical for the function of these promoters).

Single genes may be transcribed from multiple promoters and these fall into three classes: (1) Tandem promoters, i. e., two (or more) promoters are oriented in the same direction and transcribe the same gene or operon; (2) divergently oriented promoters in which two RNA polymerases can bind within a common control region and transcribe in opposite directions into seperate genes or operons; (3) convergent promoters where RNA polymerases actually oppose one another and transcribe both strands of the DNA over a common interval.

Internal promoters lie within a gene or operon.

In **eukaryotes**, several DNA sequences upstream from the initiation site are important for recognition and positioning of RNA polymerase II transcribing mRNA. In contrast to polymerase II, polymerase III recognition and positioning involve internal promoters to the 5S and tRNA genes. The controlling elements for transcription by polymerase I (rRNA genes) are not yet identified.

Comparison of the 5′ extragenic regions of eukaryotic genes have revealed two conserved sequence elements: (1) An AT-rich region with the consensus sequence $5'\text{-TATA}^{T}_{A}\text{A}^{T}_{A}\text{-}3'$ (the Goldberg-Hogness or TATA box) located about 30 bp upstream from the site of transcription initiation; (2) the CCAAT box (CAT box) located about 80 bp upstream of the mRNA cap site. For efficient transcription, both the Goldberg-Hogness box and a region between 100 and 58 bp in front of the transcription initiation site are required.

Eukaryotic promoters recognized by RNA polymerase II are thus divided into several functional elements: The initiation site (I), the TATA box, the CCAAT box, and the upstream regulatory elements. These elements are found in the 5′ untranslated region adjacent to the structural gene. Transcription usually starts at a site 30 bp downstream from the TATA box. The 5′ flanking region is not essential for transcription of genes by RNA polymerase III; these genes contain internal promoters and the coding region has a dual function: Promotion of transcription as well as encoding the sequence of the RNA transcript. The internal p. of tRNA genes is split into two regions (split promoter). The anterior p. region contains a highly conserved sequence 5′-TGGCNNAGTNGG-3′ (box A) located 11 to

18 bp downstream from the 5′ end of the gene. The posterior p. region contains the conserved sequence 5′-GGTTCGANNCC-3′ (box B) and is located 33 to 42 bp downstream from the box A sequence.

A typical eukaryotic p. includes: An AT-rich region (the TATA box) and one or more sequence elements of 8 to 12 bp called upstream p. elements (UPE's). The UPE's increase the rate of transcription initiated at the TATA box and p. strength is determined by the number and type of UPE's. The cis-acting elements bind trans-acting factors with activate transcription and dictate the control of tissue-specific, hormone-induced, viral-induced, and growth-differentiation-related gene expression.

promoter-down mutation — any mutation that decreases → promoter activity, as opposed to promoter-up mutations that increase it.

promoter modulation (Lupski et al. 1984) — a mechanism for regulation of → gene expression in prokaryotes, i.e., the use of alternative → promoters of an → operon in response to altered environmental conditions which allows for changes in relative rates of operon gene expression.

P.m. may also be responsible for developmentally regulated and tissue-specific expression of genes in eukaryotes.

promoter probe vector (West et al. 1979) — any of a number of plasmid → cloning vectors constructed for cloning and subsequent evaluation of DNA fragments with promoter activity. Such vectors carry a marker gene lacking a → promoter and a cloning site on their 5′ side. Fragments of foreign DNA able to act as promoters are recognized when, after in-phase insertion at the cloning site, they lead to expression of the marker gene.

promutagen — any chemical agent that requires metabolic activation before it can exert a → mutagenic effect, as opposed to other agents that have the ability to react directly with the genetic material (→ host-mediated assay).

pronucleus (van Beneden 1875) — the egg or sperm nucleus from maturation until → karyogamy, at which time the pronuclei fuse to produce the zygote nucleus.

proofreading activity — in genetics, any mechanism by which an incorrectly selected nucleotide or aminoacyl tRNA can be removed and replaced by the correct unit (= editing activity). (→ coding fidelity; DNA polymerase).

propagule — a uni- or multicellular agent of dissemination.

prophage (Lwoff and Gutman 1950) — a bacteriophage genome covalently integrated as a part of the bacterial chromosome (the provirus stage of a → temperate phage). Some bacteriophages are integration-site-specific, others have several distinct integration sites, still others are able to integrate nearly everywhere and in so doing interfere with or modify local gene function (→ Mu phage).

Defective prophages are those that have undergone a deletion of their genome. A region of the bacterial chromosome has been inserted into the p. genome, probably in substitution for the missing region. Defective prophages can be transmitted lineally in their host strains but cannot produce infective particles (→ prophage complementation).

Cryptic prophages lack the genes necessary for the lytic cycle or contain mutants of the relevant genes.

Phage production by a cell carrying a p. occurs spontaneously with a very low frequency (10^{-2} to 10^{-5} per cell per generation). This frequency can be raised to nearly 100% by certain agents which interfere with DNA replication (prophage → induction).

prophage complementation (Thomas 1966) — the phenomenon wherein a defective phage superinfects a bacterial strain → lysogenic for a heteroimmune but closely related phage; the yield of phage progeny is usually higher than following infection of the corresponding, nonlysogenic strain.

prophage excision — the direct reversal of the integrative recombination (→ integration system) of a temperate phage leading to excision of the prophage DNA and thus → prophage induction.

prophage induction — derepression (by inactivation of a prophage-encoded repressor) of a → prophage and initiation of a → lytic cycle. P.i. may occur spontaneously and can be induced by agents or cellular mutations that tend to block bacterial chromosome replication or damage DNA.

prophage integration — the insertion of phage DNA into the bacterial chromosome (→ integration system). Mutants that are defective in p.i. form phages indistinguishable from those of wild type since they can establish abortive → lysogeny.

prophage interference — a change in the sensitivity of → lysogenic bacteria (containing a prophage) to infection by bacteriophages unrelated to the prophage integrated in the bacterial genome. P.i. is a case of "lysogenic conversion".

prophage substitution — the replacement of a → prophage by a superinfecting phage in a small number of → lysogenic bacterial cells. The frequency of p.s. is a linear function of the multiplicity of superinfection (about 0.01—0.001 substituted cells per superinfecting phage particle).

prophase (Strasburger 1884) — → mitosis, meiosis.

prophase index (Timonen 1950) — the ratio of the frequencies of metaphases and prophases in a given tissue.

prophase poison (D'Amato 1948) — → mitotic poison.

prophasing (Matsui et al. 1972) — in an interphase-metaphase binucleate cell which originated by heterophasic → cell fusion, the transformation of the interphase nucleus to a state similar to prophase in a mononucleate cell (→ premature chromosome condensation).

proplastid (Strugger 1950) — an immature, partially differentiated → plastid found in the meristematic cells of multicellular algae or higher plants and in etiolated plants (→ prolamellar body) in which chlorophyll synthesis has been suppressed by growth in the dark. Proplastids lack chlorophyll but contain small amounts of protochlorophyllide.

propositus — = → proband.

proreduplication (Hsu and Moorhead 1956) — → interreduplication.

prosome (Schmid et al. 1984) — any of a class of ubiquitous → ribonucleoprotein particles (raspberry-shaped, sedimentation coefficient 19S). Prosomes are a subcomponent of free mRNA and occur in the nucleus and cytoplasm; the function is unknown.

prosphase (Bullough 1963) — that part of → interphase during which all the syntheses necessary before the cell enters → mitosis are completed. The p. can be divided into at least three periods: Early p., the phase of DNA synthesis (S), and antephase (G_2) (→ apophase).

protamine (Felix 1960) — any of a family of small, highly basic proteins consisting of 32 or 33 residues of which about two-thirds are arginine. By displacing the → histones, the protamines have a unique role in packaging animal sperm DNA.

protandrous (Hildebrand 1867) — → dichogamous.

proteasome — in eukaryotic cells, a 700 kDa protolytic complex (= multicatalytic endopeptidase complex) which probably functions in the rapid degradation of oxygen-damaged proteins and in the ATP-dependent proteolytic pathway.

protection — → restoration, photoprotection.

protein (Berzelius) — one of a group of nitrogenous organic compounds (molecular weight ranging between 34.000 and 200.000) of complex shape and composition. Proteins are made up of molecular subunits called "polypeptides" which are themselves composed of smaller building-blocks, called → amino acids, linked together in form of long chains. Such chains may contain 200 or more amino acids of which there are 20 different kinds.

The **primary structure** of a p. is represented by the amino acid sequence in the polypeptide chain(s) which is determined by one → cistron (or → gene) during → genetic transcription and → genetic translation of its content of → genetic information.

The **secondary structure** refers to the arrangements of the polypeptide chain in a helical or other regular or irregular manner, and the **tertiary structure** to the arrangement of the secondary structures into a threedimensional (folded and superfolded) structure. The association of more than one polypeptide chain to form a stable unit corresponds to the **quaternary structure**. Interactions between amino acid side chains largely determine the folding of the polypeptide backbone in proteins and define patterns of specificity and mechanism of action of enzymes.

Catalytic proteins or → enzymes are catalysts produced by living cells with a specific action and an optimum activity at a definite pH value. They may be either "constitutive (produced in fixed amounts independent of need), "inducible" (not synthesized unless their substrates are present), or "repressible" (the rate of their production is decreased when the intracellular concentration of certain metabolites increases).

Structural proteins are without catalytic activity and form boundaries, provide containment or protection and insulation from the environment, make connections between different tissue elements, impart rigidity, hardness, elasticity, or tensile strength, endow tissues with the capacity for movement, extension or contraction, and provide a matrix for the deposition of other components such as minerals (Seiffer and Gallop 1966).

In both bacteria and eukaryotic cells, most secreted proteins are initially synthesized as precursors with hydrophobic amino-terminal extensions (→ signal peptide) which are removed by proteolytic cleavage during or after translocation of the p. through the cell membrane or to any one of a number of different cellular compartments (protein targeting).

protein kinase — any of a diverse group of regulatory proteins involved in growth control and malignant transformation in eukaryotes. These enzymes are thought to exert their regulatory effects by phosphorylating other proteins, thereby altering the activity of the modified protein.

proteinoid — a polyamino acid molecule differing from → proteins in that the bondings between constituents are not necessarily peptide bonds between α-amine and α-carboxyl residues.

protocell — a hypothetical (prebiotic) cell-like microstructure (proteinoid microsphere) which might have evolved into the first → cells.

protogynous (Hildebrand 1867) — → dichogamous.

proto-oncogene — (c-onc); any of a class of evolutionary conserved cellular genes whose mutationally altered product or aberrant expression can lead to the malignant transformation of nontumorigenic cells (→ oncogene). P.-oncogene products play a fundamental role in the proliferation and differentiation of eucells. The products are growth factors, growth factor receptors, G-like proteins or other proteins involved in signal transduction. Some have properties of → DNA-binding proteins. When mutated, incorporated within a → retrovirus, or expressed inappropriately, proto-oncogenes can induce unrestrained growth.

protoplasm (Purkinje 1840; Mohl 1846) — collectively, the cytoplasmic and nuclear components of the eukaryotic cell.

protoplasmic incompatibility (Rizet and Esser 1952) — a mechanism protecting fungi, higher plants, and animals against cytoplasmic invasion and restricting outbreeding. P.i. is controlled by many genes that determine allelic or nonallelic systems of p.i.

protoplast (Hanstein 1880) — any cell of bacteria, fungi, or higher plants from which the rigid cell wall has been enzymatically removed to expose the cell membrane. Under appropriate conditions, two (or more) protoplasts may fuse and create hybrid cells (p. fusion) or they will take up DNA. In suitable media, protoplasts will regenerate the cell wall (→ spheroplast).

prototrophic (Ryan and Lederberg 1946) — of nutritionally independent cells, as opposed to → auxotrophic ones.

provirus (Temin 1971) — the duplex DNA intermediate (→ reverse transcription) of an RNA tumor virus integrated into the eukaryotic genome. It is flanked by → long terminal repeat (LTR) sequences generated by the duplication of segments present at both ends of the viral RNA (length varying from 330 to 1.328 bp). The primary viral transcript starts in the left LTR and terminates in the right LTR. Proviruses (→ retrovirus) are insertion mutagens, excise by recombination between terminal repeats, may undergo deletions and acquire genes (not only → oncogenes) from the host genome. Most of these characteristics are shared with → transposable genetic elements.

proximal (Navashin 1912) — of a part of a chromosome arm which is nearer to the → centromere than another (distal) part (Darlington and Mather 1949).

pseudoallele (Morgan et al. 1928; Lewis 1948) — any of two (or more) mutations which are allelic (similar or related) in a functional but not in a structural sense. Pseudoalleles map at different positions (→ genetic map) and show a low order of → genetic recombination via crossing over. Such mutations (m^1 and m^2) are phenotypically mutant in trans-heterozygotes ($m^1+/+m^2$), but wild type in cis-heterozygotes ($m^1m^2/++$), that is, exhibit a cis-trans → position effect ("position pseudoalleles").

Pseudoalleles are generally interpreted as closely linked and functionally related genes. A cluster of pseudoalleles is referred to as a "pseudoallelic series", a "complex locus" (Dunn 1954), or a "region" (Benzer 1957).

pseudoamitosis (Haecker 1910) — a type of → karyokinesis which resembles an → amitosis but originates from a mitotic type of nuclear division.

pseudoaneuploid (Rieger 1963) — of cells or individuals (pseudoaneuploids) characterized by changes in the → basic number of chromosomes within a species due to fragmentation or → centric fusion. This results in chromosome numbers, which deviate from the common one. In contrast to genuine → aneuploids, the genetic material is equivalent to that of individuals with the original chromosome complement, i.e., before the chromosomal structural changes (→ pseudopolyploid) occurred.

pseudoapogamy (Farmer and Digby 1907) — the development of the → sporophyte from fused vegetative cells instead of the fertilized egg cell [= pseudomixis of Winkler (1908) and somatogamy of Renner (1916)].

pseudoautosomal (Burgoyne 1982) — of alleles located in the short region (near the → telomeres) of mammalian → X and Y chromosomes at which meiotic pairing occurs. P. alleles undergo reciprocal exchange by → crossing over and therefore do not show strict → sex-linked inheritance.

pseudobivalent (Walters 1954) — at metaphase of → meiosis, the association of two chromosomes by stickiness rather than by → chiasmata. Associations of this type are called "quasibivalents" by Östergren and Vigfusson (1953) and may coordinate like true bivalents (→ centromere orientation).

pseudochiasma (Levan and Tjio 1948) — a chiasma-like connection between two chromosomes (in meiosis I) or chromatids (in mitosis and meiosis II) due to the adhesion of chromatids (during anaphase separation) at homologous or, rarely, nonhomologous points (= point union, point error). The points of adhesion may either be terminal or, more frequently, interstitial.

pseudo-compatibility — (PC); a quantitative and/or temporary modification of → self-incompatibility frequently occurring in normally self-incompatible plants and leading to unexpected seed set (= pseudofertility).

pseudodicentric — of a → dicentric chromosome (→ telomere fusion) with one of the two → centromeres being suppressed (similarly pseudotricentric, etc.).

pseudodominant — → dominant.

pseudo-exon (Jaworski and Piatigorsky 1989) — a silenced → exon of a split (mosaic) gene.

pseudofertility (East and Park 1917) — = → pseudocompatibility.

pseudogamy (Focke 1881) — apomictic (parthenogenetic) development of a female gamete or cell after stimulation (but not ferilization) by a male gamete or gametophyte. The progeny arising from p. are maternal in character (= gynogenesis).

pseudogene (Jacq et al. 1977) — any DNA sequence significantly homologous (75 to 80%) to a functional gene which has been altered so as to prevent any normal function. Pseudogenes are a common feature of many → multigene families in higher eukaryotes and fall into two general categories : (1) Those that show all the major structural features of functional genes (such as a promoter, exons, introns) and are supposed to be created by DNA-mediated duplication of a functional gene and its silencing **(nonprocessed pseudogenes)**; (2) those that completely lack introns, have a remnant 3′ → poly(A) tail and are often flanked by short direct repeats (**processed pseudogenes** or processed retropseudogenes). They are probably created by a mechanism involving → reverse transcription of → messenger RNA followed by integration of the resulting cDNA copy into a new chromosomal site.

Nonprocessed pseudogenes (→ orphon) are generally found within parent gene clusters; processed pseudogenes are generally dispersed and show the features of → transposable genetic elements.

As opposed to pseudogenes, those homologous sequences producing functional polypeptides or structural RNA's are called "productive genes".

pseudogene compositional assimilation (Graur et al. 1989) — a process whereby a → pseudogene loses similarity to the functional gene by molecular changes and melts into the background of its surrounding DNA (→ pseudogene length abridgement).

pseudogene length abridgement (Graur et al. 1989) — a process whereby a → pseudogene becomes progressively shorter through the accumulation of deletions (→ pseudogene compositional assimilation).

pseudogonochoristic — = → pseudohermaphroditic.

pseudohaploid (Katayama 1935; Ivanov 1938) — → haploid.

pseudohermaphroditic — of individuals ("pseudohermaphrodites") with gonads of one sex only but possessing reproductive organs with some characteristics of the opposite sex. Mammals are classified as male or female according to the type of gonad present.

pseudohomeotypic (Gustafsson 1935) — of a modified first meiotic division (→ meiosis), characterized by the presence of only unpaired chromosomes (univalents) rather than pairing configurations. The chromatids of each univalent are distributed to opposite spindle poles at anaphase I. Meiosis I is, in fact, a → mitotic division and the second meiotic division is omitted.

pseudoisochromosome (Caldecott and Smith 1952) — a chromosome whose ends are homologous (as a result of reciprocal → translocations between end segments of opposite arms of the chromosomes of one pair of homologs or due to crossing over in pericentric → inversions during meiosis). These ends pair (→ chromosome pairing) internally at meiosis like → isochromosomes, but the segments proximal ("interstitial segments") to the centromere are nonhomologous.

pseudolinkage — → affinity.

pseudolysogenic (Romig and Brodetsky 1961) — of bacteriophages which are capable of establishing a carrier state in bacterial host cells and, in several instances, able to act as generalized transducing agents (→ transduction).

pseudomixis (Winkler 1908) — = → pseudogamy.

pseudomonosomic (Hiorth 1948) — → monosomic.

pseudomonothallic (Ahmad 1954) — → heteromixis.

pseudomultivalent — an association (→ pseudobivalent) of more than two chromosomes due to sticky adhesions rather than to homologous → chromosome pairing and chiasma formation, in contrast to a true → multivalent.

pseudo-overdominant (Lerner 1958) — of a genetic locus whose apparent → overdominance (indicated by heterozygote excess) is in fact due to → linkage disequilibrium of the alleles (= associative overdominance of Frydenberg 1963).

pseudopolarity — of effects which mimic → genetic polarity but are not due to factors affecting gene function. The apparent polar effects may be due to mutations affecting enzyme activity at the level of protein-protein interaction.

pseudopolyembryony (Lebegué 1952) — → polyembryony (in plants) due to fusion of several ovules, the division of the nucellus and the production of several embryo-sacs by the multicellular archespore or by several functional macrospores.

pseudopolyploid (Battaglia 1956) — of cells or individuals ("pseudopolyploids") whose chromosome number is doubled or increased to an even higher order without a corresponding increase in the quantity of the → genetic material (→ pseudoaneuploid).

pseudoreversion — → reversion

pseudoselectivity (Sedlmayr 1956) — apparent nonrandom fusion of gametes during zygote formation due to inviability of a number of the zygotes. This may be regarded as a case of selective → fertilization.

pseudo-self-compatibility — in plants with a functional → incompatibility system, the production of seeds from self- or incompatible cross pollinations in amounts ranging from less than 1 to 100% of a compatible outcross. P.-s.-c. is due to the presence of modifier genes.

pseudo-stemline (Hughes 1968) — a → stemline whose modal cells can be characterized only by having the same chromosome number, and not by its karyotype.

pseudotrisomic (Kush 1973) — = → double telotrisomic.

pseudotype (Choppin and Compans 1970) — an enveloped RNA virus which after participation in → phenotypic mixing with a similar virus from a different group shows envelope antigens of one virus and the genome of the other.

pseudovirion (Michel et al. 1967) — any viral → capsid containing linear fragments of cellular DNA, instead of the viral genome or fragments thereof. Pseudovirions are not infectious.

pseudowild type (Mitchell et al. 1952) — of a → phenotype arising not by reversion (back mutation) of an original mutational change but rather by nearby suppressor mutations (→ revertant).

psi factor (Travers et al. 1970) — any of a class of bacterial positive control elements assumed to act as secondary specificity determinants allowing the → RNA polymerase holoenzyme (not the core enzyme) to transcribe several classes of transcription units. Each element is assumed to be specific for a given class. There are indications that psi factors are regulated by low molecular weight effectors. This provides for a positive control which is readily reversible and would allow the bacterial cell to respond to readily environmental changes.

Psi and → CAP are of approximately the same size, require → sigma function, and are functionally involved in transcriptional processes. CAP activity is activated by cyclic AMP, psi is inhibited by guanosine tetraphosphate (pp G pp).

P-site — → ribosome.

puffing (Bridges 1937) — in → polytene chromosomes, a localized cell- and stage-specific, decondensation of small chromosome regions (bands) which is a visible expression of gene activity. Puffs may play a pretranscriptional role in gene regulation such that the change of chromatin conformation triggers the subsequent transcriptional activation, or they may be a consequence of ongoing RNA synthesis and accumulation of transcription products.

pulse-chase experiment — a procedure performed by very brief incubation of cells with a radioactively labeled precursor of some metabolic pathway or macromolecule and following up the fate of the label during subsequent incubation with a nonlabeled precursor.

punctuated evolution (Eldredge and Gould 1972) — a model of biological → evolution according to which → species are relatively stable and long-lived and new species are assumed to appear during concentrated outbursts of speciation, followed by differential success of certain species, i.e., the evolutionary pattern is characterized by long periods of stasis of individual species (in the fossil record), punctuated by bursts of rapid change, as opposed to → phyletic gradualism.

pure line (Johannsen 1903) — the descendants, through self-fertilization, of a single → homozygous parent, or a highly inbred line of plants or animals obtained by long continued → inbreeding. All members of a p.l. are of the same genotype within the limits set by the probability of new → gene mutations.

pure simple DNA sequence — any of the DNA sequences composed of a single or relatively few short sequence motifs, usually in tandem. Such sequences have been implicated in a range of functions including → genetic regulation, signals for → gene conversion and → genetic recombination, and the replication of → telomeres. They are assumed to accumulate by DNA slippage and mispairing during DNA replication and recombination or by extension of single-strand ends.

pyrenoid (Schmitz 1882) — a specialized structure occurring within the chloroplasts (→ plastid) and involved in starch synthesis and storage. Pyrenoids are found primarily in certain algae and liverworts. They are composed of a granular, homogeneous, or filamentous and fibrillar matrix around which are found starch deposits.

pyrimidine cluster (Szybalski et al. 1966) — a run of pyrimidines (cytosines) in DNA assumed to serve as a recognition signal for → RNA polymerase. i.e., for initiation of → genetic transcription.

pyrimidine dimer — the major UV-induced → DNA lesion, i.e., a cis-syncyclobutane dipyrimidine. P. dimers result in growth delay, mutation and cell death. They are believed to constitute an absolute block to → DNA replication, which then resumes at a site distal to the dimer, generating a single-stranded gap within the DNA duplex. A variety of → DNA repair pathways are able to remove dimers from DNA, such as → photoenzymatic repair.

Q

quadrivalent — a → multivalent consisting of four chromosomes which are either completely (as in autopolyploids) or partially → homologous (as in → translocation heterozygotes).

quadruplex (Blakeslee et al. 1923) — → nulliplex.

quadruplochromosome — a chromosome consisting of eight chromatids held together at the centromere region (→ diplochromosome).

quantal mitosis (Dienstman and Holtzer 1975) — a type of → mitosis yielding one or two daughter cells with synthetic pathways very different from those active in the mother cell (→ cytodifferentiation).

quantasome (Park and Pon 1961) — any of the numerous membrane-bound oblate spheroids (about 10×10—20 nm) with subunit characteristics identical to the quantasome. Such particles are associated with the lamellar portion (the thylakoids) of the → chloroplast and are proteins or protein aggregates. According to recent evidence, the quantosomes do not participate in photoreduction reactions but show Ca^{2+}-dependent ATPase activities. They are probably engaged in a terminal phosphorylating step of photosynthesis.

quantitative risk assessment — a methodology for providing estimates of the risk of mutation (or cancer) to an agent or entity. Used in the regulation of → mutagenic and → carcinogenic chemicals (→ mutagenicity testing).

quantum evolution (Simpson 1944) — a mode of biological → evolution characterized by the rapid change of an organism from an ancestral to a new adaptive character state in a small population system, i.e., invasion of a new adaptive zone. Q.e. is assumed to be the normal mode of evolution in the origin of major groups. In some cases, q.e. grades into → phyletic evolution (→ punctuated evolution).

quantum speciation (Grant 1985) — the budding off of a daughter species from a large, polymorphic, out-crossing, ancestral population via the intermediate stage of a small (isolated or semi-isolated) local race founded, at the periphery of the old species area, by a few individuals (→ bottleneck).

quartet — (1) the four nuclei or cells produced in the course of → meiosis (= tetrad); (2) a cell complex consisting of four cells and produced by two → cleavage divisions which took place perpendicular to one another.

quasibivalent (Östergren and Vigfusson 1953) — → pseudobivalent.

quasidiploid (Hsu 1957) — of cells with a diploid chromosome number but → aneuploid chromosome sets (e.g., $2n + 1 - 1$).

quasi-linkage (Robinson 1971) — the preferential occurrence of nonhomologous markers together in the offspring due to exceptions (→ affinity) to the rule that loci on nonhomologous chromosomes assort independently of each other during → meiosis.

quasi-linkage equilibrium (Kimura 1965) — a state rapidly reached by a population when the selective differences are small and there is loose linkage or independent chromosome assortment. With q.-l.e., the rate of mean fitness is essentially equal to the additive gametic variance.

quasi-normals — → lethal factor.

quasispecies (Eigen 1971) — (1) a set of related but different individuals in nucleic acid clones during the early evolution of life on earth; (2) a heterogeneous mixture of related genomes as observed in most RNA viruses that do not consist of a single genome species of defined sequence.

quiescent cell — any normal cell which can survive for extended periods of time without

→ cell division. Q. cells can function as a reserve cell population which may enter the proliferating pool when needed, or as terminally differentiated cells performing specialized functions. Normally, q. cells are blocked in G1 and cannot enter into S phase before an intracellular inhibitor is removed.

R

rabbit reticulocyte system — a → cell-free translation system prepared from lysed rabbit reticulocytes; it is able faithfully to translate eukaryotic → messenger RNA's from a variety of heterologous sources.

Rabl-orientation — the maintenance to the next mitotic prophase of the anaphase orientation of the chromosomes (centromeres directed to the spindle pole of the preceding mitosis).

race — an intraspecific category, primarily a population or aggregate of populations, with characteristic gene frequencies or features of chromosome structure that distinguish a particular group of individuals from other groups of the same kind within formally recognizable → subspecies or within → species. Race differences are relative, not absolute. The term subspecies is frequently used in the same sense as race. Any race is able to interbreed freely with any other r. of the same species. Whenever different races of a cross-fertilizing species occupy geographically separate territories, they are said to be allopatric; those occupying the same territory are sympatric. Races may become distinct species (→ speciation) by the production of reproductive → isolation (with respect to the other races of the same species) and thus the formation of isolated → gene pools. (1) **Geographic races** are → subspecies occupying a geographic subdivision of the range of a species. (2) **Ecological races** are local races owing their most conspicuous attributes to the selective effect of a specific environment (→ ecotype). (3) **Physiological races** are races differing in respect to features of chromosome structure (cytotypes) or in chromosome number (polyplotypes).

rad — acronym for **r**adiation **a**bsorbed **d**ose; one rad corresponds to energy deposition by ionizing radiation of 100 erg/g (about 1.8×10^{12} ion pairs formed per gram of water or cellular material) (→ rem; Roentgen).

rad-equivalent chemical (Bridges 1974) — (rec); the dose of a chemical (concentration multiplied by time) that produces the same amount of genetic damage as one → rad of ionizing radiation under chronic irradiation conditions. It is used for purposes of quantitating the mutagenic hazards of chemicals.

radial loop chromosome model (Paulson and Laemmli 1977) — a model for DNA packing in eukaryotic chromosomes according to which a fiber radiates outward from the axis of each chromatid. Other models suggest that the 30 nm fiber either coils or loops radially into larger 200—500 nm structures, which may then coil helically to form the metaphase chromosome.

radiation chimera — an individual whose blood-forming tissue contains cells that differ genotypically from the individual's own cells. The chimeric state is due to transfusion of hemopoietic cells from a genetically dissimilar donor into an irradiated recipient. Proliferation of the transplanted cells is possible because the recipient's immune system has been depressed by irradiation.

radiation protective — of agents which protect DNA in vivo against genetic damage by ionizing radiation. Such agents may protect DNA directly by binding to DNA, or indirectly by preventing radiation-induced release of DNA-destructive agents, or by protection against radiation-induced destruction of → DNA repair systems.

radiological mapping (Herrlich and Schweiger 1971) — a technique which uses premature transcription termination as a measure of transcription and protein synthesis after irradiation (the relevant target for any gene is the distance between the → promoter and the promoter-distal end of the gene). (→ UV mapping).

radiomimetic (Dustin 1947) — of chemical agents (typically represented by alkylating agents) which mimic the most important end effects induced by ionizing radiations in living systems. These include → gene and → chromosome mutations, carcinostatic and carcinogenic effects and, in vitro, depolymerized nuclei acids.

random assortment — → assortment.

random fixation (Wright 1931) — the complete loss of one of two → alleles in a population, the other allele reaching a frequency of 100% under certain calculatable circumstances (→ gene frequency; genetic drift). Whether or not r.f. will occur in a population depends on the interrelationship between the number of breeding individuals, the selective value of the allele(s), the mutation pressure, and the gene flow.

random heterozygosity — (H); the probability that a random individual from a population is → heterozygous at a random locus.

random mating — a → mating system (panmixia) in which matings are governed entirely by chance. Departures from r.m. may be classified into four types: (1) Assortative m.; (2) disassortative m.; (3) inbreeding or consanguineous m. (genotypic assortment); (4) outbreeding or cross-breeding (genotypic disassortment).

random monomorphism — (M); the probability that a population has no gene frequency in the range 0.01 to 0.99 at a random locus.

random sample — a random collection from a population, selected in such a way that all items in the population are equally likely to be present in the sample.

rapid-start complex (Mangel and Chamberlin 1974) — → initiation complex.

rate-doubling concentration — the concentration of a chemical that produces as much genetic damage as occurs spontaneously in the same period of time in a particular system used for → mutagenicity testing (→ doubling dose; rem-equivalent chemical).

rate gene (Goldschmidt 1917) — a gene controlling the rate of a certain developmental process.

rate of disintegration — the loss of heterozygosity by random → mating for any generation.

ratio cline — a → cline in the frequency of certain → genotypes (characteristic for systems of → genetic polymorphism showing a gradual change of → genotype frequency).

RBE — acronym for → **r**elative **b**iological **e**ffectiveness.

rDNA — acronym for → ribosomal DNA.

rDNA amplification (Tartoff 1971) — in the ovaries of a wide variety of organisms, the manifold replication of the genes coding for 18S and 28S → ribosomal RNA and an associated spacer DNA in the absence of replication of the remainder of the genome (→ gene amplification).

rDNA compensation (Tartoff 1971) — in *Drosophila*, a mechanism by which the rDNA content of the → chromosome can be increased about twofold. It occurs when the X nucleolus organizer (NOR) is genetically placed opposite no sex chromosome or one deficient for the majority of X heterochromatin (Xh) including the NOR. It occurs in male and female flies, is restricted to somatic cells, and is not heritable. A specific gene locus, the compensatory response locus (cr^+), regulates rDNA c.

rDNA insertion (Glover and Hogness 1977) — in some eukaryotes, any of the DNA sequences which interrupt the coding sequences for 28S rRNA. The insertions vary in size and sequence but occur at the same position.

rDNA magnification (Ritossa 1968) — in *Drosophila* males which inherited subnormal amounts of a → ribosomal DNA (rDNA) and are phenotypically abnormal, a germ line event (distinct from → rDNA compensation) that involves the synthesis of free circular rDNA copies and their integration into the chromosomal rDNA. The magnification of rDNA is a stepwise process in male flies occurring over several generations when a rDNA-deficient X chromosome is maintained alongside a rDNA-deficient Y chromosome.

reaction norm (Woltereck 1909) — = → norm of reaction.

reactivation — → host-cell r.; UV-r.; catalase r.; multiplicity r.; cross r.; photoreactivation.

reading — in genetics, the one-way linear process by which nucleotide sequences are decoded (→ genetic code; genetic transcription; genetic translation).

reading context — = → codon context.

reading frame — one of the three possible ways of reading a nucleotide sequence as a series of triplets (→ genetic code). The r.f. defines which sets of three nucleotides are read as codons and is determined by the → initiator codon (→ open reading frame; frameshift mutation).

reading mistake — the incorrect placement of one or more amino acid residues in a polypeptide chain during → genetic translation (→ mistranslation).

read through (Reznikoff 1969) — → transcriptional read through.

read through protein — any protein synthesized by → transcriptional read through.

reannealing — → DNA renaturation.

recapitulation (Kielmeyer 1793) — a theory stating that during ontogeny the individual passes through stages whose form resembles that of adults of its presumptive ancestors; the developmental history of organisms (ontogeny) recapitulates their evolutionary history (phylogeny).

RecA protein — in *E. coli*, a DNA-dependent ATPase that catalyzes strand exchange between a variety of DNA substrates to form → heteroduplex joints and is able to mediate several steps in general → genetic recombination, including the search for homology, homologous pairing, and strand transfer. Synthesis of RecA protein is regulated by the lexA repressor and is induced along with other lexA products when damage to DNA activates a protease function of RecA protein to cleave this repressor (→ SOS response).

rec assay (Slater et al. 1971; Kada et al. 1972) — a screening method, used in → mutagenicity testing, which is based upon differential killing of wildtype and dark repair-deficient bacterial cells (→ DNA repair). It provides indirect evidence of the potential → mutagenic activity of chemicals.

rec-dependent repair (Harm 1968) — = → recombinational repair.

recessive (Mendel 1865) — of genetically controlled → characters and the → alleles responsible for their production when the phenotypic expression is associated with the → homozygous state, as opposed to → dominant characters and alleles.

The failure of a gene to express phenotypically its presence in the heterozygous genotype is called "recessivity", as opposed to "dominance".

recipient conjugal DNA synthesis — → donor conjugal DNA synthesis.

reciprocal (Mendel 1866) — (1) of a second → cross (A♀ × B♂) which is similar to a first one (B♀ × A♂) but with the sexes of the parents (the sources of male and female gametes) interchanged; (2) of two → chiasmata which involve the same two chromatids (= two-strand double crossing over); (3) of inter- or intragenic recombination in which all marked sites involved segregate 2:2, as opposed to nonreciprocal recombination (→ genetic recombination); (4) of → translocations which involve the interchange of segments between two chromosomes or chromatids.

reclex error-prone repair (Witkin 1975) — = → SOS repair.

recognition site — the nucleotide sequence in duplex DNA to which a → restriction endonuclease binds initially and within which the enzyme specifically cuts the DNA (type II restrictases). Type I enzymes bind to their r.s. and then cleave the DNA at some more or less random position outside their r.s.

recombinagenic (Holliday 1963) — ref. to agents ("recombinagens") or processes increasing → genetic recombination.

recombinant — (1) of an organism containing a different combination of alleles from either of its parents due to → crossing over or independent assortment of different chromosomes at → meiosis; (2) of a structurally rearranged chromosome with a new segmental composition resulting from meiotic crossing over between a displaced segment and its normally located counterpart in certain types of structural heterozygotes (heterokaryotypes);

(3) of a DNA molecule containing a new combination of sequences (→ recombinant DNA).

recombinant DNA — a novel DNA sequence (= hybrid DNA; chimeric DNA) formed by in vitro combination of two nonhomologous DNA molecules (→ recombinant DNA technology).

recombinant DNA technology — a set of techniques (= genetic engineering) that permits the formation of new combinations of heritable material (recombinant DNA) by the insertion of DNA into → cloning vectors so as to allow their amplification and then their incorporation into a host organism. The recombination process consists of several steps: (1) Isolation (or synthesis) of DNA fragments to be cloned; (2) ligation to a cloning vector in vitro to produce recombinant DNA molecules; (3) introduction of recombinant DNA into the host for replication (using the replication origin of the vector); (4) selection of recipient cells that have acquired the recombinant DNA.

R.DNA t. allows the selection of virtually any desired single copy genomic sequence in cloned form, provided an appropriate → probe is available. There are two principal ways of isolating DNA sequences: (1) → Shotgun cloning of fragments of the total genomic DNA, followed by colony hybridization with labeled RNA or cDNA; (2) preliminary enrichment of genes (e.g., separate purification of DNA-DNA chains by molecular hybridization with RNA or cDNA, followed by annealing of the respective chains) and subsequent bacterial cloning.

R.DNAt. has many applications, including the study of gene structure, function, and efficient expression of their products, chromosome mapping, gene amplification, and production of novel genes.

recombinant frequency — (RF); the proportion or percentage of recombinant cells or individuals (= recombinant proportion; r. fraction).

recombinant joint — the point at which two recombining molecules of duplex DNA are connected, i.e., the edge of the → heteroduplex region.

recombinant plasmid — a chimeric → plasmid containing foreign DNA sequences inserted by use of → recombinant DNA technology. R. plasmids are used as → cloning vectors; their retention is accomplished most often by selection for an antibiotic resistance marker that is carried on the plasmid.

recombinant RNA — any RNA molecule joined in vitro by RNA ligase and consisting of RNA chains from different sources.

recombinant RNA technology (Miele et al. 1983) — a technique which involves the construction of recombinant RNA and allows the production of large quantities of RNA by conferring on any RNA molecule the ability to be replicated in test tube reactions (linkage of the foreign template to $Q\beta$-like RNA sequences, so that the package is accepted for replication).

recombinase (Kozinski et al. 1967) — any of the enzymes that recognize specific DNA sequences, introduce single-strand breaks, and are involved in general → genetic recombination. Enzymes sealing single-strand breaks after recombination are called repairases.

recombination (Bridges and Morgan 1923) — any process giving rise to → recombinants.

recombinational enhancer (Kamp et al. 1978; Huber et al. 1985) — in prokaryotic site-specific inversion systems, a cis-acting sequence which efficiently stimulates recombination in vivo and in vitro. Its action is largely independent of its position and orientation with respect to the crossover sites. The r. enhancers of different site-specific inversion systems which regulate the alternate expression of genes by mediating the inversion of DNA segments are interchangeable.

recombinational switching — a mechanism (→ DNA rearrangement) which causes variable expression of particular genes in uniform environments by creating or disrupting, via inversion, a transcriptional linkage between a particular gene and its → promoter. R.s. is due to site-specific recombination in short inverted repeat sequences flanking the section in question. The recombinase catalyzing the recombination is encoded by DNA adjacent to one of the inverted repeats.

recombination bar (Solari 1980) — → recombination nodule.

recombination class (Bridges and Morgan 1923) — any of the types of recombinant gametes produced by crossing over. If the parents used for crossing differ in respect to two linked pairs of alleles (AB/AB and ab/ab), four classes of gametes are produced: AB, ab, Ab, and aB. Ab and aB represent the "recombination classes" resulting from crossing over between the loci A and B.

recombination-competent — of cells capable of undergoing → genetic recombination.

recombination-defective — of mutants which impair or abolish → genetic recombination; the wild-type alleles are necessary for normal frequencies or distributions of exchange events, or both.

recombination fraction — a measure of the frequency (= r. frequency) of exchange (crossing over) occurring between two specific loci. The r.fractions of linked loci range from just over zero to just less than one-half (i.e., the value expected for unlinked loci). The r.f. is governed by the rate of nicking and the time in which unrepaired nicks are allowed to match and to crossover. Because of multiple → crossing over and positive → interference, recombination is not additive over contiguous segments.

recombination gene (Smith 1966) — any gene (allele) that controls the frequency of → genetic recombination (allelic and/or nonallelic).

recombination hotspot — any DNA sequence preferentially involved in → genetic recombination, or stimulating recombination in adjoining DNA segments.

recombination index (Darlington 1939) — the sum of the gametic (haploid) chromosome number (i.e., the number of bivalents) and the average number of chiasmata per nucleus; it is used as an estimate of the meiotic potential for → genetic recombination via crossing over in eukaryotes. A high r.i. promotes → flexibility, a low r.i. promotes → fitness.

recombination intermediate — → genetic recombination.

recombination nodule (Carpenter 1975) — (RN); in eukaryotes, any of the strongly staining, spherical, or elongated (r. bar) structures (about 100 nm in largest dimension) associated with the central space of the → synaptonemal complexes in meiocytes. Number and location of RN's are related to the expected crossover sites; → DNA repair synthesis occurs in these RN's during pachytene (→ zygotene DNA).

Two classes of RN's are defined according to their chronology of appearance in meiotic prophase: (1) Early (ellipsoidal) RN's which are possibly involved in matching homologous sequences during → chromosome pairing and may mediate → gene conversion; (2) late (spherical) RN's whose position correlates closely with → chiasma distribution and which may be directly involved in reciprocal recombination.

recombination polarity (Coen et al. 1971) — a deviation from the 1:1 ratio of the two types of reciprocal recombinants (→ transmission polarity) observed with respect to (yeast) mitochondrial genes.

recombination-proficient (Clark and Margulies 1965) — of genotypes capable of normal (complete) recombination (symbol rec^+) as opposed to recombination-deficient (rec^-) mutants.

recombination repair (Howard-Flanders and Boyce 1966) — a repair mechanism which involves strand exchange between chromosomes and is implicated in double-strand break repair and in cross-link repair. R.r. is a mode of filling a gap in one strand of duplex DNA by retrieving a homologous single strand from another duplex. R.r. is an error-prone repair pathway. Mutants deficient in r.r. are usually depressed for → genetic recombination. Two classes of → DNA repair involve recombination functions: (1) → Postreplication repair; (2) → double-strand-break repair.

recombination system (Carson 1957) — the sum total of all factors mediating and controlling the processes of → genetic recombination; the r.s. regulates the generation of genetic variability by restricting the types and frequencies of recombinants produced. In eukaryotes with sexual reproduction, recombination is a function of three main processes: → Crossing over, random distribution of chromosomes and chromatids during meiosis, and random combination of gametes during fertilization. A high level of meiotic recombination is characteristic of organisms with a high frequency of crossing over, a high number of chromosomes and a high level of outbreeding, as opposed to organisms with a low frequency of crossing over, with a low number of chromosomes and with inbreeding.

According to the breeding and chromosome behavior of the species under observation, the r.s. of eukaryotes may be either open, relatively restricted, or closed (Carson 1957):

(1) A closed r.s. is examplified by asexual reproduction as, e.g., obligate → apomixis, where the source of new variation is → mutation and → autosegregation. (2) A restricted r.s. imposes strong but not insurmountable barriers to the realization of new gene combinations (recombinants). (3) An open r.s. lacks barriers and thus is relatively unrestricted as to the number and range of new genotypes produced; it normally permits a free flow of the generated variability.

recombinator (Holliday 1968) — any of the special, cis-acting sites that stimulate homologous → genetic recombination in their neighborhood. The crossover **h**otspot **i**nstigator Chi (5′GCTGGTGG) or its complement is a r. in the *E. coli* wild-type Rec BCD (Rec) pathway of recombination.

recombinogenic — of agents (recombinogens) able to induce → genetic recombination. Probably, all → mutagenic agents are also effective recombinogens.

recon (Benzer 1958) — the smallest unit (corresponding to a single DNA nucleotide) which is exchangeable but not divisible by intragenic recombination (→ genetic recombination) between homologous linkage structures (chromosomes), and which may be detected by the techniques of genetic fine structure analysis of the linear array of → mutational sites within the → genetic material (→ cistron; muton; polaron hybrid DNA hypothesis).

recovery — the ability of an organism to survive, in spite of the introduction of lesions (usually in DNA) by some physical or chemical agents, as a result of the removal (→ DNA repair) or circumvention of such lesions (→ reinitiation recovery).

rectangular evolution (Stanley 1975) — = → punctuated evolution.

rectification — → gene rectification.

REC unit — acronym for **r**em-**e**quivalent-**c**hemical unit, i.e., that dose (product of concentration multiplied by time) which produces an amount of genetic damage equal to that produced by one → rem of chronic irradiation.

recurrence risk — in human genetics, the risk that a woman who has had a child with a certain genetically controlled trait will have another at a later pregnancy.

R. risks will vary considerably. In autosomal dominant disorders, the r.r. after an affected child has been borne is 50% but since not all carriers of a given dominant gene will manifest the disease, the actual risk will often be less than 50%. In recessive disorders, the r.r. is 25%. In X-linked recessive inheritance, there is a 50% risk for future sons to be affected if the mother is a heterozygous carrier, and a 50% risk for future daughters to be carriers.

recurrent backcrossing — repetitive sexual crossing of hybrids to one parent in order to eliminate all but the desired characteristics of the donor parent.

recurrent parent — in → backcrosses, the parent which is crossed with the first and the subsequent generations, as opposed to the nonrecurrent parent.

recurrent selection — a cyclic scheme of plant selection by which frequencies of favorable genes are increased in plant populations. All forms of r.s. have two main steps in common: (1) Selection of genotypes from a population; (2) intercrossing the selected genotypes to form an improved population.

redifferentiation — the differentiation (after → dedifferentiation) of an already determined cell into a new cell type.

reduction (Weismann 1887) — (1) **meiotic r.:** the halving of the somatic (zygotic) chromosome number at → meiosis and, by extension, its genetical concomitant, → genetic segregation (Darlington and Mather 1949); (2) **somatic (or mitotic) r.**: reductional distribution of chromosomes in tissues other than meiotic ones, which may take place spontaneously or may be induced experimentally (→ reductional grouping).

reductional grouping (Huskins 1947) — the separation of mitotic chromosomes, prior to anaphase, into two (or more) groups containing the same or different fractions of the somatic chromosome number. Such groups may enter anaphase separately and two groups may give rise to four nuclei at the end of mitosis, each with a number of chromosomes smaller than the original (somatic or mitotic → reduction, as opposed to meiotic reduction of chromosome number). R.g. may take place spontaneously in rare cases (→ genome segregation) and can be induced experimentally.

reductional separation — separation of homologous parts of nonsister chromatids at anaphase I (prereductional s.) or anaphase II (post-reductional s.) of → meiosis, as opposed to equational separation of sister parts of chromatids (Darlington and Mather 1949).

reduction division (Weismann 1887) — = → meiosis.

redundant DNA — = → repetitious DNA (→ gene redundancy).

refractory (Hill and Holland 1967) — of bacterial mutants adsorbing → colicin but resistent to its lethal effect.

regulated gene — → constitutive gene.

regulation — → genetic regulation.

regulation reversal (Coats and Nester 1967) — end-product activation in mutants which arose from genotypes with opposite regulatory properties, i.e., show end-product inhibition.

regulatory cell — any cell that controls the fate (→ cell fate) of other cells. Any mutation which disrupts the determination or differentiation of a r.c. may lead to immediate and discrete alterations in → cell lineage.

regulatory codon (Anderson 1961) — any codon (in mRNA) which is recognized by → transfer RNA species, being present in rate limiting amounts and thus exerting → translational control.

regulatory DNA sequence — any noncoding DNA region (= regulatory element) that determines if, when, and at what level a particular gene is expressed. The r.DNA s. specifically interacts with regulatory or other proteins. Many act in cis and are believed to affect DNA topology, producing local conformations that selectively allow or restrict access of → RNA polymerase to the DNA template or that facilitate selective opening of the double helix at the site of transcriptional initiation.

Two categories of cis-acting r.D. sequences in eukaryotes may be distinguished: (1) Elements near to the initiation site, termed → promoters; (2) more remotely located r. sequences, such as → transcriptional enhancers, → terminators, → operators, → catabolite activator-protein (CAP) binding sites.

The extent to which a particular set of elements affects DNA topology may be determined by several extrinsic parameters including the extent of methylation of the DNA (→ DNA methylation), the ionic environment, the chromosomal proteins, and certain locus-specific trans-acting protein factors.

regulatory gene (Jacob and Monod 1961) — in both prokaryotes and eukaryotes, any gene whose product regulates → gene expression via transcription initiation or transcription termination. The products of such genes are → transcriptional activators or → repressors; others direct the synthesis of RNA which can directly control gene expression (→ antisense RNA).

In eukaryotes, r. genes also control the timing and tissue specificity of the expression of other genes; some control only the → structural gene that they are linked to (cis-regulation); others appear to produce diffusible effector molecules that result in trans regulation.

R. genes may be more important in the evolution of new adaptations than are changes in structural genes.

regulatory RNA — any of the small RNA molecules that bind to → messenger RNA or directly to DNA, or both and regulate gene activity (→ antisense RNA).

regulon (Maas and Clark 1964) — (1) in bacterial → gene expression, a global regulatory system (consisting of several → operons) which involves the interplay of pleiotropic regulatory domains. Individual regulons are defined by the existence of a regulatory gene that responds to a physiological imbalance or stress (→ SOS response; stress response) by coordinating the expression of a variety of genetically unlinked target loci. The mechanism of regulation involves proteins operating as repressors, inducers, or alternative → sigma factors and DNA promoter enhancing elements. These often act in concert with regulatory nucleotides (such as pp Gpp). The → stringent response was the earliest of such pleiotropic regulatory systems. (2) In eukaryotes, a group of unlinked genes regulated by one or more → regulatory genes that constitute a cascade regulatory network (→ stimulon).

reinitiation recovery (Bridges 1972; Cupido 1983) — in *E. coli* and bacteriophage T4, a type of → DNA repair (= r. repair; replication repair), besides → excision repair and → postreplication repair, which gives rise to the resumption of stalled → DNA replication by initiating new replication points. R.r. in *E. coli* is inducible and dependent on both the uvr and rec loci. Phage-encoded r.r. is presumed to consist of four steps (Cupido 1983): (1) Induction of → primer RNA opposite a lesion in the template; (2) elongation of primers by → DNA polymerase; (3) ligation of daughterstrand fragments, without removal of primer RNA; (4) replication of DNA carrying RNA sequences resulting in homogeneous DNA strands.

reinitiation site (Newton and Zipser 1967) — = → reinitiator.

reinitiator — a site for the reinitiation of polypeptide synthesis (→ genetic translation) which operates to restore translational activites impaired by polar → nonsense mutations, i.e., reverses the effects of terminator codons (UAG, UAA, UGA in mRNA). The combination of a r. and a terminator codon creates a new cistron-like boundary at the site of the nonsense codon. The further the nonsense codon is from the r., the less efficient is protein synthesis reinitiation.

reiteration frequency mapping (Zasloff and Sautos 1980) — a rapid quantitative approach to the localization of → repetitious DNA sequences used for identification of DNA sequences within a cloned (→ molecular cloning) genomic fragment that are represented in RNA populations.

reiteration mutant — a virus mutant containing a genome with tandem repetitions of small segments of the wild-type genome.

rejoining — according to the breakage-reunion hypothesis, the → reunion and restitution of broken ends of chromosomes or chromatids leading to the production of → chromosome mutations (reunion), or a restoration of the prebreakage structure (restitution).

The distance within which two separately induced breaks can rejoin to yield a rearrangement is called the r. distance. It varies with the type of rearrangement, with the geometrical arrangement of the chromosomes, and with physiological conditions such as degree of hydration.

relatedness — similarity in genetic structure among groups of individuals. R. plays a key role in the evolution of interactions that affect the → fitness of genetically related individuals.

relationship pair — in human genetics, sibs, twins, parent and offspring, etc.

relational balance (Mather 1953) — → coadaptation.

relational coiling (Darlington 1935) — → chromosome coiling.

relative biological effectiveness — (RBE); the ratio of the effect of one quality of ionizing radiation to another (usually high → LET radiation compared to low LET radiation). The RBE is often expressed as the ratio of doses required to produce the same effect.

relative sexuality (Hartmann 1923) — the ability of a gamete to act either as a male or female when mated to different gametes.

relaxed response (Borek et al. 1955) — → stringent response.

relaxed type plasmid — any of a class of → plasmids with the ability to replicate in the absence of de novo protein synthesis, as opposed to stringent type plasmids.

release factor (Ganoza 1966; Capecchi 1967) — (RF); any of the protein factors (= termination factors) that respond to the → terminator codons (UAA, UAG, and UGA) in → messenger RNA to cause release of the completed polypeptide chain and the → ribosomes from mRNA. RF-dependent hydrolysis of the → peptidyl-tRNA ester bond requires location of the peptidyl-tRNA in the ribosomal P-site, NH_4^+ or K^+ ions, and is inhibited by several antibiotics that also inhibit polypeptide bond formation.

In prokaryotes, two RF's recognize specific terminator codons: (1) RF 1 catalyzes termination at UAA and UAG codons; (2) RF 2 catalyzes termination at UGA and UAA codons.

R. factors interact with specific terminator codons in three ways (*E. coli*): (1) RF binding to ribosomes; (2) formation of RF-terminator codon-70 S ribosome intermediates; (3) release of the nascent polypeptide chain from transfer RNA at the terminator site. The r.f. RF 3 stimulates the release reaction by assisting RF 1 and RF 2 to interact with the ribosome in response to its particular terminator signals. The substrate for the termination reaction is thus a completed polypeptide chain in ester linkage with tRNA and complete with ribosome and messenger RNA. GTP is inhibitory for RF 3-stimulated release.

Codon-specific release of polypeptides in **eukaryotes** (mammals) appears to be fundamentally similar to that in bacteria. One major difference from prokaryotic termination is the requirement for GTP. In mammals, a single protein seems to embody all the requisite functions. It stimulates release with all terminator codons and is much larger than the bacterial r. factors.

relic coiling (Darlington 1935) — → chromosome coiling.

rem — acronym for **r**oentgen-**e**quivalent-for-**m**an, i.e., the dose of any type of ionizing radiation that will produce the same effect as 1 → rad of reference (usually 250 kvp) X-rays. Dose in rem = quality factor × dose in rads (quality factor is related to → LET and sometimes partly to experimental → RBE values).

rem-equivalent-chemical — (REC); the dose (product of concentration × time) of a chemical agent which produces an amount of genetic damage equal to that produced by 1 → rem of chronic irradiation in the same mutagenicity test system.

renaturation — → DNA renaturation.

Renner complex — in → complex heterozygous species (*Oenothera*), a group of chromosomes (and the genes which they contain) which are distributed as one unit from generation to generation.

Renner effect (Darlington 1932) — competition (= megaspore competition) among the four genetically different spores formed by one meiosis in regard to which shall form the embryo (Darlington and Mather 1949).

reovirus — any of a large group of animal, plant, fungal, and bacterial viruses that contain segmented double-stranded RNA genomes (→ retrovirus).

repair (Muller 1954) — → DNA repair.

repairase (Kozinski et al. 1967) — any enzyme repairing or sealing DNA single-strand breaks (= repair polymerase; antimutator polymerase). Such enzymes are usually part of an enzymatic proofreading mechanism that functions during → DNA replication to compensate for spontaneous and induced mutations and inserts nucleotides into damaged DNA.

repair-deficient — of genotypes (mutants) deficient in one or the other of the genetically determined → DNA repair pathways, as opposed to repair-proficient genotypes.

repair DNA synthesis — during → DNA repair, the nonconservative synthesis of short single-strand fragments in place of damaged and excised regions of DNA (= repair replication). The amount of r.DNA s. depends on the amount and type of → DNA damage and the manner in which the damage is repaired (e.g., patch size). R. synthesis is independent of →

DNA replication (→ unscheduled DNA synthesis) and can be catalyzed by several → DNA polymerases. Error-prone r. DNA s. requires a special type of DNA polymerase (repairase) since the "normal" polymerases are highly accurate (→ SOS repair).

repair inducer — any of the signals inducing → DNA repair. The inducing signals may be unrepaired → DNA lesions and discontinuities in the DNA duplex which may also be generated during abortive repair (→ repair inhibitor).

repair inhibitor — any agent that (if present before, during, or after mutagen treatment) at a concentration which does not affect the viability of untreated cells, enhances the degree of inactivation after mutagen exposure. R. inhibitors may be inhibitors of → DNA repair enzymes or may interfere with their access to sites of → DNA damage.

repair-mediated mutagenesis — indirect → mutagenesis via error-prone → DNA repair processes. Generally, r.-m.m. is evident when a → mutagenic agent shows a reduced or increased effectivity in certain repair-defective mutants.

repair recombination (Broker and Doerman 1975) — the increment in → genetic recombination (= repair-induced recombination) caused by any agents or conditions that damage DNA. Several types of r.r. may be distinguished: (1) Parental-parental r.r. observed in the absence of replication when single-strand regions produced by nucleases are allowed to accumulate; (2) replication-limited r.r. resulting from a reduced rate of DNA replication caused by suboptimal concentrations of deoxynucleoside triphosphate (dNTP) substrates and/or replication enzymes; (3) replication-dependent r.r. occurring at gaps or broken DNA strands left by defective replication in the vicinity of radiation or chemical damage to the DNA.

repair replication (Cleaver 1975) — = → repair DNA synthesis.

reparatose (Hirsch-Kaufmann and Schweiger 1981) — any → genetic disease due to defects in → DNA repair and characterized cytologically by an abnormally high rate of chromosomal instability (→ chromosome breakage syndrome). All reparatoses are accompanied by radiation sensitivity and a high incidence of cancer.

rep-DNA — acronym for → repetitious DNA.

repeat (Bridges 1935) — a type of → duplication where the duplicated segments lie juxtaposed (tandem), inverted or not, in the relation to each other and the rest of the chromosome. Repeats may occur at different structural levels and may involve groups of genes, single genes, or even only parts of a gene.

repeated gene — any of the members of a → multigene family.

repeated gene family — → multigene family.

repeat family — a set of mutually homologous → repetitious DNA sequence elements, as opposed to → single copy sequences.

repeating unit — in a tandem cluster of repeated DNA sequences, the length of the sequence that is repeated (appears circular on a → restriction map). R. units are → intercepts of a → chromatin strand which may include a → matrix unit plus the adjacent (preceding or subsequent) apparent → spacer unit. In the case of the DNA coding for → ribosomal RNA, the r.u. usually means the rDNA r.u., i.e., a given pre-rRNA matrix unit plus one of the two adjacent apparent spacer units.

repetition frequency — the number of copies of a given DNA sequence present in the haploid eukaryotic genome; equals 1 for nonrepetitive DNA, ⩾2 for → repetitious DNA.

repetitious DNA (Britten and Kohne 1968) — (rep-DNA); in the nuclear genomes of all multicellular eukaryotes, any of the DNA sequences occurring repeatedly in the chromosomal DNA and characterized by (1) the number of bp per sequence; (2) the specific order of bp per sequence, and (3) the number of copies of a sequence per genome. Most rep-DNA sequences are interspersed throughout the genome; some repeats are clustered in specific chromosome regions. Rep-DNA sequences comprise 20—50% of most animal genomes and more than 50% of many plant genomes. They include many structural genes constituting → multigene families and may be present either in tandem arrays (as in → satellite DNA) or interspersed within other sequences. Interspersed repeats are frequently → transposable genetic elements.

Rep-DNA sequences fall into four commonly occurring classes: (1) The repeated structural genes; (2) the interspersed middle repetitive sequences; (3) the tandemly repeated highly repetitive sequences; (4) the inverted repeat or foldback sequences.

The distribution of rep-DNA families over the eukaryotic genome is highly diverse. Some families are confined to heterochromatic regions, others occur in complex interspersion patterns with each other and with single copy DNA. **Moderately repetitious DNA** exists in two different patterns: (1) Short-period interspersion (mammals, most vertebrates, and sea urchins) in which 300 bp-long sequences are interspersed with unique sequence DNA (1000—3000 bp long); (2) long-period interspersion in which the repetitious sequences are longer than 1.5 kb and show less divergence than the short DNA sequences. The two interspersion patterns are called SINE's and LINE's, respectively; the former include the primate and rodent → Alu-like sequences, the latter proviral type and L1 sequences.

Members of a diverse array of repeated sequence families often show a high degree of homogeneity, yet differ substantially between related species. This → concerted evolution may be accounted for by some process of → molecular drive.

Highly repetitious DNA sequences can be separated into two general classes based on their complexity: (1) Simple sequence rep-DNA composed of short (7 to 20 bp) segments extensively repeated, usually with some divergence, in a tandem fashion that often gives rise to longer periodicities; (2) a longer more complex rep-DNA.

In many instances highly repetitious DNA can be isolated as a satellite to total genomic DNA by virtue of differences in buoyant density on gradients formed by cesium salts.

repetitive DNA (Britten and Kohne 1968) — = → repetitious DNA.

repetitive DNA sequence family — in eukaryotes, a set of homologous sequences of → repetitious DNA which reacts with a given cloned repetitive sequence probe.

repetitive extragenic palindromic DNA sequence (Stern et al. 1984) — (REP sequence); any of a family of repetitive palindromic DNA sequences in prokaryotes composed of hundreds of copies distributed throughout the genome. REP sequences are highly conserved and located outside structural genes and often occur in clusters (REP elements); they are assumed to be involved in the regulation of intraoperon → gene expression and can (differentially) stabilize upstream mRNA by blocking the processing action of 3′ → 5′ exonuclease. → DNA gyrase is one of the REP-binding proteins, i.e., REP's are possibly sites of action for the maintenance of DNA supercoiling and higher-order structure of bacterial chromosomes.

replacement site — in a gene, any of those sites at which mutations alter the encoded amino acid sequence. The base changes (e.g., most first- and second-position changes within a codon) are called replacement substitutions (→ silent substitution mutation).

replacement vector — a bacteriophage → cloning vector in which a portion of the phage genome is replaced by the foreign DNA fragment to be cloned.

replica plating — a means of screening populations of cells for a desired rare genetic trait in both prokaryotic and eukaryotic somatic cell genetic systems, i.e., transfer of cells from colonies on one agar plate to appropriate positions of other plates (e.g., by use of a velvet stamp).

replicase (Spiegelman and Hayashi 1963) — RNA-dependent → RNA polymerase (= RNA synthetase) which is able to initiate copying of virus RNA without an added primer. The genomic RNA of the viruses has a region (at the 3′ end) which exhibits → transfer RNA-like functions, including → aminoacylation. These regions are probably involved in replication by r. rather than in translational functions.

replicase particle (Morgan and Zweerink 1975) — any of the subviral particles of reovirus that catalyze the synthesis of double-stranded RNA in infected cells.

replicating instability (Auerbach 1967) — a specific type of premutational lesion capable of being replicated as such and reverting at an appreciable rate to the normal state as well as changing into a stable mutation.

replication — the duplication of genomic or plasmid DNA (→ DNA replication) or RNA (→ replicase) as part of the reproductive cycle of a cell or virus.

replication bubble — any replicated DNA region consisting of two daughter strands connected at two replication forks (→ DNA replication).

replication cluster (Van't Hoff and Bjerkness 1977) — in eukaryotes, a functional replication unit consisting of member → replicons which replicate their allotted DNA in unison.

replication-dependent recombination repair — in bacteriophage T4, → DNA repair (at high multiplicities of infection) by recombination of undamaged parts of phage DNA which results in an intact chromosome (= multiplicity reactivation).

replication error — any error due to damage to the DNA replication machinery after treatment with mutagens or spontaneously that might lead to base-change mutations (→ recombination error).

replication eye — a region in which DNA has been replicated within a longer, unreplicated region.

replication fidelity — → DNA replication.

replication fork — the site at which the two strands of the parental DNA duplex are separated to allow replication (→ DNA replication) of each strand. The r.f. is asymmetric, consisting of leading side where DNA synthesis is continuous, and a lagging side where discontinuous replication occurs (→ Okazaki fragment; primer RNA) and is part of either a looped structure generated by a de novo start or a → rolling circle generated by a covalent start (→ replisome).

replication form — = → replicative form.

replication granule (Nakayasu and Berezney 1989) — in eukaryotic cell nuclei, any of the structural sites consisting of replication clusters; DNA replication is sequestered within these sites which are distributed throughout the nuclear interior (nuclear matrix) and along the periphery. Numerous tandemly arranged → replicons are coordinately synthesizing DNA in the r.g. Various replicational enzymes are components of the r.g.

replication intermediate — = → Okazaki fragment.

replication mapping (Strelzoff and Ryan 1962) — (1) a method for temporal mapping by sequential mutagenesis during chromosome replication in bacteria ; the r. map is based on the doubling of specific marker mutation frequencies observed when successive aliquots of synchronous bacterial cultures are treated with a fixed dose of a nonspecific mutagen for a specified time. The distance on the map depends on the time necessary for replication of the DNA lying between the markers. This is proportional to the physical distance when the rate of replication, in terms of nucleotides added per second at a given growing point (→ replication fork), is constant. The resolving power is dependent on the degree of induced synchrony of the culture, the rate of DNA synthesis, and the duration of DNA synthesis (→ transcriptional mapping). (2) A procedure employing 5-bromodeoxyuridine (a base analog of thymidine incorporated during → DNA replication) and using the different cytological properties of substituted DNA for localization of DNA regions replicating during early, middle, and late S phase of the cell cycle.

replication origin — (ori); a unique DNA sequence of a → replicon at which → DNA replication is initiated and proceeds bidirectionally or unidirectionally. The ori sequence contains the sites where the first separation of the complementary strands occurs, a → primer RNA is synthesized, and where the switch from primer RNA to DNA synthesis takes place. Chromosomal DNA segments containing an ori sequence are able to provide autonomous replication to a variety of constructed replicons (minichromosomes or ori plasmids). The initiation of DNA replication at the origin is governed either by factors that recognize the ori sequence and direct the start of a chain de novo or by covalent extension of a nick introduced at the r.o.

replication protein — any protein (e.g., priming proteins and DNA polymerase III holoenzyme) that advances the DNA → replication fork (= elongation protein).

replication repair (Symonds et al. 1973) — = → replication-dependent recombination repair and → reinitiation recovery.

replication terminus (Louarn et al. 1977) — the site at which → DNA replication terminates; it is situated approximately opposite the replication origin in bacterial chromosomes. The clockwise replication fork is the first to arrive at the r.t. and becomes arrested; the anticlockwise fork arrives somewhat later at the r.t. and probably fuses with the clockwise fork.

replication unit (Huberman and Riggs 1968) — in eukaryotes, any of the multiple, tandemly arranged stretches of chromosomal DNA replicated by two adjacent growing points which share a common → replication origin and move in opposite directions from the origin (bidirectional replication). There are 10^4 or more such units in the genome of higher eukaryotes (most varying in length from 15 to 100 μm or 50 to 330 kb pairs). Initiation and fork progression determine the overall rate of → DNA replication on a chromosome.

replicative bypass repair (Higgins et al. 1976; Fujiwara and Tatsumi 1976) — → bypass repair.

replicative form (Sinsheimer et al. 1962) — (RF); the intracellular double-stranded circular form of the DNA of single-stranded viruses (bacteriophages) which is active in replication. DNA replication of these viruses occurs in three steps: (1) Conversion of the parental viral DNA into the double-stranded covalently closed RF; (2) replication of the RF DNA to form a pool of progeny RF molecules; (3) asymmetric replication of single-stranded viral DNA on the RF template with discontinuity in at least one strand.

replicative intermediate (Erikson et al. 1964) — (1) in the replication of single-stranded DNA and RNA viruses, a structure which consists of several nascent parental nucleic acid molecules associated with a single complementary template strand in a partially double-stranded conformation. R. intermediates are distinguished from completely double-stranded complexes called → replicative forms which may represent either a very early step in replication or one of its by-products. (2) → Okazaki fragments produced during replication of double-stranded → deoxyribonucleic acid.

replicative segregation — in cells with a mixture of wild-type and mutant → mitochondrial or → chloroplast DNA's (heteroplasmic cells), a shift of organelle DNA genotype during cellular replication due to random partitioning of organelles and organelle DNA into daughter cells during → cytokinesis. Some lineages drift toward pure wild-type or mutant organelle DNA (homoplasmic cells), others remain heteroplasmic.

replicative synthesis — the semi-conservative replication of → deoxyribonucleic acid as distinguished from → repair replication or repair synthesis.

replicative transposition (Shapiro 1979) — the main mode of → transposition of → transposable genetic elements in prokaryotes which involves movement of a copy of the element in question.

replicator (Jacob and Monod 1963) — a DNA sequence containing a → replication origin. The r. is able to promote replication of all DNA sequences attached to it (→ ARS element).

replicon (Jacob and Brenner 1963) — a segment of DNA capable of independent replication. In eukaryotic chromosomes, replicons represent a series of tandemly repeated subunits along the continuous DNA molecule. Adjacent replicons are organized into replicon clusters (~25 replicons each) that replicate as a unit at particular times in S phase, i.e., r. cluster synthesis is temporally and spatially regulated along the chromosomal DNA (→ DNA replication; mesosome; replication granule; replication origin).

repliconation (Clark 1967) — the conversion of transferred DNA into a plasmid of autonomous maintenance (→ F plasmid).

replicon fusion — the joining (→ cointegration) of two complete replicons as mediated by a → transposon.

replicon misfiring (Varshavsky 1981) — initiation of an extra (illegitimate) round of → DNA replication within any domain of DNA; it may result in extra genes (→ gene amplification) or chromatin fragments and may be increased by intra- or extracellular agents called firones.

replisome (Dressler 1975) — the hypothetical multiprotein structure that assembles at the (prokaryotic) → replication fork to undertake synthesis of DNA; it may consist of a → primosome, → DNA polymerase, → DNA helicase, as well as → DNA topoisomerase (→ replitase).

replitase (Reddy and Pardee 1980) — in eukaryotes, a multienzyme complex involved in → DNA replication; it consists of enzymes synthesizing DNA precursors juxtaposed with the replication apparatus comprising DNA polymerase, other enzymes, and structural proteins. In intact cells, the r. could be the functional component of a DNA-membrane matrix in the nucleus (→ replisome).

reporter gene — a promoterless gene whose expression is used as an indication of transcription activity of another gene after its fusion with the gene of interest (a → gene fusion marker). The expression of the r.g. should be easily detectable at many positions in a target genome, so that it comes under the control of a random selection of chromosomal → promoters. After transfer of the construct into (eukaryotic)cells, the effect of the particular sequence on the level of gene expression can be measured indirectly via the expression of the r.g.

repressible (Vogel 1957) — of enzymes (proteins) normally present in the cell (→ constitutive; inducible) whose production ceases when the intracellular concentration of certain metabolites (usually the end-products of the enzyme system) increases (→ operon). R. enzyme systems are usually anabolic.

repression (Monod and Cohen-Bazire 1953; Vogel 1957) — the alteration in the expression of a gene with the net result that specific enzyme production may fail (→ operon), as opposed to "derepression", when the level of a particular → repressible enzyme increases after removal of a specific metabolite.

R. may be due to → transcriptional control or → translational control superimposed on the former and represents a means of adaptation of enzyme production to the environment (→ induction). Coordinate r. (Ames and Garry 1959) means the simultaneous r. of many or all enzymes of a sequence controlled by scattered genes (→ regulon).

repressor (Pardee et al. 1959) — a site-specific DNA regulatory protein which exerts direct negative control over → gene expression at specific → promoters. It is the product of a → regulatory gene. The majority of prokaryotic repressors is thought to act by binding to → operator sites that overlap the RNA polymerase binding sites (promoters). When a r. binds to its operator, it sterically excludes the binding of polymerase to the promoter, i.e., repressors are competitive inhibitors of RNA polymerase binding. The efficacy of a r. as an inhibitor depends on both its intrinsic affinity for a binding site and the position of the binding site with respect to the promoter it controls.

All r. molecules can exist in both an active and an inactive form, depending on whether or not they are combined with their appropriate inducers (corepressors). Attachment of an inducer inactivates the r., binding of a corepressor changes the inactive r. into an active r.

The functional sites of a r. are (Müller-Hill 1975); (1) The operator binding site; (2) the inducer binding site; (3) the transmitter region which links the inducer binding site to the operator binding site; (4) the site involved in aggregation of r. subunits into a functional polymer (→ negative complementation).

Genetic regulation by repressors is only one of several mechanisms for gene control. Specific positive control elements are known to exist; the products of such control genes are required to turn the gene or operon on rather than to turn it off. If the control gene is deleted, the structural genes do not function.

The active forms of the r. are rotationally symmetric dimers, and their DNA binding sites are rotationally symmetric → palindromes.

reproduction — the production (self-propagation) of an organism, a cell, or cell organelle by one like itself (→ replication). (1) **Sexual reproduction:** The regular alternation (in the life cycle of → haplontic, → diplontic, and → diplohaplontic organisms) of → meiosis and → fertilization (karyogamy) which provides for the production of offspring. The main biological significance of sexual r. lies in the fact that it achieves → genetic recombination. In diploid and polyploid organisms, a large amount of → genetic variability is stored within a framework of phenotypic uniformity. Sexual r. is an effective means of pooling the → genetic information contained in the individuals of a breeding population. (2) **Parasexual reproduction** (Pontecorvo 1954): All processes of r. and genetic recombination other than a regular alternation of meiosis and fertilization (→ parasexual cycle). (3) **Asexual or agamic reproduction:** The development of a new individual from either a single cell (agamogony) or from a group of cells (vegetative r.) in the absence of any sexual process. Asexual r. may be the only mode of r. of a species, or it may take place within the sexual cycle of an organism as an essential or nonessential part of its life cycle. The capacity of an organism for either sexual or asexual r. is referred to as **"versatile r."** (Crane and Thomas 1941). (4) **Subsexual reproduction** (Darlington 1937): → Parthenogenesis following → ameiosis. There is no reduction of chromosome number but → segregation of genes occurs owing to crossing over (Darlington and Mather 1949). (5) **Differential reproduction:** In which different individuals do not contribute to the next generation in proportion to their number.

reproductive cell — any of the → germ cells (usually the gametes) and their immediate predecessors from which they arise by division.

reproductive character displacement — = → Wallace effect.

reproductive cost (Bell 1980) — = → survival cost.

reproductive death — the lose of the ability to divide and to produce progeny cells. The usual criterion for r.d. is loss of colony-forming ability.

reproductive fitness — the reproductive success of an individual over its lifetime.

reproductive effort (Fisher 1930) — the proportion of the total energy budget of an organism that is devoted (over a specified and biologically meaningful time interval) to reproductive processes. To be favorably selected, a phenotype for a given level of r.e. must enhance the → fitness of the individual exhibiting this phenotype.

reproductive isolation — any of the genetically controlled mechanisms which prevent gene exchange between two populations and preserve differences in the → gene pools of populations previously achieved by natural → selection and geographic → isolation. R.i. is essential for → speciation and for the maintenance of the genetic integrity of species.

The mechanisms of r.i. are of two types: (1) **Prezygotic mechanisms** which impede the mating between members of different populations and so prevent the formation of hybrid offspring (ecological i., temporal i., ethological i., mechanical i., gametic i.); (2) **postzygotic mechanisms**, which reduce the viability or fertility of hybrid offspring (hybrid inviability, hybrid sterility, hybrid breakdown).

R.i. may develop in two ways (Grant 1985): (1) As a by-product of evolutionary divergence from the race to the species level which allows divergent species to coexist sympatrically; (2) by re-inforcement of the existing primary r.i. with new barriers that block hybridization effectively (selection for r.i.).

reproductive potential — an organism's ability to produce surviving offspring. The r.p. includes the individual's capacity to reach the reproductive stage, its fertility during this period, and the viability of its offspring to sexual maturity.

repulsion (Bateson et al. 1905) — = → trans-configuration.

residual centric distance (Southern 1967) — → chiasma.

residual genotype — → genotype.

residual homology (Stephens 1942) — → homeologous.

residual protein — = → nonhistone chromosomal protein.

residual telomeric distance (Southern 1967) — → chiasma.

resistance plasmid (Iseki and Sakai 1953) — any of a class of bacterial → plasmids (R factors; resistance factors) which confer on their host resistance to one or more antibiotics and (in Enterobacteriaceae) can promote transfer of drug resistance to other bacteria via bacterial → conjugation (transmissible R plasmids). R. plasmids consist of closed circular duplex DNA; many are composite structures consisting of two distinguishable components: (1) A resistance transfer factor (RTF) which mediates the transfer of these plasmids during bacterial mating; (2) an r-determinant component which harbors most of the drug resistance genes of the plasmids.

Each of these components is capable of autonomous replication in certain bacterial host strains (→ transposon).

R. plasmids may be classified on the basis of their ability (fi^+) or inability (fi^-) to repress the fertility properties of an → F plasmid present in the same cell, the type of distinguishable → sex pilus that the r.f. determines, the ability of a r.f. to coexist in a cell specifically with other plasmids, and the presence of genes in the r.f. that specify → DNA modification and restriction systems.

R factors that promote their own transfer generally promote the transfer of the host chromosome, or in certain cases the transfer of r. plasmids that are nontransmissible. In certain bacterial species, the RFT and the r. determinants remain covalently attached to one another; in others they are able to reversibly dissociate and replicate independently (i.e., both are independent → replicons). Reassociation occurs by a recombinational event between DNA circles.

The molecular weight of the **RTF unit** is about 5×10^6; its major functions comprise: (1) Initiation and regulation of plasmid DNA replication; (2) maintenance of plasmids in the cell and their segregation to bacterial daughter cells at division; (3) specification of the gene products needed for bacterial conjugation; (4) transfer of DNA during conjugation.

The resistance determinants (R determinants) specify the molecules that confer the resistance. These are frequently biochemical products that destroy or inactivate the antibiotics; sometimes they prevent the antibacterial agent from reaching its target within the cell.

Bacterial strains carrying r. factors resist superinfection with the same or similar factors. In Gram-positive bacteria, transfer of r. factors occurs by → transduction.

resistance transfer factor (Watanabe and Fukasawa 1960) — → resistance plasmid.

resolvase (Reed 1981) — in certain prokaryotic transposons, any of the proteins which promote site-specific recombination of super-coiled DNA's containing directly repeated resolution (res) sites. The res site contains both the crossover point for recombination and the r.-regulated promoters. Distant sites are synapsed, broken, reorientated, and ligated (often resulting in knotted or catenated products). (→ transposase).

response element — any polynucleotide receptor sequence (= receptor; responsive e.) involved in the rate of expression of a given gene. R. elements are recognized by regulatory proteins (regulators).

restart polypeptide synthesis — = → translational reinitiation.

resting nucleus (Gregoire 1905) — the → interphase nucleus of eukaryotes.

restitution (Darlington and Upcott 1941) — → reunion.

restitution nucleus (Rosenberg 1927) — (1) meiotic r.n.: A single nucleus with unreduced chromosome number, produced in place of two nuclei owing to failure of the first or second meiotic division (→ meiosis); (2) mitotic r.n.: A single nucleus with tetraploid chromosome number, produced in place of two diploid nuclei owing to failure of → mitosis (→ C-mitosis).

restorer gene — a gene that recoups the changes brought about by a cytoplasmic induced sterility.

restriction (Dussoix and Arber 1962) — (1) the degradation (→ DNA modification-restriction system) of unmodified (foreign) DNA from various sources (bacteria, phages, plasmids) by

→ restriction endonucleases capable of recognizing particular patterns of specificity (due to DNA modification by methylation or glycosylation). (2) A reduction in the rate of → genetic translation due to mutations in a gene coding for a protein in the 30S subunit of bacterial ribosomes. These mutations generally reduce (restrict) the efficiency of various → nonsense suppressors.

restriction endonuclease (Meselson and Yuan 1968) — any of a class of site-specific endonucleases (restrictases) that recognize specific DNA sequences and cleave the DNA duplex at or near these sites to produce DNA fragments (restriction fragments) having discrete molecular weights (→ restriction fragment length polymorphism). Strain-specific restrictases (and modification methylases) are widespread in bacteria (→ DNA modification-restriction):

Three different types of restrictases may be distinguished: (1) Type I enzymes recognize a specific sequence of DNA but make cuts elsewhere; (2) type II enzymes make cuts within the recognition site; (3) type III enzymes are like type I enzymes in their activity but consist of multiple subunits encoded in different genes.

All restrictases make two single-strand breaks but these breaks have two distinct arrangements: (1) Both breaks at the center of the rotational symmetry of the recognition site generating flush or blunt ends; (2) breaks symmetrically placed around the line of symmetry (staggered breaks) generating cohesive ends.

Because of its specifity a particular r.e. generates a unique family of fragments for a particular DNA molecule. Type II restrictases are widely used in → recombinant DNA technology. Some, if not all, type II restrictases are supposed to play some roles in DNA metabolism other than restriction of foreign DNA in bacteria, such as site-specific recombination.

restriction endonuclease mapping — a method which allows a relative quick survey (mapping) of large segments of DNA as well as identification of restriction sites, individual base variations, and the sites of insertions or deletions in DNA, relative to a standard sequence, in a selected section of DNA. Interesting sequences can be cloned (→ molecular cloning) for use in the → recombinant DNA technology.

The methodology for ordering the cleavage sites and producing a cleavage map involves comparison of multiple restriction digests by gel analysis.

restriction enzyme (Linn and Arber 1968) — = → restriction endonuclease.

restriction enzyme site polymorphism — variation in the DNA sequences at the sites where → restriction endonucleases cut DNA resulting in different patterns of cutting. If there is a correlation between a particular pattern of restriction endonuclease cutting and the presence of a certain gene, r.e.s.p. may be used as a gene marker.

restriction fragment — any of the individual polynucleotide sequences produced by digestion of DNA with a → restriction endonuclease.

restriction fragment assay (Mizuuchi and Nash 1976) — a procedure which provides physical, rather than genetical, assessment of the conversion of parental (substrate) DNA to recombinant DNA during an in vitro recombination reaction. Substrate and product DNA are cleaved with a → restriction endonuclease and the resulting fragments are separated by electrophoresis in agarose gels. The substrate DNA is chosen so that recombination to be studied deletes a segment of DNA; the remaining DNA gives rise to a unique restriction fragment, as does the DNA segment that has been removed.

restriction fragment length polymorphism (Grodzicker et al. 1974) — (RFLP); variation in the length of → restriction fragments that are generated after DNA is digested with different restricition enzymes and analyzed with a variety of cloned DNA probes which detect specific homologous DNA fragments. RFLP's are due to introduction or removal of restriction sites, or sequence deletions, additions, and rearrangements which affect the length of DNA between restriction sites; they represent codominant, typically neutral, genetic markers and can be used to construct detailed genetic linkage maps (RFLP linkage maps), to follow inheritance of genetic diseases and other traits, and to examine variation between and within populations.

Once linkage of a single RFLP with a specific gene has been demonstrated, other RFLP's that straddle the gene in question can be identified and ordered by observation of recombination. This allows the definition of a stretch of DNA that contains the relevant gene which can subsequently be isolated by appropriate methods.

RFLP's are not rare, occurring in the human genome every 200 to 300 bp. Since they are inherited in simple Mendelian fashion, they can be used to study the inheritance of a relevant allele over the generations by following the inheritance of readily detectable RFLP's linked to it.

restriction gene — → modifier gene.

restriction map (Danna and Nathans 1971) — a linear array of sites on DNA cleaved by various → restriction endonucleases, as established by → restriction endonuclease mapping.

restriction-modification methylase — in bacteria, any of the enzymes that methylate DNA at a specific site and thereby protect the DNA against cleavage by the corresponding → restriction endonuclease that recognizes that site. S-adenosylmethionine is used as methyl donor and duplex DNA as acceptor. ATP and Mg^{2+} are required (→ DNA modification-restriction).

restriction-modification system — → DNA modification-restriction.

restriction site — a specific nucleotide sequence (in general 4 to 6 bp with dyad symmetry) in duplex DNA recognized and cleaved by a → restriction endonuclease. The two strands may be cut either opposite to one another (blunt ends), or in a staggered manner (sticky or cohesive ends), depending on the enzyme involved. (→ recognition site).

restrictive — (1) of environmental conditions under which a → conditional mutant shows the mutant phenotype; (2) of cells which do not allow infecting virus to multiply and produce infective progenies. (→ permissive).

reticuloplasm (Krstic 1979) — the contents of the → endoplasmic reticulum (ER). The luminal protein components of the r. are called reticuloplasmins. They all possess the same C-terminal sequence Lys-Asp-Glu-Leu and this sequence is possibly of importance for their specific retention by the ER.

reticulosome (Pollak and Shorey 1964) — a cytoplasmic component (particularly frequent in the → microsomal fraction) which can give rise to a membraneous structure in the presence of phospholipids. It is assumed to be involved in membrane formation.

retroposition — RNA-mediated → transposition, i.e., the insertion of reverse-transcribed sequences from RNA back into the DNA genome. The transposed sequence is called a → retroposon. R. is a major evolutionary force that contributes to continuous sequence duplication and dispersion resulting in rearrangements that perpetuate the dynamic character of the eukaryotic nuclear genome.

retroposon (Rogers 1983) — in eukaryotic genomes, any of at least three classes of dispersed sequences generated by → reverse transcription from RNA intermediates. Their common features include sequence boundaries corresponding to RNA species, a → poly(A) tail at the 3′side, and direct terminal repeats (TR) of 8 to 19 bp of the flanking target sequence both at the 5′ and 3′ ends. The three classes include: (1) The → processed retropseudogenes; (2) the → pseudogenes derived from → small nuclear RNA's; (3) the → Alu-like sequence family (and similar sequences) which constitutes most of the dispersed, moderately → repetitious DNA of mammalian genomes.

Another classification involves two types: (1) The **viral-type retroposons** including → retroviruses and elements such as *Drosophila* → copia, yeast → Ty elements, and others (all have LTR-like ends, encode their own → reverse transcriptases and integrases, and generate 4 to 5 bp direct repeats at the target sites upon insertion); (2) the **nonviral type retroposons** including short interspersed repeats (SINE's), such as the Alu-like sequences, long interspersed repeats (LINE's), such as the primate Kpn I and the *Drosophila* F elements, small nuclear RNA pseudogenes, processed pseudogenes (they do not code for activities related to reverse transcription or integration and transpose in a passive way). The nonviral retroposons have no LTR-type direct repeats at their ends; their 3′ shows a A-rich stretch. On insertion they generate a target site duplication of 7 to 20 bp.

retroregulation (Schindler and Echols 1981) — a regulation of a gene by a cis-acting element (the retroregulator) located distal to it. The retroregulator may negatively or positively regulate the gene's expression. Transcription of the retroregulator sequence probably leads to the incorporation of a stem and loop structure at the 3′end of the mRNA; this structure protects the mRNA from exonucleolytic degradation and enhances protein synthesis of the target gene.

retrotranscription — = → reverse transcription.

retrotransposon (Bocke et al. 1985) — any → transposable genetic element that replicates and transposes through an RNA intermediate.

retroviral vector — any of a class of → cloning vectors derived from → retroviruses and used for → gene transfer into mammalian embryos and embryonic and hematopoietic stem cells.

retrovirus (Fenner 1976) — any of a family of eukaryotic single-stranded RNA viruses which replicate by way of a duplex DNA provirus integrated into cellular DNA and synthesized by → reverse transcription. Primer-dependent reverse transcription of viral RNA (initiated by addition of a deoxynucleotide to the 3′end of a cellular tRNA molecule that is hydrogen-bonded to the primer binding site of the viral genome RNA) generates a linear DNA with long terminal repeats (LTR's) of 300 to 1200 bp which contain the transcriptional → promoter and other signals for gene expression and for integration. Some retroviruses are highly oncogenic (acute transforming retroviruses) and represent genetic recombinants between retroviral sequences and transformation-specific sequences of cellular → oncogenes. Retroviruses that cause disease after long latent periods (nonacute transforming retroviruses) have only genes for replication.

The termini of integrated retroviruses show short (3 to 6 bp) direct repeats of host DNA at the junction between the provirus and host DNA and a short (11 bp) inverted repeat at the termini of each LTR.

Retroviruses resemble → transposable genetic elements and are natural vectors for introducing foreign DNA into the vertebrate genome. Typically, recombinant particles are defective for replication and require → helper virus to be passed from cell to cell.

reunion (Darlington and Upcott 1941) — according to the "breakage-reunion hypothesis", the rejoining of broken ends of chromosomes or chromatids in such a way that chromosomal structural changes (→ chromosome mutations) arise, as opposed to "restitution", i.e., the rejoining of broken ends which restores the prebreakage structure and represents a process for repairing chromosomal damage. The proportion of recognizable breaks in the chromosomes or chromatids of a sample of nuclei which undergo recognizable reunion is referred to as the "empirical coefficient of reunion" (ECR).

reversed genetics (Weissmann 1978; Mantei et al. 1979) — a type of genetic analysis in which a cloned DNA sequence is altered at predetermined sites by site-specific mutagenesis and then assayed for changes in gene function, as opposed to the selection of deviant phenotypes to explore the genotypes (classical genetics).

reverse gyrase (Mirambeau et al. 1984) — in archebacteria, an ATP-dependent type I → DNA topoisomerase that converts relaxed or negatively supercoiled DNA (→ DNA supercoiling) into a positively supercoiled form in vitro. Positive supercoiling might be required both to stabilize DNA in high temperature growth conditions and to inactivate genes.

reverse mutation — → revertant.

reverse selection system — a selective system (→ forward s.s.) which allows the selection of nonmutant (wild type) cells from a population of mutant cells. R.s.systems are specific for the mutant type in question and allow the selection of → back mutations or hybrid cells originating from somatic → cell fusion.

reverse self-splicing (Woodson and Cech 1989) — in group I → introns, the full reversibility of the → self-splicing reaction in vitro. R.s.-s. has implications for transposition of group I introns.

reverse splicing — the phenomenon that an excised intron (→ RNA splicing), which contains all of the catalytic activity, inserts back into an unrelated RNA. R.s. would enable the same intron to be placed into different RNA's.

reverse transcript — a DNA sequence generated from an RNA sequence by means of → reverse transcription.

reverse transcriptase (Baltimore 1970; Temin and Mizitany 1970) — an RNA-directed → DNA polymerase (DNA nucleotidyl transferase) able to copy single-stranded RNA sequences into complementary double-stranded DNA. R.t. is the key enzyme in the retroviral life cycle (→ retrovirus). It uses viral RNA as template and a tRNA primer and synthesizes DNA identical to the RNA (anticomplementary DNA), using cDNA as template. R.t. is probably involved in the generation of various → transposable genetic elements (retroposons).

reverse transcription — RNA-directed DNA synthesis (→ reverse transcriptase).

reversion — a change in DNA that results in a → revertant.

reversion analysis — the analysis of susceptibility of a mutant (→ gene mutation) to reversion induction by specific mutagens as a means of examining the chemical nature of a mutational lesion.

revertant — any cell clone or individual arising from mutants by true reverse mutation at the same nucleotide as the original mutation or by pseudoreversion, i.e., as a result of a second mutation that occurred either in the same gene as the original mutation (intragenic suppression) or in a gene different from that in which the original mutation occurred (extragenic suppression).

In pseudorevertants the second mutation masks or suppresses the expression of the phenotype of the original mutation (→ suppressor mutation).

RF — acronym for → **r**eplicative **f**orm.

R factor = → resistance plasmid.

RFLP — acronym for → **r**estriction **f**ragment **l**ength **p**olymorphism.

RFLP map — a → linkage map constructed from → restriction fragment length polymorphisms in eukaryotic genomes. Such maps can be used to predict the inheritance of important phenotypic characters and assist in the → molecular cloning of linked genes which affect the phenotypes. → RFLP markers closely linked to the DNA sequence coding for relevant phenotypes can be used to monitor any portion of the genome in simple assays.

RFLP marker — any marker resulting from changes in the length of genomic DNA produced by digestion with specific → restriction endonucleases (→ restriction fragment length polymorphism).

rho factor (Roberts 1969) — in bacteria *(E.coli)* and coliphages, a multimeric transcription termination factor (product of the rho gene) that participates in various site-specific termination events, including → attenuation, termination of transcription at the end of certain genes or operons, polarity due to nonsense or insertion mutations. Rho, which has an RNA-dependent ATPase activity, probably interacts directly with the nascent RNA transcript and with RNA polymerase as well to effect the termination and release of RNA chains. (→ sigma factor).

rho-independent terminator — in bacteria *(E. coli)*, any DNA sequence that causes RNA polymerase to terminate → genetic transcription in the absence of the → rho factor.

ribonuclease — (RNase); any of a family of enzymes which hydrolyze RNA (single-strand specific and double-strand specific enzymes exist and many are involved in the in vivo processing of RNA precursor molecules).

RNase H is an enzyme that degrades specifically the RNA strand of RNA-DNA molecules.

RNase P cleaves precursor tRNA to generate the 5′terminus of the mature → transfer RNA. It recognizes the common tRNA domain and can cleave many tRNA precursors, irrespective of different primary sequences.

ribonuclease P — (RNase P); in prokaryotes and eukaryotes, one of the enzymes participating in the processing of → pretransfer RNA's. The enzyme generates the 5′ termini of mature tRNAs by cleavage of extra nucleotides from the 5′ termini of precursor tRNA's.

ribonucleic acid — (RNA); a polynucleotide (symbol RNA) consisting essentially of chains with a repeating backbone of phosphate and sugar units to which nitrogenous bases are attached. RNA is unique among biological marcromolecules in that it can encode genetic information, serve as an abundant structural component of cells, and also possesses catalytic activity (→ ribozyme).

RNA governs all protein manufacture and exists in a variety of different species. Most sequenced RNA's have the potential to form secondary structures by means of local base pairing and loop formation in otherwise single-stranded molecules. RNA metabolism of eukaryotes is more complex than that of prokaryotes and many species of RNA have been found to have functions other than those classically assigned to RNA (→ heterogeneous nuclear RNA; low molecular weight nuclear RNA; chromosomal RNA).

Three main differences distinguish → deoxyribonucleic acid (DNA) and RNA: (1) The D-ribose sugar of RNA differs from the 2-deoxy-D-ribose of DNA in having a total of four rather than three hydroxyl groups. The extra OH group is at the C 2 position. (2) Both DNA and RNA contain four principal bases but only three of these (adenine, guanine, and cytosine) are common to both nucleic acids. The RNA counterpart of thymine is uracil, the former being the 5-methyl derivative opf the latter. (3) The secondary and tertiary structures of RNA show structural and functional heterogeneities which allow one to distinguish two main classes of RNA; (a) **Genetic RNA:** The carrier of primary genetic information in most plant viruses, some animal viruses, and certain bacteriophages (→ reverse transcriptase; replicative form) which may be single- or double-stranded. (b) **Nongenetic RNA:** RNA specified by DNA in the course of → genetic transcription. It is heterogeneous and three principal types can be distinguished: → ribosomal RNA (80% of the total cell RNA), → transfer RNA (10 to 15% of the total), and messenger → RNA (5 to 10% of the total). These types of RNA are usually synthesized via the sequential processing of primary transcription units (→ transcription; RNA processing) which are longer in nucleotide sequence than the mature products (precursor molecules).

Regulation of RNA synthesis (transcription) in vivo is of two types, generalized and specific. Generalized regulation as a function of growth rate or of the presence of essential amino acids is coordinate, i.e., applies simultaneously to all species of RNA (→ relaxed; stringent). Regulation of the synthesis of specific RNA species occurs at the level of genetic transcription (→ transcriptional control; genetic regulation).

Two types of reactions leading to the formation of functional RNA molecules are known which require either DNA (DNA-dependent synthesis of cellular RNA) or RNA (RNA-dependent synthesis of viral RNA) as the primer and are catalyzed by distinct classes of enzymes.

(1)**DNA-dependent RNA synthesis:** DNA acts as the template, i.e., its nucleotide sequence determines the formation of complementary base pairs in the nucleotide sequence of the synthesized RNA molecules. The process is → genetic transcription and is catalyzed by DNA-dependent RNA polymerase which links together the monomeric precursors (the ribonucleoside triphosphates ATP, GTP, UTP, and CTP).

(2) **RNA-dependent RNA synthesis:** RNA acts as a template and the process corresponds to the → replication of viral RNA. It is catalyzed by RNA-dependent RNA polymerase (RNA replicase, RNA synthetase) capable of forming new RNA strands upon parental RNA templates. When viral RNA enters the cell and single strands attach to host → ribosomes, this enzyme is formed and ribonucleoside-5′-triphosphates are polymerized to form a

long chain of polynucleotids (with elimination of inorganic phosphate) by transfer of the activated 5′-phosphate end of a nucleotide to the free 3′-hydroxyl group of a second, adjacent nucleotide in the growing polynucleotide (→ replicative form).

ribonucleoprotein particle — (RNP particle); any of the ribonucleoprotein complexes involved in → gene expression of eukaryotes (→ messenger RNA).

ribonucleotide — a nucleotide in → ribonucleic acid.

ribonucleotide reductase (Reichard 1957) — any of a group of enzymes which catalyze the formation of deoxyribonucleotides from their corresponding ribonucleotides.

ribosomal ambiguity mutation — any mutation which gives rise to alterations of certain ribosomal proteins and results in decreased translation fidelity in vivo and in vitro.

ribosomal DNA — (rDNA); the family of chromosomal or (in the case of → rDNA amplification) extrachromosomal DNA sequences encoding → ribosomal RNA.

In **prokaryotes,** the genes coding for 16S, 23S, and 5S rRNA are linked. Due to the action of specific processing endoribonucleases during transcription and → ribosome assembly, no large primary transcript is formed.

In higher **eukaryotes,** the genes coding for 17–18S, 5.8S, and 25–28S rRNA are arranged as tandemly repeated units (7 to 50 kb) which are cotranscribed and contain, in addition to the rRNA genes, the so-called nontranscribed spacer (NTS). Each unit starts with an external spacer (ETS). The rRNA genes are separated from each other by two internal transcribed spacers (ITS1 and ITS2). The genes for 5S rRNA are unlinked and separate from the larger rDNA repeat. In some fungi and protozoans, the 5S rRNA gene is part of the rDNA unit (in yeast it divides the NTS into two parts: NTS 1 and NTS 2). Most signals for the expression of rDNA (initiation and termination, processing) are located in the spacer regions. Ribosomal rDNA is transcribed by → RNA polymerase I, the separate 5S rDNA by RNA polymerase III. (→ rDNA amplification; rDNA insertion; rDNA magnification).

The ribosomal DNA of → **chloroplasts** is usually part of an inverted repeat unit consisting of the genes coding for 16S rRNA, two tRNA's (for isoleucine and alanine), 23S rRNA, 4.5S rRNA, and 5S rRNA (which seems to be independently transcribed). Ribosomal DNA's of animal, plant, and fungal → mitochondria differ substantially.

The genes for the large rRNA may contain, in some organisms, → introns.

ribosomal frameshifting — a shift of the → reading frame within structural gene sequences during translation of → messenger RNA's. Site-specific r.f. apparently plays a role in regulating the synthesis of relatively nonabundant proteins and may allow two proteins to be produced from a single transcript.

ribosomal peptidyl transferase — an enzyme synthesizing all proteins translated from → messenger RNA. It repeatedly promotes the reaction between a peptidyl ester and an amine to form peptidylamine, i.e., a peptidyl-ester which contains one or more amino acids (→ genetic translation).

ribosomal precursor RNA (Scherrer et al. 1963) — = → pre-ribosomal RNA.

ribosomal protein — (r-protein); any of the essential proteins (70 to 80 in eukaryotes) which are synthesized individually and assembled into the → ribosome. Genes coding for r-proteins are reiterated, tend to cluster, but are found in many transcription units.

They may be regulated at various levels including RNA processing, mRNA translation, mRNA turnover, and protein turnover.

ribosomal RNA (Kurland 1960) — (rRNA); the transcription product of → ribosomal DNA.

Eukaryotic cytoplasmic rRNA's are usually transcribed from two separate transcription units: 5′ spacer-17-18 S-spacer-5.8 S–28 S 3′; the 5S rRNA is transcribed by RNA polymerase III from multiple genes. The primary product of the large transcription unit (located at one or a few → nucleolus organizers) is → pre-ribosomal RNA (about 45S) transcribed by RNA polymerase I which becomes processed through a series of processing steps (→ RNA processing) to yield the mature rRNA species. Eukaryotic 5.8S rRNA is equivalent to the 5′ end of prokaryotic 23S rRNA and base-pairs with tRNA's and 28S rRNA.

Bacterial rRNA's are cotranscribed from several transcriptional units (leader-16S-spacer-23S-spacer-5S) into 30S pre-ribosomal RNA. The monocistronic products of primary processing are subject to secondary processing and yield the mature rRNA species m16, m23, and m5. Prokaryotic 5S rRNA is essential for protein synthesis by pairing with special sequences of transfer RNA's, 16S rRNA, and 23S rRNA.

The transcription unit for **rRNA in chloroplasts** of higher plants consists of 16S rRNA, two tRNA's (isoleucine and alanine), the 23S rRNA, 4.5.S rRNA, and 5S rRNA. 4.5S rRNA arises from a 3′ segment of 23S rRNA; 5S rRNA of chloroplasts seems to be independently transcribed.

The rRNA species of plant **mitochondria** differ substantially from those of animal and fungal mitochondria (animals: 16S and 12S; yeast: 21S and 15S; plants: 26S and 18S). Plant mitochondria contain a 5S rRNA which is not present in any other mitochondria; the genes coding for 5S and 18S rRNA's are closely linked, the 26S rRNA encoding gene is located at a greater distance.

All rRNA species contain multiple hairpin regions and undergo long-distance interactions. Post-transcriptional modification of rRNA includes a number of site-specific methylations (in eukaryotes, ribose methylations; in bacteria, mostly base methylations).

ribosomal scanning model (Kozak 1978) — a model according to which the eukaryotic → ribosome (the 40S ribosomal subunit in association with methionine initiator tRNA and initiation factors) binds to the capped 5′ end of → messenger RNA and then linearly scans (5′ to 3′ direction) the mRNA for the first → initiator codon (AUG) whereupon the → initiation complex forms and the reading frame is established. Any type of direct internal initiation process which bypasses the 5′ end is excluded. The AUG codon itself serves as a basic signal for ribosomal recognition of the initiator region. When a → terminator (stop) codon is reached, the ribosome leaves the mRNA, i.e., an AUG following a stop codon cannot serve for initiation.

ribosomal suppression — suppression of nonsense and frameshift mutations in the absence of → suppressor tRNA's. R.s. is achieved by alterations in ribosomal protein components which allow a range of codon misreading and low suppression.

ribosomal translocation — during the elongation cycle of protein biosynthesis, a partial reaction dissociating the deacylated → transfer RNA from the → ribosome; the peptidyl-tRNA moves into the P site, the → messenger RNA advances by the length of one → codon, and a molecule of GTP is hydrolyzed. R.t. is catalyzed (in *E. coli*) by the → elongation factor G (EF-G).

ribosome (Roberts 1958; Dintzis et al. 1958) — any of a class of multicomponent structures (25 to 30 nm in diameter and more or less spherical in shape) found in all cells, in mitochondria and chloroplasts. Ribosomes have roles both in the → genetic translation of transcripts and in the manufacture and secretion of the proteins; they consist of → ribosomal RNA and → ribosomal proteins and serve to bring together a single → messenger RNA molecule and charged → aminoacyl-tRNA molecules in the proper position and orientation so that the base sequence of the mRNA (→ genetic code) is translated into an amino acid sequence.

Ribosomes are composed of a small and a large subunit. The sites functioning in translation are clustered into part of the r. (the translational domain) and include the → initiation factor binding sites located in the cleft of the small subunit, the messenger binding sites located on the platform of the small subunit, the peptidyltransferase and the 5S rRNA located on the central protuberance of the large subunit, and proteins mediating the GTP-dependent step of r. translocation found on the (L7/L12) stalk of the large subunit.

During protein synthesis the primary function of the small subunit is to hold the mRNA and the tRNA's, while the larger subunit catalyzes the formation of peptide bonds in the proteins being synthesized.

Ribosomal large and small subunits are organized in four structural patterns characteristic of ribosomes from eubacteria, archebacteria, eukaryotes, and a group of sulfur-dependent bacteria (eocytes), respectively. All share a common structural core, but each type also has additional independent structural features, including the eukaryotic lobes and the archebacterial bill on the smaller subunit. On the larger subunit, they include the eocytic lobe, eocytic gap, and eocytic bulge, and a modified central protuberance.

R. biogenesis is a complex process requiring production of the various rRNA's and r-proteins in roughly equimolar amounts as well as the stepwise assembly of these components to produce active particles. In eukaryotes, assembly and maturation of ribosomes occur predominantly in the → nucleolus. The biogenesis of organelle ribosomes is dependent on the expression of genes found in the nucleus and in the → organelle genomes.

Prokaryotic ribosomes: The intact particle (70S particle) consists of the 30S and 50S subunits. Both can be dissociated into RNA and protein. 30S subunit: One 16S rRNA molecule + 21 different proteins (S 1 to S 21). 50S subunit: One 5S rRNA molecule + one 23S rRNA molecule + 32 different proteins (L 1 to L 32).

Sizes of the RNA's: 5S rRNA 120 nucleotides; 16S rRNA 1541 nucleotides; 23S rRNA 2904 nucleotides. About 70% of the bases in each molecule is internally paired to form complex structures with stems and loops.

Eukaryotic ribosomes: they are larger than those of prokaryotes (80S), contain about 80 proteins and an additional RNA molecule (5.8S rRNA); the subunits are the 40S and 60S particle.

40S subunit: One 18S rRNA molecule + about 30 proteins. 60S subunit: One 5S rRNA, one 5.8S rRNA, and one 28S rRNA molecule + about 50 proteins. The bacterial counterpart to the eukaryotic 5S rRNA is probably present as part of the 23S rRNA. The initial processes of r. biogenesis, i.e., synthesis and processing of precursors to cytoplasmic rRNA molecules (pre-rRNA) and their assembly with specific ribosomal proteins, take place in the → nucleolus.

Chloroplast and **mitochondrial ribosomes:** The ribosomes are similar to, but distinct from, the major eukaryotic type of cytoplasmic ribosomes and more closely resemble the bacterial type. The structure of chloroplast ribosomes is very much like that of bacteria with respect to size and chemical composition; they contain 23S, 16S, 5S, and 4.5S rRNA. Mitochondrial ribosomes vary significantly from one organism to the next, but two distinct classes may be distinguished: (1) The 70 to 80S particles with 16S and 23S rRNA's (fungi and protozoa); (2) the miniribosomes (50 to 60S) and rRNA species of 12S and 16S (higher animals and man).

Ribosomes from *E. coli* contain in addition to the classical → A and → P sites a third tRNA binding site (E site) which is specific for deacylated tRNA which is not released from the P site but rather is translocated from the P to the E site concomitantly with the movement of peptidyl-tRNA from the A to the P site.

The ribosomes of prokaryotes are distributed throughout the cell without any apparent spatial organization. In most eukaryotes, there are two major classes of ribosomes: (1) Those bound to the → endoplasmic reticulum (attached r.); (2) free ribosomes in the cytoplasm.

ribosome cycle — the exchange of the large and small ribosomal subunits which occurs in growing bacteria and in the cytoplasm of eukaryotes. Exchange is due to dissociation of the subunits after each passage over messenger RNA and reformation from a pool of ribosomal subunits that continuously recycle through polyribosomes. Recycling requires an → initiation factor (= ribosome dissociation factor), maintains the polyribosomes, inhibits the association of ribosomal subunits, promotes the quantitative entry of the larger subunit into polysomes, allows the accumulation of ribosomal subunits (instead of single ribosomes), and is required for the binding of messenger RNA to ribosomes (→ initiation complex).

ribosome binding site (Shine-Dalgarno 1974) — (RBS); in → messenger RNA a site for binding ribosomes and required for initiation of → genetic translation. In *E. coli*, three basic elements constitute the RBS: (1) The → Shine-Dalgarno sequence; (2) the → initiator codon (usually AUG; rarely, GUG); (3) an intervening spacer region, normally 5 to 9 nucleotides long.

The Shine-Dalgarno sequence is complementary to the 3′end of 16S rRNA; base-pairing between the mRNA and rRNA possibly plays a role in ribosome recognition and binding.

The RBS of eukaryotes is characterized by considerable sequence variation in the 5′ untranslated region, the only recognizable common features are the AUG initiator codon and a 5′ terminal methylated cap structure (→ messenger RNA capping).

ribosome dissociation (Kohler et al. 1968) — → ribosome cycle.

ribosome dissociation factor — → dissociation factor; ribosome cycle.

ribosome editor hypothesis (Menninger 1977) — the hypothesis stating that an inappropriate peptidyl-tRNA (p-tRNA), whose structure does not correctly complement the mRNA, dissociates preferentially from the → ribosome during protein synthesis.

ribosome recognition site — → transfer RNA.

ribosome releasing factor (Ogawa and Kaji 1975) — in *E. coli*, a protein factor that releases ribosomes from mRNA at the → terminator codon.

ribozyme — any of the RNA molecules possessing enzymatic activity and acting as a biological catalyst. Two categories: (1) Self-splicing ribozymes (→ RNA splicing); (2) self-cleaving ribozymes.

right splicing junction — the boundary between the right end of an → intron and the left end of the adjacent → exon.

ring bivalent — a → bivalent at the first meiotic division consisting of two paired chromosomes (→ chromosome pairing) with terminalized chiasmata in both arms (→ chiasma terminalization; ring multivalent).

ring chromosome — a physically circular chromosome (→ genetic circularity). Ring chromosomes are a normal feature of chromosome organization in some prokaryotes (e.g., **E. coli** and some viruses). In eukaryotes, ring chromosomes arise as a result of chromosomal structural changes (→ chromosome mutation) and are usually meiotically unstable due to the production, by meiotic → crossing over, of double-sized dicentric chromosomes which normally are eliminated.

During mitosis, a ring chromosome usually gives rise to two daughter rings of equal size which are regularly distributed to daughter nuclei. Under these circumstances the ring chromosomes are somatically stable.

As a result of "sister strand exchange", ring chromosomes may give rise to "double sized rings" in somatic tissues, i.e., continuous dicentric rings of double the original size which may lead to a double bridge at anaphase (→ chromosome bridge). In the case of bridge rupture, two new rings may be formed which are usually of different size. This process can result in → variegation due to the formation of → duplications and → deletions.

ring multivalent — a ring-like pairing configuration at the first meiotic division (→ chromosome pairing) which consists of more than two completely (in autopolyploids) or partially homologous (e.g., in diploid → translocation heterozygotes) chromosomes held together by terminalized chiasmata up to anaphase I (→ ring bivalent).

Ri plasmid — a root-inducing → plasmid of **Agrobacterium rhizogenes** with a DNA transfer mechanism similar to that of the → Ti plasmid. The transferred DNA (Ri T-DNA) is in some plasmids composed of TR (right-hand) and TL (left-hand) segments. The roots transformed by wild-type Ri T-DNA show an altered phenotype and the plants regenerated from these roots are also phenotypically different with respect to apical dominance, physiology of flowering, and leaf morphology.

R-loop (White and Hogness 1977) — a three-stranded structure formed by the displacement and annealing of one strand of duplex DNA by complementary RNA, leaving a DNA loop with a characteristic appearance. If the displacing strand is DNA, the corresponding structure is a D-loop.

RNA — shorthand for → ribonucleic acid.

RNA binding domain (Adam et al. 1986) — in RNA binding proteins an amino acid octamer which is directly involved in their binding to specific RNA sequences (RNP consensus). The RNA b.d. may occur once or several times in the proteins which include, e.g., proteins of → heterogeneous nuclear ribonucleoprotein particles, poly(A)binding protein and some U-class small nuclear ribonucleoprotein proteins.

RNA cap — → messenger RNA capping.

RNA-dependent DNA polymerase — → DNA polymerase; reverse transcriptase.

RNA-dependent RNA synthesis — synthesis of RNA mediated by an RNA-dependent → RNA polymerase as observed in plants, animal cells, and bacteria. In most such systems, RNA-d. RNA s. is the result of cooperation between host proteins and proteins encoded by a virus.

RNA-directed DNA synthesis — the synthesis of double-stranded DNA from a single-stranded RNA template. It requires: (1) An enzyme for the synthesis of the first DNA strand from an RNA template and the second DNA strand from the first; (2) primers for each of the two strands; (3) a means for removing the RNA template after → reverse transcription, to allow synthesis of the second DNA strand. (→ retrovirus; retroposon).

RNA-driven hybridization — the use of an excess of RNA to react with all complementary sequences of single-stranded DNA. RNA-d.h. is used for estimating the sequence complexity of RNA in eukaryotic cells and tissues and for measuring the relative and absolute concentration of specific mRNA species.

RNA editing (Benne et al. 1986) — (1) a process resulting in the addition of nongenomically encoded uridine residues (Us) to mitochondrial mRNA's in the kinetoplastid protozoa. (2) Any process that results in the production of an RNA molecule which differs in nucleotide sequence in coding regions from the DNA template, provided that the difference is not the result of the removal of → introns by a splicing mechanism. RNA e. is a mechanism of modulating genetic activity and involves the precise addition and deletion of nucleotides within the coding regions of mRNAs.

RNA ligase (Silber et al. 1972) — an enzyme (polyribonucleotide synthase) in T4 infected cells, which catalyzes ATP-dependent, intra- and intermolecular phosphodiester bond formation between 3′-OH (acceptor) and 5′-P (donor) termini of RNA, i.e., an enzyme which can join RNA molecules together. Single-stranded ribonucleotide acceptors are preferred; single-stranded DNA is also joined by the enzyme.

RNA maturase — any of the enzymes encoded by self-splicing → introns and required for splicing of the intron that encodes them (→ RNA splicing).

RNA methylation — the post-transcriptional addition of methyl groups (by RNA methylases) at definite sites within → ribosomal RNA, → transfer RNA, and → messenger RNA which plays a crucial role in → RNA processing and/or the functional properties of mature RNA's.

RNA paraprocessing (Gillespie and Gallo 1975) — a process of → RNA processing by which a normal cellular RNA transcript may give rise to a → retrovirus genome. If the paraprocessed gene transcript codes for a → reverse transcriptase, replication and mutation of the RNA become possible.

RNA plasmid (Finnegan and Brown 1986) — an apparently independently replicating single- or double-stranded RNA species (as observed in maize mitochondria).

RNA polymerase — any of a number of enzymes (ribonucleoside triphosphate: RNA nucleotidyltransferase) capable of polymerizing ribonucleoside triphosphates into polyribonucleotides (RNA). DNA (DNA-dependent RNA p.) or RNA (RNA-dependent RNA polymerase or RNA replicase) is used as the template. DNA-dependent RNA p. is involved in → genetic transcription, RNA-dependent RNA p. in the replication of RNA viruses with double-stranded RNA molecules.

The overall polymerization process can be divided into three steps: (1) Binding of RNA p. to its template; (2) association of the initial nucleoside triphosphate with the enzyme-template complex; (3) polymerization (direction 5′ → 3′) of further nucleoside triphosphates.

Table 12. Functions of *E. coli* polymerase subunits. (Ishihama 1988)

Subunit	No. of amino acids	M_r	Functions
β	1342	150, 618	Catalysis of RNA synthesis (initiation and elongation) Recognition of substrate ribonucleoside 5'-triphosphates Binding of product RNA Stringent control Autogenous regulation of ββ' synthesis Binding of rifampicin and streptolydigin
β'	1407	155, 163	Binding of template DNA Binding of sigma subunits
α	329	36, 512	Association of β and β' subunits
σ^{70}	613	70, 263	Promotion of core enzyme maturation Recognition of regular promoters
σ^{32}	284	32, 381	Recognition of heat shock promoters

Prokaryotes contain multiple forms of RNA p. holoenzyme that differ only in their → sigma subunits. They are interconvertible among various forms by structural modifications (phosphorylation and ADP-ribosylation), or by interaction with various → transcription factors. Such structural changes may influence → promoter selectivity and play a role in the global regulation of → gene expression. E. coli RNA p. consists of at least four subunits (*α, β, β'* and *σ*) and is present in two major enzyme forms: Core enzyme (α_2, *β, β'*) and holoenzyme (core enzyme plus sigma subunit). The core enzyme is capable of RNA chain elongation but requires sigma for specific transcription initiation. Synthesis of all major species of RNA (rRNA, tRNA, mRNA) in prokaryotes is directed by the single DNA-dependent RNA p.

In **eukaryotes**, RNA synthesis is carried out by three classes of RNA p.: (1) RNA p. I (located in the → nucleolus) transcribes → pre-ribosomal RNA; (2) RNA p. II (located in the cytoplasm) synthesizes → pre-messenger RNA; (3) RNA p. III (located in the cytoplasm) transcribes → pre-transfer RNA, 5S → ribosomal RNA, and some families of short dispersed repetitive nucleotide sequences.

The subunit compositions of the three RNA polymerases reflect these distinct functions: Each enzyme consists of a group of subunits unique to the polymerase class in addition to a group of subunits common to all three enzyme classes. Similarly, there are different sets of transcription initation factors for each RNA p. Phosphorylation constitutes one of the regulatory mechanisms for the enzyme activities and also for proteins interacting with the polymerases.

A specific type of RNA p. synthesizing → Okazaki fragments is called → DNA primase.

Mitochondrial RNA p. (RNA p. IV) displays properties different form those of the nuclear polymerases and consists of a single polypeptide chain. Its activity is unaffected by *α*-amanitin and inhibited by rifampicin, acriflavin, and actinomycin D.

RNA-dependent RNA polymerase is a virus-dependent RNA p. (RNA replicase) necessary for the replication of double-stranded RNA viruses. It requires all four ribonucleotide triphosphates, RNA as a template, and divalent cations. The enzyme is analogous to DNA-dependent RNA p. and transcribes messenger RNA from DNA without affecting the integrity of the template (→ reverse transcriptase).

RNA primer (Zechel et al. 1975) — (pRNA); RNA which mediates → DNA replication; pRNA defines a starting point and an RNA-DNA transition point. Two signal structures on the DNA template are required (one controlling initiation and another for termination of the RNA primer and its extension into DNA; the latter controls the actual inception of DNA synthesis). (→ Okazaki fragment).

RNA processing — in prokaryotes and eukaryotes, the conversion of the primary transcription products (→ pre-mRNA, pre-tRNA, pre-rRNA) by post-transcriptional modification into functional mature RNA molecules. The alterations include: (1) Nucleolytic reactions, such as cleavages and trimming of the large precursor molecules (exception, bacterial mRNA's); (2) terminal additions of nucleotides, such as 3′-polyadenylation of mRNA, addition of CCA to the 3′-terminus of tRNA, and 5′-capping of mRNA; (3) nucleotide modifications, such as base or ribose methylation and the conversion of uridine to pseudouridine in otherwise processed molecules.

In eukaryotic → mosaic genes (split genes) encoding primary transcripts consisting of → exons and → introns, RNA p. includes the excision of introns and splicing of the exons (→ RNA splicing). Alternative RNA p. may generate alternative RNA and protein products from a gene by the selection of alternative exons for incorporation into their respective mRNA's (eukaryotic viruses, eukaryotic genes). The underlying mechanisms include use of alternative transcriptional initiation sites, the stochastic use of alternative splice donor sites, and the use of alternative poly(A) sites.

RNA replicase — → RNA polymerase.

RNase mapping — a method used to detect and quantify specific RNA's and to determine structural features of RNA molecules by hybridization with SP 6 single-stranded ^{32}-P-RNA probes. An excess of ^{32}P-RNA complementary to the test RNA is syntesized with SP 6 RNA polymerase and hybridized to the RNA sample (formation of a ^{32}-P-RNA-RNA hybrid). Ribonuclease treatment digests the unhybridized single-stranded ^{32}-P-probe; ^{32}-P-RNA-RNA hybrids are ribonuclease resistent and can be detected and quantitated by gel electrophoresis.

RNase P — the enzyme responsible for maturation of the 5′ termini of → transfer RNA precursor molecules in vivo; it consists of an RNA subunit and a protein subunit.

RNA self-splicing (Cech et al. 1981) — a set of linked cleavage-ligation reactions mediated by intron RNA itself and conserving the number of phosphodiester bonds (other forms of → RNA splicing require a minimum of two phosphodiester bond cleavages and one ligation), which lead to intron self-excision and exon ligation in the absence of enzymes, releasing a small circular RNA. RNA s.-s. exemplifies intramolecular catalysis and occurs in some nuclear → pre-ribosomal RNA's and various mitochondrial and chloroplast → premessenger RNAs. All other types of RNA splicing require interactions between cellular factors and specific recognition signals in the RNA precursors.

RNA splicing (Berget et al. 1977) — the process whereby the → introns of eukaryotic mosaic (split) genes are removed from precursor RNA (pre-mRNA, pre-rRNA, pre-tRNA) and the coding sequences (exons) are joined.

The basic mechanism of RNA s. appears to be the same in almost all systems, with all reactions proceeding via a series of transesterifications. The involvement of extrinsic factors such as proteins and other RNA's seems to vary considerably from one system to another.

RNAs catalyzed by some introns in the absence of proteins is called → RNA **self-splicing**; in these cases, splice site selection, intron excision, and exon ligation rely on elements of the RNA structure alone (→ reverse splicing). Comparisons of many nuclear pre-mRNA introns have revealed sequence similarity only near the splice junctions. Consensus sequences for the donor (5′) and acceptor (3′) splice sites are:

$\genfrac{}{}{0pt}{}{C}{A}$-A-G/g-t-$\genfrac{}{}{0pt}{}{a}{g}$-a-g-t (capital letters denote exon sequences, lower case letters denote introns,/ denotes exon-intron boundary). The acceptor consensus sequence is less specific and consists of a pyrimidine-rich region followed by the dinucleotide A-G:$\binom{c}{t}_n$-n-$\genfrac{}{}{0pt}{}{c}{t}$-a-g/G (n denotes any nucleotide; n = 11).

Four major types of RNA s. (Table 13) may be distinguished (Cech 1988): (1) Group I intron self-splicing; (2) group II intron self-splicing; (3) nuclear pre-messenger RNA splicing; (4) nuclear pre-transfer RNA splicing.

Three are two-step reactions (the 5′ splice site is cleaved in the first step, the 3′ splice site in the second). The cleavage-ligation reactions occur by transesterification mechanisms and there are similarities in the conserved bases at the splice junctions. The three systems differ with respect to the elements determining selection of the 5′ splice site. In the case of nuclear pre-mRNA's, the intron sequences near the 5′ splice junction are required for determining the 5′ splice site; in group I and II introns, flanking exon sequences play the main role.

The main difference between pre-mRNA splicing and self-splicing is that in the first case the catalyst must be assembled from splicing factors (including multiple proteins and four → small nuclear ribonucleoprotein particles termed U1, U2, U5, and U4/U6), which together with the pre-mRNA represent the spliceosome in which cleavage-ligation reactions occur. The reactions of group II intron self-splicing and nuclear pre-mRNA splicing are identical. Group I introns occur in nuclear rRNA genes, mitochondrial mRNA and rRNA genes, and in chloroplast tRNA genes. They have a conserved uridine at the 3′ end of exon 1 and a conserved guanosine at the 3′ end of the intron. Group II introns occur in structural genes of fungal and mitochondrial DNA and in structural and tRNA genes of chloroplasts. They have a conserved GU dinucleotide at the 5′ splice site, use adenosine for RNA branch formation and show a 2′, 5′ phosphodiester bond (RNA branch) between the adenosine at the branch point and the guanosine at the 5′ end of the intron (→ lariat RNA).

Alternative RNA s. is a mechanism by which a single transcript can give rise to more than one mature mRNA (most a. RNA s. events affect the region of the mRNA that encodes protein).

Deletion or alteration of the consensus GT splice donor region in eukaryotic genes may result in: (1) Complete absence of transcript; (2) accumulation of unspliced precursor RNA; (3) use of cryptic splice sites or 3′ (donor) splice sites normally used in other contexts.

RNA 3′-terminal phosphate cyclase (Filipowicz et al. 1983) — an enzyme which catalyzes conversion of a 3′-phosphate group to the 2′, 3′-cyclic phosphodiester at the 3′ end of RNA and is probably involved in → RNA splicing.

RNA transcript — any of the RNA sequences produced by → genetic transcription of DNA (→ RNA processing; RNA splicing).

RNA transcript mapping — a procedure involving hybridization of the RNA to be analyzed with a radioactively labeled and purified DNA probe under condition which favor RNA-DNA or DNA-DNA duplexes. After digestion with S 1 nuclease (or mung bean nuclease), the nuclease resistant fragments are resolved (polyacrylamide gel electrophoresis). The size of the nuclease-protected fragment serves to determine the transcription start point (S 1 nuclease protection assay).

RNA trimming — the removal of stretches of RNA at one or both end of the primary → RNA transcript (→ RNA processing).

RNA tumor virus — = → retrovirus.

RNP — acronym for → **r**ibo**n**ucleo**p**rotein.

Robertsonian — of chromosome structural changes due to → centric fusion or → centric fission.

roentgen — (r); the exposure unit for electromagnetic radiation (X- or γ -rays): 2.58×10^{-4} coulomb/kg of dry air. Exposure of dry air to one r results in energy absorption of 93 ergs/g; exposure of water or cellular material to one r results in energy absorption of about 98 ergs/g (→ Rad, RBE, Rem, LET).

Table 13. Major categories of RNA splicing. (Cech 1988)

	Splice site[a]		Required RNA structures[b]	Splicing mechanism	Catalysis
	5′	3′			
Group I	U↓	G↓	Intron folding and intron-exon pairing	Guanosine addition two-step transesterification	Self-splicing modulated by proteins
Group II	↓GUGCG	Y_nAY↓	Intron folding and intron-exon pairing	Lariat formation two-step transesterification	Self-splicing modulated by proteins
Nuclear mRNA	AG↓GURAGU	Y_nAG↓G	snRNA folding and snRNA-intron pairing	Lariat formation two-step transesterification	Spliceosome/sn RNP's
Nuclear tRNA	no conservation		Exon-exon pairing	Endonuclease + ATP-dependent ligase	Enzymatic

[a] Y, pyrimidine; R, purine.
[b] snRNA, small nuclear RNA.

roentgen equivalent dose (Crow 1973) — that dose (symbol RED) of a chemical mutagen which is equivalent to the X-ray dose necessary to induce quantitatively and qualitatively the same effect (= radequivalent).

rolling circle replication (Gilbert and Cressler 1968) — a → DNA replication mode in bacterial mating and during replication of circular phage DNA in which a → replication fork proceeds around a circular template for an indefinite number of revolutions. The nearly synthesized DNA strand in each revolution displaces the strand synthesized in the previous revolution, giving a tail containing a linear series of concatemeric sequences complementary to the circular template strand.

A variant of r.c.r. is "looped out r.c.r." which produces a progeny single-stranded circle from a double-stranded circular template (e.g., in phage ∅×174).

The main difference between r.c.r. and other modes is the asymmetry of replication: One strand remains circular, the other is cut. The circular strand may be copied several times, while the other parental strand is copied only once.

root nodule — in leguminous plants, any of the unique organs which originate by interaction between certain soil bacteria containing → plasmids (→ Ti plasmid; Ri plasmid) and in which molecular nitrogen is reduced to ammonia by bacterial nitrogenase. The nodule-specific proteins are termed nodulins.

Rot value — the product of intial RNA concentration (R_0) and time (t) of incubation in an → RNA-driven hybridization reaction.

R plasmid — = → resistance plasmid.

R-prime plasmid — a → plasmid carrying selectable bacterial genes.

r-protein — = → ribosomal protein.

r selection — → selection favoring a rapid rate of population increase; typical of species that colonize short-lived environments or of species that undergo large fluctuations in population size (r represents the intrinsic rate of increase). (→ k selection).

runaway-replication plasmid (Uhlin and Clark 1981) — a plasmid → cloning vector which exhibits temperature-dependent amplification of plasmid DNA. Such plasmids are maintained at controlled copy number at a certain temperature; after shift to higher temperatures (from 30 to 37 °C) replication becomes uncontrolled resulting in an exponential increase in the concentration of plasmid DNA. R.-r. plasmid vectors are useful for synthesis of large amounts of specific gene products.

S

S_1, S_2, S_3, etc. — shorthands for generations of self-fertilization in plants: S_1 denotes the generation obtained by selfing the parent plants, S_2 the generations obtained by selfing the S_1 plant, etc.

Sall box (Grummt et al. 1985) — a sequence element (AGGTCGACCAG$\frac{\text{AT}}{\text{TA}}$NTCCG) that directs termination in (mouse) → ribosomal DNA transcription by → RNA polymerase I. The sequence is present eight times in the spacer region downstream of the 3′ end of → preribosomal RNA. Transcription termination requires interaction with some factor(s).

S-allele — any → allele of those genes that control → incompatibility in plants. Alleles present in both style and pollen are referred to as "matched S-alleles", whereas those present only in either style or pollen are called "unmatched S-alleles" (Lewis 1943). S_F-alleles are alleles belonging to the same series of multiple alleles as S-alleles but which control self-fertility.

Salmonella-microsomal assay (Ames et al. 1973) — in → mutagenicity testing, a system which uses an exogenous microsomal activation system to convert → promutagens to their mutagenic form; the S.-m.a. is a rapid screen for chemical mutagens.

saltational speciation (Lewis 1966) — a rapid → speciation (or evolution) process characterized by the existence of pairs or groups of diploid (plant) species differentiated by little more than chromosome structural changes, often accompanied by → aneuploidy (= quantum speciation).

saltatory DNA replication (Davidson and Britten 1971) — DNA amplification resulting in organized sets of → repetitious DNA. It may be due to unequal → crossing over or repeated replications of a template progenitor sequence.

samesense mutation — a base substitution mutation in the third base of the relevant → codon which is redundant or synonymous (= synonymous mutation). Because of the degeneracy of the → genetic code, a substitution of the third base does not necessarily alter the amino acid sequence of a polypeptide chain.

Sanger method — → DNA sequencing.

SAT-chromosome (Heitz 1931) — a chromosome with a secondary → constriction which separates a → satellite from the rest of the chromosome. SAT-chromosomes whose secondary constriction is associated with the formation of the → nucleolus are referred to as "nucleolar" SAT-chromosomes and the constriction is a "nucleolar" secondary constriction.

satellite (Navashin 1912) — a chromosomal segment separated from the main body of the chromosome by one secondary → constriction if located terminally (terminal s.), or by two if intercalary (intercalary s.). Together, the s. and the secondary constriction are termed "satellite region".

satellite association (Ferguson-Smith and Handmaker 1961) — in mitosis and meiosis, the tendency of (human) acrocentric chromosomes to be located near one another. S.a. is probably the result of the involvement of these chromosomes in → nucleologenesis and occurs during telophase when a number of nucleoli fuse to form the interphase → nucleolus. Frequency and constitution of these associations at metaphase reflect the pattern of nucleolus behavior and fusion in the preceding interphase.

satellite DNA (Sueoka 1961; Kit 1961; Britten and Kohne 1968) — (sDNA); originally, nuclear DNA that can be separated from the bulk of nuclear DNA as distinct bands during isopycnic CsCl gradient centrifugation. It occurs in animals and plants (not in lower fungi) and consists of short (from several base pairs to several thousand bp in length), tandemly arranged repeats of simple sequence DNA located in → heterochromatic chromosome regions which are usually not transcribed. Amount and sequence arrangement of sDNA vary significantly between different species of the same family.

Some species have cryptic (hidden) satellites with the same density as main band DNA.

satellite virus — a defective virus (bacteriophage) which requires the presence of a → helper virus (helper phage) to complete its life cycle.

saturation density — cell density at which a cell population stops multiplying.

saturation hybridization — a method using a large excess of one component and causing all complementary sequences of the other component to enter a duplex form.

saturation mutagenesis — a procedure used for determining the nucleotice sequence requirements of various genetic elements. Each base pair within a region known to be required for a particular function is altered and the phenotypic effects are analyzed.

scaffold association region — (SAR); = → matrix association region.

scaffold attachment region — (SAR); any of those regions in nontranscribed DNA fragments anchoring DNA loops (domains) of chromosomes to the → nuclear matrix or scaffold.

scaffolding — a method which allows any gene or DNA sequence to integrate into the genome of a desired organism: (1) The sequence from the desired organism is cloned into a circular vector; (2) the vector carrying that sequence is used for integration into the genome by homologous recombination; (3) any DNA sequence cloned using the same vector can be inserted into the chromosome due to the homology between the transforming and the integrated vectors.

scaffolding protein (Paulsen and Laemmli 1977) — → chromosome scaffold.

scanning ribosome model (Kozak 1978) — = → ribosomal scanning model.

scn DNA — acronym for single copy nuclear DNA.

screen — a survey of a large number of cells or individuals for a desired activity or phenotype.

scripton (Szybalski et al. 1970) — = → transcripton.

sealase (Hurwitz et al. 1967) — → nick.

secondary pairing (Darlington 1928) — → chromosome pairing.

second division segregation — reductional ("postreductional") separation and thus → segregation of a → heterozygous pair of alleles during the second division of → meiosis (where the first meiotic division ws "equational" for the allele pair in question), as opposed to first division segregation, i.e., reductional ("prereductional") separation of a heterozygous pair of alleles during the first meiotic division (where the second meiotic division is then equational).

second messenger (Sutherland et al. 1965) — any agent responsible for the transmission of hormonal stimuli within the cell or within compartments of the cell. 3′, 5′-cyclic adenosine monophosphate (c-AMP) is a s.m. with tissue-specific action and a variety of different effects in different organs and organisms. At the molecular level it activates various protein kinases which in turn are able to regulate the activities of numerous other enzymes. cAMP also influences negatively controlled activities of genes (→ operon; catabolic repression) and thus does not act solely via activation mechanisms.

second site mutation — the second of a double mutation within a gene; in some cases, the s.s.m. suppresses the first mutation so that the double mutant has the wild-type phenotype.

second site reversion (Yanofsky et al. 1961) — → suppressor mutation.

secretion granule — in exocrine cells, the intracellular storage compartment for protein destined for export. S. granules are located in the apical cytoplasm of the exocrine cells between the Golgi apparatus region, where the granules originate, and the apical cell membrane, where the period of storage is terminated by release of the granule contents into the extracellular space by the process of → exocytosis.

secretory vesicle — a membrane-bounded container in secretory eucells (= s. granule). In the s.v. secretory products accumulate. It is generally located at the apex of the cell.

sector — a tissue area whose phenotype is different from the surrounding tissue phenotype.

sectorial (Baur 1909) — of a → chimera in which the idiotypically distinct tissues are cross-sectionally arranged as sectors of a circle.

seduction (Bulat et al. 1985) — the irregular excision of an integrated episomal → plasmid together with the linked chromosomal genes and transfer to other chromosomes or cells (e.g., in yeast). (→ sexduction).

seed incompatibility (Valentine 1960) — all forms of sterility in plants due to disturbances after fertilization or the development of tissues involved in the production of seeds.

segment (Belling 1927) — a portion of a chromosome taken as a unit under given circumstances (→ differential s.; interstitial s.; intercentric s.).

segmental allopolyploid (Stebbins 1947) — → alloploid.

segmentation gene — in eukaryotes, any of the genes that determine the number and polarity of body segments (→ homeobox).

segmented genome — in the case of some single- and double-stranded RtNA viruses, the presence of several nonidentical RNA molecules, each carrying separate genetic information.

segregation (Bateson and Saunders 1902) — = → genetic segregation.

segregational lag (Witkin 1951) — a delay in the phenotypic manifestation of a newly arisen → gene mutation in bacterial cells containing more than one → nucleoid. Since there is more than one set of genetic material, phenotypic expression of a mutation in one of the chromosomes must await the partitioning (segregation) of that chromosome into a cell that contains no nonmutant chromosomes (→ phenotypic lag).

segregational load (Crow 1958) — → genetic load.

segregational sterility (Darlington and Mather 1949) — → sterility resulting from gene segregation in structural and numerical hybrids which gives rise to genetically unbalanced gametes, as opposed to "genotypic sterility" which is due to an unbalance in the zygotic genotype itself.

segregation distortion (Sandler and Hiraizumi 1961) — any systematic deviation from equal representation of alleles or homologous chromosomes among the functional gametes (→ meiotic drive).

S.d. may be an important factor generating or maintaining → gametic disequilibrium.

selection (Darwin 1858) — any process resulting in observable changes within a generation in the means, variances, and covariances of phenotypic distributions. The evolutionary response to s. depends on inheritance, i.e., its primary consequence is change in the frequencies of genes affecting the trait upon which s. acts. Thus different genotypes contribute alleles to the → gene pool of each generation not precisely in proportion to their number, but relative to the → selective advantage they enjoy. The unit of s. is the individual organism.

Natural s. is the result of two constraints imposed on every living organism: (1) The requirement for reproduction; (2) the requirement for permanent interaction with the environment. The interaction between the two factors results in differential reproduction. It is natural s. that gives direction to changes (mutants, recombinants), orients chance, filters genotypic fitness arrays, and slowly gives rise to evolutionary changes (→ phyletic evolution; punctuated evolution).

Sexual s. is a special form of natural s. responsible for the evolution of traits that promote success in competition for mates and has two aspects: (1) Intrasexual s. focuses on → competition between members of one sex (usually male) for the access to the other sex; (2) epigamic s. or female choice focuses on the consequences of mate preferences of one sex (usually females) with regard to the other sex.

Artificial s., in contrast to natural s., is a purposeful process with definite goals set by the breeder. It normally occurs with controlled matings of a few selected genotypes, and its goal is to change specific phenotypic traits of a population.

The **modes of Darwinian selection** within populations may be classified according to their operation into stabilizing, directional, and disruptive s. (Mather 1953), but in any given situation two or even all three modes may act simultaneously: (1) **Stabilizing s.** (= centripetal s. of Simpson 1953; normalizing s.) favors a single optimum in the population. The main effect of s. is the elimination of peripheral variants which arise by mutation, immigration, or recombination. In stable environments and populations which have already achieved a high state of adaptation, a certain range of genotypes of proven fitness is preserved from generation to generation. According to Waddington (1953, 1957), two different stabilizing responses may result from s. as parent individuals near the mean; either genotypes which control the development of abnormal phenotypes ("normalizing" s.) or which render developmental pathways of the developing individual sensitive to the potentially disturbing effects of environmental stresses ("canalizing s.") will be eliminated. (2) **Directional s.** (progressive s. of Schmalhausen 1949; linear s.; dynamic s.) favors a single optimum like stabilizing s. but results in a systematic shift (in the direction of applied s.) in gene frequencies and in the population mean for the character considered. Directional s. operates in a progressively changing environment and leads to a state of adaptation. (3) **Disruptive s.** (= centrifugal s. of Simpson 1953) simultaneously favors more than one optimum in a population occupying a heterogeneous habitat, i.e., two or more different genotypes are at an advantage and intermediate types are at a disadvantage. The favored genotypes are maintained in a state of polymorphic equilibrium by the tendency of each of the several selective pressures to preserve a cetain genotypic variant which is optimally adapted to one particular feature of the environment.

In contrast to Darwinian s. within populations, **interpopulation s.** (Grant 1963) means the differential reproduction of populations of any size (from local breeding groups to non-interbreeding species).

Interpopulational s. may be subdivided into: (1) **Group s.,** i.e., competition among different populations of one and the same species; (2) **species s.,** i.e., competition among whole species, which is one of the major evolutionary forces.

selection coefficient — (s); a quantitative measure of the intensity of selection, i.e., the proportional reduction in the genetic contribution of a particular genotype compared with a standard genotype, usually the most favored, or the proportional increase or decrease of → fitness of one genotype in relation to another. The contribution of the favored genotype is taken to be 1 and the contribution of the genotype selected against is then (1-s). The s.c. can have a range of values from 0 to 1 and may, in the course of selection, undergo fluctuations and reversals. (→ selection pressure).

selection differential (Lush 1945) — (S); a measure of selection intensity, i.e., the difference between the mean of a population and the mean of the individuals selected to be parents of the next generation. The mean may shift within a generation for two reasons: (1) Because selection has acted directly on the character in question; (2) because selection has acted on a correlated character.

selection gradient (Lande 1979) — the partial regression of relative fitness of a character which measures the direct force of selection on that character.

selection index (Smith 1936) — an index designed to maximize genetic-economic merit or aggregate breeding value for multiple traits among individuals in a population.

selection limit — a population is said to be at a s.l. when the response to selection has ceased. The s.l. is reached gradually with the response to selection becoming progressively slower.

selection pressure (Wrigth 1921) — the intensity of natural → selection which is usually measured by the change of → gene frequency per generation due to the influence of selection. The degree of selective advantage of a preferred allele or genotype over the others is one of the fundamental factors determining the rate at which allelic frequencies are changed under s.p. This factor may be measured in terms of → selection coefficients.

selection progress — the difference between the mean of a population and the mean of the offspring in the next generation born to selected parents.

selection response — the product of the → heritability and the → selection differential.

selective advantage — an advantage in the competition for survival between one genotype as compared with another showing "selective disadvantage" and producing fewer viable offspring. In **natural selection,** the s.a. of a certain genotype is the property of all its phenotypic expressions and may have different causes, e.g., the production of larger progenies, better resistance of the progeny to environmental factors with deleterious effects, better resistance of the progeny to predators (including parasites), better ability of the progeny to find a suitable habitat, etc. In **artificial selection,** s.a. is conferred by the breeder on some members of a population in preference to others because they or their relatives possess certain phenotypic characters which are visible, measurable, or can be evaluated in some manner by the agency of control over the relative reproductive rate (Lerner 1958).

In the absence of other differences, the genotype showing selective disadvantage is gradually replaced by the genotype with s.a. if both are present within one habitat (→ annidation).

selective deletion — the introduction, by means of → recombinant DNA technology, of specific deletions in a genome; s.d. may be used to locate a regulatory site, to delineate the boundaries of a gene, and to generate new mutant strains.

selective disadvantage — → selective advantage.

selective fertilization — → fertilization.

selective neutrality — a situation in which different alleles of a certain gene confer equal → fitness.

selective peak (Wright 1956) — a → gene frequency system characterized by homeostatic properties (→ homeostasis). The population occupying a s.p. tends to swing back to that peak after temporary disturbances, provided these are not so drastic as to shift the group of individuals into the sphere of influence and control of another s.p. (→ adaptive peak).

selective silencing (Sager and Kitchin 1975) — any mechanism which consistently eliminates genes of one parent from the zygote.

selective system — any technique which enhances the recovery of specific, usually rare genotypes (→ sieve).

selective value — a value proportional to the probability that an individual (genotype) survives from conception to adulthood (= fitness). The mean **s.v. of a gene** is the mean of the s. values of the genotypes into which the gene enters, weighted by the frequencies of the other genes involved. Mean s. values of genes are function of the gene frequencies and, therefore, change when the gene frequencies change.

The mean **s.v. of a population** is the mean of the s. values of the various genotypes, weighted by their frequencies at the time of conception. The change of the mean s.v. of a population due to natural selection is proportional to the additive variance (→ genetic variability) of the s. values of the individuals in the population (Jacquard 1974).

The "index of the opportunity of selection" (Crow 1958) measures the maximum change in the s.v. which the population's mortality and fecundity characteristics can entail.

self — to fertilize female gametes with male gametes from the same individual.

self-assembly — the ability of certain multimeric biological structures to assemble from their component parts through random movements of the molecules and formation of weak chemical bonds between surfaces with complementary shapes.

self-catabolite repression (Katz and Engelsberg 1971) — a feedback mechanism involved in → catabolite repression which, in interaction with induction, provides a means of insuring that optimal levels of enzymes are produced for the efficient metabolism of the → inducer of an operon. This mechanism is probably common to all instances where the inducer of an operon is metabolized. S.-c. r. insures the prevention of enzyme overproduction through the induction process and is due to the production of → effectors which dampen the expression of the relevant operons.

self-compatibility — → self-incompatibility.

self-fertile — of individuals having reproductive organs and gametes which permit zygote formation by fusion of gametes derived from the same parent organism. The phenomenon is referred to as self-fertility.

self-fertilization — the union (= selfing, automixis) of male and female gametes and/or nuclei from the same haploid, diploid, or polyploid organism (→ fertilization; paedogamy; parthenogamy).

self-incompatibility (Stout 1917) — in diverse angiosperm families, the failure of self-(or cross-)fertilization, as opposed to self-compatibility. S.-i. systems operate prior to fertilization during the interaction between the male gametophyte (pollen or pollen tube) and the female sporophytic tissue of the pistil. S.-i. promotes outbreeding and heterozygosity; the specificity of the incompatible interaction is determined by one or more genetic loci, and requires that the alleles (S-alleles) carried by the male and female parents are identical. S.-i. systems are classified as heteromorphic or homomorphic according to whether or not they are associated with differences in floral structure. Homomorphic systems may be further classified according to the genetic control of pollen phenotype: (1) In sporophytic systems, pollen behavior is determined by the genotype of the plant producing it; (2) in gametophytic systems, pollen behavior is determined by the gene(s) carried by the pollen.

Pseudo-self-compatibility occurs when plants with a functional s.-i. system yield seeds following self-pollination (due to modifying genes inhibiting S gene activity).

selfish DNA (Crick 1979; Dover and Ford-Dolittle 1980) — in eukaryotes, any tandemly repeated or dispersed repetitive DNA sequence which has no obvious function but can spread and accumulate (by unequal crossing over) in the species because of its rapid replication (= parasitic DNA).

self-pollination — the transfer of pollen to stigma within the same flower or in the same → clone.

self-regulation (Smith and Magasanik 1971) — = → autoregulation.

self-splicing intron (Cech et al. 1921) — → RNA self-splicing.

self-sterility (Darwin 1876) — → sterility; self-incompatibility.

semantide (Zuckerkandl and Pauling 1965) — any → informational macromolecule.

semi-apospory (Fagerlind 1940) — → apomixis following a → pseudohomeotypic division with the gametophyte arising from an unreduced egg cell. Apospory, semiapospory, and diplospory blend into one another without clearcut distinctions.

semi-bivalent (Battaglia and Boyes 1955) — any of the units of anaphase movement in the case of "postreductional → meiosis", consisting of two chromatids paired side by side or end to end. With more than two homologues (polyploidy or polysomy) semi-trivalents, semi-quadrivalents, etc., may arise.

semi-conservative DNA replication (Delbrück and Stent 1957) — the normal mode of → DNA replication in which each of the two parental strands is conserved in a complementary daughter molecule (→ DNA polymerase).

semi-discontinuous DNA replication — a mode of → DNA replication in which the leading strand is synthesized continuously, the lagging strand discontinuously.

semi-dominant — → dominant.

semigamy (Battaglia 1963) — in plants, an abnormal mode of fertilization in which the male nucleus penetrates the egg cell but does not fuse with the egg nucleus. Both nuclei divide independently and give rise to a chimeric embryo formed of male and female (hybrid) tissues.

semi-geographic — of a mode of → speciation which consists in the partitioning of species along a line of secondary → intergradation or along lines of strong ecological contrast (*Mayr* 1963).

semi-heterotypic (Rosenberg 1926) — → nonreduction.

semi-homologous — of chromosomes or genomes which are partially → homologous (= homeologous).

semi-isolation — incomplete → isolation between two populations, allowing incidental → gene flow.

semi-lethals (Muller and Altenburg 1919) — = sublethals (→ lethal factor).

semi-species (Mayr 1940) — borderline cases between → species and → subspecies characterized by some, but not all, attributes of species rank. They may show greater morphological differences and greater discontinuity in their variation pattern than is usual for subspecies (races) without possessing the → isolating mechanisms against interbreeding that are characteristic of true species. Related s. are linked by limited amounts of → gene exchange (interbreeding is not as free as between subspecies, but freer than between species) and represent breeding units of higher order. An assembly of s. is referred to as a **superspecies or syngameon** depending on the geographical relationships between the component semi-species. If the semi-species under observation are allopatric, the assembly represents a superspecies, while a syngameon consists of sympatric semi-species (Grant 1963).

semi-sterility (Belling 1914) — sterility in genic and structural hybrids (heterozygous for → chromosome mutations such as reciprocal → translocations) characterized by inviability of approximately half of the male and female gametes.

sen DNA (Stahl et al. 1978) — an amplified section of mitochondrial DNA in senescent **Podospora** cultures which is circular and plasmid-like in nature. Sen DNA is the primary cause of senescence in **Podospora**.

S_1 endonuclease — an enzyme which specifically degrades unpaired (single-stranded) DNA sequences.

S_1 endonuclease hypersensitive site — → nuclease hypersensitive site.

S_1 endonuclease mapping (Berk and Sharp 1977) — a gel electrophoresis method of measuring and locating regions coding for RNA transcripts within a cloned DNA fragment. RNA is annealed to denatured DNA under conditions favoring RNA-DNA duplex formation. Subsequently single-strand nucleic acid is digested by S1 endonuclease and the lengths of the DNA segments protected by hybridization with RNA from digestion are determined electrophoretically. Natural gel electrophoresis allows the measurement of RNA-DNA duplex length, alkaline gel electrophoresis permits determination of the lengths of the single-strand DNA.

sense strand — = → codogenic strand.

sensitizing mutation (Koch and Drake 1970) — a → leaky mutation which is used to enhance the recovery of normally undetectable, → cryptic mutations.

sensor gene (Britten and Davidson 1971) — a hypothetical type of eukaryotic gene which is assumed to receive a stimulus (probably hormonal) and to cause → integrator genes to transcribe into activator RNA detected by receptor genes. The latter are assumed to cause → structural genes to transcribe.

sentinal phenotype — a clinical disorder or syndrome of characteristic features (Mullihill and Czeizel 1985): (1) It occurs sporadically as a consequence of a simple, highly penetrant mutant gene; (2) it is a dominant or X-linked trait of considerable frequency and low fitness; (3) it is uniformly expressed and accurately diagnosable with minimal effort at or soon after birth.

separon (Benbow et al. 1984) — any of the separated, nonrandomly localized domains of the DNA helix in eukaryotes which are stably denatured and 0.2 kb in length (in *Xenopus leavis*).

septate junction (Wood 1959) — a cell-to-cell specialization in localized areas between apposing cells which may function as an intercellular permeability barrier and in the maintenance of intercellular cohesion. (= septate desmosome).

sequenase — a → DNA polymerase (from bacteriophage T7) used for → DNA sequencing; it allows reading of sequence information beyond the point where other sequencing enzymes fail. The s. (it has no exonucleolytic activity) polymerizes thousands of nucleotides uninterrupted, efficiently incorporates nucleotide analogs, and is not impeded by template secondary structure. It is used in combination with M14 cloning vectors, primers, and either ^{35}S or ^{32}P labeled dATP. The products can be run on any standard sequencing gel apparatus.

sequence amplification — → DNA amplification; gene amplification.

sequence correction — = → sequence homogenization.

sequence-directed DNA bending (Marini et al. 1982) — in double-stranded DNA molecules from prokaryotic and eukaryotic organisms, the curvature (bent structures) found in association with sequences consisting of short oligomeric adenine residue runs, which are regularly spaced, in phase, with the helical repeat. S.-d. DNA b. is required for → DNA replication, → genetic transcription, and → genetic recombination and may provide regulatory elements. It may be due to protein-DNA interactions.

sequence divergence — the percent difference in nucleotide sequence between two related nucleic acid sequences or in amino acid sequence between two proteins.

sequence homogenization — a mechanism (= sequence correction; gene correction) whereby a relative small number of "founder" copies of DNA sequences correct alterations that collect in individual members of a sequence family (→ multigene family; concerted evolution; molecular drive; repetitious DNA).

sequence homology map — a physical → chromosome map obtained from → heteroduplex mapping and division of the heteroduplex map into fractions of → map units.

sequential enzyme induction — → induction.

sequence interspersion — → repetitious DNA; LINE; SINE.

sequence ladder — an array of bands in a gel corresponding to a DNA sequence.

sequence-specific transcriptional regulatory protein — any of a class of proteins with α-helix-turn-α-helix supersecondary structure. 24 amino acids comprise this motif and exhibit an identical three-dimensional structure as well as a homologous primary sequence. The two helix-turn-helix motifs of the → DNA-binding protein dimer interact in a twofold symmetric fashion with successive major grooves of right-handed → B DNA.

sequential mutagenesis (Cerda-Olmedo et al. 1968) — a technique which exploits the increased susceptibility of the replication point of the bacterial chromosome to mutagenesis by N-methyl-N′-nitro-nitrosoguanidine. Mutagen treatment of successive samples from a synchronized bacterial culture and the scoring of treated cells for mutants results in a maximum number of mutants for particular genes occurring at the time of gene replication.

serial genes (Grant 1964) — → gene interaction.

serum-responsive element — (SRE); a palindromic sequence, which because of its dyad symmetry is also called the dyad symmetry element (DSE); the SRE is the binding site for a ubiquitous nuclear phosphoprotein (the serum-responsive factor; SRF) which is presumed to mediate the activation of certains genes to serum or growth factors contained in it. Serum might increase the ability of the protein to activate transcription without affecting its affinity to DNA or it might promote the formation of a complex between SRF dimer and a separate protein containing a transcriptional activity domain.

serum-responsive factor — (SRF); → serum-responsive element.

service unit (Lifschytz 1971) — in eukaryotes, a DNA sequence for which the → chromomere or → band is the cytological counterpart; in the simplest case, a s.u. is assumed to be composed of a structural gene and all sequences needed for its function (such as initiation and termination sites for replication and transcription as well as recognition sequences needed for the regulation of the translation process). In certain cases, more than one structural gene can be associated with one s.u. or, conversely, some service sequences can be duplicated (→ transcription).

seven-transmembrane segment (Dohlman et al. 1987) — in receptors coupled to GTP-binding proteins, a conserved structural hydrophobic motif which probably forms membrane-spanning α-helices. Mammalian genes coding for predicted gene products with sequence similarity to these receptors are known.

Sewall Writht effect — = → genetic drift.

sex — the contrasting and complementary characteristics exhibited by male and female, + or − (if the differences are only physiological), donor and recipient (in prokaryotes) individuals or clones within the same species, as well as the capacity for performing amphimixis and/or → genetic recombination by such individuals (→ reproduction). The basis for the biological divergence of the sexes may be either inherent in the organism or impressed on it by its environment (→ sex determination).

Sex is able to affect the phenotypic expression and the genetic ratios of nuclear hereditary determinants in several ways: (1) The → inheritance of genes located in the → sex chromosomes (and showing sex-linked inheritance) follow special rules based on the pattern of homology and transmission of these chromosomes. (2) Dominance of alleles may differ in heterozygotes of the two sexes. This phenomenon is termed "sex-influenced dominance". (3) One sex may be uniform in phenotype with respect to particular characters yet transfer genes which give rise to phenotypic differences in the offspring of the other sex. This is referred to as "sex-limited gene expression". (4) Cumulatively acting genes may be required to exceed different thresholds in males and females before a phenotypic difference becomes manifest and, therefore, phenotypic segregation simply results in two classes, variously reached in the two sexes.

In the case of extranuclear hereditary determinants, transfer to the offspring may be exclusively (at least with oogamy) through the female line; this type of → inheritance is called "matrilinear".

sex cell — = → gamete.

sex character — any of the primary sex differences concerning gametes and sexual organs, as well as the secondary characters that do not concern gametes and sexual organs.

sex chromatin (Barr and Bertram 1949) — one, in special cases more than one, plano-convex, spherical, or pyramidal and Feulgen positive intranuclear body (about $0.8 \times 1.1\ \mu m$ in size), commonly situated at the periphery of the interphase nucleus just inside the nuclear envelope (= Barr body). It represents a single → X-chromosome which is positively → heteropycnotic and was originally called a "nucleolar satellite". Depending on the tissue and technical factors, 20–96% of nuclei of all females of many mammalin species contain s.c., whereas the comparable structure is rare or absent in nuclei of diploid males, with the exception of species with unusual → sex chromosomes. The s.c. of leukocytes is contained in a specific drumstick-like nuclear appendage which is usually referred to as a → "drumstick".

Depending on the number of X-chromosomes, the maximum number of sex chromatin bodies in any one cell is: none in XY and XO individuals; one in XX, XXY, XXYY; two in XXX, XXXY; and three in XXXX individuals, i.e., one less than the sum of X-chromosomes in a diploid or aneuploid individual. Cells with less than the maximum number have fewer and, accordingly, larger bodies.

Sex chromatin is derived from one of the two X-chromosomes normally present per cell in diploid females and is in a heavily condensed heteropycnotic state (→ dosage compensation).

sex chromosome (Wilson 1906) — in eukaryotes with separate sexes, a chromosome or group of chromosomes (multiple s. chromosomes) which are represented differentially in the male and female sex and are causally involved in the genotypic control of → sex determination.

In organisms that have the sexes combined in a single individual, as in hermaphroditic animals and the majority of higher plant species, there are no special s.-chromosomes in the karyotype. The formation of male and female germ cells (microspores and megaspores in plants) is accomplished in such cases by a process of histological differentiation.

In the **basic system of sex chromosomes,** in one sex a pair of microscopically indistinguishable chromosomes (referred to as the X-chromosomes) is represented in the opposite sex by a pair of visibly different chromosomes. One of these two is identical to the X-chromosome of the opposite sex, the other is structurally and functionally different, restricted to the one sex and termed the Y-chromosome. Regions in which → chromosome pairing and → crossing over between X- and Y-chromosomes can take place and which are → homologous are called "pairing segments", as opposed to "differential segments" which are nonhomologous and do not pair at meiosis.

Normally, the XX sex is female and the XY sex male. The XX and XY sexes are also called "homo-" and "heterogametic", repectively. The homogametic sex produces (during meiosis) gametes of one kind (all contain one X) while the heterogametic sex gives rise to two types of gametes with either one X or one Y chromosome. The X- and Y-chromosomes of the heterogametic sex regularly disjoin to opposite spindle poles at either the first ("prereduction" of the sex chromosomes) or the second ("postreduction" of the sex chromosomes) meiotic division, so that equal numbers of gametes with either an X- or a Y-chromosome are produced. Union of an X-bearing male gamete with an X-bearing female gamete results in a female zygote, while union of a Y-bearing male and an X-bearing female gamete gives rise to a male zygote.

In some organisms, the situation is reversed, i.e., structurally identical sex chromosomes are present in the (homogametic) male sex and structurally distinctive ones in the female (heterogametic) sex. With female heterogamety the sex chromosomes are often called Z (= X) and W (= Y) chromosomes.

Various types of **derived sex chromosome systems** differing from the basic XX♀ — XY♂ or ZZ♂ —ZW♀-systems are known. In some the Y-chromosome has been lost altogether (XX-XO-systems) or the X- and Y-chromosomes have become completely nonhomologous, that is, unable to pair and to undergo crossing over during meiosis. In others, called systems with multiple sex chromosomes, either the X- or Y-chromosome is represented by two or more chromosomes.

Five main mechanisms may give rise to multiple sex chromosome systems (Dietz 1958): (1) Reciprocal → translocations (→ centric fusions) between sex chromosomes and autosomes may bring autosomes (called neo-X- and neo-Y-chromosomes) into the sex-chromosome system and transform a system of the XX-XO-type into a neosex chromosome system of the XX-XY-type or a system of the XX-XY-type into a X_nY-, XY_n-, or X_nY_n-system. (2) Fragmentation of an X-chromosome with diffuse → centromere organization. (3) Increase due to polyploidy of the number of sex chromosomes. (4) Autonomous transformation of autosomes into X-chromosomes. (5) Increase in the number of Y-chromosomes due to → nondisjunction which gives rise to supernumerary Y-chromosomes.

The behavior in meiotic segregation of multiple sex chromosomes may be grouped into four basic types (Cooper 1946): (1) Systems with a sex chromosome → multivalent may show a definite orientation and distribution of the constituent chromosomes at meiosis. (2) Systems without physical contact between the segregating chromosomes in which alternative sets of sex chromosomes take part in distance or "touch-and-go" pairing. (3) Systems (e.g., X_1X_2OO males) characterized by multiple sex chromosomes whose components are all of the same type and are distributed together to one and the same spindle pole. (4) Systems in which one sex chromosome bivalent and one univalent, or two and more sex chromosome bivalents are regularly distributed in such a way that definite groups of sex chromosomes (not in physical contact with one another) reach opposite spindle poles.

sex composite — = → gynandromorph.

sex-controlled (Goldschmidt 1920) — of characters, whose degree of phenotypic expression is controlled by the sexual constitution of the individual. The type of → inheritance in which sex controls or changes the manifestation of characters, but not the chromosomal transmission of the hereditary determinants controlling these characters (→ sex linkage), is referred to as "sex-controlled inheritance".

sex cycle — the alternation of → karyogamy and → meiosis in eukaryotes, or (by extension) any recurrent series of processes that lead to → genetic recombination (→ parasexual) in prokaryotes.

sex determination — the determination of sex differentiation in eusexual (→ parasexual) organisms by processes which shift the so-called "bisexual potency" (Hartmann 1923) or the balance between male (M) and female (F) determining genes (sex realizors) of cells, organs, and individuals in either the male or female direction. (1) **Environmental (or phenotypic) s.d.** (Burgeff 1915): A mechanism in which zygotes differentiate as male or female depending largely upon environmental circumstances, i.e., genetic s.d. without a sex chromosome mechanism. (2) **Genotypic s.d.** (Burgeff 1915): Sex is mainly determined by the genotype of the zygote or spores. If s.d. takes place during haplophase by meiotic segregation of genes, resulting in half male and half female gametes (or spores), this is called **"haplogenotypic s.d."** in contrast to **"diplogenotypic s.d."** where the diplophase (in diplontic organism) is also sexually determined (either male or female).

There are two main cytogenetic mechanisms of diplogenotypic s.d. in which the so-called → sex chromosomes play a decisive role in the inheritance and determination of sex. The XY-system and its several variations based on the balance between X-chromosomes and autosomes occurs in higher animals and dioecious plants and represents a multigenic mechanism.

In other XY-systems, the Y-chromosome of the heterogametic sex is actively sex-determining and individuals carrying such a chromosome are of a definite sex (male in the case of male heterogamety, female in the case of female heterogamety) irrespective of their autosomal genotype. The sex of the embryo is determined at fertilization by the sex chromosomes contributed by the sperm and ovum.

Haplo-diploid s.d. is a mode of s.d. in which the males arise from a haploid egg and the females from a diploid.

Maternal s.d. is a mode of s.d. in which sex is determined by the → idiotype of the mother or egg cell.

Syngamic s.d. is a mode of s.d. in which the sex of the progeny is fixed as a result of gamete fusion and karyogamy.

Progamic s.d. is a mode of s.d. in which the sex of the progeny is determined in the egg prior to fertilization and even prior to meiosis.

Metagamic s.d. is a mode of s.d. in which the sex of the progeny is not fixed by karyogamy and, therefore, is largely subject to environmental influences. Genotypically similar individuals or cell complexes become male or female after fertilization by influences external to the genotype, or a genotypically male individual may be transformed into a phenotypically female individual by external factors and vice versa.

sex differentiation — the origin and development of organisms with different (male and female) sex organs and sex characteristics and the production of male and female, or (+) and (−) gametes, by one individual or different individuals. By extension, also the production of donor and recipient strains among bacteria (→ parasexual; sex determination; sex factor).

sex digamety — the capacity of one sex (called heterogametic) to produce male- and female-determining gametes by means of the → sex chromosome mechanisms. (→ sex monogamety; sex polygamety).

sex duction (Jacob and Wollman 1960) — = → F duction.

sex factor — = → F plasmid.

sex fimbrium (Duguid and Anderson 1967) — → pilus.

sex-influenced — → sex-controlled.

sex-limited (Morgan 1910) — of genetically controlled characters which are phenotypically manifest in only one of the two sexes due to sex-limited gene expression. The underlying genes (which may be located in → sex chromosomes as well as in → autosomes) are normally transmitted in → inheritance by the male and female parents (→ sex-controlled).

sex-linkage (Morgan 1914) — → linkage of genes located in the → sex chromosomes (X- and Y-chromosomes) of eukaryotes. Such genes and their mode of → inheritance, as well as the phenotypic characters whose manifestion these genes control, are all referred to as "sex-linked" (→ sex-limited).

(1) **Partial (incomplete) s.l.** (Darlington et al. 1934): The genes concerned have loci on both X and Y and thus are located in pairing segments of the two chromosomes and may be exchanged by → crossing over between these chromosomes. (2) **Complete s.l.** (Morgan 1914): The genes concerned have loci only on the X and thus are in the differential (nonhomologous) segments of the X- and Y-chromosomes. They are not recombinable by crossing over.

Genes located in the Y-chromosomes are termed → "holandric".

sex monogamety (Bacci 1961) — in the case of balanced hermaphoridites, labile → gonochoric and exclusively → parthenogenetic populations, the supposed capacity for producing gametes of only one kind with respect to sex determination (→ sex digamety).

sex mosaic — = → gynandromorph or gynander.

sex pilus (Meynell and Lawn 1967) — → pilus.

sex plasmid — = → F plasmid.

sex polygamety — the capacity of male and female or hermaphroditic individuals for producing gametes of many different types with respect to sex determination by a polyfactorial mechanism (Bacci 1965). (→ sex monogamety; sex digamety).

sex ratio — (1) The number of males per 100 females at zygote formation (primary s.r.), at birth (secondary s.r.), or at sexual maturity (tertiary s.r.). (2) A form of → meiotic drive (in *Drosophila*) characterized by the production of only or mostly female progeny by a male whose → X-chromosome carries a certain genetic factor.

sex realizer (v. Wettstein 1924) — any of the sex-determining or sex-deciding genes (symbolized by F for female and M for male determiners) which are assumed either to decide whether the male (A) or female (G) part of the so-called "bisexual potency" (Hartmann 1923) will become more active (→ AG complex), or to represent a balanced system with an alternative norm of reaction (Goldschmidt 1929/1930), which may be shifted in either the male or female direction.

sex reversal — a change (either natural, pathological, or experimentally induced) in the sex of an individual from male to female or vice versa.

sexual — of all processes involving both → meiosis and → fertilization by means of which sexual → reproduction and → genetic recombination are achieved in "eusexual" organisms (→ parasexual).

sexual character — any of those traits which define the sex of an individual and are produced, after → sex determination, by the processes of → sexual differentiation. Primary sexual characters are those directly involved in → reproduction (male and female reproductive organs), secondary sexual characters are those expressed differentially in males and females but which do not participate directly in reproduction.

sexual cycle — in eukaryotes, a cycle consisting in its basic form of the pairwise fusion of haploid gametic nuclei to give diploid zygotes. Zygote nuclei, or some of their mitotic products, then undergo → meiosis so that haploid nuclei are produced. These, or mitotic descendants, then serve gametic functions and the cycle is complete.

sexual differentiation — the developmental pathway realizing the genetic instructions for → sex determination.

sexual dimorphism — the presence of marked differences (in shape, size, structure, color, etc.) between male and female individuals of the same species.

sexual incompatibility — the inability of genetically similar ("homogenic s.i.") or heterogeneous ("heterogenic s.i.") gametes or nuclei to undergo karyogamy (Esser 1956). The term is used for all system restricting sexual propagation on the basis of their genetic determination (→ incompatibility).

sexual isolation — = reproductive → isolation.

sexual reproduction — → reproduction (= gamogony; gamogenesis).

sexual selection — → selection.

sexuparous (Caullery 1913) — producing male and female offspring either by → parthenogenesis or by sexual reproduction (amphigony).

sex vesicle (Sachs 1954) — a prominent heteropycnotic body formed by the X/Y chromosome pair of mammals during meiotic prophase.

shear — to fragment DNA molecules into smaller fragments.

shift — a chromosomal structural change (also called "internal → translocation") which results in the transposition of a chromosome segment to another place in the same chromosome.

shifty transfer RNA (Weiss 1984) — a normal → transfer RNA molecule that promotes frameshifting at a site of nontriplet movement called a shifty → codon. Such tRNA's can cause two-base and four-base translocation errors by mistranslating certain noncognate codons.

Shine-Dalgarno sequence (Shine and Dalgarno 1974) — (SD sequence); in prokaryotes, a group of six to eight purine-rich nucleotides (5′AAGGAGGU3′) located in → messenger RNA 4 to 15 bases upstream (5′) from the → initiator codon AUG. It is complementary to a region close to the 3′ end of 16S → ribosomal RNA (the anti-SD sequence) and involved in binding → ribosomes to mRNA (SD interaction).

The 3′ terminus of eukaryotic mRNA lacks the SD sequence (→ scanning ribosome model).

short interspersed nucleotide element (Davidson et al. 1973) — (SINE); any of the short (about 70 to 300 bp) repetitive sequence elements interspersed with longer single-copy regions in eukaryotic genomes (almost all seem to be → retroposons). Members of each SINE family generally have well-defined ends, with variations between elements occurring primarily in the characteristic simple sequence [usually oligo (A)] found at the 3′ end of the element. SINE's usually make target site duplications of 6 to 21 bp upon insertion. Most of them have a functional internal → promoter (recognized by → RNA polymerase III) which can cotranscribe adjacent chromosomal sequences and thereby mobilize them (→ long interspersed nucleotide element). The best-characterized family of SINE's is the mammalian → Alu-like sequence family.

short patch repair (Cooper and Hanawalt 1972) — → excision repair.

short period interspersion — in most eukaryotic nuclear genomes, interspersion of repetitive → short interspersed nucleotide elements with longer single-copy regions, as opposed to long-period interspersion (→ repetitious DNA).

shotgun cloning — the → molecular cloning of unselected (random) segments of the DNA of a donor cell to form a → gene library of an entire genome out of which the desired clone has to be identified and isolated (→ recombinant DNA technology). This requires hybridization of a probe to a large unordered collection of bacterial colonies, selection of colonies with DNA complementary to the probe, and destruction of the remaining colonies.

shuttle mutagenesis (Seifert et al. 1986) — a technique which uses cloned genes (in *E. coli*) as targets for → transposition and requires only that the organism has a transformation system with the ability to recombine insertions into the chromosome. It allows transposition of a bacterial → transposon containing a selectable gene (of yeast) into a cloned fragment of yeast DNA in *E. coli*; the transposition insertion can be returned to the yeast genome by homologous recombination. S.m. extends → transposon mutagenesis to eukaryotes.

shuttle transfer (Mergeay et al. 1987) — a form of gene capture operating in bacterial strains harboring some peculiar → plasmids. S.t., as opposed to transfer of chromosomal markers in the usual direction (from donor to recipient cells), mediates transfer also in the opposite direction (from the recipient to the donor).

shuttle vector — any bifunctional or multifunctional → cloning vector that can autonomously replicate in two or more different organisms and may be used to shuttle genes from one organism to another. S. vectors result from combining phage, plasmid, or viral DNA's.

sibling(s) — the progeny (= sibs) of the same parents.

sibling effect (Eaves 1976) — in human genetics, the phenomenon wherein the phenotype of one sibling influences the behavior of another, since one sibling constitutes part of the environment in which another develops (competitive or cooperative effects on development).

sibling species (Mayr 1942) — reproductively isolated populations that are morphologically similar or identical and frequently sympatric.

sib mating — brother-sister mating. In many species of plants and animals, specifically those which are self-incompatible or dioecious, s.m. is the closest possible form of → inbreeding.

sib method — in human genetics, a method of deriving the proportions of individuals with and without a character under observation from those members of a sibship other than the → propositus by whom the sibship (→ pedigree) was identified.

side-arm bridge — a type of chromosome structural change inducible only during prophase of mitosis or from pachytene to metaphase I of meiosis and observable at the anaphase immediately following treatment. S.-a. b. also occur spontaneously (→ pseudochiasma).

side-line — → stem line.

sieve (Turner 1977) — a → selective system which, given a range of mutants of roughly equal selective value in relation to some component of the environment, gives preference to those with particular properties of dominance, linkage, sex limitation, or other features of the genetic system.

Sievert — (Sv); one Sv is equal to 100 → rem (→ Gray).

sigma element (Del Rey et al. 1982) — in yeast, any of a family of → transposable genetic elements (341 bp in length) that have 8 bp inverted repeats at the ends and are flanked by 5 bp direct repeats of the target sequence. They are closely associated with genes coding for transfer RNA.

sigma factor (Burgess et al. 1969; Bautz et al. 1969) —in the bacterial → RNA polymerase holoenzyme, any of those subunits that determine the → promoter specificity. S. factors are composed of two functional domains: A core-binding domain and a DNA-binding domain. The major s.f. is responsible for transcription initiation of most genes during growth; alternative s. factors have a role in cellular responses to environmental stimuli such as heat shock or nitrogen starvation.

The s.f. is released from RNA polymerase after initiation of → genetic transcription, i.e., is free to program another polymerase molecule; the core enzyme continues to transcribe.

signal peptide (Blobel and Dobberstein 1975) — in prokaryote and eukaryote cells, an NH_2-terminal presequence (= leader sequence; signal sequence) that initiates and mediates translocation of membrane and secretory proteins across the cell membrane or the → endoplasmic reticulum (ER). Messenger RNA for such proteins contains immediately after the → initiator codon a group of signal codons which at the beginning of protein synthesis become translated into the s.p. The s.p. is recognized by a protein-RNA complex (the signal-recognition particle; SRP) that acts to temporarily arrest translocation by directly juxtaposing the protein-synthesizing apparatus with the membrane of the ER. The nascent peptide is extruded across the lipid bilayer, where a signal peptidase cleaves the signal from the protein.

The s.p. consists of three functionally distinct regions: (1) A short, positively charged N-terminal region (n-region); (2) a 7- to 15-amino-acid-residue-long central hydrophobic core (h-regions); (3) a more polar C-terminal region (c-region) defining the cleavage site.

signal recogniton particle (Walter and Blobel 1980) — (SRP); a ribonucleoprotein complex required for co-translational translocation of secretory proteins with a → signal peptide on their nascent chains. The SRP-ribosome/mRNA complex attaches to the ER membrane by associating with the SRP receptor (the docking protein) and the nascent chain is subsequently vectorially translocated across the membrane.

silencer sequence — a cis-acting DNA sequence, usually located upstream of the promoter of particular eukaryotic genes, which exerts a negative effect on the rate of → genetic transcription. S.sequences are cell-type-specific negative regulatory elements, i.e., repress promoter activity in cell types other than their target cells. A combination of positive and negative regulatory elements probably generates the correct cell-type-specific expression of the genes in question.

silent gene (Saedler et al. 1974) — a gene without → promoter which can be activated by, e.g., an insertion sequence placed near the unexpressed coding sequence. The insertion element provides a promoter which directs transcription into sequences adjacent to its insertion site.

silent mutation (Sonneborn 1965) — any → gene mutation (substitution mutation) that does not change the product of a gene, i.e., gives rise to a → codon that signifies the same amino acid (samesense m.) (→ replacement substitution).

silent section — any segment of genetic material for which no functions have been found.

simple sequence DNA (Walker 1971) — in eukaryotic genomes, any of the short DNA sequences (mostly less than 100 bp long and either AA/TT or GT/CA) which occur near genes, in some introns, in the spacers of some gene glusters as well as in the DNA regions between some genes; as opposed to → satellite DNA sequences, they are interspersed in the genome and usually transcribed into poly(A)$^+$RNA. → Transposition presumably is the means for their spreading throughout the genome. (= simple quadruplet tandem repeat).

simplex (Blakeslee et al. 1923) — → nulliplex.

SINE (Singer 1982) — acronym for → **s**hort **i**nterspersed **n**ucleotide **e**lement.

single burst — lysis of a bacterial cell after → bacteriophage infection in isolation from other cells.

single-copy sequence — a DNA sequence or gene which occurs only once per genome, as opposed to → repetitious DNA sequences.

single-copy plasmid — any of a class of → plasmids maintained in bacteria at a ratio of one plasmid for every host chromosome.

single site mutation — → multisite mutation; gene mutation.

single-strand binding protein (Bandyophadhyay and Wu 1978) — (SSB protein); any of a class of → DNA-binding proteins that bind tightly to both single-stranded DNA and to one another. SSB proteins destabilize partially double-stranded polynucleotides and stimulate the activity of homologous → DNA polymerase. They play a key role in DNA metabolism and are involved in → DNA replication, → DNA repair, and → genetic recombination (preventing the single-stranded regions from annealing or from forming intrastrand hydrogen bonds).

single-strand break — (ssb); loss of continuity of the phosphate-sugar backbone in one strand of the DNA duplex (→ double-strand break).

single-strand DNA initiation signal — (ssi signal); any nucleotide sequence on single-stranded DNA templates required for priming → DNA replication.

single-strand exchange — a reaction during which one of the strands of a DNA duplex leaves its former partner and instead pairs with the complementary strand in another molecule, displacing its homolog in the second duplex.

single-strand invasion — creation of a triple-stranded region in a circular DNA molecule. A single-stranded fragment base pairs (during → DNA breathing) with the circle. By branch migration the entire single strand can be assimilated.

sire — the male parent in mammals (→ dam).

sister chromatid — a → chromatid derived from duplication (replication) of one and the same chromosome during interphase of the → cell cycle.

sister chromatid differential staining (Latt 1973) — (SCD); differential staining with fluorochromes and conventional dyes of sister chromatids when the DNA has been substituted for two rounds of replication with 5′-bromodeoxyuridine (BrdUrd) or for one round followed by another in thymidine.

sister chromatid exchange (Taylor et al. 1957) — (SCE); the exchange, by breakage and reunion, of DNA sequences between → sister chromatids at apparently homologous sites. SCE occurs during DNA synthesis between DNA double strands having the same polarity. Analysis of SCE's is a sensitive and reproducible indicator for detecting → DNA damage induced by mutagens and carcinogens (SCE assay) and is used in → mutagenicity testing. SCE frequency may indicate the extent of a bypass type of → DNA repair occurring during S-phase of the eukaryotic → cell cycle.

site (Demerec 1956) — (1) the place in a genome or chromosome occupied by a mutation (mutational site); (2) a small (about 100 bp) region in DNA or RNA that, by interaction with cellular factors, brings about an effect on the expression of adjacent genetic material (e.g., promoters, operators, terminators, attenuators, cap sites, retroregulators, translational control sites, RNA spice sites, polyadenylation sites, enhancers, replication origins, site-specific recombination sites, switch sites, etc.).

site-directed mutagenesis (Hutchinson et al. 1978) — any of several techniques (= site-specific m.; local m.) giving rise to hereditary changes within particular regions of a given sequence of DNA (RNA) molecules or within a given genome. Systems of s.-d.m. include: (1) Local mutagenesis due to pulse incorporation of base analogs or radioactively labeled precursors of nucleic acids; (2) local m. at replication forks; (3) local m. by mutagen addressing; (4) local m. by controlled in vitro replication either of mutagenized template or by use of mutagenized primer. (→ oligonucleotide-directed m.; deletion loop m.; reversed genetics).

site-specific — of any process (→ s.-directed mutagenesis) or enzyme which acts at a defined sequence within nucleic acid molecules.

site-specific DNA inversion — in prokaryotes, any of a number of DNA inversion systems that can complement each other and result in a switch of → gene expression. Crossing over occurs within two inverted repeat sequences mediated by a recombinase (the coding gene of which usually lies adjacent to one repeat). S.-s. DNA i. controls the host range of phages P1 and Mu, and regulates the expression of flagellar antigens in *Salmonella.* Specific DNA sequences can act as enhancers of s.-s. DNA i. and several proteins participate in → inversion.

site-specific inversion system (Kamp et al. 1978) — in prokaryotes, any of the systems which regulate the alternate expression of genes by mediating the site-specific inversion of DNA segments. Such systems consist of a recombinase, two inverted cross over sites and a → recombinational enhancer.

site-specific recombination — any system of → genetic recombination that promotes genetic exchange between specific DNA sequences. S.-s.r. plays a crucial role in → DNA transposition, gene regulation, and generation of genetic diversity. Recombination between two sites that are directly repeated within a single DNA molecule (head-to-tail orientation) causes deletion of the DNA between the sites; recombination between inverted sites (head-to-head orientation) inverts the sequence between the sites.

small circular DNA — = → polydisperse nuclear DNA.

small cytoplasmic RNA — (scRNA); any of several diverse groups of small RNA molecules present in the cytoplasm of eukaryotic cells (distinct from → ribosomal RNA's and → transfer RNA's). scRNA's are complexed noncovalently with proteins (scRNP's) and seem to function in various aspects of protein synthesis and in targeting of proteins to various subcellular compartments.

small nuclear RNA (Weinberg and Penman 1968) — (snRNA); any of a class of highly conserved, low molecular weight RNA species (80 to 220 nucleotides) in the cell nucleus of eukaryotes characterized by four features: (1) They are synthesized by → RNA polymerase I from repetitive DNA sequences; (2) they show an unusual tri-methyl G cap; (3) they are associated with a specific group of RNA-binding proteins (snRNP); (4) they are uridylate-rich (hence U-snRNAs). There are six major species (U1 to U6), and several minor species (→ U-snRNA's).

Several of the snRNA's base-pair with specific regulatory sequences in pre-mRNA and participate in → RNA processing, both in RNA splicing and the generation of the 3′end of pre-mRNA.

small polydisperse circular DNA (Stanfield and Helinski 1976) — (spcDNA); in many eukaryotic cells, small DNA molecules (ranging from a few hundred to tens of thousands of bp) derived from chromosomal DNA and including a wide variety of nucleotide sequences. They occur in nuclei, organelles, and cytosol, and possibly originate by looping out of the spcDNA region and joining flanking DNA by illegitimate recombination.

snap-back sequence — → inverted repeat.

SOB response (Yasbin 1977) — a response in *Bacillus subtilis*, similar in several respects to the → SOS response in *Escherichia coli.*

social-group selection (Grant 1985) — a form of group → selection in which the → demes are social groups.

sociobiology (Wilson 1975) — a biological discipline which is the study of all evolutionary forms of social behavior and stresses that social systems evolve to increase the genetic → fitness of individuals in specific environments, i.e., emphasizes the genetic basis of social behavior.

solenoid (Finch and Klug 1976) — a shallow helical coiling of the 10 nm → chromatin fiber (the nucleofilament), with successive turns in close apposition to give the s. structure (diameter 30 nm) with about six → nucleosomes per turn and a central hole.

soluble RNA — = → transfer RNA.

solution-hybridization analysis — a method used to quantitate specific eukaryotic nucleic acids. Highly labeled, single-stranded (^{3}H) cDNA prepared form purified mRNA is incubated in solution with RNA or denatured DNA. The extent of annealing is analyzed by techniques that separate nonhybridized from hybridized (^{3}H) cDNA, including incubation with single-strand-specific nuclease, hydroxyapatite chromatography, and isopycnic centrifugation.

soma (Weismann) — the body of an organism apart from the germ line cells.

somaclonal variation (Larkin and Scowcroft 1981) — increased genetic variation observed in plants (somaclones) regenerated from in vitro cultivated somatic cells via a callus phase. The variability includes → aneuploidy, chromosome structural changes, → gene amplification and deamplification, activation of → transposable genetic elements, point mutations, organelle genome rearrangements, and changes in ploidy level. Somaclone 1 (SC_1) is the regenerant and SC_2, SC_3 represent the subsequent generations (SC_2: The family derived from selfed seed of a single SC_1 plant).

somatic — of cell types, structures, and processes in an individual other than those which are characteristic of, or give rise to, germ cells.

somatic cell fusion — homo- or heterophasic fusion between somatic cells in culture which may give rise to somatic → cell hybridization. As a normal part of growth and development, s.c.f. occurs between genetically identical cells of mammalian skeletal muscle, fungal hyphae, and true slime mold plasmodia.

somatic cell genetics — a branch of genetics which includes two distinct, though complementary and overlapping approaches: (1) The use of somatic cells in culture (very much like the use of cultures of microorganisms) for the study of general problems as, e.g., mutation, replication of DNA, differentiation, metabolism of macromolecules, selection. (2) The use of somatic cells, mainly in culture, as an alternative to germ cells for the genetic analysis of the individual from which the somatic cells were taken (Pontecorvo 1975). (→ somatic cell fusion).

somatic cell hybridization (Barski et al. 1960) — → cell hybridization.

somatic cell hybrid clone panel (Ruddle and Creagan 1975) — a panel of hybrid cell clones used for mapping of human genes. Each clone contains a unique combination of the 24 human chromosomes. By correlating the presence or absence of a particular human gene marker to be mapped with the clones in the panel, the chromosome location of that gene may be assigned. Regional mapping is possible when the panel of hybrid cell clones contains different segments of a particular human chromosome.

somatic crossing over — = mitotic → crossing over.

somatic embryogenesis — the production of embryo-like structures (embryoids) from sporophytic or somatic cells of the plant, as opposed to gametophytic or germ cells (zygotic embryogenesis).

somatic mutation — a → mutation in any cell not destined to become a → germ cell.

somatic pairing (Metz 1916) — → chromosome pairing.

somatic reduction — reduction of (normally polyploid) somatic chromosome complements (→ reductional grouping; cell hybridization) by formation of → aneuploid intermediates. S.r. in premeiotic cells suggests a possible origin for some existing diploid and tetraploid taxa from polyploids through a process of depolyploidization. S.r. may give rise to haploid cells (→ haploidization) during the parasexual cycle of certain fungi.

somatic segregation — the → segregation at → mitosis of chromosomally or extrachromosomally inherited genetic determinants (due to gene mutation, somatic recombination, aneuploidy). S.s. may give rise to → variegation or sectoring.

somatogamy (Renner 1916) — = → pseudoapogamy.

SOS-chromotest (Quillard et al. 1982) — a bacterial short-term test used as a screening assay for mutagenicity and carcinogenicity (→ mutagenicity testing). *E. coli* PQ 37, in which the structural gene for β-galactosidase lac z is placed under the control of the SOS-gene sfi A (→ SOS response), is used as indicator organism. The expression of sfi A, induced by → DNA damage, is measured indirectly by the determination of the β-galactosidase activity in a simple colorimetric assay.

SOS gene — in *E. coli*, any of a group of genes which are induced upon → DNA damage by physical and chemical genotoxic agents, leading to the inactivation of the Lex A repressor. SOS genes are involved in DNA repair (→ SOS response), mutagenesis and cell division, as well as a whole set of plasmid-encoded → colicins.

SOS mutator effect (Caillet et al. 1977) — in *E. coli*, a mutator effect generating (untargeted) mutations in undamaged DNA by a mutagenic bypass of replication blocking lesions in DNA. The perturbation of the replication fork by DNA lesions generates a new, transitory, replication complex of lower fidelity that allows error-prone trans-lesion synthesis in two steps: (1) Generation of misincorporation opposite damaged bases; (2) DNA elongation past the lesions (→ SOS response).

SOS repair (Radman 1974) — in *E. coli*, an inducible (error-prone) mutagenic repair mechanism which is controlled by the recA and lexA genes. DNA damage, which blocks the replication fork, induces a modification of the RecA protein into an activated form ($RecA^+$) that promotes the cleavage of the LexA repressor and thereby allows the expression of the → SOS response and DNA chain growth across damaged segments by error-prone filling of gaps in the course of de novo DNA synthesis.

SOS response (Radman 1974) — in *E. coli*, a group of functions induced by DNA-damaging agents. These include: (1) Error-prone → SOS repair; (2) expression of at least part of the post-replication recombination repair pathway; (3) lytic development of certain prophages; (4) massive synthesis of the RecA protein. The SOS r. requires a $RecA^+$ $LexA^+$ genotype. The LexA protein is the repressor of certain SOS operons and the RecA protein is a protease which, after DNA damage, cleaves the LexA repressor and the repressors of certain prophages.

A similar response has been observed in *Bacillus subtilis* and is called SOB response (→ adaptive response).

Southern blotting — transfer of single-stranded, restricted DNA fragments (= Southern transfer), separated in an agarose gel, to a nitrocellulose filter (or other binding matrix) which is then analyzed by hybridization to radioactive or biotinylated single-stranded DNA or RNA probes. The hybrids are detected by autoradiography or a color change, respectively. S.b. reveals details of sequence organization.

spacer sequence (Brown and Weber 1968; Birnstiel et al. 1968) — any DNA sequence separating neighboring genes. S. sequences may or may not be transcribed.

spacer tRNA gene (Lund et al. 1976) — in *E. coli*, any gene coding for → transfer RNA and located in the spacer sequence between the genes coding for 16S and 23S → ribosomal RNA or downstream from the 5S rRNA genes.

spacer unit (Franke and Scheer 1978) — any of the morphologically identified → intercepts located between → pre-ribosomal RNA matrix units and usually not covered with lateral fibrils.

spc DNA — acronym for → **s**mall **p**olydisperse **c**ircular DNA.

special-function gene — any gene which is typically expressed in only one or a few tissues or exhibits a strong, tissus-specific inductive response. Such genes are, in general, transcriptionally regulated, as opposed to housekeeping genes which are expressed in all cell types.

specialization — in evolutionary biology, any of a set of adaptations and tolerances characteristic for the adaptive zone.

special segment (Darlington and LaCour 1938) — a reversible modification of certain (probably → heterochromatic) chromosome segments of definite and species-specific location in mitotic chromosome complements.

speciation (Simpson 1944) — the evolutionary generation of → species which, in essence, occurs by the formation of barriers to gene exchange (reproductive isolation) between genetically diverging populations. Isolating mechanisms are promoted by natural → selection if gene exchange lowers the → fitness of its products.

The main modes of s. are (Mayr 1963): (1) **Sympatric speciation:** S. without geographic isolation, i.e., the acquisition of isolating mechanisms within a → deme. (2) **Allopatric speciation:** S. with geographic isolation, i.e., the acquisition of isolating mechanisms by a population during a period of geographic isolation ("geographic s."), or the partitioning of species along lines of secondary → intergradations, or along lines of strong ecological contrast ("semi-geographic s."). (3) **Stasipatric speciation** (White et al. 1967): S. whereby an ancestral species gives rise to a number of descendent species through chromosomal rearrangements, arising within its range, spreading to occupy limited areas, and acting as incipient isolating mechanisms in narrow zones of hybridization (White 1973). (4) **Instantaneous or cytocatalytic speciation:** S. by the origination of a single (e.g., → alloploid) individual which is reproductively isolated from the species to which its parents belong. It is reproductively and ecologically capable of establishing a new population. An analogous result may be obtained by multiple structural changes of the karyotype (saltational s. of Lewis 1954) which, after hybridization between the ancestral species and the resultant derivatives with structurally changed chromosomes, give rise to a high degree of hybrid sterility.

species(Ray 1670) — groups of actually or potentially interbreeding natural → populations which are reproductively isolated from other such groups. S. are the largest and most inclusive reproductive communities of sexual and cross-fertilization individuals that share in a common → gene pool (Mayr 1940; Dobshansky 1950). Species containing only a single → subspecies are called "monotypic", while those containing more than one subspecies are termed "polytypic" (→ semi-species, sibling species, superspecies).

Besides those sexually reproducing s. which are unequivocally defined by (1) morphological and physiological differences, (2) reproductive isolation, and (3) ecological differences, there are many natural populations (especially in the plant kingdom) which reproduce asexually, by apomixis or obligate autogamy, and are difficult to characterize by the three aforementioned criteria, or for which there is insufficient knowledge for definition as a species. Provided all three groups of the above-mentioned information are available, the following classification of a group of individuals under observation is proposed by Mayr (1948):

The individuals are:	Not reproductively isolated	Reproductively isolated
Identical in morphology and sympatric	1. Same population	5. Sibling species
Identical in morphology and sympatric	2. Same subspecies	6. Sibling species
Different in morphology and sympatric	3. Individual variants of the same population	7. Separate species
Different in morphology and allopatric	4. Separate subspecies	8. Separate species

species group — a group of closely related species (→ superspecies) with (usually) partially overlapping range (Mayr 1963).

species selection (Stanley 1975) — any mode of → selection in which the → species is the unit of selection: (1) Selection for ecological divergence; (2) species replacement; (3) selection for reproductive isolation.

specificity (Timoféeff-Ressovksy 1926) — the quality of gene action as defined by the location, phenotypic appearance, and mode of variation of the corresponding character resulting from the expression of the gene under observation. There is no clearcut distinction between s. and → expressivity.

specific locus test (Russell 1951) — in → mutagenicity testing, a mouse test that involves mutations transmitted from treated germ cells and recovered in first-generation offspring. It detects intragenic lesions, deletion of the marked locus, and small multi-locus deficiencies involving the marked locus and some neighboring loci. Since several loci are marked, the mutational spectrum can detect qualitative as well as quantitative parameters, e.g., mutagen specificity.

specioid (Mansfeld 1950) — a systematic category in cultivated plants which corresponds to a → species.

sperm — the male gamete in animals (= spermatozoon) and in primitive plants (= spermatozoid).

spermateleosis — the differentiation of a spermatid into a spermatozoon (→ spermatogenesis).

spermatid (La Valette St. George 1886) — → spermatogenesis.

spermatocyte (La Valette St. George 1886) — → spermatogenesis.

spermatogenesis — the development of the male animal → germ cells (spermatozoa, sperm) within the male gonad (the testis).

The testes contain the primordial germ cells **(primary spermatogonia)** which enter into a phase of rapid mitotic multiplication producing the **secondary spermatogonia**. The secondary spermatogonia transform into **primary spermatocytes** (meiocytes). Each of the latter (after increasing in size and a series of complex nuclear changes) undergoes the first meiotic division and gives rise to the two **secondary spermatocytes**, which in turn undergo the second meiotic division. As a result of these two divisions, four **spermatids** with reduced chromosome number arise. The spermatids differentiate into the mature male gametes, the **spermatozoa** or sperm (→ oogenesis; megasporogenesis; microsporogenesis).

spermatogonium (La Valette St. George 1886) — any of the immature reproductive cells in male animals after the formation of the testis. By division and differentiation spermatogonia give rise to spermatocytes (→ spermatogenesis).

(1) **Predefinitive spermatogonia** are cells that produce (through stem cell divisions) cells like themselves as well as definitive spermatogonia. (2) **Indefinitive spermatogonia** are those generations of spermatogonia occurring between the stem cell divisions and the definitive divisions (= primary spermatogonia). (3) **Definitive spermatogonia** are cells irrevocably determined through (dichotomous) division and differentiation to become the spermatocytes and eventually the spermatozoa.

spermatozoon — → spermatogenesis.

spermiogenesis — the morphogenetic period during which the spermatid (→ spermatogenesis) is changed into the shape of a mature spermatozoon (progressive nuclear condensation, changes of chromatin structure, development of the → acrosome and tail). S. occurs at the end of the second meiotid division.

sperm morphology test — in → mutagenicity testing, a test that detects positive effects of an agent as an increase in the frequency of sperm with abnormal morphology in treated mice. Such changes are due to influences of the chemical on the differentiation process during → spermatogenesis and seem to be correlated with chromosome aberrations.

S phase (Howard and Pelc 1953) — in eukaryotes, the period of chromosomal DNA replication during interphase of the → cell cycle.

S phase recovery (Lehmann 1972) — a → DNA repair system active during normal → DNA replication of eukaryotes which enables DNA to be replicated on damaged templates.

spheroid — a multicell association of in vitro grown animal cells.

spheroplast (Tulasne et al. 1959) — a microbial or plant cell after removal of most of the cell wall so that it assumes a spherical shape (→ protoplast). Bacterial spheroplasts are frequently also called L forms.

spherosome (Dangeard 1919; Perner 1952) — in plants, organelles that correspond to the → lysosomes of animal cells.

spindle — in eukaryotes, an ellipsoidal (bipolar) collection of fibers (s. fibers) visible during nuclear division and cytokinesis. The s. is responsible for the accurate segregation (→ chromosome movement) of chromatids or chromosomes to daughter cells during → mitosis and → meiosis and mainly consists of → microtubules associated with → microfilaments and → intermediate filaments. In animal cells, the s. is surrounded by a fibrous case; in meiosis of some animals, each → bivalent has a s. of its own and the → mitotic apparatus as a whole consists of a bundle of these individual spindles. In some protozoa, the → centrioles are intranuclear and the whole mitotic apparatus is formed within the → nuclear envelope (intranuclear s.), which does not break down (premitosis). In flagellates and all higher organisms, the centrioles have become extranuclear and division is by eumitosis. The spindles of higher plants are anastral, i.e., they lack asters and centrioles.

In general, most mitotic spindles are composed of various classes of microtubules or s. fibers (Bajer 1968): (1) **Centromere** (or kinetochore) **microtubules** going from the → centromere region towards the s. poles. They are present from early prometaphase of mitosis to early telophase (= chromosomal fibers). (2) **Continuous** (or interpolar) **microtubules** connecting the two polar regions of the s. observed from early prometaphase to early telophase (= continuous s. fibers). (3) **Neocentric microtubules** connected with parts of the chromosome other than the centromere (→ neocentric activity) and ending inside the chromosome. (4) **Interzonal microtubules** or interzonal connections which represent continuous microtubules persisting during anaphase in the interzonal region and found only at the anaphase between the separating groups of chromosomes. (5) **Phragmoplast microtubules** (in plants) which begin to be formed by middle anaphase and increase in number during telophase. They persist for a long time after the → cell plate has formed.

In the **direct s. type** (Schrader 1944), the chromosomes are directly connected by the centromere microtubules to the spindle poles, in the **indirect s. type**, centromere microtubules first connect with the continuous microtubules and these, in turn, make the connections to the spindle poles.

spindle plaque (Robinow and Caten 1968) — an electron-opaque structure associated with the → nuclear envelope at the poles of the intranuclear → spindle in *Ascomycetes.*

spindle poison (Dustin 1938) — any → mitotic poison affecting the formation or function of the → spindle and blocking karyokinesis at metaphase.

The prototype of a s.p. is colchicine (→ C-mitosis; C-meiosis).

spiralization (Darlington 1932) — → chromosome coiling.

splice junction (Breathnach et al. 1978) — the region of intron-exon boundaries in eukaryotic mosaic genes and their primary transcription products which aid in the process of → RNA splicing. The DNA s.j. sequence is ${}^{C}_{A}AG/GT^{A}_{G}AGT$ for exon-intron boundaries (5′donor sequence) and $({}^{T}_{G}N^{C}_{T}AG/G$ for intron-exon boundaries (3′acceptor sequence).

The long pyrimidine tract at the 3′end of introns in eukaryotic and viral genes is absent from mitochondrial DNA. No homology exists with the sequence of the s.j. in chloroplast rRNA genes.

Single base changes affecting the 5′ and 3′ s.j. consensus sequences in eukaryotic mRNA's cause aberrant splicing of pre-mRNA's. The disruption of one splice site often results in the use of an alternative cognate sequence. Such cryptic s. sites are usually found in the immediate vicinity of the disrupted site and in some cases are only a poor match with the consensus sequence.

spliceosome (Brody and Abelson 1985; Grabowski et al. 1985) — a multicomponent ribonucleoprotein complex (40S to 60S) involved in the removal of → intron sequences from nuclear → pre-messenger RNA's (→ RNA splicing). Requirements for s. formation include: (1) Functional splice junction sequences in the pre-mRNA; (2) activity of at least three → small nuclear ribonucleoprotein particles (U2, U5, and U4 + U6); (3) presence of ATP and the 5′exon and the → lariat RNA of the intron sequences linked by a normal phosphodiester bond to the 3′exon.

snRNP — snRNP interactions probably dictate the backbone of the s. A complex containing precursor RNA and U2 snRNP is a likely intermediate in s. formation.

Two catalytic reactions occur in the s.: (1) The precursor mRNA is cleaved at the 5′ exon-intron junction with the formation of a lariat structure (5′ guanosine of the intron joined by 2′, 5′,-phosphodiester bond to an adenine near the 3′ end of the intron); (2) the 3′intron-exon junction is cleaved and the exons are ligated together.

splicing — (1) gene s.: Any manipulation which attaches one DNA molecule to another (→ recombinant DNA technology). (2) RNA s.: Removal of → introns from → precursor RNA (→ RNA splicing).

splicing enzyme (Knapp et al. 1978) — any enzyme involved in → RNA splicing.

splicing homeostasis (Lazowska et al. 1980) — a process of mitochondrial → RNA processing controlled by an introncoded mitochondrial protein which controls its own synthesis by destroying the → pre-messenger RNA which encoded it. The protein is required for splicing out of the same intron.

spligase (Church and Gilbert 1980) — = → maturase.

split gene — any eukaryotic gene (most protein-coding genes, some genes coding for → transfer RNA and → ribosomal RNA) in which the coding sequences (the → exons) are interrupted by a number of usually noncoding (untranslated) regions (the → introns). S. genes occur in eukaryotic nuclear (viral), mitochondrial and chloroplast genomes.

split promoter (Hofstetter et al. 1981) — the intragenic → promoter of eukaryotic genes encoding → transfer RNA's which consists of two separated regions within the gene's coding sequence.

split spindle — a → spindle type which is split into two parts and gives rise to a double metaphase plate. In certain Hemiptera, the autosomes are contained in one, the sex-chromosomes in the other (smaller) of the two metaphase plates which are rather widely separated from one another.

sporad (Webber 1933) — any group of normally four spores (tetrad) produced by the first and second meiotic divisions from each spore mother cell in higher plants. Instead of a tetrad, sporads with aberrant numbers of cells (monads, dyads, triads, pentads, etc.) may be produced under certain conditions (e.g., in haploids of normally diploid plants) due to irregular → meiosis.

spore — a reproductive cell in plants (→ germ cells). Spores are agametes and may be produced meiotically ("meiospores", "tetraspores") or mitotically ("gonidia"). In higher plants, meiospores (→ megasporogenesis; microsporogenesis) give rise to female and male → gametophytes.

Spory, sporogenesis, and sporogony are the terms used for the process of spore formation.

sporocyte — a spore mother cell (→ spore).

sporogenesis — = → spory.

sporogony — → fision.

sporophyte — the (normally) diploid, asexual generation (in → diplo-haplontic plants having an → alternation of generations) which usually produces → spores. In higher plants, the s. bears the haploid → gametophytes; in mosses and liverworts, the s. is smaller than the gametophyte and is borne by the latter.

sporulation — (1) the formation from vegetative cells of bacteria and other organisms of dry, metabolically inactive cells with thick surface coats (spores) which can resist extreme environmental conditions (→ bacterial sporulation); (2) in yeast the induction of → meiosis resulting in four haploid spores.

spory (Battaglia 1955) — the formation of → spores (= sporogenesis; sporogony).

spot test (Ames 1971) — in → mutagenicity testing, an in vivo method for detecting mutagens and carcinogens. The s.t. involves placing the agent at the center of a Petri dish containing minimal medium that is seeded with → auxotrophic bacteria unable to grow (due to a deficiency in a metabolic pathway). Revertants from auxotrophy to prototrophy are identified as colonies in a ring around the area (spot) where the potential mutagen was deposited. By use of appropriate bacterial strains, the s.t. permits the characterization of mutagens inducing → base pair substitutions, → frameshift mutations, or large → deletions. (→ rec-assay).

spreading effect (Lewis 1950) — → position effect.

ssi signal (Nomura and Ray 1980) — any of the specific nucleotide sequences on a single-strand template for directing single-strand initiation of (plasmid) DNA synthesis. The ssi signal sequences contain possible stem and loop structures and are diverse in their structures, functions, and biological implications.

S-S recombination — a type of recombination concerning immunoglobulin heavy chain (H) genes (→ V-D-J-recombination) which mediates the H chain class switch that associates a particular antigen binding specificity and a completed V region with a series of different heavy-chain constant (C_H) regions. The rearrangement is called S-S r. since it joins two switch (S) regions which are located in the 5′flanking region of each C_H gene.

stabilizing selection — the most prevalent type of natural → selection (= centripetal s.; normalizing s.) which eliminates phenotypically extreme individuals and preserves those that are near the population mean.

stacking — the packing of the flattish nitrogen bases at the center of the DNA double helix.

staggered cuts — in duplex DNA, the cleavage of the two strands at different points near each other (→ restriction endonuclease).

stamen — the male organ of a flower (the androecium) consisting of a stalk or filament with an anther containing pollen. Flowers containing only stamens are called "staminate" (→ pistil).

standard type — an arbitrarily specified → genotype used as a basis of comparison for genetic investigations. The genes of the s.t. are usually symbolized by a (+), irrespective of their being → dominant or recessive. All genes deviating from those of the s.t. are designated by names or symbols (→ genetic nomenclature). Where possible, the most frequent → wild type of the taxonomically typical subspecies of a species of the first thoroughly analyzed wild type should be used as s.t.

stasigenesis (Huxley 1957) — in → evolution, all processes by means of which stability and persistence of types and organizational patterns (from the species up to the phylum) is achieved. S. gives rise to "grades", i.e., groups of individuals similar in their level of organization which represent persisting units of → anagenesis.

stasipatric — → speciation.

stathmokinetic — of agents which will arrest cells in mitotic metaphase (→ mitotic poison).

stem cell — any cell that can proliferate in an undifferentiated state, as well as gives rise to differentiated products (→ germ cell).

stem cell spermatogonium — any of the → spermatogonia responsible for the maintenance of the → germ line in male animals.

stem-line (Levan and Hauschka 1953) — the most frequent karyotype (with respect to chromosome number and/or chromosome structure) of a tumor cell population observed in a direct preparation or following short-term culture. All other lines are called sidelines or sublines.

sterility — any complete or partial (→ semi-sterility) failure of an individual to produce functional gametes or viable zygotes ("gametic" and "zygotic s." of Renner 1929; "haplontic" and "diplontic s." of Müntzing 1930) under a given set of environmental conditions. It may concern both the male and female sexes, or either the male ("male s.") or female ("female s.") sex.

S. may be caused by environmental effects, extranuclear hereditary determinants ("cytoplasmic s."), incompatibility between endosperm and embryo in plants ("somatoplastic s." of Bryant 1935), genotypic defects, or segregation in hybrids, polyploids and aneuploids ("genic" and "chromosomal s.").

In contrast to "true s.", parasterility or sexual → incompatibility is the inability of genetically like ("homogenic sexual incompatibility") gametes or nuclei to undergo karyogamy and thus to achieve fertilization. The same gametes may successfully unite with genotypically different partners giving rise to viable zygotes.

steroid hormone receptor — any of the receptors that activate specific gene transcription by binding as hormone-receptor complexes to DNA enhancer elements (→ transcriptional enhancer) termed hormone-responsive elements.

stickiness (Beadle 1932) — a sort of chromosomal "agglutination" of unknown nature which results in a pycnotic or sticky appearance of chromosomes. S. may give rise to → sticky adhesions between two or more chromosomes and to the formation of "sticky bridges" at anaphase when the chromosomes or chromatids associated in such adhesions move to opposite spindle poles. Sticky effects result from certain gene mutations and may be induced by the application of a wide variety of chemical and physical agents during the condensed state of chromosomes in mitotis and meiosis.

sticky association (Price 1956) — any association of two or more chromosomes at mitosis or meiosis due to nonspecific sticky adhesions rather than to homologous → chromosome pairing and chiasma formation (→ pseudobivalent; pseudomultivalent).

sticky ends — complementary single strands of DNA which protrude from opposite ends of a duplex or from ends of different duplex molecules; such ends can be generated by → staggered cuts in duplex DNA (→ restriction endonuclease).

stimulon — a set of genes that respond to a single stimulus (→ regulon, operon). Stimulons may overlap if they contain an operon with multiple promoters that have different regulatory properties.

stop codon — = → terminator codon.

strain —(1) a group of related viruses, cells, or individuals distinguished by certain characters from other such groups; (2) the progeny of a single plant irrespective of the mode of pollination.

strand transferase (Shibata et al. 1979) — any of the enzymes that promote homologous pairing by forming a complex between a double-stranded DNA molecule and a homologous single-stranded DNA molecule (→ DNA strand displacement).

stress response — a selective increase in the synthesis of a number of proteins (stress proteins) in response to various stress conditions (→ heat shock response).

stringent response (Borek et al. 1955) — in bacteria, the reduction of stable RNA synthesis in response to amino acid deprivation (relA$^+$ strains), as opposed to the relaxed response during which RNA synthesis is stimulated (relA strains). The responses are correlated with the accumulation or disappearance, respectively, of the nucleotide guanosine tetraphosphate (ppGpp), which is thought to be a general signal molecule in a control system that senses an imbalance or deficiency in amino acid supply and in response redirects the cell's economy. Possibly, ppGpp is a negative effector in the control of stable RNA synthesis.

stroma — the nonlamellar ground substance of the → chloroplast (→ plastid) and → mitochondrion.

structural gene (Jacob and Monod 1959) — any → gene coding for the primary structure (the amino acid sequence) of a polypeptide. The transcription product of s. genes is → pre-messenger RNA (→ RNA processing). S. genes may consist of a continuous coding sequence (prokaryotes), or a set of closely linked coding sequences (exons) interrupted by → introns (most eukaryotic s. genes).

In eukaryotes, the flanking DNA of s. genes (transcribed by → RNA polymerase II) contains regulatory elements for → gene expression. The canonical TATA and CCAAT boxes in the → promoter required for specific and efficient → genetic transcription are located within the 100 bases immediately upstream of the 5′end. Further upstream in the flanking DNA, there are sequences that recognize regulatory molecules such as inducers and steroid hormone-receptor complexes for genes whose expression is regulated by them. Flanking DNA may also contain → transcriptional enhancers.

structural heterozygosity — heterozygosity for a chromosomal structural change (→ chromosome mutation). Heterokaryotypes characterized by one or more → heterozygous structural changes in the chromosome complement (karyotype) are called "structural heterozygotes" or "structural hybrids". S.h. usually entails a reduction in fertility (→ hybrid sterility) due to the production of genotypically defective gametes and unbalanced zygotes and thus reduces recombination and variation (→ chromosomal polymorphism).

structural homozygosity — the structural identity of two chromosome regions, chromosomes, or chromosome sets, in contrast to → structural heterozygosity.

streuctural hybrid (Darlington 1929) — → structural heterozygosity; hybrid.

S-type position effect (Lewis 1950) — = cis-trans → position effect.

subandroecious — of primarily male plants with a few hermaphroditic or female flowers, as opposed to → subgynoecious.

subcloning — a procedure in which smaller DNA fragments from a large insert, which has already been cloned (→ molecular cloning), are cloned in the same or a different → cloning vector in order to identify a specific part of a gene (such as its → promoter), or to remove unwanted DNA segments from a clone.

subdioecious — of plant populations composed in the majority of unisexual plants but containing a few → monoecious individuals (→ dioecious).

subgynoecious — of primarily female plants with a few hermaphroditic or male flowers, as opposed to → subandroecious.

subhaploid (Kiellander 1942) — → haploid.

sublethal factor (Johannsen 1909) — → lethal factor.

subsexual (Darlington 1937) — → reproduction.

subspecies (v. Wettstein 1898; du Rietz 1930) — an aggregate of (local) breeding populations of a given → species occupying a geographic subdivision of the species range (= geographic → race). A s. usually differs from other similar breeding groups of the same species both taxonomically and with respect to certain → gene pool characteristics (such as the frequency or prevalence of certain genes: → gene frequency). Subspecies, like → semi-species and species, may be either allo- or sympatric. The small local populations of which a s. consists are called "microgeographic races" or "microsubspecies" (Huxley 1940). A s. represented by widely separated, phenotypically identical populations is said to be "polytopic".

substitution — → chromosome substitution.

substitution line — a line in which one chromosome pair is transferred from the donor variety into the recipient variety (→ alien addition line; alien substitution line).

substitutional haploid (Riley 1960) — → haploid.

substitutional load — → genetic load.

substitution mutation — a → gene mutation resulting from the substitution of one base for another.

subvital (Hadorn 1948) — of a gene or genotype which, when expressed, reduces the viability of the carrier to a level significantly below that of the → standard type (→ lethal factor).

subvital factor (Hadorn 1948) — → lethal factor.

supercoiled — of duplex DNA with a different number of coils from that expected from the double helix (= superhelical, supertwisted). S. DNA has some unique physical and chemical properties, and the closed, circular, superhelical form is required in a number of genetic processes. The number of superhelical turns per 10 bp is the superhelical density of a circular DNA molecule.

superdominant (Fisher et al. 1932) — = → overdominant.

superfemale (Bridges 1914) — = → metafemale.

supergene (Darlington and Mather 1949) — an operational unit consisting of a cluster of tightly linked, apparently different, but functionally related genes.

superinfection curing — the loss of the → prophage after infection of → lysogenic cells by heteroimmune phages having the same integration site as the resident prophage.

superinfection immunity (Watenabe 1963) — (1) in the case of transmissible bacterial → plasmids (→ sex factor), the reduction of the plasmid's ability to bring about its own transfer from one bacterial cell to another by conjugation when both members of a conjugal pair harbor isogenic or closely related plasmids. S.i. may be due to → entry exclusion and → plasmid incompatibility. (2) The inability of two viruses of the same kind to coexist stably in the same cell either due to entry exlusion or immunity proper, where either the resident or the superinfecting genome is reproduced, but not both.

supermale (Bridges 1914) — a phenotypically male, abnormal (sterile) individual with fewer → X-chromosomes than half the sets of autosomes, in organisms (e.g., *Drosophila*) with an X-chromosome-autosomes balance type of → sex-determination. (→ metafemale; superfemale).

supernumerary chromosome (Longley 1927) — = → B-chromosome.

superoxide dismutase — in all aerobic organisms any of a class of metalloproteins which catalyze the disproportionation of the superperoxide anion (O^2) to H_2O_2 and O_2 providing an important defense against oxygen toxicity.

superspecies (Mayr 1942) — a monophyletic group of entirely or essentially allopatric → semi-species that are too distinct to be considered as → subspecies of the same → species and not sufficiently distinct to be regarded as full species (Grant 1963) (→ syngameon).

super-suppressor (Hawthorne and Mortimer 1963; Gilmore 1967) — any of a class of dominant suppressors with pleiotropic action that can suppress a variety of other mutations, typically a → nonsense suppressor. Supersuppressors are caused by mutation in genes which code for transfer RNA. The tRNA's can gain activity in → suppression of nonsense codons through base substitutions occurring directly in the anticodon of the tRNA (→ suppressor tRNA) (= multisite suppressor).

supervital — a gene or genotype which, when expressed, is significantly more viable than the → standard type (→ heterosis).

suppression (Sturtevant 1920) — → genetic suppression; suppressor mutation.

suppression restriction (Gorini 1971) — a decrease of the in vivo efficiency of → genetic suppression of nonsense and missense mutations by certain mutations of genes coding for → ribosomal proteins.

suppressor (Bridges 1932) — any secondary mutation (second-site mutation) that totally or partially restores (suppression) a function lost due to a primary → nonsense, → missense, or → frameshift mutation. In contrast to → back mutation, s. mutations do not result in a reversal of the original mutation and can be separated from the primary mutation by → genetic recombination.

Suppressor mutations (s.m.) may be classified as follows (Gorini and Beckwith 1966):

(1) **From the genetic point of view** the s.m. may be located either in the same gene or cistron as the original mutation (intragenic or internal s.m.), within another gene in the same chromosome, or even in a different chromosome (intergenic, extragenic or external s.m.). A s.m. may be strictly allele-specific, i.e., it suppresses only mutations which arise at a particular codon within a cistron, or it may be able to suppress equally well several or all mutations occurring at different codons within the same cistron ("cistron-specific suppressors"), or mutations of the same codon occurring in different, unrelated cistrons.

Intragenic suppressors restore the wild-type phenotype by allelic interaction between the primary mutation and a compensatory secondary mutation at a site very closely linked to the original mutation. **Intergenic suppressors** restore the wild-type phenotype by genic interaction between linked or unlinked primary mutations and suppressor mutations at a site outside the gene in which the primary mutation took place. Such suppressors usually act on functionally unrelated genes and their mode of action must involve some common step in the steps from DNA to protein, i.e., the code translation mechanism. (2) **From the point of view of their mode of action** suppressor mutations may be grouped as follows: (a) The second mutation may be in a gene whose mutated product opens an alternative pathway of deficient metabolic synthesis. As a result of a second mutation, an enzyme activity may appear which permits the synthesis of an intermediate whose synthesis, through the usual pathway, was blocked (→ genetic block) by the first mutation (Davis and Woodward 1962). (b) The second mutation may be in a gene whose mutationally altered product may substitute directly for the function of the product of the first gene. A second enzyme (or protein) may mutate to a broader specificity (or function) encompassing that of the missing first enzyme (or protein). (c) The second mutation provides cytoplasmic conditions which affect the structure of the finished product of the first mutation. The condition may stabilize an enzyme which owing to a structural change is unstable in the wild-type cytoplasm. (d) The second mutation introduces a second amino acid substitution which reintegrates, completely or partially, the protein function, with reappearance of the wild-type or a quasi-wild-type phenotype (Suskind and Kurek 1959). (e) The second mutation reintroduces the correct code meaning by frameshift (→ frameshift mutation) within the same cistron in which the codon reading frame was altered by addition or deletion of a base pair. (f) The second mutation (informational s.) concerns one of the factors controlling the mechanism of transfer of genetic information from DNA to protein, i.e., repair of the mutational defect owing to an alteration in → genetic translation of the genetic information (Champe and Benzer 1962).

Categories (a) to (c) act indirectly to circumvent, rather than to repair directly the primary genetic lesion **("indirect suppression")**. Categories (d) and (e) represent intragenic suppressors and are characterized by the fact that (barring the extreme case of intracodon suppression) the primary and suppressor mutations are both transcribed (→ genetic transcription) into one gene product. This gives rise to two amino acid substitutions [or possibly more in case of category (e)], in the suppressed mutant protein with respect to the wild-type protein. Suppression according to category (f) is due to alterations in meaning of a codon and are not themselves transcribed in the suppressed mutant protein as a second amino acid substitution. Suppressors of this type are always intergenic and may be designated "informational suppressors" (Gorini and Beckwith 1966).

Several types of informational suppressors are known to affect → transfer RNA structure or ribosomal proteins which control the fidelity of genetic translation (→ mistranslation). Nonsense and missense suppressors have been described which involve tRNA species with altered coding specificity. In most cases this is due to changes in the anticodon which allow the → suppressor tRNA to translate a nonsense or missense codon.

After separation of intragenic suppressors from the primary mutation by → genetic recombination, such suppressors, as a rule, appear as nonfunctional mutants. Intergenic suppressors usually only partially compensate for the phenotypic effect of a primary mutation and this may be distinguished from true → back-mutation. Another means of differentiating them from back-mutations is by back-crossing to the wild type and observing the segregation ratios produced. (→ antisuppressor).

suppressor-enhancing mutation — (sue mutation); any secondary mutation that increases the efficiency of suppression, usually between three- and tenfold.

suppressor-sensitive mutation — any mutation that behaves like the wild-type allele when a → suppressor is present. The main types of s.-s. mutations are → nonsense and → missense mutations.

suppressor tRNA (Capecchi and Gussin 1965) — any → transfer RNA species which, due to mutation in a transfer RNA gene, shows altered coding specificity and is able to translate (→ genetic translation) → nonsense or → missense codons.

Nonsense and **missense suppressor tRNA** is frequently modified in the anticodon loop or in an adjacent region. Some mutant tRNA's with temperature-sensitive suppressor activity have been shown to have nucleotide substitutions in variable regions outside the anticodon.

Efficiency of suppression is determined at each site by the specific reading context of the suppressor codon. The degree of sensitivity to context effects differs among s.tRNA's. The most important role appears to be played by the base that is immediately adjacent to the codon. Possibly, context effects are produced by the whole codon following the relevant codon in the message. Translation of a nonsense codon by a s.tRNA is expected to force peptide synthesis to continue in an unproductive → reading frame.

Bacterial suppressor tRNA's are usually codon-specific rather than gene-specific.

Frameshift suppressor tRNA compensates for → frameshift mutations by reading nontriplet codons. Such tRNA molecules may carry an additional base in their anticodon, such as CCCC instead of CCC in the wild-type tRNA. This would allow them to translate frameshifts which involve the change of a GGG codon to a GGGG quadruplet. Or an additional base in the tRNA anticodon may serve as a spacer such that the tRNA occupies an unusually large space over the mRNA message and so suppresses + 1 frameshifts.

tRNA suppressor mutations are lethal if they result in the inability to translate a normal (nonmutant) codon. Under such circumstances only dispensable tRNA species (such as → isoacceptor tRNA's) can be converted to suppressor tRNA or another, nonmutated tRNA molecule must be synthesized and rescue the function abolished by the suppressor mutation.

Frequently, suppressor tRNA's, besides their major decoding specificity (including → wobble), also posses some minor decoding properties which result in mistranslation.

supraoperon control (Roth and Nester 1971) — The co-ordinate control of a → gene cluster with multiple → operator and → promoter sites which consists, in essence, of several interacting → operons controlling different biochemical pathways.

surface coat — a layer of material outside the limiting membrane (→ cell membrane) of the animal cell. It assists the cell in either adhering to its substratum or to other cells and in the binding of particles (from ions to viruses) for subsequent → pinocytosis. Furthermore, the s.c. plays a role in the determination of the antigenic identity of the cell. (→ cell wall).

surface exclusion (Lederberg et al. 1952) — = → entry exclusion.

surface-modulating assembly (Edelman 1976) — any of the submembraneous arrays of → microtubules, → microfilaments, and associated contractile and membrane proteins which control the mobility of → cell surface receptors, regulate signals from the cell surface to the cell interior and play a role in growth control.

surrogate genetics (Birnstiel and Chipchase 1977) — = → reversed genetics.

survival cost (Bell 1980) — the cost of present reproduction (= reproductive cost) on future survival and/or fecundity. Usually, any increment of fecundity is attained only at the expense of adding to the risk of mortality.

suspension arch structure (Imai et al. 1986) — a nonrandom chromosomal configuration at pachytene resulting from bonding the ends of bivalents to the nuclear envelope (= hammock structure).

switch gene — a gene which causes the → epigenotype to switch to a different → developmental pathway.

switching site — any breakpoint at which gene segments combine in gene or → DNA rearrangements.

swivel (Cairns 1963) — a point in the DNA duplex which allows for unwinding of the parental strands during → DNA replication. Enzymes capable of introducing a transient s. by nikking one strand of the DNA and subsequently rejoining it (DNA swivel enzymes; swivelases) occur in both prokaryotes and eukaryotes.

symmetrical transcription — in prokaryotes and eukaryotes, transcription of the same DNA region in duplex DNA from opposite strands. The complementary transcripts can be expressed in a tissue-specific fashion in eukaryotes. In some cases, regulatory → antisense RNA is produced by s.t.; in others, the sequence shows long → open reading frames in the antisense strand of many protein-coding genes. Transcription of → mitochondrial DNA is generally symmetrical.

sympatric (Poulton 1903; Mayr 1942) — of two or more → populations occupying the same geographic area, or of a population existing in a breeding condition within the cruising range of individuals of another population, as opposed to → allopatric. S. populations are not necessarily in physical contact, or at least reproductive contact; they may remain completely separated by different habitat preferences, or by different habits.

S. speciation is species formation in the absence of geographic → isolation, by acquiring mechanisms of reproductive isolation within a → deme (Mayr 1963).

symplast (Hanstein 1880) — a multinucleate body of protoplasm which according to its origin may be a → plasmodium or a → syncytium (→ coenocyte).

synapsis (Moore 1895) — meiotic → chromosome pairing (= syndesis).

synaptene (Winiwarter 1900) — = zygotene (→ meiosis).

synaptic (Riley and Law 1965) — ref. to the processes and forces of meiotic → chromosome pairing or synapsis. "Synaptic genes" are those which influence the extent of meiotic pairing (→ desynaptic; asynaptic).

synaptic adjustment (Moses 1977) — replacement of homologous → chromosome pairing by nonhomologous synapsis (in late pachytene) in regions heterozygous for chromosome rearrangements.

synaptomere (King 1970) — any of the specific points where the → synaptonemal complex starts forming (→ zygomere).

synapton (Whitehouse 1969) — = → synaptonemal complex.

synaptonemal complex (Moses 1958) — (SC); in eukaryotes, a regularly occurring, tripartite structure consisting of proteinaceous parallel (co-planar) strands (0.2 μm wide), coaxial to homologous chromosomes involved in meiotic chromosome pairing.

The s.c. is considered to be involved in chromosome condensation, chromosome pairing, recombination, and segregation. Formation of the s.c. involves the following steps: (1) Formation of a thin, compact proteinaceous axis along each chromosome (leptotene); (2) alignment of the axes of homologous chromosomes by the formation of transversal filaments (the axes become the lateral elements of the SCs) during zygotene; (3) formation of a central element (CE) between the LE's on the transversal filaments (zygotene and pachytene). Recombination occurs when chromosome pairing is complete. Subsequently, the transversal filaments, the CE, and later the LE's disappear, and the recombinant chromosomes of each bivalent segregate (diplotene, diakinesis, and metaphase I).

Local thickenings of the CE are called → recombination nodules; they are equal to the mean number of → chiasmata.

The SC's of most organisms terminate at the → nuclear envelope. The attachment points may be polarized or may show no discernable pattern of arrangement.

The SC is probably formed by self-assembly of specific protein molecules a major component of which is a basic, arginine-rich protein. Current ideas on structure and function of the SC propose that the SC forms a structural framework around which the chromatin is arranged in such a way that the mass of chromatin, which is not directly involved in → genetic recombination, is sequestered on the outer surface of the lateral elements and only selected stretches of chromatin penetrate the lateral element and enter the central space where they engage in molecular pairing (effective pairing) and genetic exchange.

Usually the SC is set free from the chromosome at the diplotene stage of meiosis and either becomes dispersed or aggregated into stacks, so-called **"polycomplexes"**.

synclone — → caryonide.

syncyte (Levan 1942) — a → polyploid or multinucleate cell. The production of syncytes may take place by: (1) Inhibition of → cytokinesis in mitotic divisions. The resulting syncytes may be di- or multinucleate or may contain one polyploid nucleus if a → restitution nucleus is formed. (2) Migration of the nucleus from one cell to another. The phenomenon is called "cytomictic formation of a s." (Price 1956). (3) Fusion of cells after dissolution of the cell membranes and cell walls.

syncytium (Haeckel 1894) — = → syncyte.

syndiploid (Strasburger 1907) — of cells which possess twice the diploid number of chromosomes owing to fusion of mitotically produced daughter nuclei or inhibition of the → spindle mechanism. The phenomenon is called "syndiploidy" (= tetraploidy).

syndrome — a group of specific characters (symptoms) characteristic of a certain (abnormal) genetic condition. An "inheritable" or "genetic s." may be transmitted to the offspring as a unit.

synergid — in higher plants, any of the (two) haploid nuclei found with the egg at the micropylar end of the → embryo sac (→ macrosporogenesis).

syngameon (Lotsy 1916; Grant 1957) — a group of sympatric → semi-species (→ superspecies).

syngamic (Haecker 1902) — → sex determination.

syngamodeme (Gilmour and Heslop-Harrison 1954) — = → comparium (→ deme).

syngamy — sexual → reproduction: The union (during → fertilization) of male and female gametes, resulting in karyogamy and zygote formation (→ apogamy).

syngen (Sonneborn 1957) — a reproductively isolated population representing an evolutionary unit in ciliates. The term s. formally corresponds to a → species, but since species are defined by their → gene pools and common gene pools are absent in ciliates, syngens are not species in the true sense.

syngenote (Morse et al. 1956) — a bacterial → merozygote which, in addition to its own

→ genome (referred to as the endogenote), contains an additional genome fragment (referred to as the exogenote). A s. is partially diploid (hyperploid) and may be → heterogenotic or homogenotic.

syngonic — of organisms producing male and female gametes in the same sexual organ.

synizesis (McClung 1905) — a clumping (into a more or less dense knot) of the chromosomes in early prophase of first meiotic division. S. may be a "normal" feature of → meiosis in some organisms and can be induced artificially through fixation or other kinds of cellular damage.

synkaryon — the → nucleus of a → zygote, resulting from the fusion (karyogamy) of the nuclei of the female and male gamete during fertilization.

synkaryophyte — = → sporophyte.

synkaryotic (Buller 1941) — of diploid cells in Asco- and Basidiomycetes, as opposed to dikaryotic cells (→ dikaryon). S. cells are produced by "synkaryosis" which takes place at the end of the dikaryophase and consists of the pairwise fusion of haploid nuclei of opposite → mating type..

synoecius (Correns 1928) — of haploid or diploid organisms producing both male and female gametes. The term includes → monoecious and → hermaphroditic individuals.

synonymous mutation (Sonneborn 1965) — a class of → gene mutations (samesense mutations) which represent → base-pair substitutions in DNA that do not result in the substitution of a different amino acid into the encoded protein. This is due to the degeneracy of the → genetic code. The codon produced by mutation is called a synonymous codon or synonym.

syntelic (Bauer et al. 1961) — of the → centromeres of the two chromatids of each chromosome if they are oriented to the same spindle pole at first meiotic division. Orientation to different poles is called amphitelic.

syntenic (Renwick 1971) — of two genetic loci presumed to be linked to the same chromosome, whether or not → linkage has been demonstrated, as opposed to asynthenic loci, which are linked to different chromosomes.

The term synteny may be used when referring to linkage relationships based upon analysis after somatic → cell hybridization, i.e., determination of linkage relationships by the results of correlated loss or retention of markers in hybrid cells. If concordant segregation of markers occurs, then linkage may be presumed; if discordant segregation occurs, then linkage can be denied.

synthetic gene — → gene synthesis.

synthetic lethal (Dobzhansky 1946) — → lethal factor.

syntrophic (Davis 1950) — of any two bacterial strains, → auxotrophic for different growth factors which are able to cross-feed each other and grow together on minimal medium. The phemomenon is called syntrophism or cross-feeding and is found with many bacterial mutants.

T

tachytelic (Simpson 1944) — ref. to → evolution at a rate distinctly higher than a standard or → horotelic rate (→ bradytelic). Tachytelic evolutionary rates represent exceptions characteristic of certain relatively short periods of time. The factors causing t. evolution are assumed to be environmental in nature (changes in geological environment, transposition of the evolving group of individuals into a new environment).

TACTAAC box — a DNA sequence close to the 3′ end of the → intron that is stringently required for → RNA splicing in yeast. Due to this sequence, yeast is incapable of removing introns from transcripts of higher eukaryotic genes.

tailing — nontranscriptive addition of homopolymeric nucleotide sequences to the 3′ end of DNA or of a restriction fragment by use of the enzyme → terminal transferase.

tandem duplication — → duplication.

tandem fusion (White 1957) — the fusion in the course of chromosome structural changes of two acrocentric chromosomes "end to end" with loss of one centromere, so as to give rise to a double-length acrocentric. Alternatively, the fusion of an acrocentric, which lost its centromere, with the end of one arm of a metacentric chromosome. Both types of t.f. played some role in the evolution of the karyotype of certain organisms (→ centric fusion; translocation).

tandem inversion — → inversion.

tandem repeats — multiple copies of the same DNA sequence lying in series (= tandem → duplication).

tandem satellite (Taylor 1926) — two → satellites of the same chromosome, separated from each other by a secondary → constriction.

tandon (Tartof 1975) — a multiple-copy, tandemly repeated gene (→ multigene family).

targeted gene transfer — the transfer of gene sequences directed to a specific site in the genome (→ gene targeting). The vector used to introduce the new gene into cells must carry nucleotide sequences identical to those of the DNA at the genomic site where the gene should integrate by homologous recombination. T.g.t. can be used to correct or inactivate genes.

targeted mutagenesis (Witkin and Wermundson 1978) — mutagenic processing of → DNA lesions in the same gene where the mutation occurs, as opposed to nontargeted (untargeted) mutagenesis, i.e., mutagenesis in undamaged DNA sequences (→ SOS response) due to reduced fidelity of DNA replication on intact templates.

targeting sequence — any amino acid sequence of mitochondrial, chloroplast, and secretory proteins that directs the proteins synthesized in the cytoplasm and coded for by nuclear genes to their site of destination (→ signal peptide).

target theory (Crowther 1924) — a theory according to which a reaction (e.g., a gene mutation or a chromosome structural change) is produced whenever a ionization takes place within a well-defined small volume. The target is usually in or close to the chromosome.

TATAAA sequence — a eukaryotic → promoter element which shows homology to the prokaryotic -10 region (consensus sequence TATAAT).

TATA box (Goldberg 1979) — a nucleotide sequence at the 5′ end of eukaryotic → messenger RNA encoding genes (about 30 bp long and upstream from the genomic location of the cap site) which has a critical role in the initiation of genetic transcription. TATA boxes are general promoter elements that are recognized by a common transcription factor (= Hogness-Goldberg box). The TATA box is analogous to the → Pribnow box of prokaryotic promoters found around position -30.

The consensus sequence for the TATA b. is 5′-TATA$^{A}_{T}$A$^{A}_{T}$-3′.

tautomeric shift — the spontaneous isomerization of a nitrogen base in DNA to an alternative hydrogen-bonding condition; t. shifts may result in → gene mutations.

taxon (Rickett 1958) — a taxonomic group of individuals (population) recognized as a formal unit at any level of a hierarchical classification.

T-chromosome (Kattermann 1939) — a monocentric chromosome whose terminal segments (T-ends) show → neocentric activity at meiosis.

(TC)n·(GA)n sequence element (Sures et al. 1978) — in eukaryotic (vertebrate) genomes, (dT-dC)n·(dG-dA)n repeating dinucleotide sequence elements (0.05–0.07% of all DNA sequences) which are spread throughout the genomes and map primarily within → introns and intergenic regions. Possibly they act as pause or arrest signals for DNA replication and amplification.

tc RNA (Heywood et al. 1974) — an acronym for → translational control RNA.

T-DNA — a variable length segment of the bacterial *(Agrobacterium tumefaciens)* → Ti plasmid which is transferred across bacterial and plant cell membranes and becomes stably integrated into the plant nuclear genome. In addition to the neoplastic nature of the transformed tissue unusual amino acids or sugar derivatives (opines) are frequently synthesized by enzymes coded by the T-DNA; it is a vector for introducing foreign genes into plants.

tectine (Mazia and Ruby 1968) — any of a class of similar proteins which form various eukaryotic structures, such as → microtobules, → microfilaments, and membranes. Tectines show resemblances to each other in, e.g., amino acid composition, and the molecules of each t. specifically aggregate together.

TE element (Ising 1981) — in *Drosophila*, any of a family of large (several hundred kb) → transposable genetic elements. They are composite elements; their termini consist of foldback sequences which are themselves transposable. The DNA involved in TE's can include significant components of the genome.

telocentric (Darlington 1939) — of chromosomes or chromatids with a terminal → centromere. Such chromosomes may arise by → centromere misdivision or breakage induced within the centromere region. Usually telocentrics are unstable and are either eliminated in the course of a few cell divisions or transformed into so-called → "isochromosomes".

telochromomere — any → chromomere localized at the end of a chromosome arm (→ telomere).

telochromosome — a chromosome with a terminally located → centromere.

teloisodisomic (Kimber and Sears 1968) — a cell or individual in which one chromosome pair is missing but a telocentric for one arm of the missing pair and an → isochromosome for the same arm are present (→ monotelomonoisosomic).

telolecithal — of egg cells in which the yolk (reserve material) is accumulated towards one end, referred to as the vegetal pole (→ isolecithal).

telomerase (Szostak and Blackburn 1982) — in eukaryotes, a ribonucleoprotein enzyme whose RNA and protein components are essential for activity in → telomere replication.

telomere (Muller 1940) — the special structure at the molecular end of eukaryotic linear chromosomes which provides stability to the chromosome, prevents fusion with other natural or broken ends, and allows replication without loss. With prototype semiconservative DNA replication requiring RNA primers, excision of the 5′-terminal primer would leave one of the daughter strands shorter than its parental strands. A specific terminal transferase (telomerase) synthesizes the ends of chromosomes de novo by adding the telomeric sequences in a nontemplated way back to the chromosome one base at a time. Telomeres consist of simple DNA repeat sequences (2 to 10 bp long) fitting the formula 5′ $C_n(A/T)_m/G_n$ $(T/A)_m$ 3′ (n=1 to 8; m=1 to 4) with defined orientation (C-A-rich strand running 5′ to 3′ from the end towards the interior of the DNA molecule). The C-A-rich strand usually contains nicks and the G-T-rich strand extands into a unique terminal structure. The special features which differentiate telomeres from internal chromosome regions include: (1) Nonligatable single-strand gaps; (2) blockage of the extreme end by a hairpin loop at the 3′ end in one strand (the G strand) anchored by guanine-guanine bonds (instead of the usual guanine-cytosine bonds); (3) non-nucleosomal protein complexes; (4) presence of subtelomeric middle-repetitive sequences (t.-associated sequences) in a number of species.

The telomerase adds at the 3′ end d GTP/d TTP in form of species-specific repeats (7 to 10 bp per replication) and allows complete replication of both the leading (3′) and lagging (5′) strands.

In higher eukaryotes, telomeres are often associated with blocks of constitutive → heterochromatin anchored to the → nuclear envelope; they are heterogeneous in length and consist of short, direct repeats, as in lower eukaryotes.

telomere fusion (Niebuhr 1972) — the end-to-end joining (under some specific conditions) of two chromosomes to form one pseudodicentric chromosome by terminal rearrangement. If one → centromere of the pseudodicentric is suppressed, t.f. may result in a stable chromosome.

telomere terminal transferase (Greider and Blackburn 1985) — a ribonucleoprotein enzyme which can synthesize the tandem repeat units of → telomeres to the 3′ end of telomeric primers without template. The RNA component of t.t.t. contains the complementary sequence of the telomeric repeats that it synthesizes.

telomitic (Carothers 1917) — = → telocentric.

telophase (Heidenhain 1894) — → mitosis, meiosis.

teloreduplication (Hsu and Moorhead 1956) — → interreduplication.

telosome (Endrizzi and Kohel 1966) — a → telocentric chromosome.

telotrisomic (Kimber and Sears 1968) — an individual deficient in one chromosome but which has a → telocentric chromosome and an isochromosome for the same arm of the missing chromosome (→ diiso-compensating trisomic).

temperate (Jacob et al. 1953) — of bacteriophages that can provoke one of two responses upon infection of a host cell: (1) The **lytic response** in which the bacterial cell lyses and releases new virus particles, or (2) the → **lysogenic response** in which the cell survives with the viral DNA associated with the bacterial chromosome and replicates with it in the → prophage state. Whether the lytic or lysogenic pathway is followed is influenced by various phage functions (repression of viral functions required for the lytic response) and the prevailing physiological conditions of the recipient cell (→ coimmune; heteroimmune). **A virulent phage** can provoke only the lytic response.

The relationship between a t. phage and the host bacterium is characterized by three general properties: (1) Each prophage has a specific location (attachment site) on the bacterial chromosomes which is different for different t. phages. (2) The prophage confers upon the bacterium immunity against a lytic response upon superinfection by the same phage. (3) A bacterium harboring a prophage still carries the potentiality for producing a lytic response and release of virus particles (→ induction).

Defective t. phages (Roming 1968) are those which form particles with all the morphological attributes of normal phage but are incapable of replication.

temperature-conditional (Davis 1950) — of temperature-sensitive conditional lethal mutations that can be perpetuated at the permissive temperature. The lethal defect may be studied at the nonpermissive temperature. This type of conditional lethal mutation is usually due to → missense mutation. → Nonsence and → frameshift mutations are more likely to abolish function at all temperatures and to lead to absolute lethality in the absence of a → nonsense or → frameshift suppressor. The defect in t.-c. mutations is usually attributable to a protein which is nonfunctional at a high (nonpermissive) temperature, but can function normally at lower temperatures.

template — a molecule that shapes the structure or sequence of another molecule, e.g., the polynucleotide strand (template strand) of DNA during → DNA replication or → genetic transcription.

template RNA — = → messenger RNA.

temporal gene (Paigen and Ganschow 1965) — any gene that determines the relative expression of → structural genes in a particular organ or at a particular developmental time. Two classes of t. genes may be distinguished: (1) Trans-acting t. genes which lie at a distance from the structural gene whose expression they influence; (2) cis-acting t. genes which lie in close proximity to the structural gene.

temporal mapping — → replication map.

T-end (Prakken and Müntzing 1942) — → T-chromosome.

ten-nanometer filament — = → intermediate filament.

teratogenic — of any agent which interferes with normal development (→ carcinogenic; mutagenic).

terminal — ref. to chromosome or chromatid segments terminal in position, as opposed to intercalary segments.

terminal affinity (Darlington 1932) — the forces (of unknown nature) by which chromosomes are held together end to end from diplotene to metaphasse I of → meiosis, due to the inhibition of → chiasma terminalization at the chromosome ends, or are brought together in this way at metaphase in special cases (Darlington and Mather 1949).

terminal association — nonspecific → chromosome pairing which results in nonchiasmatic end-to-end association of chromosomes.

terminal capping — → messenger RNA capping.

terminal deoxynucleotidyltransferase (Bollum 1960) — (TdT); a unique → DNA polymerase (terminal transferase) that catalyzes the polymerization of 5′-dNTP's onto the 3′-OH terminus of an oligodeoxyribonucleotide. TdT does not require a template but has an absolute requirement for an initiator molecule $(dNMP)_n$, containing at least two to three nucleotide residues and a free 3′hydroxyl end (→ telomere). All four of the common 5′-dNTP's can be polymerized by TdT.

TdT has been implicated in the postembryonic DNA synthesis required in the → immunoglobin gene recombination event.

terminalization (Darlington 1929) — → chiasma terminalization.

terminalization coefficient — → chiasma terminalization.

terminal protein — a protein bound to the 5′ end of the DNA of several virus genomes with an inverted terminal repeat of variable length. Viral replication is initiated by a protein-directed priming mechanism. At.p. has also been found in linear → mitochondrial DNA's of maize and in a linear plasmid from Streptomyces.

terminal redundancy (Streisinger et al. 1964) — in many bacteriophages, a linear DNA molecule with single-stranded complementary ends that is longer than necessary to close the DNA circle. T.r. may be of variable length, ranging from 0.5 to 3% of the phage genome.

terminal riboadenylate transferase — any enzyme (TrT) which catalyzes the transfer of adenylate residues from ATP to the 3′-OH group of certain polyribonucleotides in the presence of Mn^{2+}. Enzymes of this kind may be involved in the processing of → heterogeneous nuclear RNA to → messenger RNA in eukaryotic cells.

terminal transferase — = → terminal deoxynucleotidyltransferase.

terminase system (Campell et al. 1969) — an enzymatic maturation system of bacteriophage lambda which produces mature lambda chromosomes by cleavage of multichromosomal lengths of lambda DNA (ter system). Mature chromosomes would be generated by ter cleavage at two cohesive end sites (cos sites) to generate unit length molecules with → cohesive ends (→ integration system; prophage).

terminated leader transcript — the transcript terminated at the → attenuator.

termination — in protein biosynthesis (→ genetic translation), the incorporation of the last amino acid into a nascent polypeptide chain and the release of the complete chain from the ribosomes and messenger RNA (→ release factor; terminator codon).

termination factor (Ganoza 1966) — = → release factor.

termination sequence — a DNA sequence at the end of a → transcription unit that signals the end of → genetic transcription.

termination signal — in some phages, in which initially early genes are transcribed by a bacterial polymerase, a protein factor which halts polymerase. Subsequently an anti-termination signal (protein) is synthesized to allow transcription beyond these signals.

termination structure — the segment of the leader transcript (→ leader region) that is thought to be recognized as a termination signal by → RNA polymerase.

terminator (Reznikoff 1972) — → transcriptional terminator.

terminator codon — any → codon which, in the usual course of events, is read by → release factors that trigger the hydrolysis of the aminoacyl bond connecting the completed polypeptide to the → transfer RNA cognate to the carboxyterminal amino acid (= termination or stop codon; chain-terminating codon). T. codons (UAA, UAG, and UGA) do not specify amino acids. Occasionally t. codons can be translated as a result of the presence of → suppressor tRNA or of the action of an amino acid tRNA. The resulting extended protein is identical with a normal gene product except for its carboxyl extension. Translation of a t.c. is used to make a variety of minor gene products whose time of appearance and level of expression are coupled to the major product.

In bacteria, release factor RF1 recognizes t. codons UAG and UAA, and RF2 recognizes UGA and UAA.

Premature polypeptide chain termination may result if a mutation occurs in a gene so that a triplet in the correct phase on the mRNA becomes a t.c. (→ nonsense mutation; nonsense suppression; reinitiator).

terminator stem — in factor (rho)-independent → transcription termination *(E. coli)* an inverted repeat that encodes a hairpin in the transcript centered about 30 nucleotides before the termination site. A second sequence element is a sequence rich in adenine residues in the DNA template strand immediately preceding the stop site. These elements invariably cause termination unless specific regulatory devices intervene, e.g., antiterminators or the upstream RNA elements of → attenuation systems that base pair with the t.s. and block its formation.

terminus-site-specific recombination — in bacteria, a mode of → genetic recombination (recA-independent) which plays a role in the structural evolution and organization of bacterial → plasmids. The addition by t.-s.-s.r. of defined DNA segments on to plasmids, phages, and bacterial chromosomes, and the excision of specific DNA segments from such chromosomes may represent a principal mechanism by which the organization of prokaryotic DNA has evolved and provides a potential mechanism for the exchange of genetic information between diverse biological species.

test cross (Bridges 1934) — a → cross of a double or multiple heterozygote (hybrid) with the corresponding double or multiple recessive (nonhybrid) in order to estimate the → linkage relationships of the gene loci in question.

tetraallelic (Altwood 1944) — → monoallelic.

tetrad (Němec 1910) — (1) the four chromatids of any → bivalent in the first division of → meiosis (Němec). (2) A quartet of cells (each with reduced chromosome number) resulting from a single meiotic event (→ meiosis) which takes place in a → meiocyte or mother cell. In some organisms (certain fungi, bryophytes, and algae), all four meiotic products may be recovered and analyzed genetically (→ tetrad analysis). Among such organisms the immediate products of meiosis (referred to as gones) undergo one or two subsequent mitoses, giving rise to 8 to 16 products which represent exact replicates of the original four.

Tetrads (or their derivatives) may be ordered (in *Neurospora* and some yeasts) or unordered (in most organisms). (1) **Unordered tetrads:** With two segregating gene pairs (A/a and

B/b) and the parental combinations AB and ab, three segregation patterns are possible. Tetrads of the type AB/AB/ab/ab (no recombinants) are referred to as **"parental ditype"** (PD), tetrads of the type AB/Ab/aB/ab (two recombinants) are referred to as **"tetratype"** (TT) and tetrads of the type Ab/Ab/aB/aB (four recombinants) are called **"nonparental ditype"** (NPD).

These three types of tetrads arise (by not more than two crossovers between the loci A and B) as follows: PD due to no crossing over or two-strand double cross overs between A and B; NPD due to two crossovers of the four-strand type between A and B (this type of t. is rare for linked markers); TT due to either a single crossover or a three-strand double crossover between A and B.

By means of tetrad analysis the occurrence of linkage between two markers may be determined. If there is → linkage, the NPD's will be rare relative to the other two tetrad types. If the markers are unlinked, the NPD's and PD's will be equal in number, since both types result from random → assortment of the chromosomes (with unlinked markers, the TT's result from crossing over between either A and its → centromere, or B and its centromere). Recombination frequency between A and B equals ½ TT + NPD divided by the total number of tetrads. (2) **Ordered tetrads:** The meiotic products are ordered and the order indicates the pattern of meiotic events. The analysis of ordered tetrads may be readily used to obtain → recombination frequencies both between different genes and between genes and their centromeres. By classifying the tetrads as PD, TT, and NPD with respect to each gene pair, differentiation among two-, three-, and four-strand double crossing over can be made.

tetrad analysis — the genetic analysis of all the products of a single meiotic event in organisms with (ordinarily) cytologically observable chromosomes (eukaryotes) and a standard → meiosis. T.a. is possible in those organisms in which the four products (or their derivatives) of a single nucleus which has undergone meiosis are grouped together (→ tetrad).

tetraploid (Němec 1910) — of → auto- or → alloploid cells, tissues, and individuals with four chromosome sets in their nuclei (symbol 4x).

tetrasomic (Blakeslee 1921) — of → polysomic cells, tissues, or individuals with one chromosome represented four times instead of twice in an otherwise diploid (disomic) chromosome complement (symbol 2n + 2). Tetrasomics may result from intercrossing or selfing of → trisomic individuals (2n + 1). The frequency of tetrasomic individuals among the offspring of such matings depends upon the frequency of transmission (in the male and female line) of the extra chromosome present in trisomics. Tetrasomics rarely arise spontaneously from diploids.

Tetrasomics accentuate the phenotypes of the corresponing trisomics and their viability is generally greatly lowered.

Tetrasomic inheritance is a special pettern of inheritance due to one linkage group being present four times instead of twice. The following segregation ratios, resulting from a meiotic two-by-two distribution of the four homologous chromosomes, is theoretically expected for a tetrasomic genotype AAaa:

Gametes	1 AA	4 Aa	1 aa
1 AA	1 AAAA	4 AAAa	1 AAaa
4 Aa	4 AAAa	16 AAaa	4 Aaaa
1 aa	1 AAaa	4 Aaaa	1 aaaa

With A completely → dominant over a, a phenotypic segregation ratio of 35A:1a (→ autoploid) is expected.

Compensating nullisome-tetrasome combinations are allotetraploid karyotypes (→ alloploid) in which a certain tetrasome compensates more or less completely for a certain nullisome (→ nullisomic), indicating that the chromosomes under observation are reasonably homologous. Nullisome-tetrasome analysis is a powerful tool for genome analysis in segmental allopolyploids.

tetraspore (Renner 1916) — → germ cell.

T-factor — = → elongation factor or transfer factor.

theletoky (v. Siebold 1871) — → parthenogenesis.

thelygenous — producing exclusively or predominantly female offspring, as opposed to arrhenogenous.

thermal denaturation — → denaturation.

thermal recovery — → liquid-holding recovery.

thermotolerance — the induced capacity of cells to survive an otherwise lethal temperature after their exposure to some stressful stimulus such as a heat shock (→ heat shock response).

threshold character (Dempster and Lerner 1950) — a term used for those phenotypic → characters whose segregating distributions are phenotypically discontinuous but whose inheritance is multigenic like that of continuously varying (quantitative) characters. The discontinuous segregations of such "quasi-continuous characters" (Grüneberg 1952) result from threshold effects, i.e., the characters in question have an underlying continuity with a threshold which imposes a discontinuity on their visible expression. Threshold characters may segregate into many discontinuous phenotypic classes; those that segregate into two classes only are also called "all-or-none characters".

Thomas circle (Schachat and Hogness 1973) — a double-stranded DNA fragment from eukaryotic → chromosomes which can form circular structures after treatment with exonucleases (→ nuclease). The formation of T. circles is assumed to be due to the presence of → repetitious DNA sequences in the fragments.

thylakoid (Menke 1961) — any of the internal membranes of the → chloroplast which constitute the sites where transformation of light energy into chemical energy (photosynthesis) is catalyzed by multisubunit complexes composed of pigments, lipids, and proteins. The t. includes granal (stacked) lamellae, containing mainly the photosystem II components, and interconnecting stromal (unstacked) lamellae, which preferentially contain the photosystem I components and the ATPase complex.

In prokaryotes, the t. are of various shapes and usually free in the cytoplasm.

Light energy collection and its efficient conversion into chemical energy requires a highly organized t. membrane structure. Most t. membrane polypeptides are arranged into multiprotein complexes spanning the membrane. They include photosystem I (PS I), photosystem II (PS II), and the ATP synthase.

thymineless death (Cohen and Barner 1954) — (TLD); cell death occurring when thymine-auxotrophic bacterial (or other organisms) are deprived of a source of thymidylate. Under this condition, DNA synthesis is inhibited and DNA strand breaks accumulate. Possibly, TLD is due to attempted replication in the absence of thymine. The dUTP incorporated instead of dTTP is subject to recognition and removalby uracil glycosylase and the resulting AP sites will then be attacked by AP endonucleases.

tier system — in → mutagenicity testing, a test system in which potentially mutagenic chemical detected in a relatively simple test are subjected sequentially to more biologically complex tests until they are either eliminated from further testing or bioassayed in full-scale animal systems.

tight junction (Farquhar and Palade 1963) — in epithelial tissues, a belt-like region of contact between cells that serves as a selective barrier to small molecules and as a total barrier to large molecules (= zonula occludens; occluding junction). The main function of the t.j. is to enable a group of cells to maintain an internal environment from the external one (→ adhering junction; gap junction; desmosome).

Ti plasmid (Zaenen et al. 1974) — in *Agrobacterium tumefaciens*, a large tumor-producing → plasmid which transfers a specific segment of DNA (the T-DNA) into plant cells where it is integrated into the nuclear genome and induces crown gall tumors (in gymnospers and dicotyledonous angiosperms). The T-DNA is flanked by nearly perfect 25 bp direct repeats which act as T-DNA transfer signals. T-DNA transfer relies on products of the Ti p. virulence genes and the chromosomal virulence genes. Any sequences inserted between the terminal repeats can be integrated into plant nuclear DNA.

Ti plasmids are classified in three major groups: (1) Octopine Ti plasmids; (2) nopaline Ti plasmids; (3) agropine Ti plasmids.

tissue — a spatially ordered assembly of → cells with a particular pattern of interaction among them. A typical **mitotic t.** consists of two main groups of cells, those involved in → mitosis and the others in aging and function (→ dichophase; chalone). The balance between cell gain, cell function, and cell loss in any normal t. is determined by a homeostatic mechanism of selection between two alternative gene-controlled programs of syntheses. In a **nonmitotic t.**, composed of cells that are incapable of mitosis, no such balance exists. New cells are never created and although the t. may increase in size, any such growth is due solely to an increase in cell size.

The mass of any t. is determined by the relation between the rate of cell production by division and the rate of cell loss.

tissue culture — the growth and maintenance of cells from higher organisms in vitro, i.e., outside the tissue of which they are normally a part (→ cell hybridization).

T_m — → melting temperature.

T_n element — a bacterial → transposon.

tokozygote (Renner 1916) — the → zygote in *Thallophyta* which is immediately transformed into a gonotokont (= gonotokozygote) and gives rise to haploid → germ cells (gonospores) or vegetative cells (gonocytes).

tonofilament — in most epithelial cells of vertebrates, a meshwork of bundles of → intermediate filaments extending through the cytoplasm and often attached to the → desmosomes of the surface membrane.

tonoplast (de Vries 1885) — the membrane separating the cytoplasm and the vacuole in plant cells.

topogenesis — a process of movement in relation to → morphogenesis.

topogenic signal (Blobel 1980) — any signal which routes proteins to their final location inside or outside the cell. Transiently attached → signal peptides serve to initiate translocation across the → endoplasmatic reticulum membrane; mitochondrial and chloroplast targeting sequences direct proteins to various destinations in these organelles. Nuclear location signals ensure protein transport into the nucleus. T. signals are often found in the N′-terminal region of the protein.

topo-inhibition (Dulbecco 1970) — = → contact-inhibition.

topoisomerase (Wang and Liu 1979) — → DNA topoisomerase.

topological isomer — any of the DNA molecules that are identical except for a difference in the → topological winding number.

topological winding number (Vinograd et al. 1968) — in covalently closed circular DNA, the sum of the helical winding number and the number of superhelical turns in the native form of DNA.

toposome (Noll et al. 1985) — any of the large oligomeric glycoprotein complexes that are able to mediate cell adhesion and to express positional information (positional guidance of cells). Toposomes mediate morphogenesis and are responsible for mechanical linkage between cells and their positional guidance during embryogenesis.

torsion pairing (Darlington 1935) — nonspecific pairing between nonhomologous chromosome segments at prophase of the first meiotic division (→ chromosome pairing). T.p. does not result in → crossing over or → chiasma formation.

touch-and-go pairing (Wilson 1925) — a transitory but highly specific (end-to-end) association of chromosomes (especially of → sex-chromosomes) at the second meiotic division in the heterogametic sex of *Hemiptera.*

totipotency — the ability of a cell to proceed through all the stages of development and thus to produce a normal adult.

trailer —a nontranslated sequence at the 3′ end of a → messenger RNA transcript, following the → terminator codon.

trait — = → character.

trans-acting factor —any of the protein factors that mediate transcription of (eukaryotic) genes by transcriptional activation and bind to → cis-acting DNA sequence elements. Most → promoters for → RNA polymerase II contain binding sites of t.-a. factors. Such factors may act by: (1) Transcription activation of genes encoding t.-a. factors; (2) nuclear translocation; (3) modification of an already DNA-associated factor to acquire transcription activating competence (→ zinc finger motiv; leucine zipper; turn-helix-turn motif; GC box).

transactivation (Prell 1973) — (1) the phenomenon wherein a superinfecting → bacteriophage activates a thus far nonexpressed prophage in a lysogenic host bacterium, irrespective of the mechanism underlying this activation. (2) In *Herpes simples* virus (HSV), positive regulation (transactivation) of transcription of a second group of genes by gene products translated immediately after infection.

trans-configuration (Haldane 1942) — → cis-configuration.

transconjugant (Novick et al. 1973) — a bacterial cell which has received genetic material from another bacterium by → conjugation.

transcribed spacer — that part of a → ribosomal RNA transcription unit that is transcribed but discarded during → RNA processing.

transcript — the immediate product of → genetic transcription (= RNA transcript) which, after → RNA processing, gives rise to functional mature RNA's.

transcriptase (Spiegelman and Hayashi 1963) — = DNA-dependent → RNA polymerase (→ reverse transcriptase) responsible for the transcription of → ribosomal RNA, → transfer RNA, and → messenger RNA from a DNA template.

transcript cleavage site — a DNA sequence which is responsible for cleavage of polycistronic transcripts.

transcript heterogeneity — the phenomenon whereby a single eukaryotic gene may give rise to several → messenger RNA's. This may be due to: (1) Several transcription start points; (2) several polyadenylation sites; (3) alternative → RNA splicing.

transcript hybridization — a variant of the → in situ hybridization technique. Labeled DNA is being hybridized with nascent RNA of the locus in question.

transcription — → genetic transcription.

transcriptional activator (Brent and Ptashne 1985) — any → DNA-binding protein that stimulates transcription by interacting with DNA sequence motifs (transcriptional control regions), RNA polymerases, or both (→ transcriptional repressor; POU domain). At least some t. activators have two surfaces: (1) The DNA-binding region which positions the protein on the DNA; (2) the activating region which interacts with a target protein to promote initiation of transcription.

transcriptional antitermination — a mechanism which allows → transcription termination signals of an operon to be overcome.

transcriptional control — control of → gene expression at the level of → genetic transcription which can be exerted through regulation of → transcription initiation or → transcription termination or through selective post-transcriptional processing. It allows RNA synthesis to occur at selected loci of the genetic information and may be negative or positive (→ genetic regulation): (1) **Negative t.c.:** Genetic transcription is turned off by → repressor molecules that can bind to an → operator and prevent RNA polymerase from travelling across the operator. (2) **Positive t.c.:** Genetic transcription is turned on by a positive controller, which may or may not be a protein (→ sigma factor), and does not involve release of repression.

Transcriptional control regions of eukaryotic genes consist of three major classes of sequences: (1) Proximal control signals such as the → TATA box; (2) distal control signals such as the → CAATT box; (3) activating sequences such as → transcriptional enhancers.

transcriptional enhancer (Gruss et al. 1971) — any of a class of cis-acting DNA sequences (usually 200 bp long) that are able to activate the transcription (by RNA polymerase II) of cellular and viral genes in a manner relatively independent of position and orientation. Transcriptional activation can occur over distances as great as 6 kb from either the 5′ or 3′end of a gene. Many RNA polymerase II t. enhancer elements include an octamer sequence motif, ATTTGCATT. Enhancers are believed to be, in certain cases at least, the receptor sites of some cellular regulator molecules. Some enhancers can function in a number of different cell types, others show distinct cell-type specificity. Frequently, t. enhancers are found near regions of alternating purines and pyrimidines, a structure capable of forming → Z-DNA. In some genes, enhancer sequences have been localized in → introns. (→ transcriptional silencer).

transcriptional polarity (Kano and Imamoto 1975) — → genetic polarity.

transcriptional readthrough — prevention of → transcription termination and production of a longer than usual transcript, e.g., a r. protein. T.r. can occur in the presence of a → suppressor tRNA which allows insertion of an amino acid into a growing polypeptide chain in response to a → terminator codon. Possibly natural t.r. is occasionally used to create proteins in which two different polypeptide chains are joined together.

transcriptional repressor — any regulatory molecule that reduces the initiation of → genetic transcription by preventing RNA polymerase binding to the → promoter (→ transcriptional activator).

transcriptional silencer (Brand et al. 1985) — a regulatory DNA sequence element that suppresses enhancer-dependent transcriptional activity (→ transcriptional enhancer). T. silencers are position- and orientation-independent. A cis-dependent relationship to the transcription unit on which it acts is required.

transcription antitermination — any transcription regulatory mechanism, mediated by specific proteins, which allows transcription of genes downstream from transcription termination signals and results in increased expression of these genes (→ transcription attenuation).

transcription attenuation (Kasai 1974) — in prokaryotes, a regulatory mechanism which involves premature termination of an elongating RNA chain and is translationally coupled to specific control regions (→ attenuators) near the 5′-termini of the appropriate mRNA species. T.a. is controlled by translation of a leader peptide coding region, which in turn determines which transcript secondary structures form as → RNA polymerase traverses the leader region of the operon: (1) The terminator promotes transcription termination at the attenuator, a site immediately preceding the structural genes of the operon; (2) the antiterminator forms when a ribosome stalls at one of the tandem codons of the coding region for the leader peptide. Antiterminator formation prevents terminator formation and permits polymerase readthrough into the structural genes.

T.a. is used for early termination of transcription of an amino acid biosynthetic operon when an abundance of the corresponding charged → transfer RNA signals a sufficiency of the amino acid.

transcription bubble — a local opening (destabilization) of the DNA double helix during → genetic transcription in those regions where temporary DNA-RNA hybrids are formed as intermediates in the transcription of DNA. As transcription proceeds, the RNA chain is gradually displaced from its template DNA chain by the reunion of the two DNA chains to reform the double helix.

transcription complex (Cizewski and Soellner-Webb 1983) — a complex formed by interaction of multiple proteins with a region of DNA within a gene called the internal control region. The t.c. is remarkably stable and may account at least in part for the maintenance of the differentiated state of cells which express one set of genes and repress another set. T.

complexes have been found for genes transcribed by all three classes of eukaryotic → RNA polymerases.

transcription elongation — in → genetic transcription, the processes after t. initiation and ending with t. termination. Blocks of t.e. are important regulatory steps in prokaryotic and eukaryotic gene expression (→ attenuator; transcription attenuation).

transcription error — an error in the transcription of genetic information (→ genetic transcription). If a t.e. occurs in the transcription of a typical → structural gene, all polypeptides translated (→ genetic translation) from the mRNA will be incorrect. The particular error need not be, however, a serious one for the genes if the mRNA output is high and/or under feedback control. In the case of genes whose mRNA output is fixed and of the order of one copy per cell cycle, the t.e. may become a problem if the gene concerned is critical to cell function. If the gene subject to a t.e. codes for one of the components of the translation apparatus, a certain proportion of such errors would cause the affected compound to malfunction which, in turn, may increase its own error frequency.

transcription factor — any of the multiple ancillary DNA-binding proteins which interact with the cis-regulatory DNA sequences to control → gene expression by: (1) Commitment of the template after binding of a subset of the t. factors; (2) formation of an activated state through the action of other t. factors and RNA polymerase; (3) fulfilment of an energy requirement; (4) initiation of transcription; (5) transcription elongation. (→ POU domain; repressor; activator; sigma factor; catabolite activating protein).

transcription initiation — a multistep process which occurs as a result of recognition by → RNA polymerase of specific DNA sequences called → promoters and initiates RNA chains during → genetic transcription. T.i. involves: (1) RNA polymerase binding to a promoter in a DNA template to form a closed complex; (2) a segment of DNA in the closed complex is melted out and an open complex is formed; (3) binding of the first ribonucleotide triphosphate converts the open complex into an → initiation complex. Elongation can then start in the presence of all four ribonucleotide triphosphates.

Regulation of t.i., is a major control point in → gene expression (→ transcription factor).

transcription mapping (Bleyman and Woese 1969) — the mapping of transcripts on well-defined DNA segments. The gene must be cut once at a known site, from which it is then possible to map the transcripts toward the 3′ and 5′ends of the gene. T.m. permits the fine mapping of RNA transcripts with the same level of accuracy as the mapping of → restriction endonuclease cleavage sites on DNA.

transcription modulatory sequence (Ciliberto et al. 1982) — in certain eukaryotic genes coding for → transfer RNA, a sequence contained in the 5′-flanking regions (in addition to the internal → promoter) which has positive or negative effects on transcription efficiency.

transcription pausing — the phenomenon wherein both prokaryotic (→ transcription attenuation) and eukaryotic → RNA polymerases stop or slow down → genetic transcription at transcription pause sites when particular regions of a DNA molecular are being transcribed, i.e., chain elongation does not occur at a constant rate. T.p. may be essential for proper regulation of transcription and occurs immediately after transcription of regions with dyad symmetry. Formation of a hairpin near the 3′end of a growing transcript delays one or more of the steps required for transcript elongation.

transcription start point — the position on DNA corresponding to the first nucleotide incorporated into a transcript (= t. start site).

transcription termination — the process which terminates → genetic transcription at the 3′ends of genes, gene clusters, or operons. T.t. is thought to consist of three separate events: (1) Cessation of movement of → RNA polymerase and of chain growth at a terminator sequence; (2) release of the completed RNA chain from the enzyme; (3) dissociation of RNA polymerase from the DNA template.

Prokaryotic t.t. can be divided into several groups. It may occur directly in response to a nucleotide sequence or structure (the → t. terminator), or requires additional factors (rho-independent and rho-dependent t.t.).

Transcription terminators may be located in various positions of prokaryotic operons: (1) Distal to the last gene of an operon and defining the 3′end of primary transcripts initiated from an upstream → promoter; (2) factor-dependent terminators within or between genes of a polycistronic operon which may result in transcriptional polarity; (3) translationally controlled termination within the leader region of an operon (→ transcriptional attenuation).

The efficiency of t.t. and its modulation by → attenuator sites, antitermination factors, and processing/modulation signals, can have profound effects on → gene expression. Terminators may be constitutive or regulatable (→ rho factor).

Eukaryotic t.t. is heterogeneous in a region of AT-rich dyad symmetry and may depend on specific DNA regions located downstream from the 3′ polyadenylation site. The terminators appear to be similar to the rho-dependent terminators of prokaryotes.

transcription terminator (Reznikoff 1972) — a DNA sequence downstream of the coding segment of a gene which is recognized by → RNA polymerase, specifies → transcription termination and plays a vital role in the regulation of → gene expression. Such sites may have several purposes in the cell: (1) As efficient punctuation signals, they permit the differential regulation of adjacent gene clusters; (2) as modulating elements (→ transcriptional attenuation), they permit differential control of expression within operons; (3) as conditional abortive elements (e.g., in → genetic polarity), they prevent wasteful depletion of cellular metabolites; (4) as barriers to elongation, they minimize sequestering of important enzymes.

In **prokaryotes**, three structural domains (as in → promoters) define terminators: (1) A region of dyad symmetry (an inverted repeat) preceding the termination site; (2) uracil residues at the 3′end of the RNA transcript; (3) a GC-rich sequence of variable length surrounding the hyphenation of the dyad symmetry and preceding the termination site. This allows the formation of a stem-loop structure at the end of the DNA molecule **(rho-independent t.t.)**.

A second type of t.t. lacks either dyad symmetry or uridine-encoding segments, and extensive sequence or structural homologies are absent **(rho-dependent t.t.)**.

Eukaryotic t. terminators appear to be similar to the rho-dependent terminators of prokaryotes and termination occurs in a region of AT-rich dyad symmetry.

transcription unit — = → transcripton.

transcriptive intermediate (Dortner and Kingsbury 1972) — (TI); in RNA virus-infected cells, an RNA complex with properties analogous to the → replicative intermediate. The t.i. is metabolically active, partially double-stranded with base-paired regions and also contains substantial single-stranded structures. T. intermediates are virus-specific and absent in uninfected cells.

transcripton (Hayashi et al. 1964; Szybalski et al. 1970) — a DNA sequence defined by the sites at which → genetic transcription is initiated and terminated (= transcription unit). The RNA produced by a t. is the primary RNA transcript (→ RNA processing). The t. may include more than one gene (→ operon). A single t. may contain multiple → promoters of different strengths and different regulatory properties. Differential use of multiple promoters allows fine regulation of transcription.

An unregulated transcription unit is expressed constitutively (all of the time and at essentially constant rate). Control of a regulated t. is exerted at the promoter and/or terminator, i.e., by regulation of promoter and/or terminator strength.

Transcriptons which can produce multiple transcripts encoding different gene products are called complex transcription units; in eukaryotes they often involve alternative patterns of → RNA splicing.

trans-damage DNA synthesis — DNA synthesis by altered DNA polymerizing activity post → DNA lesions which would normally block → DNA replication (e.g., trans-dimer DNA s.).

trans-derepression (Willard and Echols 1968) — → derepression of certain bacterial → operons (gal and lac of *E. coli*) by infection with certain bacteriophages (lambda dg and phi 80 d lac, respectively). T.-d. is related to → transduction escape synthesis and suggests that phage-carried bacterial operons effect the removal of cellular → repressors, perhaps by binding of repressor directly to specific operons on DNA.

transdetermination (Hadorn 1966) — the process whereby cells determined to form one type of structure change their developmental pathway and develop, under certain conditions, into quite different structures. For any given state of → determination there exists a distinct probability or frequency of t. in a specific direction. T. is based upon activation of new sets of genes.

transdifferentiation — the transition from one differentiated state of a cell to another (= cell conversion; metaplasia).

transducing phage — a → bacteriophage containing host DNA (= t. particle). Two types may be distinguished: (1) **Generalized t. phages**, which contain only bacterial DNA and form during a lytic cycle when host DNA fragments are packaged (since fragmentation is random, all possible host sequences are represented in a heterogeneous population of g.t. phages); (2) **specialized t. phages**, which are produced during excision of an integrated → prophage and contain both host and phage DNA linked in one continuous molecule (a single t.p. can serve as a template for production of a homogeneous population of identical t. phages). Only one or at most two regions of the host DNA, specifically those flanking the prophage, are found in these phages.

Both types of t.p. can transfer host DNA from one bacterium to another by a process called → transduction. Formation of specialized, but not generalized, t. phages is the result of → lysogenization.

transduction (Zinder and Lederberg 1952) — the transfer of bacterial DNA from one cell to another by means of generalized or specialized → transducing phages and DNA injection. This is followed by DNA integration or plasmid formation in the recipient cell.

Complete t. depends on the attachment or insertion of the transducing DNA into the host chromosome (integration). In this state, the transducing phage genome, like that of a normal prophage or integrated → F plasmid, reproduces in concert with the bacterial chromosome. If integration is not achieved, the transducing DNA, unlike a normal vegetative phage or → sex factor, is unable to replicate, t. is abortive, and transmission of the transduced feature is unilinear.

Cotransduction means t.of more than one bacterial marker gene in a single event, indicating that transducing bacteriophages can sometimes transport a segment of the bacterial chromosome large enough for neighboring (closely linked) genes to be transferred together from donor to recipient bacteria. Cotransduction permits the inference of relative → linkage of two genetic markers on the bacterial chromosome from the relative frequency with which they are cotransduced. (→ mini-muduction).

transfection (Földes and Trautner 1964) – the transfer of DNA into living cells by several methods: (1) Biochemically mediated absorption of viral DNA; (2) DEAE dextran-mediated DNA transfer; (3) directed microinjection with micropipets; (4) calcium phosphate precipitation; (5) cell or liposome mediated microinjection; (6) electroporation. (→ gene transfer; cotransfection).

transferase – any of the soluble protein factors (aminoacyl transferases) required, besides GTP and aminoacyl-tRNA's, for the elongation of polypeptide chains on mammalian ribosomes.

Transferase I is the aminoacyl-tRNA binding factor required for the interaction of the substrates with ribosomes. If peptidyl-tRNA is present on the ribosomal peptidyl site, a ribosome-associated peptidyl-transferase activity brings about the formation of a peptide bond between the peptidyl-tRNA and the aminoacyl-tRNA. Transferase I binds to the small ribosomal subunit.

Transferase II is a factor needed to transport peptidyl-tRNA from the acceptor (aminoacyl) site to the donor (peptidyl) site on the mammalian ribosome. The activity of the enzyme depends on GDP and intact sulfhydryl groups. A complex is first formed between t. II and GTP, even in the absence of ribosomes. Binding of the complex requires both ribosomal subunits, reduced sulfhydryl groups on the enzyme and the ribosome, and occurs at the acceptor site.

transfer DNA – any of the cistrons in DNA coding for → transfer RNA (acronym tDNA).

transfer element – = → sex factor.

transfer enzyme (Conway and Lipman 1964) – any of the enzymes that are responsible for the transfer of amino acids from → transfer RNA to the growing peptide chain on the → ribosome (→ genetic translation).

transfer factor (Nathans and Lipman 1962) – (1) = → elongation factor; (2) = → sex factor.

transfer origin – a specific site on the → plasmid DNA molecule at which the transfer of plasmid DNA during bacterial → conjugation begins (oriT). DNA transfer is initiated by a strand- and site-specific nick at oriT.

transfer-proficient (Achtman et al. 1971) – of bacterial cells (symbol tra^+) carrying a → sex factor and acting as efficient donors in bacterial → conjugation.

transfer replication – the synthesis of DNA associated with DNA transfer during bacterial → conjugation (mating).

transfer RNA (Hoagland et al. 1957) – any of a class of ribonucleic acid molecules (= amino acid acceptor RNA; adaptor RNA; soluble RNA; symbols tRNA or sRNA) each with a molecular weight of about 25.000 (4S) consisting of 73 to 90 nucleotides. Transfer RNA molecules are involved in → genetic translation of → messenger RNA into the amino acid sequences of protein molecules (→ aminoacyl tRNA), in → nonsense and → frameshift suppression (→ suppressor tRNA), and (in bacteria) in the regulation of a number of amino acid biosynthetic → operons.

In genetic translation, tRNA's are able to recognize specific amino acids and corresponding genetic code words in → messenger RNA. In doing so, they translate the information encoded in nucleic acid messages into protein products. Multiple steps in this process include amino acid binding (→ aminoacyl-tRNA-synthetase), transport to the → ribosome, binding to one ribosomal site (the A site), peptide bond formation, translocation to another ribosomal site (the P site), and release from the ribosome.

The transcription products of the genes coding for tRNA are → pre-tRNA molecules which undergo a variety of modification prior to or at specific times during the maturation process (→ RNA processing).

The → genetic code could be translated by 31 tRNA species (assuming no other restrictions on codon-anticodon recognition than those due to → wobble). However, among prokaryotes 40 to 80 tRNA genes have been found, and for eukaryotic cells the number is even larger: 320 to 1400 depending on the organism. The redundancy (→ gene reiteration) of tRNA genes paralleles that of the major → ribosomal RNA cistrons so that the ratio of tRNA genes: rRNA genes is between 4 and 10 in bacteria (*E. coli*). The genes for some of the tRNA molecules are clustered in groups of 2 or 3 on the chromosome.

Prokaryotic tRNA genes are distributed throughout the genome. Some tRNA's are synthesized as part of a → ribosomal RNA operon or a polycistronic transcript of tRNA and mRNA cotranscribed from a distant promoter.

Transcription of **eukaryotic tRNA genes** involves, as a primary event, the stable binding of a protein factor to the intragenic → promoter (recognized by RNA polymerase III). In higher eukaryotes, tRNA genes constitute a moderately repetitive and dispersed → multigene family. Some tRNA genes contain → introns and all lack the common – CCA sequence found at the 3′ terminus of mature tRNA, i.e., these residues must be added post-transcriptionally during processing of tRNA precursors by tRNA nucleotidyltransferase.

tRNA makes up about 10–15% of the total RNA content of a cell; its main role in genetic translation is the acceptance of specific amino acids, the binding of aminoacyl tRNA to the messenger RNA-ribosome → initiation complex and the release of the amino acids to the growing polypeptide chain. Every amino acid has its specific tRNA for activation during protein synthesis. tRNA species which accept the same amino acid but differ in their code word recognition patterns are called **isoaccepting tRNA's** (→ isoacceptor tRNA). The amount of redundant tRNA species varies widely. The biological significance of this phenomenon may be seen in the action of isoaccepting tRNA as a safeguard against mutations. They also offer an explanation for the origin of suppressor tRNA's. The terms **heterogeneric** and **homogeneric** tRNA's (White and Tener 1973) describe tRNA molecules that are transcribed from different DNA cistrons or from the same cistron, respectively. Although being the products of the same gene, homogeneric tRNA's are chromatographically distinct because of different degrees of post-transcriptional modification. The rather high degree of polymorphism is compatible with a molecule whose surface is largely repetitive and densely populated with charged acidic groups.

The rules of pairing between tRNA and mRNA result in the → genetic code. tRNA molecules contain up to some five percent of modified nucleosides. All modifications occur post-transcriptionally and comprise methylation, thiolation, isopentenylation, 2-thiomethyl-6-isopentenyl adenosine formation, pseudouridine formation, and esterification of carboxymethyluridine.

All tRNA's can be folded into the **cloverleaf structure** and share a set of absolutely conserved nucleotides. The constant nucleotides are involved in stabilizing the tertiary structure of the tRNA molecule. A small number of positions in a tRNA molecule are important for specific recognition by the cognate → aminoacyl-tRNA synthetase; these may include the fourth nucleotide from the 3′end (the discriminator) and the anticodon. Many tRNA's contain a modified wobble base at the 5′end of the anticodon or a modified purine adjacent to the 3′end of the anticodon.

The cloverleaf structure has double-stranded regions of constant lengths: Seven base pairs near the 5′-end, at least three bases in the dihydrouridine arm, five base pairs in the anticodon loop, and five base pairs in the T ψ arm. The remaining single-stranded regions in the loop show much similarity. There are only a few variable places in the T loop and in the dihydrouridine loop. The finger or extra loop region varies in size between four and 14 bases. The length of the dihydrouridine region may also differ.

The particular positions of 2′-O-methylation of the sugar unit are the middle of the dihydrouridine loop, the turning points from the double-stranded regions to the looped single-stranded regions, and the wobble base position of the anticodon.

tRNA's can be classified into two families according to the length of the variable region: (1) Class I tRNA's with short variable loops of 4–5 nucleotides; (2) class II tRNA's (specific for leucine, serine or tyrosine) possess a variable region of 10 nucleotides or more. This region is generally supposed to fold in a short stem (3 to 7 bp) and a mini-loop of 3 to 5 nucleotides.

The cloverleaf secondary sturcture of tRNA is further folded to give rise to a tertiary three-dimensional structure characteristic of the tRNA molecule.

This structure consists essentially of three major interlinked substructures orientated to each other in the shape of a T: The long, almost perfect double helix is formed by stacking the acceptor and T ψ C stems. A so-called "D system" is attached to the long helical segment almost at the middle part and is augmented by several unusual base pairs and postulated triplets. This forms the T junction. The anticodon stem extends outwards from the augmented D helix, with the helices separate and noncolinear. The anticodon helix with its loop out at the far end of the molecule allows for mobility, possibly during genetic translation. It is assumed that bases susceptible to modification are those not involved in tertiary interactions, i.e., double-helical residues are unaffected by modification. The bases reactive in modification are on the surface of the molecule.

The tRNA molcule must be able to recognize its → aminoacyl-tRNA synthetase, the tRNA binding site on the ribosomes, the appropriate ribosomal factors, and its matching codon in the messenger RNA. These recognition functions are fulfilled by four specific sites present in each tRNA molecule:

(1) **The amino-acid attachment site** represented by the ... CCA sequence at one end of the molecule. (2) **A site that recognizes the specific amino-activating enzyme** which is specific for a given amino acid and varies among different tRNA types. (3) **The ribosome-recognition site** necessary for the interaction of tRNA with the → ribosomes. (4) The **coding-recognition site** (anticodon, nodoc) which is specific for a particular tRNA molecule and responsible for the correct recognition (by complementary base pairing) of the corresponding codon in messenger RNA and, thus, for the correct insertion of the amino acid attached to the tRNA molecule into the growing polypeptide. Present evidence indicates that a tRNA species can recognize multiple codons differing in the third letter (→ genetic code). The pattern of multiple recognition is consistent with the postulates of the so-called "wobble hypothesis" (Crick 1966).

According to the **"wobble hypothesis"**, the standard base pairs may be used rather strictly in the first two positions of the triplet, but there may be some "wobble" in the pairing of the third base. U (uracil; the first base, assuming antiparallel base-pairing) on the anticodon of tRNA can recognize both A (adenine) and G (guanine) in the third place on the messenger RNA codon; the base C (cytosine) on the anticodon can recognize only G in the third place; the base A can only recognize U; the base G can recognize both U and C; and the base hypoxanthine can recognize U, C, and A in the third place.

The tRNA sites with special function, noted above, are assumed to be located at the unpaired nucleotide sequences in the bends or loops of the tRNA structure where the bases are exposed and not in portions of the molecules that are DNA-like in structure.

Mutational alterations of the primary structure of tRNA may result in functional changes of the relevant tRNA species. Such alterations may change codon-anticodon pairing and amino acid recognition. The latter seems to be determined by various components of the tRNA molecule, perhaps including its tertiary structure. In addition, changes have been obtained which lead to the absence of function, temperature-sensitive function and altered recognition by the activating enzyme.

tRNA species which ensure the initiation of genetic translation are called → initiator tRNA.

Chloroplast tRNA's show both prokaryotic and eukaryotic features. They contain sequences similar to prokaryotic tRNAs in their structural parts, except for the 3′ CCA ends, and to prokaryotic → Pribnow boxes and −35 regions in their upstream regions. Some chloroplast tRNA genes contain rather long introns (451 to 949 bp) compared with those of nuclear tRNA genes (16 to 60 bp).

Transfer RNA's participate in several reactions other than protein synthesis. One tRNA is a primer for → reverse transcriptase, another participates in the formation of the pentapeptide bridges in bacterial cell walls; still another tRNA is a component of the ubiquitin and ATP-dependent proteolytic system of mammalian cells. Additionally, tRNA has catalytic properties, such as, e.g., in the → self-splicing of some introns.

transfer RNA activator (Markman and Cautoni 1966) – a factor which increases the capability of transfer RNA to accept certain amino acids via → aminoacyl tRNA synthetases. The tRNA a. acts catalytically.

transfer RNA binding site – → A site; E site; P site; genetic translation.

transfer RNA ligase — any of the ligases which bind the amino acids to their respective transfer RNA's.

transfer RNA methylase — any of the enzymes that post-transcriptionally modify → transfer RNA species by catalyzing the transfer of methyl groups from S-adenosylmethionine to specific sites in the tRNA molecule. These enzymes are unique in being species-specific, organ-specific, and base-specific. In biological systems undergoing developmental changes, large alterations occur in the levels of tRNA m. activity.

transfer RNA mischarging — any process that results in aminoacylation of tRNA with the wrong amino acid. tRNA m. occurs after single-base substitutions in tRNA that result in altered amino acid acceptor specificity. The substitutions are termed mischarging mutations.

transfer RNA nucleotidyltransferase (Canellakis 1957) — an enzyme which catalyzes the incorporation of AMP and CMP residues into transfer RNA molecules to form the complete 3′-terminal-CCA sequence. The enzyme repairs tRNA's with incomplete 3′ends and is concerned with tRNA maturation.

transfer RNA pseudogene — any of the eukaryotic interspersed repetitive sequences (→ SINE's) derived from transfer RNA sequences and sharing extensive sequence homology with specific tRNA molecules. They can be folded into cloverleaf tRNA-like structures, though with less base-pairing than in tRNA.

transfer RNA recognition — the recognition and interaction of → transfer RNA molecules with the enzyme → aminoacyl-tRNA synthetase resulting in the attachment of the appropriate amino acid to a specific tRNA.

transfer RNA selection — selection of a transfer RNA molecule from among diverse competing cognate species (→ isoacceptor tRNA) by trial and error, with numerous abortive and noncognate interactions occurring for each tRNA molecule ultimately selected. The relative abundance and concentration of competing tRNA species affects their selection.

transformant — any cell which has either stably integrated transferred DNA (→ gene transfer) into its chromosomal DNA or harbors the transferred DNA transiently (= transferent; transgenote).

transformation — → genetic transformation; cell transformation.

transformosome (Kahn et al. 1983) — any of the membraneous extensions (20 nm in diameter) within which *Haemophilus influenzae* DNA is in a protected state resistant to external DNase and cellular restriction enzymes, although remaining unmodified (→ DNA modification-restriction) and double-stranded. The t. is supposed to be the organelle responsible for DNA binding and uptake during → genetic transformation. The organelles extend about 35 nm from the cell, are located at points of fusion between the inner and outer membrane, and are composed of a lipid bilayer.

transgenic (Gordon and Ruddle 1981) — of individuals containing a foreign gene (a transgene), introduced by means of → gene transfer techniques, in all of their cells. The new information is passed on through the germ line of the adult individual.

The site of integration of transgenes in the host genome can affect levels of expression and occasionally confer ectopic patterns of expression on tissue-specific genes.

transgenome (Althwal and McBride 1977) — a genetic element transferred by → chromosome-mediated gene transfer.

transgenosis (Doy et al. 1973) — the transfer, maintenance, and expression of genes in completely unrelated organisms (such as, e.g., the expression of bacterial genes in plant cells). T. is recommended as an appropriate term in those cases where donor and recipient cells are widely separated by evolution.

trans-Golgi network (Griffiths and Simons 1986) — an extensive tubular reticulum on one side of the Golgi stack which represents the last Golgi compartment (→ Golgi apparatur) through which proteins destined for the plasma membrane, lysosomes, and secretion pass.

transgression — the appearance in a segregating generation (F_2, backcross, etc.) of one or more genotypes (individuals) which fall outside the limits of variation defined by the parents and the F_1 of the cross in respect to one or more characters (Darlington and Mather 1949).

transition (Freese 1959) —the substitution of one pyrimidine for another (e.g., replacement of deoxycytosine by thymidine) or of one purine for another at any base location (site) in DNA (or RNA if this is the primary carrier of genetic information, as in some viruses). At the next replication of the altered region, a similar t. will be induced in the daughter strand, i.e., pyrimidine t. will result in a purine t. in the newly synthesized complementary strand. With this type of → base pair substitution the purine-pyrimidine orientation is preserved. By means of spontaneous or induced transitions → gene mutations (t. mutations) may arise. Any transition mutation can revert spontaneously or by any transitional agent (→ transversion).

transit peptide (Chua and Schmidt 1979) — in the case of post-translationally imported mitochondrial and chloroplast proteins (synthesized in the cytoplasm), an amino terminal sequence which contains targeting information and is removed during or shortly after import into the organelles (→ signal peptide).

transition zone — in → speciation and population evolution, a hybrid, intergradation and contact zone which may be due to the differentiation of the population in situ (primary t.z.) as in clinal and stasipatric speciation, or due to the seconcary contact of populations (secondary t.z.) as in allopatric speciation.

translation — → genetic translation.

translational ambiguity (v. Ehrenstein 1966) — the fact that a given → codon in mRNA can be translated in more than one way. T.a. may be due to genetic factors (mutations) or external agents (e.g., streptomycin). (→ genetic translation; mistranslation).

translational control — the regulation of protein synthesis operating either by controlling the overall level of protein synthesis or by causing preferential translation (→ genetic translation) of particular → messenger RNA's. Several mechanisms can differentially regulate the translation of mRNA (Bergman and Lodish 1979): (1) mRNA can be masked or inactivated to withdraw certain species from the pool of translatable mRNA's; (2) different messenger-specific → initiation factors may affect the efficient initiation of translation on certain mRNA's; (3) mRNA's can differ in their ability to compete for common initiation factors; (4) the levels of → isoaccepting tRNA species can affect the time needed by a ribosome to translate a mRNA and thus the translation efficiency of a mRNA; (5) the destruction of transcripts may occur at nonuniform rates. (→ transcriptional control).

translational control RNA (Heywood et al. 1974) — (tc RNA); in eukaryotes, any of a class of RNA molecules involved in the utilization of long-lived (masked, or stored) → messenger RNA's and in the maintenance of these molecules in the inactive form. Such RNA's are effective in the regulation of protein synthesis within the cell from which they are derived.

translational polarity — in prokaryotes, elimination of → genetic translation of a distal mRNA segment caused by a lesion (such as a nonsense mutation, deletion, frameshift mutation or insertion sequence) in a more promoter-proximal gene of the operon in question. T.p. is a consequence of the coupling between transcription and translation in prokaryotic systems and can be alleviated by nonsense suppressors that permit translation to continue through to the end of the affected gene.

translational reinitiation — restart of → genetic translation, after premature termination (triggered by a nonsense mutation), at an internal initiation site (AUG, GUG, or UUG) that functions only in the presence of a nearby terminator codon. The product of t.r. is a restart polypeptide that lacks some of the amino acids normally present at the N-terminus of the protein.

Mutations which cause t.r. are called translation-restart mutations (Sarabhal and Brenner 1967).

translational repression (McLellan and Vogel 1970) — a translational regulatory mechanism in bacteria causing a reduction in genetic translation under repressive conditions.

translational repressor (Maxwell et al. 1971) — = → interference factor.

translation coupling (Lodish and Zinder 1966) — in polycistronic transcripts of bacteria and phages, the phenomenon that some ribosome binding sites function efficiently only when the preceding cistrons are translated.

translation error — → genetic translation; mistranslation; transfer RNA mischarging.
translation factor — any protein factor, such as → initiation factors, → elongation factors and → release factors, which transiently reside on the → ribosomes during protein synthesis. Association with GTP favors their binding to the ribosome-mRNA complex; hydrolysis of GTP into GDP and inorganic phosphate induces their release from the ribosome.
translation fidelity — the accuracy of → genetic translation determined by anticodon-codon interaction. Errors occur only about one time per 10^4 amino acids incorporated. Error frequencies differ for different → transfer RNA's.
translation initiation — → genetic translation; ribosomal scanning model.
translation initiation region — = → ribosome binding site.
translation termination — → genetic translation.
translesion synthesis — a type of → DNA repair in which a modified replication apparatus synthesizes through a damaged DNA region (with high error frequency at the sites of noncoding → DNA lesions).
translocatable genetic element — = → transposable genetic element.
translocation — I. a chromosomal structural change characterized by the change in position of chromosome segments (and the gene sequences they contain) within the chromosome complement (→ chromosome mutation).

In many organisms, translocations result in specific patterns of → genetic segregation, such as unusual → linkages and reduced meiotic recombination, reduced viability of a fraction of the offspring, familial patterns of trisomy, etc. These can be used for the detection and mapping of translocations in genetically well-analyzed organisms. Generally, in higher organisms cytological investigations are used for their identification or confirmation.

Simple translocations may be classified as follows:

(1) **Intrachromosomal (internal) translocations or shifts:** Change in position of a chromosome segment within the same chromosome either from one chromosome arm to the other **(extraradial)** or within the same chromosome arm **(intraradial)**.

(2) **Interchromosomal translocations:** Transfer of a chromosome segment from one chromosome into another chromosome **(transposition)** or reciprocal exchange of segments between chromosomes **(reciprocal t., mutual t.** or **interchange**). An interchange of segments between homologous chromosomes is called "fraternal", while one between nonhomologous chromosomes is called "external".

In the case of **eucentric** t., the sequence of loci contained in the transposed segments is the same as that in the original segment position with respect to the → centromere, while the sequence is reversed in case of a **"dyscentric"** t.

Intercalary incorporation of a chromosome segment due to transposition is called an **"intercalation"** or **"insertion"** ("insertional translocation"). Fraternal transpositions result in an intrachromosomal → duplication in one, and the corresponding → deletion in the other chromosome of a pair of homologs.

Terminal transpositions of chromosome segments to the natural ends (→ telomere) of another chromosome are not possible.

In those cases where a chromosome segment is laterally transposed to another chromosome, a three-armed configuration arises which is said to be a **"triradial"** due to lateral translocation.

The unit of translocation may be the chromosomes **(chromosome t.)** or the single chromatid **(chromatid t.)**. The sites of translocation are called **translocation points**.

Reciprocal translocations (interchanges) may be either **asymmetrical** (= aneucentric) or **symmetrical** (= eucentric):

In the former, one → dicentric and one acentric product of t. arise; in the latter the translocation products are both monocentric. Asymmetrical interchanges may lead to the production of a → chromosome bridge at anaphase if the two centromeres of the dicentric product are distributed to opposite spindle poles.

Whole-arm translocations (Muller 1940) or whole arm transfers are translocations in which whole (or nearly whole) arms are transposed or interchanged. By this process one acrocentric (·A) and one metacentric (B·C) chromosome may give rise to one telocentric (·C) and one new metacentric (A·B) chromosome, or from two metacentrics (A·B and C·D) two new ones (A·C and B·D) may be produced.

Special cases of "whole arm" translocations are the "centric fusion" (Robertson 1916), the „dissociation", and the "tandem fusion" (White 1957). With **centric fusion** the long arms of two acrocentric chromosomes are united to form a metacentric chromosome. The reverse process by which one metacentric with long arms and one with short arms may give rise to two acrocentric chromosomes is referred to as **"dissociation"**. **"Tandem fusions"** result from one "break" in the vicinity of the centromere in one chromosome, and another "break" near the chromosome end in a second. If the chromosomes engaged in a subsequent interchange are two acrocentrics, then one large and one very small metacentric chromosome arise. If one of the chromosomes exchanging segments is metacentric, the other acrocentric, then two acrocentrics (one short, the second large) may be produced.

Both centric and tandem fusions may lead to a decrease in standard chromosome number if the small chromosome which frequently consists only of heterochromatin is eliminated. Under these circumstances the number of large chromosome arms in the complement (fundamental number) remains unchanged. Centric fusions in which → sex chromosomes and → autosomes take part may represent the origin of "multiple sex chromosome" systems.

In **individuals heterozygous for a symmetrical reciprocal translocation** (structural hybrids), four chromosomes share a partial homology but no two are identical. Consequently, → chromosome pairing at prophase of first division of → meiosis results in a cross-configuration involving the four chromosomes.

The subsequent behavior of this cross-configuration depends upon the frequency and location of the → chiasmata and the mode of → centromere orientation (which is influenced by the location of chiasmata and the morphology of the chromosomes).

If crossing over and chiasma formation takes place in all four "pairing segments" (→ interstitial segment), a ring of four chromosomes results. If chiasma formation is absent in one of the four pairing segments a chain of four chromosomes results. A chain of three and one univalent, or two bivalents, results as a consequence of chiasma formation in two adjacent or alternate pairing segments. Rings and chains contain structurally normal and structurally changed chromosomes in alternating sequence.

The distribution of the four chromosomes in a ring or chain configuration at anaphase I of meiosis is determined by their centromere orientation, which is either concordant or discordant. In the case of → **alternative distribution**, chromosomes located alternately in the pairing configuration are distributed to the same spindle pole, while in → **adjacent distribution** neighboring chromosomes reach the same pole. If neighboring chromosomes with nonhomologous centromeres reach the same pole, this mode of distribution is called "adjacent-I distribution". In the case of "adjacent-II distribution", those neighboring chromosomes in the ring or chain which are distributed to the same pole have homologous centromeres.

In the alternative distribution, a complete genome reaches each spindle pole, but with the two types of adjacent distribution the polar groups are characterized by → duplications and deletions and the resulting gametes are genetically unbalanced. This unbalance may give rise to gametic → sterility in plants and to zygotic sterility in animals. The percent sterility is usually about 50% (between 30 and 70%) and the phenomenon is called "semisterility".

In the case of discordant centromere orientation, gametes with $n-1$ and $n+1$ chromosomes may arise. Disomic $(n+1)$ gametes may contain one or both chromosomes with interchanged segments and may produce → trisomics by the union with normal (n) gametes.

These are referred to as tertiary trisomics and translocation trisomics if they contain one or both chromosomes which participated in the interchange.

If crossing over and chiasma formation occurs in the interstitial segments (located between the centromere and the translocation points) of the pairing configuration, there results either alternative or nonhomologous adjacent distribution with (theoretically) 50% genetically balanced and 50% genetically unbalanced gametes.

When one chromosome or a pair of homologs is engaged in two or more reciprocal translocations, pairing configurations consisting of six or more chromosomes arise. In such pairing configurations, pairing-, interstitial-, and → differential segments may be distinguished. **Differential segments** are the segments which lie between the two points of interchange where a chromosome pair is involved in two interchanges. As a consequence of crossing over within differential segments, 50% genetically balanced and 50% genetically unbalanced gametes are (theoretically) expected, irrespective of whether centromere orientation and chromossome distribution are alternative or adjacent.

In extreme cases, whole sets of chromosomes may be involved in a series of interchanges (e.g., *Oenothera, Rhoeo*). Heterozygotes of this type are said to be → **complex heterozygous**.

Heterozygosity for interchanges like that for → inversions may affect the **crossing over frequency** both intra- and interchromosomally.

In animals, centric fusions and whole-arm translocations have been shown to be important factors in **karyotype evolution**, and of evolutionary changes in the → sex-chromosome system, while in plants, in addition to interchange homozygosity, complex heterozygosity due to serial translocations has played a major role in the evolution of certain groups.

II. The transfer (→ transpeptidation) of the growing peptide chain from the A site to the P site in the → ribosome during → genetic translation.

translocation factor — = → G factor.

translocation test — in → mutagenicity testing, a test devised to measure sterility and heritable semi sterility in the F_1 progeny of rodents exposed to a test agent. Males or females are treated with the test agent, mated with untreated animals, and F_1 individuals (usually males) are selected for further breeding and examination of the litter size produced by the F_1 animals. In general, any animal showing greater than 50% fetal death is considered to be semi-sterile due to an induced chromosomal translocations. The translocation can then be detected cytogenetically.

transmembrane receptor — any receptor sensing stimuli (such as light, hormones, or neurotransmitters) and activating particular second messenger systems. These second messengers are responsible for producing the appropriate response to the stimulus and coordinating the response to different stimuli.

transmission genetics — the study of how genes are transmitted from cell to cell and generation to generation, and how they recombine and segregate.

transpeptidation — in → genetic translation, the movement of peptidyl-tRNA from the → A site to the P site of the → ribosome, expulsion of previously deaminoacylated tRNA from the P site and concurrent movement of mRNA by one codon relative to the A and P sites of the ribosome (= translocation). → Elongation factor EF-2 is essential for t.

transposable bacteriophage — any type of phages (such as the → Mu phage) which use → transposition as part of their normal mode of production of phage progeny. Early in infection they integrate their DNA into the bacterial chromosome, creating a typical transposon sequence, i.e., phage DNA is flanked by a duplicated target sequence. Packaging of phage DNA ultimately occurs from these chromosomally located units and the amount of DNA packaged is greater than the length of the coding sequence of the phage since the termini of the DNA within the phage particles always contain bacterial DNA.

transposable genetic element (McClintock 1952; Cohen 1976) — in prokaryotes and eukaryotes, any of a class of diverse DNA segments that can insert into nonhomologous DNA (chromosomes, plasmids, virus DNA, mitochondrial and plastid DNA), exit, and relocate in a reaction which is independent of the general recombination function of the host (= mobile genetic elements; translocatable elements). Different elements differ in the nature of the final product of → DNA transposition but usually encode two types of transposition information: (1) DNA sequences at the ends (direct or inverted repeats) which participate directly in transposition; (2) proteins (→ transposase) which promote transposition and interact directly with the cis-acting sites at the ends of the element.

T.g. elements are able to move to numerous sites in a genome (replicon), to alter → gene expression and to induce genome rearrangements (→ insertion mutagenesis). According to their structure t.g. elements may be subdivided into four classes: (1) Elements with long terminal direct repeats; (2) elements with long terminal inverted repeats; (3) elements with short inverted repeats; (4) elements without terminal repeats.

The integration event of a t.g.e. is usually accompanied by the duplication of a short (3 to 13 bp) oligonucleotide sequence of the target DNA; most of the t.g. elements can induce the fusion of replicons (the resulting structure is called a → cointegrate and contains one copy of the t.g.e. at each interreplicon boundary in direct orientation).

Prokaryotic (bacterial) t.g. elements are classified into three groups: (1) → Insertion sequences (IS elements) of a usual size between 0.8 and 1.5 kb, which encode only transposition determinants; (2) → transposons (TN) which are larger (usually more than 5 kb), encode additional determinants, such as antibiotic resistance, besides the information for replicative transposition, and are often bracketed by IS elements; (3) → transposable phages carrying out site-specific recombination reactions which neither require DNA replication nor degradation or resynthesis of DNA.

Eukaryotic t.g. elements fall into three major classes: (1) Classical transposons; (2) retrovirus or retroviral-like → retroposons; (3) retroposons. The latter two classes transpose by a process involving the → reverse transcription of RNA's and the generation of free extrachromosomal species.

The following properties are shared by all eukaryotic t.g. elements and integrated provirus retroviruses: (1) Their sequences consist of an internal DNA segment (several kb long) flanked by a pair of identical DNA segments of several hundred bp which are both arranged in the same direct orientation. Each of these repeats is bounded by small inverted repeats (several nucleotide pairs long). (2) The complete element is flanked by an identical pair of host DNA sequences (usually 4 to 12 bp long) that are present once in the target site for integration. (3) No detectable homology between the DNA sequences at the insertion site and the end of the element. (4) The terminal dinucleotides of the elements are 5′TG ... CA3′ and in some cases this sequence homology extends further into the elements.

Many of the eukaryotic t.g. elements are transcribed to give abundant poly(A)$^+$RNA's.

T.g. elements mediate a variety of genetic phenomena such as gene rearrangement, plasmid-chromosome integration, and formation of plasmid cointegrates.

(→ copia-like element; FB element; P element; Ty element).

transposagenic (Datta et al. 1983) — of any agent (transposagen) which promotes → transposition.

transposase (Berg 1977) — in some families of (prokaryotic) → transposable genetic elements a trans-acting, element-specific protein required for → transposition. Short distinctive sequences at the element's ends are the sites of t. action which probably also participates in insertion site selection.

transposition (Cohen 1976) — → DNA transposition.

transposition immunity (Robinson et al. 1977) — the ability of a → transposable genetic element present in a target DNA molecule to reduce substantially the frequency of subsequent insertion of another copy of the element into this target.

transposition mutagenesis — = → insertion mutagenesis.

transposon (Hedges and Jacob 1974) — any of a group of composite bacterial → transposable genetic elements consisting of a (bacterial) gene or group of genes, functionally unrelated to → transposition, bracketed by two identical, or nearly identical, → insertion sequences (IS elements) arranged as direct, or inverted repeats. Others contain either one or no complete IS element.

transposon-induced cointegrate — during → DNA transposition, a → cointegrate structure in which donor and recipient replicons are fused with a directly repeated copy of a → transposon at each juncture point. Transposition is completed by recombination between the direct repeats of the transposon.

transposon tagging (Bingham et al. 1981) — a technique for isolating genes which encode unknown gene products by use of a → transposable genetic element as a molecular tag (→ transposon mutagenesis; insertion mutagenesis).

trans-regulation — the phenomenon whereby specific genes are recognized and turned on by trans-acting factors (proteins) encoded by → regulatory genes.

trans-splicing — intermolecular splicing of → exons on different precursors, as opposed to cis-splicing (intramolecular splicing). T.-s. results in the splicing together of normally unassociated exons and is dependent on close alignment of two → pre-messenger RNA's.

transvection (Lewis 1954) — the ability of a gene to influence the activity of an allele on a homologous chromosome when the two homologs are paired (synapsis-dependent genetic complementation).

transversion (Freese 1959) — a base pair substitution (→ transition) that involves the substitution of one purine by a pyrimidine or vice versa. The substituted pyrimidine would base-pair with a purine, giving rise to an inversion of the pyrimidine-purine base pair with respect to the two DNA strands containing it. Transition agents are unable to revert t. mutants and t. agents do not revert transition mutants.

tribe — a group of human beings which share a common gene pool and exhibit a more or less common culture and language.

trihybrid — a → hybrid heterozygous for three pairs of alleles (genes).

triisosomic (Kimber and Sears 1968) — of a cell or individual which lacks one chromosome pair but has three homologous isochromosomes for the same arm of the missing pair (→ monoisosomic).

trimonoecious (Errera and Gevaert 1878) — of plants with hermaphroditic, male, and female flowers on the same individual (= coenomoneocious).

trioecious — of plants having male, female, and hermaphroditic flowers on different individuals (→ dioecious).

triparental cross (Hershey and Chase 1951) — in virus genetics, the simultaneous infection of bacteria with three parent bacteriophages which differ from each other in three of their genetically controlled markers (e.g., $ab^+c^+ \times a^+b^+c \times a^+bc^+$). From such triply infected cells a triparental recombinant abc (one genetic marker from each of the three parents) may be recovered.

triplex (Blakeslee et al. 1923) — → nulliplex.

triploid (Němec 1910) — of → auto- or → alloploid cells, tissues, and individuals with three chromosome sets in their nuclei (symbol 3n or 3x).

triradial — a chromosome with three arms arising as a result of a "lateral" → translocation.

trisomic (Blakeslee 1921) — of → polysomic cells, tissues, or individuals with one or more (doubly t., etc.) extra chromosomes in an otherwise diploid (disomic) chromosome complement. This phenomenon is referred to as trisomy and is symbolized by $2n + 1$, $2n + 1 + 1$, etc. The extra chromosome is either completely (primary trisomics) or partially homologous (secondary and tertiary trisomics) to certain chromosomes of the standard complement.

(1) **Primary trisomy** (Blakeslee 1924; Rhoades 1933; Goodspeed and Avery 1939): The extra chromosome is completely homologous to one pair of homologs in the chromosome complement and, therefore, one chromosome is represented three times in an otherwise diploid complement. Such homology permits trivalent formation at the first meiotic division.

The principal sources of p. trisomics are (Khush 1973): (a) Normal disomics (spontaneous occurrence or by treatment with physical and chemical agents); (b) asynaptic and desynaptic disomics; (c) polyploids (e.g., haploids, triploids, and tetraploids); (d) translocation heterozygotes; (e) primary trisomics, tetrasomics, multiple trisomics, and secondary, tertiary, and compensating trisomics; (f) monosomics; and (g) other chromosomal abnormalities.

Frequently, primary trisomics differ from normal diploids and from each other in a number of phenotypic characters due to changes in the genic balance brought about by the extra chromosome. Their viability and fertility is usually lower than that of the diploids from which they arose.

(2) **Secondary trisomy** (Blakeslee 1924; Rhoades 1933; Goodspeed and Avery 1939): The extra chromosome has two identical arms and is an → isochromosome. For each primary two types of secondaries are possible. Thus, in secondary trisomics, one chromosome segment (arm) is represented four times in an otherwise diploid chromosome complement.

The diagnostic pairing configuration at the first meiotic division is either a ring trivalent (which only rarely occurs in primaries), or a bivalent and one U-shaped ring resulting from the internal pairing of the univalent with identical arms.

Secondary trisomics may occur among the progeny of normal diploid individuals, but their best source is among the progeny of individuals with one or more univalent chromosomes in which → centromere mis-division takes place.

(3) **Telosomic trisomy** (Burnham 1962): The extra chromosome is a telocentric fragment chromosome homologous to one arm of a chromosome pair in the standard complement (= telotrisomy).

(4) **Tertiary trisomy** (Belling and Blakeslee 1926; Rhoades 1933): The extra chromosome is the result of a reciprocal → translocation (interchange between two standard chromosomes). Its ends are homologous with the ends of two different chromosomes. Tertiary trisomics occur regularly (and possibly exclusively) in the progenies of translocation heterozygotes by 3:1 segregation of the four chromosomes which are partly homologous in such heterozygotes and may pair meiotically as a quadrivalent.

The meiotic pairing configuration diagnostic for a tertiary trisomy is a chain of five chromosomes or two pairs of chromosomes connected by a single chromosome. A tertiary trisomic individual has two chromosome segments represented three times each in an otherwise diploid complement.

(5) **Compensating trisomy** (Blakeslee 1927): one chromosome of the diploid standard complement is missing but is compensated for by the presence of two other chromosomes which together are equivalent to the missing one. The missing chromosome in a compensating trisomic may be compensated for by: (a) Two tertiary chromosomes, (b) two isochromosomes, (c) one tertiary and one telocentric chromosome, (d) one iso- and one telocentric chromosome, and (e) one iso- and one tertiary chromosome (Khush 1973). In the case of compensating trisomy, a five-membered ring pairing configuration may occur at the first meiotic division.

Segregation of genes included in a primary trisomic association differs from that of genes on any of the other chromosomes of the complement. For a particular locus (A/a), a trisomic may have the genotypes AAA (triplex), AAa (duplex), Aaa (simplex), or aaa (nulliplex). If one of the three chromosomes is not eliminated, selfing or intercrossing of trisomics would lead to 25% tetrasomics, 50% trisomics, and 25% disomics (diploids) in the progeny. This expectation is rarely realized because of meiotic irregularities and/or nonfunctioning of $n+1$ gametes.

For the genotypes AAa and Aaa and a regular chromosome distribution (at meiosis I two to one pole, one to the other), a gametic ratio of 1AA:2Aa:2A:1a (dominant: recessive = 5:1) and 2Aa:1aa:1A:2A (dominant: recessive = 1:1), respectively, are expected. If male gametes possessing the additional chromosome are nonfunctional (as in many plants) and → chromatid segregation is absent, the following genotypic proportions are expected in the progenies of selfed or intercrossed trisomics of the genotypes AAa and Aaa:

Genotype	AAa		Genotype	Aaa	
Gametes	2 A	1 a	Gametes	1 A	2 a
1 AA	2 AAA	1 AAa	2 Aa	2 AAa	4 Aaa
2 Aa	4 AAa	2 Aaa	1 aa	1 Aaa	2 aaa
2 A	4 AA	2 Aa	1 a	1 Aa	2 Aa
1 a	2 Aa	1 aa	2 a	2 Aa	4 aa

The observed genotypic ratio in the progenies may deviate considerably from this theoretical expectation. If instead of → chromosome segregation, chromatid segregation takes place (and the locus in question is so far from the centromere that crossing over takes place regularly between centromere and locus) then the expected gametic ratio for the genotype AAa will be 6AA:8Aa:1aa:10A:5a instead of 1AA:2Aa:2A:1a.

tritelosomic (Kimber and Sears 1968) — of a cell or individual which lacks one chromosome pair but has three homologous telocentric chromosomes (→ ditelosomic).

trivalent — a → multivalent consisting of three meiotically paired chromosomes.

tRNA — acronym for → transfer RNA.

tRNA releasing factor (Ishitsuka and Kaji 1970) — a factor (TR) catalyzing the release of → transfer RNA from ribosomes. TR is distinct from the → G factor in that it loses its activity upon heating (50 °C) and does not require GTP for its action. TR may play a role in removing tRNA from the ribosome after the polypeptide chain termination factor has released the completed chain from the ribosome-bound tRNA.

trophic resource stability theory (Ayala and Valentine 1974) — a theory which predicts a negative correlation between the amount of temporal variation in the food resources of a species and the amount of genetic variation within this species.

true breeding — of genotypes or certain of their gene loci which do not segregate (→ segregation) because they are → homozygous.

truncation selection — (1) natural → selection that eliminates all individuals beyond a given phenotypic value (the t. point). (2) A breeding technique in which individuals above or below a certain phenotypic value (the truncation point) are selected as parents for the next generation.

tube nucleus — the → nucleus of the pollen tube which controls the growth and behavior of the pollen tube in higher plants.

tubulin (Boris and Taylor 1967) — the major protein constituent of → microtubules which is a heterodimer composed of α- and β-subunits (each with a molecular weight of about 55.000). In addition to t., microtubules contain a heterogeneous class of proteins, microtubule-associated proteins (MAP's), which copolymerize with t. and are potent promoters of t. assembly in vitro. The t. molecule has two guanine nucleotide binding sites of different affinities (E site and N site) which are possibly involved in the control of polymerization and energy transduction.

tumorigenesis — a multistep process in which a normal cell progresses in a step-by-step fashion to a fully malignant tumor cell (→ oncogene).

tumor suppressor gene — any of a category of genes that can suppress transformation or tumorigenicity.

twin meiosis (Gutz 1967) — after copulation of diploid yeast cells, separate meiosis in both nuclei in a common cytoplasm in those cases where karyogamy does not occur immediately after copulation. By t.m. asci with eight haploid spores are formed.

twin method — in human genetics, a method using identical and fraternal twins for studying the influences of heredity and environment on the phenotype.

twins — two embryos developed simultaneously within one uterus (similarly triplets, quadruplets, etc.). T. may be fraternal (dizygotic) if they originate from two fertilized eggs, or identical (monozygotic) if they arise from only one egg. Monozygotic t. are genetically identical.

twin species — two → species which are phenotypically very similar but separated from each other by reproductive → isolation (→ sibling species).

twin spot — two adjacent cell groups which differ in genotype **and** phenotype due to mitotic → crossing over or segregation within the somatic cells of a → heterozygous individual.

two-factor cross — a genetic recombination experiment involving two → genetic markers (e.g., $a^+b^+ \times ab$).

two micron plasmid — (2 μm plasmid); in yeast, any of a family of → plasmids of similar size and structure (30 to 200 copies per cell). The *S. cerevisiae* plasmid is 6318 bp long and comprises unique stretches of 2774 bp and 2346 bp separated by two identical 599 bp inverted repeats across which the p. undergoes site-specific recombination resulting in two isomers of the plasmid. The 2 μm p. encodes four genes, whose products are involved in plasmid maintenance, and has two additional cis-acting loci (ORI and STB); ORI is the origin of replication containing an → ARS sequence, STB is required for plasmid partitioning.

two-out-of-three reading frame (Lagerkvist 1978) — the hypothesis according to which → codons may be read by relying mainly on the Watson-Crick base pairs formed with the first two codon positions; mispaired nucleotides in the third codon and anticodon → wobble positions make only a marginal contribution to the total stability of the reading interaction (→ genetic code).

two-plane theory (Sharp 1924) — → chiasmatype theory.

Ty element (Cameron et al. 1979) — in yeast, any of three related families of dispersed → transposable genetic elements (Ty = transposon yeast) which share considerable structural and functional features with → retroviruses. They contain terminal direct repeated sequences (~ 335 bp), share extensive homology at the nucleotide sequence level, and transpose by a reverse transcriptase-mediated process (→ retroposon). Most are about 5.9 kb long and encode several protein species. Ty elements transpose preferentially into the regulatory regions at the 5′ends of yeast genes.

type number (Harvey 1917) — the most frequently occurring chromosome number in a particular taxonomic group (= modal number). The term t.n. is often used with the implication of its being the ancestral number of the group in question. Other numbers are then assumed to be derived from the type number.

U

UACUAAC box — in yeast → introns, a conserved sequence which specifies the branch point for the → lariat and to which U2 sn RNA binds by complementary base pairing during → RNA splicing.

UAS — acronym for the upstream transcription activation sequence mediating → gene expression.

ubiquitin (Schlesinger et al. 1975) — (Ub); a conserved protein (76 amino acid residues) that is present in any prokaryotic and eukaryotic cell, both as a free molecule and conjugated to other proteins via a peptide or isopeptide bond between its COOH-terminal residue and NH_2-groups of the target proteins. It may play a role in chromatin organization and regulation and in the cellular response to stress.

ultraviolet promoter mapping — a method used for the determination of the lengths of individual → transcription units in prokaryotes and eukaryotes after UV-induction of random transcription-terminating lesions in DNA. The probability that a given gene will be expressed is determined by the UV dose and the distance of the gene from the → promoter. Therefore, by monitoring the decrease in synthesis of either RNA or the protein product of the gene with the UV dose, it is possible to estimate the distance of the gene from its promoter.

unassigned reading frame — (URF); a gene-like nucleotide sequence with proper → initiator and → terminator codons but with no known function (= unidentified r.f.). Intronic URF's in fungal mitochondrial genes probably encode proteins required for splicing of these → introns from presursors of the mRNAs (→ RNA splicing; open reading frame).

under-replication — slower replication of certain → heterochromatic chromosome regions and of → ribosomal DNA in relation to the rest of the DNA.

undifferentiated — of cells which are descendants of the fertilized ovum and have not yet developed recognizable heterogeneity (→ cytodifferentiation).

undulipodium (Frey-Wyssling and Mühlethaler 1965) — any of the long, slender → tubulin containing intracellular organelles of motility (diameter ~2000 nm) in eukaryotes, which are intrinsically mobile throughout their length (cilia and 9 + 2 flagella). The → axoneme is the shaft of the u.

unilateral inheritance (Winge 1927) — = → uniparental inheritance.

uniparental inheritance — in → organelle transmission genetics, the exclusive transmission of organelle genes from only one parent, as opposed to biparental i. of (most) nuclear genes.

unipolar — of a → spindle with only one spindle pole. Under these circumstances (= monaster) the chromosomes usually lack a polarized movement (→ chromosome movement).

unique DNA — any specifically ordered sequence of nucleotides in DNA (= single-copy DNA; unique sequence DNA) which occurs only once in a genome (chromosome set), as opposed to → repetitious DNA.

unisexual — of organisms having separate sexes (dioecious, gonochoric) and producing only one kind (male or female) of gametes. Such organismus are thus dimorphic (male and female).

unit character (Castle 1905) — a character or trait controlled by one pair of → alleles, as opposed to quantitative or polygenic characters.

unit-copy plasmid — any of a class of → plasmids (such as the → F plasmid) which are maintained in approximate numerical parity with the chromosomes of the (bacterial) host cell.

unit membrane (Robertson 1959) — any lipoprotein (about 40% lipid and 60% protein) membrane, irrespective of cellular location (→ cell membrane; endomembrane system; nonunit membrane), which can be resolved as a triple-layered strucutre (6 to 10 nm wide) consisting of two electron-dense outer layers and a less dense central layer (all about 2.5 nm in diameter).

Different membranes may show considerable modifications of the basic trilayered structure. Their total thickness may be greater than 6 to 10 nm, and the two outer dense layers are often of unequal thickness. In some cell types, an additional layer of filamentous material may be present on the outer surface. A wide spectrum of variation also exists with respect to chemical content, enzyme activity, permeability characteristics, metabolism and function.

Cellular and organelle membranes frequently consist of two u. membranes in paired arrangement.

unit of replication — = → replicon; replication unit.

unit of transcription — = → transcripton.

univalent — an unpaired chromosome (→ chromosome pairing) at the first meiotic division which corresponds to a particular chromosome of the mitotic complement (→ bivalent; multivalent). They are observed at metaphase I in haploids, triploids, and aneuploids, in interspecific hybrids, and in diploids showing → asynapsis or → desynapsis. Centromere orientation of univalent at anaphase I may be of two main types: (1) Syntelic, i.e., the centromeres of the two chromatids orient to the same pole (reductional distribution); (2) amphitelic, i.e., the centromeres of the two chromatids orient to opposite poles (equational distribution).

Some univalents may not be included in one of the daughter nuclei and are then lost in the cytoplasm (formation of micronuclei). Furthermore, the presence of several univalents in one meiocyte may completely hinder the separation of the chromosomes into two daughter groups and result in a → restitution nucleus.

"False univalents" (Upcott 1938) is the term proposed for univalents which result from indifferent co-orientation (→ centromere orientation) of the chromosomes contained in multivalents rather than from asynapsis or desynapsis.

univalent shift (Person 1956) — after crossing disomic (diploid) and → monosomic individuals (which lack a particular chromosome) the occurrence, among the progeny, of monosomic individuals which lack a chromosome other than that which is lacking in the monosomic parent.

unscheduled DNA synthesis (Rasmussen and Painter 1964) — (USD); in eukaryotes, DNA synthesis outside the S phase of the → cell cycle that maintains the integrity of the genetic material by repairing → DNA lesions induced through natural cellular events or through interaction with DNA-damaging agents (→ DNA repair). USD is of nonsemi-conservative nature and its measurement provides a method for detecting potentially hazardous agents that may have mutagenic or carcinogenic properties (USD assay in → mutagenicity testing).

In addition to → DNA repair, USD also occurs during → genetic recombination, → gene amplification, and the formation of → DNA puffs.

unstable allele — any mutant allele (= hypermutable allele) characterized by frequent reversion to the wild type allele or by frequent changes to other mutant alleles (changes at rates as high as 50%). (→ transposable genetic element).

untargeted mutagenesis (Witkin and Wermundson 1976) — the phenomenon (= nontargeted m.) wherein mutations do not always occur in DNA at the site where a → DNA lesion is formed, as opposed to targeted m. U.m. is often attributed to the action of error-prone → DNA repair mechanisms. (→ gene mutation).

untranslated region — any region of → messenger RNA not coding for protein. All the mRNA's that have been purified are known to contain more nucleotides than are required for coding. These extra residues, presumably present on each side of the coding region, and defined as the 5′ and 3′ untranslated regions are of unknown function and probably serve in → translational control, the termination of transcription, the addition of poly(A), the binding of transport proteins, and mRNA degradation.

up promoter mutation — a mutation that increases the frequency of initiation of → genetic transcription (→ promoter).

upstream — on the 5′ side of any given site in DNA or RNA and on the N-terminal side within a polypeptide (→ downstream). With respect to its transcriptional activity, upstream regions of a gene are on the 3′ side of any given region in the coding (sense) strand.

upstream activation sequence (Guarente 1984; Siliciano and Tätchell 1986) — (UAS); in eukaryotic genes, any nucleotide sequence which is part of the → promoter and located some distance upstream from the transcriptional start site and the → TATA box. The UAS affects the extent of transcription from a given promoter without affecting the site of transcription initiation and may represent the site at which the → RNA polymerase enters the chromatin complex.

upstream activation site (Buther and McConnell 1988), (UAS); an activating sequence located in cis upstream to a structural gene and mediating its transcription (e.g., in response to metabolites).

upstream regulatory element — any of the regulatory sequence elements which confer tissue-specific or developmental patterns of expression on their attached (eukaryotic) genes.

upstream repressing sequence — (URS); a DNA sequence located in cis upstream to a structural gene repressing → gene expression.

upstream RNA — in eukaryotes, → messenger RNA sequences linked to sequences transcribed from the 5′-flanking region of the gene; they are usually polyadenylated and contain 700 to 800 nucleotides; their 5′ regions are heterogeneous.

upstream stimulatory factor (Sawadogo and Roeder 1985) — (USF); in eukaryotes, a gene-specific transcription initiation factor that binds at a palindromic sequence relative to the + 1 start site.

uracil-DNA glycosylase (Lindahl 1974) — an ubiquitous enzyme [dUra(DNA) glycosylase] which specifically hydrolyses glycosyl base-sugar bonds of dUMP in polymeric form and removes uracil residues introduced into DNA, either by deamination of cytosine bases or by incorporation of dUMP by → DNA polymerase. The reaction products are free uracil and an apyrimidinic site in DNA (→ AP site). Ura (DNA) g. activity is necessary to preserve the integrity of DNA; mutants deficient in dUra (DNA) g. are mutators and in vivo deamination of cytosines is the source of these mutations (→ DNA repair).

URF — acronym for → unassigned (unidentified) reading frame.

UTR — acronym for the 3′ transcribed but untranslated region of a gene which may play a regulatory role in specific gene expression.

U-type small nuclear RNA (Lerner and Steitz 1981) — (U-sn RNA); in nuclei of (higher) eukaryotes, a class of uridine-rich → small nuclear RNA molecules (90 to 400 nucleotides) which are complexed with seven to eight proteins to form small nuclear ribonucleoprotein particles (snurps). U-sn RNA's contain several types of modified bases, have no 3′ poly (A) tail, and (with the exception of U6) contain a 5′ trimethylguanosine cap structure. In higher eukaryotes, at least 11 U-sn RNA species are known. Some of these (U1–U7 sn RNA's; perhaps U11 sn RNA) are directly involved in the maturation and processing of cellular RNA's (→ RNA splicing). U-type sn RNA's are classified according to their relative abundance in nuclei into (1) the major U-type sn RNA's (U1–U6; 2×10^5–10^6 copies) and (2) minor U-type sn RNA's (U7–U11; 2.5×10^4 copies). Multigene families code for the major U-type sn RNA's. With the exception of U6 sn RNA genes (transcribed by RNA polymerase III), all the major U-type sn RNA genes are transcribed by RNA polymerase II. All U-type sn RNA's function through specific base-paired hydrogen bonding with other RNA's.

UV endonuclease — any enzyme which initiates repair of pyrimidine dimers, introduced into DNA by UV irradiation, by a mechanism of nucleotide → excision repair. Such enzymes act in → DNA repair of only a few unusually radiation-resistant organisms or virus-infected cells, in which they presumably provide an accessory repair pathway. After production of a single-strand break, the dimer is removed by an exonuclease and repair can be completed by DNA polymerase and DNA ligase activities.

UV protection (Mortelmans and Stocker 1976) — the reduction of the lethal effect of UV irradiation in bacteria carrying certain → plasmids at the cost of enhanced mutagenesis. The plasmids increase the host's ability ot repair UV-damaged DNA by some error-prone process of → DNA repair.

UV reactivation (Weigle 1953) — = → Weigle reactivation.

UV transcription mapping (Sauerbier et al. 1970) — a technique applied in prokaryotes and viruses which allows the mapping of gene positions. Pyrimidine dimers, induced by UV irradiation, cause premature chain termination without affecting initiation, i.e., promoter-proximal transcription is affected less than promoter-distal transcription, and within a given transcription unit there is an exponentially increasing chance of a UV lesion in proportion to the distance from the promoter. In this way, the distance between a given gene and its promoter can be measured by UV t.m.

V

vacuole (Dujardin 1841) — in mature plant cells, the large vesicle (limited by a membrane referred to as the → tonoplast) which occupies the majority of the intracellular volume. The v. is a water-filled space whose main function is the accumulation of solutes responsible for the generation of turgor pressure. Provacuolar structures (transition elements) derive from the → endoplasmic reticulum which then coalesce or enlarge to form the v.

vacuolar system — in eukaryotic cells, the elements of the exo- and endocytotic pathways; they include the → Golgi complex, the secondary vacuoles, the → endosomes, and the → lysosomes as well as other organelles involved in biosynthesis, processing, transport, storage, release, and degradation of soluble and membrane-bound macromolecules (= vacuolar apparatus).

vacuome (Dangeard 1919) — all the membrane-bounded spaces of the eukaryotic cell with the exception of the → mitochondria and → plastids which make up the chondriome and the plastome, respectively.

variability — the state or quality of being variable, or subject to → variation, i.e., of having the tendency to vary in form, nature, substance, etc. (→ genetic variability; variation).

variance — (V); a statistical measure of the variation of individual values about the mean, given in terms of the square of the units used to measure the values. In genetics, V_T refers to the total v. of a character, while V_G and V_E refer to the variances due to genotype and environment, respectively. V_1 refers to the v. resulting from genotype-environment interaction.

variant — a cell or individual that is recognizably different from an arbitrary standard. The proportion of variants already existent in a population is the v. frequency.

variation — the occurrence of heritable or nonheritable differences in the permanent structures of cells (intra-individual v.), among individuals of one population (individual v.), or among populations (group v.).

The chief sources of v. among the characteristics of related organisms are genic differences (due to → mutation or new assortment of genes) or environmentally induced differences which cause only temporary changes of the phenotype.

Primarily, biological v. may be subdivided into three catgories: (1) **Phenotypic variation:** The total biological v. of a given → character (symbol V_p). V. without natural discontininuities is called **continuous v.** and characters that exhibit it are quantitative or metric characters, as opposed to **discontinuous v.** and qualitative characters.

The components of continuous and discontinuous phenotypic v. in a segregating population may be grouped into two major classes referred to as genetic (symbol V_G) and nongenetic or environmental (symbol V_E) variation. By definition, phenotypic v. is then $V_P = V_G + V_E$. (2) **Environmental variation:** V. due to all intra- and extracellular factors which influence the expression of the → genotype. This part of v. is expressed in terms of environmental variance (symbol V_E) and consists of two main components. One is represented by the intangible statistical residue known as "error" and certain genotype-environment interactions; the other is the controllable environment. (3) **Genetic variation:** V. due to the contribution of segregating genes and → gene interactions (symbol V_G) represented as the proportion of the total phenotypic variance that is exclusively genetic. This proportion is called the → heritability (symbol H) and is commonly expressed as a percentage (H = 100% if all the variation is due to genetic causes or there is no environmental variance; as the environmental component of variance increases, H decreases). The main components of genetic variance are "additive genetic variance" (due to any two alleles that have different quantitative effects contributing to this component), "dominance deviations" (deviations from additive variance due to dominance), and "interaction" or "epistatic deviations" (deviations from additive variance due to → epistasis and other nonallelic interactions).

Genetic v. in a population is controlled by three main factors (Mayr 1963): The input of new genetic information due to mutation and → gene flow; the erosion of this variation by selection and sampling errors; and the protection of the stored variability by cytophysiological devices and ecological factors (→ genetic variability).

variegation (Schultz 1936) — the occurrence, within one tissue, organ, or individual, or a mosaic phenotype (→ mosaicism) with respect to pigmentation (e.g., diversity in color due to the deprivation of a part of the normally green cells of a plant or their chlorophyll content) or any other easily observable phenotypic character. V. is a widespread phenomenon caused by several (observed or inferred) mechanisms (Ar-Rushdi 1957): (1) Plastid variation and the behavior of plastid variants in somatic divisions of plants; (2) genic instability and the recurrence of somatic mutations; (3) instability of the phenotypic expression of genes as a result of → position effect; (4) genetically controlled → stickiness of chromosomes; (5) somatic → crossing over; (6) somatic behavior of → ring chromosomes and → dicentric chromosomes (→ bridge-breakage-fusion-bridge cycle); (7) infection.

variety — in classical taxonomy, a heterogeneous grouping including nongenetic variations of the phenotype, morphs, domestic breeds, and geographic races (Mayr 1963).

vector — → cloning vector.

vesicle — in eukaryotic cells, any of the closed structures completely surrounded by a → unit membrane (→ lamella). As opposed to vacuoles, the contents of vesicles are not in the liquid state.

vesicular conglomerate (Yamada et al. 1957) — = → multivesicular body.

v-gene reactivation (Streisinger 1956) — (vR); a phage-specific → reactivation system, due to the presence in the genome of phage T 4 of gene v^+ which makes the phage more than twice as resistant to UV-irradiation as the related phages T2 and T6 and the mutants of T4 carrying the recessive allele v (Rupert and Harm 1966) (→ x-gene reactivation).

The product of the v^+ gene (endonuclease V) is essential for the formation of the single-strand breaks which initiate → excision repair.

viability — the capability for living or continuing to develop (→ lethal factor).

viability polymorphism (Brues 1969) — a term proposed to substitute for → genetic load (with the exception of mutational load) which is used with a variety of meanings.

viral oncogene — → oncogene.

viral transformation — = → cell transformation.

virion (Lwoff et al. 1959) — a complete, infectious virus particle, consisting of a nucleic acid core embedded in the protein → capsid (→ pseudovirion).

virogeny (Koprowski 1964) — in a nonlysing mammalian cell culture, the induction or active appearance of an infectious virus, the induction of an infectious nucleic acid, the presence of virus antigen, and the resistance of the cell to superinfection.

viroid (Diener 1971) — any of the small (246 to 375 nucleotides), unencapsidated, covalently closed, single-stranded RNA molecules capable of autonomous replication and able to incite disease in a variety of plants. They to not code for proteins and their biological properties are the consequence of direct v.-host interactions. Disease induction might result from v. interference with host gene expression. Viroids probably replicate by a → rolling circle in which oligomeric (+) strand molecules are cleaved to monomeric units. These circularize to mature viroids. Because of extensive regions of intramolecular configuration, viroids can assume a rod-like quasidouble-stranded configuration, in which short base-paired regions alternate with small internal loops (→ virusoid).

virulent — of a → virus or bacteriophage capable only of → lytic growth, as opposed to temperate phages establishing the → lysogenic response.

virus (Iwanowsky 1892; Beijerinck 1899) — any of the infectious, subcellular, and ultramicroscopic particles (usually nucleoprotein, more rarely exclusively DNA or RNA in nature) representing potentially pathogenic agents, whose replication in the cell and transmission by infection results in characteristic reactions of the host cells and host individuals. Viruses are independent genetic systems (endowed with genetic continuity and mutability) which possess an evolutionary history of their own. They use the synthetic machinery of living cells to direct the synthesis of specialized particles, the → virions, which contain the viral genome and transfer it to other cells. Major groups of viruses are bacteriophages, animal viruses, and plant viruses.

Nucleic acid molecules from mature → virions can be put into four classes: linear or circular and these either single- or double-stranded (→ chromosome).

Some viruses have segmented genomes which consist of separate and unique nucleotide sequences each of which is transcribed into messenger RNA. The life cycle of all virulent viruses consists of the following stages: (1) Adsorption; (2) entry of nucleic acid into the cell; (3) transcription, translation, and replication; (4) maturation of particles; (5) release of particles.

Genomes of endogenous viruses are integrated into the host cell genome (→ retrovirus).

virus maturation — a complex process which results in the formation of infective progeny particles inside the cell. It includes the formation of mature viral nucleic acid in the form found in the parent particles, the encoating or packaging of viral nucleic acid within a capsule, and the assembly of packaged nucleic acid with the accessory structures necessary for infectivity.

virusoid (Randles et al. 1981) — in certain plant viruses with a bipartite genome, any of a class of single-stranded, small circular RNA's which exist as highly base-paired, rod-like structures. The virusoids depend on a linear viral RNA (about 4500 residues) for their replication and encapsidation, as opposed to the unencapsidated → viroids which are replicated by a host polymerase.

visibles — mutations recognizable by their phenotypic expression, in contrast to → vitality mutations and → lethal factors.

vitality mutation — any mutation the effect of which, in contrast to → visibles and → lethal factors, is difficult to determine, but which changes the vitality (viability) of the carrier genotype if present in an effective dosage.

Hadorn (1949) classified vitality mutations as follows: (1) **Semilethal mutations** (viability of the carrier less than 50% as compared with a standard type). (2) **Subvital mutations** (viability of the carrier higher than 50%, but lower than normal when compared with a standard type). (3) **Supervital mutations** (viability of the carrier higher than that of the standard type).

VSP repair (Lieb 1983) — in *E. coli*, a repair mechanism (very short patch repair) that repairs T/G mismatches regenerating the original CC (AT) GG sites. The mechanism has probably evolved to minimize C → T transitions that result from deamination of methylated cytosines (→ DNA methylation).

V-type position effect (Lewis 1950) — → position effect.

W

Wahlund effect (Wahlund 1928) — a deficiency of heterozygotes, compared to expectations based on a → Hardy-Weinberg equilibrium, when two divergent populations (previously isolated and subject to forces differentiating allele frequencies between them) occur sympatrically and are sampled as a single population.

Wallace-effect (Grant 1966) — selection for reproductive → isolation in areas of sympatry reinforcing previously existing barriers and thereby reducing genetic wastage, hybridization, and disruptive → gene flow.

W-chromosome — the → sex chromosome which, in cases of female heterogamety, is limited to the female sex. In male heterogamety this is the Y-chromosome (→ Z-chromosome).

Weigle reactivation (Radman 1974) — (W-R); the increased capacity (→ SOS response) of UV irradiated bacteria to reactivate irradiated phage; coupled with increased mutagenesis (W mutagenesis). W-R also occurs in mammalian cells when UV irradiated nuclear replicating DNA viruses are plated on UV irradiated host cells. W-R is an inducible mode of repair of → DNA damage in bacteria (= → SOS repair; UV reactivation).

Western blotting — analogous to → Southern blotting, the transfer of proteins from a polyacrylamide gel onto a suitable immobilizing matrix (e.g., a nitrocellulose sheet). The proteins may then be probed with, e.g., a specific antibody to identify a particular protein species.

whole-arm transfer (Muller 1940) — → translocation.

whole-arm transposition — = whole-arm transfer (→ translocation).

wide hybridization — interspecific and intergeneric hybridization which is a first step to introduce alien variation and to transfer desirable genes and traits from wild species into cultivated species. Barriers to w.h. are: (1) Incompatibility between the parent species; (2) inviability of the F_1 hybrids; (3) sterility of the F_1 hybrids or their progeny.

wild type — a gene, genotype, or phenotype predominating in the wild population or in the standard laboratory stock for a given organism (→ standard type; genetic nomenclature).

wobble pairing (Crick 1966) — the ability of certain bases at the third position (5′end) of an → anticodon in transfer RNA to form hydrogen bonds in various ways, causing alignment with several possible → codons (at the 3′end). The third position deviations consist of pairing between:

5′end of anticodon (third position)	3′end of codon
G	U or C
C	G only
A	U only
U	A or G
I	U, C, or A.

According to the w. rules, every tRNA reads either one, two, or three codons, the number depending on whether the first base of the anticodon (the wobble base) is C or A (one codon), U or G (2 codons), or a modified base such as I (3 codons). The modified bases have probably evolved in the wobble position because they maximize the number of codons that can be read by a single tRNA molecule (→ two-out-of-three reading frame).

W reactivation (Weigle 1953) — = → Weigle reactivation.

writing number — the number of times a duplex axis of DNA crosses over itself in space.

X

x — a symbol for the → basic number of eukaryotic chromosomes, i.e., the smallest number of chromosomes in a → chromosome set or → genome.

X-chromosome (Henking 1891; McClung 1902) — in organisms with male heterogamety and diploid sex differentiation ("diplogenotypic → sex determination"), the → sex chromosome which is present in both sexes (→ sex linkage). One sex (the homogametic, female sex) is homozygous (XX) with respect to this chromosome, the other (the heterogametic, male sex) is heterozygous (XO or XY). In organisms with haploid sex differentiation ("haplogenotypic sex determination"), the sex-chromosome of the female.

In **"multiple sex chromosome systems"** the X-chromosome may be represented more than twice in the female and more than once in the male sex (e.g., X_nO, X_nY, X_nY_n).

The **"inactive-X hypothesis"** (Lyon 1961) states that in XY-males the single X-chromosome is active in all cells, while in each cell of the early female mammalian embryo (XX), one of the two X-chromosomes becomes inactivated. (→ dosage compensation).

xenia (Focke 1881) — seed whose hybrid nature is recognizable by particular phenotypic characters (color, shape, size, etc.) owing to direct influences exerted by the genotype of the pollen on the embryo or on the maternal tissue (endosperm) of the fruit.

xenobiotic (Mason et al. 1965) — ref. to components of the chemical environment which are foreign to the metabolic network of an organism.

xenotropic (Levy 1973) — of any endogenous → virus which productively infects only cells from animals which are foreign to the host species, as opposed to ecotropic viruses which preferentially infect and grow in cells from their own host species.

xerograft — = → heterograft.

x-gene reactivation (Harm 1963) — (xR); a phage-specific → reactivation system which increases the UV-resistance of T-even phages (absent in mutants T 4x and T 4vx) due to the presence of gene v^+ in the genomes. The mutation from x^+ to x is coupled with a decrease in the frequency of genetic recombination in the mutants. The x^+ gene is probably involved in a → post-replication (recombinational) repair system (→ v gene reactivation; UV-reactivation).

X-inactivation — → dosage compensation.

Y

Y-chromosome (Wilson 1909) — in organisms with male heterogamety and diploid sex differentiation ("diplogenotypic → sex determination"), the → sex chromosome limited to one sex (the heterogametic, male sex) which pairs with the → X-chromosome (→ sex linkage). In organisms with haploid sex differentiation ("haplogenotypic sex determination"), the sex chromosome of the male sex. In mammals, a DNA sequence encoding a → zinc finger protein resides in the region of the Y c. that causes male development (a gene for testis determination).

In **"multiple sex chromosome systems"** the Y-chromosome may be represented more than once (e.g., XY_n, X_nY_n) in the chromosome complement.

yeast centromeric plasmid — (YCp); a *Saccharomyces cerevisiae* → cloning vector containing a cloned → centromere sequence, a chromosomal replicator ARS-sequence, a selectable yeast gene and (frequently) an *E. coli* origin to permit replication in that organism (→ shuttle vector).

yeast episomal plasmid — (YEp); a *Saccharomyces cerevisiae* → cloning vector which relies for replication on the 2 μm plasmid origin and contains a selectable yeast gene and (frequently) an *E. coli* origin of replication.

yeast integrative plasmid — (YIp); a *Saccharomyces cerevisiae* → cloning vector consisting of a selectable yeast gene inserted into an *E. coli* plasmid (e.g., pBR322). The vector is without a yeast replicator and thus can only be maintained if integrated into a chromosome.

yeast replicative plasmid — (YRp); a *Saccharomyces cerevisiae* → cloning vector which uses a chromosomal replicator (ARS) to replicate in yeast, a selectable yeast gene and an *E. coli* origin.

y gene reactivation (Boyle and Symonds 1969) — a phage T4 gene that affects radiation sensitivity in a manner similar to → gene reactivation.

yolk — a complex collection of macromolecules and smaller nutrient molecules with which the oocyte is preloaded prior to → fertilization. Synthesis and deposition of y. constitute one of the major events of → oogenesis.

yolk nucleus — an intensely osmiophilic body situated in the vicinity of the oocyte nucleus in invertebrates and vertebrates. The y.n. is a conglomeration of lamellae and vesicles as observed with the electron microscope (= Balbiani body).

yolk platelet (Anderson 1968) — in mature unfertilized animal eggs, any of a distinct species of subcellular organelles which are ensembled from derivatives of the → Golgi apparatus during → oogenesis. Yolk platelets are storage granules for metabolites required during embryogenesis. They and the → cortical granules exhibit many structural and functional features usually associated with → lysosomes in adult tissues and have been termed "lysosome-like" organelles. Mucopolysaccharides and other glycoproteins as well as acid phosphates are biochemical constituents of y. platelets.

Z

Z-chromosome — the → sex chromosome which, in cases of female heterogamety, is present in both sexes, in contrast to the → W-chromosomes.

Z-DNA (Wang et al. 1979) — DNA characterized by a left-handed (zigzag) double helix and only a single groove, as opposed to the → B-DNA duplex which is right-handed and has two grooves (one major and one minor). The backbones of Z-DNA produce a zigzag pattern around the helix; the orientation of the bases in Z-DNA is quite different from that in B-DNA. Supercoiling at densities known to exist under physiological conditions induces the reversible B to Z transition. Alternating purine-pyrimidine sequences having a strong potential to adapt stable left-handed Z-conformation are repetitive elements in prokaryotic and eukaryotic DNA genomes. Z-DNA-binding proteins are known and may be involved in the control of gene activity. Left-handed stretches in DNA may be recognition signals in processes such as regulation of genetic transcription and DNA-replication. Z-DNA forming sequences (→ CA repeat) cause elevated recombination frequencies, may play a role in → gene conversion events, and might cause genetic instability (→ Z-RNA).

zero-time binding DNA — in eukaryotes, a fraction of DNA which renatures almost instantaneously after denaturation (= → foldback DNA).

zinc finger motiv (Bogenhagen et al. 1980) — any of the tandemly repeated protein sequence motifs (28 to 30 amino acids each) containing two cysteines and two histidines at invariant positions. Z.f. motifs are common to a wide variety of eukaryotic → DNA-binding proteins; each of these units folds into a finger-like loop as an independent domain, centered on a zinc ion, and binds to DNA in a sequence-specific manner. Z.f. motifs are encoded by various → regulatory genes (→ helix-turn-helix motif). Proteins containing z.-f. motifs participate in the regulation of → genetic transcription.

zona pellucida — (ZP); a thick extracellular glycoprotein matrix (coat) that is assembled during growth of the mammalian oocyte and surrounds seggs and preimplantation embryos. The ZP performs different function: (1) It becomes a selectively permeable barrier while funtioning as a structural lattice providing support for the oocyte and follicular cells; (2) it acts as a sperm receptor during early events of → fertilization; (3) it assists in prevention of polyspermy and protects preimplantation-stage embryos as they travel toward the uterus.

zona occludens (Farquhar and Palade 1963) — = → tight junction.

zonula nucleum limitans (Patrizi and Poger 1967) — = → nuclear lamina.

Z-RNA (Hall et al. 1984) — a left-handed RNA double helix; the conformational transition from the A-form to the Z-form RNA sequences requires more extreme conditions than the transition to → Z-DNA and may be of relevance in those cases in which double-stranded RNA occurs (e.g., in → ribosomes and in some viruses).

zygomere (Sybenga 1966) — any of the specialized sites of DNA (the synaptic initiation sites) responsible for the initiation of intimate eukaryotic → chromosome pairing during meiosis.

zygonema (Gregoire 1907) — = zygotene (→ meiosis).

zygote (Bateson 1902) — (1) **Holozygote:** In eukaryotic organisms with sexual → reproduction, a diploid cell formed by the fusion of two haploid gametes during fertilization and (usually) containing two **complete** → genomes (hence holozygote); also the individual derived from it. For those genes which were represented by different → alleles in the genomes of the fusing gametes, the z. is said to be → **heterozygous** or hybrid, while it is **homozygous** with respect to genes represented by identical alleles in the gametes. (2) **Merozygote** (Wollman et al. 1956): In prokaryotes (bacteria), an **incomplete** "zygote" formed by parasexual mechanisms of genetic transfer ("meromixis") due to bacterial → conjugation mediated by → F plasmids, → transduction, and → transformation), being diploid only for part of its genetic material and haploid for the remainder. A merozygote contains the complete genome of the recipient cell (the endogenote) and a → merogenote (or exogenote) derived from a donor cell. Merozygotes are usually very unstable and may be → **hetero- or homogenotic** in respect to their genetic constitution. The potentiality for → genetic recombination in a merozygote is limited to the diploid part of its genetic material.

zygotene (v. Winiwarter 1900) — → meiosis.

zygotene DNA (Hotta et al. 1984) — (zyg DNA); a fraction of the eukaryotic nuclear DNA which fails to replicate during the premeiotic S-phase and undergoes replication during zygotene in coordination with → chromosome pairing. Replication of zyg DNA is incomplete; small gaps remain at the ends of newly synthesized strands and these are not filled until after termination of meiotic prophase. Suppression of zyg DNA replication is due to a protein (L-protein). The poly(A)$^+$RNA molecules transcribed from zyg DNA on initiation of chromosome pairing is called zygotene RNA (zyg RNA).

zygotene RNA (Hotta et al. 1985) — (zyg RNA); → zygotene DNA.

zygotic embryogenesis — → somatic embryogenesis.

zygotic induction (Jacob and Wollman 1956) — → induction.

zygotic mortality — one of the post-mating mechanisms (in addition to gametic m., hybrid inviability, and hybrid sterility) that prevent the success of an interspecific cross even though mating has taken place. The egg is fertilized but the zygote does not develop into an organism. In the case of **gametic m.** either the male or female gametes are destroyed because of the interspecific cross.

zygotonucleus — the nucleus of a holozygote formed by fusion (karyogamy) of two gametic nuclei.

References

Aaronson RP, Blobel G (1974) J Cell Biol 62: 746
Aaronson RP, Blobel G (1975) Proc Natl Acad Sci USA 72: 1007
Abdel-Monem M, Laupp H-F, Kartenbeck J, Dürwald A, Hoffmann-Berlin H (1977) J Mol Biol 110: 667
Abercrombie M, Heaysman JEM (1953) Exp Cell Res 5: 111
Abercrombie M, Heaysman JEM (1953) Exp Cell Res 6: 293
Abercrombie M, Heaysman JEM, Pegrum SM (1971) Exp Cell Res 67: 359
Achtman MA (1975) J Bact 123: 505
Achtman M, Willetts N, Clark AJ (1971) J Bact 106: 529
Adam SA, Nakagawa T, Swanson MS, Woodruff TK, Dreyfuss G (1986) Mol Cell Biol 6: 2932
Adhya S, Gottesman M, De Combrugghe B (1974) Proc Natl Acad Sci USA 71: 2534
Adler HI, Fischer WD, Cohen A, Hardigree AA (1967) Proc Natl Acad Sci USA 57: 321
Adolph KW, Cheng SM, Paulson JR, Laemmli UK (1977) Proc Natl Acad Sci USA 74: 4937
Agar WE (1911) Quartl J Micr Sci 67: 1
Ahlström CG (1951) Acta Path Microbiol Scand (Suppl 91): 52
Ahmad M (1954) Pakist J Sci 5: 59
Air GM, Blackburn EH, Coulson AR, Galibert F, Sanger F, Sedal JW, Ziff EB (1976) J Mol Biol 107: 445
Akaboshi E, Inouye M, Tsugita A (1976) Mol Gen Genet 149: 1
Alberts BM, Amodio FJ, Jenkins M, Gutmann ED, Ferris FL (1968) Cold Spring Harb Symp 33: 289
Alberts BM, Frey L (1970) Nature 227: 1313
Alberts B, Worcel A, Weintraub H (1977) In: Bradbury R, Jahaveriam K (eds) Proceedings of the Int Symp on the Eukaryotic Genome, Acad Press, New York
Alexeiff A (1917) Compt Rend Soc Biol, Paris 80: 512
Alföldi L, Jacob F, Wollman EL (1957) Compt Rend Acad Sci, Paris 244: 2974
Allen CE (1912) Arch Zellforsch 8: 121
Allen JS (1967) Genetics 55: 797
Allen RD, Hagström BE (1955) Exp Cell Res (Suppl) 3: 1
Allen SL (1963) J Protozool 10: 413
Alpers JB, Paulus H (1971) Nature 233: 478
Altmann R (1889) Arch Anat Physiol, Lpz, Physiol Abt 524
Ames BN (1971) The detection of chemical mutagens with enteric bacteria. In: Hollaender A "Chemical Mutagens" Vol 1, p 267 Plenum Press, New York
Ames BN, Garry B (1959) Proc Natl Acad Sci USA 45: 1453
Ames BN, Hartman PE (1963) Cold Spr Harb Symp 28: 439
Ames BN, Durston WE, Yamasaki E, Lee FD (1973) Proc Natl Acad Sci USA 70: 2281
Anderson E (1949) Introgressive Hybridization. Wiley, New York
Anderson E (1953) Biol Rev 28: 280
Anderson E (1968) J Cell Biol 37: 514
Anderson, E, Hubricht L (1938) Am J Bot 25: 396
Anderson TF (1958) Cold Spr Harb Symp 23: 47
Angel T, Austin B, Catcheside DG (1970) Aust J Biol Sci 23: 1229
Anker P, Strouth M, Greppin H, Fredy M (1971) Nature New Biol 234: 184
Arber W, Dussoix D (1962) J Mol Biol 5: 18
Arber W, Linn S (1969) Ann Rev Biochem 38: 467

Ar-Rushdi AH (1957) Genetics 42: 312
Astbury WT (1950) Harvey Lectures 46: 3
Athwal RS, McBride OW (1977) Proc Natl Acad Sci USA 74: 2943
Atwood SS (1944) Proc Natl Acad Sci USA 30: 69
Auerbach C (1967) Heritage from Mendel. Proc of the Mendel Centennial Symp 1965. Brink RA (ed) Univ Wisconsin Press, Madison, Milwaukee, London
Auerbach L (1874) Organologische Studien, Breslau
Austin CR (1960) J Cell Comp Physiol (Symp on Mammalian Genetics and Reproduction (Suppl 1) 56: 1
Avery DT, Macleod CM, McCarthy M (1944) J Exp Med 79: 137

Baccarini I (1908) G Bot Ital 15: 180
Bacci G (1950) Boll Zool 17: (Suppl) 193
Bacci G (1961) Boll Zool 28: 469
Bacci G (1965) Sexdetermination. Pergamon Press, Oxford
Bacharova R, Davidson EH, Allfrey VG, Mirsky AE (1966) Proc Natl Acad Sci USA 55: 359
Backman K, Ptashne M, Gilbert W (1976) Proc Natl Acad Sci USA 73: 4174
Bade EG, Gonzales NS, Algranati ID (1969) Proc Natl Acad Sci USA 64: 654
Bainbridge BW, Roper JA (1966) J Gen Microbiol 42: 417
Bajer A (1968) Chromosoma 25: 249
Baker JR (1951) Nature 168: 1089
Balfour FM (1880) Comparative Embryology I
Ball O (1925) Wien Klin Wschr 34: 447
Baltimore D (1970) Nature 226: 1209
Bandyophadhyay PK, Wu CW (1978) Biochemistry 17: 4078
Bangham AD, Staudish MM, Watkins JC (1965) J Mol Biol 13: 238
Barber HN (1940) Proc Roy Soc B 128: 170
Barber HN, Callan HG (1943) Proc Roy Soc B 131: 258
Barr ML, Bertram EG (1949) Nature 163: 676
Barrell BG, Air GM, Hutchinson CA (1976) Nature 264: 33
Barski G, Sorieul S, Cornefert F (1960) Compt Rend Acad Sci, Paris 251: 1825
Bary de A (1877) Vergleichende Anatomie der Vegetationsorgane der Phanerogamen und Farne. Leipzig
Bateman AJ (1958) Heredity 12: 213
Bateson W (1894) Materials for the Study of Variation. MacMillan, London
Bateson W (1902) Mendels Principles of Heredity. Cambridge
Bateson W (1905) in a letter to Sedgewick A, from Bateson W (1928) Essays and Adresses, edited by Bateson B, Cambridge Univ Press, Cambridge
Bateson W (1906) Progr Rei Botanicae 1: 368
Bateson W, Saunders ER (1902) Rep Evolut Comm Roy Soc, Rep I: 160
Bateson W, Saunders ER, Punnett RC (1905) Rep Evolut Comm Roy Soc, Rep II: 1 and 80
Bateson W, Saunders ER, Punnett RC (1908) Rep Evolut Comm Roy Soc, Rep IV: 60
Battaglia E (1945) Nuovo G Bot Ital 52: 42
Battaglia E (1947) Nuovo G Bot Ital 54: 674
Battaglia E (1952) Atti Della Soc Toscana 59: 166
Battaglia E (1955) Bull Torrey Bot Cl 82: 383
Battaglia E (1955) Phytomorphology 5: 171
Battaglia E (1955) Phytomorphology 5: 173
Battaglia E (1956) Caryologia 8: 214
Battaglia E (1963) Apomixis. In: Recent Adv in the Embryology of Angiosperms. Maheshwari P (ed): 249
Battaglia E, Boyes JW (1955) Caryologia 8: 87

Bauch R (1947) Naturwiss 34: 346
Bauer W, Vinograd J (1968) J Mol Biol 33: 141
Bauer H (1943) Fortschr Zool 7: 256
Bauer H (1952) Zool Anz (Suppl Bd) 17: 252
Bauer H, Dietz R, Röbbelen C (1961) Chromosoma 12: 116
Baur E (1909) Z Indukt Abstamm- u. Vererb.-Lehre 1: 330
Bautz EK, Bautz FA, Dunn JJ (1969) Nature 223: 1022
Beadle GW (1931) Cornell Univ Agr Exp Stat Mem 135: 12
Beadle GW (1932) Z Indukt Abstamm- u. Vererbungslehre 63: 195
Beadle GW (1932) Proc Natl Acad Sci USA 18: 160
Beadle GW (1933) Cytologia 4: 269
Beadle GW (1957) The role of the nucleus in heredity. In: McElroy WD, Glass B (eds) "The Chemical Bais of Heredity". John Hopkins Press, Baltimore
Beadle GW, Tatum EL (1941) Proc Natl Acad Sci USA 27: 499
Beatty RA (1957) Parthenogenesis and Polyploidy in Mammalian Development. University Press, Cambridge
Becker MM, Wang IC (1984) Nature 309: 795
Beckwith J (1963) Biochem Biophys Acta 76: 162
Beer de GR (1951) Embryos and Ancestors. Univ Press, Oxford
Beermann W (1952) Chromosoma 5: 139
Beermann W, Bahr GF (1954) Exp Cell Res 6: 195
Beijerinck MW (1899) Zentrbl Bakt Parasit 5: 27
Belar K (1928) Die zytologischen Grundlagen der Vererbung Hdb Vererb I. Borntraeger Berlin
Bell G (1980) Am Nat 116: 45
Belling J (1914) J Hered 5: 65
Belling J (1914) Z indukt Abstamm- u. VererbLehre 12: 303
Belling J (1924) Bot Gaz 78: 458
Belling J (1925) J Genet 15: 245
Belling J (1927) J Genet 18: 177
Belling J (1928) Univ California Publicat Botany 14: 335
Belling J, Blakeslee AF (1923) Proc Natl Acad Sci USA 9: 106
Belling J, Blakeslee AF (1926) Proc Natl Acad Sci USA 12: 7
Benbow RA, Gandette MF, Hines PJ, Hiriynna (1984) J Cell Sci 99: 10a
Benda C (1891) Verhandl Berliner physiol Gesellsch Arch Anat Physiol 1891: 549
Benda C (1898) Arch Anat Physiol, Physiol Abtlg, Leipzig: 393
Benda C (1902) Ergebn Anat 12: 743
Bender W, Spierer P, Hogness DS (1983) J Mol Biol 168: 17
Beneden E van (1870) Mém Cour Acad R Belg 34: 11
Beneden E van (1875) Bull Acad Belg Cl Sci XL II. Serie, 686
Beneden E van (1883) Arch Biol 4: 265
Beneden E van, Bessels E (1868) Mém Cour Acad R Belg 34
Benne R, van den Burg J, Brakenhoff JPL, Slooz P, van Boom JH, Tromp ML (1986) Cell 46: 819
Bennet HS (1963) J Histochem Cytochem 11: 2
Bennet HS (1969) The cell surface: Components and configurations. In: Lima-de-Faria A (ed) "Handbook of Molecular Cytology" North-Holland Publ Comp, Amsterdam: 1261
Bennett MD (1972) Proc Roy Soc B 191: 109
Benoist C,O'Hare K, Breathnack R, Chambon P (1980) Nucleic Acids Res 8: 127
Benzer S (1955) Proc Natl Acad Sci USA 41: 344
Benzer S (1957) The elementary units of heredity. In: McElroy WD, Glass B (eds) "The Chemical Basis of Heredity", John Hopkins Press, Baltimore

Benzer S (1961) Proc Natl Acad Sci USA 47: 403
Berezney R, Coffey DS (1974) Biochem Biophys Res Comm 60: 1410
Berezney R, Coffey DS (1975) Science 189: 291
Berg D (1977) In: Bukhari AI et al. (eds) "DNA Insertion Elements, Plasmids and Episomes" Cold Spr Harb Lab, New York, 205
Bergel SM, Moore C, Sharop PA (1977) Proc Natl Acad Sci USA 74: 3171
Bergmann JE, Codish HF (1979) J Biol Chem 254: 11927
Berk AJ, Sharp PA (1977) Cell 12: 721
Bernardi G, Bernardi G (1985) J Mol Evol 22: 363
Bernhard W, Frayssinet C, Lafarge C, Leberton E (1965) Compt Rend Acad Sci, Paris, 261: 1785
Bernhard W, Gautier A, Rouiller C (1954) Arch Anat Microsc 35: 236
Bernhard W, Granboulan N (1968) In: Dalton AJ, Haguenau F (eds) "The Nucleus" Acad Press, New York, London: 81
Bernhard W, Haguenau F, Oberling U (1952) Experientia 8: 58
Bernheimer HP, Wermundsen IE, Austrian R (1967) J Bact 93: 320
Berns MW, Rattner JB, Brenner S, Mercaith S (1977) J Cell Biol 72: 351
Bertani G, Weigle JJ (1954) J Bact 65: 113
Bessis M, Breton-Gorius J, Thiery JP (1958) Rev Hématol 13: 363
Bhalla SC, Sokal RR (1964) Evolution 18: 312
Bingham PM, Levis R, Rubin GM (1981) Cell 25: 693
Binns A, Meins F (1973) Proc Natl Acad Sci USA 70: 2660
Bird AP, Taggart MH, Frommer M, Miller OJ, MacLeod D (1985) Cell 40: 91
Birdsell JB (1950) Cold Spr Harb Symp 15: 259
Birky CW, Acton AR, Dietrich R (1982) In: Alleverdi G, Porst P, Slonimsky PP (eds) Cold Spr Harbor Lab
Birnboium HD, Mitchell REJ, Straus NA (1973) Proc Natl Acad Sci USA 70: 2189
Birnstiel M, Chipchase M (1977) Trends Biochem Sci 2: 149
Birnstiel M, Speirs J, Purdom J, Jones K, Loening UE (1968) Nature 219: 454
Bischoff GW (1835) Nova Acta Acad Caes Leop Carol Nat Cur 17: 911
Bizzozero G (1871) Zentr Med Wochenschr 9: 482
Blakeslee AF (1904) Proc Am Acad Arts Sci 40: 205
Blakeslee AF (1921) Am Nat 55: 254
Blakeslee AF (1924) Proc Natl Acad Sci USA 10: 109
Blakeslee AF (1927) Ann N Y Acad Sci 30: 1
Blakeslee AF, Belling J, Farnham M (1923) Bot Gaz 76: 329
Blattner FR, Dahlberg JE, Doettiger JK, Fiandt M, Szybalski W (1972) Nature New Biol 237: 232
Bleyman M, Woese C (1969) Proc Natl Acad Sci USA 63: 532
Blobel G (1980) Proc Natl Acad Sci USA 77: 1496
Blobel G, Dobberstein B (1975) J Cell Biol 67: 835
Bloom W, Leider RJ (1962) J Cell Biol 13: 269
Bocke JD, Garfinkel DJ, Styles CA, Fink GR (1985) Cell 40: 491
Börner T (1984) Biol Zbl 103: 485
Bogenhagen DF, Sakonju S, Brown DD (1980) Cell 19: 27
Boller K, Schmid W (1970) Humangenetik 11: 35
Bollum FJ (1960) J biol Chem 235: PC 18
Böök JA (1945) Hereditas 31: 177
Bordet IC (1925) Ann Inst Pasteur 84: 273
Borek E, Ryan FJ (1958) Proc Natl Acad Sci USA 44: 374
Borek E, Ryan FJ, Rockenbach J (1955) J Bact 69: 460
Borisy GG, Taylor EW (1967) J Cell Biol 34: 525

Boveri T (1887) S B Ges Morph Physiol, München, 3: 71
Boveri T (1888) Jena Z Naturw 22: 685
Boveri T (1891) Anatomische Hefte 1, Abt II Ergebnisse
Boveri T (1895) Verh phys med Ges Würzb 29: 1
Boveri T (1901) Spengels Zool Jahrb 14: 630
Boveri T (1905) Jena Z Naturw 39: 445
Boveri T (1907) Jena Z Naturw 43: 1
Bower FO (1886) J Linn Soc Bot 21: 360
Boyce RP, Howard-Flanders P (1964) Proc Natl Acad Sci USA 51: 293
Boycott AE, Diver C (1923) Proc Roy Soc B 195: 143
Boyle JM, Symonds N (1969) Mut Res 8: 431
Bradshaw AD (1965) Adv Genet 13: 115
Brand A, Breeden L, Abraham J, Sternglans R, Nasmyth K (1985) Cell 41: 41
Braun A (1851) Betrachtungen über die Erscheinung der Verjüngung in der Natur, insbesondere in der Lebens- und Bildungsgeschichte der Pflanze. Engelmann W, Leipzig
Braun W (1965) Bacterial Genetics (2 ed) Saunders, Philadelphia
Breathnach R, Mandel JC, Chambon P (1977) Nature 270: 314
Breathnach R, Benoist C, O'Hare K, Cannon F, Chambon P (1978) Prcp Natl Acad Sci USA 75: 4853
Brehme KS (1939) Genetics 24: 131
Breidenbach RW, Beevers H (1967) Biochem Biophys Res Comm 27: 462
Brenner S, Beckwith JR (1965) J Mol Biol 13: 629
Brenner S, Jacob F, Meselson M (1961) Nature 190: 576
Brenner S, Stretton AOW, Kaplan S (1965) Nature 206: 994
Brenner S, Barnett L, Crick FHC, Orgel A (1961) J Mol Biol 3: 121
Brenner S, Barnett L, Katz ER, Crick FHC (1967) Nature 213: 449
Brent R, Ptashne M (1985) Cell 43: 729
Breuer ME, Pavan CK (1955) Chromosoma 7: 371
Brewer GJ (1971) Am J Hum Genet 23: 92
Bridges BA (1972) Nature New Biol 240: 32
Bridges BA (1974) Mut Res 26: 335
Bridges CB (1913) J exp Zool 15: 587
Bridges CB (1914) Science 40: 107
Bridges CB (1916) Genetics 1: 1 and 107
Bridges CB (1917) Genetics 2: 445
Bridges CB (1919) J exp Zool 28: 337
Bridges CB (1922) Am Nat 56: 51
Bridges CB (1923) Aberrations in chromosoma materials. Scient Pap 2nd Int Congr Eugenics 1: 76
Bridges CB (1932) Z indukt Abstamm-u.VererbLehre 60: 207
Bridges CB (1932) Am Nat 66: 571
Bridges CB (1934) J Hered 25: 18
Bridges CB (1935) J Hered 26: 60
Bridges CB (1936) Am Nat 70: 41
Bridges CB (1937) Cytologia Fujii Jub Vol 745
Bridges CB, Morgan TH (1923) Publ Carneg Inst Nr. 327
Brieger F (1928) Z indukt Abstamm-u.VererbLehre 47: 1
Brieger F (1958) Populationsgenetik. In: Roemer T, Rudorf W (eds) "Handbuch der Pflanzenzüchtung" (2. Aufl) Parey Berlin, Bd I, 176
Brink RA (1932) Am Nat 66: 444
Brink RA (1958) Cold Spr Harb Symp 23: 379
Brink RA (1962) Quart Rev Biol 37: 1

Brinkley BR, Stubblefield E (1966) Chromosoma 19: 28
Brinkley BR, Stubblefield E (1970) Ultrastructure and interaction of the kinetochore and centriole in mitosis and meiosis. In: Prescott DM et al (eds) "Advances in Cell Biology" 1: 119
Brinton C (1972) Epiviruses. In: Raspe G (ed) Workshop on Mechanism and Prospects of Genetic Exchange. Adv in Biosciences 8: 75
Brinton C, Buzzell A, Lauffer MA (1954) Biochim Biophys Acta 15: 533
Brinton C, Gemski P, Carnaham J (1964) Proc Natl Acad Sci USA 52: 776
Britten RJ, Davidson (1969) Science 165: 349
Britten RJ, Kohne DE (1968) Science 161: 529
Britten RJ, Smith J (1970) Carn Inst Wash Yearbook 68: 378
Britten RJ, Davidson EH (1971) Quart Rev Biol 46: 111
Broach JR, Guarascio VR, Jaygram M (1982) Cell 29: 227
Brodsky VY, Urgvaeva IV (1977) Int Rev Cytol 50: 275
Brody S, Yanofsky C (1963) Proc Natl Acad Sci USA 50: 9
Brody E, Abelson I (1985) Science 228: 963
Broker TR, Doerman AH (1975) Ann Rev Genet 9: 213
Brongniart A (1834) Ann Sci Nat 1: 65
Bronson MJ, Squires C, Yanofsky C (1973) Proc Natl Acad Sci USA 70: 2335
Brookes P, Druckrey H, Ehrenberg L, Lagerlöf B, Litwin J, Williams GM (1973) In: Ramel C (ed) "Evaluation of Genetic Risks of Environmental chemicals" Ambio Special Report No 3: 15
Brown DD, David IB (1968) Science 160: 272
Brown DD, Weber CS (1968) J Mol Biol 34: 681
Brown DD, Wensink PC, Jordan E (1972) J Mol Biol 63: 57
Brown R (1831) Trans Linn Soc, London (Bot)
Brown SW (1963) Chromosoma 14: 360
Brown SW (1964) Genetics 49: 797
Brown SW (1965) Science 151: 417
Brown SW, Cleveland C (1968) Chromosoma 24: 210
Brown WK, Wilson EO (1956) Systemat Zool 5: 49
Brues AM (1969) Science 164: 1130
Bütschli O (1876) Abh Senckenb Naturf Ges 10
Bulat SA, Peshekhonov VT, Chepurnaya OV, Zakharov IA (1985) Curr Genetics 9: 119
Buller AHR (1931) Proc 5th Int bot Congr pp 357
Buller AHR (1941) Bot Rev 7: 335
Bullough WS (1963) Nature 199: 859
Bullough WS (1965) Cancer Res 25: 1683
Bullough WS (1967) The Evolution of Differentiation. Acad Press, London, New York
Bunn CL, Wallace DC, Eisenstadt JM (1974) Proc Natl Acad Sci USA 71: 1681
Burgeff H (1915) Flora 108: 353
Burgess RR, Travers AA, Dunn JJ, Bautz EKF (1969) Nature 221: 43
Burgoyne PS (1982) Hum Genet 61: 85
Burnett IH (1956) New Phytol 55: 50
Burnham CR (1962) Discussions in Cytogenetics. Burgess, Minneapolis
Buther G, McConnell DJ (1980) Curr Genet 14: 405
Butler EG (1933) J exp Zool 65: 271
Buttin G (1960) J Mol Biol 7: 164

Cahn RD (1969) Factors affecting inheritance and expression of differentiation: Some methods of analysis. In: Beermann W et al. (eds) "Results and Problems in Cell Differentiation", Vol 1: 58
Caillet-Fauquet P, Defais M, Radman M (1977) J Mol Biol 117: 95

Cairns J (1963) Cold Spr Harb Symp 28: 43
Caldecott RS, Smith L (1952) Cytologia 17: 224
Calhoun DH, Hatfield GW (1973) Proc Natl Acad Sci USA 70: 757
Callan HG, Lloyd L (1960) Phil Trans Roy Soc B 243: 135
Callan HG, Tomlin SG (1950) Proc Roy Soc B 137: 367
Cameron JR, Coh EY, Davis RW (1979) Cell 16: 739
Camp WH, Gilly CL (1942) Brittonia, N Y 4: 323
Campbell AM (1962) Adv Genet 11: 101
Canellakis ES (1957) Biochim Biophys Acta 25: 217
Cannon WB (1929) Phys Rev 9: 399
Cantoni GL, Ishikura H, Richards HH, Tanaka K (1963) Cold Spr Harb Symp 28: 123
Capecchi MR (1967) Proc Natl Acad Sci USA 58: 1144
Capecchi MR, Gussin GN (1965) Science 149: 417
Carothers EE (1917) J Morph 28: 445
Carpenter ATC (1975) Proc Natl Acad Sci USA 72: 3186
Carrano AV, Gray JW, Langlois RG, Burghardt-Schultz KJ, Dithe MA v (1979) Proc Natl Acad Sci USA 76: 1382
Carraway M, Rewinsky C, Wu TH, Marinus MG (1988) Gene 74: 157
Carson HL (1955) Cold Spr Harb Symp 20: 276
Carson HL (1957) The species as a field for gene recombination. In: Mayr E (ed) "The species problem", Am Ass Adv of Sci Publ 50
Carson HL (1959) Cold Spr Harb Symp 24: 87
Carter TC, Robertson A (1952) Proc Roy Soc B 139: 410
Castle W (1905) Pop Sci Mon 66: 193
Castle W (1906) Science 24: 275
Catcheside DG (1958) Proc Roy Soc B 148: 285
Caullery M (1913) Le Problems de la Sexualite. Flammarion, Paris
Cavalier-Smith T (1950) J Cell Sci 34: 247
Cech TR (1985) Int Rev Cytol 93: 3
Cech TR (1988) Gene 73: 259
Cech TR, Bass BL (1986) Ann Rev Biochem 55: 599
Cech TR, Zang AJ, Grambowski PJ (1981) Cell 27: 487
Cerda-Olmedo EP, Hanawalt PC, Guerola N (1968) J Mol Biol 33: 705
Chakrabarsky R, Weiss KM (1988) Proc Natl Acad Sci USA 85: 9119
Chakrabarty AM (1973) Proc Natl Acad Sci USA 70: 1641
Chambers R (1940) J Roy Soc Microsc 60: 113
Chambon P, Weill JD, Mandel P (1963) Biochem Biophys Res Comm 11: 39
Chambon P, Weill JD, Doly J, Strosser MT, Mandel P (1966) Biochem Biophys Res Comm 25: 1966
Champe SP, Benzer S (1962) Proc Natl Acad Sci USA 48: 532
Champoux JJ, Dulbecco R (1972) Proc Natl Acad Sci USA 69: 143
Chandra HS, Brown SW (1975) Nature 253: 165
Chang S, Ho D, McLaughlin JR, Chang SY (1984) Gene 29: 255
Changeux JP (1964) Brookh Symp Biol 17: 232
Charlier D, Severne Y, Zafaull AM, Glansdorff N (1983) Genetics 105: 469
Charpak M, Dedonder R (1965) Compt Rend Hebd Acad Sci, Paris, 260: 5638
Chase M, Doerman AH (1958) Genetics 43: 332
Chatton E (1920) Arch Zool Exp Gén 59: 1
Chatton E (1925) Ann Sci Natur Zool 8: 5
Chiarugi A (1933) Atti Soc Ital par il Prog della Scienze 3: 1
Childs G, Maxson R, Cohn RH, Kedes L (1981) Cell 23: 651
Choppin PW, Compans RW (1970) J Virol 12: 609

Chouinard LA (1960) J Cell Sci 6: 73
Chu EHY (1978) J Cell Physiol 95: 365
Chu EHY, Thuline HC, Norby DE (1964) Cytogenetics 3: 1
Chua NH, Schmidt GW (1979) J Cell Biol 81: 461
Chun EHY, Vaughan MH, Rich A (1963) J Mol Biol 7: 130
Church GM, Gilbert W (1984) Proc Natl Acad Sci USA 81: 1991
Church GM, Gilbert W (1980) In: Joseph DR (ed) "Mobilization and Reassembly of Genetic Information". Acad Press, New York: 379
Church GM, Slonimsky PP, Gilbert W (1979) Cell 18: 1 209
Ciliberto G, Castagnoli L, Melton DA, Cortese R (1982) Proc Natl Acad Sci USA 79: 1 195
Cizewski V, Soellner-Webb B (1983) Nucleic Acids Res 11: 7043
Clark AJ (1967) J Cell Physiol 70 (Suppl 1): 165
Clark AJ, Adelberg EA (1962) Ann Rev Microbiol 16: 289
Clark AJ, Margulies AD (1965) Proc Natl Acad Sci USA 53: 451
Clark MA, Shay JW (1982) Nature 295: 605
Clausen J, Hiesey WM (1958) Publ Carnegie Inst Wash: 615
Clausen J, Keck DD, Hiesey WM (1945) Publ Carnegie Inst Wash 564: pp 174
Clayton DA Vinograd J (1967) Nature 216: 652
Cleaver JE (1974) Adv Rad Biol 4: 1
Cleaver JE (1975) Methods Cancer Res 11: 123
Cleland RE (1922) Am J Bot 9: 391
Cleland RE (1957) Cytologia (Proc Int Genet Symp) 1956, 5
Cleland RE (1962) Adv Genet 11: 147
Cleveland IR (1947) Science 105: 287
Clusius C (1601) Rariorum plantarum historia. Antwerpen
Coats JH, Nester EN (1967) J Biol Chem 242: 4948
Coen ES, Dover GA (1983) Cell 33: 849
Coen D, Deutsch J, Netter P, Petrochilo E, Slonimski PP (1970) Mitochrondrial Genetics. I Methodology and Phenomenology. In: Miller PL (ed) "Control of Organelle Development", Cambridge Univ Press, Cambridge: 449
Coetzes JN, Datta N, Hedges RW (1972) J Gen Microbiol 72: 543
Cohen D (1983) Proc Natl Acad Sci USA 80: 6289
Cohen SN (1976) Nature 263: 731
Cohen SN, Chang ACY (1974) Mol Gen Genet 134: 133
Cohen SN, Boyer HW, Helling RB (1973) Proc Natl Acad Sci USA 70: 3240
Cohen SS, Barner HD (1954) Proc Natl Acad Sci USA 40: 885
Cole RD (1972) zit nach Elgin SCR, Bonner J: "Isolated chromatin in the study of gene expression". In: Pollak IK, Wilson JL (eds) "The Biochemistry of Gene Expression in Higher Organisms"
Coleclough C, Erlitz EL (1985) Gene 34: 305
Collins GN, Kempton JH (1916) J Hered 7: 106
Collins J, Hohn B (1978) Proc Natl Acad Sci USA 75: 4242
Comings DE (1972) Exp Cell Res 70: 259
Cone CD (1969) J Theoret Biol 22: 365
Contents S, Dubnau D (1979) Plasmid 2: 555
Conway TW, Lipmann F (1964) Proc Natl Acad Sci USA 52: 1462
Cook SA (1965) Reproduction, Heredity, and Sexuality. MacMillan, London
Cooper KW (1941) Proc Natl Acad Sci USA 27: 109
Cooper KW (1944) Genetics 29: 537
Cooper KW (1946) Genetics 31: 181
Cooper KW (1959) Chromosoma 10: 535
Cooper PK, Hanawald PC (1972) Proc Natl Acad Sci USA 69: 1156

Correns C (1904) Ber Dtsch Bot Ges 22: 506
Correns C (1928) Bestimmung, Vererbung und Verteilung des Geschlechts bei höheren Pflanzen. Hdb Vererb Wiss 2, Borntraeger, Berlin
Corsaro CM, Migeon BR (1975) Exp Cell Res 95: 39
Craig-Cameron T, Jones GH (1970) Heredity 25: 223
Craig-Holmes AP, Shaw MW (1971) Science 174: 702
Crana MB, Thomas PT (1949) Heredity 3: 99
Crick FHC (1958) Symp Soc Exp Biol 12: 138
Crick FHC (1963) Progr Nucleic Acid Res Mol Biol 1: 163
Crick FHC (1966) J Mol Biol 19: 458
Crick FHC (1976) Proc Natl Acad Sci USA 73: 2639
Crick FHC (1979) Science 204: 264
Crick FHC, Orgel LE (1964) J Mol Biol 8: 161
Crick FHC, Klug A (1975) Nature 255: 530
Crick FHC, Barnett L, Brenner S, Watts-Tobin RJ (1961) Nature 192: 1227
Cronen JE, Navasimhan ML, Rawlings M (1988) Gene 70: 161
Crouse HV (1960) Genetics 45: 1429
Crow JF (1948) Genetics 33: 477
Crow JE (1958) Hum Biol 30: 1
Crow JF (1970) Genetic loads and the cost of natural selection. In: Kojima K (ed) Biomathematics, Vol 1 "Mathematical Topics in Population Genetics". Springer, Berlin, Heidelberg, New York
Crow JF (1973) Env Health Persp 6: 1
Crowe JK (1964) Heredity 19: 435
Crowther JA (1924) Proc Roy Soc B 96: 207
Cupido M (1983) Mut Res 109: 1

Dahmus ME, Bonner J (1970) Fed Proc 29: 1255
D'Amato F (1948) Hereditas 24: 83
D'Amato F (1954) Caryologia 6: 160
Dangeard PA (1919) Compt Rend Acad Sci Paris 169: 1005
Dangeard PA (1920) Compt Rend Acad Sci Paris 170: 301
Dangeard PA (1947) Cytologie Végétale et Cytologie Générale, Lechevalier, Paris
Danna K, Nathans D (1971) Proc Natl Acad Sci USA 68: 2913
Danser BH (1929) Genetics 11: 399
Darlington CD (1926) J Genet 16: 237
Darlington CD (1928) J Genet 19: 213
Darlington CD (1929) J Genet 21: 207
Darlington CD (1929) John Innes Hort Inst Conf on Polyploidy, p 42
Darlington CD (1930) Proc Roy Soc B 107: 50
Darlington CD (1931) Proc II Int Bot Congr, p 189
Darlington CD (1931) Biol Rev 6: 221
Darlington CD (1931) Science 73: 561
Darlington CD (1932) Recent Advances in Cytology. Churchill, London
Darlington CD (1932) Am Nat. 66: 25
Darlington CD (1935) J Genet 31: 185
Darlington CD (1935) Proc Roy Soc B 118: 33
Darlington CD (1935) Proc Roy Soc B 118: 59
Darlington CD (1935) Proc Roy Soc B 118: 74
Darlington CD (1936) J Genet 33: 465
Darlington CD (1936) Proc Roy Soc B 121: 264
Darlington CD (1937) J Hered 28: 308

Darlington CD (1937) Recent Advances in Cytology. 2nd Ed Churchill, London
Darlington CD (1939) J Genet 37: 341
Darlington CD (1939) The Evolution of Genetic Systems. Univ Press, Cambridge
Darlington CD (1940) J Genet 39: 351
Darlington CD (1941) Ann Bot 5: 203
Darlington CD (1944) Nature 154: 164
Darlington CD (1957) Conference on Chromosomes. Willink, Zwolle, Wageningen
Darlington CD (1958) The Evolution of Genetic Systems. 2nd Ed Oliver & Boyd, Edingburgh
Darlington CD (1963) Chromosome Botany and the Origins of Cultivated Plants. 2nd ed Allen and Unwin, London
Darlington CD (1965) Cytology. Churchill, London
Darlington CD, Gairdner AE (1937) J Genet 35: 97
Darlington CD, Janaki-Ammal EK (1945) Chromosome Atlas of Cultivated Plants. Allen & Unwin, London
Darlington CD, LaCour LF (1938) Ann Bot 2: 615
Darlington CD, LaCour LF (1940) J Genet 40: 85
Darlington CD, Mather K (1932) Cytologia 4: 1
Darlington CD, Mather K (1949) The Elements of Genetics. Allen & Unwin, London
Darlington CD, Moffet AA (1930) J Genet 22: 129
Darlington CD, Thomas PF (1941) Proc Roy Soc B 130: 127
Darlington CD, Upcott MB (1939) Chromosoma 1: 23
Darlington CD, Upcott MB (1941) J Genet 41: 275
Darlington CD, Wylie AP (1953) Heredity 6 (Suppl Vol) 197
Darlington CD, Haldane JBS, Koller PC (1934) Nature 133: 417
Darlington PJ (1972) Proc Natl Acad Sci USA 69: 293
Darwin Ch (1858/59) Journ of the Proc of the Linnean Soc, Zoology 3: Nr. 9, 45
Darwin Ch (1865) J Linn Soc 8: 169
Darwin Ch (1876) The Effects of Cross and Self-fertilization in the Vegetable Kingdom. Murray, London
Darwin Ch (1877) The Different Forms of Flowers on Plants of the Same Species. Murray, London
Datta P, Gest H (1964) Nature 203: 1259
Datta N, Hedges RW (1971) Nature 234: 222
Datta AR, Randolph BW, Rosner JL (1983) Mol Gen Genet 189: 245
Davidson EH, Britton RJ (1971) Quart Rev Biol 46: 111
Davidson EH, Hough BR, Amenson CS, Britten RJ (1973) J Mol Biol 77: 1
Davidson WM, Smith DR (1954) Brit Med J 2: 6
Davis BD (1949) Proc Natl Acad Sci USA 35: 1
Davis BD (1950) Experientia 6: 41
Davis HM (1908) Ann Bot 22: 1
Davis RH, Woodward VW (1962) Genetics 47: 1075
Davis RW, Hyman RW (1971) J Mol Biol 62: 287
Davison J, Pilarski LM, Echols H (1969) Proc Natl Acad Sci USA 63: 168
Dawid IB, Long EO, DiNocera PP, Pardue ML (1981) Cell 25: 399
De Duve C (1957) Symp Soc Exp Biol 10: 50
De Duve C (1963) The Lysosome Concept. In: Reuck AVS, Cameron MP (eds) "Ciba Foundation Symposium on Lysosomes". Little, Brown & Comp, Boston, Mass: 1
De Duve C (1965) J Cell Biol 27: 25
De Duve C, Baudhuin P (1966) Physiol Rev 46: 323
De Duve C, Wattiaux E (1966) Ann Rev Physiol 28: 435
De Duve C, Pressman BC, Gianetto R, Wattiaux R, Appelmans F (1955) Biochem J 60: 604
Deflandre G (1934) Compt Rend Acad Sci Paris 198: 497

Degnen GE, Miller IL, Eisenstadt JM, Adelberg EA (1976) Proc Natl Acad Sci USA 73: 2838
Delage H (1899) Arch Zool Exp 7: 383
Delbrück M (1945) J Bac 50: 151
Delbrück M (1963) Spektrum 6: 74
Delbrück M, Stent GS (1957) On the mechanism of DNA replication. In: McElroy WD, Glass B "The Chemical Basis of Heredity". John Hopkins Press, Baltimore
Del Rey FJ, Donahne TF, Fink GR (1982) Proc Natl Acad Sci USA 79: 4138
Demarly Y (1979) Proc Conf Broadening Genet Base Crops, Wageningen 1978
Demerec M (1937) Genetics 22: 469
Demerec M (1946) Proc Natl Acad Sci USA 32: 36
Demerec M (1956) Terminology and nomenclature. In: "Genetic Studies with Bacteria" Publ Carneg Inst Wash 612: 1
Demerec M (1962) Proc Natl Acad Sci USA 48: 1696
Demerec M (1964) Proc Natl Acad Sci USA 51: 1057
Demerec M, Cahn E (1953) J Bact 65: 27
Demerec M, Hartman PE (1959) Ann Rev Microbiol 13: 377
Demerec M, Hoover ME (1936) J Hered 27: 207
Demerec M, Adelberg EA, Clark AJ, Hartman PE (1966) Genetics 54: 61
Dempster ER, Lerner IM (1950) Genetics 35: 212
Dermen H (1947) Am Bot 34: 32
Dermen H (1953) J Hered 44: 31
Dewey MM, Barr L (1962) Science 137: 670
Diamelidis T (1951) Portug Acta Biol (A) 3: 151
Diaz de la Espina SM, Franke WW, Krohne G, Trendelenburg MF, Grund C, Scheer U (1982) Eur J Cell Biol 27: 14 1
Diener TO (1971) Virology 45: 411
Dienstman SR, Holtzer H (1979) In: Reinert J, Holtzer H (eds) Cell Cycle and Cell Differentiation 7. Springer, Berlin, Heidelberg, New York
Dietz R (1958) Chromosoma 9: 359
Digby L (1912) Ann Bot 26: 357
Dilla M van, Steinmetz L, Davis D, Gray JW (1974) IEEF Trans Nucl Sci NS 21: 714
Dintzis HM, Borsook H, Vinograd J (1958) Microsomal Particles and Protein Synthesis. Roberts RB (ed), Pergamon NY
Dobzhansky T (1935) Phil Sci 2: 344
Dobzhansky T (1937) Genetics and the Origin of Species. (1st ed) Columbia Univ Press, New York
Dobzhansky T (1946) Genetics 31: 269
Dobzhansky T (1950) Am Nat 84: 401
Dobzhansky T (1951) Mendelian populations and their evolution. In: Dunn LC (ed) "Genetics in the 20th Century". MacMillan, New York: 573
Dobzhansky T (1951) Genetics and the Origin of Species (3rd ed) Columbia Univ Press, New York
Dobzhansky T (1952) Nature and origin of heterosis. In: *Gowen* JW (ed) "Heterosis". Iowa State College Press, Ames: 218
Dobzhansky T (1956) Am Nat 90: 337
Dobzhansky T (1962) Mankind Evolving, Yale Univ Press, New Haven
Dobzhansky T (1965) Genetic diversity and fitness. In: Proc XI Int Congr Genet, The Hague 3: 541
Dobzhansky T (1967) Science 155: 409
Dobzhansky T (1968) Evol Biol 2: 1
Dobzhansky T (1970) Genetics of the Evolution Process. Columbia Univ Press, New York, London

Dodge BO (1945) Mycologia 37: 629
Dodge ID (1964) Arch Mikrobiol 48: 66
Dohlman WG, Caron MG, Lefkowitz R (1987) Biochemistry 26: 2657
Donachie WD (1968) Nature 219: 1077
Doudney CO, Haas FL (1958) Proc Natl Acad Sci USA 44: 390
Dougherty EC (1957) J Protozool 4 (Suppl): 14
Dover G (1982) Nature 299: 111
Dover G, Coen E (1981) Nature 290: 731
Dover GA, Ford-Dolittle W (1980) Nature 288: 646
Dover GA, Tautz D (1986) Philos Trans Roy Soc Biol 312: 275
Doy CH, Gresshoff PM, Rolfe BG (1973) Proc Natl Acad Sci USA 70: 723
Drake JW, Allen EF (1968) Cold Spr Harb Symp 33: 339
Dressler D (1975) Ann Rev Microbiol 29: 525
Druery CT (1886) J Linn Soc Bot 21: 354
Dubinin NP (1948) J Gen Biol Moscow (Zhurn Obsc Biol) 9: 203
Dubinin NP, Sidorov BN (1935) Biol Zh 4: 555
Duguid JP, Anderson ES (1967) Nature 215: 89
Duguid JP, Smith IW, Dempster G, Edwards PN (1955) J Path Bact 70: 335
Dujardin F (1841) Histoire Naturelle des Zoophytes. Roret, Paris
Dujon B (1980) Cell 20: 185
Dulbecco R (1970) Nature 227: 802
Dunkel VC, Williams GM (1981) In: Proc Third Life Sciences Symp on Health Risk Analysis. Franklin Press, Chagrin, p 249
Dunn LC (1954) Proc 9th Int Congr Genet (Caryologia Suppl 6) 1: 156
Dunsmuir P, Brorein jr WJ, Simon MA, Rubin GM (1980) Cell 21: 575
Du Praw EJ (1970) DNA and Chromosomes. Rinehard & Winston, New York
Du Rietz (1930) Svensk Bot Tidskr 24: 333
Durrant A (1962) Heredity 17: 27
Dussoix D, Arber W (1962) J Mol Biol 5: 37
Dustin AP (1934) Bull Acad R Med Belg 14: 487
Dustin AP (1938) Compt Rend Ass Anat 33
Dustin AP (1947) Nature 159: 794
Dvorak J, McGuire PE (1981) Genetics 97: 391
Dyer A, Jong FK, Ratter JA (1970) Notes Roy Bot Garden, Edinburgh 30: 177

Earle (1952) J Nat Cancer Inst 12: 133
Earnshaw WC, Honda BH, Laskey RA (1980) Cell 21: 373
East EM, Park JB (1917) Genetics 2: 505
Eaves LJ (1976) Heredity 36: 205
Echols H (1963) J Bact 85: 262
Edelman GM (1976) Science 192: 218
Edgar RS, Lielausis I (1964) Genetics 49: 649
Edmond M, Abrams R (1960) J Biol Chem 235: 1142
Edström J-E (1968) Nature 220: 1196
Edwards MR, Stevens RW (1963) J Bact 86: 414
Egawa R, Hirota Y (1962) Jap J Genet 37: 66
Ege T, Ringertz NR (1974) Exp Cell Res 87: 378
Ehrenberg L, Brookes P, Druckrey H, Lagerlöf B, Litwin J, Williams G (1973) The relation of cancer induction and genetic damage. In: "Evaluation of Genetic Risks of Environmental Chemicals". AMBIO Special Report No 3, p 15, Roy Swed Acad Sci, Universitels Forlarget
Ehrenstein v G (1966) Cold Spr Harb Symp 31: 705
Ehrman L (1962) Quart Rev Biol 37: 279

Eicher EM, Washburn LL (1978) Proc Natl Acad Sci USA 75: 946
Eigen M (1971) Naturwiss 58: 465
Eisen G (1900) J Morph 17: 1
Eldredge N, Gould SJ (1972) Punctuated Equilibria: An Alternative to Phyletic Gradualism. In: Schopf TJM (Ed) "Models in Paleobiology". Freeman Copper & Co, San Francisco
Ellis J (1982) Nature 299: 678
Elton CH (1927) Animal Exology. Sidgewich & Jackson, London
Endrizzi JE, Kohel RJ (1966) Genetics 54: 535
Engelhardt P, Pusa K (1972) Nature New Biol 240: 163
Engler A, Prantl K (1897) Die natürlichen Pflanzenfamilien. Engelmann, Leipzig
Englesberg E (1961) J Bact 71: 996
Englesberg E, Ingraham L (1957) Proc Natl Acad Sci USA 43: 369
Enriques P (1922) Genetics 7: 583
Ephrussi-Taylor H (1951) Cold Spr Harb Symp 16: 445
Epps H, Gale EF (1942) Biochem J 36: 619
Epstein RH, Bolle A, Steinberg CM, Kellenberger E, Boy de la Tour E, Chevalley R, Edgar RS, Susman M, Denhard G, Leilausis A (1963) Cold Spr Harb Symp 28: 375
Epstein W, Beckwith JR (1968) Ann Rev Biochem 37: 411
Erikson RL, Fenwick ML, Franklin RM (1964) J Mol Biol 10: 519
Erlanger RV (1897) Biol Zbl 17: 745
Ernst-Schwarzenbach M (1939) Arch Klaus-Stift VererbForsch 14: 361
Errera L (1888) Bot Zbl 34: 395
Errera L, Gevaert G (1878) Bull Soc Bot Belg 17
Esposito MS (1968) Genetics 58: 507
Esser K (1956) Z Indukt Abstamm-u VererbLehre 87: 595
Esser K (1971) Mol Gen Genet 110: 86
Esser K, Kuenen R (1965) Genetik der Pilze. Springer Berlin, Heidelberg, New York
Estable C, Sotelo JR (1950) Inst Ciencias Biologicas Montevideo 1: 105

Faelen M, Toussaint A, Resibois A (1979) Mol Gen Genet 176: 191
Fagerlind F (1940) Hereditas 26: 1
Fagerlind F (1944) Hereditas 30: 590
Fahrig R (1977) Arch Toxicol 38: 87
Falconer DS (1960) An Introduction to Population Genetics. Oliver & Boyd London
Farmer JB, Digba L (1907) Ann Bot 21: 161
Farmer JB, Moore JES (1905) Quart J Micr Sci 48: 489
Farquhar MG, Palade GE (1963) J Cell Biol 17: 375
Fawcett DW (1961) Cilia and Flagella. In: Brachet J, Mirsky ME (eds) "The Cell", Vol 2, 217, Academic Press, New York
Fawcett DW 1965) J Histochem Cytochem 13: 75
Feagin IE, Abraham JM, Stuart K (1988) Cell 53: 413
Felix K (1960) Adv Prot Chem 15: 1
Fenner (1976) Virology 71: 371
Feodoroff S (1967) J Nat Cancer Inst 38: 607
Ferenczy L, Kevei F, Szegedi M, Franko A, Rojik I (1976) Experientia 32: 1156
Ferguson-Smith MA, Handmaker SD (1961) Lancet I: 638
Fernandez-Morán H (1963) Science 140: 381
Filipowicz W, Konovska M, Groß HJ, Shatkin AJ (1983) Nucleic Acids Res 11: 1405
Finch JT, Klug A (1976) Proc Natl Acad Sci USA 73: 1897
Finch JT, Lutter LC, Rhodes D, Brown RS, Rushton B, Levitt M, Klug A (1977) Nature 269: 29
Fincham JRS (1966) Genetic Complementation. Benjamin WA, New York

Finnegan P, Brown G (1986) Proc Natl Acad Sci USA 83: 5175
Fischer E (1939) Z Indukt Abstamm-u.VererbLehre 76: 47
Fisher RA (1928) Am Nat 62: 115
Fisher RA (1930) The Genetical Theory of Natural Selection. Clarendon Press, Oxford
Fisher RA, Immer FR, Tedin O (1932) Genetics 17: 107
Fitch WM (1977) Genetics 86: 623
Fitch WM, Markowitz E (1970) Biochem Genet 4: 579
Fitz-James PC (1960) J Biophys Biochem Cytol 8: 507
Flavell RB, Rimpau J, Smit DB (1977) Chromosoma 63: 205
Flemming W (1879) Beiträge zur Kenntnis der Zelle und ihrer Lebenserscheinungen. I.Arch Mikr Anat 16: 302; II. Arch Mikr Anat 18: 151; III. Arch Micr Anat 20: 1
Flemming W (1882) Zellsubstanz, Kern und Zellteilung. Vogel, Leipzig
Flemming W (1887) Arch Mikr Anat 29: 389
Flock JI (1983) Mol Gen Genet 189: 304
Focke WO (1881) Die Pflanzenmischlinge, ein Beitrag zur Biologie der Gewächse. Borntraeger, Berlin
Fogel S, Mortimer R, Lusnak K, Tavares F (1978) Cold Spr Harb Symp 43: 1 325
Földes J, Trautner TA (1964) Z VererbLehre 95: 57
Fol H (1877) Arch Sci Nat et Phys Genève 58
Fol H (1891) Arch Sci Phys Nat 25: 393
Fol H (1896) Lehrbuch der vergleichenden mikroskopischen Anatomie mit Einschluß der vergleichenden Histologie und Histogenie. Die Zelle. Engelmann, Leipzig
Ford EB (1940) Polymorphism and Taxonomy. In: Huxley JS (ed) "New Systematics", Clarendon Press, Oxford, 493
Fournier REK, Ruddle FH (1977) Proc Natl Acad Sci USA 74: 319
Fox AS, Yoon SB, Gelbart WM (1971) Proc Natl Acad Sci USA 68: 342
Franke WW, Scheer U (1978) "Structure of Eukaryotic chromosomes" Phil Transakt Roy Soc Lond, p 333
Franke WW, Scheer U, Trendelenburg M, Zentgraf H, Spring H (1977) Cold Spr Harb Symp 42: 755
Franklin NC, Luria SE (1961) Virology 15: 299
Fredericq P, Netz-Bareau M (1953) Compt Rend Soc Biol 147: 1653
Freese E (1959) Proc Natl Acad Sci USA 45: 622
Freese E, Rhaese HJ, Bautz-Freese E (1969) Jap J Genet 44, Suppl 1: 94
Frey-Wyssling A, Mühlethaler K (1965) Ultrastructural Plant Cytology. Elsevier, Amsterdam
Frydenberg D (1963) Hereditas 50: 89
Fujiwara Y, Tatsumi M (1976) Mut Res 37: 91
Fuller FB (1971) Proc Natl Acad Sci USA 68: 815
Fulton C (1971) Centrioles. In: Reinert J, Ursprung H (eds) "Origin and Continuity of Cell Organelles". Results and Problems in Cell Differentiation 2: 170, Springer, Berlin, Heidelberg, New York

Gabridge MG, Legator MS (1969) Proc Soc Exp Biol Med 130: 831
Gaertner FH, Cole KW (1977) Biochem Biophys Res Comm 75: 259
Gall JG (1961) J Biophys Biochem Cytol 10: 163
Gall JG, Pardue ML (1969) Proc Natl Acad Sci USA 63: 378
Galton F (1883) Inquires into Human Faculty and its Development. McMillan, London
Ganoza MC (1966) Cold Spr Harb Symp 31: 273
Garber RL, Kuroiwe A, Gehring WJ (1983) EMBO J 2: 2027
Garen A, Zinder ND (1955) Virology 1: 347
Garnier C (1897) Bibliogr Anat 5: 278
Garnjobst L, Wilson JF (1956) Proc Natl Acad Sci USA 42: 613

Garrod AE (1902) Lancet 2: 1616
Garrod AE (1908) Lancet 175: 2, 1, 73, 142, 214
Gasser M, Laemmli KK (1986) Cell 46: 521
Gates RR (1911) Ann Bot 25: 909
Gavaudan P (1943) Compt Rend Soc Biol Paris 137: 281
Gedda L, Brenci G (1973) Acta Genet Med Gemellol 22: 3
Geitler L (1939) Chromosoma 1: 1
Geitler L (1953) Endomitose und endomitotische Polyploidisierung. Protoplasmatologia 60, Springer, Wien
Gellert M, Mizuuchi K, O'dea MH, Nash HA (1976) Proc Natl Acad Sci USA 73: 3872
Generoso WM, Cain KT, Huff SW, Gosslee DG (1978) Chemical Mutagens Vol 5: 55
Georgiev GP (1961) Biokhimia 26: 1095
Georgiev GP, Mantieva VL (1962) Biochem Biophys Acta 61: 153
German J (1969) Birth Defects 5: 117
Giardini A (1901) Int Monatsschrift Anat Physiol 18: 418
Gibson J, Beale GH (1961) Genet Res 2: 82
Gierer A (1977) Curr Top Dev Biol 11: 17
Gilbert W (1978) Nature 271: 5645
Gilbert W, Dressler D (1968) Cold Spr Harb Symp 33: 473
Gillespie D, Gallo RC (1975) Science 188: 802
Gilmore RA (1967) Genetics 56: 641
Gilmour JSL, Gregor JW (1939) Nature 144: 333
Gilmour JSL, Heslop-Harrison J (1954) Genetica 27: 147
Gilpin RW, Patterson SK (1976) J Bact 125: 845
Ginsburg J (1954) Zoologica, N.Y. 39: 31
Glover DM, Hogness DS (1977) Cell 10: 167
Goebel K (1880) Bot Ztg 38: 545
Goldberg M (1979) Ph D thesis, Stanford University 1979
Goldberger RF (1974) Science 183: 810
Goldman RD, Lazarides E, Pollack R, Weber K (1975) Exp Cell Res 90: 333
Goldschmidt R (1915) Biol Zbl 35: 565
Goldschmidt R (1917) J Exp Zool 22: 593
Goldschmidt R (1917) Endocrinology 1: 433
Goldschmidt R (1920) Z Indukt Abstamm-u.VererbLehre 23: 1
Goldschmidt R (1929) Biol Zbl 49: 641
Goldschmidt R (1935) Z Indukt Abstamm-u.VererbLehre 69: 38
Goldschmidt R (1935) Z Indukt Abstamm-u.VererbLehre 69: 70
Goldschmidt R (1935) Biol Zbl 55: 534
Goldschmidt R (1937) Am Nat 71: 83
Goldschmidt R (1938) Physiological Genetics. McGraw Hill, New York
Goldschmidt R (1940) The Material Basis of Evolution. Yale Univ. Press, New Haven
Goldschmidt R (1945) J Morph 77: 71
Goldschmidt R (1949) Heterochromatic heredity. Proc VIII Int Congr Genet Stockholm (Suppl Vol Hereditas) 244
Goldschmidt R (1955) Theoretical Genetics. Calif Univ Press, Berkeley
Golgi C (1898) Arch Ital Biol 30: 60
Gonda DK, Radding CM (1983) Cell 34: 647
Gondos B, Bhiralens B (1970) Z Zellforschg Mikrosk Anat 111: 149
Gonzales A (1967) Genet Iberica 19: 1
Goodspeed TH, Avery P (1939) J Genet 38: 381
Goodwin GH, Sanden C, Johns EW (1973) Eur J Biochem 38: 14
Gordon JW, Ruddle FH (1981) Science 214: 1244

Gorini L (1971) Nature 234: 261
Gorini L (1974) In: Nomura M et al (eds) Ribosomes Cold Spr Harb Lab, New York, 791
Gorini L, Beckwith JR (1966) Ann Rev Microbiol 20: 401
Gorini L, Rosset R, Zimmermann RA (1967) Science 157: 1314
Grabowski PJ, Seiler SR, Sharp PA (1985) Cell 42: 345
Graia A (1925) Compt Rend Acad Sci, Paris 93: 1040
Grant V (1953) The plant species in theory and practice. In: MayrE (ed) "The Species Problem", Am Ass Adv of Sci Publ 50
Grant V (1963) The Origin of Adaptations. Columbia Univ Press, New York
Grant V (1964) The Architecture of the Germplasm. Wiley, New York
Grant V (1966) Am Nat 100: 99
Grant V (1971) Evolution 25: 144
Grant V (1979) Bot Gaz 140: 443
Grant V (1985) The Evolutionary Process. Columbia Univ Press, New York
Graur D, Shuali Y, Wen-Hsiung Li (1989) J Mol Evol 28: 279
Gray JW, Guillery RW (1963) Ztschr Zellforschg 59: 738
Gray JW, Carrano AV, Moore DH, Steinmetz LL, Minkler J, Mayall BH, Mendelsohn ML, van Dilla MA (1975) Clin Chem 21: 1258
Green MM (1959) Genetica 29: 1
Gregoire V (1905) Cellule 21: 297
Gregoire V (1907) Ann Soc Zool Belg 42: 267
Gregoire V (1907) Cellule 24: 369
Gregoire V (1932) Bull Acad Belg Cl Sci, V. Ser 17: 1435
Greider CW, Blackburn EH (1985) Cell 43: 405
Grell RF (1962) Genetics 47: 1737
Griffing B (1956) Austral J Biol Sci 9: 463
Griffith F (1928) J Hyg 27: 113
Griffith JD (1975) Science 187: 1202
Griffith JD, Kornberg A (1974) Virology 59: 139
Griffiths G, Simons K (1986) Science 234: 438
Grinnell A (1904) Auk 21: 364
Grobstein C (1959) Differentiation of Vertebrate Cells. In: Brachet J & Mirsky AE (eds) "The Cell". Academic Press, New York and London, p 437
Grodzickev T, Williams J, Sharp P, Sambrook J (1974) Cold Spr Harb Symp 39: 439
Groner Y, Pollack Y, Berissi H, Revel M (1972) Nature 239: 16
Grosjean H, Fiers W (1982) Gene 18: 199
Grossman L, Braun A, Feldberg R, Mahler I (1975) Ann Rev Biochem 44: 19
Groudine M et al (1981) Mol Cell Biol 1: 281
Grüneberg H (1938) Proc Roy Soc B 125: 123
Grüneberg H (1952) J Genet 51: 95
Grula JW, Hall TJ, Hunt JA, Giugni TD, Graham GJ, Davidson EH, Britten (1982) Evolution 36: 665
Grummt I, Maier U, Öhrlein A, Hassouna N, Bachellerie JP (1985) Cell 43: 801
Gruss P, Dhar R, Khoury G (1981) Proc Natl Acad Sci USA 78: 943
Guarante L (1984) Cell 36: 799
Guerola, N, Ingraham JL, Cerdá-Olmedo E (1971) Nature New Biol 230: 122
Guerry P, Falkow S, Datta N (1974) J Bact 199: 144
Guilliermond A (1910) Bull Sci Fr Belg 44
Gunning BES (1982) In: Lloyd CW (ed) The Cytoskeleton in Plant Growth and Development p 229 Acad Press, London, New York
Gurdon JB (1974) The Control of Gene Expression in Animal Development. Clarendon Press, Oxford

Gurdon JB, Lana CD, Woodland HR, Marbaix G (1971) Nature 233: 177
Gustafsson A (1935) Hereditas 21: 1
Gutherz S (1907) Arch Mikr Anat 69: 491
Gutz H (1967) Science 158: 796

Haapala OK, Nienstedt I (1976) Hereditas 84: 49
Haapala OK, Nokkala S (1982) Hereditas 96: 215
Haas D, Holloway BW (1977) Mol Gen Genet 158: 229
Hadorn E (1945) Arch Klaus-Stift VererbForsch, Ergänzungsbd zu Bd 20: 82
Hadorn E (1948) Symp Soc Exp Biol 2: 179
Hadorn E (1949) Arch Klaus-Stift VererbForsch 24: 105
Hadorn E (1955) Letalfaktoren in ihrer Bedeutung für Erbpathologie und Genphysiologie der Entwicklung. Thieme, Stuttgart
Hadorn E (1965) Brookh Symp Biol 18: 148
Hadorn E (1966) Develop Biol 13: 424
Haeckel E (1866) Generelle Morphologie der Organismen. Reimer, Berlin
Haeckel E (1874) Anthropogenie und Entwicklungsgeschichte des Menschen, Leipzig
Haeckel E (1891) Anthropogenie und Entwicklungsgeschichte des Menschen (2. Aufl), Leipzig
Haeckel E (1894) Systematische Phylogenie, Reimer, Berlin
Haecker V (1892) Ber Naturf Ges Freiburg in Br 6: 160
Haecker V (1895) Arch Mikr Anat 45: 200
Haecker V (1895) Arch Mikr Anat 46: 579
Haecker V (1897) Biol Zbl 17: 689
Haecker V (1910) Zeitschr Ind Abstamm-u.Vererbl 3: 181
Haecker V (1918) Entwicklungsgeschichtliche Eigenschaftsanalyse. (Phänogenetik), Fischer, Jena
Haecker V (1925) Bibl Genet Lpz 1: 93
Haefner K (1965) Intern J Rad Biol 9: 545
Hagemann R (1958) Z VererbLehre 89: 587
Hakansson A, Levan A (1957) Hereditas 43: 179
Haldane JBS (1919) J Genet 8: 299
Haldane JBS (1930) J Genet 22: 359
Haldane JBS (1942) New Paths in Genetics. Harper & Brothers, New York
Haldane JBS (1957) J Genet 55: 511
Hall JB, Sinsheimer RL (1963) J Mol Biol 6: 115
Hall K, Cruz P, Tinoco jr I, Jovin TM, van de Sande JH (1984) Nature 311: 584
Hamer DH (1977) Science 196: 220
Hamilton WD (1963) Am Nat 97: 354
Hanahan D, Lane D, Lipsich L, Wigler M, Botchan M (1980) Cell 21: 127
Hansen HN, SmithRE (1932) Phytopathology 22: 953
Hanstein J v. (1880) Bot Abh (Bonn) 4, H 2
Harada K, Kameda M, Suzuki M, Kakinuma Y, Mitsuhashi S (1961) Gunma J Med Sci 10: 201
Hardin G (1960) Science 131: 1292
Hardy GH (1908) Science 28: 49
Harm W (1963) Virology 19: 66
Harm W (1968) Mut Res 6: 25
Harris H (1970) Cell Fusion, Harvard Univ Press, Cambridge, Mass
Harris H (1963) Progr Nucleic Acid Res Mol Biol 2: 19
Harris P, James T (1952) Experientia 8: 384
Hartmann M (1904) Biol Zbl 24: 18

Hartmann M (1923) Über sexuelle Differenzierung und relative Sexualität. Studia Mendeliana Brünn 203.
Hartmann M (1939) Geschlecht und Geschlechtsbestimmung im Tier- und Pflanzenreich. De Gruyter, Berlin
Harven E de, Bernhard W (1956) Z Zellforsch 45: 378
Harvey EB (1917) J Morphol 28: 1
Haselkorn R, Fried VA (1964) Proc Natl Acad Sci USA 51: 308
Hashimoto H, Hirota Y (1966) J Bact 91: 51
Hastings PJ, Quash S-K, v. Borstel RC (1976) Nature 264: 719
Hawthorne DC, Mortimer RK (1963) Genetics 48: 617
Hawthorne EDC, Friis J (1964) Genetics 50: 829
Hayashi M, Hayashi MN, Spiegelman S (1964) Proc Natl Acad Sci USA 51: 351
Haynes RH, Eckardt F, Kunz BA (1984) Brit J Cancer Suppl VI: 81
Hedges RW, Jacob AE (1974) Mol Gen Genet 132: 31
Hedrick PW, Levin DA (1984) Am Nat 124: 789
Heffron F, So M, McCarty BJ (1978) Proc Natl Acad Sci USA 75: 6012
Heidenhain M (1894) Arch Mikr Anat 43: 423
Heilmeyer L (1950) Naturwiss 37: 58
Heitz E (1928) Pringheims Jb Wiss Bot 69: 762
Heitz E (1929) Ber Dtsch Bot Ges 47: 274
Heitz E (1931) Planta 12: 775
Heitz E (1935) Z Indukt Abstamm-u.VererbLehre 70: 402
Helmstetter C, Cooper S, Pierucci O, Revelas E (1968) Cold Spr Harb Symp 33: 809
Henikoff S, Keene MA, Fechtel K, Fristrom JW (1986) Cell 44: 33
Henking H (1891) Z Wiss Zool 51: 685
Hennig W, Huijser P, Vogt P, Jäckle H, Edstroem JE (1983) EMBO J 2: 1741
Hentschel CC (1982) Nature 295: 714
Heppel LA (1969) J Gen Physiol 54: 95
Hermann F (1891) Arch Mikr Anat
Herr W, Sturm RA, Clerc RG, Corcovan LM, Baltimore D, Sharp PA, Ingraham HA, Rosenfeld MG, Finney M, Ruvkun G, Horvitz HR (1988) Genes Dev 2: 1513
Herrlich P, Schweiger M (1971) Proc First Eur Biophys Congr Vol 1: 489, Wiener Med Akad
Hershey AA (1971) Carnegie Inst Wash Yearb 69: 717
Hershey AA, Chase M (1952) Cold Spr Harb Symp 16: 471
Hershey AA, Burgi E, Ingraham L (1963) Proc Natl Acad Sci USA 49: 748
Hershey AA, Roesel C, Chase M, Forman S (1951) Carnegie Inst Wash Yearb 50: 195
Herskowitz IH (1987) Nature 329: 219
Herskowitz IH, Muller HJ (1953) Genetics 38: 669
Hertwig O (1893) Die Zelle und die Gewebe. Fischer, Jena
Hertwig R (1902) Sitz Ber Bayer Akad Wiss 32: 57
Hertwig R (1903) Biol Zbl 23: 108
Heywood SM (1970) Proc Natl Acad Sci USA 67: 1782
Heywood SM, Kennedy DS, BesterAJ (1974) Proc Natl Acad Sci USA 71: 2428
Hicks JB, Herskowitz I (1977) Genetics 85: 373
Higgins NP, Kato K, Strauss B (1976) J Mol Biol 101: 417
Hildebrand F (1867) Die Geschlechter-Verteilung bei den Pflanzen und das Gesetz der vermiedenen und unvorteilhaften stetigen Selbstbefruchtung, Engelmann, Leipzig
Hill C, Holland IB (1967) J Bact 99: 677
Hill J (1967) Genetics 55: 735
Hiorth G (1948) Z Indukt Abstamm-u.VererbLehre 82: 1
Hiraizumi Y (1979) Genetics 93: 449
Hirsch HJ, Starlinger P, Brachet P (1972) Mol Gen Genet 119: 191

Hirsch-Kaufmann M, Schweiger M (1981) Mol Gen Genet 184: 17
Hoagland MB (1959) Brookh Symp in Biol 12: 40
Hoagland MB, Zamecnick PC, Stephenson ML (1957) Biochim Biophys Acta 24: 215
Hodge AJ, McCleyn JD, Mercer FV (1956) J Biophysic Biochem Cytol 2: 597
Höfler K (1957) Protoplasma 48: 167
Hofmeister W (1849) Die Entstehung des Embryo der Phanerogamen. Fr Hofmeister, Leipzig
Hofmeister W (1851) Vergleichende Untersuchungen über die Keimung, Entwicklung und Befruchtung der höheren Kryptogamen und über die Befruchtung der Koniferen, Engelmann, Leipzig
Hofstetter H, Kressmann A, Birnstiel MC (1982) Cell 24: 573
Holliday R (1963) The genetic effect of mitomycin C. In: *Geerts* SJ (ed): "Genetics Today", Proc 11th Int Congr Genet 1: 13, Pergamon Press, Oxford, London, New York, Paris
Holliday R (1968) Genetic Control of Recombination in Fungi. In: Peacock WJ, Brock RD (eds) Replication and Recombination of Genetic Material. Austral Acad Sci, Canberra p 157
Holliday R (1964) Genet Res 5: 282
Hollis GF, Hieter PA, McBride OW, Swan D, Leder P (1982) Nature 296: 321
Holmquist GP, Dancis B (1979) Proc Natl Acad Sci USA 76: 4566
Hong J-S, Ames BN (1971) Proc Natl Acad Sci USA 68: 3158
Hood L (1972) Fed Proc 31: 177
Hood L, Campbell JH, Elgin SCR (1975) Ann Rev Genet 9: 305
Hooke R (1665) Micrographia, or some physiological descriptions on minute bodies by magnifying glasses. Facsimile ed publ by Gunther RT in: "Early Science in Oxford" XIII: The Life and Works of Hooke R, Oxford 1938
Hopfield JJ (1974) Proc Natl Acad Sci USA 71: 4135
Hotta Y, Bassel A (1965) Proc Natl Acad Sci USA 53: 356
Hotta Y, Stern H (1970) J Cell Biol 47: 91
Hotta Y, Stern H (1971) Nature 234: 83
Hotta Y, Stern H (1971) J Mol Biol 55: 337
Hotta Y, Stern H (1974) Chromosoma 46: 279
Hotta Y, Stern H (1981) Cell 27: 309
Hotta Y, Stern H (1984) Chromosoma 89: 127
Hotta Y, Tabata S, Stern H (1984) Chromosoma 90: 243
Hotta Y, Tabata S, Stern H (1985) Cell 40: 785
Howard A, Pelc SR (1953) Heredity 6 (Suppl): 261
Howard-Flanders P, Boyce RP (1966) Radiat Res (Suppl) 6: 156
Howell WH (1891) J Morph 4: 97
Howell WM, Hsu TC (1979) Chromosoma 73: 61
Hsu TC (1957) Numerical variation of chromosomes in higher animals. In: Rudnick D (ed) "Developmental Cytology", Ronald Press, New York, 47
Hsu TC, Moorhead PS (1956) Ann N Y Acad Sci 63: 1083
Huang WM, Ao SZ, Casjens S, Orlandi R, Zeikus R, Weiss R, Winge D, Fang M (1988) Science 239: 1005
Huber HE, Iida S, Arber W, Bickle TA (1985) Proc Natl Sci USA 82: 3776
Huberman JA, Riggs AD (1968) J Mol Biol 32: 327
Huberman JA, Kornberg A, Alberts BM (1971) J Mol Biol 62: 39
Huebner RI, Todaro GI (1969) Proc Natl Acad Sci USA 64: 1087
Hughes DT (1968) Nature 217: 518
Hughes-Schrader S, Ris H (1941) J exp Zool 87: 429
Hull FH (1946) Genetics 31: 219
Hurwitz J, Becker A, Gefter ML, Gold M (1967) J Cell Comp Physiol 70: 181
Huskins CL (1931) J Genet 25: 113

Huskins CL (1947) Am Nat 81: 401
Hutchinson GE (1958) Cold Spr Harb Symp 22: 415
Hutchinson CA, Philips S, Edgell MK, Gillam S, Jahnke P, Smith M (1978) J Biol Chem 253: 6551
Hutt FB (1969) Genetic aspects of infertility. In: Bernischke K "Comparative Mammalian Cytogenetics" p. 146
Huxley JS (1932) Problems of Relative Growth. Methuen, London
Huxley JS (1935) Biol Rev 10: 427
Huxley JS (1939) Bijdr Dierk 27: 491
Huxley JS (1940) Towards the new systematics. In: Huxley JS (ed) "New Systematics". Clarendon Press, Oxford
Huxley JS (1942) Evolution, the Modern Synthesis. Allen & Unwin, London
Huxley JS (1955) Heredity 9: 1
Huxley JS (1955) Proc Roy Soc B 144: 215
Huxley JS (1957) Nature 180: 454
Huxley JS (1958) Univ Arsskr Uppsala 21
Hynes RO (1976) Biochem Biophys Acta 458: 73
Hynes RO (1987) Cell 48: 549

Iino T, Katsubabe K (1983) In: Mizokuchi K, Watanabe I, Watson JD (eds) Nucleic Acids Research, Acad Press, New York p. 395
Ilyin YV, Chmeliaus Kaite VG, Ananiev EV, Georgiev GP (1980) Chromosoma 81: 27
Imai HT, Maruyama T, Gojobors T, Inone Y, Crozier RH (1986) Am Nat 128: 900
Imai Y (1936) J Genet 33: 169
Imai Y (1936) Genetics 21: 752
Inge-Vechtomov SG, Andrianova VM (1970) Genetika 6: 111 (Russian)
Inman RB (1966) J Mol Biol 18: 464
Isaacs A, Lindemann J (1957) Proc Roy Soc B 147: 258
Iseki S, Sakai T (1953) Proc Jap Acad 29: 121
Ishihama A (1988) Trends in Genetics 4: 282
Ishikawa H, Bischoff R, Holtzer H (1968) J Cell Biol 38: 538
Ishitsuka H, Kaji A (1970) Proc Natl Acad Sci USA 66: 168
Ising G, Block K (1981) Cold Spr Harb Symp 45: 527
Ito J, Crawford IP (1965) Genetics 52: 1303
Itoh T, Tomizawa J (1980) Proc Natl Acad Sci USA 77: 2450
Ivanov MA (1938) Genetika, Leningrad 20: 295
Ivanovsky D (1892) Bull Acad Sci St Petersburg 3: 67
Iyengar NK (1945) Indian J Genet 5: 32
Izant JG, Weintraub H (1985) Science 229: 345

Jackson DA, McCready SJ, Cook PR (1981) Nature 292: 552
Jackson EN, Yanovsky C (1973) J Mol Biol 76: 89
Jackson EN, Yanovsky C (1974) J Bact 117: 502
Jacob F, Brenner S (1963) Compt Rend Acad Sci, Paris 256: 298
Jacob F, Monod J (1959) Compt Rend Acad Sci, Paris 249: 1282
Jacob F, Monod J (1961) J Mol Biol 3: 318
Jacob F, Monod J (1961) Cold Spr Harb Symp 26: 193
Jacob F, Wollman EL (1956) Ann Inst Pasteur 91: 486
Jacob F, Wollman EL (1958) Compt Rend Acad Sci, Paris 247: 154
Jacob F, Brenner S, Cuzin F (1964) Cold Spr Harb Symp 28: 329
Jacob F, Schaeffer P, Wollman EL (1960) Symp Soc Gen Microbiol 10: 67
Jacob F, Ullman A, Monod J (1965) J Mol Biol 13: 704

Jacob F, Lwoff A, Simonovitch L, Wollman EL (1953) Ann Inst Pasteur 84: 222
Jacob F, Perrin D, Sanchez C, Monod J (1960) Compt Rend Acad Sci Paris 250: 1727
Jacq C, Miller JR, Brownlee GG (1977) Cell 13: 109
Jagger J, Stafford RS (1965) B Biophys J 5: 75
James AP, Werner MW (1966) Rad Res 29: 523
Janssens FA (1905) Cellule 22: 377
Janssens FA (1909) Cellule 25: 389
Janssens FA (1924) Cellule 34: 135
Jaquard A (1974) The Genetic Structure of Populations. Springer Berlin, Heidelberg, New York.
Jaworski CJ, Piatigorsky J (1989) Nature 337: 752
Jeffreys AJ, Flavell RB (1977) Cell 12: 1027
Jeffreys AJ, Wilson V, Thein SL (1985) Nature 314: 67
Jeggo P, Defais M, Samson L, Schendel P (1977) Mol Gen Genet 157: 1
Jinks JL (1958) Proc Roy Soc B 148: 314
Joachimiak A (1983) Proc Natl Acad Sci USA 80: 668
Johannsen W (1903) Über Erblichkeit in Populationen und reinen Linien, Fischer, Jena
Johannsen W (1909) Elemente der exakten Erblichkeitslehre. Fischer, Jena
Johannsen W (1926) Elemente der exakten Erblichkeitslehre. (3. Aufl) Fischer, Jena
John B, Lewis KR (1965) The Meiotic System. Protoplasmatologia 6 (F 1), Springer, Berlin
John HA, Birnstiel ML, Jones KW (1969) Nature 223: 582
Johns EW (1964) Biochem J 92: 55
Johnson RT, Rao PN (1970) Nature 226: 717
Johnson SM, Bangham AD, Hill MW, Korn ED (1973) Biochem Biophys Acta 233: 820
Jokelainen PT (1967) J Ultrastr Res 19: 19
Jollos V (1921) Arch Protistenk 43: 1
Jollos V (1939) Grundbegriffe der Vererbungslehre, insbesondere Mutation, Dauermodifikation, Modifikation. In: Baur E, Hartmann M(eds) "Handbuch der Vererbungswissenschaft" Bd IV Borntraeger, Berlin
Jolly S (1905) Compt Rend Soc Biol Paris 58: 593
Jordan E, Saedler H, Starlinger P (1968) Mol Gen Genet 102: 353
Jorgensen CA (1928) J Genet 19: 133
Jorgensen CA, Crane MB (1927) J Genet 18: 247
Jukes TH (1963) Adv Biol Med Phys 9: 1
Jukes TH (1973) Nature 246: 22

Kada T (1974) Ann Rep Natl Inst Genet 24: 43
Kada T et al (1972) Mut Res 16: 165
Kada T, Inoue T, Namiki M (1981) In: Kletowski EJ (ed). "Environmental Desmutagens and Antimutagens" Praeger, pp 134
Kadonaga JT, Jones KA, Tjian R (1986) Trends Biochem 11: 20
Kahn ME, Helinski DR (1978) Proc Natl Acad Sci USA 75: 2200
Kahn ME, Borany F, Smith HO (1983) Proc Natl Acad Sci USA 80: 6927
Kaiser D, Masuda T (1973) Proc Natl Acad Sci USA 70: 260
Kalderon D, Oostra BA, Ely BK, Smith AE (1982) Nucleic Acids Res 10: 5161
Kalmus H, Maynard Smith S (1966) Am Nat 100: 619
Kamp D, Chow LT, Broker TR, Kwoh D, Zipser D, Krahmann R (1978) Cold Spr Harb Symp 43: 1159
Kano Y, Imamoto F (1975) Genetics 80: 651
Karpechenko GD (1935) Die Theorie der Kreuzung entfernter Formen. In: Vavilov NI (ed) "Theoretische Grundlagen der Pflanzenzüchtung" Bd I, Kap 14
Kasai T (1974) Nature 249: 523

Katayama Y (1934) Cytologia 5: 235
Katayama Y (1935) Jap J Bot 7: 349
Kettermann G (1938) Ztschr Indukt Abstamm-u.VererbLehre 74: 354
Kattermann G (1939) Chromosoma 1: 284
Katz L, Engelsberg E (1971) J Bact 107: 34
Kaufmann BP (1925) Am Nat 59: 190
Kaufmann BP (1933) Proc Natl Acad Sci USA 19: 830
Kellenberger E (1962) The study of nature and artificial DNA plasms by thin sections. In: "The Interpretation of Ultrastructure", 1. Symp Int Soc Cell Biol, Acad Press, London, New York, p 233
Kellenberger E, Huber L (1953) Experientia 9: 289
Kelly RB, Cozzarelli NR, Deutscher MP, Lehmann IR, Kornberg A (1970) J Biol Chem 245: 39
Kelner A (1949) J Bact 58: 511
Keosian J (1965) The Origin of Life. Chapman & Hall, London
Kerner A (1881) Pflanzenleben. Bd I. Bibl Inst, Leipzig, Wien
Khoury G, Gruss P (1983) Cell 33: 313
Khush GS (1973) Cytogenetics of Aneuploids. Academic Press, New York, London
Kidwell MG, Kidwell JF (1976) Genetics 84: 333
Kiellander CL (1942) Svensk bot Tidskr 36: 200
Kielmeyer KF (1793) Über die Verhältnisse der organischen Kräfte. Stuttgart
Kihara H (1930) Cytologia 1: 263
Kihara H, Ono T (1926) Z Zellforsch 4: 475
Kimber F, Riley R (1963) Bot Rev 29: 480
Kimber F, Sears ER (1968) Int Wheat Genet Symp Canberra, 468
Kimura M (1956) Evolution 10: 278
Kimura M (1960) J Genet 57: 21
Kimura M (1965) Genetics 52: 875
Kimura M (1968) Nature 217: 624
King JL, Jukes TH (1969) Science 164: 788
King JP, Ohta T (1975) Genetics 79: 681
King RC (1970) Int Rev Cytol 28: 125
Kingman JFC (1982) Stochastic Processes and their Applications 13: 235
Kirkpatrick FH (1979) BioSystems 11: 93
Kiryanov G, Manamshjan TA, Polyakov VY, Fais D, Chentsov YS (1976) FEBS Letters 67: 323
Kit S (1961) J Mol Biol 3: 711
Klysik J, Stirdivant SM, Larson JE, Hart PA, Wells RD (1981) Nature 290: 672
Knapp G, Beckmann JS, Johnson PF, Fuhrman SA, Abelson J (1978) Cell 14: 221
Kniep H (1920) Verh Physik-Med Ges Würzburg 46: 1
Kniep H (1928) Die Sexualität der niederen Pflanzen. Fischer, Jena
Koch RE, Drake JW (1970) Genetics 65: 379
Kohler RE, Ron EZ, Davis BD (1968) J Mol Biol 36: 71
Kohn KW, Spears CL, Doty P (1966) J Mol Biol 19: 266
Kojima K-I, Schaffer HE (1967) Evolution 21: 518
Koltsov NK (1929) Sov Med Enzycl
Kopecko DJ, Brevert J, Cohen SN (1976) J Mol Biol 108: 333
Koprowski H (1964) Criteria for the proof of virogeny in mammalian cells. In: Krooth RS (Ed) "Somatic Cell Genetics", Univ of Michigan Press, Ann Arbor, 274
Kornberg RD (1974) Science 184: 868
Korschelt E, Heider K (1903) Eireifung, Samenreifung und Befruchtung. Lehrbuch d. vergl. Entw.-Gesch. Allg.Teil, Fischer, Jena

Kossel A (1884) Z physiol Chemie 8: 511
Kostoff D (1939) Genetics 21: 285
Kourilsky P, Gros F (1974) In "Regulation of gene expression in cultured cells"; Fogarty Int Conf Center Proc Nr 22, U.S. Government Printing Office, Washington D.C.
Kozak M (1978) Cell 15: 1109
Kozinski AW, Kozinski PB, James R (1967) J Virol 1: 758
Kressmann A, Hofstetter H, DiCapua E, Grosschedl R, Birnstiel ML (1979) Nucleic Acids Res 7: 1749
Krichevskaya AA, Georgiev GP (1969) Biochem Biophys Acta 164: 619
Krooth RS (1969) Gene action in human diploid cell strains. In: Bernischke K (ed) "Comparative Mammalian Cytogenetics" p 154
Krstic RV (1984) Ultrastructure of the Mammalian Cell. Springer Berlin, Heidelberg, New York
Kühn A (1927) Nachr Ges Wiss Göttingen, Math.-Phys. Klasse, 407
Kühne W (1878) Unters Physiol Inst Heidelberg 1: 291
Kuhn B, Abdel-Monem M, Hoffmann-Berling H (1978) Cold Spr Harb Symp 43: 63
Kumar CC, Novick RP (1985) Proc Natl Acad Sci USA 82: 638
Kupffer C (1875) Schr naturw Ver Schl.-Holst 1
Kurland CG (1960) J Mol Biol 2: 83
Küster E (1935) Die Pflanzenzelle. Fischer, Jena
Kuusela P, Ruoslahti E, Engvall E, Vahen A (1976) Immunochem 13: 639
Kyryanov GI, Manamshjan TA, Polyakov VY, Fais D, Chentsov YS (1976) FEBS Letters 67: 323

La Bella FS, Krass ME (1968) Nature 220: 1141
La Cour LF, Rutishauser A (1953) Nature 172: 501
Lacadena JL (1968) Genet Iber 20: 195
Lacy PE (1961) Am J Med 31: 851
Lagerkvist U (1978) Proc Natl Acad Sci USA 75: 1759
Lam ST, Stahl MM, Milin KD, Stahl FW (1979) Genetics 77: 425
Lamb BC (1975) Mol Gen Genet 137: 305
Landauer W (1958) Am Nat 92: 201
Lande R (1979) Evolution 33: 402
Landschutz HW, Johnson PF, McKnight SL (1988) Science 240: 1759
Lange M (1952) Dansk bot Ark 14: 6
Langlet OF (1927) Svensk bot Tidskr 21: 1
Lark KG (1972) Nature New Biol 240: 237
Larkin PJ, Scowcroft WR (1981) Theor Appl Genet 60: 197
Laskey RA, Honda BM, Mills AD, Finch JT (1978) Nature 275: 416
Latarjet R, Gray LH (1954) Acta Radiol Stockh 41: 61
La Valette St George (1886) Arch Mikr Anat 15: 261
Lawrence WJ (1931) Cytologia 2: 352
Lazowska J, Jacq C, Slonimsky PP (1980) Cell 22: 333
Lea D, Catcheside DG (1942) I Genet 44: 216
Lebégue A (1952) Bull Soc bot Fr 99: 329
Leblond CP (1964) J Nat Cancer Inst 14: 119
Ledbetter MC, Porter KR (1963) J Cell Biol 19: 239
Lederberg J (1952) Physiol Rev 32: 403
Lederberg J (1954) Genetics 39: 978 (Abstr)
Lederberg J (1955) J Cell Comp Physiol (Suppl II) 45: 75
Lederberg J (1966) Am Nat 100: 519
Lederberg J (1966) Current Topics in Dev Biol 1: IX
Lederberg J, Cavalli LL, Lederberg EM (1952) Genetics 37: 720

Lee CS, Thomas CA (1973) J Mol Biol 77: 25
Lee CS, Davis RW, Davidson N (1970) J Mol Biol 48: 1
Lee-Huang S, Ochoa S (1972) Biochem Biophys Res Com 49: 371
Lefevre G (1955) Genetics 40: 374
Lehmann AR (1972) J Mol Biol 66: 319
Lehmann FE (1947) Rev Suisse Zool 54: 246
Leighty CE, Taylor IW (1924) J Agr Res 28: 567
Lenz F (1938) Erbarzt 5: 83
Lerman LS (1961) J Mol Biol 3: 18
Lerner JM (1950) Population Genetics and Animal Improvement. Univ Press, Cambridge
Lerner JM (1954) Caryologia (Suppl I) 6: 124
Lerner JM (1954) Genetic Homeostasis. Oliver & Boyd, London
Lerner JM (1958) The Genetic Basis of Selection. Wiley, New York
Lerner JM (1968) Heredity, Evolution and Society. Freeman & Comp., San Francisco
Lerner MR, Steitz JA (1981) Cell 25: 298
Leung J, Addison WR, Delaney AD, Mackay RM, Miller jr RC, Spiegelmann GB, Grighatti TA, Tener GM (1985) Gene 34: 207
Levan A (1937) Hereditas 23: 317
Levan A (1938) Hereditas 24: 471
Levan A (1939) Hereditas 25: 9
Levan, A (1942) Hereditas 28: 177
Levan A, Hauschka TS (1953) J Nat Cancer Inst 14: 1
Levan A, Müntzing A (1963) Portug Acta Biol Ser A 7: 1
Levan A, Tjio JH (1948) Hereditas 34: 453
Levan A, Fredga K, Sandberg AA (1964) Hereditas 52: 201
Levens PA (1909) Ber Ot Chem Ges 42: 2474
Levin, DH, Kyner D, Acs G (1973) J Biol Chem 248: 6416
Levinthal C (1959) Bacteriophage genetics. In: Burnet FM, Stanley WM (Ed) "The Viruses", Vol 2, 281, Academic Press, New York, London
Levitan M (1954) Am Nat 88: 419
Levitsky GA (1924) Materielle Grundlagen der Vererbung (russ) Kiew, Staatsverlag
Levitsky GA (1925) Planta 1: 301
Levitsky GA (1931) Bull Appl Bot, St.Petersb. 27: 19
Levitsky GA (1931) Bull Appl Bot, St.Petersb. 27: 187
Levy JA (1973) Science 182: 1151
Levy JA (1974) Am J Clin Pathol 62: 258
Lewis D (1943) J Genet 45: 171
Lewis D (1947) Heredity 1: 85
Lewis D (1954) Adv Genet 6: 235
Lewis D (1954) Heredity 8: 333
Lewis EB (1948) Genetics 33: 113
Lewis EB (1950) Adv Genet 3: 75
Lewis EB (1951) Cold Spr Harb Symp 16: 159
Lewis EB (1955) Am Nat 89: 73
Lewis H (1966) Science 152: 167
Lewis KR, John B (1963) Chromosome Marker, Churchill, London
Lewis KR, John B (1964) The Matter of Mendelian Heredity. Churchill, London
Lewis KR, John B (1970) The Organisation of Heredity. E Arnold, London
Lewis WH (1931) Johns Hopk Hosp Bull 49: 17
Lewis WH (1939) Arch Exp Zellforsch 23: 1
Lewontin RC (1955) Evolution 9: 27
Li HW, Pao WK, Li CH (1945) Am J Bot 32: 92

Lichter P, Cremer T, Borden J, Manuelidis L, Ward DC (1988) Hum Genet 80: 224
Lieb M (1966) J Mol Biol 16: 149
Lifschytz E (1971) Mut Res 12: 35
Lilienfeld FA (1951) Cytologia Tokyo 16: 101
Lima-de-Faria A (1949) Hereditas 35: 77
Lima-de-Faria A (1949) Hereditas 35: 422
Lima-de-Faria A (1952) Hereditas 38: 246
Lima-de-Faria A (1952) Chromosoma 5: 1
Lima-de-Faria A (1954) Chromosoma 6: 330
Lima-de-Faria A (1956) Hereditas 42: 85
Lima-de-Faria A (1958) Int Rev Cytol 7: 123
Lima-de-Faria A, Daskaloff S, Nell AE (1973) Hereditas 73: 99
Lindahl T (1974) Proc Natl Acad Sci USA 71: 3649
Lindahl T (1979) Progr in Nucleid Acids Res Mol Biol 22: 135
Lindegren CC (1949) Proc 8th Int Congr Genet, Hereditas (Suppl) 338
Lindsley, DI, Sandler L (1972) Genetics 71: 157
Linn S, Arber W (1968) Proc Natl Acad Sci USA 59: 1300
Linn S, Kuhnlein U, Deutsch WA (1979) In: "DNA Repair Mechanisms" Hanawalt PC, Friedberg EC, Fox CF (eds) Academic Press, New York
Lissouba P, Rizet G (1960) Compt Rend Acad Sci, Paris 250: 3408
Lissouba P, Mousseau J, Rizet G, Rossignol JL (1962) Adv Genet 11: 343
Litardiére R de (1925) Cellule 35: 19
Livneh Z, Elad D, Sperling J (1979) Proc Natl Acad Sci USA 76: 1089
Ljungdahl H (1924) Svensk bot Tidskr 18: 279
Lodish HF, Zinder ND (1966) J Mol Biol 19: 333
Löve A, Löve D (1949) Portug Acta Biol Goldschmidt Vol 273
Loew E (1895) Einführung in die Blütenbiologie. Dümmler, Berlin
Longley AE (1927) Genetics 30: 100
Longley AE (1937) J Agric Res 54: 835
Lotsy JP (1904) Flora 93: 65
Lotsy JP (1916) Evolution by Means of Hybridization. Nijhoff, The Hague
Louarn J, Patte J, Louarn JM (1977) J Mol Biol 115: 295
Loveless A, Revell S (1949) Nature 164: 938
Low RL, Arai KI, Kornberg A (1981) Proc Natl Acad Sci USA 78: 1436
Lucas-Lenard J, Lipman F (1966) Proc Natl Acad Sci USA 55: 1562
Luchnik N, Porjadkova A (1977) Chromosoma 63: 11
Luck DJL, Reich E (1964) Proc Natl Acad Sci USA 52: 931
Ludwig W (1948) Darwins Zuchtwahllehre in moderner Fassung. Aufs Senckb Naturf Ges
Ludwig W (1954) Die Selektionstheorie. In: Heberer G (ed) "Die Evolution der Organismen" (2. Aufl), Fischer, Stuttgart, 662
Lueders, KK Kuff EL (1977) Cell 12: 963
Lund et al (1976) Cell 7: 165
Lundegardh H (1912) Arch Mikr Anat 80: 223
Lupski JR, Ruiz AA, Godson N (1984) Mol Gen Genet 195: 391
Luria SE (1945) Genetics 30: 84
Luria SE (1947) Proc Natl Acad Sci USA 33: 253
Luria SE (1950) Science 111: 507
Luria SE (1963) Rec Progr Microbiol 8: 604
Luria SE, Delbrück M (1942) Arch Biochem 1: 207
Luria SE, Human MC (1952) J Bact 64: 557
Lush JL (1941) Methods of Measuring the Heritability of Individual Differences among Farm Animals. Proc of 7th Intern Congr Genetics 1939. J Genetics (Suppl Vol): 199

Lush JL (1945) Animal Breeding in Plants. 3rd ed. Iowa State College Press, Ames
Luyks P (1974) The organization of meiotic chromosomes. In "The Cell Nucleus" (Busch H, ed), Acad, Press, New York, London, Vol 2, p 163
Lwoff A (1949) Edition du Centre Nat de la Rech Sci Paris, 7
Lwoff A (1953) Bact Rev 17: 269
Lwoff A (1953) Ann Inst Pasteur 84: 225
Lwoff A, Gutmann A (1949) Compt Rend Acad Sci Paris 229/679
Lwoff A, Gutmann A (1950) Ann Inst Pasteur 78: 711
Lwoff A, Anderson TF, Jacob F (1959) Ann Inst Pasteur 97: 281
Lyon MF (1961) Nature 190: 372
Lyon MF (1972) Biol Rev 47: 1

Maas WK, Clark AJ (1964) J Mol Biol 8: 365
Mac Arthur RH, Wilson ED (1967) "The Theory of Island Biogeography" Princeton Univ Press, Princeton, NJ
Magasanik B (1961) Cold Spr Harb Symp 26: 249
Magni GE (1964) J Cell Comp Physiol (Suppl 1) 64: 165
Magnus P v (1954) Adv Virus-Res 2: 59
Maio JJ (1971) J Mol Biol 56: 579
Maire R (1902) Bull Soc Mycol Fr 18: 1
Malamy MH, Fiandt M, Szybalski W (1972) Mol Gen Genet 119: 207
Malecot G (1948) Les Mathematique de L'hérédité. Masson & Cie, Paris
Malheiros-Gardé N, Gardé A (1950) Genet iber 2: 257
Mangel WF, Champerlin MJ (1974) J Biol Chem 249: 2995
Mansfeld R (1950) Züchter 20: 8
Mantei N, Boll W, Weissmann O (1979) Nature 281: 5726
Marcker KA, Sanger F (1964) J Mol Biol 8: 835
Margolin P (1965) Science 147: 1456
Margolin P, Bauerle RH (1966) Cold Spr Harb Symp 31: 311
Margulis L (1970) Origin of Eukaryotic Cell. Yale Univ Press, New Haven Conn
Marini JC, Levéne SD, Crothers DM, Englund PT (1982) Proc Natl Acad Sci USA 79: 7664
Mark J (1967) Hereditas 57: 1
Markman, Cautoni (1976) Biochemistry 5: 2246
Markower M, Bevan EA (1963) Proc Int Congr Genet XI, 1: 202
Marlor RL, Parkhurst SM, Corces VG (1986) Mol Cell Biol 6: 1129
Marquardt H (1952) Ber Dtsch Bot Ges 65: 197
Martin GM, Spargue CA (1973) Exp Mol Pathol 18: 125
Mason HS, North JC, Vaneste M (1965) Fed Proc 24: 1172
Mather K (1932) J Genet 26: 129
Mather K (1933) Am Nat 67: 476
Mather K (1933) J Genet 27: 243
Mather K (1936) J Genet 32: 287
Mather K (1936) J Genet 33: 207
Mather K (1936) Proc Roy Soc B 120: 208
Mather K (1939) Genetics 24: 413
Mather K (1941) J Genet 41: 159
Mather K (1943) Biol Rev 18: 32
Mather K (1948) Symp Soc Exp Biol 2: 196
Mather K (1953) Heredity 7: 297
Matsui S, Weinfeld H, Sandberg AA (1972) J Nat Cancer Inst 49: 1621
Matsuura H (1937) Cytologia Tokyo 8: 142
Matthey R (1945) Experientia 1: 50 and 78

Matthey R (1949) Les Chromosomes des Vertébrés. F Rouge, Lausanne
Maxam AM, Gilbert W (1977) Proc Natl Acad Sci USA 74: 560
Maxwell CR, Kamper CS, Rabinovitz M (1971) J Mol Biol 58: 317
Mayfield JE, Ellison JR (1975) Chromosoma 52: 37
Maynard Smith J (1976) Am Nat 110: 331
Mayr E (1940) Am Nat 74: 249
Mayr E (1942) Systematics and the Origin of Species. Columbia Univ Press, New York
Mayr E (1948) Adv Genet 2: 205
Mary E (1954) Change of Genetic Environment and Evolution. In: Huxley J, Hardy AC, Ford EB (eds), Allen & Unwin, London
Mayr E (1963) Animal Species and Evolution. Belknap Press of Havard Univ Press, Cambridge, Mass
Mayzel W (1875) Zbl med Wiss 13: 849
McAthey, P, Kilbey JB (1976) Biol Zbl 95: 415
McBride OW, Ozer HL (1973) Proc Natl Acad Sci USA 70: 1258
McCarthy BJ, Britten RJ, Roberts RB (1962) Biophys J 2: 83
McClintock B (1934) Z Zellforsch 21: 294
McClintock B (1945) Am J Bot 32: 671
McClintock B (1952) Cold Spr Harb Symp 16: 13
McClintock B (1978) Stadler Symp 10: 25
McClung CE (1900) Kans Univ Quart A 9: 73
McClung CE (1902) Biol Bull 3: 43
McClung CE (1902) Kansas Univ Sci Bull 1: 185
McClung CE (1905) Biol Bull Wood's Hole 9: 304
McCready SJ, Cox BS (1973) Mol Gen Genet 124: 305
McDaniel RG, Sarkissian JV (1966) Science 152: 1640
McElroy WD, Deluca M, Travis J (1967) Science 157: 150
McGhee, JD, Felsenfeld G (1980) Ann Rev Biochem 49: 1115
McKnight SL, Kingsburg R (1982) Science 217: 316
McKusick VA, Ruddle FH (1977) Science 196: 390
McLellan WL, Vogel HJ (1970) Proc Natl Acad Sci USA 67: 1703
McMorris et al (1973) Neurosci Res Progr Bull 11: 411
McPherson I, Montagnier L (1964) Virology 23: 291
Meacock PA, Cohen SN (1980) Cell 20: 529
Mecuse (1957) Protoplasmatologia II H 1c: 1
Mendel JG (1866) Versuche über Pflanzenhybriden. Verh Naturf Ver Brünn 4 (Abh)
Menke W (1961) Z Naturf. 16b: 334
Menninger JR (1977) Mech Aging Dev 6: 131
Menninger JR, Mulholland MC, Stirewalt WS (1970) Biochem Biophys Acta 217: 496
Mergeay M, Lejeune P, Sadonk A, Gerits J, Fabry L (1987) Mol Gen Genet 209: 61
Merrell DJ (1950) Evolution 4: 326
Mertz DB, Craig DM, Wade MJ, Boyer JF (1984) Evolution 38: 560
Meselson M, Yuan R (1968) Nature 217: 1110
Messing J (1983) In: Wu R, Grossman L, Moldave K (eds) "Methods in Enzymology" 101: 20
Metz CW (1916) J Exp Zool 21: 213
Metzenberg RL (1968) Arch Biochem Biophys 125: 532
Metzenberg RL (1972) Ann Rev Genet 6: 111
Meves F (1907) Anat Anz 31: 399
Meves F (1908) Arch mikr Anat 72: 816
Meyer A (1883) Das Chlorophyllkorn in chemischer, morphologischer und biologischer Beziehung. A Felix, Leipzig
Meyer A (1896) Bot Ztg 54: 187

Meynell GG, Lawn AM (1967) Genet Res 9: 359
Meynell E, Meynell GG, Datta N (1968) Bact Rev 32: 55
Michaelis P (1955) Plasma-Vererbung. In: Kappert H, Rudorf W (Ed), "Handbuch d. Pflanzenzüchtung", Parey, Berlin 140
Michaelis P (1957) Planta 50: 60
Michel MR, Hirt B, Weil R (1967) Proc Natl Acad Sci USA 58: 1381
Michie D, Wallace ME (1953) Nature 171: 26
Miele EA, Mills DR, Kramer FR (1983) J Mol Biol 171: 281
Miller J, McLachlan AD, Klug A (1985) EMBO J 4: 1604
Milne A (1961) Symp Soc Exp Biol 15: 40
Minot CS (1908) The Problems of Age, Growth and Death. A Study of Cytomorphosis. Putmans's Sons, New York, London
Mintz B (1962) Am Zool 2: 541
Mintz B (1967) Proc Natl Acad Sci USA 58: 344
Mirambean G, Dugnet M, Forterre P (1984) J Mol Biol 179: 559
Mitchell MB, Pittenger TH, Mitchell HK (1952) Proc Natl Acad Sci USA 38: 442
Mitra SK, Lustig F, Äkesson B, Lagerkvist U (1977) J Biol Chem 252: 471
Mitra SK, Lustig F, Äkesson B, Axberg B, Elias P, Lagerkvist U (1979) J Biol Chem 254: 6397
Mittwoch U (1954) J Genet 52: 547
Mizuno T, Chou M, Inouye M (1983) Proc Jap Acad Sci 59: 335
Mizuno T, Chou MY, Inouye M (1984) Proc Natl Acad Sci USA 81: 1966
Mizuuchi K, Nash HA (1976) Proc Natl Acad Sci USA 73: 3524
Mizuuchi K, Kamper B, Hays J, Weisberg RA (1982) Cell 29: 357
Möbius K (1877) Die Auster und die Austernwirtschaft, Berlin
Mohl H v (1846) Bot Ztg 4: 73
Mohn GR, Kerklaan PRM, ten Bokkum-Coenradi WPC, ten Hulscher TEM (1983) Mut Res 113: 403
Mohr OL (1923) Z indukt Abstamm-u. VererbLehre 30: 279
Mohr OL (1926) Z indukt Abstamm- u. VererbLehre 41: 59
Mok DWS, Peloquin SJ (1975) Can J Genet Cytol 17: 217
Monné L (1948) Advanc Enzymol 8: 1
Monneron A, Bernhard W (1969) J Ultrastruct Res 27: 266
Monod J (1942) Recherches sur la Croissance des Cultures bacteriénnes. Hermann, Paris
Monod J, Cohen-Bazire G (1953) Compt Rend Acad Sci Paris 236: 530
Monod J, Jacob F (1961) Cold Spr Harb Symp 26: 389
Monod J, Changeux JP, Jacob G (1963) J Mol Biol 6: 306
Monod J, Cohen-Bazire G, Cohn M (1951) Biochim Biophys Acta 7: 585
Monod J, Torriani AM, Jolit M (1949) Compt Rend Acad Sci Paris 229: 557
Montgomery TH (1904) Biol Bull 6: 137
Montgomery TH (1906) Trans Am Phil Soc 21: 97
Moore GW, Goodman M, Calahan C, Holmquist R, Moise H (1976) J Mol Biol 105: 15
Moore JES (1859) Quart J Micr Sci 38
Moore, RT, McAlear JH (1961) Mycologia 53: 194
Morgan TH (1910) Am Nat 44: 449
Morgan TH (1911) J Exp Zool 11: 365
Morgan TH (1911) Science 34: 636
Morgan TH (1914) Heredity and Sex (2 Ed), Columbia Univ Press, New York
Morgan TH, Cattell E (1912) J Exp Zool 13: 79
Morgan TH, Bridges CB, Sturtevant AH (1928) Yearb Carneg Inst 27: 1
Morgan TH, Sturtevant AH, Muller HJ, Bridges CB (1915) The Mechanism of Mendelian Heredity. Holt H & Comp, New York
Morgan EM, Zweerink HJ (1975) Virology 68: 455

Morikawa H, Iida A, Matsui G, Ikegami M, Yamada Y (1986) Gene 41: 121
Mormeneo S, Knott R, Perlman D (1987) Gene 61: 21
Morré DJ, Mollenhauer HH, Braeth CE (1971) Origin and continuity of Golgi apparatus. In: Reinart J, Ursprung H (eds) "Results and Problems of Cell Differentiation", Vol 2, p 82, Springer Berlin, Heidelberg, New York
Morrison JG (1966) Functional Organelles. Reinhold Publ Corp, New York
Morse ML, Lederberg EM, Lederberg J (1956) Genetics 41: 758
Mortelmans KE, Stocker BAD (1976) J Bact 128: 271
Morton NE, Crow JF, Muller HJ (1956) Proc Natl Acad Sci USA 42: 855
Morton NE et al (1977) Am J Hum Genetics 29: 52
Moscona AA (1968) Devel Biol 18: 250
Moses MJ (1958) J Biophys Biochem Cytol 4: 633
Moses MJ (1968) Ann Rev Genet 2: 363
Moses MJ (1977) Chromosomes Today 6: 71
Müller-Hill B (1975) Progr in Biophysics and Molecular Biol 30: 227
Müller-Hill B, Crapo L, Gilpert W (1968) Proc Natl Acd Sci USA 59: 1259
Müntzing A (1953) Ärftlighets Forskning. LTs Förlag, Stockholm
Mukherjee AS, Beermann W (1965) Nature 207: 785
Muller HJ (1916) Am Nat 50: 193
Muller HJ (1917) Proc Natl Acad Sci USA 3: 619
Muller HJ (1918) Genetics 3: 422
Muller HJ (1927) V. Int Kongr Vererbgsl 234
Muller HJ (1927) Anat Rec 37: 174
Muller HJ (1928) Genetics 13: 279
Muller HJ (1930) J Genet 22: 299
Muller HJ (1930) Am Nat 64: 220
Muller HJ (1932) Proc 6. Int Congr Genet 1: 213
Muller HJ (1934) The effects of X-rays upon the hereditary material. In: "Science of Radiology", 305, Baillière, Tindall & Cox, London
Muller HJ (1940) J Genet 40: 1
Muller JH (1940) Bearings of the *Drosophila* work on systematics. In: "The New Systematics", Clarendon Press, Oxford, p 185
Muller HJ (1941) Cold Spr Harb Symp 9: 151
Muller HJ (1948) Symp Radiol Genet Biol Div Oakridge Nat Lab 9
Muller HJ (1950) Am J hum Genet 2: 111
Muller HJ (1954) The manner of production of mutations by radiation. In: Hollaender A, (ed.) "Radiation Biology" 1: 475, McGraw-Hill Book Comp, New York
Muller HJ (1954) Acta Radiol, Stockh 41: 5
Muller HJ, Altenburg E (1919) Proc Soc Exptl Biol NY 17: 10
Muller HJ, Painter TS (1932) Z VererbLehre 62: 316
Muller HJ, League BB, Offerman CA (1931) Anat Rec 51 (Suppl): 110
Mulvihill JJ, Czeizel A (1985) Biol Zbl 104: 457
Murphy EA (1970) Ann of Human Genet 34: 73

Nachtsheim H (1957) Experientia 13: 57
Naegeli C (1884) Mechanisch-physiologische Theorie der Abstammungslehre. Oldenbourg R, München, Leipzig
Nakayasu H, Berczney R (1989) J Cell Biol 108: 1
Nanney DL (1958) Proc Natl Acad Sci USA 44: 712
Nanney DL (1963) Genetics 48: 737
Nass, MK, Nass S (1962) Exptl Cell Res 26: 424
Nathans D, Lipman F (1961) Proc Natl Acad Sci USA 47: 497

Navashin MS (1926) Z Zellforsch 4: 171
Navashin MS (1927) Z Zellforsch 6: 195
Navashin MS (1934) Cytologia 5: 169
Navashin SG (1912) Bull Acad Sci, St Petersb 22: 273 (russ)
Nazarenko IA, Potapou A, Romaschenko AG, Salganik RI (1978) FEBS Lett 86: 201
Needham J, Waddington CH, Needham DM (1934) Proc Roy Soc B 114: 393
Nei M, Roychoudhury AK (1974) Genetics 76: 379
Nemec B (1910) Das Problem der Befruchtungsvorgänge und andere zytologische Fragen. Borntraeger, Berlin
Newcombe HB (1953) Genetics 38: 134
Nichols WW, Levan A, Adula P, Norrby E (1964) Hereditas 51: 380
Nicolosi-Roncati F (1912) Boll Orto bot Napoli 2: 531
Niebuhr E (1972) Hum Genet 16: 217
Nilsen-Hamilton M, Shapiro JM, Massoglia SL, Hamilton RT (1980) Cell 20: 19
Nilsson-Ehle H (1908) Bot Notiser 257
Nilsson-Ehle H (1911) Lund Univ Årsskr 7 Nr 6
Nishimura Y, Caro L, Berg CM, Hirota Y (1971) J Mol Biol 55: 441
Nishizuka Y, Lipmann F (1966) Proc Natl Acad Sci USA 55: 212
Noll H, Matranga V, Cervello M, Humphreys T, Kuwasaki B, Adelson D (1985) Pro Natl Acad Sci USA 82: 8062
Nomura N, Ray DS (1980) Proc Natl Acad Sci USA 77: 6566
Normanly R, Ogden C, Horvath SJ, Abelson J (1986) Nature 321: 213
Nose K, Okamoto H (1983) Biochem Biophys Res Comm 111: 383
Novick A (1956) Brookhaven Symp Biol 8: 201
Novick A, Szilard L (1951) Science 113: 34
Novick RP, Clowes RC, Cohen SN (1976) Bact Rev 40: 168
Novick RP, Clowes, RC, Cohen SN, Datta N, Falkow S (1973) Uniform nomenclature for bacterial plasmids: A proposal, Unpublished draft.
Novikoff AB (1973) Lysosomes and storage disease. Hers G, van Hoof F (eds) Acad Press, New York
Novikoff, AB, Novikoff PM (1972) J Cell Biol 53: 532
Novikoff, AB, Novikoff PM (1973) J Histochem Cytochem 21: 963
Novikoff AB, Essner E, Quintana N (1964) Fed Proc 23, 3: 1010
Novitski E (1954) Genetics 39: 127

Ochman, H, Gerber AS, Harte L (1988) Genetics 120: 621
Odor L (1956) J Biophysic Biochem Cytol 2 (Suppl): 105
Oehlkers F (1956) Das Leben der Gewächse. Vol I Die Pflanze als Individuum. Springer, Berlin
Östergren G (1947) Bot Notiser 1947: 176
Östergren G (1950) Hereditas 36: 1
Östergren G, Vigfusson E (1953) Hereditas 29: 33
Östergren G, Morris R, Wakonig T (1958) Hereditas 44: 1
Ogawa K, Kaji A (1975) Eur J Biochem 58: 411
Ohashi M, Dove WF (1976) Virology 72: 299
Ohno S (1970) Evolution by Gene Duplication. Springer, Berlin, Heidelberg, New York, pp 160
Oksala T (1952) Hereditas 38: 449
Olins AL, Olins DE (1973) J Cell Biol 59: 252a
Oliver DA, Manis JJ, Whitfield HJ (1974) J Bact 119: 192
O'Mara JG (1940) Genetics 25: 401
Oppenheim AB, Newbauer Z, Calef E (1970) Nature 226: 31

Orrick LR, Olson MJ, Busch H (1973) Proc Natl Acad Sci USA 70: 1316
Orr-Weaver T, Szostak J, Rothstein R (1981) Proc Natl Acad Sci USA 78: 6364
Oshimura M, Barrett JC (1986) Env Mutagenesis 8: 129
Osiewacz HD, Esser K (1984) Current Genetics 8: 299
Oudet P, Gross-Bellard M, Chambon P (1975) Cell 4: 281
Owen FV (1945) J Agric Res 71: 423
Ozeki H, Howarth S (1961) Nature 190: 986

Padgett RA, Konarska MM, Grabowski PJ, Hardy SF, Sharp PA (1984) Science 225: 898
Paetau K (1941) Chromosoma 2: 36
Paigen K, Ganschow RS (1965) Brookh Symp Biol 18: 99
Painter TS (1939) Am Nat 73: 315
Painter TS, Muller HJ (1929) J Hered 20: 287
Palade GE (1953) Histochem Cytochem 1: 188
Palade GE, Siekevitz P (1955) Fed Proc 14: 262
Pandey KK (1975) Nature 256: 310
Pardee AB, Jacob F, Monod J (1959) J Mol Biol 1: 165
Park RB, Pon NG (1961) J Mol Biol 3: 1
Parry JM (1972) Mol Gen Genet 118: 33
Patrick MH, Haynes RH, Uretz RB (1964) Rad Res 21: 144
Patrizi G, Poger M (1967) J Ultrastruct Res 17: 127
Pauling L (1964) Bull N Y Acad Med 40: 334
Pauling L, Itano HA, Singer SJ, Wells IC (1949) Science 110: 543
Paulson JR, Laemmli UK (1977) Cell 12: 817
Paulus H, Alpers JB (1971) Enzyme 12: 385
Paulus H, Gray E (1964) J Biol Chem 239: PC 4008
Payton BW, Bennett MVL, Pappas GD (1969) Science 166: 1641
Pelc SR (1968) Acta Histochem 8 (Suppl): 41
Percival J (1932) Ann Bot 46: 479
Perkins FO (1970) J Cell Sci 6: 639
Perner ES (1952) Ber dtsch bot Ges 65: 52
Pernis BG, Chiappino G, Kelns AS, Gell PGH (1965) J Exp Med 122: 853
Perroncito A (1910) Arch ital Biol 54
Perry P, Evans HJ (1975) Nature 258: 121
Person C (1956) Canad J Bot 34: 60
Perucho M, Hanahan D, Wigler M (1980) Cell 22: 309
Peterson PA (1976) Mol Gen Genet 149: 5
Pfeffer W (1877) Osmotische Untersuchungen. Engelmann, Leipzig
Philiptschenko J (1927) Variabilität und Variation. Borntraeger, Berlin, pp 101
Piatigorsky J, Wistow GJ (1989) Cell 57: 197
Pickett-Heaps J (1969) Cytobios 3: 257
Piekarski G (1937) Arch Mikrobiol 8: 428
Piperno R, Carere A, Sermonti G (1966) Ann Ist Super Sanità 2: 393
Pirie NW (1937) The meaningless of the terms life and living. In: Needham J, Green F (eds) "Perspectives in Biochemistry", Univ Press, Cambridge
Pitelka DR (1969) Centriole replication. In: Lima-de-Faria A (ed) "Handbook of Molecular Cytology", North-Holland Publ Comp, Amsterdam, London
Plate L (1910) Vererbungslehre und Deszendenztheorie. Festschr f Hertwig R II, Fischer, Jena, 537
Plate L (1913) Vererbungslehre. Engelmann, Leipzig
Plowe JQ (1931) Protoplasma 12: 196
Poessnerova V, Smetana K (1966) Folia Morphol (Praha) 14: 240

Policard A, Bessis M (1956) Compt Rend Acad Sci Paris 242: 2496
Pollak KJ, Shorey CD (1964) Biochem J 93: 360
Pontecorvo G (1949) J General Microbiol 3: 122
Pontecorvo G (1953) Advanc Genet 5: 141
Pontecorvo G (1954) Proc 9.Int Congr Genet., Caryologia 6, Suppl 1: 192
Pontecorvo G (1955) Proc Roy Soc B 144: 171
Pontecorvo G (1958) Trends in Genetic Analysis. Columbia Univ Press, New York
Pontecorvo G (1975) The Bateson Lecture 1974: "Alternatives to sex" – genetics by means of somatic cells. In: Markham R et al (eds) "Modification of the Information Content in Plant Cells", Second John Innes Symp, North-Holland, American Elsevier, p 1
Pontecorvo G, Gloor ET, Forbes E (1954) J Genetics 52: 226
Pontecorvo G, Riddle PN, Hales A (1977) Nature 265: 257
Porter KR, Caulfield JB (1958) Proc 4th Int Congr Electron Microscopy 2: 503
Porter KR, Claude A, Fullmann EF (1945) J Exp Med 81: 233
Portin P, Ruohonen M (1972) Dros Inf Service 49: 70
Potter H, Dressler D (1976) Proc Natl Acad Sci USA 73: 3000
Potter H, Dressler D (1980) Proc Natl Acad Sci USA 77: 2390
Potter SS (1982) Nature 297: 201
Potter S, Truett M, Phillips M, Maher A (1980) Cell 20: 639
Poulton EB (1903) Proc R Ent Soc Lond 1903: 77
Prakken R, Müntzing A (1942) Hereditas 28: 441
Breer JR, Siegel RW, Stark PS (1953) Proc Natl Acad Sci USA 39: 1228
Prell HH (1973) Mol Gen Genet 127: 327
Price S (1956) Cytologia Tokyo 21: 21
Primrose SB, Ehrlich SD (1981) Plasmid 6: 193
Pring DR, Levings CS, Hu WWL, Timothy DH (1977) Proc Natl Acad Sci USA 74: 2904
Prusinger SB (1982) Science 216: 136
Punnett RC (1905) Mendelism. MacMillan, Cambridge
Purkinje JE (1825) Subjectae sunt symbolae ad ovi avium historiam ante incubationem (ed 1), Vratislaviae (ed II Lipsiae 1830)
Purkinje JE (1840) Über die Analogien in den Strukturelementen des thierischen und pflanzlichen Organismus. In: Übers Arb Veränder schles Ges vaterl Cultur 1839: 81 (=Opera omina II)

Quillardet P, Huisman O, D'Ari R, Hofnung M (1982) Proc Natl Acad Sci USA 79: 5971

Radman M (1974) in "Molecular and Environmental Aspects of Mutagenesis", Prakash L, Sherman F, Miller MW, Lawrence CW, Taber HW (eds), Thomas, Springfield Ill, p 128
Radman M (1975) Phenomenology of an inducible mutagenic DNA repair pathway in *E coli:* SOS repair hypothesis. In: Miller M (Ed) "Molecular and Environmental Aspects of Mutagenesis", Thomas, Springfield
Radmann M, Villani G, Boiteux S, Defais M, Laillet-Fouquet P, Spaden S (1977) In: Watson JD, Hiatt I, Winston H "Origins of Human Cancer" Cold Spring Harb Lab New York, pp 903
Ramakrishnan T, Adelberg EA (1965) J Bact 89: 661; 90: 295
Ramon y Cajal S (1908) Histologie du système nerveux de l'homme et des vertébrés. Neudruck: Instituto Ramon y Cajal, Madrid 1952
Randles JW, Davies C, Hatta T, Gould AR, Franki RIB (1981) Virology 108: 111
Randolph LF (1928) Cornell Univ Agr Exp Mem 117
Raper JR, Esser K (1964) The fungi. In: Brachet J, Mirsky AE (ed) "The Cell" 6: 139, Acad Press, New York
Raper JR, SanAntonio JP (1954) Am J Bot 41: 69
Rasmussen RE, Painter RB (1964) Nature 203: 1360

Raven CP (1961) The Storage of Developmental Information. Pergamon Press, Oxford
Ravin AW (1963) Am Nat 97: 307
Razin A, Hirose T, Itakura K, Riggs AD (1978) Proc Natl Acad Sci USA 75: 4268
Rechler MM, Martin RG (1970) Nature 226: 908
Reddy PV, Pardee AP (1980) Proc Natl Acad Sci USA 77: 3312
Reed RR (1981) Cell 25: 713
Reichard P (1957) The Biosynthesis of Deoxyribose. Ciba Lectures in Biochem. Wiley & Sons, New York
Reichard P, Eliasson R, Söderman G (1974) Proc Natl Acad Sci USA 71: 4901
Reif HJ, Saedler H (1975) Mol Gen Genet 137: 17
Reinig WG (1938) Elimination und Selektion. Fischer, Jena
Renner O (1916) Biol Zbl 36: 337
Renner O (1917) Z Indukt Abstamm-u. VererbLehre 18: 121
Renner O (1918) Flora 111–112: 641
Renner O (1921) Z Bot 131: 609
Renner O (1924) Z Indukt Abstamm-u. VererbLehre 33: 317
Renner O (1929) Artbastarde bei Pflanzen, Handb Vererbungswissenschaft Bd II A, Borntraeger, Berlin
Renner O (1934) Ber Mat-Phys Kl Sächs Akad Wiss, Leipzig 86: 241
Renner O (1937) Z Indukt Abstamm-u. VererbLehre 74: 91
Renner O (1941) Flora 35: 201
Renner O (1949) Z Indukt Abstamm-u. VererbLehre 83: 1
Renner O, Kupper, W (1921) Ber Dtsch Bot Ges 39: 201
Rensch B (1947) Neuere Probleme der Abstammungslehre. Die transspezifische Evolution. Enke, Stuttgart
Renwick JH (1971) Ann Human Genet 35: 79
Renwick JH (1971) Ann Rev Genet 5: 84
Resch A (1952) Chromosoma 5: 296
Resende F (1940) Chromosoma 1: 486
Revel M, Pollack Y, Groner Y, Scheps R, Inouye H, Berissi H, Zeller H (1973) Biochemie 55: 41
Reznikoff WS (1972) Ann Rev Genet 6: 133
Rheinberger HJ, Sternbach H, Nierhaus KH (1981) Proc Natl Acad Sci USA 78: 5310
Rhoades MM (1933) Proc Natl Acad Sci USA 19: 1031
Rhoades MM (1942) Genetics 27: 395
Rhoades MM (1952) Preferential segregation in maize. In: Gowen JW (ed) "Heterosis", Iowa State Coll Press, Ames Iowa 66
Rhoades MM, Kerr WE (1949) Proc Natl Acad Sci USA 35: 129
Rhodin J (1954) Correlation of ultrastructural organization and function in normal and experimentally changed proximal convoluted tubule cells of the mouse kidney. Doctorale thesis Karolinska Inst, Stockholm
Richardson CC (1965) Proc Natl Acad Sci USA 54: 158
Rich A, Rajbhandary UL (1976) Ann Rev Biochem 45: 805
Rick CM (1943) Bot Gaz 104: 528
Rickett HW (1958) Taxon 7: 37
Ricklefs R (1973) Ecology. Chiron, Newton, Mass 861 pp
Rieger R (1963) Die Genommutationen (Ploidiemutationen). Fischer, Jena
Riley R (1960) Heredity 15: 407
Riley R, Chapman V (1958) Nature 182: 713
Riley R, Law CN (1965) Adv Genet 13: 57
Ris H (1961) Canad J Genet Cytol 3: 95
Ris H, Chandler BL (1963) Cold Spr Harb Symp 28: 1

Ritossa F (1968) Proc Natl Acad Sci USA 60: 509
Ritossa F (1962) Experientia 18: 571
Ritossa F, Scala G (1969) Genetics (Suppl 1) 61: 305
Rizet G, Esser K (1952) Compt Rend hebd 237: 760
Robards AW (1968) Planta 82: 200
Robberson DL, Kasamatsu H, Vinograd J (1972) Proc Natl Acad Sci USA 69: 737
Roberts HS (1961) Quart Rev Biol 36: 155
Robert JW (1969) Nature 224: 1168
Roberts JW (1970) Cold Spring Harb Symp 35: 121
Roberts RB (1958) Introduction to "Microsomal Particles and Protein Synthesis", Pergamon, New York
Roberts RB, Aldous E (1949) J Bact 57: 363
Robertson JD (1959) Biochem Soc Symp 16: 3
Robertson JD (1963) J Cell Biol 19: 201
Robertson W (1916) J Morph 27: 179
Robinow CF, Caten CE (1968) J Cell Sci 5: 403
Robinson R (1971) Gene Mapping in Laboratory Mammals, Plenum Press, London, New York, Parts A and B
Robinson MK, Bennett PM, Richmond MH (1977) J Bact 129: 407
Rogers J (1983) Nature 301: 460
Roman HL (1956) Cold Spr Harb Symp 21: 175
Romig WR (1968) Bact Rev 32: 349
Romig WR, Brodetsky AM (1961) J Bact 82: 135
Rosenberg O (1926) Ber Dtsch Bot Ges 44: 455
Rosenberg O (1926) Hereditas 8: 305
Rosenberg O (1927) Hereditas 9: 285
Roth CW, Nester EW (1971) J Mol Biol 62: 577
Roth TF, Porter KR (1964) J Cell Biol 20: 313
Rottman F, Shatkin AJ, Perry RP (1974) Cell 3: 197
Roux W (1905) Die Entwicklungsmechanik, ein neuer Zweig der biologischen Wissenschaft. In: Roux W (ed) „Vorträge und Aufsätze über Entwicklungsmechanik der Organismen" 1, Engelmann, Leipzig
Rowen L, Kornberg A (1978) J Biol Chem 253: 758
Rubin GM, Finnegan DJ, Hogness DS (1976) Progr Nucleic Acid Res Mol Biol 19: 221
Rubin GM, Kidwell MG, Bingham PM (1982) Cell 29: 987
Rückert J (1882) Anat Anz 7: 107
Ruddle FH, Creagan P (1975) Ann Rev Genet 9: 407
Ruddle F, Ricciuti FA, McMorris J, Tischfield R, Creagan R, Darlington G, Chen T (1972) Science 176: 1429
Rupers CS (1960) J Gen Physiol 43: 573
Rupert C (1961) J Cell Comp Physiol 58 (Suppl 1) 57
Rupert C, Harm W (1966) Adv Rad Biol 2: 1
Rupp WD, Howard-Flanders P (1968) J Mol Biol 31: 291
Russell WL (1951) Cold Spr Harb Symp 16: 327
Ryan FJ (1954) Proc Natl Acad Sci USA 40: 178
Ryan FJ, Lederberg J (1946) Proc Natl Acad Sci USA 32: 163

Sachs L (1954) Ann Eugen 18: 255
Saedler H, Reif HJ, Hu S, Davidson N (1974) Mol Gen Genet 132: 2265
Sagan L (1967) J Theoret Biol 14: 225
Sager R (1973) Cytoplasmic Genes and Organelles. Academic Press London, New York
Sager, R, Ishida MR (1963) Proc Natl Acad Sci USA 50: 725

Sager R, Kitchin R (1975) Science 189: 426
Saito I, Stark GR (1987) Proc Natl Acad Sci USA 83: 8664
Sakano H, Hüpi K, Heinrich G, Tonegawa S (1979) Nature 280: 288
Salisbury EJ (1940) Ecological aspects of plant taxonomy. In: Huxley Js (ed) "The New Systematics", Univ Press, Oxford, 329
Sallei J, Zuckerkandl E (1975) Biochemie 57: 343
Samarina OP, Lukanidin EM, Georgiev GP (1967) Biochem Biophys Acta 142: 561
Samarina OP, Lukanidin EM, Molnar J, Georgiev GP (1968) J Mol Biol 33: 251
Samson L, Cairns J (1977) Nature 267: 281
Sandler L, Hiraizumi Y (1961) Canad J Genet Cytol 3: 34
Sandler L, Novitski E (1957) Am Nat 91: 105
Sandler L, Lindsley DL, Nicoletti B, Trippa G (1968) Genetics 60: 525
Sanger F, Nicklen S, Caulson AR (1977) Proc Natl Acad Sci USA 74: 5463
Sansome ER (1933) Cytologia 5: 15
Sauerbier W, Millette R, Hackett P (1970) Biochim Biophys Acta 209: 368
Sawadogo M, Roeder RG (1985) Cell 43: 165
Sax K (1932) J Arnold Abor 12: 3
Sax K (1935) J Arnold Arbor 16: 216
Sax K (1941) Cold Spr Harb Symp 9: 93
Scaife J, Gross JD (1962) Biochem Biophys Res Comm 7: 403
Schaal BA, Levin DA (1976) Am Nat 110: 191
Schachat FH, Hogness DS (1973) Cold Spr Harb Symp 38: 371
Schäfer R, Zillig W, Zechel K (1973) Eur J Biochem 33: 207
Schell J (1979) Proc Roy Soc Lond B 204: 251
Scherrer K (1971) FEBS Lett 17: 68
Scherrer K (1973) 6th Karolinska Symp "Protein Synthesis in Reproductive Tissue", Geneva, p 95
Scherrer K, Darnell JE (1962) Biochem Biophys Res Comm 7: 486
Scherrer K, Marcaud L (1965) Bull Soc Chim Biol 47: 1697
Scherrer K, Marcaud L (1968) J Cell Physiol (Suppl 1) 72: 181
Scherzinger E, Lanka E, Hillenbrand G (1977) Nucleic Acids Res 4: 4151
Schibler U, Sierra F (1987) Ann Rev Genet 21: 237
Schimper AJW (1885) Jahrb Wiss Bot 16: 1
Schindler D, Echols H (1981) Proc Natl Acad Sci USA 78: 4475
Schindler CW, Rush MG (1985) Nucl Acids Res 13: 8242
Schleicher W (1878) Arch Mikr Anat 16: 248
Schlesinger DH, Goldstein G, Nian HD (1975) Biochemistry 14: 2214
Schlesinger MJ, Levinthal C (1963) J Mol Biol 7: 1
Schmalhausen II (1949) Factors of Evolution. The Theory of Stabilizing Selection. Blakiston, Philadelphia
Schmieger H (1972) Mol Gen Genet 119: 75
Schmid HP, Khayato A, Martins de Sa C, Purion F, Koehler K, Scherrer K (1984) EMBO J 3: 29
Schmitz F (1882) Die Chromatophoren der Algen. Bonn
Schrader F (1932) Z Wiss Zool 142: 520
Schrader F (1936) Biol Bull 70: 484
Schrader F (1944) Mitosis. Columbia Univ Press, New York
Schrantz JP (1970) Rev Cytol Biol Vég 33: 1
Schreier MH, Erni B, Staehelin T (1977) J Mol Biol 116: 727
Schultz J (1936) Proc Natl Acad Sci USA 22: 27
Schultz J, Redfield H (1951) Cold Spr Harb Symp 16: 175
Schvartzman JB, Van't Hof J (1982) Nucleic Acids Res 10:

Schwann T (1838) Mikroskopische Untersuchungen über die Übereinstimmung in Struktur und dem Wachsthum der Thiere und Pflanzen. Ostwalds Klassiker Nr. 176, Engelmann, Leipzig
Scott FM (1950) Bot Gaz 111: 252
Scott WA, Wigmore DJ (1978) Cell 15: 1511
Sears ER (1941) Genetics 26: 167
Sears ER (1972) Stadler Symp, Univ of Missouri, Columbia USA 4: 23
Sears ER, Camara A (1950) Genet Iber 2: 239
Sears ER, Camara A (1952) Genetics 37: 125
Sedliakova M, Brozmanova J, Slezankova V, Mazek F, Fandlova E (1975) Neoplasma 22: 361
Sedlmayer K (1956) S B Dtsch Akad Ldwtsch Wiss 5: 2
Sedlmayer K (1956) Wiss Z Univ Leipzig 4: 258
Seeberg E, Steinum AL (1982) Proc Natl Acad Sci USA 79: 988
Seifert HS, Chen EY, So M, Heffron F (1986) Proc Natl Acad Sci USA 83: 735
Seiffer S, Gallop PM (1966) The structure proteins. In: Neurath H (ed) "The Proteins" 4: 153, Academic Press, New York
Seifriz W (1928) Arch Exp Zellforsch 6: 341
Seiler J (1914) Arch Zellforsch 13: 159
Seiler J, Haniel CB (1921) Z Indukt Abstamm-u.VererbLehre 27: 81
Selleck SB, Majors J (1987) Nature 325: 173
Semancik JS, Weathers LG (1972) Nature New Biol 237: 242
Sen D, Gilbert W (1988) Nature 334: 364
Sen KN (1952) Genetics 37: 227
Senger F, Nicklen S, Coulsen AR (1977) Proc Natl Acad Sci USA 74: 5463
Sermonti G, Carere A (1968) Mol Gen Genet 103: 141
Serra JA (1942) Bol Soc Brot 16: 83
Serra JA (1948) Portug Acta Biol (A) 2: 257
Serra JA (1949) Portug Acta Biol (A) Goldschmidt-vol, 401
Setlow R, Carrier W (1964) Proc Natl Acad Sci USA 51: 226
Setlow JK, Setlow RB (1963) Nature 197: 560
Shapiro JA (1969) J Mol Biol 40: 93
Shapiro JA (1979) Proc Natl Acad Sci USA 76: 1933
Shapiro JA, Chargaff E (1964) Biochem Biophys Acta 91: 262
Sharp LW (1914) Bot Gaz 58: 419
Sharp LW (1929) Bot Gaz 88: 349
Sharp LW (1934) Introduction to Cytology. McGraw Hill Comp, New York
Shaw MW (1970) Ann Rev Med 21: 409
Shenkin A, Burdon RH (1974) J Mol Biol 85: 19
Shibata T, Das Gupta RP, Cunningham RM, Radding CM (1979) Proc Natl Acad Sci USA 76: 1638
Shine J, Dalgarno L (1974) Proc Natl Acad Sci USA 71: 4734
Shull GH (1911) Bot Gaz 52: 480
Shull GH (1914) Z indukt Abstamm-u.VererbLehre 12: 97
Sibatani A, Hiai S (1964) J theoret Biol 7: 393
Siegel RW (1953) Genetics 38: 550
Sigal N, Alberts B (1972) J Mol Biol 71: 789
Silber R, Malathi V, Hurwitz J (1972) Proc Natl Acad Sci USA 69: 3009
Siliciano PG, Tätchell K (1986) Proc Natl Acad Sci USA 83: 2320
Simons RW, Kleckner N (1983) Cell 34: 683
Simpson GG (1944) Tempo and Mode in Evolution. Columbia Univ Press, New York
Simpson GG (1953) The Major Features of Evolution. Columbia Univ Press, New York
Simpson GG (1953) Evolution 7: 110

Simpson L (1972) Int Rev Cytol 32: 139
Simpson RT (1978) Biochemistry 17: 5524
Singer MF (1982) Cell 28: 433
Siniscalco M (1972) Adv in Biosciences 8: 307
Sinoto Y, Sato D (1940) Cytologia, Tokyo 10: 529
Sinsheimer RL, Starman B, Nagler C, Guthrie S (1962) J Mol Biol 4: 142
Sitte P (1961) Ber Dtsch Bot Ges 74: 177
Sjöstrand FS (1955) Compt Rend Colloque Intern. "Les Tech Rec Microscopie Electronique Corpusc", Toulouse 151
Slater E, Anderson MD, Rosenkranz HS (1971) Cancer Res 31: 570
Slautterback DB (1963) J Cell Biol 18: 367
Slightom JL, Blechl AE, Smithies O (1980) Cell 21: 627
Slizinski BM (1945) Proc Roy Soc Edingburgh 62: 114
Slizinski BM (1955) J Genet 53: 597
Smith GP (1973) Cold Spr Harb Symp 38: 507
Smith GR, Magasanik B (1971) Proc Natl Acad Sci USA 68: 1493
Smith HF (1936) Am Engenics 7: 240
Smith HM (1955) Turtox News 33: 74
Smith HM (1965) Systemat Zool 14: 57
Smith SG (1935) J Genet 30: 227
Solari AJ (1980) Chromosoma 81: 315
Solioz M, Yen M C, Marrs B (1975) J of Bacteriol 123: 651
Soll DR, Waddell DR (1975) Devel Biol 47: 292
Sonenberg N, Morgan MA, Merrick WC, Shatkin AJ (1978) Proc Natl Acad Sci USA 75: 4843
Sonneborn TM (1937) Proc Natl Acad Sci USA 23: 378
Sonneborn TM (1938) Proc Am Phil Soc 79: 411
Sonneborn TM (1943) Proc Natl Acad Sci USA 29: 329
Sonneborn TM (1957) Breeding systems, reproductive methods, and species problems in Protozoa. In: Mayr E (ed) "The Species Problem", Amer Ass for the Advancement of Science (AAAS), Washington D C 50: 155
Sonneborn TM (1963) Does preformed cell structure play an essential role in cell heredity? In: Allen JM (ed) "The Nature of Biological Diversity", McGraw Hill, New York 165
Sonneborn TM (1964) Proc Natl Acad Sci USA 51: 915
Sonneborn TM (1965) Am Nat 99: 279
Sonneborn TM (1965) Degeneracy of the genetic code. Extent, nature, and genetic implications. In: Bryson & Vogel (ed) "Evolving Genes and Proteins", New York, Acad Press, p 377
Sonneborn TM (1967) The evolutionary integration of the genetic material into genetic systems. In: Brink RA (ed) "Heritage from Mendel", p 375, The University of Wisconsin Press, Madison
Sorokin H (1929) Am J Bot 16: 407
Sotelo JR, Porter KR (1959) J Biophysic Biochem Cytol 5: 327
Southern DJ (1967) Chromosoma 22: 164
Spemann H (1918) Roux Arch Entwicklungsmech 43: 448
Spencer H (1863) The Principles of Biology. Williams, London
Spiegelman S, Hayashi M (1963) Cold Spr Harb Symp 28: 161
Spirin AS, Belitsina NV, Ajtkhozin MA (1964)8 J Gen Biol, Moscow, 25: 321
Spraque GF (1932) Genetics 17: 358
Sprague GF, McKinney HH (1966) Genetics 54: 1287
Sprengel CK (1793) Das entdeckte Geheimnis der Natur im Bau und in der Befruchtung der Blumen, Berlin
Staden R (1980) Nucleic Acids Res 8: 3673
Stadler D, Kariya B (1973) Genetics 75: 423

Stahl V, Lemke PA, Tudzynski P, Küch U, Esser K (1978) Mol Gen Genet 162: 341
Stanfield S, Helinsky DR (1976) Cell 9: 333
Stanley SM (1975) Proc Natl Acad Sci USA 72: 646
Starling E (1905) In: Croonian Lectures to the College of Physicians. Lancet 11: 339
Stebbins GL (1941) Bot Rev 7: 507
Stebbins GL (1945) Bot Rev 11: 463
Stebbins GL (1947) Adv Genet 1: 403
Stebbins GL (1950) Variation and Evolution in Plants. Columbia Univ Press, New York
Stebbins GL (1966) Processes of Organic Evolution. Prentice-Hall, Englewood Cliffs, New Jersey
Stedman E, Stedman E (1943) Nature 152: 267
Steiner RF (1965) The Chemical Foundations of Molecular Biology. Van Nostrand Comp, Princeton
Steitz JA (1969) Nature 224: 957
Steitz TA, Ohlendorf DH, McKay DB, Anderson WF, Mathews BW (1982) Proc Natl Acad Sci USA 79: 3097
Stent GS (1963) Molecular Biology of Bacterial Viruses. Freeman, San Francisco
Stent GS (1964) Science 144: 816
Stephens SG (1942) J Genet 44: 272
Stephens JC, Artz SW, Ames BN (1975) Proc Natl Acad Sci USA 2: 4389
Stern C (1929) Biol Zbl 49: 261
Stern C (1933) Faktorenkopplung und Faktorenaustausch. In: Baur H, Hartmann M (ed) "Hdb Vererbungswissensch" 1, Abt. H; Borntraeger, Berlin
Stern C (1958) Am Nat 92: 313
Stern C (1959) The Lancet 12: 1088
Stern C (1960) Principles of Human Genetics. Freeman, San Francisco
Stern C, Kodani M (1944) Genetics 40: 343
Stern C, Schaeffer EW (1943) Proc Natl Acad Sci USA 29: 361
Stern MJ, Ames G, Ferro-Luzzi G, Smith NA, Robinson EC, Higgins CF (1984) Cell 37: 1015
Stout AB (1917) Am J Bot 4: 395
Stout AB (1918) J Genet 8: 71
Strandskov HH (1950) Cold Spr Harb Symp 15: 1
Strasburger E (1877) Jena Z Naturw 11: 435
Strasburger E (1878) Jena Z Naturw 12: 647
Strasburger E (1880) Zellbildung und Zellteilung, Fischer, Jena
Strasburger E (1882) Arch mikr Anat 21: 476
Strasburger E (1884) Neue Untersuchungen über den Befruchtungsvorgang bei den Phanerogamen als Grundlage für eine Theorie der Zeugung. Fischer, Jena
Strasburger E (1893) Anat Anz 8: 177
Strasburger E (1901) Über Plasmaverbindungen pflanzlicher Zellen Jb Wiss Bot 36: 493
Strasburger E (1905) Typische und allotypische Kernteilung. Ergebnisse und Erörterungen. Pringsheims Jb Wiss Bot 42: 1
Strasburger E (1907) Flora 97: 123
Strasburger E (1910) Flora 100: 398
Strauss W (1958) J Biophysic Biochem Cytol 4: 541
Streisinger G (1956) Virology 2: 1
Streisinger G, Edgar RS, Denhards GH (1964) Proc Natl Acad Sci USA 51: 775
Streisinger G, Emrich J, Stahl MM (1967) Proc Natl Acad Sci USA 57: 292
Strelzoff E, Ryan FJ (1962) Biophys Biochem Res Comm 7: 471
Strugger S (1950) Naturwiss 37: 166
Struhl K, Stinchcomb DT, Scherer S, Dawis RW (1974) Proc Natl Acad Sci USA 76: 1035
Stubbe H (1940) Biol Zbl 60: 113

Sturtevant AH (1914) Am Nat 48: 535
Sturtevant AH (1920) Proc Soc Exp Biol 17: 70
Sturtevant AH (1925) Genetics 10: 117
Sturtevant AH (1926) Biol Zbl 46: 697
Sturtevant AH, Beadle GW (1936) Genetics 21: 554
Subak-Sharpe H, Bürk RR, Pitts JD (1966) Heredity 21: 342
Sueoka N (1961) J Mol Biol 3: 31
Sueoka N (1962) Proc Natl Acad Sci USA 48: 166
Sueoka N, Kano-Soeoka T (1964) Proc Natl Acad Sci USA 52: 1535
Sunshine MG, Kelly B (1971) J Bact 108: 695
Suomaleinen E (1950) Adv Genet 3: 193
Surani MAH, Barton SC, Norris ML (1984) Nature 308: 548
Sures I, Lowry J, Kades LH (1978) Cell 15: 1033
Suskind SR, Kurek LJ (1959) Proc Natl Acad Sci USA 45: 193
Susman M (1965) Ann Rev Microbiol 19: 59
Sutcliffe JG, Milner RJ, Bloom FE, Lerner RA (1982) Proc Natl Acad Sci USA 79: 4942
Sutherland GR (1979) Am J Hum Genet 31: 125
Sutherland EW, Oye I, Butcher RW (1965) Rec Progr Hormone Res 21: 623
Sutton WS (1903) Biol Bull Wood's Hole 4: 231
Sved JA (1976) Aust J Biol Sci 29: 375
Svedelius N (1915) Nova Acta Regiae Soc Scient Upsalensis 4: 1
Swift H (1950) Physiol Zool 23: 169
Swift H (1956) J Biophys Biochem Cytol 2: Suppl 2, 415
Sybenga J (1965) Genetica 36: 243
Sybenga J (1966) Genetica 37: 186
Symonds N, Heindl H, White P (1973) Mol Gen Genet 120: 253
Szostak JW, Blackburn EH (1982) Cell 29: 245
Szybalski W, Bovre K, Fiandt M, Hayes S, Hradecna Z, Kumar S, Lozeron NA, Nijkamp HJJ, Stevens WF (1970) Cold Spr Harb Symp 35: 341
Szybalski W, Kubinsky H, Sheldrick P (1966) Cold Spr Harb Symp 31: 123

Täckholm G (1922) Acta Hort Berg 7
Tamkun JW, Schwarzbauer JE, Hynes RO (1984) Proc Natl Acad Sci USA 81: 5140
Tangl E (1879) Jb Wiss Bot 12: 170
Tartoff KD (1971) Science 171: 294
Tartoff KD (1975) Ann Rev Genet: 9: 355
Taschenberg O (1892) Abh Naturf Ges Halle 17: 365
Tata JR, Baker B (1978) J Mol Biol 118: 249
Taylor AC (1966) J Cell Biol 28: 155
Taylor ALj (1963) Proc Natl Acad Sci USA 50: 1043
Taylor JH, Woods PS, Hughes WL (1957) Proc Natl Acad Sci USA 43: 122
Taylor WR (1926) Am J Bot 13: 179
Tecott LH, Barchas JD, Eberwine JH (1988) Science 240: 1661
Temin HM (1971) J Nat Cancer Inst 46: III
Temin HM, Mizutani S (1970) Nature 226: 1211
Templeton AR (1981) Evolution 34: 719
Thach SS, Thach RE (1971) Proc Natl Acad Sci USA 68: 1791
Thoday JM (1953) Symp Soc Exp Biol 7: 96
Thoday JM (1972) Proc Roy Soc London B 182: 109
Thoday JM (1975) Nature 255: 675
Thomas CA (1966) Progr in Nucleic Acid Res & Mol Biol 5: 315
Thomas CA, Kelly TJ, Rhoades M (1968) Cold Spr Harb Symp 33: 417

Thomas KR, Capecchi MR (1986) Nature 324: 34
Thomas R (1955) Biochim Biophys Acta 18: 467
Thomas R (1966) J Mol Biol 22: 79
Thompson DH (1931) Genetics 16: 267
Thorell B (1944) Cold Spr Harb Symp 12: 247
Thornton RM, Thimann KV (1964) J Cell Biol 20: 345
Tilney-Basset RAE (1963) Heredity 18: 265
Timmis K, Cabello F, Cohen SN (1975) Proc Natl Acad Sci USA 72: 2242
Timoféeff-Ressovsky HA, Timoféeff-Ressovsky NW (1926) Roux Arch EntwMech 108: 146
Timonen S (1950) Acta Obstetr Skand 31 (Suppl): 1
Tischler G (1920) Biol Zbl 40: 15
Tischler G (1937) J Ind Bot Soc 16: 165
Tobias P v (1953) S Afr J Sci 50: 134
Togby HA (1943) J Genetics 45: 67
Tournefort JP (1700) Institutiones Rei Herbarii. Paris
Trautner TA, Spatz HC (1973) Curr Topics in Microbiology and Immunology 62: 61
Travers A, Kamen R, Cashel M (1970) Cold Spr Harb Symp 35: 415
Traverse KL, Pardue ML (1989) Chromosoma 97: 261
Trelstad RL, Hay ED, Revell JP (1967) Develop Biol 16: 78
Trivers RL (1972) Parental investment and sexual selection. In: Campbell B (ed) "Sexual Selection and the Descent of Man". Aldine-Atherton, Chicago
Trow AH (1895) Ann Bot 9: 609
Tsai RL, Green H (1973) Nature New Biol 243: 168
Tsanew R (1975) In: Results and Problems in Cell Differentiation Vol 7: 197
Tsuchida N, Nonoyama M, Ikeda Y (1966) J Mol Biol 20: 575
Tulasne R, Minck R, Kirn A, Krembel J (1959) Ann Inst Pasteur 99: 859
Turesson G (1922) Hereditas 3: 211
Turesson G (1923) Hereditas 4: 171
Turesson G (1926) Hereditas 8: 161
Turesson G (1929) Hereditas 12: 323
Turing AM (1952) Phil Trans Roy Soc B 237: 37
Turner JRG (1977) Evol Biol 10: 163
Tyson GE, Bulger RE (1973) Z Zellforsch Mikrosk Anatomie 141: 44

Uexküll-Syllenband E v (1901) Bibl Bot 52: 5
Uhlin BE, Clark AJ (1981) J Bact 148: 386
Umbarger HE (1956) Science 123: 848
Unrau J, Person C, Kuspira J (1956) Can J Bot 34: 629
Upcott M (1938) Cytologia 8: 398

Vaarama A (1954) Ann Bot Soc Zool-Bot Fenn Vanamo 28: 1
Valentine DN (1960) Nature 185: 778
Valentine GG (1836) Repertorium für Anatomie und Physiologie, I, Berlin
Vandel A (1938) III Bull Biol 72: 147
Vandel A (1945) IX Bull Biol 79: 168
Van't Hoff J, Bjerkness CA (1977) Chromosoma 64: 287
Varshavsky A (1981) Proc Natl Acad Sci USA 78: 3673
Varshavsky AJ, Sundine OH, Bohn MJ (1978) Nucleic Acids Res 5: 3469
Vavilov NJ (1922) J Genet 12: 47
Vejdovsky F (1912) Zum Problem der Vererbungsträger. Böhm, Prag
Verly, Paquette (1970) Proc Can Fed Biol Soc 48: 685

Verworn M (1891) Pflüg Arch Ges Physiol 51: 1
Vickerman K (1978) Nature 273: 613
Vinograd J, Lebowitz J, Watson R (1968) J Mol Biol 33: 173
Virkki N, Kimura M (1978) Biosystems 10: 213
Vogel F (1958) Verh Dtsch Ges Inn Med 64
Vogel HJ (1957) Repression and induction as control mechanisms of enzyme biogenesis: the adaptive formation of acetylornithinase. In: McElroy WD & Glass B(eds) "The Chemical Basis of Heredity", John Hopkins Press, Baltimore, 276
Vogel HJ, Bryson V, Lampen JO (Eds) (1963) Informational Macromolecules. Acad Press, New York
Vogt O (1926) Z Ges Neurol Psychiat 101: 805
Vries H de (1885) Pringsheims Jb Wiss Bot 16: 464
Vries H de (1900) Ber Dtsch Bot Ges 18: 83
Vries H de (1900) Compt Rend Acad Sci Paris 130: 845
Vries H de (1901) Die Mutationstheorie Bd I Veit & Comp, Leipzig
Vries H de (1911) Biol Zbl 31: 97
Vries H de (1917) Ber Dtsch Bot Ges 35: 128
Vosberg, Vinograd J (1975) In: Goulian M, Hanowalt P (eds), Benjamin WA, Menlo Park, CA, Vol 3: 94

Waardenburg PJ (1932) Bibliogr Genet 7
Waddington CH (1932) Phil Trans B 221: 179
Waddington CH (1939) An Introduction to Modern Genetics. Allen & Unwin, London
Waddington CH (1940) J Genet 41: 75
Waddington CH (1942) Nature 150: 563
Waddington CH (1953) Symp Soc Exp Biol 7: 186
Waddington CH (1953) Endeavour 12: 134
Waddington CH (1953) Evolution 7: 386
Waddington CH (1953) J Genet 51: 243
Waddington CH (1956) Principles of Embryology. Allen & Unwin, London
Waddington CH (1957) The Strategy of the Genes. Allen & Unwin, London
Waddington CH (1961) Adv Genet 10: 257
Waddington CH (1962) New Patterns in Genetics and Development. Columbia Univ Press, New York
Waddington CH, Schmidt GA (1933) Roux Arch EntwMech Organ 128: 522
Wagner R, Meselson M (1976) Proc Natl Acad Sci USA 73: 4135
Wahlund S (1928) Hereditas 11: 65
Wakker JH (1888) Jb Wiss Bot 19: 423
Waldeyer W (1888) Arch Mikr Anat 32: 1
Walker PMB (1971) Progr Biophys Mol Biol 23: 145
Wallace B (1963) Cand J Genet Cytol 5: 239
Wallace RB, Johnson MJ, Hirose T, Miyahe T, Kawashima EH, Ithakura K (1981) Nucleic Acids Res 9: 879
Waller AE (1917) Ohio J Sci 17: 273
Walter P, Blobel G (1980) Proc Natl Acad Sci USA 77: 7112
Walters MS (1954) Am J Bot 41: 160
Wang A, Quigley GJ, Crawford FL, van Boom JH, van der Mavel G, Rich A (1979) Nature 282: 680
Wang JC (1971) J Mol Biol 55: 523
Wang JC, Liu LF (1979) In: Molecular Genetics. Taylor JH (ed) Academic Press, New York, Part 3: 65
Warburton FE (1955) Am Nat 89: 129

Warmke HE, Blakeslee AF (1939) J Hered 30: 419
Warner JR, Roeiro R (1967) Proc Natl Acad Sci USA 59: 1984
Warner JR, Rich A, Hall CE (1962) Science 138: 1399
Washington WJ (1971) Can J Genet Cytol 13: 169
Wassermann F (1926) Z Ges Anat I Z Anat EntwGesch 80: 344
Watanabe T (1963) Bact Rev 27: 87
Watanabe T, Nishida H, Ogalta C, Arai T, Sato S (1964) J Bact 88: 716
Watson JD (1964) Bull Soc Chim Biol 46: 1399
Watson JD, Crick FCH (1953) Nature 171: 964
Watson MD, Scaife JG (1980) J Bact 142: 462
Watson ML (1955) J Biophysic Biochem Cytol 1: 257
Watson ML (1959) J Biophys Biochem Cytol 6: 147
Weber AF, Frommes SP (1963) Science 141: 912
Webber HJ (1897) Bot Gaz 24: 225
Webber HJ (1903) Science 18: 501
Webber JM (1933) J Agric Res 47: 845
Weigle JJ (1953) Proc Natl Acad Sci USA 39: 628
Weinberg RA, Penman SP (1968) J Mol Biol 38: 289
Weinberg W (1908) Jhrb Ver Vaterl Naturk Württemberg 64: 368
Weinstein A (1936) Genetics 21: 155
Weir BS, Cockerham CC (1969) Genetics 61: 923
Weismann A (1883) Über die Vererbung. In: Aufsätze über Vererbung und verwandte biologische Fragen. Fischer, Jena 1892, 73
Weismann A (1885) Die Kontinuität des Keimplasmas als Grundlage einer Theorie der Vererbung. In: Aufsätze über Vererbung und verwandte biologische Fragen. Fischer, Jena 1892, 191
Weismann A (1887) Über die Zahl der Richtungskörper und ihre Bedeutung für die Vererbung. Fischer, Jena
Weismann A (1891) Amphimixis oder die Vermischung der Individuen. Fischer, Jena
Weismann A (1895) Neue Gedanken zur Vererbungsfrage. Eine Antwort an Herrn Spencer. Fischer, Jena
Weismann C, Nagata S, Taniguchi T, Weber H, Meyer F (1979) In: Genetic Engineering, Principles and Methods. Setlow JK, Hollander A (eds) 1 133 Plenum, New York
Weiss B, Richardson CC (1967) Proc Natl Acad Sci USA 57: 1021
Weiss P (1945) Growth 5 Suppl: 163
Weiss P (1959) Rev Modern Phys 31: 11
Weiss MC, Green H (1967) Proc Natl Acad Sci USA 58: 1104
Weiss RB (1984) Proc Natl Acad Sci USA 81: 5797
Welch JE, Grimball EL (1947) Science 106: 594
Wells JA, Vassev M, Powers DB (1985) Gene 34: 315
Wenrich DN (1916) Bull Mus Comp Zool Harv 60: 57
Went HA (1966) Protoplasmatologia 6, G 1, Springer, Berlin
West RW, jr. Neve RL, Rodriguez RL (1979) Gene 7: 271
Wettstein F v (1924) Ztschr Ind Abst. u Vererbgsl 33: 1
Wettstein F v (1927) Nachr Ges Wiss Göttingen, Math Phys Kl: 250
Wettstein F v (1932) Naturw 20: 981
Wettstein FO, Staehelin T, Noll H (1963) Nature 197: 430
Wettstein R v (1898) Grundzüge der geographisch-morphologischen Methode der Pflanzensystematik. Fischer, Jena
Whaley (1975) The Golgi Apparatus, Springer Berlin, Heidelberg, New York
White BN, Tener GM (1973) J Mol Biol 74: 635
White MJD (1935) Proc Roy Soc B 119: 61

White MJD (1935) Naturwiss 23: 390
White MJD (1941) J Genet 42: 143
White MJD (1943) Nature 152: 536
White MJD (1945) Animal Cytology and Evolution. Univ Press, Cambridge
White MJD (1949) Genetics 34: 537
White MJD (1951) Adv Genet 4: 267
White MJC (1954) Animal Cytology and Evolution (2 ed) Univ Press, Cambridge
White MJD (1957) Surv Biol Progr 3: 109
White MJD (1965) Chromosoma 16: 271
White MJD (1973) The Chromosomes (6th ed) Chapman & Hall, London
White MJD (1973) Animal Cytology and Evolution. Third Ed, Cambridge Univ Press, London
White MJD, Blackith RE, Blackith RM, Cheney J (1967) Austral J Zool 15: 263
White RL, Hogness DS (1967) Cell 10: 177
Whitehouse HLK (1965) Towards an Understanding of the Mechanism of Heredity. Arnold E, London
Whitehouse HLK (1969) Towards an Understanding of the Mechanism of Heredity, 2nd Ed, Arnold E. London
Whiting PW, Greb RJ, Speicher BR (1934) Biol Bull, Wood's Hole 66: 152
Whitman CO (1887) J Morph
Whittaker RH (1975) The design and stability of plant communities. In: van Dobben WH, Lowe-McConnel RH (eds), Unifying Concepts in Ecology, Junk, The Hague
Wichtermann R (1939) Nature 144: 123
Wichtermann R (1940) J Morph 66: 423
Wilson DA, Thomas CA (1974) J Mol Biol 84: 115
Wilson EB (1882) J Morph 6
Wilson EB (1896) The Cell in Development and Heredity. MacMillan, New York (1st ed)
Wilson EB (1901) Roux Arch EntwMech Organ 13: 352
Wilson EB (1906) J Exp Zool 3: 1
Wilson EB (1909) Science 29: 53
Wilson EB (1909) J Exp Zool 6: 69
Wilson EB (1910) Science 31: 788
Wilson EB (1911) Arch Mikr Anat 77: 249
Wilson EB (1925) The Cell in Development and Heredity. (3rd ed), MacMillan, New York
Wilson EO (1975) Sociobiology. Harvard Univ Press, Cambridge
Winge O (1927) Hereditas 9: 274
Winiwarter H v (1900) Arch Biol, Paris 17: 33
Winkler H (1906) Ann Jard Bot Buitenzorg 20: 208
Winkler H (1907) Ber Dtsch Bot Ges 25: 568
Winkler H (1908) Über Parthenogenesis und Apogamie im Pflanzenreich. Fischer, Jena
Winkler H (1916) Z Bot 8: 417
Winkler H (1920) Verbreitung und Ursache der Parthenogenese im Pflanzen- und Tierreiche. Fischer, Jena
Winkler H (1930) Die Konversion der Gene. Fischer, Jena
Witkin EM (1951) Cold Spr Harb Symp 16: 357
Witkin EM (1956) Cold Spr Harb Symp 21: 123
Witkin EM (1969) Ann Rev of Genet 3: 525
Witkin EM (1975) Relationsships among repair mutagenesis and survival: Overview. In: Hanawalt PC, Setlow RB (eds) "Molecular Mechanisms for Repair of DNA", Part A, Plenum Press, New York p 347
Witkin EM, Wermundson IE (1978) Cold Spr Harb Symp 43: 881
Woese CR, Fox GE (1977) Proc Natl Acad Sci USA 74: 5088
Wollman EL, Jacob F, Hayes W (1956) Cold Spr Harb Symp 21: 141

Wolosewick JJ, Porter KR (1976) Am J Anat 147: 303
Wolpert L (1969) J Theoret Biol 25: 1
Woltereck R (1909) Verh Dtsch Zool Ges 19: 110
Wood RL (1959) J Biophys Biochem Cytol 6: 343
Woodruff RC, Thompson JN jr, Lyman RF (1979) Nature 278: 277
Woodson SA, Cedi TP (1989) Cell 57: 335
Wright S (1921) Genetics 6: 111
Wright S (1922) Am Nat 56: 330
Wright S (1929) Am Nat 63: 247
Wright S (1931) Genetics 16: 97
Wright S (1932) Proc 6th Int Congr Genetics 1: 356
Wright S (1943) Genetics 28: 114
Wright S (1949) Adaption and selection. In: Jepsen GL, Mayr E, Simpson GG "Genetics, Paleontology, and Evolution", Univ Press, Princeton, 365
Wright S (1955) Cold Spr Harb Symp 20: 16
Wright S (1956) Am Nat 90: 5
Wyatt GR (1951) Biochem J 48: 584
Wynne-Edwards VD (1962) Animal Dispersion in Relation to Social Behaviour. Oliver & Boyd, Edinburgh, London

Yagura t, Tanaka ST, Seno T, Korn D (1938) J Biol Chem 258: 6698
Yamada E (1955) J Biophys Biochem Cytol 1: 551
Yamamoto Y (1938) Mem Coll Agric Kyoto, 43
Yamamoto H, Shiraiwa S, Ashida T, Ito Y (1969) J Electron Microsc 18: 57
Yanofsky C, Helsinski D, Malling B (1961) Cold Spr Harb Symp 26: 11
Yasbin RE (1977) Mol Gen Genet 153: 211
Yasuda S, Hiruta Y (1977) Proc Natl Acad Sci USA 74: 5458
Yasuzumi G (1951) Chromosoma 4: 222
Yates RA, Pardee AB (1956) J Biol Chem 221: 757
Yee T, Furuichi T, Inouye S, Inouye M (1984) Cell 38: 203
Young MW (1979) Proc Natl Acad Sci USA 76: 6274
Young CSH, Cox BS (1971) Heredity 26: 413
Young ND, Tanksley SD (1989) Theor Appl Genet 77: 95
Yura T, Ishihama A (1979) Ann Rev Genet 13: 59
Yourno J, Barr D, Tanemura S (1959) J Bac 100: 453

Zaenen I, Larebeke N van, Tenchy H, Montagu M van, Schell J (1974) J Mol Biol 86: 109
Zamenhof S (1967) Progr Nucleic Acid Res and Mol Biol 6: 1
Zang KD, Back E (1968) Cytogenetics 7: 455
Zasloff M, Santos T (1980) Proc Natl Acad Sci USA 77: 5668
Zechel K, Bouché JP, Kornberg A (1975) J Biol Chem 250: 4684
Zimmer EA, Martin SL, Beverley SM, Kan YM, Wilson AC (1980) Proc Natl Acad Sci USA 77: 2158
Zimmermann W (1943) Methoden der Phylogenetik. In: Heberer G, "Die Evolution der Organismen" (1. Aufl), 20, Fischer, Jena
Zimmermann W (1948) Grundfragen der Evolution. Klostermann, Frankf.
Zimmermann W (1953) Evolution, Geschichte ihrer Probleme und Erkenntnisse. Alber, Freyburg-München
Zinder ND, Lederberg J (1953) J Bact 64: 679
Zissler Y (1967) Virology 31: 189
Zubay G, Schwartz D, Beckwith J (1970) Proc Natl Acad Sci USA 66: 104
Zuckerkandl E (1978) J Mol Evol 12: 57

Zuckerkandl E, Pauling L (1965) J Theoret Biol 8: 357
Zuckerkandl E, Pauling L (1965) Evolutionary divergence and convergence in proteins. In: Bryson V, Vogel HJ (eds), "Evolving Genes and Proteins". Acad Press, New York
Zukov NI (1941) Promyslennosh (Krasnodar) 143: 142
Zurawski G, Elseviers D, Stauffer GV, Yanofsky C (1978) Proc Natl Acad Sci USA 75: 5988